ENCYCLOPEDIA OF 20TH-CENTURY

TECHNOLOGY

ENCYCLOPEDIA OF 20TH-CENTURY

TECHNOLOGY

Volume 1
A–L

Colin A. Hempstead, Editor

William E. Worthington, Jr., Associate Editor

ROUTLEDGE

NEW YORK AND LONDON

Published in 2005 by
Routledge
270 Madison Avenue
New York, NY 10016
www.routledge-ny.com

Published in Great Britain by
Routledge
2 Park Square
Milton Park, Abingdon
Oxon OX14 4RN, UK
www.routledge.co.uk

10 9 8 7 6 5 4 3 2 1

Library of Congress Cataloging-in-Publication Data

Encyclopedia of 20th-century technology / Colin A. Hempstead, editor; William E. Worthington, associate editor.
   p. cm.
Includes bibliographical references and index.
ISBN 1-57958-386-5 (set : alk. paper)—ISBN 1-57958-463-2 (vol. 1 : alk. paper)—
ISBN 1-57958-464-0 (vol. 2 alk. paper)
  1. Technology—Encyclopedias. I. Hempstead, Colin. II. Worthington, William E., 1948–
  T9.E462 2005
  603—dc22

Advisers

Contents

List of Entries

LIST OF ENTRIES

Thematic List of Entries

Biotechnology
Antibacterial Chemotherapy
Artificial Insemination and *in Vitro* Fertilization
Biopolymers
Biotechnology
Breeding, Animal: Genetic Methods
Breeding, Plant: Genetic Methods
Cloning, Testing and Treatment Methods
Gene Therapy
Genetic Engineering, Methods
Genetic Engineering, Applications
Genetic Screening and Testing
Tissue Culturing
See also **Food and Agriculture; Health and Medicine**

Chemistry
Biopolymers
Boranes
Chemicals
Chemical Process Engineering
Chromatography
Coatings, Pigments, and Paints
Combinatorial Chemistry
Cracking
Detergents
Dyes
Electrochemistry
Electrophoresis
Environmental Monitoring
Explosives
Feedstocks
Green Chemistry
Industrial Gases
Isotopic Analysis
Nitrogen Fixation
Oil from Coal Process
Radioactive Dating
Reppe Chemistry
Solvents
Synthetic Resins
Synthetic Rubber
Warfare, Chemical

See also **Materials; Scientific Research/ Measurement**

Communications
Communications
Electronic Communications
Fax Machine
Mobile (Cell) Telephones
Radio-Frequency Electronics
Satellites, Communications
Telecommunications
Telephony, Automatic Systems
Telephony, Digital
Telephony, Long Distance

Computers
Artificial Intelligence
Computer and Video Games
Computer Displays
Computer Memory, Early
Computer Memory, Personal Computers
Computer Modeling
Computer Networks
Computer Science
Computer-Aided Design and Manufacture
Computers, Analog
Computers, Early Digital
Computers, Hybrid
Computers, Mainframe
Computers, Personal
Computers, Supercomputers
Computers, Uses and Consequences
Computer–User Interface
Control Technology, Computer-Aided
Control Technology, Electronic Signals
Error Checking and Correction
Encryption and Code Breaking
Global Positioning System (GPS)
Gyrocompass and Inertial Guidance
Information Theory
Internet
Packet Switching

Editor's Preface

All editors of encyclopedias are faced with the problem of what to include. Even if the title is agreed and the numbers of volumes and pages have been decided, the sum of possible entries could be very large. In the case of the *Encyclopedia of 20th-Century Technology*, the editor decided that in order to construct a logical and consistent set of entries it was necessary to adopt what could be described as an analytic framework. During the 20th century a plethora of manufactured articles have appeared for which the real costs have continuously fallen. The products in industrialized societies have become universal, and many of the good ones are within the reach of a large proportion of humanity. In keeping with this democratic trend of the century it was decided that people and their experiences with technology should be central to the encyclopedia. Readers are urged to read the entries in the light of the humanistic core.

An examination of people and their lives led to six broad, related areas of society from which the four hundred entries that comprise these volumes could be derived. The type of analysis carried out is indicated in the diagrams on the next page. The first shows the six basic areas; the second diagram is an outline of the detailed application for the category FOOD. Five or six levels of analysis allowed the definition of headers that provided the individual entries. Of course, entries could be found in two or more basic areas or could be related to others: entries in refrigerating in the domestic situation as found in food preservation would lead to entries in the technology of refrigeration *per se*. Thus the contents were defined.

The encyclopedia contains two types of entries. The greatest number of entries are of 1000 words, and as far as possible these standard entries are devoid of interpretation. Nevertheless, it is recognized that all history is redolent of the era in which it is constructed, and this encyclopedia is of its own particular society, that of Western industrial. The factual nature of the standard entries is leavened by longer essays in which historical and interpretative themes are explored. Among other things, these essays describe and analyze the relationship between society and technology, touch on the modern debates on the nature of the history of technology of history, and relate what people expect of the products of modern industrial civilisation.

The encyclopedia is concerned with 20th-century technology but not with 20th-century inventions. The technologies included are those that had an impact on the mass of the population in industrial societies. So many technologies invented in the 19th century did not begin to impinge markedly on many lives until the middle of the 20th century, so they are considered to be of the 20th century. Similarly, many products in the constructed world are old conceptions, transformed by modern materials or production methods. They have found a place in the encyclopedia. The inclusion of pre-20th-century products compares with the exclusion of recently developed technologies that have yet to have any effect on the mass of the public. However, the encyclopedia is not intended to be futuristic. In the 20th century, scientific engineering came to majority, and many if not all the products of modern technology can be seen to be the results of science. However, there are no entries that discuss science itself. Within the essays, however, science as science related to each subject is described.

Even with four hundred entries, the encyclopedia is not canonical, and gaps will be noted. However, the standard entries, the interpretative essays, and the lists of references and further reading suggestions allow readers to appreciate the breadth and depth of the technology of the 20th century.

Colin Hempstead

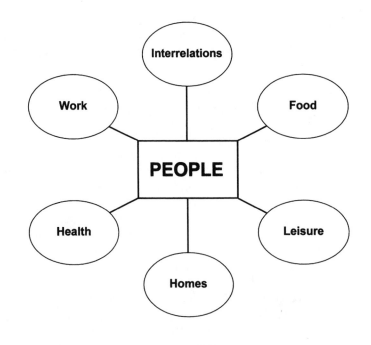

FOOD

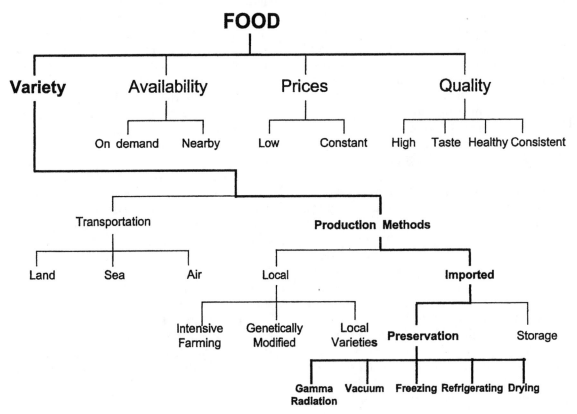

Associate Editor's Preface

Technology is a vital subject. It grows continuously. New technologies are introduced, existing technologies evolve, and the outmoded are abandoned. Looking dispassionately at technology, it is always exciting, for it is the product of human ingenuity. For the purposes of this encyclopedia, we felt it could not and should not be discussed devoid of its human element. It is breathtaking to consider the panoply of developments which occurred during the last century, but it is necessary to recall that these developments did not take place in isolation. It was our desire to see that events, where possible, were described in context. Thus, you will find names, places, dates, and events critical to the development of a particular technology. The reader will note that some entries contain a surprising amount of information on 19th-century events. This was appropriate, for some 20th-century technologies were firmly rooted in that earlier time and can be best understood in light of the past. To avoid a deadly dull recitation of formulae and regurgitation of dry facts, we sought to give the reader the broadest possible picture.

The encyclopedia was created for the lay reader and students as well as for historians of science and technology. In light of this, we attempted to minimize the use of the jargon that tends to grow around some technologies. Although many of the subjects are highly technical, our belief was that even complicated subjects could be rendered in such a way as to make them comprehensible to a wide audience. In the same way that an electrical engineer might need explanations when encountering genetic terminology, students and non-specialists will also appreciate the clarification. Because of the pervasiveness of the subjects in all facets of our lives, the encyclopedia should be a handy reference tool for a broad range of readers. Our aim was to make the subjects, which many of us deal with daily and do not necessarily grasp completely, readily understood with a minimum need for additional reference. However, should the reader wish to delve further into any particular subject, our expert authors have provided a selection of further bibliographic readings with which to begin.

The scope of the encyclopedia is intended to be international. Discussions were to be as inclusive as possible and avoid focus solely on the events of any one country. Nonetheless, some skewing was unavoidable due simply to the prodigious number of developments that have taken place in some countries.

William E. Worthington, Jr.

Acknowledgments

A host of workers and authors contributed to this encyclopedia, and I wish to extend my thanks to every person without whom these volumes would be stillborn. My particular thanks are offered to Gillian Lindsey of Routledge. Gillian conceived the idea of an encyclopedia of 20th-century technology, and appointed me the editor of the work in 2000. Her energy and ideas were legion, although she glossed over the amount of work for me! However, the editorship was rewarding, offering the possibility of producing a worthwhile publication with academic colleagues from around the globe. The selection of technologies and of particular subjects suggested by Gillian and me were critiqued and extended by our advisers. Their contributions, drawn from their specialist knowledge and scholarship, were invaluable. When circumstances forced my withdrawal from the active editorship, William Worthington, then with the National Museum of American History in Washington, stepped into the hot seat. To William I give my heartfelt thanks.

Finally I acknowledge the publishers and the 20th century which presented all of us with the opportunity to examine and extol some of the content and effects of modern technology. Nevertheless, the encyclopedia is partial, and any omissions and shortcomings are mine.

Colin Hempstead

My thanks go to Gillian Lindsey for presenting me with the challenge of filling the void left by Colin's departure. However, the prospect of assuming a role in a project already well under way and natural differences in approach and style were concerns. Nonetheless, the final third of the encyclopedia was crafted in such a way that it blends seamlessly with the sections completed under Colin's careful guidance. This was due in no small part to the untiring efforts of Sally Barhydt, and to her I extend sincere thanks.

William E. Worthington, Jr.

Contributors

W.A. Addis, Buro Happold, Middlesex, United Kingdom.

Aaron Alcorn, Cleveland Heights, Ohio, USA.

K.W. Allen, Joining Technology Research Center, Oxford Brookes University, United Kingdom.

Paul Anastas, White House Office of Science and Technology, National Security and International Activities Division, USA.

Joe Anderson, Agricultural History and Rural Studies Program, Iowa State University, USA.

Stuart Anderson, Department of Public Health and Policy, London School of Hygiene and Tropical Medicine, United Kingdom.

Rachel A. Ankeny, Unit for History and Philosophy of Science, University of Sydney, Australia.

Peter Asaro, Departments of Computer Science and Philosophy, University of Illinois at Champaign, USA.

Glen Asner, Department of History, Carnegie Mellon University, Pittsburgh, Pennsylvania, USA.

Janet Bainbridge, School of Science and Technology, University of Teesside, and Chief Executive of EPICC (European Process Industries Competitiveness Centre), United Kingdom.

Guy D. Ball, Product Information, Unisys, Tustin, California, USA.

Susan B. Barnes, Department of Communication and Media Studies, Fordham University, New York, USA.

Keith Bartle, Department of Chemistry, University of Leeds, United Kingdom.

Donald R Baucom, Department of Defense, Ballistic Missile Defense Organization, USA.

Reinhold Bauer, Universität der Bundeswehr Hamburg, Germany.

Joyce E. Bedi, National Museum of American History, Lemelson Center for the Study of Invention and Innovation, USA.

Randal Beeman, Archives Director, Bakersfield College, California, USA.

Hal Berghel, Department of Computer Science, University of Nevada at Las Vegas, USA.

Beverly Biderman, Adaptive Technology Resource Centre, University of Toronto, Canada.

David I. Bleiwas, U.S. Geological Survey, Reston, Virginia, USA.

F.K. Boersma, Department of Culture, Organization and Management, Vrije Universiteit Amsterdam, Netherlands.

James Bohning, Center for Emeritus Scientists in Academic Research (CESAR), Lehigh University, Bethlehem, Pennsylvania, USA.

Brian Bowers, Engineering Historian and Writer, Retired Senior Curator, Science Museum, London, United Kingdom.

Hans-Joachim Braun, Universität der Bundeswehr Hamburg, Germany.

Catherine Brosnan, United Kingdom.

David J. Brown, Ove Arup & Partners, London, United Kingdom.

Louis Brown, Department of Terrestrial Magnetism, Carnegie Institution of Washington, USA.

Nik Brown, Science and Technology Studies Unit, University of York, United Kingdom.

Timothy S. Brown, Department of History, Pomona College, California, USA.

Robert Bud, Head of Research (Collections), Science Museum, London, United Kingdom.

William L. Budde, Office of Research and Development, U.S. Environmental Protection Agency.

Ian Burdon, Head of Sustainable Energy Developments, Energy and Utility Consulting, PB Power Ltd., Newcastle, United Kingdom.

Larry Burke, Carnegie Mellon University, Pittsburgh, Pennsylvania, USA.

Russell W. Burns, Retired Professor, Nottingham, United Kingdom.

Michael Bussell, London, United Kingdom.

J. Stewart Cameron, St. Thomas' and Guy's Hospital, London, United Kingdom.

CONTRIBUTORS

Rodney P. Carlisle, History Associates Incorporated, Rockville, Maryland, USA.

Stéphane Castonguay, Département des sciences humaines, Université du Québec á Trois-Rivières, Canada.

Carol A. Cheek, Rensselaer Polytechnic Institute, Troy, New York, USA.

Dennis W. Cheek, John Templeton Foundation, Radnor, Pennsylvania, USA.

Mark Clark, General Studies Department, Oregon Institute of Technology, USA.

Noel G. Coley, Department of the History of Science, Technology and Medicine, Open University, Milton Keynes, United Kingdom.

Paul Collins, Ironbridge Institute, University of Birmingham, United Kingdom.

Yonina Cooper, University of Nevada at Las Vegas, USA.

Peter Copeland, National Sound Archives, British Library, United Kingdom.

Anthony Coulls, Formerly of the Museum of Science and Industry in Manchester, United Kingdom.

Jennifer Cousineau, Charlottesville, Virginia, USA.

Trevor Cox, Acoustics Research Center, University of Salford, Greater Manchester, United Kingdom.

Jennifer Croissant, Program on Culture, Science, Technology, and Society, University of Arizona, USA.

Donard de Cogan, School of Information Systems, University of East Anglia, Norwich, United Kingdom.

Guillaume de Syon, Department of History, Albright College, USA.

Marc J. de Vries, Philosophy and Social Sciences, Eindhoven University of Technology, Netherlands.

Andrew Dequasie, Pownal, Vermont, USA.

Maggie Dennis, National Museum of American History, Smithsonian Institution, Washington, D.C., USA.

Panos Diamantopoulos, School of Engineering, University of Sussex, United Kingdom.

John Dolan, Retired, Nobel Division ICI, United Kingdom.

Michael Duffy, Lancashire, United Kingdom; formerly of the Department of Engineering, University of Sunderland.

Charles Duvall, Bandwidth 9, Duluth, Georgia, USA.

Matthew Eisler, Department of History and Classics, University of Alberta, Edmonton, Canada.

Boelie Elzen, University of Twente, The Netherlands.

Linda Endersby, Thomas A. Edison Papers, Rutgers University, New Jersey, USA.

Ted Everson, Institute for the History and Philosophy of Science and Technology, University of Toronto, Canada.

Jim Farmer, Chief Technical Officer, Wave 7 Optics, Alpharetta, Georgia, USA.

David L. Ferro, Computer Science Department, Weber State University, Ogden, Utah, USA.

Mark Finlay, Department of History, Armstrong Atlantic State University, Savannah, Georgia, USA.

Gerard J. Fitzgerald, Department of History, Carnegie Mellon University, Pittsburgh, Pennsylvania, USA.

Amy Foster, Ph.D. Candidate, Auburn University, Alabama, USA.

Philip L. Frana, Charles Babbage Institute, University of Minnesota, USA.

Philip J. Gibbon, Temple University, Philadelphia, Pennsylvania, USA.

Bruce Gillespie, Braamfontein, South Africa.

Julia Chenot GoodFox, University of Kansas, USA.

David Grier, Department of History, Center for History of Recent Science, George Washington University, Washington, D.C., USA.

Reese Griffin, 2 Stroke International Engine Co., Beaufort, South Carolina, USA.

Eric Grove, Center for Security Studies, University of Hull, United Kingdom.

David Haberstich, National Museum of American History, Smithsonian Institution, Washington, D.C., USA.

John Hamblin, Medical Doctor, Essex, United Kingdom.

Jennifer Harrison, Department of Information Technology, College of William & Mary, Williamsburg, Virginia, USA.

Ulf Hashagen, Deutsches Museum, Germany.

Peter Hawkes, CEMES-CNRS, Toulouse, France.

Stephen Healy, School of Science and Technology Studies, University of New South Wales, Sydney, Australia.

David Healy, North Wales Department of Psychological Medicine, University of Wales College of Medicine, Bangor, United Kingdom.

Colin Hempstead, Darlington, United Kingdom; formerly Reader in History of Science and Technology, University of Teesside.

Martin Hempstead, Corning Photonics Technology, Corning Inc., New York, USA.

Klaus Hentschel, Institut für Philosophie, University of Berne, Switzerland.

Arne Hessenbruch, Dibner Institute, Massachusetts Institute of Technology, USA.

Robert D. Hicks, Chemical Heritage Foundation, Philadelphia, Pennsylvania, USA.

Roger Howkins, Ove Arup & Partners, London, United Kingdom.

Peter J. Hugill, Department of Geography, Texas A&M University, USA.

Merritt Ierley, Sussex, New Jersey, USA.

Mary Ingram, Department of Sociology, University of California, Santa Barbara, USA.

Muzaffar Iqbal, Center for Islam and Science, Canada.

Ann Johnson, Department of History, Fordham University, New York, USA.

Sean Johnston, Science Studies, University of Glasgow, United Kingdom.

Suzanne W. Junod, History Office, U.S. Food and Drug Administration.

David Kaplan, Biomedical Engineering, Tufts University, Boston, Massachusetts, USA.

Christine Keiner, Science, Technology, and Society Department, Rochester Institute of Technology, New York, USA.

Karen D. Kelley, U.S. Geological Survey, Reston, Virginia, USA.

David Kirsch, Smith School of Business, University of Maryland, USA.

Timothy Kneeland, Department of History and Political Science, Nazareth College, Rochester, New York, USA.

Ramunas A. Kondratas, Division of Science, Medicine, and Society, National Museum of American History, Smithsonian Institution, Washington, D.C., USA.

Helge Kragh, History of Science Department, University of Aarhus, Denmark.

John Krige, School of History, Technology, and Society, Georgia Institute of Technology, USA.

Alex Law, Department of Sociology, University of Abertay, Dundee, United Kingdom.

Michal Lebl, Illumina, Inc. & Spyder Instruments, Inc., San Diego, California, USA.

Tim LeCain, Department of History, Montana State University, USA.

Trudy Levine, Computer Science Department, Fairleigh Dickinson University, Teaneck, New Jersey, USA.

Gillian Lindsey, Writer and Editor, New York, USA.

Eric Livo, U.S. Geological Survey, Reston, Virginia, USA.

Henry Lowood, Curator, History of Science and Technology Collections, Stanford University, California, USA.

Juan Lucena, Liberal Arts and International Studies, Colorado School of Mines, USA.

Harro Maat, Technology & Agrarian Development (TAO), Wageningen University, Netherlands.

Alex Magoun, Executive Director, David Sarnoff Library, Princeton, New Jersey, USA.

A.M. Mannion, Department of Geography, University of Reading, United Kingdom.

J. Rosser Matthews, History Office, National Institutes of Health, Bethesda, Maryland, USA.

W. Patrick McCray, Center for History of Physics, American Institute of Physics, College Park, Maryland, USA.

Ian C. McKay, Department of Immunology and Bacteriology, University of Glasgow, United Kingdom.

Shelley McKellar, Department of History, University of Western Ontario, London, Canada.

Dennis McMullan, London, United Kingdom.

Kenneth Mernitz, History and Social Studies Department, Buffalo State College, New York, USA.

Lolly Merrell, Paonia, Colorado, USA.

Andre Millard, Department of History, University of Alabama at Birmingham, USA.

Carl Mitcham, Liberal Arts and International Studies, Colorado School of Mines, USA.

Susan Molyneux-Hodgson, Department of Sociological Studies, University of Sheffield, United Kingdom.

Gijs Mom, Foundation for the History of Technology, Technical University of Eindhoven, Netherlands.

John Morello, Department of General Education, DeVry Institute of Technology, Addison, Illinois, USA.

Peter Morris, Science Museum, London, United Kingdom.

Robin Morris, Retired Lecturer, West Malvern, United Kingdom.

David L. Morton, Tucker, Georgia, USA.

Susan Mossman, Senior Curator, Science Museum, London, United Kingdom.

Karel Mulder, Technology Assessment Group, Delft University of Technology, Netherlands.

Peter Myers, Department of Chemistry, University of Leeds, United Kingdom.

Francis Neary, Centre for the History of Science, Technology and Medicine, University of Manchester, United Kingdom.

CONTRIBUTORS

Caryn E. Neumann, Department of History, The Ohio State University, USA.

William O'Neil, CNA Corporation, Alexandria, Virginia, USA.

Andrew Panay, Department of Sociology, University of Abertay, Dundee, United Kingdom.

Kayhan Parsi, Neiswanger Institute for Bioethics and Health Policy, Stritch School of Medicine, Loyola University of Chicago, USA.

Mike Pavelec, Department of History, The Ohio State University, USA.

Niocola Perrin, Nuffield Council on Bioethics, London, United Kingdom.

James Perry, Formerly Strategic Assessment Center, SAIC, USA.

John Pfotenhauer, Applied Superconductivity Center, University of Wisconsin-Madison, USA.

Robert Raikes, Meko Ltd., United Kingdom.

Thomas W. Redpath, Department of Biomedical Physics and Bioengineering, University of Aberdeen, United Kingdom.

Antoni Rogalski, Institute of Applied Physics, Military University of Technology, Warsaw, Poland.

David Rose, Retired Otorhinolaryngology Consultancy, Stockport NHS Trust, Manchester, United Kingdom.

Paul Rosen, Department of Sociology, University of York, United Kingdom.

Robin Roy, Department of Design and Innovation, Open University, United Kingdom.

Pedro Ruiz-Castell, St. Cross College, University of Oxford, Valencia, Spain.

Robert W. Rydell, Department of History, Montana State University, USA.

Nicholas Saunders, London, United Kingdom.

Roger Scantlebury, Integra SP, London, United Kingdom.

Jessica R. Schaap, Policy and Communications, NewMediaBC, Canada.

Elizabeth Schafer, Loachapoka, Alabama, USA.

Thomas Schlich, Institut fuer Geschichte der Medizin, Albert-Ludwigs-Universitaet, Freiburg, Germany.

Jeff Schramm, History Department, University of Missouri-Rolla, USA.

Stuart Shapiro, Senior Information Security Scientist, The MITRE Corporation, Bedford, Massachusetts, USA.

G. Terry Sharrer, National Museum of American History, Smithsonian Institution, Washington, D.C., USA.

Duncan Shepherd, School of Engineering, University of Birmingham, United Kingdom.

John K. Smith, Department of History, Lehigh University, Bethlehem, Pennsylvania, USA.

John S. Sobolewski, Computer & Information Research and Technology, University of New Mexico, USA.

Lawrence Souder, Department of Culture and Communication, Drexel University, Philadelphia, Pennsylvania, USA.

James Steele, National Physical Laboratory, United Kingdom.

Carlene Stephens, National Museum of American History, Smithsonian Institution, Washington, D.C., USA.

Christopher Sterling, George Washington University, Washington, D.C., USA.

Jack Stilgoe, University College London, United Kingdom.

Anthony Stranges, Department of History, Texas A & M University, USA.

James Streckfuss, College of Evening and Continuing Education, University of Cincinnati, USA.

Rick Sturdevant, HQ AFSPC/HO, Peterson, Colorado, USA.

Eric G. Swedin, Computer Science Department, Weber State University, Ogden, Utah, USA.

Derek Taylor, Altechnica, Milton Keynes, United Kingdom.

Ernie Teagarden, Professor Emeritus, Dakota State University, Madison, South Dakota, USA.

Jessica Teisch, Department of Geography, University of California, Berkeley, USA.

Thom Thomas, Halmstad University, Sweden.

Lana Thompson, Florida Atlantic University, Boca Raton, USA.

James E. Tomayko, School of Computer Science, Carnegie Mellon University, Pittsburgh, Pennsylvania, USA.

Anthony S. Travis, Sidney M. Edelstein Center for History and Philosophy, Hebrew University, Jerusalem, Israel.

Simone Turchetti, Centre for History of Science, Technology and Medicine, University of Manchester, United Kingdom.

Steven Turner, National Museum of American History, Smithsonian Institution, Washington, D.C., USA.

Aristotle Tympas, Department of Philosophy and History of Science, University of Athens, Panepistimioupolis, Greece.

Eric v.d. Luft, Historical Collections, Health Sciences Library, State University of New York Upstate Medical University, Syracuse, USA.

Helen Valier, Centre for History of Science, Technology and Medicine, University of Manchester, United Kingdom.

Peter Van Olinda, Power Systems Engineer, New York, USA.

Colin Walsh, Medical Physics and Bioengineering Department, St. James's Hospital, Dublin, Ireland.

John Ward, Senior Research Fellow, Science Museum, London, United Kingdom.

Frank Watson, Reynolds Metals Company, Richmond, Virginia, USA.

David R. Wilburn, U.S. Geological Survey, Reston, Virginia, USA.

Mark Williamson, Kirkby Thore, The Glebe House, United Kingdom.

Duncan Wilson, Centre for History of Science, Technology and Medicine, University of Manchester, United Kingdom.

Frank Winter, Curator, Rocketry, National Air and Space Museum, Smithsonian Institution, Washington D.C., USA.

Bob Wintermute, Scholar in Residence, Army Heritage Center Foundation; Doctoral Candidate, Department of History, Temple University, Philadelphia, Pennsylvania, USA.

Stewart Wolpin, New York, New York, USA.

William E. Worthington, Jr., National Museum of American History, Smithsonian Institution, Washington, D.C., USA.

Jeffrey C. Wynn, U.S. Geological Survey, Reston, Virginia, USA.

Yeang, Chen Pang, Program in Science, Technology and Society, Massachusetts Institute of Technology, USA.

Ronald Young, History Department, Georgia Southern University, USA.

Thomas Zeller, Department of History, University of Maryland, USA.

Stelios Zimeras, Medcom GmbH, Germany.

A

Absorbent Materials

For thousands of years, plant-derived materials have served as the primary ingredient of absorbent materials. Jute, flax, silk, hemp, potatoes, and primarily cotton, have been employed since pre-Roman times. These simple plant-based fibers demonstrated molecular properties such as surface tension and colloid attraction, but it wasn't until the development of the ultramicroscope in 1903 that the size and structure of molecules was better understood and the actual chemical process of absorption grasped. The late nineteenth century inspired a new wave of design for the specialized applications of absorbent material—as sanitary napkins and diapers—and eventually helped drive innovative applications for the burgeoning fields of organic and polymer science in the twentieth century.

The need for sterile bandages in medicine precipitated the design of mass-producible, absorbent materials. In 1886, the medical supply company Johnson & Johnson developed surgical wound dressings made of heated, sterilized absorbent cotton with a gauze overlay to prevent fibers sticking to wounds. This design for sterile wound dressing became a fixed part of medical treatment, although it was still unavailable to the general public. However, as women changed their clothing styles and became more independent, demand increased for transportable absorbent menstrual napkins, as well as disposable diapers. In 1887 an American, Maria Allen, created a cotton textile diaper covered with a perforated layer of paper, to draw blood away from the skin, with a gauze layer stitched around it. It was an improvement over the usual washable cotton "rag" that was extremely leaky (as both a sanitary napkin and a diaper). However, it was too expensive for mass production.

Johnson & Johnson continued to improve on the absorption capacity of their original bandage. They discovered that heating and compressing several layers of cotton together provided higher absorption, less leakage, and less bulk in their dressings. When the Lister Towel, as it was named, became widespread in 1896, menstrual products such as the German-manufactured Hartman's Pads and bolts of "sanitary" cotton cloth appeared in catalogs for women. However, the Johnson & Johnson product was expensive. Cotton, while readily available, still had to be hand picked, processed and sterilized. So, in 1915, an American paper supply company called Kimberly–Clark developed Cellucotton, a bandage material that combined sterile cotton with wood pulp-derived cellulose. During World War I, nurses working in Europe began to use both the Lister Towel and Cellucotton as menstrual pads. By 1921, propelled by this innovative application, Kimberly–Clark manufactured Cellucotton-based disposable pads called Kotex. Thick, with a gauze overlay, they employed several different securing devices. Used in diapers, Cellucotton was sometimes covered by a thick rubber pant, which inhibited evaporation and could exacerbate diaper rash and urinary tract infections in babies. "Breathability" would become one of the challenges in the decades to come.

After the turn of the twentieth century, the molecular properties of most fibers were thoroughly understood. Protein fiber-based materials, such as wool, are made up of long, parallel, molecular chains connected by cross-linkages.

While able to absorb 30 percent of its weight, it would also expel liquid readily when squeezed, making it an unattractive menstrual or diaper material. Plant-based material such as cotton was made up of long chains of cellulose molecules arranged in a collapsed tube-like fiber. Cotton could easily absorb water by holding the water molecules within the tubes and between the fibers. In addition, the shape of the fibers meant that cotton could be easily manipulated by surfactants and additives. The rate of absorption depended largely on the surface tension between the absorbent material, and the fluid it was absorbing. Manipulating surface tension would become an element of future products.

For the first half of the twentieth century, absorbent materials varied little, but design changed dramatically. Tampons, available for millennia, now incorporated the new cotton-hybrid materials and by 1930 appeared widely on the market. In 1936, Dr. Earle C. Haas, an American physician, created and earned a patent for a cardboard tampon applicator. Soon thereafter, his product became the first Tampax brand tampon and was sold by Tambrands.

By 1938, American chemist Wallace Hume Carothers of the DuPont Company had helped create nylon, the first polymer textile, and it was soon included as a barrier to prevent leakage. In 1950, American housewife Marion Donovan created a plastic envelope from a nylon shower curtain that was perforated on one side and filled with absorbent cotton gauze. By 1973, scientists working at the Illinois-based National Center for Agricultural Utilization Research invented H-Span. They combined synthetic chemicals with cornstarch to create a uniquely absorbent polymer of hydrolyzed starch called polyacrylonitrile. The "Super Slurper," as it became known, was capable of absorbing up to 5,000 times its weight in water. In a dry powdered state, the polymer chains are coiled and then treated with carboxylate to initiate a faster colloid transfer of water molecules to the starch.

Soon afterwards, "superthirsty" fibers appeared in absorbent products around the world. By the late 1970s, disposable diapers included a layer of some sort of highly absorbent fibers, covered with a lightweight plastic or nylon shell that allowed for more evaporation without leakage. The American-based company Procter & Gamble introduced a "superthirsty" synthetic material, made up of carboxymethylcellulose and polyester, into their tampons. The product, named Rely, far surpassed the absorbency of other competing tampons.

Under competitive pressure, Tambrands and Playtex both produced versions of superthirsty tampons using derivatives of polyacrylate fibers. Diaper designs began to include convenience features such as refastenable tapes, elastic legs, barrier leg cuffs, elasticized waistbands, and "fit" guides to guarantee less leakage. The popular creped cotton tissue interior was replaced with denser cellulose-fiber mats, utilizing a highly absorbent cotton treated with a surfactant to encourage rapid absorption by increasing the surface tension between water molecules and cotton.

Research continued and resulted in a new wave of polymer-manipulated superabsorbers, namely hydrophilic cross-linked polymers. Incorporating a three-dimensional polymeric structure, this material did not dissolve in water and could absorb in three dimensions. By 1980, Japanese scientists created the first disposable diaper incorporating a superabsorbent polymer. Procter & Gamble soon developed "ultra thin" pads using a crystalline polymer layer that would gel when it absorbed water. This design also included a "Dri-Weave" top sheet, separating the wearer from the absorbent layer and using a capillary-like, nonwoven material to inhibit a reverse flow.

In the late 1970s, a dramatic increase in cases of toxic shock syndrome appeared among users of superabsorbent tampons. Eventually, the "superthirsty" absorbent was found to encourage growth of the bacteria *Staphyloccocus aureus*. In the early 1980s more health problems seemed to be linked to improvements in absorption, and by 1986 Tambrands and Playtex had removed their polyacrylate tampons from the market. Six years later the U.S. Food and Drug Administration reported that trace amounts of dioxin used to bleach and sterilize cotton components of pads, tampons, and diapers could cause birth defects and possibly cancer.

At the beginning of the twenty-first century, pads were comprised of anything from an absorbent, compressed cotton and cellulose-pulp core, a plastic moisture-proof liner, a soft nonwoven textile for drawing moisture away from the skin (like viscose rayon and cotton blend), and chemicals such as polyacrylates to prevent leakage and keep the product from falling apart. Scientists working for the U.S. Department of Agriculture had discovered that the cellulose properties of ground chicken feathers could be manipulated and used as an absorbent material, utilizing billions of tons of discarded poultry-plant waste. The fibers are straight polymer chains—like cotton—making them highly absorbent. Internationally, the use of tampons, disposable diapers, and sanitary napkins is still largely reserved for developed countries. However, as more innovative techniques reduce the reliance on expensive imported products (e.g., bird feathers), the convenience of absorbent technology may stretch beyond current economic, cultural, and geographic borders.

See also **Fibers; Synthetic; Semi-Synthetic**

LOLLY MERRELL

Further Reading

Asimov, I. *New Guide to Science*. New York, 1994, 533–550.

Gutcho, M. *Tampons and Other Catamenial Receptors*. Noyes Data Corporation, Park Ridge, NJ, 1979.

Hall, A. *Cotton-Cellulose: Its Chemistry and Technology*. E. Benn, London, 1924.

Park, S. The Modern and Postmodern Marketing of Menstrual Products. *J. Pop. Cult.*, 30, 149, 1996.

Swasy, A. *Soap Opera*. Times Books, New York, 1993.

Useful Websites

U.S. Environmental Protection Agency fact sheet on acrylamide: http://www.epa.gov/ttn/uatw/hlthef/acrylami.html

Activated Carbon

Activated carbon is made from any substance with a high carbon content, and activation refers to the development of the property of adsorption. Activated carbon is important in purification processes, in which molecules of various contaminants are concentrated on and adhere to the solid surface of the carbon. Through physical adsorption, activated carbon removes taste and odor-causing organic compounds, volatile organic compounds, and many organic compounds that do not undergo biological degradation from the atmosphere and from water, including potable supplies, process streams, and waste streams. The action can be compared to precipitation. Activated carbon is generally nonpolar, and because of this it adsorbs other nonpolar, mainly organic, substances. Extensive porosity (pore volume) and large available internal surface area of the pores are responsible for adsorption.

Processes used to produce activated carbons with defined properties became available only after 1900. Steam activation was patented by R. von Ostreijko in Britain, France, Germany, and the U.S. from 1900 to 1903. When made from wood, the activated carbon product was called Eponite (1909); when made from peat, it was called Norit (1911). Activated carbon processes began in Holland, Germany, and the U.S., and the products were in all cases a powdered form of activated carbon mainly used for decolorizing sugar solutions. This remained an important use, requiring some 1800 tons each year, into the twenty-first century.

In the U.S., coconut char activated by steam was developed for use in gas masks during World War I. The advantage of using coconut shell was that it was a waste product that could be converted to charcoal in primitive kilns at little cost. By 1923, activated carbon was available from black ash, paper pulp waste residue, and lignite. In 1919, the U.S. Public Health Service conducted experiments on filtration of surface water contaminated with industrial waste through activated carbon. At first, cost considerations militated against the widespread use of activated carbon for water treatment. It was employed at some British works before 1930, and at Hackensack in New Jersey. From that time there was an interest in the application of granular activated carbon in water treatment, and its subsequent use for this purpose grew rapidly. As improved forms became available, activated carbon often replaced sand in water treatment where potable supplies were required.

Coal-based processes for high-grade adsorbent required for use in gas masks originally involved prior pulverization and briquetting under pressure, followed by carbonization, and activation. The process was simplified after 1933 when the British Fuel Research Station in East Greenwich, at the request of the Chemical Research Defence Establishment, began experiments on direct production from coke activated by steam at elevated temperatures. In 1940, Pittsburgh Coke & Iron Company, developed a process for producing granular activated carbon from bituminous coal for use in military gas masks. During World War II, this replaced the coconut char previously obtained from India and the Philippines. The large surface area created by the pores and its mechanical hardness made this new material particularly useful in continuous decolorization processes. The Pittsburgh processes developed by the Pittsburgh Activated Carbon Company were acquired in 1965 by the Calgon Company. In late twentieth century processes, carbon was crushed, mixed with binder, sized and processed in low-temperature bakers, and subjected to high temperatures in furnaces where the pore structure of the carbon is developed. The activation process can be adjusted to create pores of the required size for a particular application. Activation normally takes place at 800–900°C with steam or carbon dioxide. Powdered activated carbon is suitable for liquid and flue gas applications—the granulated form for the liquid and gas phases, and pelleted activated carbon for the gas phase. Granulated activated carbon is used as a filter medium for contaminated water or air, while the powdered form is mixed into wastewater where it adsorbs the contaminants and is later filtered or settled from the mixture. Activated carbon has also been used in chemical analysis for prior removal and concentration of contaminants in water. Trade names for activated carbon used in these processes are Nuchar and Darco.

Activated carbon has been used in the large-scale treatment of liquid waste, of which the effluent from the synthetic dye industry is a good example. Synthetic dye manufacture involves reactions of aromatic chemicals, and the reactants and products are sometimes toxic. In addition to an unpleasant taste and odor imparted to water, this waste is also highly colored, complex, and invariably very difficult to degrade. Fortunately, many of the refractory aromatic compounds are nonpolar, the property that permits adsorption onto activated carbon. In the 1970s, three large dye-making works in New Jersey used activated carbon to remove aromatics and even trace metals such as toxic lead and cadmium from liquid waste. In two cases, powdered activated carbon was added to the activated sludge treatment process to enhance removal of contaminants. In a third case, following biological treatment, the liquid effluent was adsorbed during upward passage in towers packed with granular activated carbon. The spent carbon from this continuous process was regenerated in a furnace, and at the same time the adsorbed waste solute was destroyed.

In 1962, Calgon utilized activated granular carbon for treating drinking water, and at the end of the twentieth century, municipal water purification had become the largest market for activated carbon. The older methods that involved disposal of spent carbon after use were replaced by the continuous processes using granulated activated carbon. By continuous reuse of the regenerated activated carbon, the process is ecologically more desirable. Apart from the inability to remove soluble contaminants (since they are polar) and the need for low concentrations of both organic and inorganic contaminants, the cost of the carbon is the greatest limitation in the continuous process.

Activated carbon also found wide application in the pharmaceutical, alcoholic beverage, and electroplating industries; in the removal of pesticides and waste of pesticide manufacture; for treatment of wastewater from petroleum refineries and textile factories; and for remediation of polluted groundwater. Although activated carbons are manufactured for specific uses, it is difficult to characterize them quantitatively. As a result, laboratory trials and pilot plant experiments on a specific waste type normally precede installation of activated carbon facilities.

See also **Green Chemistry; Technology, Society, and the Environment**

ANTHONY S. TRAVIS

Further Reading

Cheremisinoff, P.N. and Ellersbusch, E. *Carbon Adsorption Handbook*. Ann Arbor, MI, 1978.
Hassler, J.W. *Activated Carbon*. Chemical Publishing, New York, 1963. Includes a historical survey on pp. 1–14.

Useful Websites

A Brief History of Activated Carbon and a Summary of its Uses: http://www.cee.vt.edu/program_areas/environmental/teach/gwprimer/group23/achistory.html
Calgon Carbon, Company History: http://www.calgoncarbon.com/calgon/calgonhistory.html
Chemviron Carbon: http://www.chemvironcarbon.com/activity/what/history/menu.htm

Adhesives

Adhesives have been used for about six millennia, but it was only from the first decade of the twentieth century that any significant development took place, with the introduction of synthetic materials to augment earlier natural materials. The driving force for development has been the needs of particular industries rather than technological advances themselves. The introduction of synthetic resins began in about 1909, but although the growth in plywood manufacture was accelerated by World War I, little innovation was involved. Significant advances began with World War II and the development of epoxy and urea/formaldehyde adhesives for the construction of wooden aircraft, followed by the phenol/formaldehyde/polyvinyl formal adhesives for bonding aluminum, which cannot generally be welded. Later, adhesive bonding in conjunction with riveting was applied to automobile construction, initially to high-performance models but increasingly to mass-produced vehicles. The fastening of composite materials is, with few exceptions, accomplished by use of adhesives.

If the forces of adhesion are to be effective, intimate contact must be established between two components, one of them a liquid that will wet and flow across the other before solidifying so that the bond can resist and transmit any applied force. This change of phase from liquid to solid is achieved in a variety of ways.

Solution-Based Adhesives

The earliest adhesives were all natural products such as starch and animal protein solutions in water. These are still in use for applications where only low strength is required (e.g., woodworking or attaching paper and similar materials).In these cases, the cost has to be low because the uses are high volume. Until about 1930 these were the main adhesives used in all carpentry and furniture. Polyvinyl acetate adhesives are now probably the most important range of water-based adhesives. The base polymer is dispersed in water to give an emulsion that has to be stabilized, usually with approximately 5 percent polyvinyl alcohol.

Solutions in organic solvents were first introduced in 1928, and they are now perhaps the most widely used adhesives both for manufacturing and for do-it-yourself purposes. Based on solutions of polychloroprene as base polymer dissolved in organic solvents, they provide a fairly strong "quick-stick" bond. Particular grades are extensively used in the footwear industry. Because of the toxic, environmentally unfavorable properties of the solvents, considerable efforts are being devoted to replacing these with water-based products, but these have not yet been entirely satisfactory.

Hot-Melt Adhesives

One of the oldest types of adhesive is sealing wax. Since about 1960, these hot-melt adhesives have been introduced initially for large-scale industrial use and more recently for small-scale and do-it-yourself uses. Polyethylene is extensively used as the base for hot-melt adhesives since it is widely available in a range of grades and at low cost. Ethylene vinyl acetate is similarly a useful base, and the two are commonly used in combination to give effective adhesives with application temperatures in the range of 160–190°C. This means that the adhesives have an upper limit of service use of perhaps 140°C, and the materials being joined must be able to withstand the higher temperature. These adhesives are quite widely used in large-scale manufacturing. However there are a considerable number of applications where the temperature involved for normal hot-melt adhesives is exces-

sive. Consequently, in the 1990s a group of special formulations evolved that have an application temperature in the range of 90 to 120°C without any loss of adhesive strength. The most recent developments are adhesives that are applied as hot-melts and are then "cured" by various means. They have all the advantages of ease of application and quick achievement of useful strength supplemented by a much higher service temperature. Curing may be achieved either by heating to a higher temperature than that of application or by irradiation with an electron beam.

Reactive Adhesives

Reactive adhesives include epoxides, urethanes, phenolics, silicones, and acrylates.

Epoxides. Introduced in the early 1940s, these depend on three-membered epoxy or oxirane rings at the end of carbon chains with pendant hydroxyl groups, all of which react with various second components to produce thermoset polymers. The second components are principally amines or acid anhydrides. Generally the epoxides give bonds of considerable strength and durability, but until recently they tended to be too brittle for many purposes. Developments beginning in the 1970s have enhanced the toughness of these and other structural adhesives.

Urethanes. These involve the reaction of an isocyanate with an organic compound containing a hydroxyl group. Like the epoxides, variation of the properties of the final polymer can readily be controlled with two ingredients to give a product that may be an elastomer, a foam, or one that is stiff and bristle-like. Urethanes are increasingly used in a wide variety of situations.

Phenolics. The phenolics group of adhesives includes two that are somewhat different in their uses. The first, urea/formaldehyde formulations, were developed in the 1920s and 1930s and are mainly significant in the manufacture of plywood and similar products. The second group is phenol/polyvinyl formal formulations mainly used in aircraft construction for bonding aluminum and developed during World War II. Phenolics all involve curing under considerable pressure at an elevated temperature, typically 1500°C for 30 minutes at a pressure of 10 atmospheres for an aircraft adhesive. The bonds are of considerable strength and durability, suitable for primary aircraft structures.

Silicones. Silicones, generally silicone (or siloxane) rubbers, are largely used as sealants that combine adhesion with their gap-filling characteristics. Commonly used for sealing around baths and similar fittings, they cure by reaction with moisture from the environment. Industrially, particularly in automobile construction, there are many situations where providing a bond of moderate strength together with filling a gap between parts, which may amount to several millimeters, is required.

Acrylates. The acrylates include four types of adhesives.

1. Anaerobic adhesives (c. 1950) are formulations in which polymerization is prevented by the presence of oxygen. If oxygen is removed and ferrous ions are present, the liquid very quickly polymerizes to a hard, rather brittle solid. The main use for this is in thread locking in machinery and in the securing of coaxial joints.

2. Cyanoacrylates, or "super glues," were developed in 1957. They are colorless, very mobile liquids derived from cyanoacrylic acid. They readily polymerize, particularly in conjunction with the imperceptible film of moisture that is invariably present on surfaces. The bonds are very susceptible to attack by water and are only stable below about 80°C. Nevertheless, they are extensively used in product assembly in the electronics industry where they are likely to be exposed to only benign conditions.

3. Reactive acrylics (sometimes called "second generation" acrylates, developed in 1975) depend upon a polymerization reaction that follows a free radical path. This means that the ratio of the two components is relatively unimportant, so careful control of quantities is unnecessary. In parallel with the development of this system, a technique was perfected for increasing the toughness of the cured adhesive by incorporating minute particles of rubber. The adhesive is in two parts: a viscous gel and a mobile liquid. These two are spread one on each side of the joint. When the two are brought together, they react quickly to give a strong bond, which is handleable in 2 to 3 minutes, with full working strength in 1 hour and ultimate strength in 24 hours. These adhesives not only give a strong bond of high toughness very quickly, they are also able to contend with oily surfaces. They provide an exceedingly satisfactory product that meets a number of requirements in advanced assembly, especially within the automobile industry.

4. A series of acrylic adhesives has been produced which are cured by irradiation with ultraviolet light. Clearly they can only be used where the radiation can reach the adhesive; for example, where one component is transparent to the UV wavelength. While a considerable range of these products has been developed, very little information has been released about their composition.

High-Temperature Adhesives

All the adhesives considered so far can only provide useful bonds up to very limited temperatures, commonly 100°C or perhaps 150°C. There are demands, mainly military, for bonds that can withstand up to 300°C. To meet these needs, some adhesive base polymers have been developed that are based on carbon and nitrogen ring systems with a limited service life at these high temperatures.

Pressure-Sensitive Adhesives

Pressure-sensitive adhesives (e.g., Scotch Tape, first sold in 1940) are totally different from any others. These adhesives depend on an exceedingly high-viscosity liquid that retains this state throughout its life and never cross-links or cures. The strength of the bond is dependent on the pressure applied to it as the bond is made. The useful life of pressure-sensitive adhesives is generally limited to perhaps one or two years.

KEITH W. ALLEN

Further Reading

Adams, R.D., Comyn, J. and Wake, W.C. *Structural Adhesive Joints in Engineering,* 2nd edn. Chapman & Hall, London, 1997.

Comyn, J. *Adhesion Science.* Royal Society of Chemistry, Cambridge, 1997.

Kinloch, A.J. *Adhesion and Adhesives: Science and Technology.* Chapman & Hall, London, 1987.

Packham, D.E. *Handbook of Adhesion,* Longman, Harlow, 1992.

Pocius, A.V. *Adhesion and Adhesives Technology: An Introduction,* 2nd edn. Carl Hanser, Munich, 2002.

Wake, W.C. *Adhesion and the Formulation of Adhesives,* 2nd edn. Applied Science Publishers, London, 1982.

Agriculture and Food

In late-twentieth century Western societies, food was available in abundance. Shops and supermarkets offered a wide choice in products and brands. The fast-food industry had outlets in every neighborhood and village. For those in search of something more exclusive, there were smart restaurants and classy catering services. People chose what they ate and drank with little awareness of the sources or processes involved as long as the food was tasty, nutritious, safe, and sufficient for everyone. These conditions have not always been met over the last century when food shortages caused by economic crises, drought, or armed conflicts and war, occurred in various places. During the second half of the twentieth century, food deficiency was a feature of countries outside the Western world, especially in Africa. The twentieth century also witnessed a different sort of food crisis in the form of a widespread concern over the quality and safety of food that mainly resulted from major changes in production processes, products, composition, or preferences. Technology plays a key role in both types of crises, as both cause and cure, and it is the character of technological development in food and agriculture that will be discussed. The first section examines the roots of technological developments of modern times. The second is an overview of three patterns of agricultural technology. The final two sections cover developments according to geographical differences.

Before we can assess technological developments in agriculture and food, we must define the terms and concepts. A very broad description of agriculture is the manipulation of plants and animals in a way that is functional to a wide range of societal needs. Manipulation hints at technology in a broad sense; covering knowledge, skills, and tools applied for production and consumption of (parts or extractions of) plants and animals. Societal needs include the basic human need for food. Many agricultural products are food products or end up as such. However, crops such as rubber or flax and animals raised for their skin are only a few examples of agricultural products that do not end up in the food chain. Conversely, not all food stems from agricultural production. Some food is collected directly from natural sources, like fish, and there are borderline cases such as beekeeping. Some food products and many food ingredients are artificially made through complicated biochemical processes. This relates to a narrow segment of technology, namely science-based food technology.

Both broad and narrow descriptions of agriculture are relevant to consider. In sugar production for example, from the cultivation of cane or beets to the extraction of sugar crystals, both traditional and science-based technologies are applied. Moreover, chemical research and development resulted in sugar replacements such as saccharin and aspartame. Consequently, a randomly chosen soft drink might consist of only water, artificial sweeteners, artificial colorings and flavorings, and although no agriculture is needed to produce such products, there is still a relationship to it. One can imagine that a structural replacement of sugar by artificial sweeteners will affect world sugar prices and therewith the income of cane and beet sugar producers. Such global food chains exemplify the complex nature of technological development in food and agriculture.

The Roots of Technological Development

Science-based technologies were exceptional in agriculture until the mid-nineteenth century. Innovations in agriculture were developed and applied by the people cultivating the land, and the innovations related to the interaction between crops, soils, and cattle. Such innovation is exemplified by farmers in Northern Europe who confronted particular difficulties caused by the climate. Low temperatures meant slow decomposition of organic material, and the short growing season meant a limited production of organic material to be decomposed. Both factors resulted in slow recuperation of the soil's natural fertility after exploitation. The short growing season also meant that farmers had to produce enough for the entire year in less than a year. Farmers therefore developed systems in which cattle and other livestock played a pivotal role as manure producers for fertilizer. Changes in the feed crop could allow an increase in livestock, which produced more manure to be used for fertilizing the arable land, resulting in higher yields. Through the ages, farmers in Northern Europe intensified this cycle. From about the 1820s the purchase of external supplies increased the productivity of farming in the temperate zones. Technological improvements made increases in productivity not only possible but also attractive, as nearby markets grew and distant markets came within reach as a result of the nineteenth century transportation revolution.

An important development at mid-nineteenth century was the growing interest in applying science to agricultural development. The two disciplines with the largest impact were chemistry and biology. The name attached to agricultural chemistry is Justus von Liebig, a German chemist who in the 1840s formulated a theory on the processes underlying soil fertility and plant growth. He propagated his organic chemistry as the key to the application of the right type and amount of fertilizer. Liebig launched his ideas at a time when farmers were organizing themselves based on a common interest in cheap supplies. The synergy of these developments resulted in the creation of many laboratories for experimentation with these products, primarily fertilizers. During the second half of the nineteenth century, agricultural experiment stations were opened all over Europe and North America.

Sometime later, experimental biology became entangled with agriculture. Inspired by the ideas of the British naturalist Charles Darwin, biologists became interested in the reproduction and growth of agricultural crops and animals. Botany and, to a lesser extent, zoology became important disciplines at the experimental stations or provided reasons to create new research laboratories. Research into the reproductive systems of different species, investigating patterns of inheritance and growth of plant and animal species, and experimentation in cross-breeding and selection by farmers and scientists together lay the foundations of genetic modification techniques in the twentieth century.

By the turn of the century, about 600 agricultural experiment stations were spread around the Western world, often operating in conjunction with universities or agricultural schools. Moreover, technologies that were not specifically developed for agriculture and food had a clear impact on the sector. Large ocean-going steamships, telegraphy, railways, and refrigeration, reduced time and increased loads between farms and markets. Key trade routes brought supplies of grain and other products to Europe from North America and the British dominions, resulting in a severe economic crisis in the 1880s for European agriculture. Heat and power from steam engines industrialized food production by taking over farm activities like cheese making or by expanding and intensifying existing industrial production such as sugar extraction. The development of synthetic dyes made crop-based colorants redundant, strongly reducing or even eliminating cultivation of the herb madder or indigo plants. These developments formed the basis of major technological changes in agriculture and food through the twentieth century.

Patterns of Technology Development

The twentieth century brought an enormous amount of technology developed for and applied to agriculture. These developments may be examined by highlighting the patterns of technology in three areas—infrastructure, public sector, and commercial factory—as if they were seen in cross section. The patterns are based on combined material and institutional forces that shaped technology.

A major development related to infrastructure concerns mechanization and transport. The combustion engine had a significant effect on agriculture and food. Not only did tractors replace animal and manual labor, but trucks and buses also connected farmers, traders, and markets. The development of cooling technology increased storage life and the distribution range for fresh products. Developments in packaging in general were very important. It was said that World War I would have been impossible without canned food. Storage and packaging is closely related to hygiene. Knowledge about sources and causes of decay and contamination initiated new methods of safe handling of food, affecting products and trade as well as initiating other innovations. In the dairy sector, for example, expanding markets led to the growth and mergers of dairy factories. That changed the logistics of milk collection, resulting in the development of on-farm storage tanks. These were mostly introduced together with compression and tube systems for machine milking, which increased milking capacity and improved hygiene conditions. A different area of infrastructure development is related to water management. Over the twentieth century, technologies for irrigation and drainage had implications for improved "carrying capacity" of the land, allowing the use of heavy machinery. Improved drainage also meant greater water discharge, which in turn required wider ditches and canals. Water control also had implications for shipping and for supplies of drinking water that required contractual arrangements between farmers, governing bodies, and other agencies.

During the twentieth century, most governments supported their agricultural and food sectors. The overall interest in food security and food safety moved governments to invest in technologies that increased productivity and maintained or improved quality. Public education and extension services informed farmers about the latest methods and techniques. Governments also became directly involved in technological development, most notably crop improvement. Seed is a difficult product to exploit commercially. Farmers can easily put aside part of the harvest as seed for the next season. Public institutes for plant breeding were set up to improve food crops—primarily wheat, rice, and maize—and governments looked for ways to attract private investment in this area. Regulatory and control mechanisms were introduced to protect commercial seed production, multiplication, and trade. Private companies in turn looked for methods to make seed reproduction less attractive to farmers, and they were successful in the case of so-called hybrid maize. The genetic make-up of hybrid maize is such that seeds give very high yields in the first year but much less in the following years. To maintain productivity levels, farmers have to purchase new seed every season. Developments in genetic engineering increased the options for companies to commercially exploit seed production.

Most private companies that became involved in genetic engineering and plant breeding over the last three decades of the twentieth century started as chemical companies. Genetic engineering allowed for commercially attractive combinations of crops and chemicals. A classic example is the herbicide Roundup, developed by the chemical company Monsanto. Several crops, most prominently soy, are made resistant to the powerful chemical. Buying the resistant seed in combination with the chemical makes weed control an easy job for farmers. This type of commercial development of chemical technologies and products dominated the agricultural and food sector over the twentieth century. Artificially made nitrogen fertilizers are one such development that had a worldwide impact. In 1908, Fritz Haber, chemist at the Technische Hochschule in Karlsruhe, fixed nitrogen to hydrogen under high pressure in a laboratory setting. To exploit the process, Haber needed equipment and knowledge to deal with high pressures in a factory setting, and he approached the chemical company BASF. Haber and BASF engineer Carl Bosch built a crude version of a reactor, further developed by a range of specialists BASF assigned to the project. The result was a range of nitrogen fertilizer products made in a capital and knowledge-intensive factory environment. This type of development was also applied to creating chemicals such as DDT for control of various pests (dichloro-diphenyl-trichloroethane), developed in 1939 by Geigy researcher Paul Müller and his team. DDT may exemplify the reverse side of the generally positive large-scale application of chemicals in agricultural production—the unpredictable and detrimental effects on the environment and human health.

The commercial factory setting for technology development was omnipresent in the food sector. The combination of knowledge of chemical processes and mechanical engineering determined the introduction of entirely new products: artificial flavorings, products, and brands of products based on particular food combinations, or new processes such as drying and freezing, and storing and packaging methods.

Patterns of Technology Development in the Western World

Technological developments in agriculture and food differ with regard to geography and diverging social and economic factors. In regions with large stretches of relatively flat lands, where soil conditions are rather similar and population is low, a rise in productivity is best realized by technologies that work on the economies of scale. The introduction of mechanical technologies was most intensive in regions with these characteristics. Beginning early in the twentieth century, widespread mechanization was a common feature of Western agriculture, but it took different forms. In the Netherlands, for example, average farm size was relatively small and labor was not particularly scarce. Consequently, the use of tractors was limited for the first half of the twentieth century as emphasis was placed on improved cultivation methods. Tractors became widely used only after the 1950s when equipment became lighter and more cost-effective and labor costs rose sharply. The result was an overall increase of farm size in these regions as well. The Dutch government changed the countryside with a land policy of connecting and merging individual parcels as much as possible. This huge operation created favorable conditions for expansion; but where the land was already under cultivation, the only way to expand was to buy up neighboring farms. The effect was a considerable reduction in the number of farm units. An exception to this process was the Dutch greenhouse sector, in which improvements in construction, climate regulation, and introduction of hydroponic cultivation, increased production without considerable growth of land per farm unit.

The Dutch greenhouse sector is also an exemplary case of technological support in decision making and farm management. In Western countries a vast service sector emerged around agriculture and food. This process in fact started early in the twentieth century with the rise of extension services, set up as government agencies or private companies. Experimental methods based on multivariate statistics, developed by the British mathematician Karl Fisher, are the major tool in turning results of field experiments into general advisories. In keeping with the development of modern computers, digital models of crop growth and farming systems became more effective. Computer programs help farmers perform certain actions and monitor other equipment and machinery; yet even in the most technologically advanced greenhouses, the skilled eye of the farmer is a factor that makes a considerable difference in the quality and quantity of the final product.

The means by which agriculture in the West raised productivity have been questioned. Doubts about the safety of food products and worries over the restoration of nature's capacity became recurrent issues in public debate. Moreover, technological advances in tandem with subsidies resulted in overproduction, confronting national and international governing bodies with problems in trade and distribution, and a public resistance against intensive agriculture, sometimes called agribusiness. Technology is neither good nor bad; much of the knowledge underlying technologies with a detrimental effect also helps detect polluting factors and health hazards. Although a substantial part of research and technological efforts are aimed at replacing and avoiding harmful factors, many such "clean" technologies are commercially less interesting to farmers and companies. Subsidies and other financial arrangements are again being used to steer technology development, this time in the direction of environmentally friendly and safe forms of production.

Patterns of Technology Development in Less Developed Countries

From the beginning of the twentieth century, scientific and technological developments in the agricultural and food sector were introduced to less developed countries either by Western colonizing powers or by other forms of global interaction. The search for improved farming methods and new technology were mostly institutionalized at existing botanical gardens and established in previous centuries. Plant transfer and economic botany were a major modality of twentieth century technological improvement in less developed countries.

The early decades of the century featured an emphasis on technological improvement for plantation agriculture. Plantation owners invested in scientific research for agriculture, often supported by colonial administrations. The gradual abolition of slavery during the nineteenth century, increasing labor costs, was a reason to invest in technology. Other factors were more specific to particular sectors; for example, the rise of European beet sugar production encouraging cane sugar manufacturers to invest in technological improvement. Another example was the emergence of the automobile industry, which initiated a boom in rubber production.

Most colonial administrations launched programs, based on the combination of botanical and chemical research, to improve food crop production in the first decades of the twentieth century. It was recognized that dispersion of new technologies to a small number of plantation owners was different from initiating change among a vast group of local food crop producers. The major differences concerned the ecology of farming (crop patterns and soil conditions) and the socioeconomic conditions (organization of labor or available capital). Agronomists had to be familiar with local farming systems, occasionally resulting in pleas for a technology transfer that would better meet the complexity of local production. The overall approach, however, was an emphasis on improvement of fertilization and crop varieties. Transfer of the Western model gained momentum in the decades after World War II. Food shortages in the immediate postwar years encouraged European colonial powers to open up large tropical areas for mechanized farming. Unfortunately, the result was largely either a short-lived disaster, as in the case of the British-run groundnut scheme in Tanzania, or a more enduring problem, as in case of the Dutch-run mechanized rice-farming schemes in Surinam. The 1940s also saw the beginnings of a movement that came to be known as the "green revolution." Driven by the idea that hunger is a breeding ground for communism, American agencies initiated a research program for crop improvement, primarily by breeding fertilizer-responsive varieties of wheat and rice. Agencies were put together in a Consultative Group on International Agricultural Research (CGIAR). Technological progress was realized by bringing together experts and plant material from various parts of the world. Modified breeding techniques and a wide availability of parent material resulted in high-yielding varieties of wheat and rice. Encouraged by lucrative credit facilities, farmers, especially in Asia, quickly adopted the new varieties and the required chemicals for fertilization and pest control. Research on the adoption process of these varieties made clear that many farmers modified the seed technology based on specific conditions of the farming systems. In areas where such modifications could not be achieved—primarily rice growing regions in Africa—green revolution varieties were not very successful. Based on these findings, CGIAR researchers began to readdress issues of variation in ecology and farming systems. This type of research is very similar to that done by colonial experts several decades earlier. However, because of decolonization and anti-imperialist sentiments among Western nations, much of this earlier expertise has been neglected. This is just one of the opportunities for further research in the domain of agriculture and food technology.

See also **Biotechnology; Breeding: Animal, Genetic Methods; Breeding: Plant, Genetic Methods; Dairy Farming; Farming, Agricultural Methods; Fertilizers; Food Preservation; Irrigation; Pesticides; Transport, Foodstuffs**

HARRO MAAT

Further Reading

Baum, W.C. and Lejeune, M.L. *Partners Against Hunger: The Consultative Group on International Agricultural Research*. World Bank, Washington, 1986.

Beckett, J.V. *The Agricultural Revolution*. Blackwell, Oxford, 1990.

Bray, F. *The Rice Economies: Technology and Development in Asian Societies*. Blackwell, Oxford, 1986.

Brockway, L.H. *Science and Colonial Expansion: the Role of the British Royal Botanic Gardens*. Academic Press, New York, 1979.

Bonano, A., Busch L. and Friedland W.H., Eds. *From Columbus to ConAgra: the Globalization of Agriculture and Food*. University Press of Kansas, Lawrence, Kansas, 1994.

Busch, L., Ed. *Science and Agricultural Development*. Allenheld Osmun, New Jersey, 1981.

Cittadino, E. *Nature as the Laboratory: Darwinian Plant Ecology in the German Empire, 1800–1900*. Cambridge University Press, Cambridge, 1990.

Fitzgerald, D. *The Business of Breeding: Hybrid Corn in Illinois, 1890–1940*. Cornell University Press, London, 1990.

Gigerenzer, G. *The Empire of Chance: How Probability Changed Science and Everyday Life*. Cambridge University Press, Cambridge, 1989.

Hurt, R. D. and M.E. Hurt. *The History of Agricultural Science and Technology: An International Annotated Bibliography*. Garland, New York, 1994.

Headrick, D.R. *The Tentacles of Progress: Technology Transfer in the Age of Imperialism, 1850–1940*. Oxford University Press, Oxford, 1988.

Kimmelman, B.A. *A Progressive Era Disciple: Genetics at American Agricultural Colleges and Experiment Stations, 1900–1920*. UMI, Ann Arbor, MI, 1987.

Kiple, K.F. and Ornelas, K.C., Eds. *The Cambridge World History of Food*. Cambridge University Press, Cambridge, 2000.

Maat, H. *Science Cultivating Practice: A History of Agricultural Science in the Netherlands and its Colonies*. Kluwer, Dordrecht, 2001.

Perkens, J.H. *Geopolitics and the Green Revolution: Wheat, Genes and the Cold War*. Oxford University Press, New York, 1997.

Richards, P. *Indigenous Agricultural Revolution: Ecology and Food Production in West Africa*. Unwin Hyman, London, 1985.

Rossiter, M.W. *The Emergence of Agricultural Science: Justus Liebig and the Americans, 1840–1880*. Yale University Press, New Haven, 1975.

Slicher van Bath, B.H. *The Agrarian History of Western Europe A.D. 500–1850*. Edward Arnold, London, 1963.

Zanden, J.L. *The Transformation of European Agriculture in the Nineteenth Century: the Case of the Netherlands*. Free University Press, Amsterdam, 1994.

Air Conditioning

Air conditioning is a mechanical means of controlling the interior environment of buildings through the circulation, filtration, refrigeration and dehumidification of air. Although commonly thought of as a method of *cooling* interiors, it treats the interrelated factors of temperature, humidity, purity, and movement of air and is closely linked to developments in building electrification. The history of the modification of building air by human beings goes back millennia, and the major components of what we now call air conditioning, such as forced ventilation and mechanical refrigeration, were well developed by the nineteenth century. Fifty years before the term air conditioning was used to describe a mechanical system, the new Houses of Parliament in London were built to David Boswell Reid's comprehensive and controversial scheme for cooling, heating, purifying, humidifying, and circulating air. Outside air was drawn into a duct, passed through filtering sheets, heated by a hot water heater or cooled over blocks of ice, then drawn up through holes in the floor into the House of Commons. The used air was drawn up by the heat of a ventilating fire through raised panels in a glass ceiling and discharged.

The term "air conditioning" was introduced as a technical term in the twentieth century by an American textile engineer, Stuart W. Cramer. He used it to describe a process by which textile mills could be humidified and ventilated so as to "condition" the yarn produced there. Both his wall and ceiling units drew air into their casings, where it was pulled through a water spray and a cloth filter and discharged. Cramer's hygrometer measured the heat and relative humidity of the air and controlled the entire system automatically.

Cramer's contemporary Willis Carrier was among the most relentless researchers and promoters of "man-made weather," and while he is known as the father of air conditioning, the range of processes and products involved in air conditioning cannot be attributed to a single author. Carrier's "Apparatus for Treating Air" was only used to cleanse air that was already circulating through existing ventilation systems. Technologies that would culminate in room cooling were being developed in the refrigeration industry contemporaneously with Carrier's work in humidification. For centuries, ice and water had been manipulated to cool air that circulated in theaters, hospitals, factories, and other large public spaces. The New York Stock Exchange was air-cooled in 1904 using a toxic ammonia refrigerant. Air was channeled from roof level to the basement where it was filtered through layers of hanging cheesecloth and brine-chilled coils, and then blown into the building by a fan.

The development of industrial air conditioning, also called process air conditioning, dominated the newly created industry at the beginning of the twentieth century. Each air conditioning system was custom designed by engineers for the buildings into which they were installed. Human comfort was a byproduct of technologies aimed at improving industrial production. Comfort began to be promoted as a luxury in the 1920s when thousands of "evaporating–cooling stations" were installed in movie theaters where they relieved crowds of heat and smells. The J.L. Hudson Department Store in Detroit, Michigan, was the first commercial space to benefit from a "centrifugal chiller" installed by the Carrier Corporation in 1924.

Although Alfred Wolff had installed air conditioning systems in elite houses in the last decade of the nineteenth century, significant challenges remained in the twentieth century to the mass production of domestic room or central air conditioners: size, weight, cost, safety, and limitations on electrical service capacity for domestic buildings. One of the earliest room coolers, the Frigidaire Room Cooler of 1928, weighed 270 kilograms, required 2.7 kilograms of a toxic refrigerant, and was available in a single output size of 5500 Btu per hour. A room cooler of this size could only be (and often was) used in violation of electrical codes. Early room coolers cost in the thousands of dollars. It is not surprising that in 1930, General Electric could sell and install only thirty of the DH-5 models, the casing of which resembled a Victrola phonograph cabinet. Air conditioning began to be marketed as a comfort device for domestic consumption during the 1930s as manufacturers and mass production techniques helped democratize a product that was expensive, cumbersome, and custom designed and installed. The deprivation of the Great Depression followed by the post-World War II housing boom (particularly in the United States) facilitated the mass production and installation of domestic air conditioning across the class spectrum.

The development of chlorofluorocarbon gas (CFC or Freon) by Thomas Midgely in the 1920s, its manufacture by DuPont, and its successful application in the refrigeration industry galvanized the development of air conditioning as both a cooling device and mass-market product. Freon was considered the first nontoxic refrigerant and a replacement for sulfur dioxide, carbon dioxide, and ammonia. Together with development of the hermetically sealed motor compressor and lightweight finned coils it provided the foundation for air conditioning in its current form.

Air conditioners manufactured after the early 1930s, whether placed in a window or installed as part of the internal ducting of a building (central air conditioning), have included five basic mechanical components: compressor, fan (often two), condenser coil, evaporator coil, and chemical refrigerant. In central, or "split," systems there is a hot side, located outside the building, and a cold side, located inside. On the cold side, hot indoor air is drawn into the furnace and blown over an evaporator coil in which the refrigerant liquid is located. The refrigerant absorbs the heat of the air and evaporates, and the cooled air is then circulated throughout the building in internal ducts. The evaporated refrigerant is then pumped through a compressor and over condenser coils in contact with outside air. Once the heat of the refrigerant is transferred to the outside air, it liquefies and recirculates through an expansion valve and back into the cold side of the system. The hot air is expelled to the outside by a fan. A window air conditioner, which is powered by electricity and can weigh in the 25 kilograms range, also contains a hot and cold side, located on the outside and inside of the window, respectively. Other than the fact that the hot and cold sides of the central system are split and that the central system has a much higher capacity, these two systems function on essentially identical principles. Large buildings, taxed by extensive piping and larger capacities, often employ chilled water systems and cooling towers.

By 1970, over five million room air conditioners were being produced per year. By the end of the twentieth century, over 80 percent of single-family houses in the United States had air conditioning.

Since the 1970s, developments in air conditioning have focused on efficiency and environmental concerns. Freon was discovered to be destroying the ozone layer, and restrictions on its use and manufacture were imposed. It has been replaced in air conditioners by safer coolants such as hydrofluorocarbons and hydrochlorocarbons. A movement to replace thousands of Freon or CFC air conditioners with higher efficiency, non-CFC models was underway at the end of the twentieth century.

See also **Refrigeration, Mechanical**

JENNIFER COUSINEAU

Further Reading

Banham, R. *The Architecture of the Well Tempered Environment*. Architectural Press, London, 1969.

Bruegman, R. Central heating and forced ventilation: origins and effects on architectural design. *J. Soc. Archit. Hist.*, October 1978, 143–160.

Cooper, G. *Air-Conditioning America: Engineers and the Controlled Environment, 1900–1960*. Johns Hopkins University Press, Baltimore, 1998.

Elliot, C.D. *Technics and Architecture: The Development of Materials and Systems for Buildings*. MIT Press, Cambridge, MA, 1992.

Faust, F.H. The early development of self-contained and packaged air conditioners. *J. ASHRAE*, 1986, 353–369.

Fitch, J. M. and Bobenhausen, W. *American Building: The Environmental Forces That Shape It*. Oxford University Press, New York, 1999.

Ingels, M. *Father of Air Conditioning*. Country Life Press, Garden City, NY, 1952.

Air Traffic Control Systems

When a technology is fully assimilated, it is relatively invisible. Such is the way with air traffic control. It seems to go on all around us without our caring much about it, even if we fly.

Air traffic control began via the mail service. The airmail had to be delivered, and many new routes in strange places were pioneered by the need to fly mail. Mail pilots used writing on building roofs below to guide them during the day and various lights at night. However, most of the congestion was at the airports themselves. Controllers would be stationed in a tower or at the end of a runway and would use light signals (still the same ones used in a communications failure today) to guide landings or to hold aircraft for a break in the traffic.

When airplanes were rare, the threat of collision, even in cloudy weather, was low. As the number of commercial passenger airlines grew in the 1930s, a method of separating them from one another and a way of proceeding as safely in the clouds was necessary. Radio was not in widespread aerial use, and radar had not yet been invented. In this depression era, the U.S. government had little money, so it required the airlines to establish Air Traffic Control Units (ATCUs) to separate and space the traffic. Markers, called "shrimp boats," which could carry information about an airplane's intended altitude and air speed, represented an aircraft on a flight plan. The controllers moved the shrimp boats on maps following the approximate motion of the planes. In 1937, the federal government had enough money to incorporate the ATCUs. Units were renamed Air Traffic Control Stations (ATCS) and most controllers became civil service employees.

Cleveland had the first radio installed in a tower in the late 1930s, but there were few other developments until after World War II. The U.S. became crisscrossed by very high frequency (VHF) omnidirectional range transmitters (VORs), which were used for navigation. The rest of the world, including the U.K., which had many former bomber bases, used radio beacons that were much less accurate. With the VORs, it was possible to proceed on an exact path in the sky.

It was not until 1956 that the first radar dedicated to air traffic control was installed in Indianapolis, Indiana. This development was slow in coming because radar used for interceptions was different from radar used for spacing air traffic. Controllers still used the shrimp boats, but the radar could tell if a flight was progressing as planned.

Finally, computers came to be used. They would store the flight plans and drive displays that showed the controllers the identification, altitude, and airspeed of airplanes. Each aircraft was required to carry a transponder that could broadcast a four-digit code to enable flights to be paired with their radar track. Some of these devices had nearly a quarter-century of service when they were retired. The airspace system founded with their help enabled separation of only 300 meters vertically and 8 kilometers horizontally. The rest of the world based their air traffic control on the American system, with some differences where there is intermittent radar coverage, like the 80-kilometer separation in South America.

In the modern air traffic system, preparation begins when either the pilot or the dispatch office of an airline files a flight plan with a preferred route, altitude, and airspeed. Before departing, the pilot calls Clearance Delivery to obtain a flight clearance, which may include changes in the route due to expected traffic or to weather problems, and which also includes a transponder identifier number. The pilot sets this number into the transponder, calls Ground Control for taxi clearance, and taxies the plane to the appropriate runway for take-off.

After arriving at the runway entrance and performing any last-minute checks, the pilot calls the tower air traffic controller. The controller tells the pilot when to enter the runway to avoid take-off and landing traffic and makes sure that there is enough separation to avoid invisible wake turbulence from airplanes that have already left. Cleared for take-off, the aircraft does its ground roll down the runway to rotation speed (the speed at which it can lift off), climbs, and retracts the landing gear.

Once the wheels have entered their wells, the tower tells the pilot to call Departure Control. This frequency was given as part of the clearance, which also gives an initial altitude. The departure controller has a flight strip for the plane printed out by a computer before the aircraft takes off. It contains the desired altitude. When traffic allows, the controller clears the plane to climb to that altitude.

Airplanes going west are at even-numbered altitudes, those going east at odd ones; thus the vertical separation is maintained by pilots' requests for appropriate altitudes. Horizontal separation is the controller's job, as well as monitoring vertical separation. For most of the flight, a Center controller is watching the airplane. The U.S. is divided into only 22 Centers, each with sectors assigned to controllers at displays with flight strips.

Approaching the destination airport, the process is essentially reversed. Within about 50 kilometers of the airport, the pilot speaks with Approach Control and is then handed off to the tower controller. When the controller gives the pilot the order "cleared to land," the field is the property of the pilot. At that point pilots can do anything they need to get their airplanes down. After landing, taxi instructions are given by Ground Control; at large airports, airplanes are transferred to a ramp controller for "parking."

Today's relatively close separation and ability to handle many thousands of flights make air traffic control one of the twentieth century's most ubiquitous technologies. It is also one of its most successful.

See also **Radar Aboard Aircraft; Civil Aircraft, Jet Driven; Civil Aircraft, Propeller Driven.**

JAMES E. TOMAYKO

Further Reading

National Research Council. *The Future of Air Traffic Control: Human Operators and Automation*, Wickens, C.D. and Parasuraman, R., Eds. National Academy Press, Washington D.C., 1998.

National Research Council. *Flight to the Future: Human Factors in Air Traffic Control*. Wickens, C.D., Mayor, A.S. and McGee J.P., Eds. National Academy Press, Washington D.C., 1997.

Nolan, M.S. *Fundamentals of Air Traffic Control*. Wadsworth, Belmont, CA, 1990.

Smolensky, M.W. and Stein, E.S., Eds. *Human Factors in Air Traffic Control*. Academic Press, San Diego, 1998.

Aircraft Carriers

Three nations built fleets of aircraft carriers—Britain, Japan and the United States—and each contributed to carrier design trends. Experiments began before World War I when, in November 1910, Eugene Ely flew a Curtiss biplane from a specially built forward deck of the cruiser *USS Birmingham* moored off Hampton Roads, Virginia. Two months later he accomplished the more difficult task of *landing* on a deck built over the stern of the cruiser *Pennsylvania*. Sandbags were used to anchor ropes stretched across the deck to help stop the airplane, which trailed a crude hook to catch the ropes.

Late in World War I, the British Royal Navy was the first to develop crude carriers when they adapted several merchant or naval vessels to carry seaplanes. The light cruiser *Furious* was converted in 1917 to carry eight aircraft, using a canted forward deck to allow them to take off; they then landed at sea or at nearby airstrips. By 1918 the ship's modified aft deck allowed separate take-offs and landings. *Furious* was thus the world's first true aircraft carrier. She was modified in the early 1920s to remove her bridge structure and extend the flying deck for nearly the length of the vessel, and she was rebuilt again in the 1930s. *Furious* and her two sister ships helped to pioneer the use of powered elevators (to raise or lower aircraft between the lower hangar and the landing and take-off deck) and of aircraft catapults, both of which would become standard carrier equipment.

In July 1917 the Royal Navy ordered the 10,000-ton *Hermes,* the world's first carrier designed as such. She set the design model of an "island" bridge on the right-hand side of the deck that was followed for years. In April 1942 she also became the first carrier lost to aircraft flown from another carrier. The 1938 *Ark Royal* was more than twice the size and was fitted with the enclosed "weather" bow and stern that became hallmarks of British carriers, but it lacked the radar already being employed by American vessels. The postwar 37,000-ton *Eagle* and a second *Ark Royal* saw extensive modification over time, including the British-developed angle-deck extension allowing for simultaneous take-off and landing operations. The three ships of the 16,000-ton *Invincible* class, commissioned in the early 1980s, were soon fitted with "ski jump" bows to assist jet fighter take-off from their relatively short decks.

Given its Pacific location, the Japanese Imperial Navy was an early convert to the importance of aircraft carriers. In 1922 a converted oiler became *Hosho,* the country's first carrier, designed with British help. The 1927 *Akagi* was the country's first large (30,000 ton) fleet carrier, converted from an unfinished cruiser. At the time the Americans were making similar conversions. As with American (but not British) carriers of the time, the flight decks of Japanese vessels were made of wood to save weight and allow rapid repair when damaged. The best Japanese carriers were the two *Shokakus* of 1941. Displacing 26,000 tons and carrying up to 84 aircraft, they were more heavily armored and armed than their predecessors. The mid-1942 battles of the Coral Sea and Midway were the first naval confrontations where combat was conducted solely by carrier-based airplanes. In 1944 the 30,000-ton *Taiho* was the first Japanese carrier to feature an armored deck as well as enclosed bows for better sea keeping, as was common with British carriers of the period. Paralleling British and American wartime practice, many smaller "escort" carriers were created using military or merchant ship hulls. Japan constructed no aircraft carriers after 1945.

America's carrier experience began in 1922 with the small (13,000 ton) *Langley,* converted from a naval collier. She featured a compressed-air catapult to launch aircraft. The *Saratoga* and *Lexington* were converted from battle cruiser hulls and displaced nearly 40,000 tons when launched in 1927. Both vessels featured hydraulic aircraft elevators. The *Ranger* of 1934 was the first keel-up American carrier, though she proved to be too small for practical application. The three-ship *Yorktown* class launched between 1937 and 1941 featured light wooden decks, open bow and stern structures, multiple airplane elevators, and the ability to carry nearly 100 aircraft. No matter their size, however, carriers never had sufficient space. This led to such expedients as parking aircraft with only their front wheels on deck, and their tails hanging over the sea, supported by

special extended bars. Biplane fighters were suspended from hangar ceilings to store them out of the way. More important was the folded wing, first developed by the British in the late 1930s, and adopted by the U.S. Navy early in 1941. Folding up one third or one half of each wing allowed many more aircraft on crowded decks and in hangar spaces.

American wartime carrier production featured the 24-ship *Essex* class of 27,000-ton vessels. The largest class of big carriers built by any nation, they could carry 91 aircraft. Ship antitorpedo protection was much improved, as were the elevators (more and better located) and anti-aircraft armament. Several of these ships were vastly modified in postwar years to take angle decks and jet aircraft. Completed only as the war ended, three vessels of the *Midway* class were, at more than 48,000 tons, the largest carriers to enter service until 1955. They were also the first American carriers to adopt the British-developed armored flight deck, which had saved several bombed British carriers during the war. They featured huge aviation gasoline storage capacity, large hangars, and enough space that they were substantially rebuilt to accommodate jet aircraft in the postwar years.

The *Forrestal* class of 62,000-ton carriers in the late 1950s became the basic model for all subsequent American vessels. Built for jet operations, the *Forrestals* featured an armored and canted deck to allow for simultaneous take-off and landing operations. Each of the four ships carried 90 aircraft and a crew of more than 4,000. Early in their service, anti-aircraft guns were removed and replaced with surface-to-air missile batteries that would figure on future carriers. Electronics and other features were updated continually. Four "improved" carriers of about the same size were added to the fleet in the 1960s with larger flight decks, improved electronics, and missile rather than gun defense.

The *Enterprise*, America's first atomic-powered carrier, entered service in 1961 with a range of 320,000 kilometers, or capable of four years' cruising. She was similar to the *Forrestal* carriers except for her small square island structure that originally featured "billboard" radar installations. Despite a huge cost increase (about 70 percent more than the *Forrestals*), she became the prototype for the ultimate *Nimitz* class of nuclear carriers that began to enter fleet service in the mid-1970s. Displacing nearly 95,000 tons, each had a crew of some 6,500 men. Driven by concerns about the growing expense of building and operating the huge American fleet carriers and their vulnerability, research into smaller carrier designs continued.

See also **Battleships**

CHRISTOPHER H. STERLING

Further Reading

Blundell, W.G.D. *British Aircraft Carriers.* Model & Allied Publications, Hemel Hempstead, 1969.

Chesneau, R. *Aircraft Carriers of the World, 1914 to the Present: An Illustrated Encyclopedia.* Naval Institute Press, Annapolis, MD, 1984.

Friedman, N. *U.S. Aircraft Carriers: An Illustrated Design History.* Naval Institute Press, Annapolis, MD, 1983.

Musciano, W.A. *Warbirds of the Sea: A History of Aircraft Carriers and Carrier-Based Aircraft.* Schiffer, Atglen, PA, 1994.

Polmar, N. *Aircraft Carriers: A Graphic History of Carrier Aviation and Its Influence on World Events.* Doubleday, Garden City, NY, 1969.

Terzibaschitsch. S. *Aircraft Carriers of the U.S. Navy,* 2nd edn. Naval Institute Press, Annapolis, MD, 1989.

Aircraft Design

No innovation is more distinctive of the twentieth century, or more influential on its life or imagination, than the airplane. While the root technologies were established in the nineteenth century, it was the Wright brothers in 1903 who first synthesized them into a machine capable of sustained, controlled flight. Heavier-than-air flight stretched the limits of engine power and structural strength per unit weight and gave great impetus to technology, with influence far beyond aeronautics. It also largely initiated the engineering science of aerodynamics, again with far-reaching implications. The scope of aircraft design through the twentieth century can be considered by viewing the developments of four types of aircraft:

1. Biplane and externally braced wings
2. Streamlined monoplanes
3. Transonic aircraft
4. Supersonic and hypersonic aircraft.

Biplane and Externally Braced Wings

For their first four decades, successful airplanes were almost exclusively powered by spark-ignition internal combustion piston engines. Early engines followed automotive practice but gave greater attention to weight reduction and utilized lightweight materials such as aluminum for the block. Take-off power was typically about 40 kW, and weight per unit power was about 3 kilograms per kilowatt (kg/kW). Both overhaul life and mean

time between in-flight failure were no more than a few hours. The theory of the piston engine was understood with reasonable clarity, but lack of experience and refinement in mechanical detail imposed severe limits.

The most distinctive feature of airplanes was the wing. Designers relied on avian models to frame their ideas for airfoil sections and on primitive wind tunnels, developed since the 1870s, to test them. A thin airfoil relative to its chord length was thought necessary for efficiency. To make a thin wing strong enough with the existing materials (generally wood, tension-braced with steel wire) external bracing was necessary. The Wrights joined two wings with a wood and wire truss structure in a biplane configuration, and most others followed them, although some monoplanes with external bracing were also seen as well as a few triplanes.

It had long been appreciated that an unstabilized wing would pitch down. This could be corrected by an aft-mounted horizontal tailplane surface rigged to provide negative lift, or a forward surface providing positive lift, the so-called "canard," as employed by the Wrights. Horizontal tailplanes were usually combined with a vertical stabilizer—a tail-aft configuration—because this lent itself to the most practical distribution of masses along the length of the aircraft. In particular, a tail-aft airplane could mount its engine forward for best cooling. A truss fuselage carried the tail at one end and the engine at the other, with the pilot accommodated amidships. Airfoil surfaces were covered with sewn and varnished fabric, and increasingly so was the fuselage. Other configurations were tried, but this one quickly became dominant for most applications because it offered the lowest weight and reasonable drag.

As the Wrights were the first to clearly recognize, control was crucial. Their scheme was adopted as the standard for control surfaces, but with modifications. Hinged rudders and elevators were fitted to fixed stabilizers, thus adding static stability while preserving control. Wing warping was replaced with simpler hinged ailerons. As aircraft grew from being purely sporting vehicles to a practical form of transport, designers moved from the Wright practice of negative static stability to positively stable designs that were able to fly steadily without constant control inputs.

By 1910 airplanes were typically capable of carrying one or two people at speeds of about 125 km/h for one hour, with applications chiefly for sport and military reconnaissance and observation. Their value in World War I led to considerable pressure for improvement. Engine outputs as great as 200 kilowatts allowed larger and more robust structures carrying loads of 400 kilograms or more at speeds of up to 200 km/h for several hours. Frameworks of welded steel tubing began to appear, and a few aircraft were sheathed in aluminum.

Before World War I, a novel type of airplane engine, the rotary, was developed. The cylinders were disposed radially, like spokes of a wheel, and the crankcase revolved with the propeller about a crankshaft fixed to the airframe. Both lubrication (using a once-through total-loss system) and cooling were thus improved, and rotaries were relatively lightweight (less than 2 kilograms per kilowatt) and reliable. The inertial forces of the whirling engine hindered application of more powerful rotaries, but this led to interest in the fixed radial configuration, with a stationary crankcase and rotating crankshaft. Once vibration problems were ameliorated, the radial became one of the two major airplane engine types later in the 1920s. The other was the water-cooled inline engine, often with two banks in a V8 or V12 configuration. By the end of the 1920s, outputs as great as 425 kW with weights as low as 1 kg/kW were becoming available.

Increasing engine power and a clearer appreciation of the demands of flight led to general advances in performance and to the beginnings of commercial air service in the 1920s. Externally braced monoplane wings became common, as did two or three engines for larger models. However, flight remained hazardous and limited to distances of less than 2,000 kilometers and speeds of less than 260 km/h, racers and stunts aside.

Although physicists had built an impressive body of theory about fluid flows in the eighteenth and nineteenth centuries, little of this was useful to early airplane designers. More relevant discoveries were quick in coming, particularly in Germany, under the leadership of Ludwig Prandtl. However, this work did not become generally known and accepted among designers until after World War I when it led to the adoption of thicker airfoil sections that provided better performance while allowing for stronger structures needing less drag-inducing bracing (Figure 1). However, overall flight efficiency, as measured by the ratio of lift to drag (L:D), improved little.

Achievements of the 1920s did not represent an ultimate limit on externally braced airplanes with fabric-covered tube structures, but development of the streamlined monoplane gradually led to the virtual extinction of the earlier aircraft. By the end

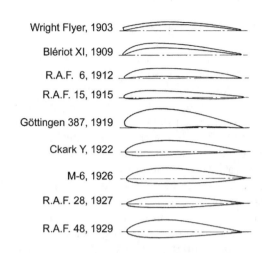

Figure 1. Evolution of airfoil sections, showing the trend to thicker airfoils as aerodynamic knowledge grew.

of the twentieth century such construction was limited to light sport and special purpose aircraft.

Streamlined Monoplanes

Before 1930, airplane development was little affected by science, and few designers knew much about the achievements of Ludwig Prandtl and his followers in understanding the physical mechanisms of lift and drag. Physics-based theories of structural strength, imported from architectural practice, found some use, but they were not readily applied to more elaborate aeronautical structures. In any event, the lack of knowledge of aerodynamic loads limited their use. The formula of the fabric-covered steel-tube structure with externally braced biplane or monoplane wings had become a comfortable one for many designers, and they saw the needs in terms of incremental improvement, not radical change.

As aerodynamicists drew on theory and experimental data to show a great gap between actual and potential performance, more and more designers became determined to close it. Particularly thought-provoking were findings that small struts and even bracing wires produced far more drag than had been supposed. Systematic scientific experimentation with wind tunnels, guided by the developing aerodynamic theory, stimulated many improvements in the late 1920s, of which cowlings that dramatically reduced the drag of radial engines while providing good cooling were among the most significant. Flaps and slats were devised to allow wing surfaces to be made smaller while still providing sufficient lift for take-off and landing at reasonable speeds.

Mechanisms were developed to permit undercarriages to be retracted.

Full implementation of aerodynamic improvements depended on structural advances. With new theories and extensive experimentation, designers learned to build airplane structures as smooth, integrally stiffened shells out of easily worked sheet aluminum fastened with rivets. They gained the knowledge and confidence to make wings of higher aspect ratio (the ratio of span to average chord length, from leading edge to trailing), without external bracing, thus improving lift efficiency. As speeds rose to 300 km/h and beyond, small details began to become quite significant. Engineers developed techniques of flush riveting in order to make even smoother structures.

These developments tended to make aircraft heavier, but weight was offset with power. By 1933, 800 kW was available with a weight about 0.8 kg/kW. By 1940, 1500 kW was available for 0.7 kg/kW. Radial and inline engines developed in parallel, with radials generally preferred for transports, bombers, and all naval aircraft, and inlines used for land-based fighters (with some exceptions in both directions). Pressurized cooling systems allowed reductions in radiator size and drag. Engine supercharging became widely used, either in mechanically driven or exhaust-turbine form, permitting aircraft to operate at higher altitudes where decreased atmospheric density reduced drag. In the U.S., development of high octane fuels permitted increased compression ratios.

The overall result was a remarkable transformation in the shape and performance of aircraft in less than a decade. Although much of the basic knowledge behind the transformation came from Germany, it was America that took the lead, reflecting favorable economic and geographic conditions and extensive government investment in practical research. British industry initially was somewhat slow to take up the new technologies, but it was quick to close the gap as rearmament began later in the 1930s.

The externally braced Ford AT-5 Trimotor transport of 1930, weighing 6100 kg and carrying 15 passengers at less than 215 km/h, had yielded by 1936 to the Douglas DC-3 of 11,400 kg, carrying 21 passengers in much greater comfort and capable of 370 km/h (although normal cruising speed was about 275 km/h). In Britain, fighters ranged from the 1930 Bristol Bulldog, with speeds reaching 280 km/h, to the 415 km/h Gloster Gladiator biplane that entered service as late as 1937, to the 585 km/h Supermarine Spitfire of 1939. If World War II had started in 1929, the planes that would have fought in it would have been only marginally superior to those of World War I. But by the time the war actually began in 1939, aircraft had moved far ahead and correspondingly played a much larger role.

Even more than in World War I, the pressures of conflict prompted great advances. By 1948, piston engine outputs had reached as high as 2600 kW, and some units could achieve 0.5 kg/kW. The North American P-51H fighter of 1945 could reach 780 km/h, while the 25,000 kg Lockheed Constellation airliner of 1946 carried 51 passengers at speeds up to 530 km/h, cruising at 440 km/h. The Constellation incorporated cabin pressurization, allowing comfortable operation at altitudes of 6000 m and more. Most of the performance gains of this period simply reflected increased engine power with incremental aerodynamic refinement.

The fundamental engineering science that underlay these achievements had all been in place by the mid-1930s. As early as the 1840s, physicists had formulated the exact equations for fluid flow (the Navier–Stokes equations). They were far too complex to be solved for any realistic cases, and even workable approximate solutions were very difficult to obtain. Efforts were made to analyze flow over airfoils with the assumption of incompressible, irrotational, and inviscid (friction-free) flow, but the results agreed poorly with measurements, especially for drag.

Prandtl and those who followed him showed that viscosity was in fact of crucial importance but that its effects were largely confined to a very thin layer of air, the "boundary layer," which normally lay immediately adjacent to the airplane's surface. Understanding of boundary layer phenomena was complicated by the existence of two quite different kinds of flow: smooth, relatively simple "laminar" flow and highly complex turbulent flow. Full understanding and prediction of turbulent flows remained elusive even at the end of the nineteenth century. Nevertheless, by making allowance for the boundary layer in their mathematics, aerodynamicists were able to reach approximate solutions to flow problems that were useful in practical cases, guiding designers to aerodynamic forms which could be refined with reasonable time and effort through experiments in wind tunnels.

The main problem was to gain lift by favorable distributions of air pressure and to avoid drag caused by adverse pressures. Pressure is constant for air at rest, but it is modified by the flow of air past the airplane in ways predicted by the approximate solutions of the Navier–Stokes equations. Lift is generated by excess pressure on the underside of the airfoil relative to that on top. Some drag comes from viscous friction, but much is a result of the excess pressure on the front of airfoils and other parts of the plane over that acting on the rear. Particularly to be avoided are situations in which the flow separates entirely from the surface, leaving a relative void filled by irregular eddies; the low pressures thus generated result in high drag. Intensive application of this knowledge, involving large-scale testing and analysis in specialized facilities, directly supported aircraft and engine designers.

By 1950, the streamlined metal propeller-driven monoplane with piston engine power had nearly reached its inherent technological limits. Its triumph had been magnificent but brief. Gas-turbine power plants had now appeared, bringing new opportunities and new challenges. The classic subsonic airplane, with piston or turboprop engines, continued to have important market niches at the end of the twentieth century but only in subsidiary roles.

Transonic Aircraft

The aerodynamic theory that had been so valuable in the 1930s depended on the important simplifying assumption that compression of the air by the aircraft could be ignored. This assumption of incompressible flow broke down by 1940, as speeds approached Mach 0.7. (The Mach number is the ratio of the speed of the flow of air to that of sound. All moderate pressure disturbances travel at Mach 1 = 1225 km/h = 340 meters per second under standard conditions at sea level.)

The aircraft itself need not approach Mach 1 very closely to encounter compressibility, since the air accelerates in flowing past the airplane. Any advance beyond the flight conditions achieved by the end of World War II in 1945 could come only by penetrating this new "transonic" region of flight.

Before this became possible, new means of propulsion were necessary. The speed of a propeller blade through the air is, of course, higher than that of the aircraft itself. Therefore propellers were the first thing to be affected as sonic speeds were approached, with consequent falloff of performance. As the 1930s advanced, engineers in several countries turned to a new application of an old idea: a gas turbine to generate a hot gas stream for propulsion by jet reaction using atmospheric air, rather than carrying oxidizer on board the aircraft

as in a rocket. These ideas were pursued in Britain and Nazi Germany, and late in World War II they were put into practice by both countries.

The stage was set for transonic flight if its problems could be overcome. With German aeronautical research and development prostrate following World War II, the lead passed principally to the U.S., Britain, and the Soviet Union, with the U.S. taking the most prominent role owing to its great resources. Quite rapid progress was made both in theoretical understanding of transonic flight (building to a significant extent on foundations laid down in Germany) and in the difficult problems of measurement. There were many important discoveries, but three deserve special mention. First was the use of wing sweepback. It is principally the component of airflow at right angles to the wing's leading edge that determines lift and drag, so sweep increases the wing's critical Mach number (the point at which its drag begins to rise very sharply due to compressibility). Second was the use of very thin sections with quite sharp leading edges, generally with little sweepback and low aspect ratio. Finally, area rules governed the overall design of transonic aircraft; that is, the aircraft is designed to ensure that the total cross section of all areas varies in as smooth a way as possible along its length, with no bumps or hollows, thus minimizing wave drag caused by formation of shock fronts. Through these and other means it was possible to raise the lift-to-drag ratios of large aircraft to 20:1 or more, offsetting the high fuel consumption of early jet engines and ultimately giving transonic aircraft unprecedented range–payload performance as engine efficiencies were improved.

High flight speeds increased the loads on the structures of transonic aircraft, as these vary according to the square of speed. The aerodynamic innovations posed significant additional challenges as they dictated greater deviations from structurally efficient forms. The problems were most notable in the wings, where depth had to be sacrificed and where sweepback increased effective structural aspect ratio. Designers met these challenges with structures employing complex internal geometries to optimize stiffness and strength. Highly stressed parts were frequently made of stainless steel or titanium, often in the form of massive forgings machined to net shape. These innovations raised the cost of aircraft construction significantly, but the advantages of transonic flight justified the expense for many applications.

The theoretical and experimental understanding of transonic aircraft aerodynamics grew to meet the need for loads data on which to base structural design. It became clear that practical aircraft structures could not be even approximately rigid under transonic flight and maneuver loads, fostering much greater development of the discipline of aeroelasticity—analysis of the deformation of the plane's structure under aerodynamic loads and of the effect on aerodynamics (and hence on loads) of the change in its shape.

With the ratio of flight speed to stalling speed now reaching 4:1 or more, control problems took on new complexity. Controls that were effective and necessary at low speeds could become entirely ineffective or even counter-effective at transonic speeds. The classic example is the aileron on a long, swept wing, which had to be supplemented or supplanted with spoilers or other controls. Manual control circuits were replaced with powered irreversible controls.

In several cases, problems—particularly those associated with control and aeroelastic issues—were first recognized through catastrophic accidents, some of them in airliners in service. In earlier days it had been assumed that aviation was quite dangerous. However, as aircraft became increasingly employed for transport on a large scale, tolerance for accidents declined sharply. By the 1960s, even the military services had come to reject the human and economic costs of high loss rates. This emphasis on safety as well as reliability combined with the rigors of transonic flight to change the design of the airplane from something accomplished by ten to 25 engineers over a period of a year into a massive engineering project involving hundreds and even thousands of specialists over a number of years.

Transonic aircraft reached relative maturity by the 1960s, and most of the development in the last third of the century was in the nature of incremental improvements. Three significant developments do deserve mention however. First was the replacement of direct manual control with computer-mediated electronic "fly-by-wire" controls. Second was the introduction of fiber-reinforced materials, especially those incorporating carbon fibers in a thermosetting plastic matrix. After a lengthy gestation, occasioned both by concern for proven safety and the complexities of designing for a quite different kind of material, carbon-fiber composites finally began to see service in quantity in the final decade of the century, bringing significant benefits of weight reduction and promises of possible future reductions in fabrication costs (balanced against higher material costs). The third significant development was increased computerization of the design process, ranging from paperless drawings to complex calculations of aerodynamic flows. This has not, as once hoped, gone far in cutting the time or cost of design, but it has permitted unprecedented thoroughness, resulting in a better and ultimately more economical aircraft.

At the end of the twentieth century, the jet-propelled transonic aircraft was the dominant type for the great majority of transport services and for a wide range of military applications. There was no immediate prospect that it might be supplanted by supersonic or hypersonic aircraft in most roles, suggesting that transonic aircraft would continue to see wide use and evolve into the twenty-first century.

Supersonic and Hypersonic Aircraft

As the airflow over an aircraft reaches and exceeds Mach 1, drag begins to rise very steeply. Most transonic aircraft lack propulsion power to push very far into the drag rise in level flight, but by 1950 it had been verified that transonic aircraft could exceed flight speeds of Mach 1 in a dive.

Fighter aircraft need excess power for maneuver, generally achieved by adding afterburning to the jet engine. The next step, begun early in the 1950s, was to refine the aerodynamics and controls of fighters to permit them to fly at supersonic (over Mach 1) speeds in afterburning. The most notable changes involved adoption of greater sweep angle, highly swept delta wings, or thin, low-aspect wings with sharp leading edges. By 1960, most fighters entering service were capable of exceeding Mach 2. This was possible only at high altitudes (limited by aerodynamic heating and forces) and for brief periods (limited by high afterburning fuel consumption), but the speed was tactically useful. These were transonic airplanes that were capable of supersonic sprints.

As transonic aircraft entered service in substantial numbers for military and commercial purposes in the 1950s, it was generally anticipated that they would soon be supplanted or at least widely supplemented by truly supersonic aircraft that normally flew at over Mach 1. However, by 2000, only one type of aircraft regularly spent more than half of its time aloft in supersonic flight, the Anglo–French Concorde airliner, and only about a dozen Concordes remained in service.

Obstacles to wider supersonic flight included weight, cost, and environmental impact. Theory and experiment quickly led to the conclusion that the best shape for supersonic flight was slender and arrow-like and that suitable slender aircraft could cruise supersonically with efficiency generally matching that of transonic aircraft. Slender airplanes were not inherently suited to the relatively low speeds needed for landing and take-off, however. Compromises and adaptations were necessary for controllable and efficient flight over a range of speeds that varied by 10:1 or more from maximum to stalling, leading to extra weight and expense. Moreover, supersonic flight presented even greater structural challenges than transonic flight, and this also brought cost and weight penalties. These arose in part from the high dynamic pressures involved in flight at very high speeds, but even more so from aerodynamic heating, representing the sum both of friction and of air compression in the supersonic flow. For sustained flight at more than Mach 2.5, aluminum loses too much strength due to heating to be used as a structural material unless it is cooled or protected. Steel or titanium may be used instead.

In aircraft, any increase in weight brings cascading problems. This is especially true for supersonic aircraft, which tend to be most attractive for long-range missions requiring large fuel loads. High weight allied with the need for special materials and structures pushed costs up for supersonic aircraft. Moreover the supersonic shock wave reaches disturbing and even destructive levels on the ground below the path of the supersonic plane even when it flies at altitudes of 20 kilometers or more. These problems combined to drastically slow acceptance of supersonic flight. Indeed, one supersonic type that did see successful service, the U.S. Lockheed SR-71 Mach 3+ strategic reconnaissance aircraft, was ultimately withdrawn from operations because its functions could be performed more economically by other means.

At over Mach 4, a series of changes in aerodynamic phenomena led to the application of the label "hypersonic." In principle, hypersonic flight presents attractive opportunities. In the 1960s there was a belief that supersonic aircraft might be supplemented relatively rapidly by hypersonic types. In practice, the problems of weight, cost, and environmental effects proved to be even more intractable. At hypersonic speed, heating is so intense that even steel and titanium lose strength. A number of research programs relating to hypersonic flight, stimulated in part by the demands of reentry from space, led to the accumulation of considerable knowledge of many of the issues. Progress on development of air-breathing propulsion systems was halting however, and several efforts aimed at construction of a prototype hypersonic aircraft collapsed owing to cost and technology issues. Thus at the end of the twentieth century, the promise of supersonic flight seemed just out of reach and that of hypersonic flight not yet clearly in view.

See also **Civil Aircraft, Jet-Driven; Civil Aircraft, Propeller-Driven; Civil Aircraft, Supersonic; Rocket Planes; Wright Flyers**

WILLIAM D. O'NEIL

Further Reading

Anderson, J.D. Jr. *A History of Aerodynamics and its Impact on Flying Machines.* Cambridge University Press, Cambridge, 1998.

Boot, R. *From Spitfire to Eurofighter: 45 Years of Combat Aircraft Design.* Airlife Publishing, Shrewsbury, 1990. A personal story of important designer of 1950s through 1970s illustrating design process and setting.

Boyne, W.J. *Technical Progress in World War I Aviation.* AIAA, Washington, 1978, 78–3002.

Boyne, W.J. and Donald S. Lopez, Eds. *The Jet Age: Forty Years of Jet Aviation.* National Air & Space Museum, Washington, 1979.

Brooks, P.W. *The Modern Airliner: Its Origins and Development.* Putnam, London, 1961.

Covert, E.E. Introduction, in *Thrust and Drag: Its Prediction and Verification,* Covert E.E., Ed. AIAA, Washington, 1985. Extended historical survey.

Dale, H. *Early Flying Machines.* British Library, London, 1992.

Fielding, J.P. *Introduction to Aircraft Design.* Cambridge University Press, Cambridge, 1999. Accessible modern survey requiring minimal technical preparation.

Foss, R.L and Blay, R. From Propellers to Jets in Fighter Design. *Lockheed Horizons,* 23, 2–24, 1987.

Gunston, B. *The Development of Pisto AeroEngines.* Patrick Stephens, Sparkford, 1993.

Howard, H.B. Aircraft Structures. *J. Aeronaut. Soc.,* 70, 54–66, 1966. Summary of development by a pioneer structural engineer.

Jarrett, P., Ed. *Aircraft of the Second World War: The Development of the Warplane, 1939–45.* Putnam, London, 1977.

Jarrett, P.P. *Biplane to Monoplane: Aircraft Development, 1919–39.* Putnam, London, 1997.

Kermode, A.C. *The Aeroplane Structure.* Pitman, London, 1964. Nontechnical survey.

Küchemann, D. *The Aerodynamic Design of Aircraft: A Detailed Introduction to the Current Aerodynamic Knowledge and Practical Guide to the Solution of Aircraft Design Problems.* Pergamon Press, Oxford, 1978. Authoritative survey looking ahead to supersonic and hypersonic aircraft.

Langley, M. *Metal Aircraft Construction: A Review for Aeronautical Engineers of the Modern International Practice in Metal Construction of Aircraft.* Pitman, London, 1937. Richly illustrated contemporary survey of metal aircraft development.

Loening, G.C. *Military Aeroplanes: An Explanatory Consideration of Their Characteristics, Performances, Construction, Maintenance, and Operation, for the Use of Aviators.* Best, Boston, 1917. Contemporary text summarizing state of art c. 1916. There are several editions.

Loftin, L.K. Jr. *Quest for Performance: The Evolution of Modern Aircraft*. NASA, Washington, (SP-486), 1985.

Miller, R. and Sawers D. *The Technical Development of Modern Aviation*. Praeger, New York, 1970.

Murphy, A.J. Materials in aircraft structures. *J. R. Aeronaut. Soc.*, 70, 114–119, Jan 1966. Historical survey by a pioneer.

Raymer, D.P. *Aircraft Design: A Conceptual Approach*, 3rd edn. AIAA, Washington, 1999. Comprehensive standard basic text with extensive bibliography.

Rhodes, T. History of the Airframe: Part 1. *Aerospace Engineering*, May 1989, 11–15.

Rhodes, T. History of the Airframe: Part 2. *Aerospace Engineering*, June 1989, 11–17.

Rhodes, T. History of the Airframe: Part 3. *Aerospace Engineering*, September 1989, 27–32.

Rhodes, T. History of the Airframe: Part 4. *Aerospace Engineering*, October 1989, 16–21.

Shenstone, B.S. Hindsight is always one hundred percent. *J. R. Aeronaut. Soc.*, 70 131–134, Jan 1966. Story of the streamlined monoplane revolution by an important designer of the period.

Shute, N. *Slide Rule: The Autobiography of an Engineer*. Heinemann, London, 1954. Personal story of important designer of the 1920s and 1930s.

Stinton, D. *The Anatomy of the Aeroplane*. Blackwell, Oxford, 1985. Another accessible modern survey with minimal technical demands.

Taylor, C.F. *Aircraft Propulsion: A Review of the Evolution of Aircraft Piston Engines*. National Air & Space Museum, Washington, 1971.

Warner, E.P. *Airplane Design: Aerodynamics*, 1st edn. McGraw-Hill, New York, 1927.

Warner, E.P. *Airplane Design: Performance*, 2nd edn. McGraw-Hill, New York, 1936. Two editions of a contemporary text illustrating the rapid progress between the late 1920s and 1930s.

Whitford, R. *Design for Air Combat*. Jane's, Coulsdon, Surrey, 1989. Deals extensively with aerodynamic issues of transonic and supersonic airplanes on a relatively nontechnical level; not restricted to fighters.

Whitford, R. *Fundamentals of Fighter Design*. U.K. Airlife Publishing, Shrewsbury, 2000. Surveys all aspects of fighter design from historical perspective on relatively nontechnical level; much of what is said applies to aircraft generally.

Aircraft Instrumentation

When Wilbur Wright took off from Kitty Hawk, North Carolina in December 1903, the only instruments on board the first successful powered heavier-than-air machine were a stopwatch and a revolution counter. The historical first flight began the quest to improve aircraft as machines to conquer the air, and instrumentation evolved accordingly.

Aircraft instrumentation can be divided into four categories: engine performance, communications, ground-based instruments, and air-based radar.

Engine Performance

Instruments to determine the performance of the plane were first to develop. The Wrights thought it was important to monitor engine revolutions to maintain engine performance. The stopwatch—which failed on the first flight—was intended to record the actual flight time for speed and endurance computations. Continued evolution of performance instrumentation record and report on the engine(s), on the airframe, and on the aircraft's attitude in the air. The first such instruments were timers and revolution counters (from the outset of heavier-than-air flight), airspeed indicators and compasses (from before World War I), as well as turn-and-bank indicators and artificial horizon instruments (during and after World War I). The six most important—and most prevalent—aircraft instruments are: altimeter (to gauge altitude), airspeed indicator (to gauge speed), the artificial horizon (to determine the attitude of the aircraft relative to the ground), the compass (for heading and direction finding), angle-of-attack indicator (to gauge the angle that the aircraft is climbing or diving), and the turn-and-bank indicator (to indicate the stability of the aircraft in a turn). Throughout their evolution, aircraft have included all of these in-plane instruments; the significant difference is that today most of these instruments are digital and computer enhanced rather than analog.

Communications Technology

Communications technology was the second system that developed in aircraft instrumentation. Initially too heavy for the primitive machines, wireless transmission sets (W/T or radio) were not installed in aircraft on a regular basis until well into World War I. The need was foreseen, but technological development had to catch up with requirements. When radios were installed, pilots could communicate with other planes and ground stations transmitting information back and forth. This had many important effects including military observation and coordination as well as increased safety in the air. Radio beams were also used introduced in the 1940s as directional beacons to pinpoint specific locations. They were used initially to guide planes to landing sites (the Lorenz system). Later, radio direction finding was adapted to guide bombers to distant targets. Both the Germans (*Knickebein* and *X-Gerät*) and the British (*Gee* and *Oboe*) used radio beams extensively to aid in targeting enemy cities during World War II. Radio direction finding is still used extensively to guide planes to their destinations through instrument flight rules (IRF).

Ground-Based Instruments

The third interconnected evolution in aircraft instrumentation was the ability of ground-based instruments to locate and identify aircraft. In the 1930s the British and Germans developed ground-based radio-direction finding (radar) to be able to locate aircraft from a distance. The British also added the component of IFF (identification, friend or foe) beacons in aircraft to be able to identify British Royal Air Force planes from their German Luftwaffe attackers. The "Battle of Britain" in 1940 was won by the British through the use of Fighter Command: an integrated system of ground-based radar, landline communications, a central command system, and fighter squadrons. Ground-based radar has evolved to the point where air traffic controllers the world over maintain close contact with all aircraft in order to ensure safety in the air. Ground-based radar, although not specifically an aircraft instrument, is important for the operation of large number of aircraft in the skies today.

Air-Based Radar

The final example of important aircraft instrumentation also made its debut in wartime. Air-based radar—radar instruments in aircraft—was first used by the British RAF Bomber Command to locate German cities at night. The British were able to find and bomb cities using airborne radar. To counter Bomber Command, the German Luftwaffe devised *Lichtenstein*, airborne radar used to find RAF bombers in the air. To this day, modern military air forces use airborne radar to locate and identify other aircraft. However, not all uses are so nefarious. Airborne radar has been adapted to not only "see" other planes in the air, but also to find weather fronts in dense cloud so that bad weather can be avoided. Airborne radar has made commercial aviation safer than ever.

Targeting Instruments

In the military sphere, one additional evolution is in targeting instruments. From the early iron gunsights and dead-reckoning bombsights, instruments have been developed for military aircraft to deliver weapons payloads with ever-increasing efficiency. During World War I, pilots had to rely on skill alone to deliver bombs and bullets against enemy targets. The first computerized gunsights

were developed during World War II, as well as the American Norden bombsight—the bombsight developed for the American "precision" bombing campaign against Germany. Heads-up-displays (or HUDs) in the cockpit have increased the accuracy of weapons delivery into the modern era with the addition of laser-guided munitions and "smart" bombs. The modern HUD and complementary bombsights can track a number of targets, which adds to the destructive efficiency of modern-day military aircraft.

During World War II Albert Rose, Paul K. Weimer and Harold B. Law at RCA developed a small experimental compact image orthicon camera system to guide GB-4 glide bombs to their target (see Figure 2). A television camera carried under the bomb's fuselage transmitted a radio signal that was picked up by an antenna on the controlling plane, before being displayed as an image on a cathode ray tube. The image was displayed to the bombardier, who could then send radio commands to correct the glide bomb's course.

Additional Instrumentation

Additional instrumentation has been added to aircraft at different times to record high altitude,

Figure 2. Guided missile cam, WWII.
[*Courtesy of the David Sarnoff Library.*]

high-speed flight, atmospheric conditions, and global navigation. In the modern era, pilots can rely on global positioning satellites (GPS) to maintain precise flight paths. However, even to this day, pilots of both fixed-wing and rotary-wing aircraft rely on simple instruments—sometimes disguised as complicated computer enhanced systems—such as the aircraft compass, engine(s) gauges, artificial horizon, turn-and-bank indicator, flight clock, radio and direction finding aids. The most important instruments that pilots possess, however, may still be instinct, common sense, and seat-of-the-pants daring.

See also **Fly-by-Wire Systems; Global Positioning System (GPS); Gyro-Compass and Inertial Guidance; Radar Aboard Aircraft; Radionavigation**
MIKE PAVELEC

Further Reading

Beij, K.H. *Air Speed Instruments.* Government Printing Office, Washington, D.C., 1932.

Dwiggins, D. *The Complete Book of Cockpits.* Tab Books, Blue Ridge Summit, PA, 1982.

Gracey, W. *Measurement of Aircraft Speed and Altitude.* Wiley, New York, 1981.

Hersey, M. *Aeronautic Instruments, General Classification of Instruments and Problems.* Government Printing Office, Washington, D.C., 1922.

Hersey, M. *Aeronautic Instruments, General Principles of Construction, Testing, and Use.* MIT Press, Cambridge, MA, 1922.

Hunt, F. *Aeronautic Instruments.* Government Printing Office, Washington, D.C., 1923.

Irvin, G. *Aircraft Instruments.* McGraw-Hill, New York, 1944.

Molloy, E. *Aeroplane Instruments.* Chemical Publishing, New York, 1940.

Nijboer, D. *Cockpit.* Boston Mills Press, Toronto, 1998.

Pallett, E.H.J. *Aircraft Instruments, Principles and Applications.* Pitman, London, 1981.

Paterson, J. *Aircraft Compass Characteristics.* National Advisory Committee for Aeronautics (NACA), Washington, D.C., 1936.

Aircraft Instruments, United States Naval Air Technical Training Command, Washington, D.C., 1954.

Alloys, Light and Ferrous

Although alloys have been part of industrial practice since antiquity (bronze, an alloy of copper and tin, has been in use for thousands of years), the systematic development of alloys dates to the middle of the nineteenth century. Due to improvements in techniques for chemical analysis and the rise of systematic testing of material properties, the basic theoretical principles of alloys were developed in the late 1800s.

Broadly speaking, the development and use of alloys in the twentieth century was essentially an extension of the discoveries made in the nineteenth century. Refinements in the purity of source materials, the systematic testing of different alloy combinations and methods of heat treatment, and improvements in manufacturing techniques led to significant improvements in material properties. However, no new fundamental principles that led to radical breakthroughs were discovered during the twentieth century.

During the twentieth century, steel was the dominant engineering material in the industrialized world due to its low cost and versatility. Much of that versatility is due to a class of steels known as alloy steels. By adding other metals to the basic mix of iron and carbon found in steel, the properties of alloy steels can be varied over a wide range. Alloy steels offer the potential of increased strength, hardness, and corrosion resistance as compared to plain carbon steel. The main limitation on the use of alloy steels was that they typically cost more than plain carbon steels, though that price differential declined over the course of the twentieth century.

Although steel was the dominant engineering material in the twentieth century, a number of other alloys developed during the twentieth century found widespread use in particular applications. Higher cost limited the use of specialty alloys to particular applications where their material properties were essential for engineering reasons. This entry covers the use of alloys in mechanical applications. Alloys used in electrical applications are discussed in the entry Alloys, Magnetic.

Definitions

An alloy is a mixture of two or more metallic elements or metallic and nonmetallic elements fused together or dissolving into one another when molten. The mixture is physical, and does not involve the formation of molecular bonds. Strictly speaking, steel is an alloy, since it is a mixture of iron and carbon, but is not normally referred to in that way. Rather, when one speaks of alloy steel, one is referring to steel (iron plus carbon) with other elements added to it.

The formal definition of alloy steel is a steel where the maximum range of alloying elements content exceeds one or more of the following limits: 1.6 percent manganese, 0.6 percent silicon, or 0.6 percent copper. In addition, alloy steels are recognized as containing specific (minimum or otherwise) quantities of aluminum, boron, chromium (up to 3.99 percent), cobalt, nickel, titanium, tungsten, vanadium, zirconium, or any other alloying element that is added in order to obtain a desired alloying effect.

Somewhat confusingly, a number of alloys that are commonly referred to as alloy steels actually contain no carbon at all. For example, maraging steel is a carbon-free alloy of iron and nickel, additionally alloyed with cobalt, molybdenum, titanium and some other elements.

Another commonly used industry term is "special" (or in the U.S. "specialty") steel. Most, though not all, special steels are alloy steels, and the two terms are often used interchangeably. Other industry terms refer to the properties of the steel rather than a specific material composition. For example, "high strength" steel refers to any steel that can withstand loads of over 1241 MPa, while "tool-and-die" steel refers to any steel hard enough to be used for cutting tools, stamping dies, or similar applications.

The names of nonsteel alloys are usually defined by the names of their primary constituent metals. For example, nickel–chromium alloy consists of a mix of approximately 80 percent nickel and 20 percent chromium. Titanium alloys are primarily titanium mixed with aluminum, vanadium, molybdenum, manganese, iron or chromium. However, some alloys are referred to by trade names that have become part of the standard engineering vocabulary. A good example is Invar, an alloy of 64 percent iron and 36 percent nickel. The name is a contraction of the word "invariable," reflecting Invar's very low rate of thermal expansion.

Alloys are useful for industrial purposes because they often possess properties that pure metals do not. For example, titanium alloys have yield strengths up to five times as high as pure titanium, yet are still very low in density. Even when alloys have the same properties as pure materials, alloy materials—particularly alloy steels—are often cheaper than a pure material for a given purpose.

The differences in properties between a pure material and its alloys are due to changes in atomic microstructure brought about by the mixture of two or more types of atoms. The addition of even small amounts of an alloying element can have a major impact on the arrangement of atoms in a material and their degree of orderly arrangement. In particular, alloying elements affect the way dislocations are formed within microstructures. These changes in microstructure lead to large-scale changes in the properties of the material, and often change the way a material responds to heat treatment.

It is important to note that the addition of alloying elements can have both positive and negative effects on the properties of a material from an engineering point of view. In the manufacture of alloys, it is often just as important to avoid or remove certain chemical elements as it is to add them. Careful control of the chemical composition of raw materials and various processing techniques are used to minimize the presence of undesirable elements.

Alloy Steel

The development of alloy steel has its origins in the crucible process, perfected by Benjamin Huntsman in England around 1740. By melting bar iron and carbon in clay pots and then pouring ingots, Huntsman created superior steel with carbon uniformly dispersed throughout the metal. Used for cutlery, die stamps and metal-cutting tools; crucible steel was the first specialty steel.

In 1868, Robert F. Mushet, the son of a Scottish ironmaster, found that the addition of finely powdered tungsten to crucible steel while it was melted made for much harder steel. Suitable for metal cutting tools that could operate at high speed, Mushet tungsten tool steel was the first commercial alloy steel. The English metallurgist and steelmaker Sir Robert Hadfeld is generally considered to be the founder of modern alloy steel practice, with his invention of manganese steel in 1882. This steel, containing 12 percent manganese, has the property of becoming harder as it is worked. This made it ideal for certain types of machinery, such as digging equipment. Hadfeld also invented silicon steel, which has electrical properties that make it useful for building transformers. His work showed conclusively that the controlled addition of alloying elements to steel could lead to significant new specialty products. Hadfeld's discoveries, which were well publicized, led many other engineers and steelmakers to experiment with the use of alloying elements, and the period between about 1890 and 1930 was a very active one for the development of new alloys.

The first highly systematic investigation of alloy steels was carried out by Frederick W. Taylor and Maunsel White at the Bethlehem Steel Works in the 1890s. In addition to testing various alloy compositions, the two men also compared the impact of different types of heat treatment. The experiments they conducted led to the development of high-speed steel, an alloy steel where tungsten and chromium are the major alloying elements, along with molybdenum, vanadium and cobalt in varying amounts. These steels allowed the development of metal cutting tools that could operate at speeds three times faster than previous tools. The primary application of high-speed steel during the twentieth century was for the manufacture of drill bits.

Military applications were also a major factor in the development of alloy steels. The demand for better armor plate, stronger gun barrels, and harder shells capable of penetrating armor led to the establishment of research laboratories at many leading steel firms. This played a significant role in the development of the science of metallurgy, with major firms like Vickers in the U.K. and Krupps in Germany funding metallurgical research. The most notable discovery that came out of this work was the use of nickel as an alloying element. Nickel in quantities between 0.5 and 5.0 percent increases the toughness of steel, especially when alloyed with chromium and molybdenum. Nickel also slows the hardening process and so allows larger sections to be heat-treated successfully.

The young science of metallurgy gradually began to play a greater role in nonmilitary fields, most notably in automotive engineering. Vanadium steel, independently discovered by the metallurgists Kent Smith and John Oliver Arnold of the U.K. and Léon Guillet of France just after the beginning of the twentieth century, allowed the construction of lighter car frames. Research showed that the addition of as little as 0.2 percent vanadium considerably increased the steel's resistance to dynamic stress, crucial for car components subject to the shocks caused by bad roads. By 1905, British and French automobile manufacturers were using vanadium steel in their products. More significantly, Henry Ford learned of the properties of vanadium from Kent Smith and used vanadium alloy steel in the construction of the Model T. Vanadium steel was cheaper than other steels with equivalent properties, and could be easily heat-treated and machined. As a result, roughly 50 percent of all the steel used in the original Model T was vanadium alloy. As the price of vanadium increased after World War I, Ford and other automobile manufacturers replaced it with other alloys, but vanadium had established the precedent of using alloy steel. By 1923, for example, the automobile industry consumed over 90 percent of the alloy steel output of the U.S., and the average passenger car used some 320 kilograms of alloy steel.

The extensive use of alloy steels by the automobile industry led to the establishment of standards for steel composition. First developed by the Society of Automotive Engineers (SAE) in 1911 and refined over the following decade, these standards for the description of steel were widely adopted and used industry-wide by the 1920s, and continued to be used for the rest of the century. The system imposes a numerical code, where the initial numbers described the alloy composition of the steel and the final numbers the percentage of carbon in the steel. The specifications also described the physical properties that could be expected from the steel, and so made the specification and use of alloys steels much easier for steel consumers.

One of the goals of automotive engineers in the 1910s and 1920s was the development of so-called "universal" alloy steel, by which they meant a steel that would have broad applications for engineering purposes. While no one alloy steel could serve all needs, the search for a universal steel led to the widespread adoption of steel alloyed with chromium and molybdenum, or "chrome–moly" steel. This alloy combines high strength, toughness, and is relatively easy to machine and stamp, making it the default choice for many applications.

The final major class of alloy steel to be discovered was stainless steel. The invention of stainless steel is claimed for some ten different candidates in both Europe and the U.S. in the years around 1910. These various individuals all found that high levels of chromium (12 percent or more) gave exceptional levels of corrosion resistance. The terms "stainless" is a bit of an exaggeration—stainless steel alloys will corrode under extreme conditions, though at a far slower rate than other steels. It is this resistance to corrosion, combined with strength and toughness, that made stainless steels so commercially important in the twentieth century. The first commercial stainless steels were being sold by 1914 for use in cutlery and turbine blades, and by the 1920s the material was commonly used in the chemical industry for reactor vessels and piping. Stainless steel later found widespread application in the food processing industry, particularly in dairy processing and beer making. By the end of the twentieth century, stainless steel was the most widely produced alloy steel.

After the 1920s, the development of alloys steels was largely a matter of refinement rather than of significant new discoveries. Systematic experimentation led to changes in the mix of various alloys and the substitution of one alloy for another over time. The most significant factor has been the cost and availability of alloying elements, some of which are available in limited quantities from only a few locations. For example, wartime shortages of particular elements put pressure on researchers to develop alternatives. During World War II, metallurgists found that the addition of very small amounts of boron (as little as 0.0005 percent) allowed the reduction of other alloying elements by as much as half in a variety of low- and medium-carbon steels. This started a trend that continued after the war of attempts to minimize the use of alloying elements for cost reasons and to more exactly regulate heat treatment to produce more consistent results.

The manufacture of alloy steels changed significantly over the period 1900–1925. The widespread introduction of electrical steel making replaced the use of crucible furnaces for alloy steel processing. Electrical furnaces increased the scale of alloy steel manufacture, and allowed the easy addition of alloying elements during the steel melt. As a result, steel produced electrically had a uniform composition and could be easily tailored to specific requirements. In particular, electric steel-making made the mass production of stainless steel possible, and the material became cheap enough in the interwar period that it could be used for large-scale applications like the production of railway cars and the cladding of the Chrysler and Empire State skyscrapers in New York.

A major refinement in steel manufacture, vacuum degassing, was introduced in the 1950s and became widespread by the 1970s. By subjecting molten steel to a strong vacuum, undesirable gases and volatile elements could be removed from the steel. This improved the quality of alloy steel, or alternatively allowed lower levels of alloy materials for the same physical properties.

As a result of manufacturing innovations, alloy steel gradually became cheaper and more widely used over the twentieth century. As early as the 1960s, the distinction between bulk and special steel became blurred, since bulk steels were being produced to more rigid standards and specialty steels were being produced in larger quantities. By the end of the twentieth century, nearly half of all steel production consisted of special steels.

Other Alloys

A variety of nonsteel alloy materials were developed during the twentieth century for particular engineering applications. The most commercially significant of these were nickel alloys and titanium alloys. Nickel alloys, particularly nickel–chromium alloys, are particularly useful in high temperature applications. Titanium alloys are light in weight and very strong, making them useful for aviation and space applications. The application of both materials was constrained largely by cost, and in the case of titanium, processing difficulties.

Nickel–chromium alloy was significant in the development of the gas turbine engine in the 1930s. This alloy—roughly 80 percent nickel and 20 percent chromium—resists oxidation, maintains strength at high temperatures, and resists fatigue, particularly from embrittlement. It was later found that the addition of small amounts of aluminum and titanium added strength through precipitation hardening. The primary application of these alloys later in the twentieth century was in heating elements and exhaust components such as exhaust valves and diesel glow-plugs, as well as in turbine blades in gas turbines.

Pure titanium is about as strong as steel yet nearly 50 percent lighter. When alloyed, its strength is dramatically increased, making it particularly suitable for applications where weight is critical. Titanium was discovered by the Reverend William Gregor of Cornwall, U.K., in 1791. However, the pure elemental metal was not made until 1910 by New Zealand born American metallurgist Matthew A. Hunter. The metal remained a laboratory curiosity until 1946, when William Justin Kroll of Luxembourg showed that titanium could be produced commercially by reducing titanium tetrachloride (TiCl4) with magnesium. Titanium metal production through the end of the twentieth century was based on this method.

After the World War II, U.S. Air Force studies concluded that titanium-based alloys were of potentially great importance. The emerging need for higher strength-to-weight ratios in jet aircraft structures and engines could not be satisfied efficiently by either steel or aluminum. As a result, the American government subsidized the development of the titanium industry. Once military needs were satisfied, the ready availability of the metal gave rise to opportunities in other industries, most notably chemical processing, medicine, and power generation.

Titanium's strength-to-weight ratio and resistance to most forms of corrosion were the primary incentives for utilizing titanium in industry, replacing stainless steels, copper alloys, and other metals. The main alloy used in the aerospace industry was Titanium 6.4. It is composed of 90 percent titanium, 6 percent aluminum and 4 percent vanadium. Titanium 6.4 was developed in the 1950s and is known as aircraft-grade titanium. Aircraft-grade titanium has a tensile strength of up to 1030 MPa and a Brinell hardness value of 330. But the low ductility of 6.4's made it difficult to draw into tubing, so a leaner alloy called 3-2.5 (3 percent aluminum, 2.5 percent vanadium, 94.5 percent titanium) was created, which could be processed by special tube-making machinery. As a result, virtually all the titanium tubing in aircraft and aerospace consists of 3-2.5 alloy. Its use spread in the 1970s to sports products such as golf shafts, and in the 1980s to wheelchairs, ski poles, pool cues, bicycle frames, and tennis rackets.

Titanium is expensive, but not because it is rare. In fact, it is the fourth most abundant structural metallic element in the earth's crust after aluminum, iron, and magnesium. High refining costs, high tooling costs, and the need to provide an oxygen-free atmosphere for heat-treating and annealing explain why titanium has historically been much more expensive than other structural metals.

As a result of these high costs, titanium has historically been used in applications were its low weight justified the extra expense. At the end of the twentieth century the aerospace industry continued to be the primary consumer of titanium alloys. For example, one Boeing 747 uses over 43,000 kg of titanium.

See also Alloys, Magnetic; Iron and Steel Manufacture

MARK CLARK

Further Reading

Freytag D.A. and Burton L.A., Eds. *The History, Making, and Modeling of Steel*. William K. Walthers, Milwaukee, 1996.
Misa, T.J. *A Nation of Steel: The Making of Modern America, 1865—1925*. Johns Hopkins University Press, Baltimore, 1995.
Seely B., Ed. *Iron and Steel in the Twentieth Century*. Facts on File, 1994
Smith, W.F. *Structure and Properties of Engineering Alloys*. McGraw-Hill, New York, 1993.
Williams, J.C. *et al. Titanium*. Springer-Verlag, 2003.

Alloys, Magnetic

The development and application of magnetic alloys in the twentieth century was driven largely by changes in the electronics industry. Magnetic alloys are used in three major applications. First, permanent magnets are required for devices like electric motors and generators, loudspeakers, and television tubes that use constant magnetic fields. Second, electrical steels are used to make electromagnets, solenoids, transformers and other devices

where changing magnetic fields are involved. Finally, they are used in magnetic storage media, which require magnetic materials that retain the impression of external magnetic fields.

The most important types of magnetic materials are the ferromagnets, of which the most commonly used are iron, cobalt, and nickel. Ferromagnets have high magnetic ability, which allows high magnetic inductions to be created using magnetic fields. They also retain magnetism so they can be used as a source of field in electric motors or for recording information. Ferromagnets are used in the manufacture of electrical steels and magnetic media.

In the early twentieth century, the most commonly used magnetic materials were steel alloys containing tungsten or chromium. Chromium steel came to dominate the market due to lower cost. In 1917, Honda and Takai found that the addition of cobalt doubled the coercivity of chromium steel. Cobalt–chromium steels were commercialized in the early 1920s. The first major permanent magnet alloy, Alnico, was discovered in 1930 by Mishima and commercially introduced in the late 1930s. An alloy of steel, aluminum, nickel, and cobalt, Alnico has magnetic properties roughly eight times better than chromium–cobalt steels.

Introduced in the 1950s, ferrites—ceramic ferromagnetic materials—are a class of magnets made from a mixture of iron oxide with other oxides such as nickel or zinc. Ferrites have greatly increased resistivity because they are oxides rather than alloys of metals. They are also very hard, which is useful in applications where wear is a factor, such as magnetic recorder heads. Unlike bulk metals, ceramic magnets can be molded directly. Although not as strong on a unit weight basis as Alnico, ferrite magnets are much cheaper, and account for the vast majority of magnets used in industry in the late twentieth century—roughly 90 percent by weight in the 1990s, for example.

The strongest magnetic materials are the "rare-earth" magnets, produced using alloys containing the rare earth elements samarium and neodymium. Samarium–cobalt magnets were introduced in the early 1970s, but increases in the price of cobalt due to unrest in Zaire limited their use. In 1983, magnets based on a neodymium–iron–boron alloy were introduced. Neodymium is cheaper and more widely available than samarium, and neodymium–iron–boron magnets were the strongest magnetic materials available at the end of the twentieth century.

However, not all applications require the strongest magnetic field possible. Throughout the twentieth century, electromagnets and electromagnetic relays were constructed almost exclusively from soft iron. This material responds rapidly to magnetic fields and is easily saturated. It also has low remnance (a measure of how strong a remaining magnetic field is), so there is little residual field when the external magnetic field is removed.

For transformers, the material properties desired are similar but not identical to those for electromagnets. The primary additional property desired is low conductivity, which limits eddy current losses. First developed just after 1900, the primary material used for power transformers is thus a silicon–iron alloy, with silicon accounting for approximately 3 to 4 percent by weight. The alloy is heat-treated and worked to orient the grain structure to increase permeability in a preferred direction. Aluminum–iron alloy is also a suitable material for this application, although it is less used due to its higher cost.

Transformers for applications that involve audio and higher frequencies make use of nickel–iron alloys, with a nickel content of 30 percent or more. Common trade names for such alloys include Permalloy, Mumetal, and Supermalloy. These alloys were developed in the early twentieth century and were first manufactured in quantity for use as submarine cable shielding. The decline in cable production in the 1930s led to their use in transformers and related applications.

Magnetic recording was first developed by the Danish inventor Valdemar Poulsen at the beginning of the twentieth century and used for sound recording. The first material used for magnetic recording, solid steel wire or tape, was originally developed for other applications. For example, steel piano wire was used as the recording media for early wire recorders. However, the property that makes particular steel alloys suitable for magnetic recording, strong remnant magnetism, is associated with increased stiffness. Thus, a recording tape or wire made from magnetic alloy steel is highly resistant to bending. This creates difficulties in the design of a mechanism for moving the recording media past the recording head.

As a result, by the late 1930s most recording media were divided into two parts. The first part was a suitable substrate, such as brass wire or plastic tape that could be fed easily through a reel or cassette mechanism. The second part was a coating that had suitable magnetic properties for recording. By the late 1940s, it was clear that the cheapest and most easily used recording media for sound recording was plastic tape coated with particles of a type of iron oxide (gamma ferric oxide). This type of tape continued in use through the end of the twentieth century due to its low cost. During the 1970s, new tape particles of chromium dioxide and cobalt-doped ferric oxide were introduced because of their superior magnetic properties, but their higher cost meant that they were used only for more specialized audio recording applications.

Magnetic media based on ferric oxide particles were used for the recording of computer data beginning in the 1950s. Initial computer recording applications used coated plastic tape. Metal disks coated with iron oxide were introduced in the late 1950s for use in computer disk drives. In the 1990s, thin metal films of cobalt alloy largely replaced metal oxide as the recording media for hard disks.

In addition to ferromagnetic materials, two additional classes of magnetic alloys exist: paramagnets and diamagnets. Aside from use in the scientific study of magnetism, paramagnets have limited uses. One limited application is the production of very low temperatures. Paramagnetic salts can be cooled conventionally and then demagnetized, producing temperatures in the millikelvin range.

Superconducting materials are a subclass of diamagnets. When cooled to a sufficiently low temperature, superconducting materials experience a significant drop in resistance. Associated with this transition is the exclusion of magnetic flux from the conductor, with the flux moving to the surface of the conductor. These properties allow for the production of very high magnetic fields when using niobium–tin alloys as the conducting material. These materials were also used in the development of magnetic resonance imaging (MRI) devices, although in the 1990s superconducting magnets were being replaced in this application by neodymium–iron–boron permanent magnet systems.

See also **Audio Recording, Compact Disk; Computer Memory; Materials and Industrial Processes**

MARK CLARK

Further Reading

Daniel, E.D., Mee, C.D. and Clark M.H. *Magnetic Recording: The First 100 Years*. IEEE Press, New York, 1999.

Jiles, D. *Introduction to Magnetism and Magnetic Materials*. Chapman & Hall, London, 1991.

Livingston J.D. *Driving Force: The Natural Force of Magnets*. Harvard University Press, Cambridge, MA, 1996.

Sowter G.A.V. Soft magnetic materials for audio transformers: history, production, and applications. *J. Aud. Eng. Soc.*, 35, 10, October, 1987.

Analgesics

Analgesics are drugs that relieve pain selectively, without affecting consciousness or sensory perception. This is the important distinction between analgesics and anesthetics. Intensive research on the biochemistry of analgesia from the 1970s has shown that there are two main classes of analgesics. The first, opioids, combine chemically with molecular "receptors" in the brain to block pain impulses in the central nervous system. Nonsteroidal anti-inflammatory drugs (NSAIDs) alleviate pain by inhibiting the production of prostaglandins, hormone-like substances that cause local inflammation and pain at the site of an injury or infection. Opioid analgesics can be used for short-term or long-term relief of severe pain; NSAIDs are used to relieve moderate pain, such as headaches, superficial injuries, or muscle strain.

Opium, the dried, powdered sap of unripe poppy seedpods, has been used as an analgesic since antiquity. Over 20 different alkaloids are found in dry opium, of which the most important is morphine. The correct chemical structure of morphine was proposed in 1925 and confirmed by total synthesis in 1955. Morphine, administered by subcutaneous injection, effectively relieves pain, but there are various side-effects such as drowsiness, respiratory depression, nausea, and vomiting. The analgesic effect peaks at about 1 hour after administration and lasts 4–5 hours. Morphine users develop physical dependence on the drug and become tolerant to it. Increasing doses are needed to maintain its effectiveness, and serious withdrawal symptoms accompany the cessation of morphine medication. Codeine, another naturally occurring opium alkaloid, is a less potent analgesic than morphine, but it is also less addictive and produces less nausea than morphine. It can be administered orally and is often used in conjunction with aspirin. Meperidine (Demerol), a synthetic morphine analog identified in 1939, was originally thought to provide short-lasting analgesia without addiction, but this proved false and the drug was widely abused. It is still the most common drug used for pain relief in childbirth and has superseded morphine. The side effects of meperidine are similar to those of morphine but are less severe and are further reduced when given with the antihistamine drug promethazine. In prolonged use meperidine may cross the placental barrier, and the drug has been found in newborn infants. Heroin, introduced in 1898, was also falsely heralded as a nonaddictive alternative to morphine. It is about ten times more potent than morphine when administered intravenously and is still used clinically for its analgesic properties in some countries, though it is not generally available for therapeutic purposes due to its highly addictive propensities. Methadone, a synthetic opioid analgesic discovered in Germany during World War II, has long-lasting analgesic effects when taken orally, and, as it alleviates the euphoria-producing effects of heroin, it is used to control the withdrawal symptoms of heroin addiction. The opioid analgesics were formally called narcotic drugs as they induce sleep and cause physiological dependence and addiction. The term is less commonly used in medicine because many drugs other than opioids also show these effects.

In 1973 complex compounds called opioid receptors, which can combine with opioid molecules, were discovered in the brain, hypothalamus, and spinal cord. At least eight such substances are known, though only four are thought to be important to the central nervous system. The best understood is the μ receptor, which affects euphoria, respiratory depression, tolerance, and analgesia. The κ receptor is also involved in analgesia as well as diuresis, sedation, and physical dependence. Soon after discovery of these receptors, peptide-like compounds consisting of chains of amino-acid residues and showing opioid properties were found in the pituitary gland. Three groups of endogenous opioid peptides known as endorphins, enkephalins, and dynorphins were discovered around 1975. Found to be neurotransmitters, one of the most important is β-endorphin, consisting of a chain of 30 amino acid residues. It is synthesized, stored, and released from cells in the pituitary gland, and it can also be prepared synthetically. Injected intravenously, β-endorphin is three times more potent than morphine. Interest in these drugs intensified when two potent analgesic pentapeptides, each containing five linked amino acids, were found in extracts of pig brain. Named enkephalins, they are derived from endorphins. They can be prepared synthetically and injected intravenously to induce analgesia by combining with receptors in the brain in the manner of the opioids.

Several chemically unrelated groups of organic compounds also show mild analgesic properties. They include derivatives of salicylic acid, pyrazolone, and phenacetin. These nonopioid drugs are often self-prescribed, though continued use can lead to adverse effects including drug abuse and addiction, allergic reactions, gastrointestinal irritation, and fatal overdose. The oldest and most widely used nonopioid analgesic is acetyl salicylic acid, or aspirin, first developed and marketed in 1899 by the German chemical company, Bayer. In addition to its analgesic properties, aspirin reduces fever and inflammation by inhibiting the synthesis of prostaglandins. The irritant effect of large doses of aspirin on the stomach lining may result in gastric ulcers. Hypersensitivity to aspirin and related drugs is thought to be due to the accumulation of prostaglandins after the pathways that break them down are blocked. All aspirin-like analgesics inhibit prostaglandin synthesis, and their potency depends on the degree to which they do so. Many share similar side effects, some of which can be serious. However, the inhibition of prostaglandins also reduces the ability of blood platelets to form clots and this effect has given aspirin added value as an antithrombotic drug.

As the mechanisms of analgesic action began to be understood, alternatives to aspirin were introduced. Acetaminophen, a derivative of phenacetin introduced in 1956, is a popular alternative drug that avoids severe symptoms of stomach irritation. This mild analgesic and antipyretic however has much weaker anti-inflammatory properties, and overdoses can cause liver and kidney damage. Pyrazolone analgesics such as phenylbutazone show similar properties to aspirin and have been used to treat rheumatoid arthritis. Recently potent NSAIDs, sometimes called "super-aspirins," are widely used to replace aspirin itself. These include the propionic acid derivatives, naproxen (first made by Syntex in 1979), ketoprofen (by Wyeth in 1986) and ibuprofen. The latter, manufactured by Upjohn in 1984, is much more potent than aspirin, causes fewer side effects, and is better tolerated by most individuals. In large doses or prolonged use however, it can cause all the symptoms of other inhibitors of prostaglandin synthesis including gastric irritation and renal toxicity.

Further Reading

Atkinson, R.S., Rushman, G.B. and Lee J.A. *A Synopsis of Anaesthesia*, 10th edn. IOP Publications, Bristol, 1987. Covers all aspects of anesthesia and analgesia as practiced in the late twentieth century.

Bonica, J.J. *et al.,* Eds. *Advances in Pain Research and Therapy*. Raven, New York, 1983, 5, 199–208 and 411–459.

Fairley, P. *The Conquest of Pain*. Michael Joseph, London, 1978

Holden, A.V. and Winlow, W., Eds. *The Neurobiology of Pain*. Manchester University Press, Manchester, 1984.

Kirk-Othmer. *Encyclopaedia of Chemical Technology*, 4th edn. Wiley, 2, 1992, pp. 729–748.

Sternbach, R.A. *Pain Patients: Traits and Treatment*. Academic Press, New York, 1974.

Anesthetics

An anesthetic is a drug that causes total loss of sensory perception and thus enables complex surgical procedures to be carried out painlessly. The first surgical uses of ether and chloroform as inhalation anesthetics began in the 1840s, but they were administered crudely by holding a sponge or cloth saturated with the drug over the patient's nose and mouth. By 1910, mixtures of nitrous oxide with ether had begun to replace chloroform, though anesthetic ether and chloroform are still manufactured. Other compounds were soon introduced; and as surgeons demanded deeper, more controlled levels of anesthesia, the professional anesthetist became an essential member of every surgical team.

There are many factors to be considered in choosing a suitable anesthetic agent for each patient. The ideal anesthetic should allow rapid induction followed by ease of control and the possibility of rapid reversal. It should give good muscle relaxation, have few toxic or adverse effects, and be stable in soda lime, used in anesthesia equipment to absorb expired carbon dioxide. Inhalation anesthetics are administered in mixtures with oxygen alone or with a 30/70 mixture of nitrous oxide and oxygen. They are also often combined with drugs that relax muscles and block neuromuscular impulse transmission, making surgical operations easier. Artificial respiration may be required to maintain proper levels of oxygen and carbon dioxide in the blood as deep anesthesia can bring the patient close to death.

Most modern inhalation anesthetics are synthetic halogen-substituted hydrocarbons or ethers. Fluorine, the most common halogen substitute, decreases flammability and boiling point and increases the stability of the molecule. It also reduces fluctuations in heart rate. Fluroxene, introduced in 1960, gives rapid onset and recovery and is stable in soda lime; however, it is unstable in light and readily metabolized in the body. It was replaced in 1962 by methoxyflurane and in 1963 by enflurane. Methoxyflurane is the most potent of the inhaled anesthetics, but like fluroxene it is metabolized. Because fluoride ions cause renal damage, the duration of methoxyflurane anesthesia must be limited. Enflurane is the least potent of the inhaled anesthetics. Later inhalation anesthetics include Sevoflurane, launched in Japan in 1989, and Desflurane. Both are less subject to metabolism in the body, but Sevoflurane is unstable in soda lime. Modern research has been concentrated on the discovery of safer anesthetic agents, and new compounds steadily replace older, less satisfactory ones. However, all halogen-substituted anesthetics tend to trigger hypermetabolic reactions accompanied by rapid temperature rise, increased oxygen use, and carbon dioxide production.

In addition to inhalation techniques, anesthesia may also be produced by intravenous injection. There are two types of injection anesthetics, those used only to induce anesthesia and those used both to induce and maintain anesthesia. Since the ideal injection agent is yet to be discovered, a "balanced" anesthesia is frequently employed with one drug injected for rapid induction followed by an inhalation agent to maintain anesthesia.

French surgeon Pierre Cyrpien Oré first attempted to produce anesthesia by intravenous injection of chloral hydrate in 1874. Improved safety was achieved with hedonal (methylpropylurethane) in 1899, and this was followed during the early 1900s by the intravenous injection of diluted chloroform and ether, as well as hedonal. Toward the end of World War I, barbiturates manufactured by the Swiss firm Hofmann–La Roche were introduced. In the 1920s chemical research revealed a large number of barbiturates, including thiopental sodium (Pentothal) and hexobarbitone, which is said to have been used in up to 10 million cases by the end of World War II. Pentothal is still widely used, but another group of drugs, the benzodiazepines, including diazepam (Valium) and chlordiazepoxide hydrochloride (Librium), were introduced in the early 1960s as muscle relaxants and tranquilizers. Several other drugs in this range are used in neurosurgery and cardiovascular surgery. Among the opiates, morphine is the most common and most potent drug. It is used in high doses as an anesthetic, despite its propensity to cause nausea and vomiting, respiratory and cardiovascular depression, and hypotension (lowered blood pressure). Fentanyl citrate, a synthetic opioid 50 to 100 times more potent than morphine, was introduced in the 1960s. It is used in cardiac surgery in very large doses to produce profound analgesia and suppress cardiovascular reflexes. Since the 1980s other synthetic opioids have been approved for clinical use.

An important difference between inhalation and injection anesthetics is that the former exert physical effects on the respiratory system, whereas the latter function by combining chemically with receptor molecules in the cells. There are two kinds of receptors, the GABA (γ-aminobutyric acid)

receptor and the opiate receptor. As the injected anesthetic is chemically bound to its receptor, removal from the system is slow and other drugs (antagonists) are required to reverse the anesthetic effects.

A third type, local anesthetics, produces loss of sensation in limited areas without loss of consciousness. They are usually administered by subcutaneous injection around sensory nerve endings to block the passage of nervous impulses. Some local anesthetics also block the motor nerves enabling operations to be carried out while the patient remains conscious. The choice of local anesthetic depends on the type and length of the dental or surgical procedure for which it is to be used. The ideal local anesthetic should have rapid onset and long duration, and it should be useful in situations requiring differential blockage of sensory and motor nerve fibers. Obstetric and postoperative pain relief requires a powerful sensory block accompanied by minimal motor block, whereas limb surgery requires both sensory and motor block. A special form of regional nerve block may be caused by injecting a local anesthetic into the spinal cord, either into the space between the membranes that surround the cord (epidural anesthesia) or into the cerebrospinal fluid (spinal anesthesia). In these cases the anesthetic can be adjusted to block conduction in nerves entering and leaving the cord at the desired level.

New compounds are constantly being investigated for anesthetic properties, though the anesthetist now has a wide range of choice. It seems likely that future advances in anesthesia will depend upon developments in the computerization of monitoring and control of the patient's physiological status in response to these agents rather than on the discovery of new anesthetic compounds.

NOEL G. COLEY

Further Reading

Adams, R.C. *Intravenous Anesthesia*. Hoeber, New York, 1944.

Armstrong Davison, M.H. *The Evolution of Anesthesia*, revised edn. John Sherrat, Altrincham, 1970.

Atkinson, R.S., Rushman, G.B. and Lee J.A. *A Synopsis of Anesthesia*, 10th edn. IOP Publications, Bristol, 1987. Surveys late twentieth century anesthesia and analgesia practices.

Duncum, B.M. *The Development of Inhalation Anesthesia*. Royal Society of Medicine Press, London, 1994

Eriksson, E. *Illustrated Handbook in Local Anesthesia*. W.B. Saunders, Philadelphia, 1979.

Holden, A.V. and Winlow, W. *The Neurobiology of Pain*. Manchester University Press, Manchester, 1984.

Kirk-Othmer. *Encyclopaedia of Chemical Technology*, 4th edn. Wiley, New York, 1992, 2, 778–800.

Smith, E.B. and Daniels, S. Eds. *Gases in Medicine: Anesthesia*. Proceedings of the 8th BOC Priestley Conference, London, Royal Society of Chemistry, 1998.

Angiography

Angiography is a radiographic procedure that employs x-rays of the blood vessels in the body to assist in diagnosing and treating a variety of ailments. The x-ray image, or angiogram, allows diagnosis of pathologies of the blood vessels such as blockages, stenosis (narrowing of the vessel), and other aberrations so that they can be treated. Angiography can be purely investigative, but it is commonly employed in combination with minimally invasive surgery or catheterization.

In the angiography procedure, a venous "contrast," or dye, is administered orally, anally, or by intravenous injection. This dye is a radiopaque substance that highlights the blood vessels in the x-ray.

Magnetic resonance angiography (MRA) does not rely on radioactive dye; rather it employs a specific sequence of radio waves to create the angiogram. The MRA is able to provide a detailed map of the patient's vasculature without an enhancer, although enhancers such as the rare earth element gadolinium are sometimes used to make the images clearer and bolder. In therapeutic or interventional angiography, the possible treatment runs from surgery to less invasive processes such as angioplasty or catheterization. In angioplasty, a catheter containing a small balloon is guided through the blood vessels to the site of the obstruction. Once in place, the balloon is inflated in order to expand the constricted area of the vessel. The catheter can also be used to guide into place surgical "stents," cylindrical mesh-like supports that keep the vessel open to the desired width. Over one million angioplasty procedures were performed in the U.S. in 2000 to prevent or treat myocardial infarction (heart attack).

Rudimentary angiography was developed not long after the x-ray came into clinical use. In 1896 in Vienna, Eduard Haschek and Otto Lindenthal took x-rays of the blood vessels in an amputated hand injected with a radiopaque substance. The radiopaque contrast agent of choice has changed over time, owing in part to the high toxicity of the earliest agents used. Common formulations of contrast agents used today include various acetrizoic acids, diatrizoic acids, iodamides, and methanesulfonic acids.

Angiography is employed today in a number of clinical situations. The most common procedures performed are cerebral angiography, thoracic aortography, pulmonary and bronchial arteriography, coronary arteriography, and angiography of the extremities. In celial and mesenteric angiography, these arteries in the abdomen are examined to diagnose gastrointestinal bleeding, aneurysm, or ischemia. In cerebral angiography the blood vessels of the brain are investigated in order to locate and treat of blood clots, aneurysms, tumors, and migraines. In coronary angiography the coronary arteries are investigated to detect vascular diseases such as heart disease, heart attack, and acute stroke. Renal angiography is of value in diagnosing kidney disease and renal failure. In the latter case, the physician uses hemodialysis catheters to divert blood from the neck and filter it through a hemodialysis machine. One problem with kidney angiography is that the contrast agent can be harmful to the kidneys if the patient already suffers from renal ailments. In ocular angiography either fluorescein or indocyanine green contrast helps to view various problems in the eye such as retinal failure.

Complications surrounding angiographic procedures can arise either from reactions to the contrast agent or from problems with the performance of the catheterization or surgery. Reactions to the contrast agent are generally mild, such as nausea, but serious allergic reactions do occasionally occur. Renal damage can occur regardless of previous problems because of the mild toxicity of most opaque media contrasts. Some contrasts have also caused minor drops in blood pressure and vasodilatation of the arteries. When a catheter is placed in the body, blood clots can form and block the vessel or a small particle can break loose and lead to embolization, a potentially deadly complication. Catheters can also tear or puncture blood vessels causing internal bleeding and an exacerbation of the already existing problem. Hematomas and hemorrhages may occur if there are any complications in the catheterization process.

As with other medical tests and procedures, the risks of angiography are generally outweighed by the benefits, as the angiogram provides specific and detailed clinical information that is invaluable in clinical diagnosis and intervention.

See also **Cardiovascular Disease, Diagnostic Methods; Medicine; Nuclear Magnetic Resonance (NMR/MRI); X-rays in Diagnostic Medicine**

TIMOTHY S. BROWN

Further Reading

Abrams, H.L. *Angiography*. Little, Brown & Company, Boston, 1971.

Grossman, W. and Bain, D.S. *Cardiac Catheterization, Angiography and Intervention*. Lea & Febiger, Philadelphia, 1991.

Silverman, J.F. *Coronary Angiography: An Introduction to Interpretation and Technique*. Addison-Wesley, Menlo Park, CA, 1984.

Tortorici, M.R. *Advanced Radiographic and Angiographic Procedures*. F.A. Davis, Philadelphia, 1995.

Antibacterial Chemotherapy

In the early years of the twentieth century, the search for agents that would be effective against internal infections proceeded along two main routes. The first was a search for naturally occurring substances that were effective against microorganisms (antibiosis). The second was a search for chemicals that would have the same effect (chemotherapy).

Despite the success of penicillin in the 1940s, the major early advances in the treatment of infection occurred not through antibiosis but through chemotherapy. The principle behind chemotherapy was that there was a relationship between chemical structure and pharmacological action. The founder of this concept was Paul Erhlich (1854–1915). An early success came in 1905 when atoxyl (an organic arsenic compound) was shown to destroy trypanosomes, the microbes that caused sleeping sickness. Unfortunately, atoxyl also damaged the optic nerve. Subsequently, Erhlich and his co-workers synthesized and tested hundreds of related arsenic compounds. Ehrlich was a co-recipient (with Ilya Ilyich Mechnikov) of the Nobel Prize in medicine in 1908 for his work on immunity.

Clinical trials in 1910 showed that another of these compounds was effective in treating not only sleeping sickness but also syphilis. It was called arsphenamine and given the trade name Salvarsan. The success of Salvarsan was evidence that synthetic drugs could destroy microbes in patients, but again there were severe limitations. A course of treatment involved an injection a week for 10 to 12 weeks. This had to be repeated two or three times, with intervals of one month or so to minimize toxic effects. Treating early syphilis took a year or more.

Some progress in antibacterial chemotherapy was made between 1920 and 1932 in the laboratories of the Bayer Company in Elberfeld, Germany. Ehrlich had developed simple methods for mass screening. He selected a suitable test system (typically a group of animals infected with a particular organism) and used it to test many substances, most of which were newly synthesized. But real progress occurred only after the appointment by Bayer of Gerhard Domagk as director of research in experimental pathology in 1927. In collaboration with two chemists, Fritz Mietzsch and Joseph Klarer, Domagk immediately began a program to develop antibacterial drugs.

Domagk and his team investigated dyestuffs as a possible source of potential drugs. In 1932 they experimented on two groups of mice, all of which had been infected with streptococci. One group was treated with a red dye called prontosil, and the other was not. The untreated group all died within four days; the treated group all survived at least a week. For commercial reasons nothing was published for over two years, during which time Domagk treated his own daughter successfully for a streptococcal infection. By 1935 a large number of clinical trials were under way, and the drug rapidly became widely available. Domagk received the Nobel Prize in medicine in 1939.

Later in 1935 Trefouel and his colleagues at the Pasteur Institute in Paris showed that prontosil was broken down *in vivo* (in the body) to a much simpler compound, sulfanilamide, which was the active disease-fighting agent. Prontosil is not active *in vitro* (outside the body), and this nearly led to the failure of its discovery. The two chemists Mietzsch and Klarer were uneasy about performing initial tests in mice and wanted to screen compounds on bacterial culture in test tubes first. If this had been done, the action of prontosil would have been missed, not for the first but for a *second* time, because sulfanilamide was not a new substance. It had been synthesized and described in a publication in 1908 as part of research on dyes. However, a major obstacle to progress in the development of antibiotics in the 1920s was the prevalence of contemporary ideas. Sulfanilamide had actually been tested as an antibacterial in about 1921 by Michael Heidelberger at the Rockefeller Institute in New York, but he failed to discover its activity. It was believed at that time that if an antibacterial drug were to be effective at all, it would work immediately. In fact, sulfanilamide takes several hours to produce its effects, and Heidelberger stood no chance of observing them under the conditions he used.

It took some time to work out the mode of action of sulfanilamide. It was a powerful and effective drug, but it did not work under certain conditions; for example, when the target streptococci were surrounded by dead tissue or pus. Eventually the process of *competitive antagonism* was understood. Microbes that are sensitive to sulfanilamide need a substance known as *p*-aminobenzoic acid (or PABA) as a building block for their growth. Sulfanilamide is a false building block shaped like PABA and is able to take its place. Once in position it prevents any further building and cell growth ends. The cells are then dealt with by the body's normal defensive mechanisms, and the patient recovers.

Soon after this discovery, a large number of related compounds, collectively known as the sulfonamides, were synthesized by modification of the parent compound. They were shown to be effective against various bacterial infections. Sulfanilamide was very effective against streptococcal infections, including puerperal fever, erysipelas, and scarlet fever. It was, however, not active against pneumococci, which cause pneumonia. A pharmaceutical company in the U.K. (May and Baker) soon developed a sulfonamide that was. Called sulfapyridine, it was marketed as M&B 693. The first trials were carried out at the Middlesex Hospital in London in 1938, under the supervision of Sir Lionel Whitby.

Other sulfonamides were being developed during this time. Trials indicating the range of antibacterial activity of sulfathiazole were reported in 1940. Sulfadiazine followed in 1941 after trials on 446 adult patients at Boston City Hospital. In 1942 the results of early trials on a sulfonamide developed by Imperial Chemical Industries, sulfadimidine, were published, indicating its value in the treatment of pneumonia, meningitis, and gonorrhoea. Sulfaguanidine was used successfully during World War II to treat bacillary dysentery in troops in the Middle and Far East.

Success in discovering a range of effective antibacterial drugs had three important consequences: it brought a range of important diseases under control for the first time; it provided a tremendous stimulus to research workers and opened up new avenues of research; and in the resulting commercial optimism, it led to heavy postwar investment in the pharmaceutical industry. The therapeutic revolution had begun.

See also **Antibiotics: Developments through 1945; Antibiotics: Use after 1945.**

STUART ANDERSON

Further Reading

Bickel, M.H. The development of sulphonamides, 1932–1938. *Gesnerus*, 45, 67–86, 1988.
Lesch, J. The Invention of the Sulfa Drugs, in *Chemical Sciences in the Modern World*, Mauskopf, S.H., Ed. University of Pennsylvania Press, Philadelphia, 1993.
Lesch, J. The Discovery of M&B 693 (Sulfapyridine), in *The Inside Story of Medicines: A Symposium,* Higby, G. and Stroud, E., Eds. American Institute of the History of Pharmacy, Madison, 1997.
Liebenau, J., Higby, G. and Stroud, E., Eds. *Pill Peddlers: Essays on the History of the Pharmaceutical Industry.* American Institute of the History of Pharmacy, Madison, 1990.
Mann, R.D. *Modern Drug Use: An Enquiry on Historical Principles.* MTP Press, the Hague, 556–57, 1984.
Slinn, J. *A History of May and Baker 1834–1984.* Hobsons, Cambridge, 1984.
Tweedale, G. *At the Sign of the Plough: 275 Years of Allen & Hanburys and the British Pharmaceutical Industry 1715–1990.* John Murray, London, 1990.
Weatherall, M. *In Search of a Cure: A History of Pharmaceutical Discovery.* Oxford University Press, Oxford, 1990.

Antibiotics, Developments through 1945

The term antibiotic was first used in 1899 by the botanist Marshall Hall, following the coining of the term antibiosis ten years earlier by French scientist Paul Vuillemin. Antibiosis referred to the idea that living organisms might produce substances that were antagonistic to one another, an idea first suggested by Louis Pasteur in 1877. In the late nineteenth century, many observations of antibiotic action among microorganisms were reported, but unfortunately the significance of these observations was not appreciated at the time.

In the history of antibiosis, an observation by Alexander Fleming (1881–1955), a bacteriologist at St. Mary's Hospital in London, proved to be the turning point. In the late summer of 1928, Fleming went to his laboratory during the holiday to inspect culture plates awaiting disposal. He noticed one that had been accidentally contaminated with a mold, or fungus, around which bacterial growth had been inhibited. The bacteria were staphylococci and the fungus was *Penicillium notatum*. Fleming cultured it, fixed the plate in formalin vapor, and gave it the name penicillin.

Fleming's tests showed that the fungus could kill or inhibit the growth of a number of other organisms harmful to man. In a paper in 1929, Fleming mentioned the possible use of penicillin as an antiseptic in surgical dressings. There was further work in other laboratories interested in Fleming's observation, but progress was limited because of the great instability of the material. The goal of isolating an effective preparation from the organism remained for the future.

In 1939, three other British scientists based in Oxford made the discovery that penicillin was effective inside the body as well as on its surface. Ernst Chain, Howard Florey, and Norman Heatley, embarked on its detailed investigation, urged on by the necessities of war. In 1940 Florey and Chain were able to make a dry, stable extract of penicillin, and the first trial was carried out on four mice on May 25 1940.

The first human received an injection of penicillin on January 27 1941, at the Radcliffe Infirmary in Oxford. The patient did not have an infection, but suffered a sharp rise in temperature after 3 hours. After removal of the pyrogen that caused the fever, enough active material was isolated to treat one ill patient. On February 12 1941, a London policeman who had a severe staphylococcal infection was the first sick patient to receive penicillin. After treatment for five days, the stock of penicillin was exhausted, and the patient relapsed and died. Over the following days further seriously ill patients were treated with varying degrees of success, and the results were published in August 1941.

By early 1942, small-scale production was being carried out in the U.K. by Imperial Chemical Industries Limited. However the investment needed to produce penicillin on a commercial scale was considerable and could not easily be found in a country at war. Florey and Heatley went to America to seek support. Penicillin was first made in the United States by the pharmaceutical company Upjohn in March 1942, using a culture supplied by the U.S. Department of Agriculture. Initial small-scale production was achieved using a surface culture method in bottles. On May 25 1943, the firm was asked to make penicillin for the military, and it began brewing the fungus in 120,000 bottles in a basement. By July 1943 the company had shipped its first batch of 100 vials, each containing 100,000 units of penicillin.

The process of surface culture in bottles had obvious limitations, and work began to develop a more efficient process that used deep culture in vat fermenters. The principle of the submerged culture method is to grow the fungus in large steel containers in a medium that is constantly aerated and agitated. Under these circumstances, the fungus grows throughout the body of the medium. Early problems in the operation of the method were overcome, and by 1949 fermenters with capacities up to 12,000 gallons were in use.

Tens of thousands of species and mutants of *Penicillium* were examined in the search for ones that produced the highest yields of penicillin. Eventually the highest yields were obtained from a strain of *Penicillium chrysogenum,* originally found growing naturally on a moldy melon. As a result yields have increased 2000 times since the original small-scale manufacture.

Solving the complex microbiological, chemical, and engineering problems involved in the large-scale manufacture of penicillin required a collaborative effort. By 1944, 20 American and 16 British academic and industrial groups were working on the problem. The first large delivery, consisting of 550 vials of 100,000 units of penicillin each, was made to the U.S. Army in February 1944. Its initial use was in the treatment of gonorrhea; widespread use of penicillin in the treatment of life-threatening conditions such as pneumonia only occurred later when sufficient supplies became available. With large-scale manufacture the price dropped dramatically from $20.05 per 100,000-unit vial in July 1943 to just $0.60 per vial by October 1945.

Full understanding of the mode of action of penicillin only emerged slowly and awaited developments in both biochemistry and bacteriology. The involvement of the cell wall was recognized early on, but even in 1949 Chain and Florey concluded, "No complete picture of how penicillin acts *in vivo* can be drawn on the evidence available."

Penicillin inhibits an enzyme that catalyzes one of the biochemical steps in the synthesis of mucopeptide, the rigid component of the cell wall. In the absence of the enzyme inhibited by penicillin, links between the peptide molecules fail to occur. When penicillin in sufficient concentration comes into contact with penicillin-sensitive organisms, it binds to the cell wall. As cell division takes place, defects occur in the rigid component of the wall. Coupled with high internal osmotic pressure, the cell then bursts, or lyses. Penicillin is therefore only active against growing bacteria and only those sensitive to it. Some bacteria contain plasmids that make them resistant to penicillin.

In subsequent years a large number of variants of penicillin were developed. Some of these were naturally produced, some were semisynthetic, and others were entirely synthetic. The search for effective antibacterial substances was widened, and the stage was set for the discovery of a wide range of new antibacterial substances in the post-World War II years.

See also **Antibacterial Chemotherapy; Antibiotics: Use after 1945**

STUART ANDERSON

Further Reading

Carlyle, R.D.B. *A Century of Caring: The Upjohn Story.* Benjamin, New York, 1987, pp. 89–95.

Elder, A., Ed. *The History of Penicillin Production.* Livingstone, New York, 1970.

Florey, H.W., Chain, E., Heatley, N.G., Jennings, M.A., Sanders, A.G., Abraham, E.P. and Flory, M.E. *Antibiotics: A Survey of Penicillin, Streptomycin, and other Antimicrobial Substances from Fungi, Actinomycetes, Bacteria and Plants.* Oxford University Press, Oxford, 1949, 1, pp. 1–73.

Florey, H.W., Chain, E., Heatley, N.G., Jennings, M.A., Sanders, A.G., Abraham, E.P. and Flory, M.E. *Antibiotics: A Survey of Penicillin, Streptomycin, and other Antimicrobial Substances from Fungi, Actinomycetes, Bacteria and Plants.* Oxford University Press, Oxford, 1949, 2, pp. 631–671.

Hare, R. *The Birth of Penicillin.* Allen & Unwin, London, 1970

Lazell, H.G. *From Pills to Penicillin: The Beecham Story.* Heinemann, London, 1975.

Maurois, A. *The Life of Alexander Fleming: Discoverer of Penicillin,* translated by Hopkins, G. Heinemann, London, 1952.

Weatherall, M. *In Search of a Cure: A History of Pharmaceutical Discovery.* Oxford University Press, Oxford, 1990.

Antibiotics: Use after 1945

Antibiotics, a category of antiinfective drugs, are chemical substances produced by microorganisms such as bacteria and fungi that at low concentrations kill or inhibit other microorganisms. The development of antibiotic drugs was one of the most important advances in twentieth century medicine. Penicillin, the first antibiotic, was first produced in the U.S. in 1942. About the same time new antibiotics, including streptomycin, were discovered in a mixture of microbial substances from a soil bacillus. Many of these drugs were found to be specific in their action but in 1947, chloramphenicol, the first so-called broad spectrum antibiotic was discovered; followed in 1948 by aureomycin, the first broad spectrum tetracycline antibiotic. Chloramphenicol, a nitrobenzene derivative of dichloroacetic acid, is manufactured synthetically and used to treat typhoid fever and severe infections caused by penicillin-resistant microorganisms.

As antibiotics destroyed the more common pathogens, highly resistant strains that escape destruction began to pose problems, and new antibiotics were required to eradicate them. By the end of the twentieth century, there were well over 10,000 known antibiotics ranging from simple to extremely complex substances. This great expansion in new antibiotics was achieved through successful synthetic modifications, and many antibiotics are now manufactured in quantity. This approach has been especially successful with antibiotics whose molecular structure contains the four-member β-lactam ring, including the penicillins and cephalosporins that currently account for over 60 percent of world production of antibiotics (see Table 1 for major types of antibiotics).

The modes of action of different antibiotics vary, though many destroy bacteria by inhibiting cell-wall biosynthesis. Bacterial cells, unlike animal cells, have a cell wall surrounding a cytoplasmic membrane. The cell wall is produced from components within the cell that are transported through the cell membrane and built into the cell wall, and antibiotics that interfere with this process inhibit cell growth. For example, the β-lactam ring in penicillin and cephalosporin molecules interferes with the final stage in the assembly of the cell wall. Other antibiotics inhibit protein synthesis in bacteria, while others act by combining with phos-

Table 1 Important Groups of Antibiotics.

Classification	Generic Drug
Aminoglycoside	Gentamicin sulfate
Antifungal	Fluconazole
β-Lactam	Imipenim-Cilastatin sodium
Cephalosporin	
First Generation	Cephalothin sodium
Second Generation	Cefonicid sodium
Third Generation	Cefotaxime sodium
Clindamycin	Clindamycin hydrochloride
Macrolide	Erythromycin
Penicillin	
Aminopenicillin	Ampicillin
Antipseudomonal Penicillin	Mezlocillin sodium
Natural Penicillin	Penicillin G potassium
Tetracycline	Tetracycline hydrochloride

pholipids in the cell membrane to interfere with its function as a selective barrier and allow essential materials in the cell to leak out. This results in the death of the bacterial cell; but since similar phospholipids are found in mammalian cells, this type of antibiotic is toxic and must be used with care. One antibiotic, rifampin, disrupts RNA synthesis in bacterial cells by combining with enzymes in the cell; as its affinity for bacterial enzymes is greater than for mammalian ones, the latter remain largely unaffected at therapeutic dosages.

Penicillin is the safest of all antibiotics, although some patients show hypersensitivity toward it, and this results in adverse reactions. Moreover some microorganisms, notably the staphylococci, develop resistance to naturally occurring penicillin, and this has led to the production of new synthetic modifications. Thus there are two groups of penicillins, those that occur naturally and the semisynthetic penicillins made by growing the *Penicillium* mold in the presence of certain chemicals. To increase the usefulness of the penicillins, the broad-spectrum penicillins were developed to treat typhoid and enteric fevers and certain infections of the urinary tract. Nevertheless, the naturally occurring penicillins are still the drugs chosen for treating many bacterial infections.

The cephalosporins, discovered in the 1950s by Sir Edward Abraham, are relatively nontoxic β-lactam antibiotics. Like penicillin they were first isolated from a fungus, but later modifications of the β-lactam ring have resulted in over 20 variations grouped according to their activity. The first generation cephalosporins were used for patients who had developed sensitivity to penicillin. They were active against many bacteria including *Escherichia coli*, but they had to be replaced by second and third generation drugs as resistance began to develop. They have been used for treating pulmonary infections, gonorrhea, and meningitis.

The aminoglycosides, which include streptomycin (discovered in 1944), inhibit protein biosynthesis. They are poorly absorbed from the gastrointestinal tract and are administered by intramuscular injection. Streptomycin was among the first of the aminoglycosides to be discovered and is still used together with penicillin for treating infections of the heart valves. Other aminoglycosides are used for treating meningitis, septicemia, and urinary tract infections, but the narrow margin between a therapeutic and a toxic dose poses difficult problems, and the risks increase with age. The antimicrobial activity of the tetracyclines,

another group of synthetic antibiotic drugs, depends on the fact that although they inhibit protein biosynthesis in both bacterial and animal cells, bacteria allow the tetracyclines to penetrate the cell, whereas animal cells do not. Tetracyclines are absorbed from the gastrointestinal tract and can be given orally.

In addition to their uses in medicine, antibiotics have also had important veterinary applications and are used as animal feed supplements to promote growth in livestock. Tetracyclines make up about half the sales of antibiotics as supplements, which surpasses all other agricultural applications, but many other antibiotics are also used for this purpose. It is thought that feed antibiotics may promote growth by preventing disease. Another important agricultural use of antibiotics is in their use as antiparasitic agents against both worms and other parasites in the gastrointestinal tract and against ectoparasites such as mites and ticks.

In addition to the biochemical antibiotics, the sulfonamides, synthetic chemotherapeutic agents, are also used in treating bacterial diseases. The first sulfonamide was prontosil, introduced in 1932 to combat streptococcal infections. The sulfonamides are broad-spectrum agents that were widely used prior to antibiotics. They work by preventing the production of folic acid, which is essential for the synthesis of nucleic acids. The reaction is reversible causing the inhibition but not the death of the microorganisms involved. Their use has diminished due to the availability of better and safer antibiotics, but they are still used effectively to treat urinary tract infections and malaria and to prevent infection after burns. Related to the sulfonamides, the sulfones are also inhibitors of folic acid biosynthesis. They tend to accumulate in the skin and inflamed tissue, and as they are retained in the tissue for long periods, they are useful in treating leprosy. There are also some other chemical synthesized drugs that show antibacterial properties and find specific clinical uses.

See also **Antibacterial Chemotherapy; Antibiotics: Developments through 1945**

Noel G. Coley

Further Reading

Bottcher, H. *Miracle Drugs: a History of Antibiotics.* Heinemann, London, 1959.
Cannon, G. Superbug: *Nature's Revenge. Why Antibiotics Can Breed Disease.* Virgin, London, 1995.
Halliday, J., Ed. *Antibiotics: The Comprehensive Guide.* Bloomsbury, London, 1990.

Kirk-Othmer. *Encyclopaedia of Chemical Technology*, 4th edn. Wiley, New York, 1992, 2, pp. 893–1018 and 3, pp. 1–346.

MacFarlane, M. *Alexander Fleming: The Man and the Myth*. Chatto and Windus, Hogarth Press, London, 1984.

Pharmaceutical Society of Great Britain. *Antibiotics: A Survey of Their Properties and Uses*. Pharmaceutical Press, London, 1952.

Reese R.E., Betts R.F. and Gumustop B. *Handbook of Antibiotics*, 3rd edn. Lippincott, Williams & Wilkins, New York, 2000.

Architecture, *see* **Constructed World**

Artificial Insemination and *In Vitro* Fertilization (Farming)

Artificial insemination (AI) involves the extraction and collection of semen together with techniques for depositing semen in the uterus in order to achieve successful fertilization and pregnancy. Throughout the twentieth century, the approach has offered animal breeders the advantage of being able to utilize the best available breeding stock and at the correct time within the female reproductive cycle, but without the limitations of having the animals in the same location. AI has been applied most intensively within the dairy and beef cattle industries and to a lesser extent horse breeding and numerous other domesticated species.

There is some anecdotal historical evidence to indicate the ancient use of AI in mammalian reproduction, dating as far back as fourteenth century Arabia, telling of how rival tribes would obtain sperm from the mated mares or from the male horses of their opponents and use it to inseminate their own mares. Based on the first visible identification of spermatozoa in the seventeenth century, the earliest documented insemination is believed to have been carried out by Italian physiologist Lazzaro Spallanzani in 1784 on a domestic dog, resulting in a successful pregnancy. This was followed a century later with the work in the U.K. by Cambridge reproductive scientist Walter Heape who in 1897 documented the importance and variation of reproductive cycles and seasonality across a number of species. However, it was not until the opening decades of the twentieth century with the work of Russian reproductive scientist E. I. Ivanow that AI was more widely investigated, particularly in Scandinavia, though still largely within the confines of research rather than general animal husbandry.

The emergence of an AI research community takes firmer shape with the first International Congress on AI and Animal Reproduction in Milan in 1948. The 1940s can be understood as the decade in which the technique progresses from a relatively small-scale research activity to becoming one of the most routine of procedures in domestic animal reproduction. The establishment of the New York Artificial Breeders Cooperative in the early 1940s resulted in the insemination of hundreds of thousands of cattle and the refinement of many of the methods on which AI would subsequently come to depend. Progress on techniques relevant to male aspects of AI includes sire selection, semen collection, and evaluation methods, as well as techniques for safely preserving and storing semen. In relation to female aspects of AI, innovation has focused on characterizing species-specific patterns of estrus detection and synchronization in order to best judge the timing of insemination. As the twentieth century drew to a close, in addition to becoming one of the most widely used techniques within animal breeding, AI has become ever more prominent in the preservation of rare and endangered species. In 1986, the American Zoo and Aquarium Association established a Species Survival Plan, which has been successful in reviving such nearly extinct animals as black-footed ferrets.

In Vitro *Fertilization*

Many of the techniques involved in artificial insemination would lay the foundation for *in vitro* fertilization (IVF) in the latter half of the twentieth century. IVF refers to the group of technologies that allow fertilization to take place outside the body involving the retrieval of ova or eggs from the female and sperm from the male, which are then combined in artificial, or "test tube," conditions leading to fertilization. The fertilized eggs then continue to develop for several days "in culture" until being transferred to the female recipient to continue developing within the uterus.

The first reported attempts to fertilize mammalian ova outside the body date back as far as 1878 and the reported attempt to fertilize guinea pig eggs on a solution of mucus taken from the uterus. However, the first verified successful application of IVF did not occur until 1959 and resulted in the live derivation of rabbit offspring. This was followed in the late 1960s with mice and coincided with similar but unsuccessful attempts at IVF in humans by Robert Edwards, Barry Bavister, and others. However, it was events in 1978 that decisively put the technique on the research agenda for reproductive science in both humans and other mammals. The birth of Louise Brown in the U.K. in 1978, the first successful application of the technique in humans, represented the point at which IVF became an intense focus of reproductive activity throughout the latter years of the twentieth century. By the early twentieth century IVF was an established reproductive technology in the production of highly prized genetic source animals and, as with AI, the preservation of endangered species.

See also **Agriculture and Food; Fertility, Human**

NIK BROWN

Further Reading

Cupps, P.T. *Reproduction in Domestic Animals*, 4th edn. Academic Press, San Diego, 1991.

Foote, R.H. *The History of Artificial Insemination. Selected Notes and Notables.* American Society of Animal Science, 2002.

Foote, R.H. Artificial Insemination from its origins up to today, in Russon, V., Dall'Olio, S. and Fontanesi, L., Eds. *Proceedings of the Spallanzani International Symposium,* Reggio Emilia, Italy, 1999, pp 23–67.

Foote, R.H. The artificial insemination industry, in *New Technologies in Animal Breeding*, Brackett, B.G., Seidel, Jr., G.E. and Seidel, S.M., Eds. Academic Press, New York, 1981, pp 13–39.

Herman, H.A. *Improving Cattle by the Millions: NAAB and the Development and Worldwide Application of Artificial Insemination.* University of Missouri Press, Columbia, 1981.

Perry, E.J., Ed. *The Artificial Insemination of Farm Animals*, 4th edn. Rutgers University Press, New Brunswick, 1968.

Sipher, E. *The Gene Revolution: The History of Cooperative Artificial Breeding in New York and New England 1938–1940.* Eastern AI Cooperative, Ithica, 1991.

Trounson, A.O. *et al. Handbook of In Vitro Fertilization.* CRC Press, London, 2000.

Artificial Intelligence

Artificial intelligence (AI) is the field of software engineering that builds computer systems and occasionally robots to perform tasks that require intelligence. The term "artificial intelligence" was coined by John McCarthy in 1958, then a graduate student at Princeton, at a summer workshop held at Dartmouth in 1956. This two month workshop marks the official birth of AI, which brought together young researchers who would nurture the field as it grew over the next several decades: Marvin Minsky, Claude Shannon, Arthur Samuel, Ray Solomonoff, Oliver Selfridge, Allen Newell, and Herbert Simon.

During the 1940s many researchers, under the guise of cybernetics, had worked out much of the theoretical groundwork for AI and had even designed the first computers. Among the most significant contributions upon which AI depended were Alan Turing's theory of computation and ACE computer, John Von Neumann's ENIAC computer, Claude Shannon's theory of communication, Norbert Weiner's theory of information and negative feedback, Warren McCulloch and Walter Pitts' neuronal logic networks, W. Ross Ashby's theory of learning and adaptive mechanisms, and W. Grey Walter's autonomous robotic tortoises. The young Dartmouth group differed from the earlier work of the cyberneticians in that they concerned themselves primarily with writing digital computer programs that performed tasks deemed to require intelligence for humans, rather than building machines or modeling brains.

AI research focused around the key aspects of intelligent behavior including automated reasoning, decision making, machine learning, machine vision, natural language processing, pattern recognition, automated planning, problem solving, and robot control. This field of research set itself ambitious goals, seeking to build machines that could "out-think" humans in particular domains of skill and knowledge, and achieving some success in this. Some researchers even speculated that it would be possible to build machines that could imitate human behavior in general by the end of the twentieth century, but most researchers now consider this goal to be unattainable by the end of the twenty-first century.

The first AI program, Logic Theorist, was presented by Newell and Simon at the Dartmouth workshop in 1956. Logic Theorist proved theorems of mathematical logic from a given set of axioms and a set of rules for deducing new axioms from those it already had. Given a theorem, Logic Theorist would attempt to build a proof by trying various chains of deductive inference until it arrived at the desired theorem. Logic Theorist was followed by General Problem Solver in 1961. This program demonstrated that the technique of proving theorems could be applied to all sorts of problems by defining a "goal" and conducting a search to find a series of valid moves that led from what is already known to the goal that is sought. This technique can work well for simple problems, but the total number of alternative moves that are possible can grow exponentially in the number of steps to a solution. Since the program has to keep backtracking and trying all the alternate routes, the technique quickly breaks down for problems with many steps. These challenges led Newell and Simon to suggest that AI research on problem solving ought to focus on finding good *heuristics,* search strategies or rules-of-thumb, to use when searching. A good heuristic helps one find a solution faster by reducing the number of dead ends encountered during a search.

On June 27, 1963, AI research was catapulted forward by a huge grant from the Defense Advanced Research Projects Agency (DARPA) of the U.S. Department of Defense to the Massachusetts Institute of Technology AI Laboratory. The grant was partly motivated by U.S. fears following the Soviet launch of Sputnik and partly motivated by extreme enthusiasm on the part of AI researchers that computers would soon have human-like intelligence. By the early 1970s, actual research at the Massachusetts Institute of Technology was limiting search spaces by studying very simplified application domains, or *micro worlds* as they came to be called. The most famous program, SHRDLU, planned manipulations in a world consisting only of wooden blocks sitting on a table called the *blocks world.* While these systems did what their designers intended, often led to theoretically interesting results, and eventually developed into useful technologies; they did not live up to either the public or military expectations for intelligent machines. By the end of the 1970s, DARPA became deeply concerned that AI would fail to deliver on its promises and eventually cut its research funding.

The insights gained from the micro worlds research eventually did find application in the area of automated planning. Planning generally starts from knowing the current state of the "world," the desired state of the world, and a set of actions called *operators* that can be taken to transform the world. The Stanford Research Institute Problem Solver (STRIPS) was an early planner that used a language to describe actions that is still widely used and enhanced. The STRIPS operators consist of three components:

1. The action description
2. The *preconditions* of the action, or the way the world must be before the action can be taken
3. The *effect*, or how the world has been changed since the action was taken. To develop a plan, the system searches for a reasonably short or cost-efficient sequence of operators that will achieve the goal. Planning systems have been used widely to generate production schedules in factories, to find the most efficient ways to lay out circuits on microchips or to machine metal

parts, and to plan and coordinate complex projects involving many people and organizations, such as space shuttle launches.

One AI technology that had early real-world applications was the expert system. Expert systems utilize a large amount of knowledge about a small area of expertise in order to solve problems in that domain. The first such system was DENDRAL, which could logically infer the structure of a molecule if given its chemical formula and information from a mass spectrogram of the molecule. This difficult task was achieved by DENDRAL because it was provided with rules-of-thumb and tricks for recognizing common patterns in the spectrograms, developed in collaboration with Joshua Lederberg, a Nobel prize-winning chemist. The next generation expert system MYCIN used rules that incorporated uncertainty as probability weights on inferences. MYCIN used some 450 such rules to diagnose infectious blood diseases. Expert systems have proven to be one of the most successful applications of AI so far. Thousands of expert systems are currently in use for medical diagnoses, servicing and troubleshooting complex mechanical devices, and aiding information searches.

Another successful AI technology has been machine learning, which develops techniques for machines to actually learn from experience and improve over time. Machine learning was first conceived by Ashby in 1940, while the first successful program for learning was Samuel's 1959 checkers-playing program. Most forms of machine learning use statistical induction techniques to infer rules and discover relationships in sample or training data. Machine learning is useful in solving problems in which the rules governing the domain are difficult to discover and a large amount of data is available for analysis.

Pattern recognition is the most common type of problem for machine learning applications. The most popular pattern recognition systems, and perhaps the most popular single AI technology, are neural networks that learn from experience by adjusting weighted connections in a network. A typical feed-forward neural network performs some version of statistical pattern classification; that is, it induces statistical patterns from training data to learn a representative function, and then applies this function to classify future examples. A classification is simply a mapping function from inputs to outputs, and so a neural network just maps the objects to be classified into their types or classes.

Consider, for example, the problem of classifying some two-dimensional geometric shapes into one type of the set (square, circle, triangle, other). A total mapping function would assign every member of the set to one of these four types. There are many possible mapping functions however, and only a few of these will classify the inputs in a desirable way. Good techniques for neural networks will find these mappings efficiently and avoid getting stuck in statistical dead ends, called "local minima." Other examples of pattern recognition include speech recognition, face recognition, handwritten letter recognition, and robotic vision and scene analysis, in which the program must match audio or visual patterns to words, faces, letters, objects, or scenes, respectively.

Another important area of AI research has been natural language processing (NLP). NLP attempts to provide computers with the ability to understand natural human languages, such as English or Russian. Work in this area has drawn heavily on theories of grammar and syntax borrowed from computational linguistics and has attempted to decompose sentences into their grammatical structures, assign the correct meanings to each word, and interpret the overall meaning of the sentence. This task turned out to be very difficult because of the possible variations of language and the many kinds of ambiguity that exist. The applications of successful NLP programs have included machine translation from one natural language to another and natural language computer interfaces. A great deal of success has been achieved in the related areas of optical character recognition and speech recognition, which employ machine learning techniques to translate text and sound inputs into words but stop short of interpreting the meaning of those words.

Game playing programs have done much to popularize AI. Programs to play simple games like tic-tack-toe (noughts and crosses) are trivial, but games such as checkers (draughts) and chess are more difficult. At IBM, Samuel began working in 1952 on the program that would be the first to play tournament level checkers, a feat it achieved by learning from its own mistakes. The first computer to beat a human grandmaster in a chess match was HITECH in 1989. And in May of 1997, IBM's Deep Blue computer beat the top-ranked chess player in the world, Gary Kasparov. Unfortunately, success in a single domain such as chess does not translate into general intelligence, but it does demonstrate that seemingly intelligent behaviors can be automated at a level of performance that exceeds human capabilities. One of the most common consumer AI applications is probably the computer opponent in video games that "plays against" the human user. These applications use techniques of varying sophistication to challenge human opponents and often allow the human to select the skill level of their opponent.

It would be difficult to argue that the technologies derived from AI research had a profound effect on our way of life by the end of the twentieth century. However, AI technologies have been successfully applied in many industrial settings, medicine and health care, and video games. Programming techniques developed in AI research were incorporated into more widespread programming practices, such as high-level programming languages and time-sharing operating systems. While AI did not succeed in constructing a computer which displays the general mental capabilities of a typical human, such as the HAL computer in Arthur C. Clarke and Stanley Kubrick's film *2001: A Space Odyssey*, it has produced programs that perform some apparently intelligent tasks, often at a much greater level of skill and reliability than humans. More than this, AI has provided a powerful and defining image of what computer technology might someday be capable of achieving.

Peter M. Asaro

Further Reading

Anderson, J.A., and Rosenfeld E., Eds. *Talking Nets: An Oral History of Neural Networks*. MIT Press, Cambridge, MA, 1998.

Barr, A., Feigenbaum, E.A. and Cohen, P.R., Eds. *The Handbook of Artificial Intelligence,* vols. 1–4. Heuris Tech Press and Kaufmann, Stanford, (1981–89).

Crevier, D. *AI: The Tumultuous History of the Search for Artificial Intelligence*. Basic Books, New York, 1993.

Dreyfus, H. *What Computers Can't Do: A Critique of Artificial Reason*. Harper & Row, New York, 1979.

Haykin, S. *Neural Networks: A Comprehensive Foundation*. Prentice Hall, Englewood Cliffs, NJ, 1998.

Jurafsky, D. *et al. Speech and Language Processing: An Introduction to Natural Language Processing, Computational Linguistics and Speech Recognition*. Prentice Hall, Englewood Cliffs, NJ, 2000.

McCarthy, J. Programs with common sense. *Proceedings of a Symposium on Mechanical Thought Processes,* vol. 1. London: Her Majesty's Stationery Office, London, 1958, pp. 77–84. Reprinted in Minsky, M.L., Ed. *Semantic Information Processing*. MIT Press, Cambridge, MA, pp. 403–418, 1968.

McCorduck, P. *Machines Who Think*. W. Freeman, San Francisco, 1979.

Mitchell, T. *Machine Learning*. McGraw-Hill, New York, 1997.

Newborn, M. and Newborn, M. *Kasparov Versus Deep Blue: Computer Chess Comes of Age*. 1996.

Russell, S.J. and Norvig P. *Artificial Intelligence: A Modern Approach*. Prentice Hall, Englewood Cliffs, NJ, 1995.

Shapiro, S.C., Ed. *Encyclopedia of Artificial Intelligence*, 2nd edn. Wiley, New York, 1992.

Stork, D.G., Ed. *HAL's Legacy: 2001's Computer as Dream and Reality*. MIT Press, Cambridge, MA, 1997.

Webber, B.L. and Nilsson, N.J., Eds. *Readings in Artificial Intelligence*. Morgan Kaufmann, San Mateo, CA, 1981.

Audio Recording, Compact Disk

The compact disk (CD) brought digital recording and playback technology within the reach of the average household. It offered a quality of sound reproduction good enough to satisfy the audiophile, unprecedented recording length within a compact format, and a system that did not damage sound quality with use. These qualities made it far superior to the micro-grooved vinyl disk for the reproduction of sound, and it rendered this long-lived technology obsolete within a decade of its introduction. The technology employed in this product has been applied to a variety of uses, from data storage and retrieval systems to an integrated audio and video (play only) disk for the home user.

Methods of saving and reproducing data have been a major area of research for engineers and scientists from the nineteenth century onwards. Many of the technologies developed have found applications in home audio—from Edison's revolving wax cylinders in the 1880s to magnetic recording tape in the 1940s. At every step, the accuracy of reproduction, the range of sound frequencies captured, and the durability of the recording medium have been progressively improved to reach its zenith with the compact disk.

In the 1960s and 1970s technologies from several different areas of this research were combined into a system of digital encoding and retrieval using optical readers to retrieve the data and computer sampling to turn analog sound into digital code. Much of the research was aimed at video recording, but it was soon applied to audio. Many scientists and business executives involved in this corporate research were audiophiles who wanted a superior method of saving and reproducing sound. The revolving vinyl disk and the system of stamping copies of master recordings provided important paradigms in the research effort. The compact disk was first introduced as a commercial product in 1982 by Sony Corporation of Japan and Philips electronics of the Netherlands.

The research that led to the CD was carried out in the U.S., Japan and Europe. James T. Russell, a physicist at the Battelle Institute, invented a system of using light to read binary code in the 1960s. By 1970 he had a digital-to-optical recording system using tiny bits of light and dark—each about 1 micron in diameter—embedded on a photosensitive platter. A laser read the tiny patterns, and a computer converted the binary code into an electrical signal that could be converted into audio or video streams. I.S. Reed and G. Solomon published a multiple error correction code in 1960 that would be employed in the encoding of data on CDs to detect and correct errors.

In Japan the NHK Technical Research Institute exhibited a pulse code modulation (PCM) digital audio recorder in 1967 that sampled sound and saved the binary data to videotape. Two years later Sony introduced a PCM recorder, and in 1978 it offered a digital audio processor and editor, the PCM-1600, to record companies and radio stations, which used them to make master recordings.

Credit for the idea of the compact disk is usually given to Klaas Compaan, a physicist working for the Philips Company. In 1969 he realized that an RCA system of stamping copies of holograms could be used to reproduce disks holding video images. With his colleague Piet Kramer he devised a glass disk on which they recorded video signals, along with a track of dimples to record the analog sound signal, that were read with a laser beam. They then moved to recording a digital code on the disk and used a digital-to-analog converter to reproduce sound from the encoded binary stream.

In the 1970s Philips, Sony, and several other companies introduced digital systems to save video and audio. In 1978 35 manufacturers of digital recorders met in Tokyo. Philips took the lead in establishing standards for the format of the audio system: the diameter of the disk was finally set at 120 millimeters; the sampling rate was to be 44.1 kHz; and a 16-bit standard was adopted for the encoding of the audio signal. This enabled 74 minutes of sound to be recorded on the disk. Manufacturers agreed to run the data track from the inside to the outside of the disk and use a polycarbonate material (developed by Polygram, a subsidiary of Philips) for the disk substrate.

Philips and Sony collaborated in the development of prototypes, and in 1980 they proposed a set of standards whose worldwide adoption was an important factor in the success of the compact disk. In 1983 CDs were first introduced in the U.S., and 800,000 disks were sold; in 1986 53,000,000 were sold. By 1990 an estimated 1 billion CDs were sold globally. Within a few years after the introduction of the compact disk player, smaller units were available as car stereos and personal audio systems.

Several other products were quickly developed from the technologies used in compact disks. Compact video disks ranging from 120–300 mm in diameter were introduced in the 1980s. CD-ROM (Read Only Memory) units were developed as high-capacity data storage systems for computers. In the 1990s CD-I (Interactive) technology was introduced to merge interactive combinations of sound, pictures, computer texts, and graphics in one format. A combined audio and video 120 mm disk system (Digital Versatile Disk) was launched in 1996, which went on to successfully challenge the VHS video tape cassette recorder.

The compact disk quickly proved to be far superior to its vinyl distant cousin. Not only did it sound better and hold much more music, it was also free of the scratches and background noise that had become an accepted part of sound recording and reproduction. Yet it failed to dislodge the compact magnetic tape cassette from its preeminent position in home audio because it was a play-only format. The union of Sony and Philips was broken in 1981 and the two companies went their separate ways to develop compact disk technology and devise a suitable recorder. Several different systems were introduced in the 1990s ranging from Sony's Mini Disk recorder to the recordable CD-R used in home computers. The popularity of downloading MP3 files from the Internet and burning the sound onto recordable CD-Rs has made this format the most likely successor to the cassette tape.

See also **Audio Recording, Disks; Computer Memory, Personal Computer; Lasers, Applications; Personal Stereo**

ANDRE MILLARD

Further Reading

Everest, F.A., and Streicher, R. *New Stereo Sound Book*, Audio Engineer, New York, 1998.

Guterl, F. Compact Disc. *IEEE Spectrum*, 25 102–108, 1988.

Millard, A. *America on Record: A History of Recorded Sound*. Cambridge University Press, New York, 1995.

Pohlmann, K. *The Compact Disc Handbook*. A-R Editions, Madison, 1992.

Reed, I.S., and Solomon, G. Polynomial Codes Over Certain Finite Fields. *J. Soc. Appl. Math.*, 8, 300–304, 1960.

Read, O., and Welch, W.L., *From Tin Foil to Stereo: Evolution of the Phonograph*. SAMS, Indianapolis, 1976.

Audio Recording, Electronic Methods

The mechanical method of recording sound was invented by Thomas A. Edison in 1877. With the help of mass production of recorded copies for entertainment, cylinder phonographs, and then disk phonographs developed into a major industry during the first quarter of the twentieth century. However, electronic amplification was not available. Modern research has shown a few examples of *pneumatic* amplification, and there were also experiments using electronic approaches. The first published record known to have been made with the help of an electronic amplifier was Guest and Merriman's recordings at the Burial Service of the Unknown Warrior in Westminster Abbey on November 11, 1920 (specifically for the Abbey Restoration Fund and not a commercial issue).

Several components were needed by experimenters: a microphone, an electronic amplifier, a loudspeaker system to hear what was being recorded and to avoid overloading, and an electro-mechanical device which would faithfully transmute the amplified signal into mechanical vibrations for cutting a groove in wax with negligible added background noise. Cutting records and the means for pressing sturdy disk records were commercial trade secrets at that time.

The vital breakthroughs occurred at the American Telephone and Telegraph Company (AT&T) after World War I, when their research section Bell Laboratories began studies for improving transcontinental telephone communication. E.C. Wente developed a microphone in 1916, and in 1924 Henry C. Harrison developed an elaborate theory of "matched impedance" for sending speech long distances without losses. He realized that the same principles could be used for the faithful reproduction and recording of mechanically recorded sound. His team translated these electrical studies using "analogies." For example, an electrical capacitance might be regarded as analogous to a mechanical spring, or the springiness of air in a confined space. Based on these analogies, the team designed a complete system, including microphone, amplifier, loudspeaker, and cutter; *and* an improved mechanical reproducer for judging the results.

The ideas were marketed by AT&T's manufacturing and licensing company Western Electric. As far as can be established, the Western Electric recording system was first used commercially in New York on February 25, 1925. A few earlier examples are now known, and some have been published.

The Western Electric amplifier was originally developed for public-address purposes, and it used electronic valves, or vacuum tubes. Several other recording systems with electronic amplification immediately appeared, demonstrating that parallel research had occurred; but a relatively small amount of electronic amplification can often compensate for classical thermodynamic inefficiencies. Among the new systems were the inventions of P.G.A.H. Voigt for the British Edison Bell Company and Captain H. Round for the British Marconi Company, both of whom developed alternative microphones.

A microphone changes sound into alternating electricity and should introduce little distortion to the frequencies picked up. It should have the same frequency response in all directions, and its linearity to acoustic waveforms should be uniform. However, by the end of the twentieth century, electronic amplification had not enabled extremely faint sounds (detectable by a healthy human ear) to be recorded. There was always added random noise arising from acoustic causes, electronic causes, and inefficiencies in acoustical–mechanical transducers. Microphone users had to select an instrument suited to the proposed application.

Although the earliest optical film sound experiments probably did not use electronic amplification, both Western Electric and the merged RCA-Victor Company developed independent methods of recording sound on optical film with the help of electronics. Here, the two methods had to be "compatible" so that any film could be shown in any theater, and this situation remained true with only minor exceptions until the 1980s.

Film studios provided two new electronic techniques widely understood today but much less so before the 1960s. "Automatic volume limiting" protected the fragile "light valves," with the side effect of improving speech intelligibility for cinema reproduction. The difficulties of recording foreign languages led to what is now called "multitrack recording," in which two or more sounds recorded at different times could be kept in synchronism and modified as necessary. Hollywood evolved the principle of three synchronous soundtracks for music, sound effects, and dialog, to facilitate the adaptation of films for foreign markets.

The second technique was "negative feedback." This principle enabled a high degree of amplification to be traded for other purposes by feeding some of the output (reversed in phase) back to a previous stage. In this process, nonlinearity and deviations in frequency response may be reduced. Mechanical transducers could be built with "motional feedback" with the same results. In sound recording, motional feedback was first used in disk equipment for the National Association of Broadcasters in America. This was a different system from that used in commercial recording, and because of the outbreak of World War II, the commercial implementation of this principle was delayed until 1949. Cutting stereophonic disks would have been impossible, however, without motional feedback.

Classical analog information theory stated that frequency range could be traded against the need for amplification. At the start of the 1930s, both films and mechanical disk records were restricted to an upper frequency range of about 5 kHz. This range was gradually extended by trading amplification for electromechanical efficiency. It is generally accepted that the full audio frequency range was first achieved by the English Decca Company during 1944 (as a spinoff from war research). An upper limit of 14 kHz was obtained for microphones, amplifiers, and disk pickups.

During World War II, German engineers rediscovered a patented American invention for reducing harmonic distortion on magnetic media due to hysteresis. The principle of ultrasonic alternating current bias greatly improved the linearity and reduced the background noise of magnetic tape, causing much debate about precisely how the principle worked, which still has not received a complete explanation. In analog sound recording, a useful element of feedback happens with this process. If a small particle of dirt should get between the tape and the recording head, the bass goes down and the treble rises. When the electronics are correctly set up, these effects cancel each other; and as a result, analog magnetic tape became the preferred mastering format for professionals. This format required even more amplification; but by the mid-1950s, magnetic tape was much simpler and cheaper than mechanical or optical recording, so it remained the favorite technology for analog recording until the end of the century. Digital audio recording using magnetic tape and optical methods began to overtake it after about 1980, but no fully digital microphone was developed.

See also **Audio Recording, Compact Disk; Audio Recording, Disks; Audio Recording, Stereo and Surround Sound; Audio Recording, Tape; Film and Cinema, Early Sound Films; Film and Cinema, High Fidelity to Surround Sound; Valves/Vacuum Tubes.**

PETER COPELAND

Further Reading

Borwick, J., Ed. *Sound Recording Practice*, 4th edn. Association of Professional Recording Studios, Oxford University Press, Oxford, 1994.

Copeland, P. *Sound Recordings*. British Library, London, 1991.

Langford-Smith, F. *Radio Designer's Handbook*, 4th edn. RCA, U.S., 1953. Chapters 12 to 21 cover all aspects of "Audio Frequencies."

Millard, A. *America on Record: A History of Recorded Sound*. Cambridge University Press, Cambridge, 1995

Morton, D. Off The Record: The Technology and Culture of Sound Recording in America. Rutgers University Press, New Brunswick, 2000.

Audio Recording, Mechanical

Mechanical recording and reproduction of sound on disk followed ten years after Thomas Edison's invention of these techniques on cylinder. In 1887, Emile Berliner filed for and received his patent on the gramophone recording and reproduction system. Shortly after 1900, technical developments and commercial agreements in the U. S. established disk records as an independent medium. By 1910 they dominated global markets for recorded sound. Mechanical recording remained the technique of choice for the storage of sound for the next 40 to 50 years, and disk "records" continued to be a significant medium for consumers of music and other sounds into the 1980s.

Disk recording requires that sound waves be converted into mechanical force that makes a stylus engrave or otherwise reproduce the waves as a spiral groove on a spinning disk composed of some pliant material. To play back the sound wave on the same disk or on a copy stamped from a matrix, a stylus tracks the groove representing the wave. This energy is then converted into mechanical force that drives one or more loudspeakers. The development of this technology during the twentieth century is one of incremental improvements punctuated by revolutionary changes in technique and materials.

Recording changed very little in form from 1901, when Eldridge Johnson's synthesis of earlier inventors' work with his own insights became the Victor Talking Machine Company, to 1925. Performers directed their voices or instruments at a horn that channeled sound waves to a diaphragm. The membrane was linked to a stylus driven by a feed screw along the radius of a metallic wax disk spinning between 74 and 82 revolutions per minute (rpm). Miniature flyball governors controlled the speed of the turntable so that the record was rotated at a constant speed. The stylus cut the wax vertically ("hill and dale") or laterally.

Once recorded, the disk was electroplated, creating a master from which records or secondary stampers could be made. Victor and its partner companies made records from mineral powder, shellac, and carbon black; Columbia Records either copied Victor's technique or used pure shellac laminated over a paper core. Edison had also moved from cylinders to records, and the "Diamond Disks" made from 1912 to 1929 used a wood flour base coated with a varnish called condensite. Consumers played records on spring-wound gramophones or talking machines where the process for recording was reversed through steel stylus, diaphragm, tone arm, and horn.

Improvements in diaphragms, styluses, room acoustics, and horn placement enabled increases in the signal-to-noise ratio and frequency range while reducing distortion in the reproduced sound. By the late 1910s, companies had standardized record diameters at 10 and 12 inches (254 and 305 mm), playing three and five minutes respectively, and progressed from recording individual performers of selected instruments to symphony orchestras.

The revolution of electronic recording with microphones and amplifiers took place after World War I. Joseph Maxfield and Henry Harrison of Bell Telephone Laboratories applied scientific techniques of research and development to extend the undistorted frequency range of recording from 220–4,000 Hz to 100–6,000 Hz, and to raise the signal–noise ratio to 32 dB. To meet competition from wireless broadcasting, record companies began licensing the Bell system in 1925.

Electronic recording stimulated systems improvements and standard speeds in reproduction. For the consumer and home market, the "78" record (78.26 rpm) was standard, and the broadcast and film industries used 33.33 rpm. Electronic amplification reduced the need for abrasive fillers that resisted the stylus pressure of mechanical reproduction, while the higher frequencies aggravated the sound of needle "scratch" generated at the interface of stylus and the record groove.

Despite the global economic depression, the 1930s were a fertile period for recording innovations. Alan D. Blumlein of Electric & Musical Industries and Arthur C. Keller of Bell Laboratories invented single-groove stereophonic recording systems, for which their corporations could not find markets. Record companies began recording on metal disks coated with man-made lacquers and stamping disks of man-made plastics.

Researchers learned more about acoustics, recording and plating techniques, and the science of the stylus–groove interface. In the latter field, Frederic V. Hunt and John Pierce of Harvard University developed a stylus and pickup ten times lighter than contemporary devices.

Hunt and Pierce also anticipated the advantages of the narrower microgrooves that RCA Victor engineers began applying in 1939 to their vinyl chloride–vinyl acetate copolymer, 7 inch, 45 rpm, record system. World War II and rising sales deferred replacement of the 78 until 1949, a year after Columbia Records introduced vinyl compound, 10- and 12-inch, 33.33 rpm long-playing records (LPs) using RCA's microgrooves. The jukebox industry quickly adopted the 45, which survived to the end of the century in the pop singles market. The new records gave less distorted sound reproduction up to 10,000 Hz with a 60 dB signal–noise ratio.

At the time, all music was monophonically recorded and reproduced. The postwar growth of magnetic recording techniques enabled multitrack recording, which led to the international standard for stereophonic disk reproduction in 1958. During this time, engineers refined the art of microphone placement and recording techniques to produce what has become known as the "Golden Age" of recording.

Efforts to put four channels on disk in the early 1970s failed because of the lack of a standard and the rise of magnetic-tape cassettes. Cassettes surpassed LP sales in the United States in the late 1970s, and this was followed by the introduction of the compact disk (CD) ten years later. Major labels abandoned vinyl records in 1990, although some production continued in the U.K. and on independent record labels.

The main refuge for disk records became audiophile and reissue pressings and discotheques, where disk jockeys had been playing extended versions of popular songs on 12-inch records since the early 1970s. The playing time of 5 to 12 minutes permitted increased amplitude of the groove and a concomitant rise in dynamic range. Ironically, in the middle of the digital revolution of CDs, the cult success of vinyl disks forced the major companies to continue pressing LPs and 45s for top performers through the end of the century.

See also **Audio Recording, Compact Disk; Audio Recording, Electronic Methods; Audio Recording, Stereo and Surround Sound; Audio Recording, Tape; Personal Stereo**

ALEXANDER B. MAGOUN

Further Reading

Gelatt, R. *The Fabulous Phonograph, 1877–1977.* Macmillan, New York, 1977. While somewhat dated, still the best single narrative.

Gronow, P. and Ilpo, S., translated by Moseley, C. *An International History of the Recording Industry.* Cassell, London, 1998. A broader survey, focusing on performers.

The phonograph and sound recording after one-hundred years. *J. Aud. Eng. Soc.,* 25, 10/11, October/November 1977. With survey articles on aspects of the phonograph system edited by the former chief engineer for RCA Records.

Guy M. and Andrews F., Eds. *Encyclopedia of Recorded Sound in the United States.* Garland, New York, 1993. An ambitious one-volume compendium that covers the art and science of the subject.

Read, O. and Welch, W.L. *From Tin Foil to Stereo: Evolution of the Phonograph,* 2nd edn. SAMS, Indianapolis, 1976. Considered the best history of the phonograph.

Audio Recording, Stereophonic and Surround Sound

It is perhaps difficult to understand the problems experienced by early listeners when they first heard "music coming out of a hole." It was an effect that annoyed some people much more than others, although few had even encountered it before the twentieth century. Together with the difficulties of low amplification and high background noise, this effect proved significant and was a further problem in sound recording and reproduction.

In the late twentieth century, the words "stereophonic" and "stereo" refer to a sound along a horizontal line, usually between two loudspeakers along one wall of a room. Good stereo can simulate a live performance, but the expression is restricted to loudspeaker listening. Listening on headphones is quite different, because our hearing has evolved to take advantage of two different phenomena. For natural sounds (and for loudspeakers), our heads form a "sound shadow," so high-pitched sounds from the left are picked up by the left ear very much louder than the right ear. For low-pitched sounds, the shadow-effect is practically nonexistent. What little directionality we possess depends upon differences in the *time* the sound arrives at the two ears. If recordings were made with microphones the same distance apart as our ears (about 17 centimeters), and something the same general shape and size as a human head were put between the microphones, we might have much more faithful stereo reproduction than any loudspeakers could give. This is called "dummy head" or "binaural" stereo. We learn directional hearing in the first year of life, correlating the senses of sight and stereo sound with feelings of balance and the sensations our muscles give us as they move. With headphone listening, the acoustic fidelity is certainly better; but when the listener moves his head, the earpieces move too and the sound remains the same, yet his other senses indicate the exact opposite.

The first publicized experiment into spatial sound reproduction was performed by Clement Ader in Paris in 1881. He placed a row of microphones along the footlights of the Paris Opera House and sent signals by telephone lines to earpieces at an electrical exhibition 3 kilometers away. A spatial effect was reproduced. American workers were first to achieve stereo sound recording. Closed-circuit experiments established that just one pair of microphones and loudspeakers would reduce the disadvantage of "sound coming out of a hole;" but attempts to reproduce physical movement showed that if the microphones were more than a few feet apart, a new "hole in the middle" might appear. A third central microphone-and-loudspeaker arrangement reduced this "hole in the middle," demonstrating that three were much better in practice.

In 1932 a significant experiment was performed using landline connections in Philadelphia, which seemed to show that for orchestral music, three channels would be sufficient. While significant, it was not a definitive experiment because there were apparently no comparisons with more channels and the vertical element was not featured. Nevertheless this established the benchmark against which future systems would be compared. This experiment led to the first stereo recordings in 1932 of the Philadelphia Orchestra conducted by Leopold Stokowski, recorded on a two-track disk at Bell Laboratories in New Jersey. Because there was only one experimental stereo disk cutter, continuous performances could not be recorded. In 1979 Bell Labs reissued the surviving material on two charity stereo long-playing records (LPs) with the missing sections in monophonic sound, clearly showing the advantage of stereo.

On the other side of the Atlantic, work started from different basic principles. It was noticed that when mono sound was reproduced on two identical loudspeakers equidistant from the listener, the sound appeared to come from a point midway between the two speakers. By dividing the same mono sound into suitable proportions and sending it to two speakers, it could appear to come from anywhere between the two. This was the claim in Alan Blumlein's British patent in 1931. Blumlein

was an employee of the record company EMI, but he considered the principal application to be soundtracks for films, unlike in the U.S. where the initial work was for sound alone. Blumlein's work showed how suitable signals might be captured by two bidirectional microphones at the same point in space. It also showed how the correct balance for monophonic listeners would result from mixing the left and right in equal proportions, so downward compatibility would occur. It also discussed how stereo sounds might be recorded on disk or film, including some remarkable predictions of what occurred decades later. The important point was that Blumlein showed that only two discrete channels could be sufficient.

Blumlein's work also mentioned a device to "steer" a mono sound between two loudspeakers, now called a "panpot." In 1934 the EMI Amateur Dramatic Society made a short stereo film called *Move The Orchestra*, in which a single restaurant scene achieved exactly that.

The public heard stereophonic sound for the first time in 1939 in the Walt Disney film *Fantasia*. In the "Fantasound" system, three discrete channels were printed onto a separate optical film that held the three soundtracks side-by-side, together with a fourth track that controlled the volumes of the other three, giving wider dynamic range. During the last musical piece (*Ave Maria*), loudspeakers at the back of the auditorium could be switched on for the choral effect. This was not only the first time the public had heard stereo, but it was the first time they encountered "surround sound." The outbreak of World War II brought all these advances to a halt.

After LPs became established in the 1950s, techniques were researched to get two channels of sound into one groove. For commercial master sound recording in the U.S., the three traditional discrete channels were initially recorded on $\frac{1}{2}$-inch magnetic tape. In Europe, EMI issued the first British "stereosonic" $\frac{1}{4}$-inch magnetic tapes in 1955 using Blumlein's principles. They were intended for loudspeaker listening; Blumlein's "coincident microphone" technique did not capture the time information needed for headphones. The standard form of stereo disk that carried two sounds in one v-shaped groove was launched in 1958. A single cutter vibrating in two directions mutually at right angles imparted two sounds onto the two groove walls, which gave compatibility between mono and stereo disk reproduction.

Although human eyes see an almost circular field of vision, most practical action takes place on a narrower plane. Widescreen filmmakers followed

American thinking: a different microphone for each loudspeaker behind the screen. But dialog was found to give problems since it was upset both by performance considerations and the acoustic properties of studios. Subsequent stereo films used sound picked up on mono microphones and "panned" into place. Ultimately, mono became standard for dialog, with only music and effects in stereo.

During the first half of the 1970s, the high fidelity, or hi-fi, market stimulated the idea of "quadraphonic" sound reproduction. The idea was a commendable one, to reproduce the entire acoustic environment of a concert hall rather than just the stage. The concept failed for several reasons, but it can be argued that the sophisticated engineering research needed for quadraphonic LPs helped prolong the life of mechanical disk records until the end of the century.

Until digital methods came to the rescue, the shortage of analog channels continued to hamper surround sound. Dolby Laboratories had invented methods of improving the perceived signal-to-noise ratio of analog recording media, and they also found ways to encode more channels of audio with extra information for cinema audiences, while retaining downward compatibility with stereo and mono. The first "Dolby Surround" films date from 1976.

When video recording became suitable for domestic use, the same principles became available to domestic listeners. At the end of the twentieth century, there were several ways of encoding surround sound. The dominant one used 5.5 channels (the "half" being low-pitched sounds such as aircraft, guns, and bombs normally requiring a large loudspeaker, which could be placed behind the viewer). It seems certain that further developments will occur in the twenty-first century. The *vertical* element of surround sound, for example, is not yet featured, although it is an increasingly important part of our environment.

See also **Audio Recording, Electronic Methods; Film and Cinema, High Fidelity to Surround Sound; Loudspeakers and Earphones; Personal Stereo; Television Recording, Tape**

PETER COPELAND

Further Reading

Copeland, P. *Sound Recordings.* British Library, London, 1991.
Hadden, H.B. *Practical Stereophony.* Iliffe Books, London, 1964.
Millard, A. *America on Record: A History of Recorded Sound.* Cambridge University Press, Cambridge, 1995.
Rumsey, F. *Spatial Audio,* Focal Press, 2001.

Audio Recording, Tape

Tape recorders were the dominant technology for sound recording in the second half of the twentieth century. By combining low cost, high fidelity, and flexibility in editing, they came to dominate both consumer and professional recording during this period. In addition to sound, tape recorders were used for video and data recording, which are discussed in separate entries.

Tape recorders are characterized by the use of a recording medium with a flat cross section. The tape is moved rapidly past a recording head, which generates a magnetic field in response to an electrical input, normally from a microphone. The magnetic field leaves the tape permanently magnetized. The field on the tape is then played back by moving the tape past a reproducing head, which senses the field and produces a signal that is amplified to recreate the original sound.

Valdemar Poulsen, inventor of the first wire recorder in 1898, also experimented with tape recorders around 1900. His machines used a solid metal tape, which provided superior performance since the recording medium always had the same geometrical relationship to the reproducing head as the recording head (wire could twist about its axis, resulting in variable sound quality). However, solid metal tape was much more expensive than wire, making tape recorders too costly for most applications.

As a result, solid metal tape recorders were built only for highly specialized applications during the 1920s and 1930s. Lorenz in Germany and Marconi in the U.K. built radio broadcast recorders using solid metal tape, and AT&T in the U.S. built telephone announcing systems, but their high cost meant that only a handful of each type of machine entered service. The key to more widespread use of tape recorders was the development of particle-coated tape by the Austrian inventor Fritz Pfleumer in 1928. By coating paper tape with small magnetic particles, he dramatically reduced the cost of magnetic tape. Coated tape was also physically much more flexible than solid metal tape, and this made the design of transport mechanisms for the tape much simpler.

Pfleumer sold his invention to the German firm AEG, which then partnered with the BASF division of I. G. Farben to develop a tape recorder. AEG developed and manufactured the machine itself, while BASF developed and manufactured the tape. BASF soon replaced paper with plastic as the backing material, further lowering costs. The AEG machine was first marketed as the Magnetophone in 1935. The firm developed a variety of models for civilian, government, and military use over the next ten years. The Nazi government made extensive use of the Magnetophone for both security and radio broadcast purposes, and the demand for higher quality reproduction drove the development of a number of innovations. The most significant of these were the invention and widespread use of the ring head, which produced more intense and defined magnetic fields on the tape, and AC-bias, an electronic noise reduction technique. AC-bias was the addition of a very high frequency tone to the input signal during recording, which dramatically reduced noise and made tape recording suitable for music as well as voice recording.

The developments at AEG and BASF were copied in the laboratory in the U.S. and the Soviet Union during World War II, but it was not until the general dissemination of knowledge about the Magnetophone after Germany was occupied by Allied forces that there was widespread interest in tape recording. The Brush Development Company in the U.S., under the technical leadership of Dr. Semi Joseph Begun, produced the first American tape recorder, the Soundmirror, in 1947. Brush was soon followed by a number of other companies, most notably Ampex, which produced a high-quality recorder suitable for radio broadcast use. Using a combination of German technology and American wartime research, these and other firms in the U.S., Europe, and Japan soon offered a variety of products whose performance matched or exceeded competing recording technologies.

By the early 1950s, tape recorders were the dominant method for making professional sound recordings. They had rapidly replaced other forms of recording in the motion picture industry and were increasingly used in radio broadcasting and the music recording industry. The success of tape recording was due not only to lower cost but also to the ease of editing. Unlike phonograph recordings, magnetic tape could easily be cut and pasted together, much like motion picture film, to make a high quality product. Such editing had been possible with sound-on-film systems, but magnetic recordings could be edited immediately rather than waiting for film to be developed.

The popularity of tape recording for professional use increased further in the 1960s with the development and widespread use of multitrack tape recorders. In addition to the basic ability to separate vocal and instrumental tracks, these recorders provided much easier sound editing in the studio and also allowed artists to experiment with new forms of musical expression. Analog multitrack recorders began to be replaced in the mid-1970s by digital multitrack recorders for studio use. By the end of the twentieth century, most professional recording was digital and increasingly recorded on hard disks or other computer-related formats rather than on tape recorders.

The use of tape recorders by consumers lagged behind that of the professional world. Reel-to-reel tape recorders, the primary consumer versions of the 1950s, were relatively inconvenient to use compared with phonograph record players and other consumer electronics and were usually limited to fixed installations in the home. A variety of cassette formats were developed to address this problem, but it was not until the introduction of the Phillips Compact Cassette (so common by the end of the twentieth century that it was simply referred to as "the cassette" as if it were the only one) and the 8-Track Cartridge in the 1960s that cassettes were widely adopted.

The 8-Track was the first successful high fidelity cassette system, and it was installed in many vehicles in the 1960s and 1970s. Because of the greater physical size and complexity, it eventually lost out to the Phillips cassette. The Phillips format was originally intended for voice recording only, but the development of new magnetic particle formulations in the 1970s, along with the introduction of Dolby® noise reduction, improved sound quality of the Phillips cassette to the point that it matched the performance of the 8-Track and the LP phonograph record and so gained consumer acceptance. At the end of the twentieth century, tape recorders and especially the Phillips cassette were still the dominant form of consumer recording technology, but they were increasingly being supplemented by newer digital recording formats primarily derived from the computer industry.

See also **Audio Recording, Wire; Computer Memory; Television Recording, Tape**

MARK H. CLARK

Further Reading

Daniel, E.D. and Mee, C.D. *Magnetic Storage Technology,* 2nd edn. McGraw-Hill, New York, 1996.

Daniel, E.D., Mee, C.D. and Clark, M.H. *Magnetic Recording: The First 100 Years*. IEEE Press, 1999.

Morton D. *America on Record*. Rutgers University Press, New Brunswick, 2000.

Audio Recording, Wire

The recording of sound using the principle of remnant magnetism (the residual magnetic fields present in some materials after they are exposed to magnetic fields of sufficient strength) was first proposed in 1878 by the American inventor Oberlin Smith. His ideas, later recorded in a patent caveat, are the intellectual starting point for the development of all modern magnetic recording systems.

Wire recorders are characterized by the use of a solid recording medium with a small circular cross section. The wire is moved rapidly past a recording head, which generates a magnetic field in response to an electrical input, normally from a microphone. The magnetic field leaves the wire permanently magnetized. The field on the wire is then played by moving the wire past a reproducing head, which senses the field and produces a recorded signal that is amplified and played back to recreate the original sound.

The primary advantage of the wire-recording medium is its low cost as compared with solid metal tape; machinery used in piano wire manufacture was easily used to produce wire for magnetic recording. Wire has many disadvantages however. In contrast to tape, which always presents the same orientation to the reproducing head, wire can twist in the transport mechanism, leading to variations in reproduction volume. As a result, wire recorders can record only a single track, in contrast to the multiple track possibilities of tape recorders. Wire also is harder to splice than tape, and it easily becomes snarled if it escapes from the reels on which it is wound. Finally, steel with ideal magnetic characteristics is very stiff and difficult to pull through transport mechanisms. This problem can be dealt with by using coated wire (developed in the early 1940s), although the higher cost reduces wire's economic advantages.

The primary application of wire recorders is sound recording. Recordings are analog recordings; that is, the magnetic pattern produced on the wire is an analog copy of the original sound. Wire recorders were also occasionally used for the digital recording of telegraph and radio telegraph signals early in the twentieth century.

The first successful magnetic recorder of any kind was a wire recorder constructed by the Danish inventor Valdemar Poulsen in 1898. Poulsen had been attempting to develop a telephone answering machine, and so he named his invention the Telegraphone (roughly meaning "distant sound recorder)." With the assistance of Danish investors led by the entrepreneur Lemvig Fog, Poulsen began to offer commercial machines in 1903. By that time, Poulsen had abandoned magnetic recording for research on radio transmission, and he and his collaborators turned further development of the telegraphone over to the American Telegraphone Company in the U.S.

After delays caused by financial and technical problems, wire recorders were offered in small numbers to the public by American Telegraphone during the 1910s and 1920s. Total sales amounted to fewer than 300 machines due to high prices, poor sound reproduction quality, and worse quality control. American Telegraphone entered bankruptcy proceedings in 1919 and ceased manufacturing in 1928, though the firm was not formally out of business until 1941.

During the late 1920s and 1930s, the primary manufacturer of wire recorders was the Echophone Company in Germany, later sold to Lorenz. The Textophone (later renamed the Dailygraph) was similar in design to the machine produced by the American Telegraphone Company, but incorporated an electronic tube amplifier to improve sound quality. The quality of the machines was higher than American Telegraphone products, but prices were just as high. As a result, the device was a niche product, selling primarily to wealthy businessmen interested in cutting-edge technology until 1933. With the rise of the Nazi Party to power after that date, Lorenz sold an increasing number of machines to the Gestapo and other state security agencies in Germany for wiretaps and the recording of interrogations. However, by 1939 competition from tape recorders produced by AEG led to the end of wire recorder production in Germany.

Although several organizations in the Soviet Union and firms in the U.S., most notably AT&T through its Bell Telephone Laboratories subsidiary, experimented with magnetic recording during the 1930s, production of wire recorders for consumer use did not take place until World War II in either country. Thus, at the same time wire recorder production was ending in Germany, it was increasing dramatically in the USSR and the U.S. However the military provided the major market for wire recorders in both countries. The primary application was in situations where vibration rendered other forms of recording impossible. Wire recorders were used primarily by reconnaissance pilots to record their comments while in flight and by journalists reporting from frontline locations.

Most military recorders used cassette designs to eliminate problems with threading the fine wire through a recording head. Cassettes also minimized the chance of wire snarls because of wire coming off the reels. Improved recording heads and the development of coated wire (an outer magnetic layer electroplated on a more flexible nonmagnetic inner core) meant that the sound quality of wartime wire recorders was much better than before. However, the most notable improvement in sound quality resulted from the introduction of alternating current (AC) bias, an electronic noise reduction technique originally developed for tape recorders.

The Armor Research Foundation was the leading wartime developer of wire recorders. Armor patented much of its work and then licensed its patents to dozens of firms in the U.S. and abroad in the late 1940s. These firms produced essentially all the wire recorders aimed at the consumer market in the U. S. and Europe after World War II. Annual production of consumer wire recorders peaked at several million around 1950.

In contrast to the high quality of military recorders built during World War II, postwar consumer wire recorders were largely inexpensive units built as cheaply as possible. In particular, very few of them incorporated cassette mechanisms. As a result, snarls and other malfunctions were a constant problem. Consumers put up with these difficulties since wire recorders sold for considerably less than tape recorders in the immediate postwar period. Wire recorders vanished from the consumer market by the mid-1950s as the price of tape recorders using coated plastic tape dropped. Although they continued to be used in special applications through the 1960s, wire recorders have essentially disappeared with the development of improved tape materials.

See also **Alloys, Magnetic; Audio Recording, Tape**

MARK H. CLARK

Further Reading

Daniel, E.D. and Mee, C.D. *Magnetic Storage Technology*, 2nd edn. McGraw-Hill, New York, 1996.

Morton D. *America on Record*. Rutgers University Press, New Brunswick, 2000.

Daniel, E.D., Mee, C.D. and Clark, M.H. *Magnetic Recording: The First 100 Years*. IEEE Press, 1999.

Audio Systems

At the start of the twentieth century, mechanical recording and reproduction of sound by cylinder or disk phonographs was dominant (Edison's cylinder, Edison's Diamond Disk, Emile Berliner's gramophone, and the Victor Talking Machine Company's Victrola). Playback quality was limited by the lack of amplification methods to increase the volume of the recorded performance above the background noise of the mechanical record. Thus inexperienced performers were rarely used. Electronic recording methods with microphones and amplifiers were developed after World War II, and valve amplifiers and the transistor, which created more efficient signal amplification, caused a revolution in audio playback.

Throughout the twentieth century, turntables or decks for playing disk records were normally kept horizontal, so the downward pressure of the stylus would be constant across the disk's surface. Eldridge Johnson, founder of the Victor Talking Machine Company in 1901, devised a spring motor to rotate the turntable at an even speed in 1896, and battery motors were also used prior to the electrical drives developed in 1925. Although disk records were mastered on lathe-type mechanisms that moved the cutter in a straight line across the disk, records were usually played with a pivoted "pickup arm" that made a curved rather than straight track across the record, introducing distortion as well as wearing the record. Careful geometric design and motors that moved the rear end of the arm reduced tracking error and improved audio quality.

Early playback was acoustomechanical, as was sound recording. Following the development of electronic sound recording, the first fully electrical playback device, the Brunswick "Panatrope" in 1926, used magnetic playback pickups (cartridges) to convert the motion of the stylus in the groove to an electrical signal. Brunswick's player also incorporated an electronic amplifier using the new vacuum tubes, and the electrical signal could be played through a loudspeaker.

The earliest material for disk records was called "shellac." Actually a mixture of materials, shellac was an organic binding material for which no synthetic substitute was found. Before the first electronic version, sounds were reproduced and amplified mechanically. The acoustic energy was emitted by a "soundbox" and fed into a conical horn providing a suitable acoustical load. Ultimately, this energy came from the rotation of the disk itself. Therefore the disks had to resist wear; and to this day, shellac has proven to be one of the longest lasting sound storage media. Shellac was used until about 1955 in the U.S., 1959 in the U.K, and even later in developing countries. The major drawback to shellac is that it is very prone to breakage.

During World War II the Allied forces made effective use of music for maintaining the morale of the troops, and music had to be distributed without risk of breakage. Shellac was gradually replaced with "vinyl" (polyvinyl chloride), which was flexible and had much less "grain," meaning lower background noise. Until then, heavy steel needles, or pickups, were used for playing disks, typically with downward pressures of 100–150 grams. When long-playing disks (LPs) were introduced after the war, the same vinyl was used. It was much more expensive than shellac, but the longer playing time offset this. Pressures from 10 to 20 grams became normal, with reduced stylus tip size. Vinyl disks also featured extended frequency range, and the extended range of good pickups also reduced wear. Also, because of vinyl's inherent flexibility, turntable designers were forced to introduce a pickup with a carefully crafted playback stylus made of jewel such as sapphire to resist wear. A number of mathematical studies showed how to optimize various dimensions of quality sound to achieve faithful reproduction, low distortion, and simultaneously low damage to disks. Vinyl records were still used by enthusiasts at the end of the twentieth century.

The faithful capturing of sound performed under "normal" conditions was led by radio broadcasting in the early 1920s. The technology of the electronic amplifier aided both the recording and the reproduction of sound. The first electronic amplifiers were based on thermionic valves, or vacuum tubes. Since vacuum tube technology was familiar to radio engineers, the principal developments in both radio and audio work tended to be described in books and magazines for the radio market. The demobilization of armed forces after World War II and expansion of the consumer market seems to have led directly to pure audio research and literature.

From 1950 onward, the pursuit of innovative, if not always faithful, sound reproduction was fed by specialist manufacturers, dealers, exhibitions, and advertising claiming "high fidelity," or hi-fi, playback. A disk deck tended to form the centerpiece of a practical domestic hi-fi system. Tape decks could also used for playing reels of magnetic tape, with cassette decks added for analog audiocassettes after 1970. A tuner, primarily for receiving analog frequency-modulated (FM) radio broadcasts, was a normal component.

The actual amplification took place in two or three separate units, depending on whether monophonic or stereophonic reproduction was intended. The preamplifier brought the various sources of audio to similar electronic voltages, allowed the bass and treble ranges to be controlled for individual preference and, in the case of the disk deck, applied equalization. This compensated for deliberate changes in frequency response to improve the capacity of a mechanical disk record, called a recording characteristic, internationally standardized in 1955.

In vacuum tube days, the main amplifier (or amplifiers for stereo) tended to be a heavy, delicate, and hot unit, so it was usually placed on the floor out of the way. It had enough output power to feed a separate loudspeaker enclosure, or two for stereo. Hi-fi loudspeakers might require between 5 and 50 watts apiece. Most 1950s designs were no more efficient than those at the end of the century, the electroacoustic efficiency being only about 1 to 3 percent. Quality transistorized amplifiers became dominant toward the end of the 1960s.

Each loudspeaker cabinet often contained several electrical-to-acoustical transducers covering different parts of the frequency range, with the audio being suitably divided by a "crossover unit." Compared with anything else in the high-fidelity system, loudspeakers had the most radical effects upon the reproduced sound quality. They were relatively cheap and simple for enthusiasts to assemble, and many did.

As the actual hardware became cheaper, with greater appeal to nonspecialists, the hi-fi craze became democratized. No longer was it confined to the wealthy or dedicated enthusiast. By the end of the twentieth century, everyone could enjoy high fidelity sound very comparable to leading technology of the 1950s.

See also **Audio Recording, Disks; Audio Recording, Electronic Methods; Audio Recording, Tape; Audio Recording, Wire; Loudspeakers and Earphones; Personal Stereo**

PETER COPELAND

Further Reading

Copeland, P. *Sound Recordings.* British Library, London, 1991
Crabbe, J. *Hi-Fi In The Home,* Blandford Press, 1968.
Day, T. *A Century of Recorded Music: Listening to Musical History.* Yale University Press, New Haven, CT, 2001.

Read, O. and Welch, W.L. *From Tin Foil to Stereo: Evolution of the Phonograph*. SAMS, Indianapolis, 1976.

Wilson, P. and Webb, G.W. *Modern Gramophones and Electrical Reproducers*. Cassell & Company, 1929.

Audio Systems, Personal Stereo, *see* **Personal Stereo**

Audiology, Hearing Aids

Treatment and testing for hearing disorders can be traced back as far as the first century BC. However, nothing prior to the sixteenth century indicates that any consistent use or application of hearing assistance devices existed.

The fundamental function of early hearing aids was to amplify sound. Before electricity, the only way to do that was to filter out other noise by directing the desired sound straight into the ear with some kind of tube or trumpet. Ear trumpets were first used by sailors and others who needed to communicate over long distances, and later on were adopted for use by the hearing impaired. Most early trumpets were custom made, and the real business of manufacturing and selling hearing aids only began around 1800. By the end of the nineteenth century a large variety of ear tubes and trumpets were available in many styles and designs, ranging from cheap devices made of tin or hard rubber to the more expensive ones constructed of more valuable materials. The more expensive models were often treated like jewellery but did not necessarily work better. Cupping the hand behind the ear makes sounds 5 to 10 decibels (dB) louder; and depending on their size and shape, ear trumpets could amplify by about 10 to 20 dB, with most of this in the range of 500 to 1000 Hertz (Hz). As the range of human speech is 300 to 3000 Hz, ear trumpets could only help people with mild hearing impairments.

Auricles and cornets were developed as an alternative to the ear trumpet. It was hoped that the device would be less observable on the user. Other hearing assistance devices used through the early part of the 1800s were bone conduction aids. Sounds are transmitted to the ear by vibrations in the air, but also by vibration of the bones in the skull. Bone conduction devices had been tested since the sixteenth century and were typically designed in two ways. The first design consisted of an apparatus held to the speaker's mouth or throat while the opposite end was held in the listener's teeth. The primary constraint of this model was the restriction on speaking distance and the number of persons involved in conversa-tion. Allowing a greater distance from the speaker, the second design involved an instrument that collected sound energy by means of a flat transmitting surface from the surrounding air, something comparable to an acoustic fan. Inventors and physicians however were not satisfied with these devices, and by the end of the 1800s efforts were being made to produce a hearing aid more powerful and effective than the ear trumpet or bone conduction aids.

Electrical hearing aids were introduced just after 1900 and brought useful amplification to a wider audience. It is uncertain who invented the first electric hearing aid, but it may have been Miller Reese Hutchinson in 1898. This first electric hearing aid came to market in 1901 and used the transmitting potential of carbon in order to amplify the sound. There existed, however, substantial static noise and distortion caused by the carbon hearing aid. Such aids were also very large and impractical, but their effectiveness in amplifying sound surpassed any of the prior devices used. They offered the same amplification as ear trumpets, but covered a wider frequency range of 500 to 1800 Hz. All such devices consisted of the carbon microphone (technology borrowed from the telephone), processing unit, battery box and headpiece. Batteries often did not last more than a few hours and were very expensive. Later models with multiple microphones provided 25 to 30 dB of amplification, and the introduction of amplifiers in the 1920s increased the range to 45 to 50 dB.

The development of the vacuum tube hearing aids in the 1920s offered a device with an even greater amount of power and reduced slightly the size of the processing components. Wearable multipart hearing aids were then developed and used in the 1930s and 1940s. The only drawback with these aids was that batteries remained large and still far from invisible. A separate battery pack was needed to warm the vacuum tubes. In addition, most of the earlier large-size batteries did not last more than a day and had to be carried in special cases. This problem was resolved in 1947 with the invention of the transistor. As transistors got smaller, so did hearing aids, and concealment became an achievable and important goal.

By the late 1950s and early 1960s, models that were "at-the-ear" or "over-the-ear" combined a microphone with a battery and transistor in one unit and could amplify sounds within the range 400 to 4,000 Hz. These models were molded with custom-made ear tubes, and were easy to conceal behind the ear or under the hair. In the 1970s, batteries became even smaller, allowing "in-the-canal" aids to fill the ear canal without anything worn outside the ear. By the late 1980s, advanced circuitry and lithium batteries made possible "in-the-ear-canal" units that could be concealed completely in the ear canal.

Since the 1990s, most manufacturers have produced four basic styles of hearing aids:

Behind-the-Ear (BTE)

The components are held in a case worn behind the ear and connected to a plastic earmold that fits inside the outer ear. Sound travels through the earmold into the ear. BTE aids are used by people of all ages for mild to profound hearing loss. Poorly fitting BTE earmolds may cause feedback, a whistle sound caused by the fit of the hearing aid or by the build up of earwax or fluid. However, BTE aids can be as sophisticated as smaller hearing aids. In fact, they can hold more circuitry and amplify sounds to a greater degree than in-the-ear types. BTE aids can be more durable than other types and a few are even waterproof.

In-the-Ear (ITE)

These devices house components in a custom-formed earmold that fits within the outer portion of the ear. ITE aids can accommodate added technical mechanisms such as a telecoil, a small magnetic coil contained in the hearing aid that improves sound transmission during telephone calls. ITE aids are used for mild to severe hearing loss but can be damaged by earwax and ear drainage. Their small size can cause adjustment problems and feedback. Usually, children do not wear them because the casings need to be replaced as the ear grows. However, its size and easy-to-use controls may be helpful for those with limited manual dexterity.

Canal Aids

These fit into the ear canal and are available in two sizes. In-the-canal (ITC) hearing aids are smaller still, with an earmold that fits down into the ear canal, and a smaller portion facing out into the outer ear. They are discreet, yet still visible within the outer ear. The ITC hearing aid can also be customized to fit the size and shape of the ear canal and is used for mild or moderately severe hearing loss. The newest generation of such hearing aids is those that fit completely in the canal (CIC). A CIC hearing aid is largely concealed in the ear canal and is used for mild to moderately severe hearing loss. In general, because of their small size, canal aids may be difficult for the user to adjust and remove, and they may not be able to hold additional devices such as a telecoil. Canal aids can also be damaged by earwax and ear drainage. They are not typically recommended for children.

Body Aids

Body aids are used by people with profound hearing loss. The aid is attached to a belt or a pocket and connected to the ear by a wire. Because of its large size, it can incorporate many signal processing options, but it is usually used only when other types of aids cannot be used.

Mechanisms of Action

The inside mechanisms of hearing aids vary among the different devices, even if they are of the same style. In general, every hearing aid is a miniature electronic circuitry encased in plastic. Every hearing aid consists of a microphone that picks up sound, an amplifier that boosts the sound, and a receiver that delivers the amplified sound into the ear. All the parts are powered by replaceable batteries. However, due to microprocessor or computer chip technology, late twentieth century hearing aids far surpass the simplicity of this description. All hearing aid styles described above use three basic types of circuitry:

1. An analog hearing aid works much the same way as traditional high-fidelity audio systems. Sound is picked up by the microphone and is then converted to electrical signals. Once sound is turned from acoustic to electrical signal, it is fed into the amplifier of the hearing aid. The sound is then amplified overall and sent to the receiver of the hearing aid, and finally to the user's ear. This type of sound processing has now been used over many years. The biggest drawback of analog processing is that the amplified sound is over the full frequency range of hearing, so low frequency (background noise) would "mask" high frequency (speech) sounds. To alleviate this problem, "potentiometers," which provide the ability to reduce or enhance the sounds needed by the end user, have been introduced to hearing aids to restore hearing to as "normal" a sound as possible. Analog circuitry is generally the least expensive.

2. Programmable hearing aids offer different (often customizable for the individual's audiogram) settings for different listening situations such as the office or home, or a noisy rock concert. Programmable devices enable the audiologist to fine tune the hearing aid using a computer. Potentiometers are built onto the circuit within the programmable hearing aid, rather than sitting externally on the hearing aid. The circuitry of analog and programmable hearing aids accommodates more than one setting. If the aid is equipped with a remote control device, the user can change the program to accommodate a given listening environment.

3. Hearing instruments incorporating digital signal processing (DSP) are widely known as digital hearing aids. The real difference between analog and digital hearing instruments is the ability of the DSP instruments to process more complex signal processing. Sound is still received into the DSP instrument by microphone and converted into "bits" of data. The circuitry within the "digital" hearing aid now acts like a very tiny computer. The computer can sample the data and far more accurately fine tune to each individual's requirements. These hearing aids are wearable minicomputers capable of monitoring the audiological environment. Digital hearing instruments compensate for different listening situations in a more flexible, accurate, and complex way than any analog circuit. Digital hearing aids use a microphone, receiver, battery, and computer chip. Digital circuitry can be used in all types of hearing aids and is typically the most expensive.

Implantable Hearing Aids

The latest twentieth century development in hearing aids is the implantable type. This type of hearing aid is placed under the skin surgically. Regarded as an extension of conventional hearing aid technology, an implantable hearing device is any electronic device completely or partially implanted to improve hearing. This allows for the hearing aid to be worn at or in the ear and still provide the user with a signal quality equal or superior to that of prior hearing aids.

Included in the category of implantable hearing aids is the cochlear implant, a prosthetic replacement for the inner ear or cochlea. In its most basic form, the cochlear implant is a transducer, which changes acoustic signals into electrical signals in order to stimulate the auditory nerve. The device is surgically implanted in the skull behind the ear, and it electronically stimulates the auditory nerve with small wires touching the cochlea. External parts of the device include a microphone, a speech processor (for converting sounds into electrical impulses), connecting cables, and a battery. Unlike a hearing aid, which just makes sounds louder, this device selects information in the speech signal and then produces a pattern of electrical pulses in the user's ear. It is impossible, however, to make sounds completely natural because a limited amount of electrodes are replacing the function of tens of thousands of hair cells in a normal hearing ear.

Hearing aid technology advanced greatly in the last few years of the twentieth century, thanks to the computer microchip and to digital circuitry. The majority of efforts in the research and further development of hearing aid technology is concentrated in the area of implantable hearing devices.

See also **Audiology, Implants and Surgery; Audiology, Testing**

PANOS DIAMANTOPOULOS

Further Reading

Berger, K.W. *The Hearing Aid: Its Operation and Development*, U.S. National Hearing Aid Society, 1984.

Dillon H. *Hearing AIDS*. Thieme, New York, 2001

Stephens, S.D.G. and Goodwin, J.C. Non-electric hearing aids to hearing: a short history. *Audiology*, 23, 215–240, 1984.

Valente, V. *Hearing AIDS: Standards, Options, and Limitations*. Thieme, New York, 2002

Useful Websites

Hearing Aid Manufacturers:
A&M Hearing: http://www.hearing-am.com
Amplifon: http://www.amplifon.it
Audio Controle: http://www.audiocontrole.com
Aurilink: http://www.aurilink.com
AVR: http://www.avrsono.com
Beltone: http://www.beltone.com
Bernafon: http://www.bernafon.ch
Cochlear: http://www.cochlear.com
Danavox: http://www.danovox.com
EarCraft Technologies: http://www.earcraft.com
Frye: http://www.frye.com
General Hearing Instruments: http://www.generalhearing.com
Magnatone: http://www.magnatone-hearing-aids.com
Micro-Tech: http://www.hearing-aid.com
Miracle-Ear: http://www.miracle-ear.com/main.asp
Oticon: http://www.oticon.com
Phonak: http://www.phonak.com
Phonic Ear: http://www.phonicear.com

Audiology, Implants and Surgery

The cochlear implant is the first effective artificial sensory organ developed for humans. The modern cochlear implant has its origin in the late eighteenth century. Alessandro Volta, the Italian physicist whose work on electricity gave his name to a measure of electric current (volt), bravely conducted an experiment on his own ears to see what would happen. He attached two metal rods to activate an electric circuit and then inserted one rod into each ear. Volta reported that he felt something like a blow to his head, followed by the sound of boiling liquid. He did not repeat his experiment.

One of the first documented cases of an implant that successfully stimulated the auditory nerve took place in France in 1957. A French electrophysiologist named Andre Djourno teamed up with otolaryngologist Dr. Charles Eyries to implant a single electrode directly on the auditory nerve of a deaf man and connect it to a crude signal generator. The man described sounds like crickets and was able to use his improved awareness of speech rhythm to lip-read more easily.

In the 1960s, Dr. William House in Los Angeles started the movement toward commercial devel-opment with a single-channel, single-electrode device that was later known as the 3M/House device. In 1964, Dr. Blair Simmons at Stanford University placed the first multichannel implant in a human. When he submitted a paper on the subject for a presentation at the American Otological Society in the following year, however, it was rejected as too controversial. By the 1980s and 1990s, cochlear implants had become generally accepted as a safe and effective method of treating sensorineural deafness in adults and children.

A 1995 U.S. National Institutes of Health Consensus statement on cochlear implants in adults and children concluded that "a majority of those [adult] individuals with the latest speech processors for their implants will score above 80 percent correct on high-context sentences even without visual cues." The development of this technology represents one of the most rapid advances ever seen in medical technology; by the end of the twentieth century, there were about 45,000 users of cochlear implants around the world, almost half of them children. Deaf children who receive implants early enough can now be expected to acquire spoken language and the ability to function well in a hearing milieu.

While hearing aids can be implanted or non-implanted and simply amplify incoming sound, a cochlear implant is always surgically implanted and directly stimulates the auditory nerve. Candidates for cochlear implants are typically those with profound or severe sensorineural deafness in both ears. They normally perform signifi-

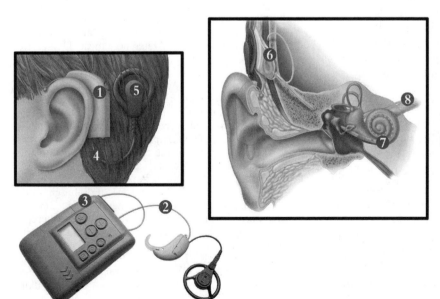

Figure 3. Nucleus® 24 Contour™ cochlear implant system. 1. Sounds are picked up by the small, directional microphone located in the headset at the ear. 2. A thin cord carries the sound from the microphone to the speech processor, a powerful miniaturized computer. 3. The speech processor filters, analyzes and digitizes the sound into coded signals. 4. The coded signals are sent from the speech processor to the transmitting coil. 5. The transmitting coil sends the coded signals as FM radio signals to the cochlear implant under the skin. 6. The cochlear implant delivers the appropriate electrical energy to the array of electrodes which has been inserted into the cochlea. 7. The electrodes along the array stimulate the remaining auditory nerve fibers in the cochlea. 8. The resulting electrical sound information is sent through the auditory system to the brain for interpretation.

cantly better with a cochlear implant than they would with a hearing aid.

The internal parts of a cochlear implant consist of two parts—a surgically inserted electrode array threaded through a hole drilled in the inner ear and inserted into the cochlea, and a receiver/stimulator magnet that is placed over the mastoid bone beneath the skin and behind the ear. The external parts consist of a digital processor, a microphone, and a transmitter. The transmitter sits on the head behind the ear and is held securely in place by its magnet, which is attracted to the magnet of the receiver/stimulator under the skin. The appearance, size, and attributes of components vary depending on the manufacturer and the date of manufacture. However, the general way in which the systems work is the same in all devices.

The cochlear implant works when the microphone sitting on the wearer's head picks up sounds in the environment and sends them to the digital processor by way of a thin cable. The processor uses an internal program that has been customized to the thresholds and comfort levels of the patient to rapidly convert the sounds into electrical codes. These codes indicate which electrodes should be stimulated and how (for example, at what intensity). The coded signals pass back along the cable to the flat transmitter resting on the wearer's head. The transmitter then sends the information across the skin as radio frequency signals to the receiver/stimulator under the skin. There, the signals are decoded and sent on to the electrode array within the inner ear to stimulate the appropriate electrodes and thence the adjacent auditory nerve endings. The signals are passed along the auditory nerve to the brain, where they are interpreted finally as sound. The entire cycle from receipt of sound by the microphone to deciphering it in the brain is rapid: in the slowest systems, the entire cycle takes place 250 times per second; in the fastest systems, the cycle can take place up to 1,000,000 times per second.

Candidates for cochlear implants require an intact auditory nerve. For those who do not have an intact auditory nerve, an auditory brain stem implant (ABI) is possible. This was first developed in the 1970s at the House Ear Institute in the U.S. The ABI is based on the technology of the cochlear implant and works in the same way, except the implanted portion is placed directly over the cochlear nucleus of the brain stem instead of within the inner ear, thus bypassing the ear completely. By 2000, approximately 300 individuals had received an ABI. People generally do not hear as well with ABIs as with cochlear implants.

There is a learning period involved in making sense of the new sounds delivered by either type of implant. Some people learn more quickly than others, and some achieve greater success in understanding speech without lip reading than others. Not all the factors affecting success are well understood, although the length of time the person has been deaf is considered the main factor.

There are three major companies developing the technology of cochlear and auditory brainstem implants: Cochlear Limited (based in Australia), Advanced Bionics (based in the U.S.), and MED-EL (based in Austria). The oldest and largest is the Australian company.

The adaptation of individuals to this technology, which was deemed quackery in the 1960s, is now teaching researchers much about how normal hearing works in those who are not deaf. Two major challenges faced developers of the technology at the beginning of the twenty-first century. First, there is a need to improve the ability of patients to hear with the devices in noisy environments. Second, the programming strategies used in the processors, as well as the fine-tuning techniques to customize them, need to become more sophisticated to maximize the individual's potential success and reduce the current variability in performance.

See also **Audiology, Hearing Aids; Audiology, Testing.**

BEVERLY BIDERMAN

Further Reading

Biderman, B. *Wired for Sound: A Journey into Hearing.* Trifolium Books, Toronto, 1998. An inside account of learning to hear with a cochlear implant. Appendices include extensive endnote references to the literature on the subject, as well as listings of resources on deafness and cochlear implants.

Clark, G.M., Cowan, R.S.C. and Dowell, R.C., Eds. *Cochlear Implantation for Infants and Children: Advances.* Singular Publishing, San Diego, 1997. Articles by the Melbourne Australia team of researchers and surgeons. The chapter, "Ethical Issues" is a thorough treatment of the ethics of cochlear implantation in children, a controversial issue.

Clark, G.M., Tong, Y.C. and Patrick, J.F., Eds. *Cochlear Prostheses.* Churchill Livingstone, Edinburgh, 1990. Excellent material on the development of the Melbourne prototype which was later commercially developed as the Nucleus device by Cochlear Limited in Australia.

Niparko, J., Kirk, K., Mellon, N., Robbins, A., Tucci, D. and Wilson, B. *Cochlear Implants: Principles and Practices.* Lippincott, Williams & Wilkins, Philadelphia, 1999. The clinical practices related to cochlear implantation from selection of candidates to techniques of device placement and activation.

Salvi, R.J., Henderson, D., Fiorino, F. and Colletti, V., Eds. *Auditory System Plasticity and Regeneration.* Thieme, New York, 1996.A collection of papers discussing the status of research into auditory plasticity, including papers on ear hair cell regeneration, the effects of deafness and electrical stimulation on the auditory system, and a discussion of cochlear implant performance over time as an indicator of auditory system plasticity.

Schindler, R.A. and Merzenich, M.A., Eds. *Cochlear Implants.* Raven, New York, 1985. Of great historical importance, this publication is based on the proceedings of the Tenth Anniversary Conference on Cochlear Implants, June 22–24, 1983 at the University of California, San Francisco. The papers by Drs. House, Simmons, and Michelson, early pioneers of the technology, are especially interesting.

Audiology, Testing

The *Archives of Paediatrics* of 1896 reported that 30 percent of the large number of children with poor educational and mental development in Germany and America had deficient hearing power. At the close of the nineteenth century, methods for measuring hearing included observing the patient's ability to detect sounds of a ticking watch, Galton's whistle, the spoken voice, and tuning forks. Half a century earlier in Germany, Hermann Ludwig Ferdinand von Helmholtz investigated hearing physiology and Adolf Rinne used tuning forks to diagnose types of deafness. Galton's whistle emitted accurate variations in tones without loudness control and allowed exploration of hearing ability; Adam Politzer's acoumeter produced loud controlled clicks for detecting severe deafness. These principles of intensity and frequency measurement have become integral features of modern audiometers.

Induction coil technology used in Alexander Graham Bell's telephone (1877) formed the basis of many sound generators. In 1914, A. Stefanini (Italy) produced an alternating current generator with a complete range of tones, and five years later Dr. Lee W. Dean and C. C. Bunch of the U.S. produced a clinical "pitch range audiometer." Prerecorded speech testing using the Edison phonograph was developed in the first decade of the twentieth century and was employed in mass testing of school children in 1927.

World War I brought about major advances in electro-acoustics in the U.S., Britain, and Germany with a revolution in audiometer construction. In 1922 the Western Electric Company (U.S.) manufactured the 1A Fowler machine using the new thermionic vacuum tubes to deliver a full range of pure tones and intensities. Later models restricted the frequencies to the speech range, reduced their size, and added bone conduction.

In 1933 the Hearing Tests Committee in the U.K. debated the problems of quantifying hearing loss with tuning forks because their frequency and amplitude depended on the mode of use. Despite the availability of electric audiometers at the time, standardization of tuning fork manufacture and strict method of use was recommended, along with the introduction of graphic documentation audiography.

Concern abroad about a standard of intensity of sound, or loudness, led to the American Standards Association adopting the bel, named after Alexander G. Bell and already used as a measurement in the acoustic industry. The decibel (dB) was adopted as a unit of measurement of the relative intensity of sounds in the U.S., Britain, and then Germany. International standards were established in 1964.

Subjective Hearing Testing

Several tests for hearing involve the subject's response to types of sounds in a controlled setting. Tests are designed to detect conductive loss, which results from obstruction and disease of the outer or middle ear, and sensorineural (perceptive) loss, caused by lesions of the inner ear (cochlea) or auditory nerve (nerve of hearing) and its connections to the brain stem. Tuning fork tests compare responses to sound transmitted through the normal route of air conduction (AC) as well as bone conduction (BC) through the skull bone to the cochlear shell. Conductive loss reduces air conduction, and perceptive disorders reduce bone conduction. The Rinne test (Figure 4) is positive when AC > BC and negative when AC < BC.

Pure tone audiometry (PTA) is a subjective measurement of AC/BC thresholds using headphones. Responses are displayed graphically. Frequencies cover 250–8,000 Hz, the range for speech understanding, and are plotted against intensities of –10 to 120 decibels, hearing level (dBHL). The internationally agreed normal threshold for hearing is 0 dB and is calibrated from healthy young adults. Classification of the degree of hearing loss is based on pure tone audiometry of each ear.

Speech audiometry tests the subject's response to phonetically balanced words at different intensities. The test is used to evaluate hearing discrimination for fitting hearing aids and for cochlear implantation.

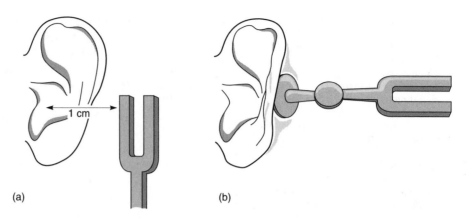

Figure 4. Rinne test using a tuning fork: (a) air conduction; (b) bone conduction. [*Source: Ludman H. and Kemp, D. T. Basic Acoustics and Hearing Tests, in Diseases of the Ear, 6th edn., Ludman, H. and Wright, T., Eds. Edward Arnold, London, 1998. Reprinted by permission of Hodder Arnold.*]

(a)

(b)

1 cm

Behavioral Testing

Around 1900, neurophysiology gained importance with the work of Charles Scott Sherrington in the U.K. and Ivan Petrovich Pavlov in Russia who were exploring innate and conditioned reflexes. Over the next few decades, interest spread beyond Europe to Japan and the U.S. Clinicians began to use these subjective behavioral responses along with various noise makers to identify hearing loss in infants and children. The year 1944 was an important landmark for the introduction of a developmental distraction test researched by Irene R. Ewing and Alexander W. G. Ewing at Manchester University. This first standard pre-school screen became a national procedure in Britain in 1950 and was later adopted in Europe, Israel, and the U.S. Observations of head turns to meaningful low- and high-frequency sounds in infants about 8 months old were made with one person distracting and another testing. This screening method, although still in use, failed to detect moderate deafness.

Automated behavioral screening to detect body movements in response to sounds represented a transition period in the mid-1970s. Two tests, the American Crib-o-gram (1973) and Bennett's Auditory Cradle (London 1979), both proved inefficient.

Objective Testing

In 1924 German psychiatrist Dr. Hans Berger recorded brain waves with surface cranial electrodes. Electroencephalography (EEG) was used fifteen years later by P. A. Davis in the U.S. to record electrical potentials from the brain with sound stimulation. In the journal *Science* in 1930, Ernest Weaver and Charles Bray, also in the U.S., confirmed the production of electrical potentials in the inner ear to sound stimulation in cats. In New York in 1960, action potentials were recorded with electrode placement through a human middle ear directly onto the inner ear. Electrocochleography (ECoG) was used for diagnosis but required anesthesia, especially for children, and was replaced by noninvasive auditory brainstem response audiometry (ABR).

Signal-averaging computerization in the 1950s enhanced the recordings of synchronous neural discharges generated within the auditory nerve pathway with EEG on repetitive acoustic stimulation. Research in the later 1960s in Israel and the U.S. identified amplified electrical potentials more easily in cats and humans with improved computerization to extract background noises. Peaked waves I to VII represented various component activities along the auditory route, wave V being the most diagnostic (Figure 5). ABR was thoroughly investigated in the U.S. and was validated in 1974 for testing high-risk infants. An automated model, automated auditory brain stem response (AABR) was manufactured in the U.S. during the next ten years and proved to be an affordable, noninvasive screening tool for babies. The device used short broadband clicks to elicit responses in the 2000 to 4000 Hz speech recognition range and compared the results with a template algorithm of age-related normal responses.

In 1978 David Kemp in London researched earlier reports of sound reemissions from the ear and redefined outer hair-cell function (inner ear workings). These sensors "amplified" and converted the mechanical sound waves traveling along the cochlear fluid duct, which sharpened and strengthened each individual sound to stimulate a specific anatomical site along the duct. The excess energy produced by this process was reflected as "echoes" and could be collected with Kemp's miniaturized ear canal microphone. These otoacoustic emissions (OAE) were ana-

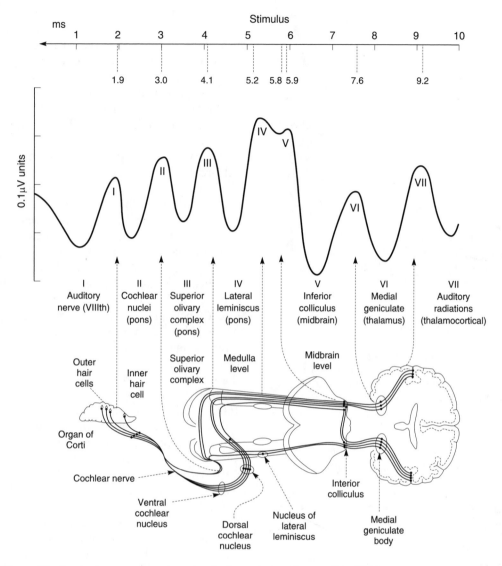

Figure 5. Normal brain stem responses and anatomical correlates of waves I to VII in auditory brain stem response (ABR).
[*Source: Ludman H. and Kemp, D. T. Basic Acoustics and Hearing Tests, in Diseases of the Ear, 6th edn., Ludman, H. and Wright, T., Eds. Edward Arnold, London, 1998. Reprinted by permission of Hodder Arnold.*]

lyzed by computerization according to specific frequency responses, thus providing an extremely sensitive apparatus for detecting cochlea hair-cell damage (Figure 6). After trials in France it was widely accepted as an efficient and safe testing tool.

Infant Screening

The OAE and AABR physiologic tests have proved successful in detecting permanent deafness in infants and consequently in encouraging early intervention with hearing aids and communication strategies. Universal newborn hearing screening programs have been established internationally.

See also **Audiology, Implants and Surgery; Audiology, Hearing Aids; Electronics; Valves or Vacuum Tubes**

DAVID ROSE

Further Reading

Bender, R.E. *The Conquest of Deafness*. Interstate Printers & Publishers, Denville, IL, 1981.
Campbell, K. *Essential Audiology for Physicians*. Singular Publishing Group, San Diego, 1988.

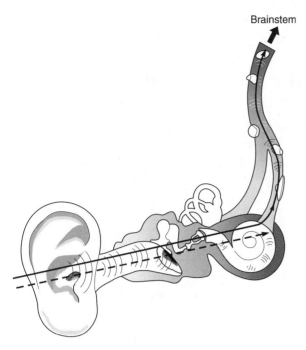

Brainstem

Figure 6. Screeners use automated auditory brain stem response (AABR) technology to test the entire hearing pathway from the ear to the brain stem. Otoacoustic emissions (OAE) test hearing from the outer ear to the cochlea only.
[*Used with permission of Natus U.K.*]

Ewing, I.R., and Ewing, A.W.G. *The Handicap of Deafness.* Longmans, Green & Co., London, 1938.

Ewing, A.W.G. The education of the deaf: history of the Department of Education of the Deaf, University of Manchester, 1919–1955. *Br. J. Educ. Stud.* 4, 2, 1956.

McGrew R.E. *Encyclopaedia of Medical History.* Macmillan, London, 1985.

Sebastion A. *A Dictionary of the History of Medicine.* Parthenon, Carnforth, 1999.

Automobiles

An automobile is commonly defined as a self-propelled, minimum three-wheeled vehicle for passenger or freight transport on ordinary roads. Today's cars are commonly driven by an internal combustion engine using volatile fuel.

Without any doubt, the motor vehicle is among the technological developments of the twentieth century with the most far-reaching social and economic consequences. Automobile production became a major industry in many industrialized countries and had a decisive influence on the development of modern mass production technologies. Furthermore, by deeply transforming the living habits of people, automobiles became an integral part of modern western culture.

At the turn of the twentieth century, these developments were not at all foreseeable. Many prototype solutions for the various subsystems of automobiles had already been developed, but all in all there was remarkable uncertainty about how a proper automobile should be designed. Commercial car production was well established, but the automobiles were still individually crafted and very expensive.

The twentieth century began with a real technical break in 1901 when the first Mercedes racing-car of the German Daimler–Motoren–Gesellschaft put an end to the "horseless carriage" period of the late nineteenth century. The new vehicle incorporated a host of improved features in a general concept that was exemplary: honeycomb radiator mounted right at the front of the car with directly attached engine hood (bonnet); longitudinal four-cylinder in-line engine with the gearbox placed directly behind it; drive to the rear axle; lightweight pressed-steel frame with large wheelbase. The international diffusion of this future standard layout characterized the technical development in the first decade of the twentieth century.

Even the most famous car of these years, the 1908 Model T of the Ford Motor Company, followed this concept. With the Model T, Henry Ford succeeded in building a reliable, high quality, and relatively low-cost car with an initial price of $950. However, even in the U.S., with its comparatively high incomes, a further price reduction was needed to create a real mass market for automobiles. Ford and his staff accomplished this through a fundamental modification of car production technology and organization. Continuous flow production, widespread use of single-purpose machine tools and, from 1913 onward, a step-by-step changeover to assembly line production revolutionized car manufacturing. With Ford's system of rigid mass production, automobiles became affordable consumer goods for middle class Americans. In 1916 the price of the Model T touring car was $360; in 1927 it was $290.

In the years before World War I, cars in general became more reliable, comfortable, and suitable for everyday purposes. In Europe, unlike the U.S., automobile purchases remained limited to a small group of wealthy customers. Only the first steps were taken toward the professional use of cars, for example by physicians.

After 1918 in the U.S., increasing mass motoring had a strong influence on automobile design. The general change from open "touring" cars to larger and more powerful closed limousines for day-to-day use typified the development. In the 1930s,

U.S. cars became remarkably more comfortable with new features such as elastic engine support, automatic transmission, and overdrive, but the conventional drive layout (front engine, rear wheel drive, rigid axles with leaf springs) remained unchanged.

The European automobile industry recovered only slowly after World War I. The still comparatively low incomes and high operating costs prevented the emergence of a real mass market for automobiles. European companies therefore took over the American mass production system only piecemeal and with hesitation. On the other hand the higher flexibility of the relative small European manufacturers allowed them to be much more innovative. In the 1930s, the increasing road speed of European cars forced the industry to overcome the restraints of conventional car design: swing and floating axles, front wheel drive, and aerodynamic bodies entered series production.

During World War II, the development of new models came to a halt in all car-producing countries. After the war, production resumed on the basis of prewar models. In the 1950s, manufacturers and customers in the U.S. favored comfortable, powerful, "gas guzzling" cars with prestigious chromium-plated bodies still designed on the conventional drive layout. Outside the U.S. these expensive models could hardly be sold.

In Europe, the 1950s were years of rapid change. All enveloping bodies and automatic transmissions were adopted from the U.S., and self-supporting bodies, fuel injection, radial-ply tires, hydropneumatic suspension systems, and servo components, improved the performance, handling, and safety of automobiles. According to the market demand, European producers focused on small and medium sized cars. Based on models such as the Morris Minor, the Citroen 2 CV, the Fiat 600, and the Volkswagen Beetle, Western Europe now finally experienced its own mass production boom in automobiles (see Table 2.)

In the 1960s the dark side of automobile traffic became more and more obvious and began, first in the U.S., to generate public debate on safety and environmental risks. The first energy crisis of 1973–1974 intensified the debate, shifting the focus to fuel consumption. Air pollution, safety, and fuel economy as public policy issues and the resulting governmental regulations exerted great pressure on the international automobile industry. A new standard type of smaller cars with front wheel drive, front mounted transverse engine, McPherson front suspension, and fastback design was developed and became dominant on the

Table 2 Automobiles per 1,000 Inhabitants in Selected Countries 1950–1995.

	1950	1960	1970	1979	1986	1995
U.S.	260	340	430	527	681	723
France	40	110	240	355	435	521
U.K.	50	110	210	256	393	472
Germany[1]	40	90	230	357	456	540
Italy	15	30	190	303	411	569
Japan	–	5	80	185	381	520

[1] 1950–1990 West Germany, 1995 united Germany.

market. The BLMC-Mini of 1959 was already a decisive ancestor of this new style, but it was the 1974 Volkswagen Rabbit (Golf) that was its first proper representative. Fuel-saving Otto and especially diesel engines gained an increasing market share. Safety components such as disk brakes, seat belts, and airbags were introduced and became standard features. With the introduction of catalytic converters to treat exhaust emissions, the ecological damage through growing car traffic was at least limited.

In the 1970s Japan became a new global competitor on the international scene. With their comparatively small, inexpensive, and fuel-saving models, Japanese auto manufacturers met the changed consumer priorities and got a firm footing on the North American and Western European market. The international adoption of Japanese production methods (lean production, just-in-time system, robot-based flexible mass production) caused a second revolution of car manufacturing.

In the 1980s and 90s environmental, fuel economy, and safety concerns still exerted a major influence, and electronic components became a key technology in automobile engineering. Microelectronic controls have not only optimized the motor management (ignition and fuel injection) but also the transmission and suspension, steering, and braking systems. The trend now is toward central electronic control units that harmonize the various subsystems of which a car consists.

Influenced by the pollution debate and with a view to the fact that the global mineral oil deposits will inevitably be drained one day, different alternatives to the conventional car have been developed in the last 30 years. At present, vehicles with hydrogen-powered Otto engines and particularly electric-powered cars with fuel cells are regarded as the most promising solutions. However, at least in the medium term, the automobile with fuel-powered Otto or diesel engine will remain prevalent. The fuel economy of the individual car will be further reduced by consequent lightweight design and an even broader application of electronic controls. Due to the ongoing process of increasing automobile use globally however, overall emissions will nevertheless continue rising.

See also **Automobiles, Electric; Automobiles, Hybrid; Automobiles, Internal Combustion; Internal Combustion Piston Engine**

REINHOLD BAUER

Further Reading

Bardue, J.P. *et al. The Automobile Revolution: The Impact of an Industry.* University of North Carolina Press, Chapel Hill, 1982.

Flink, J.J., *The Automobile Age.* MIT Press, Cambridge, MA, 1988.

Mondt, J.R. *Cleaner Cars, The History and Technology of Emission Control since the 1960s.* Society of Automotive Engineers, Warrendale, 2000.

Newcomb, T.P. and Spurr, R.T. *A Technical History of the Motor Car.* Adam Hilger, Bristol, 1989.

Rae, J.B., *The American Automobile Industry.* MIT Press, Cambridge, MA, 1990.

Rubinstein, J.M. *Making and Selling Cars: Innovation and Change in the U.S. Automotive Industry.* Johns Hopkins University Press, Baltimore, 2001.

Volti, R.A. Century of automobility. *Technol. Cult.,* 37, 663–685, 1996.

Automobiles, Electric

Since its development in the 1890s, the electric automobile has always been the car of tomorrow. Many of the earliest motor vehicles employed electric technology, and thousands of electric vehicles provided satisfactory transport service in the U.S. and Europe over the course of the twentieth century. Never as incapable as its critics claimed, neither was the electric automobile ever destined to become the "universal" vehicle of choice for the typical driver. In specific places and times and for select applications, the electric vehicle excelled, but it has never managed to live up to its lofty expectations.

Prior to 1903, the very soul of the automobile was at stake. Three factors led to the initial failure of the electric vehicle. First, early consumers played an outsized role in defining the overall shape of the technology. Wealthy elite males in search of the "sporting life" sought a technology that offered excitement and enabled exurban touring. Although many wealthy families also "stabled" an electric automobile along with their other cars and horses and used it for local transport, the electric vehicle was too practical and domesticated to satisfy these deeper symbolic needs. Successful applications of electric vehicles focused on commercial needs. Electric taxicabs operated continuously in New York City from 1897 to 1912, but these operated on a traditional, horse-based livery model of transport service. The vehicles were leased and not owner operated.

Second, exurban touring required infrastructure that favored internal combustion automobiles. In addition to roads and service facilities, which were decidedly less conducive to electric touring, access to electricity was inherently problematic. Regardless of the absolute range and the technical challenge of charging batteries in remote locations, symbolically the electric car was always "tethered to a wire." The presence of electric service was evidence of civilization and therefore the very thing that early automobilists sought to escape.

Finally, electric technology suffered from unmet expectations, while internal combustion managed to greatly exceed its initial promise. Following the electrification of lighting and street railways, many observers expected electricity to soon make it "off the rails," whereas few saw the internal combustion engine as capable of powering a transportation revolution. By 1906, the tables had turned. Rapid advances in materials and machining of gasoline engines enabled higher compression and resulted in dramatic increases in both power-to-weight ratio and reliability. The universal electric automobile had failed, and mass motorization via internal combustion had begun.

Until the end of World War I, electric vehicles continued to prosper, but always in niche markets or as parts of larger vehicle fleets. Where fire danger limited the operation of combustion engines, electric stevedores replaced horses on docks and train platforms. Electric materials handling vehicles were used inside factories and storehouses where delicate merchandise could not be exposed to dirty exhaust fumes. Local delivery service was the most durable and persistent niche market. Through the 1920s in the U.S. and well into the post-World War II era in Britain, fleets of electric vehicles delivered everything from mail to milk. The unique attributes of the electric car— quiet, reliable, economical, and capable of frequent starts and stops—could not be matched.

During the middle decades of the century, electric vehicles were limited to materials handling and other niche applications and all but disappeared from public view. In the mid-1960s, the electric automobile again emerged as the car of the future. Growing public awareness of the increasing costs of unbridled expansion of the gasoline automobile system and industry inertia prompted policy makers to look anew at the general purpose electric car. In the United States, after 1973, energy independence provided an additional argument in support of the electric vehicle, ultimately leading to the passage of the Electric and Hybrid Vehicle Research Development and Demonstration Act (1976). However, internal combustion had a 75-year head start, and expectations of range and performance had coevolved with suburbanization. Despite an infusion of government research funds, the electric vehicle proved incapable of displacing the established internal combustion standard.

This cycle repeated itself again in the 1990s. In January 1990, General Motors unveiled the Impact, a prototype electric automobile that offered greatly increased performance and range through the use of advanced materials and design. Later that year, the California Air Resources Board passed regulations requiring that a fraction of all vehicles sold in the state must be Zero Emissions Vehicles (ZEVs). Only electric vehicles met the initial criteria for ZEVs, and a decade-long battle ensued between the industry and state and federal regulators. As a byproduct, small numbers of electric cars and trucks were sold or leased by mainstream manufacturers like Chrysler, Ford, General Motors, Honda, Nissan, and Toyota. Enthusiasts adored the vehicles, but overall sales were disappointing, and industry support for the vehicles was tepid at best. Similar stories unfolded in various Western European settings. Tiny minorities were well served by the electric option, but not surprisingly the technology was still incapable of competing head-to-head with internal combustion. Gradually, researchers and policy makers again abandoned the electric vehicle, instead pinning hopes on other technological options, including hybrid electric vehicles (reintroduced by Toyota in Japan in late 1996 and in the U.S. in 2000) and the oft-promised hydrogen fuel cell.

Throughout the history of the electric automobile, one popular misconception stands out: Michael Schiffer has called it the "better battery bugaboo," the idea that the electric vehicle failed entirely on account of the limited range of its batteries. While storage batteries have their limitations, range only became a binding constraint after internal combustion emerged as the leading propulsion technology. Before 1903, electric vehicle

taxicab operators like the Electric Vehicle Company in New York City struggled with the cost of operation rather than range *per se*. Batteries were short-lived and therefore expensive, as were tires, but exchangeable batteries enabled vehicles to be refueled quickly, often while patrons waited in a lounge at the cab station. As noted herein, factors beyond the battery were at least as important, if not more so, in determining the ultimate fate of the electric car in the twentieth century.

See also **Automobiles; Automobiles, Hybrid; Automobiles, Internal Combustion; Batteries, Primary and Secondary; Fuel Cells**

DAVID A. KIRSCH

Further Reading

Kemp, R. *et al. Experimenting for Sustainable Transport: The Approach of Strategic Niche Management*. Spon Press, London, 2002.

Kirsch, D.A., *The Electric Vehicle and the Burden of History*. Rutgers University Press, New Brunswick, NJ, 2000.

Kirsch, D.A. and Mom, G. Visions of transportation: the EVC and the transition from service to product based mobility. *Bus. Hist. Rev.*, 76, 1, 75–110, 2002.

Mom, G. *The Electric Vehicle: Technology and Expectations in the Automobile Age*. Johns Hopkins University Press, Baltimore, 2004.

Mom, G. and Kirsch, D.A. Technologies in tension: horses, electric trucks and the motorization of American cities, 1900–1925. *Technol. and Cult.*, 42, 3, 489–518, 2001.

Schiffer, M. *Taking Charger: The Electric Automobile in America*. Smithsonian Institution Press, Washington D.C., 1994.

Volti, R. Why internal combustion? *Am. Herit. of Invent. Technol.*, 6, 2, 42–47, 1990.

Automobiles, Hybrid

In technology, as in biology, a hybrid is the result of "cross-fertilization," in this case referring to the application of technologies to produce a similar yet slightly different entity. Recent research in the history of automotive technology shows that hybridization has been much more common than previously thought. Thus, the automobile itself can be viewed as a hybrid with a century-long history of crossover phenomena from electrical engineering to mechanical engineering that resulted in an "electrified gasoline car."

The term hybrid, however, is generally reserved for combinations of *propulsion systems* in automobiles. Most common in the history of the automobile is the thermoelectric hybrid, mainly a combination of the internal combustion engine (gasoline or diesel) and an electric propulsion system (electric motor, battery set). Thermomechanical hybrids are possible when a combustion engine is combined with a flywheel system in which part of the kinetic energy during braking can be stored and released the moment this energy is needed, for instance for acceleration from standstill. Similarly, thermohydraulic hybrids combine combustion engines with a hydraulic energy storage system (a pump and a hydraulic accumulator). Electroelectric hybrids are also known; in these cases the actual propulsion is done by one electric motor, but the energy supply is a combination of battery storage and a supply from overhead trolley wires. Combinations of trolley systems and mechanical flywheel storage systems have also been built. Viewed from this perspective, the automobile as we know it at the end of the twentieth century is but one case among many possibilities.

Thermoelectric hybrids are nearly as old as automotive technology. Before 1900, the Belgian automobile producer Henri Pieper developed a car that was equipped with an electromagetically controlled carburetor. His patents were later bought by car manufacturers like Siemens–Schuckert (Germany), Daimler (Coventry, U.K.) and the French Société Générale d'Automobiles électro-Mécaniques (GEM). In 1908 the latter company proposed a Pieper-like hybrid called "Automobile Synthesis." At about the same time, German battery producer AFA (now Varta AG) bought a Pieper to develop a special battery for hybrid car applications. Another famous hybrid vehicle builder was the French electrical engineer Louis Kriéger. He started hybrid development in 1902 and produced a car he drove during the rallye from Paris to Florence a year later. In 1904 his hybrid was the sensation of the Paris automobile show. In 1906 he conceived a drive train based on an electric propulsion system and a gas turbine, and in the same year he developed a hybrid taxicab, 100 of which were intended to be built.

In Austria, Lohner built 52 hybrids between 1898 and 1910, designed by electrical engineer Ferdinand Porsche. These cars were later sold by Daimler, Germany, which founded a separate company for this purpose, Société Mercedes Mixte. In Germany several local fire companies built thermoelectric fire engines, some of these a combination of an electric motor with batteries and a steam engine. In this configuration, the electric drive system was meant for quick starting and for use during the first few kilometers of the trip. After ten minutes, when kettle pressure had built up, the steam engine took over to propel the truck to the fire location. All in all however, no more than a hundred or so hybrids were sold in Europe before World War I. In the U.S., there was even less hybrid construction activity during this period, the most famous being the Woods Dual Power, which was produced during the war.

Hybrids were supposed to combine the advantages of two systems while avoiding their disadvantages. For instance, because the thermal element in the hybrid system was often used (in combination with the electric motor, which for this purpose had to be repolarized to become an electricity generator) to supply a part of the stored electricity, the battery set in a hybrid tended to be smaller. It was lighter than that in a full-blown electric motor where all the energy for a trip had to be stored in the batteries before the start of the trip. In most cases the combination of systems led to a more complex and expensive construction, jeopardizing state-of-the-art reliability standards, and complicated control problems, which would only be overcome with the emergence of postwar automotive electronics. Also, despite the lighter battery, the total drive train became heavier. For this reason hybrid alternatives were especially popular among producers of heavy vehicles such as buses and trucks in which the relative importance of the drive train weight is less. Well-known examples in this respect are the brands Fisher (U.S.), Thornicroft (U.K.) and Faun (Germany).

The popularity of hybrid propulsion systems among engineers was not only, and according to some analysts not primarily, the result of technical considerations. During the first quarter century of automotive history, when the struggle between proponents of steam, electric, and gasoline engine propulsion was not yet over, hybridization often functioned as a strategic and social compromise as well. This was very clear in the case of the German fire engine community before World War I. A fierce controversy raged over the apparent unreliability of the gas combustion engine, but the proponents of electric drive trains, who boasted that electric drive trains guaranteed quick starting, high acceleration, high reliability, and no danger of fuel explosions in the neighborhood of fires, were not strong enough to monopolize the field. Several fire officials then opted for a hybrid drive, combining the advantages of electric with the advantages of the combustion engine (primarily a greater driving range), but they encountered heavy resistance from a combination of both other fire officials and the established automobile industry. Nevertheless, in 1910 the German fire engine fleet included about 15 heavy hybrids.

As with the electric alternative, hybrid automobiles experienced a revival during the last quarter of the twentieth century. This resulted in at least one commercially available hybrid automobile, the Toyota Prius. During this period, the issue of energy consumption played a role as well. Heavily subsidized by local, regional, and federal governments in Europe, Japan, and the U.S., hybrid projects used new light materials such as magnesium, plastics, and carbon fibers; and sophisticated electronic control systems (borrowed from related industries such as aerospace and information and communication technology) to enable very energy efficient solutions, initially to the surprise of many engineers. For example, a Dutch–Italian hybrid bus project resulted in exhaust emissions that were barely measurable and demonstrated very low energy consumption rates. Similar results in other experimental areas have been possible because of sophisticated combinations of small engines, flywheel systems with continuously variable transmissions, and even engine concepts that were considered obsolete, such as Sterling engines, micro gas turbines, and two-stroke engines. By now, the field of possible alternatives is so vast that several classification schemes have been proposed. The most common classification is that which distinguishes between "series hybrids," where the electric element is positioned between the thermal element and the drive wheels, and "parallel hybrids," where both the thermal and the electric element can be used separately to propel the vehicle. At the beginning of the twenty-first century, the "mild hybrid" was the latest development, in which the electric system is so small that it resembles the electric starter motor. If this development materializes, automotive history will have come full circle, producing a true compromise of an electrified gasoline car.

See also **Automobiles; Automobiles, Electric**

GIJS MOM

Further Reading

Dietmar, A. *Die Erklärung der Technikgenese des Elektroautomobils*. Frankfurt am Main, 1998.

Kirsch, D.A. Technological hybrids and the automobile system: historical considerations of the "electrified gas car" and the electrification of the automobile, in *Electric Vehicles: Socio-economic Prospects and Technological Challenges*, Cowan, R. and Hultén, S., Eds. Aldershot, 2000, pp. 74–100.

Mom, G. *History of Tomorrow's Car: Culture and Technology of the Electric Automobile*. Baltimore, forthcoming.

Mom, G. and Vincent van der Vinne, V. *De elektro-auto: een paard van Troje?* Deventer, 1995.

Sperling, D. (with contributions from Delucchi, M.A., Davis, P.M. and Burke, A.F.) *Future Drive: Electric Vehicles and Sustainable Transportation*. Covelo, CA, 1995.

Wakefield, E.H. *History of the Electric Automobile: Hybrid Electric Vehicles*. Society of Automotive Engineers, Warrendale, 1998.

Automobiles, Internal Combustion

Modern standard automobiles are driven by engines working according to the principle of internal combustion; that is, the pressure resulting from the combustion of an air and fuel mixture that is directly transformed into mechanical work. Otto engines, first built in 1877 by the German engineer Nikolaus Otto, are the standard engines for gasoline-powered cars. The only other significant and commercially successful internal combustion engine has been the diesel, developed by the German engineer Rudolf Diesel. These engines have become the standard because they are smaller, lighter, and more efficient when compared with other combustion engines of the same power.

The Otto engine operates in a continuous repetitive cycle: (1) intake of the fuel and air mixture; (2) compression, in which pressure and temperature increase slowly; (3) ignition and combustion of fuel; pressure and temperature increase rapidly; the expansion of the combustion gases forces the piston to move; pressure and temperature drop and thermal energy is converted into mechanical energy; and (4) exhaust, in which remaining pressure, or thermal energy, is released from the engine with the exhaust gases.

Today's automobile power units are highly perfected Otto and diesel engines that provide a long service life, a favorable power to mass ratio, and good start and control characteristics. If operated with conventional fuel, these cars can be driven in the range 500 to 700 kilometers on a tank of gas.

In the twentieth century, the general development of automobiles with internal combustion engines was influenced by a complex interaction between fundamental technical goals, the given technical potential, and the economic, political, and social conditions. Particularly in the early years, engine design was characterized by a host of different constructive solutions to various problems, making it difficult to detect an underlying common path of innovation. Furthermore, progress was based not only on the development of the engines themselves but also on the development of ancillary components such as the materials, the production technology, and the fuels and lubricants.

In the late nineteenth century, there were a number of basic innovations in automobile engines. The years between the turn of the century and World War I saw a consolidation of engine design. Front mounted, water-cooled, longitudinal in-line four-cylinder four-stroke internal combustion engines became the international standard. The main goal in the early years of the century was to increase engine power, at first primarily by increasing the cubic capacity. The zenith of this tendency was reached with the famous "Lightning Benz" of 1909, which achieved 200 horsepower (148 kW) from a displacement of 21.5 liters of fuel. The poor efficiency, the low power density, and the difficulty of handling the inertial forces resulting from massive pistons and rods made these gigantic engines a technical dead end.

Since the end of the first decade of the twentieth century, a higher speed of rotation therefore became the major path to more powerful engines. Prerequisites for this strategy were improved lubrication (forced-feed lubrication instead of splash lubrication), a more effective charge-changing process (mechanically operated instead of vacuum operated inlet valves), and especially more efficient ignition and carburetion (high-instead of low-voltage ignition and multijet carburetors). The standard automobile engine on the eve of World War I already reached a maximum rotation speed of 2,000 revolutions per minute (rpm) and a power density of about 15 hp/l (or 11 kW/l). In comparison, Maybach's and Daimler's first automobile engine of 1885 had a maximum rotation speed of 900 rpm and 1.5 hp/l (1.1 kW/l); the average automobile engine of 1905 had a maximum 1,200 rpm and 7 hp/l (5 kW/l). Due to the limited antiknock quality of the fuel used at that time, the compression ratio of the common engines remained low, approximately 4:1.

World War I interrupted automobile but not at all engine development. To the contrary, the enormous efforts spent on the development of airplane engines led to great strides forward. New designs and materials (light alloys, nickel–steel, and chromium–nickel steel) influenced the postwar automobile engines that, in general, became more powerful and durable but nevertheless smaller and lighter. New leaded fuels allowed higher compression ratios. Until the early 1930s, the maximum rotation speed of standard engines rose to approximately 2,900 rpm, the compression ratio to about 6:1, and, as a result, the power density up to 20 hp/l (15 kW/l).

In the period between the wars, different standard designs of automobile engines emerged in Europe and the U.S. In Europe, small, fast running, four-cylinder engines prevailed due to high fuel costs and displacement dependent taxation. In the U.S., manufacturers and customers preferred huge, smooth running, high torque (but not at all fuel-efficient) six- or eight-cylinder models. American mass production technology guaranteed low production costs but, on the other hand, forced the manufacturers to produce engines principally unchanged for as long as up to three decades.

In general, the 1930s were not a time of radical change but of continuous improvement. Engines became more reliable, the average power density rose to between 25 and 30 hp/l (18 and 22 kW/l), and the engine speed increased to a maximum 4,000 rpm. The 1930s also saw the introduction of the diesel engine for passenger cars. The Daimler–Benz Company produced the first standard car with a diesel engine in 1936, and Citroen followed in 1937–1938. In those days the use of the noisy, low torque, and expensive compression ignition engines was limited to taxis.

World War II interrupted automobile development again, and in the U.S., postwar car engines were basically the same as before the war. In the 1950s however, popular demand for more power was met by further increasing the size of the engine, the compression ratio, and the rotation speed, even if these changes were at the expense of fuel economy. At the end of the 1950s, the most powerful standard engines of average medium-sized U.S. cars reached 240 hp (175 kW) and 4,500 rpm.

In Europe, postwar engine development also followed more or less familiar lines, but the manufacturers did show a higher innovative potential in details. Consequently, lightweight design (light alloy cylinder blocks, heads, and pistons) and improved carburetion and air-to-gas intake (fuel injection systems standard since 1955—bigger valves and more lift) made the still rather small engines more powerful. In the 1960s, high-performance engines with a compression ratio of about 8:1, a power density of almost 50 hp/l (37 kW), and a rotation speed of more than 5,000 rpm were found in the standard everyday car.

Since the 1960s and especially since the energy crisis of 1973 and 1974, regulatory requirements and changing consumer priorities exerted a growing influence on engine design, making emission control and fuel economy the decisive objectives of further development. What followed was a period of rapid change primarily characterized by the introduction of microelectronic components in engine design. At first, new transistorized ignition systems allowed a fuel-saving firing. In the early 1980s, combined electronic ignition and injection systems improved engine control and, by more precise carburetion, paved the way for catalytic exhaust treatment. Today, digital engine control systems make it possible to mechanically uncouple the various components of a car engine and establish a cylinder-specific control of carburetion and ignition. Highly efficient direct fuel injection is regarded as the future of the internal combustion Otto engine in automobiles.

The 1970s energy crisis also led to an increased market share for the diesel engine automobile. Its running characteristics, performance, and fuel economy were improved, primarily by electronic control of the injection process. Furthermore, turbocharging became a common feature of standard cars with modern diesel engines. Other alternative engines with internal combustion have played a very limited role within the automobile industry. In the 1960s particularly, the Wankel engine was seen as a promising development due to specific advantages such as immediate generation of a rotary motion and comparatively simple structure. Its poor fuel economy, problematic emission levels, and the high investment requirements that a shift of production toward the rotary engine would have caused have combined to prevent its broader introduction.

Up until the early 1980s, the automobile gas turbine also seemed to be a possible alternative. However, because of further development of the conventional reciprocating engine, the principle advantages of the gas turbine (low-level raw exhaust emissions and good fuel economy at the operating point) lost relevance, whereas its specific disadvantages (poor fuel economy at part-throttle operation and high production costs) could not be overcome. Again, the high investment costs of converting mass production plants for gas turbines significantly hindered the broader introduction of this technology.

Because of their specific advantages, their high development, and their well-established production technology, gasoline and diesel internal combustion engines seemed likely to remain, at least in the medium term, the prevalent power plants of the standard car.

See also **Automobiles; Automobiles, Electric; Automobiles, Hybrid, Internal Combustion Piston Engine; Turbines, Gas in Land Vehicles**

REINHOLD BAUER

Further Reading

Caton, J.A. *History of Engine Research and Development*. American Society of Mechanical Engineers, New York, 1996.

Cummins, C.L. *Internal Fire: The Internal Combustion Engine*. Society of Automotive Engineers, Warrendale, 1989.

Cummins, C.L. *Diesel's Engine*. Carnot Press, Wilsonville, 1993.

King, D. *Computerized Engine Controls*. Delmar, Albany, 2001.

Mondt, J.R. *Cleaner Cars, The History and Technology of Emission Control Since the 1960s*. Society of Automotive Engineers, Warrendale, 2000.

Mowery, D.C. and Rosenberg, N. *Paths of Innovation: Technological Change in 20th-Century America*. Cambridge University Press, Cambridge, 1999.

Newcomb, T.P. and Spurr, R.T. *A Technical History of the Motor Car*. Adam Hilger, Bristol, 1989.

Somerscales, E.F.C. and Zagotta, A.A., Eds. *History of the Internal Combustion Engine*. American Society of Mechanical Engineers, New York, 1989.

B

Batteries, Primary and Secondary

The battery is a device that converts chemical energy into electrical energy and generally consists of two or more connected cells. A cell consists of two electrodes, one positive and one negative, and an electrolyte that works chemically on the electrodes by functioning as a conductor transferring electrons between the electrodes.

Primary cells, most often "dry cells," are exhausted (i.e., one or both of the electrodes are consumed) when they convert the chemical energy into electrical energy. These battery types are widely used in flashlights and similar devices. They generally contain carbon and zinc electrodes and an electrolyte solution of ammonium chloride and zinc chloride. Another form of primary cell, often called the mercury battery, has zinc and mercuric oxide electrodes and an electrolyte of potassium hydroxide. The mercury battery is suitable for use in electronic wristwatches and similar devices.

In 1800, Italian physicist Alessandro Volta began a new era of electrical experimentation with the voltaic pile and his discovery of the means for generating a continuous flow of electricity by chemical action for dissimilar metals separated by an electrolyte. In the nineteenth century, many different primary cells were developed and tested based on Volta's discoveries. The growth of the telegraph and railroad industries pushed battery development with their needs for various batteries with specific characteristics (e.g., sustained operation over long periods and under a wide range of temperatures). The standard form of the primary cell in the twentieth century is quite similar to the cell that Georges Leclanché invented

in France in the 1860s. Another Frenchman, Felix de Lalande, also developed cells, and Thomas A. Edison improved them in the U.S. The use of batteries for electric vehicles particularly spurred Edison's interest in battery development. By 1900, the Edison–Lalande battery had displaced most of the earlier batteries, such as the Leclanché, in telegraph, telephone, and railroad signaling. Eventually, a more foolproof and maintenance-free battery, which resembled the Edison–Lalande one, displaced it in terms of number of users. In the twentieth century, household, nontechnical users of batteries rapidly outgrew the number of other users, and they preferred less maintenance to higher current strengths available in the original battery. The personal electronics revolution in the later part of the twentieth century would spur further improvements in current strengths and capacity in the maintenance-free batteries.

Secondary cells convert chemical energy into electrical energy through a chemical reaction that is essentially reversible. In "charging," the cell is forced to operate in reverse of its discharging operation by pushing a current through in the opposite direction of the one normal in discharge. Energy is thus "stored" in these cells as chemical, not electrical, energy. They may be "recharged" by an electrical current passing through them in the opposite direction of their discharge. Secondary, or storage, cells are generally wet cells, which use a liquid electrolyte.

The lead–acid storage battery and the nickel–iron, nickel–cadmium or alkaline batteries were widely used in the twentieth century. The lead–acid storage battery, mainly used in motor vehicles, consists of alternate plates of lead and lead coated with lead oxide, with an electrolyte of sulfuric acid.

The nickel–iron ("Edison") battery has nickel-plated steel grids containing tubes or pockets to hold the active materials of nickel and iron oxides and with an electrolyte of potassium hydroxide. While industry used the nickel–iron battery widely (particularly for emergency back-ups), the similar nickel-cadmium battery became the most widely used household rechargeable battery.

The development of the storage battery began with the work of French physicist Raymond Gaston Planté, who discovered the principle of the lead–acid rechargeable battery around 1859. The lead–acid battery, the most popular storage battery, is made of materials that are relatively cheap and easily manufactured, although heavy. Every alternative either requires materials that are more expensive or requires more complicated and more expensive manufacturing techniques. The storage battery industry bloomed with the advent of improved electrical generators for rapid recharging in the late nineteenth century. Until 1880, the battery was generally an apparatus for use in the laboratory because of high labor and expense for preparing and charging plates. Edison developed the nickel–iron or alkaline battery around the turn of the twentieth century. The overall structure of the storage batteries did not change much in the twentieth century. Development focused on reducing maintenance needs, reducing weight, or reducing manufacturing cost. For example, in the 1930s, battery companies began to use a new type of lead oxide produced by the companies themselves rather than provided by companies in the lead smelting, refining, and oxide-manufacturing business.

While the battery industry started in the nineteenth century with several small firms and manufacturers, by 1900 a single, large firm emerged in each of the main countries of battery development: the U.S., the U.K., Germany, and France. Companies in each country enjoyed virtual monopolies in the early part of the twentieth century as the technology of the storage battery was essentially standardized by 1920. Small differences that existed between batteries and countries depended mainly on the costs of alternate materials of construction.

Major developments in the battery industry in the twentieth century were the growing number of batteries in use by nontechnical people and the more portable structure of batteries. The automotive industry was the main push for battery development in the first half of the twentieth century. The growth of the automobile and its use of a battery for self-starting and lighting brought many people into contact with batteries.

The transportation industry also required batteries for electric vehicles. Storage batteries were also widely used for emergency backups because of their relatively high capacity as a source of reserve power. New batteries developed had three basic requirements at mid-century: maximum electrical capacity per unit volume, maximum storage life under varying temperatures, and constant discharge voltage. In the second half of the twentieth century, a growing number of small, complex devices led to a great need for more compact, portable batteries. Hearing aids, pacemakers, electronic wristwatches, satellites, and personal entertainment devices spurred the number of batteries manufactured. Efforts at finding a practical battery for electric vehicles continued throughout the twentieth century. Each different need produced changes in the materials used and thus the characteristics, but not the overall design and structure. Battery development in the late twentieth century focused on changes in the electrochemical system that would operate most effectively for new and particular uses.

See also **Automobiles, Electric; Automobiles, Hybrid; Fuel Cells**

LINDA EIKMEIER ENDERSBY

Further Reading

Ruben, S. *The Evolution of Electric Batteries in Response to Industrial Needs.* Dorrance & Company, Philadelphia, 1978.

Schallenberg, R.H. *Bottled Energy: Electrical Engineering and the Evolution of Chemical Energy Storage.* American Philosophical Society, Philadelphia, 1982.

Vinal, G.W. *Primary Batteries.* Wiley, New York, 1950.

Vinal, G.W. *Storage Batteries: A General Treatise on the Physics and Chemistry of Secondary Batteries and Their Engineering Applications*, 4th edn. Wiley, New York, 1955.

Battleships

The modern battleship dates back to the final decade of the nineteenth century when the term came into general use in English for the most powerfully armed and armored surface warships. Material improvements allowed the construction of ships with high freeboard and good sea keeping capable of effectively fighting similar ships at sea, like the line of battleships of the sailing era. British battleships were the archetypes of the era. They displaced around 13,000 to 15,000 tons and their most useful armament was a battery of six 6-inch quick-firing guns on each side. These stood the best chance of successful hitting given the primitive fire control techniques of the day, although skilled gunnery officers might use them to gain the range for accurate shooting by the slow-firing 12-inch guns, two of which were mounted in covered barbette turrets (armored structures to protect the guns) at each end.

Increasing torpedo range forced fire control improvements that emphasized longer-ranged fire with increasingly rapid-firing big guns, and the result was the entirely big gun battleship pioneered by *HMS Dreadnought* completed in 1906. Displacing 18,000 tons and armed with ten 12-inch guns in five twin turrets, she was powered by steam turbines that gave a maximum speed of 21 knots compared to 18 in the earlier ships. She made all existing pre-*Dreadnought* obsolete, and subsequent battleships were commonly known as "dreadnoughts."

Parallel with the battleship, a new type of armored cruiser had been developed in the 1890s similar in displacement to the battleship, although longer and thinner. Their larger batteries of quick-firing guns and higher speed of 23 knots made them arguably more powerful than contemporary battleships. Admiral Fisher, the dynamic British First Sea Lord usually associated with *HMS Dreadnought*, thought the all big gun armored cruiser the better type and the next ships after *Dreadnought* were the *Invincible* class armored cruisers, soon rated battle cruisers. These ships had 6-inch belt armor compared to the 11-inch maximum in *Dreadnought* but at expected combat ranges, armor gave little protection from 12-inch guns and the higher speed of 25.5 knots gave considerable tactical advantages.

Rather than replacing battleships however, battle cruisers were built alongside battleships in both the Royal Navy and its emerging rival, the Imperial German Navy. British vessels were superior in size of armament; 13.5-inch guns were adopted in the 'super dreadnoughts' commissioned in 1912 and 15-inch guns in the battleships laid down the same year. German ships were more lightly armed with 11-inch and 12-inch guns but were more heavily protected which stood them in good stead in combat. It now seems the decisive weakness in British ships was defective ammunition handling rather than protection, but the catastrophic loss of three British battle cruisers at the battle of Jutland seemed to confirm the need for armor as well as speed, especially as still longer ranges made protection more useful. New battle cruisers like *HMS Hood* were huge ships of over 40,000 tons that allowed speed to be combined with protection.

After World War I the term capital ship came into use to cover both battleships and battle cruisers. The U.S. was outbuilding Britain both in quantity and quality, and Japan also had a capital ship program that threatened both. A 5:5:3 ratio of capital ships was agreed at the Washington Conference in 1922 as was a capital ship building "holiday." The only exception to the latter was Britain's allowance of two slow 16-inch gun battleships built to a tonnage limitation of 35,000 tons "standard" displacement. There was no need for speed as the battle cruisers of this caliber armament planned by Japan and the U.S. were canceled or converted into aircraft carriers.

The Washington system broke down in the mid-1930s. Attempts to maintain qualitative restrictions limited to 14-inch guns the first class of new British capital ship, five of which were laid down as soon as was possible in 1937. The U.S.'s new capital ships mounted 16-inch guns. Japan began building four monsters of over 62,000 tons armed with nine 18-inch guns. Two, *Yamato* and *Musashi,* were completed in 1941–1942. Germany, Italy, and France, not bound by the Washington system, laid down new vessels armed with 15-inch guns. All these ships were quite fast (27–30 knots) but because of their levels of protection were known as "battleships." The term "battle cruiser" was used unofficially for German and French types of more lightly gunned battleships, respectively developments of and answers for the heavily gunned cruisers known as "pocket battleships" built by the Germans under the restrictions of the Versailles peace treaty. The fastest battleships of the new generation were the American *Iowa* class laid down in 1940–1941 which combined an armament of nine 16-inch guns, and heavy armor of up to 19.7 inches with a speed of 32.5 knots and a standard displacement of over 48,000 tons.

The speed of the *Iowas* was to allow them to operate with fast aircraft carriers. The aircraft was becoming a major challenge to the battleship. In 1940–1941 battleships were sunk at their moorings at Taranto and Pearl Harbor. Then, on 10 December 1941, the British battleship *Prince of Wales* and battle cruiser *Repulse* were sunk at sea by Japanese land-based torpedo bombers. Both *Yamato* and *Musashi* would later succumb to torpedo bombers at sea in 1944–1945, and the Italian battleship *Roma* was sunk by a German guided bomb in 1943.

Yet vulnerability was not the battleship's major problem. Sometimes the only answer to one battleship was another, as when the *Scharnhorst* was sunk by the *Duke of York* in the Arctic darkness of December 1943. Battleships also added significantly to the anti-aircraft protection of carrier forces. Battleships were however, hungry for manpower at the same time as their supremacy as the main fleet striking unit was challenged and they thus became uneconomical. Britain and France commissioned battleships after World War II and the USSR planned abortive new units, but only the *Iowa* class survived the 1950s. All four were recommissioned in the 1980s, their armament enhanced by long-range cruise missiles. They were decommissioned in 1991–1992 but two remained in reserve ten years later. The battleship era was still not quite over.

See also **Aircraft Carriers; Radar, Defensive Systems in World War II; Ships, Bulk Carriers and Tankers; Warplanes, Fighters and Fighter Bombers**

ERIC J. GROVE

Further Reading

Breyer, S. *Battleships and Battlecruisers.* Macdonald & Janes, London, 1973.
Burt, R.A. *British Battleships 1889–1904.* Arms & Armour Press, London, 1988.
Burt, R.A. *British Battleships of World War One.* Arms & Armour Press, London, 1986.
Campbell, N.J.M. *Battlecruisers.* Conway Maritime Press, London, 1978.
Conway's *All the Worlds Fighting Ships 1860–1905, 1906–21 and 1922–46.* Conway Maritime Press, London, 1979, 1985, and 1980.
Dulin, R.O. and Garzke, W.H. *Battleships,* 3 vols. Naval Institute Press, Annapolis, 1976, 1980, and 1985.
Friedman, N. *Battleship Design and Development 1905–1945.* Conway Maritime Press, London, 1978.
Friedman, N. *US Battleships, An Illustrated Design History.* Naval Institute Press, Annapolis, 1985.
Johnston, I. and McAuley, R. *The Battleships.* Channel 4 Books, London, 2000.
Raven A. and Roberts, J. *British Battleships of World War Two.* Arms & Armour Press, London, 1976.

Bicycles and Tricycles, *see* **Transport, Human Power**

Biomass Power Generation

Biomass, or biofuels, are essentially clean fuels in that they contain no sulfur and the burning of them does not increase the long-term carbon dioxide (CO_2) levels in the atmosphere, since they are the product of recent photosynthesis (note that peat is not a biofuel in this sense). This is by no means an unimportant attribute when seen in the context of the growing awareness across the globe of the pollution and environmental problems caused by current energy production methods, and the demand for renewable energy technologies.

Biomass can be used to provide heat, make fuels, and generate electricity. The major sources of biomass include:

- Standing forests
- Wood-bark and logging residues
- Crop residues
- Short rotation coppice timber or plants
- Wood-bark mill residues
- Manures from confined livestock
- Agricultural process residues
- Seaweed
- Freshwater weed
- Algae

A few facts and figures might help to put the land-based biomass sources in perspective. The first three of the above list produce in the U.S. approximately the equivalent of 4 million barrels of oil per day in usable form. If all crop residues were collected and utilized to the full, almost 10 percent of the total U.S. energy consumption could be provided for. Although the other land-based sources of biomass are perhaps not on the same scale as this, the combined resource represents a huge untapped reservoir of potential energy. An interesting point to note is that current practices in forestry and food crop farming are aimed directly at optimizing the production of specific parts of a plant. Since biomass used for energy would make use of the whole plant, some significant advantage might be gained by growing specifically adapted crops designed to maximize the energy yield rate. It is from this origin that the energy farm concept is born.

In addition to land-based biomass, there is potential in aquatic biomass, and there are various methods by which to approach this resource. The first is by direct farming of methane as a byproduct of photosynthesis in marine plants. An example of this would be to farm huge cultivated kelp beds at great depth off the coast in suitable sea areas.

It should be noted that of the solar energy incident on the earth's surface, only 0.1 percent is harnessed through photosynthesis. Since about 2×10^{12} tons of vegetable matter grows worldwide each year, it would require only a small increase in the percentage of solar energy used in plant processes to yield a large increase in potential biomass fuel. The conversion of human and animal waste product to useful fuels has long been an interesting prospect. One method for doing just that is to employ microbial processes. It has been demonstrated that a practical regenerative system can be developed in which waste materials are used as the feedstock upon which to grow algae. Methane can be produced by fermenting the algae and the remaining nutrient-rich waste can be recycled to grow further algae.

In a biomass farm, which combines elements of both the above techniques, algae is grown in open ponds of water in the presence of carbon dioxide and recycled inorganic nutrients. Gas lift pumps introduce CO_2 to the system and growing algae. After an incubation period, algae are collected in a trough by clumping, sedimentation, or floatation techniques, and then dewatered. The harvested biomass is deposited in a biophotolytic reactor where, under a carefully controlled environment, the algae cells use sunlight to split water molecules, forming hydrogen and oxygen.

The processes for producing energy from biomass can be divided into four areas:

1. Digestion of vegetable matter
2. Thermal processing
3. Combustion of biofuels
4. Anaerobic digestion of animal waste

The first, digestion of vegetable materials, has a resource size of 3 to 4 million tons of coal equivalent per year (Mtce per year). The economics are critically sensitive both on costs of collection and the digester equipment. The resource is characterized also by the seasonal nature of the raw material. Vegetable matter can also be converted directly into liquid fuels for transportation. The two most common biofuels are ethanol and biodiesel. Ethanol, an alcohol, is made by fermenting any biomass high in carbohydrates such as corn. It is mostly used as a fuel additive to cut down a vehicle's carbon monoxide and other smog-causing emissions. Biodiesel, an ester, is made using vegetable oils, animal fats, or algae. It can be used as a diesel additive or as a pure fuel for automobiles.

The second area, possibly more promising than the former, goes under the general heading of thermal processing. This includes the gasification, the direct liquefaction, and the pyrolysis (thermal decomposition in the absence of oxygen) of low moisture content biomass. By these means, about 5 Mtce per year of methanol alone could be produced in the U.K., 10 to 15 percent of current U.K. gasoline annual energy requirements. It is unlikely that the resource will become economically viable in the short term or even medium term.

The last two technologies are significantly better prospects and indeed have been demonstrated as commercially viable even at current fuel prices.

Combustion of biofuels is said to represent a significant potential energy resource, and with most schemes having a payback period of three to five years, it presents itself as a most inviting investment. The second of these more attractive technologies is anaerobic digestion of animal wastes. The size of this resource, although still significant, is not on the same scale as the combustion of biofuels, being on the order of 1 Mtce of economically viable potential. Although much work on this resource has been carried out among farming cooperatives in Denmark, there are a number of uncertainties that significantly affect the rate of development of this resource.

The product of this process is methane, and it is most likely that the exploitation of the resource will be carried out on a local scale, perhaps at farm level. Thus the marketability of surplus methane; that is, that which is in excess of the farmer's needs, is not certain. Although the technology for digester construction is well established, the actual processes that occur during operation are still poorly understood. Consequently there are uncertainties as to the design performance and flexibility in adapting to any fuel variations that may occur.

Despite these drawbacks and the general unsuitability of high technology to the agricultural environment, research is continuing along a number of promising lines that could lead to increased controllability and the reduction of costs on less economic fuels such as dairy cattle waste. Although the manufacture of digesters for the more attractive fuels such as pig and poultry waste is well established, it remains to be seen how quickly the technology will be taken up and perform in a working environment.

See also **Electricity Generation and the Environment; Energy and Power; Power Generation, Recycling; Solar Power Generation; Wind Power Generation**

IAN BURDON

Further Reading

Andreae, M.O. Biomass burning: its history, use, and distribution and its impact on environmental quality and global climate, in *Global Biomass Burning: Atmospheric, Climatic, and Biospheric Implications*, J.S. Levine, Ed. MIT Press, Cambridge, MA, 1991, pp. 1–21.

Goldemberg, J. *Energy, Environment and Development*, Earthscan, London, 1996.

Kolar, J. Renewable energy: biomass power and biofuels as alternative sustainable energy sources. *Environ. Qual. Manag.*, 8, 2, 35–42, 1998.

Leggett, J. *The Carbon War. Global Warming at the End of the Oil Era*. Penguin, London, 1999.

MacNeill, J.R., Ed. *Something New Under the Sun: An Environmental History of the Twentieth Century.* Penguin, London, 2001.

Biopolymers

Biopolymers are natural polymers, long-chained molecules (macromolecules) consisting mostly of a repeated composition of building blocks or monomers that are formed and utilized by living organisms. Each group of biopolymers is composed of different building blocks, for example chains of sugar molecules form starch (a polysaccharide), chains of amino acids form proteins and peptides, and chains of nucleic acid form DNA and RNA (polynucleotides). Biopolymers can form gels, fibers, coatings, and films depending on the specific polymer, and serve a variety of critical functions for cells and organisms. Proteins including collagens, keratins, silks, tubulins, and actin usually form structural composites or scaffolding, or protective materials in biological systems (e.g., spider silk). Polysaccharides function in molecular recognition at cell membrane surfaces, form capsular barrier layers around cells, act as emulsifiers and adhesives, and serve as skeletal or architectural materials in plants. In many cases these polymers occur in combination with proteins to form novel composite structures such as invertebrate exoskeletons or microbial cell walls, or with lignin in the case of plant cell walls.

Natural biopolymers have a huge diversity in functions, yet only a small number of this diverse group of polymers has been extensively studied and commercially exploited. However, the impact of even these limited numbers in the twentieth century has been substantive. Biopolymers from plants (e.g., starch, cellulose in cotton and flax, natural rubber) and animals (e.g., collagen or gelatin) have traditionally been gathered and utilized for centuries as food and materials. In the twentieth century, additional biopolymers have been processed or extracted from plants (e.g., alginates from seaweeds) and exploited in many important materials-related applications, and new polymers have been designed or engineered (see below). Alginate from seaweed was developed in the 1930s when the Kelco Company commercialized the biopolymer as a food stabilizer in ice cream. Biopolymers have since transformed food industries as gums, stabilizers, and emulsifiers, the oil extraction industry in the use of xanthan to enhance oil recovery, medicine (e.g., collagen and silk biomaterials), and cellulose and starch used extensively in the pulp, paper, and textile industries. The rheological properties of biopolymers such as xanthan give them their useful properties: adding xanthan to water-based drilling fluids enhances the viscosity.

Xanthan is also used in food and pharmaceutical formulations due to its stability over a broad range of pH and salt concentrations. Alginates are used for gelation, emulsification, and stabilization in foods and ceramic formulations, as coatings for paper, and in pharmaceutical formations. Biodegradable starch-based products and bacterial polyhydroxyalkanoates generated by thermal extrusion or solution processing have been extensively studied in recent years as lower cost options for commodity plastics and for biomedical material applications.

Traditional modes of generating biopolymers have included farming, as in the case of cellulose from cotton and starch from corn. New methods of generating polymeric materials from plants may involve:

1. Improved modes of production via tissue culturing in bioreactors
2. Degrading the plant material chemically or by microbial fermentation and subsequently synthesizing new polymers
3. Tailoring the original polymeric structure of the plant material by enzymatic synthesis or genetic engineering. A variety of microbial polysaccharides produced outside of cells are produced commercially; probably the most important is xanthan gum, produced by enzymes from corn syrup. Advances in the *in vitro* synthesis of biopolymers, via cell culture systems or enzymatic catalysis, suggest future opportunities to further expand on the suite of monomers amenable to direct incorporation into biopolymers during biological synthesis. New options are being actively explored to expand the building blocks (monomers such as amino acids, sugars and fatty acids) used in these types of polymers by marrying chemistry and biology. The incorporation of nonnative building blocks, such as modified amino acids into proteins, alternative sugars into polysaccharides, and alternative fatty acids into polyesters, are examples of the expanding range of monomers that can be utilized in biopolymers that can be generated by biological synthesis, an approach that avoids the historical limitations imposed by biology and evolution or selection. For example, fluorinated amino acids and fluorinated fatty

acids have been successfully incorporated into proteins and into polysaccharides like emulsan, a biopolymer synthesized naturally by a bacterium, and which is amenable to structural tailoring.

Extensive chemical derivatization of polysaccharides is carried out industrially, particularly with starches (mostly derived from corn) and cellulose, as a means to alter solubility and properties. Genetic engineering is being pursued via transgenic animals or plants, by inserting bacterial genes that create new biosynthetic pathways. For example, transgenic collagens and silks are being pursued as a route to more cost-effective sources of these materials for biomedical and commodity materials. Genes isolated from spider species could be used in mammalian cells to produce silk.

Polyhydroxyalkanoates (PHAs) are a large family of structurally related polyesters synthesized by many bacteria or transgenic plants and accumulated as granules within the cell. These natural thermoplastics were investigated in the 1970s following the oil crisis, and developed by ICI in the U.K. These polymers are composed of intracellular homo- or copolymers of [R]-β-hydroxyalkanoic acids. PHAs have been pursued for biomedical materials as well as replacements for petrochemically derived polymers since they biodegrade naturally and completely, and have diverse options for monomer chemistries that can be incorporated biologically into the polymer. Once extracted from the cells and processed into plastics, some PHAs exhibit material properties similar to polypropylene. PHAs have been produced on an industrial scale (one polymer produced under the trade name Biopol is used as a packaging material, for example for shampoo bottles), although commercial applications for PHAs have focused in recent years on biomedical materials applications due to the high costs of bacterially synthesized polymer production.

An important feature of biological synthesis is the template-directed process used in the case of proteins and nucleic acid biosynthesis. In comparison to synthetic approaches to polymer synthesis, this method provides direct control over monomer sequence (and thus chemistry) and size of the polymers. Therefore, biosynthesis methods are carefully orchestrated processes aimed at optimizing structures for molecular recognition to drive macromolecular assembly, while also designed to conserve resources to promote survival of the organism. These polymers are also recycled back into natural geochemical and biological cycles for

reuse and serve as models for "green chemistry" approaches.

Advances in metabolic engineering, environmental considerations about renewable polymers from nonpetroleum feedstocks, and the expansion in molecular biology and protein engineering tools in general are taking biopolymer synthesis and production in new directions. The opportunity to enhance, alter, or direct the structural features of biopolymers through genetic manipulation, physiological controls, or enzymatic processes provides new routes to novel polymers with specialty functions. The use of biopolymers in commodity and specialty materials, as well as biomedical applications, can be expected to continue to increase with respect to petrochemical-derived materials. The benefit in tailoring structural features is a plus for generating higher performance properties or more specialized functional performance. Biosynthesis and disposal of biopolymers can be considered within a renewable resource loop, reducing environmental burdens associated with synthetic polymers derived from petrochemicals that often require hundreds of years to degrade. In addition, biopolymers can often be produced from low cost agricultural feedstocks versus petroleum supplies and thereby generate value-added products.

See also **Biotechnology; Synthetic Resins**

DAVID KAPLAN

Further Reading

Byrom, D. Polymer synthesis by micro-organisms: technology and economics. *Trends Biotechnol.* 5, 246–250, 1987.

Kaplan, D.L. Introduction to Biopolymers from Renewable Resources, in *Biopolymers from Renewable Resources*, Kaplan, D.L., Ed. Springer Verlag, Berlin, 1998, pp. 1–26.

Shogrun, R.L. 1998. Starch properties and material applications, in *Biopolymers from Renewable Resources*, Kaplan, D.L., Ed. Springer Verlag, Berlin, 1998, pp. 30-46

Steinbuchel, A. and Valentin. H.E. Diversity of bacterial polyhydroxyalkanoic acids. *FEMS Microbiol. Lett.* 128, 219–228, 1995.

Biotechnology

The term biotechnology came into popular use around 1980 and was understood to mean the industrial use of microorganisms to make goods and services (Commission of the European Communities, 1979). Although biotechnology is often associated with the application of genetics, that is too narrow an interpretation. Rather the

Figure 1. Different genetic sources of alfalfa are evaluated to identify plant traits that would increase growth and enhance the conversion of plant tissues into biofuel. [*Photo by Keith Weller. ARS/USDA*].

word has been used for almost a century to reflect a changing combination of the manipulation of organisms, the means of multiplying them using fermentation, and the extraction of useful products. Moreover, while the technology of the 1980s was new, claims that the introduction of biotechnology would mark a new industrial revolution had been made with conviction and vision since the time of World War I.

Biotechnology grew out of the technology of fermentation, which was called zymotechnology. This was different from the ancient craft of brewing because of its thought-out relationships to science. These were most famously conceptualized by the Prussian chemist Georg Ernst Stahl (1659–1734) in his 1697 treatise *Zymotechnia Fundamentalis*, in

which he introduced the term zymotechnology. Carl Balling, long-serving professor in Prague, the world center of brewing, drew on the work of Stahl when he published his *Bericht über die Fortschritte der zymotechnische Wissenschaften und Gewerbe* (Account of the Progress of the Zymotechnic Sciences and Arts) in the mid-nineteenth century. He used the idea of zymotechnics to compete with his German contemporary Justus Liebig for whom chemistry was the underpinning of all processes.

By the end of the nineteenth century, there were attempts to develop a new scientific study of fermentation. It was an aspect of the "second" Industrial Revolution during the period from 1870 to 1914. The emergence of the chemical industry is widely taken as emblematic of the formal research

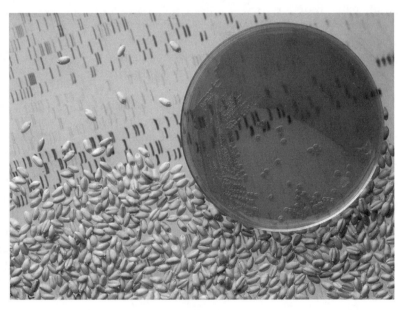

Figure 2. Wheat seeds coated with genetically modified bacteria like those colonized in this petri dish are nearly immune to wheat take-all, a root-destroying fungal disease. The sequencing gel in the background bears the genetic code for bacterial enzymes that synthesize natural antibiotics. [*Photo by Jack Dykinga. ARS/USDA*].

and development taking place at the time. The development of microbiological industries is another example. For the first time, Louis Pasteur's germ theory made it possible to provide convincing explanations of brewing and other fermentation processes.

Pasteur had published on brewing in the wake of France's humiliation in the Franco–Prussian war (1870–1871) to assert his country's superiority in an industry traditionally associated with Germany. Yet the science and technology of fermentation had a wide range of applications including the manufacture of foods (cheese, yogurt, wine, vinegar, and tea), of commodities (tobacco and leather), and of chemicals (lactic acid, citric acid, and the enzyme takaminase). The concept of zymotechnology associated principally with the brewing of beer began to appear too limited to its principal exponents. At the time, Denmark was the world leader in creating high-value agricultural produce. Cooperative farms pioneered intensive pig fattening as well as the mass production of bacon, butter, and beer. It was here that the systems of science and technology were integrated and reintegrated, conceptualized and reconceptualized.

The Dane Emil Christian Hansen discovered that infection from wild yeasts was responsible for numerous failed brews. His contemporary Alfred Jørgensen, a Copenhagen consultant closely associated with the Tuborg brewery, published a widely used textbook on zymotechnology. *Microorganisms and Fermentation* first appeared in Danish 1889 and would be translated, reedited, and reissued for the next 60 years.

The scarcity of resources on both sides during World War I brought together science and technology, further development of zymotechnology, and formulation of the concept of biotechnology. Impending and then actual war accelerated the use of fermentation technologies to make strategic materials. In Britain a variant of a process to ferment starch to make butadiene for synthetic rubber production was adapted to make acetone needed in the manufacture of explosives. The process was technically important as the first industrial sterile fermentation and was strategically important for munitions supplies. The developer, chemist Chaim Weizmann, later became well known as the first president of Israel in 1949.

In Germany scarce oil-based lubricants were replaced by glycerol made by fermentation. Animal feed was derived from yeast grown with the aid of the new synthetic ammonia in another wartime development that inspired the coining of the word biotechnology. Hungary was the agricultural base of the Austro–Hungarian empire and aspired to Danish levels of efficiency. The economist Karl Ereky (1878–1952) planned to go further and build the largest industrial pig-processing factory. He envisioned a site that would fatten 50,000 swine at a time while railroad cars of sugar beet arrived and fat, hides, and meat departed. In this forerunner of the Soviet collective farm, peasants (in any case now falling prey to the temptations of urban society) would be completely superseded by the industrialization of the biological process in large factory-like animal processing units. Ereky went further in his ruminations over the meaning of his innovation. He suggested that it presaged an industrial revolution that would follow the transformation of chemical technology. In his book entitled *Biotechnologie,* he linked specific technical injunctions to wide-ranging philosophy. Ereky was neither isolated nor obscure. He had been trained in the mainstream of reflection on the meaning of the applied sciences in Hungary, which would be remarkably productive across the sciences. After World War I, Ereky served as Hungary's minister of food in the short-lived right wing regime that succeeded the fall of the communist government of Bela Kun.

Nonetheless it was not through Ereky's direct action that his ideas seem to have spread. Rather, his book was reviewed by the influential Paul Lindner, head of botany at the Institut für Gärungsgewerbe in Berlin, who suggested that microorganisms could also be seen as biotechnological machines. This concept was already found in the production of yeast and in Weizmann's work with strategic materials, which was widely publicized at that very time. It was with this meaning that the word "Biotechnologie" entered German dictionaries in the 1920s.

Biotechnology represented more than the manipulation of existing organisms. From the beginning it was concerned with their improvement as well, and this meant the enhancement of all living creatures. Most dramatically this would include humanity itself; more mundanely it would include plants and animals of agricultural importance. The enhancement of people was called eugenics by the Victorian polymath and cousin of Charles Darwin, Francis Galton. Two strains of eugenics emerged: negative eugenics associated with weeding out the weak and positive eugenics associated with enhancing strength. In the early twentieth century, many eugenics proponents believed that the weak could be made strong. People had after all progressed beyond their biological limits by means of technology.

Jean-Jacques Virey, a follower of the French naturalist Jean-Baptiste de Monet de Lamarck, had coined the term "biotechnie" in 1828 to describe man's ability to make technology do the work of biology, but it was not till a century later that the term entered widespread use. The Scottish biologist and town planner Patrick Geddes made biotechnics popular in the English-speaking world. Geddes, too, sought to link life and technology. Before World War I he had characterized the technological evolution of mankind as a move from the paleotechnic era of coal and iron to the neotechnic era of chemicals, electricity, and steel. After the war, he detected a new era based on biology—the biotechnic era. Through his friend, writer Lewis Mumford, Geddes would have great influence. Mumford's book *Technics and Civilization,* itself a founding volume of the modern historiography of technology, promoted his vision of the Geddesian evolution.

A younger generation of English experimental biologists with a special interest in genetics, including J. B. S. Haldane, Julian Huxley, and Lancelot Hogben, also promoted a concept of biotechnology in the period between the world wars. Because they wrote popular works, they were among Britain's best-known scientists. Haldane wrote about biological invention in his far-seeing work *Daedalus.* Huxley looked forward to a blend of social and eugenics-based biological engineering. Hogben, following Geddes, was more interested in engineering plants through breeding. He tied the progressivism of biology to the advance of socialism.

The improvement of the human race, genetic manipulation of bacteria, and the development of fermentation technology were brought together by the development of penicillin during World War II. This drug was successfully extracted from the juice exuded by a strain of the *Penicillium* fungus. Although discovered by accident and then developed further for purely scientific reasons, the scarce and unstable "antibiotic" called penicillin was transformed during World War II into a powerful and widely used drug. Large networks of academic and government laboratories and pharmaceutical manufacturers in Britain and the U.S. were coordinated by agencies of the two governments. An unanticipated combination of genetics, biochemistry, chemistry, and chemical engineering skills had been required. When the natural mold was bombarded with high-frequency radiation, far more productive mutants were produced, and subsequently all the medicine was made using the product of these man-made cells. By the 1950s penicillin was cheap to produce and globally available.

The new technology of cultivating and processing large quantities of microorganisms led to calls for a new scientific discipline. Biochemical engineering was one term, and applied microbiology another. The Swedish biologist, Carl-Goran Heden, possibly influenced by German precedents, favored the term "Biotechnologi" and persuaded his friend Elmer Gaden to relabel his new journal *Biotechnology and Biochemical Engineering.* From 1962 major international conferences were held under the banner of the *Global Impact of Applied Microbiology.* During the 1960s food based on single-cell protein grown in fermenters on oil or glucose seemed, to visionary engineers and microbiologists and to major companies, to offer an immediate solution to world hunger. Tropical countries rich in biomass that could be used as raw material for fermentation were also the world's poorest. Alcohol could be manufactured by fermenting such starch or sugar rich crops as sugar cane and corn. Brazil introduced a national program of replacing oil-based petrol with alcohol in the 1970s.

It was not, however, just the developing countries that hoped to benefit. The Soviet Union developed fermentation-based protein as a major source of animal feed through the 1980s. In the U.S. it seemed that oil from surplus corn would solve the problem of low farm prices aggravated by the country's boycott of the USSR in 1979, and the term "gasohol" came into currency. Above all, the decline of established industries made the discovery of a new wealth maker an urgent priority for Western governments. Policy makers in both Germany and Japan during the 1970s were driven by a sense of the inadequacy of the last generation of technologies. These were apparently maturing, and the succession was far from clear. Even if electronics or space travel offered routes to the bright industrial future, these fields seemed to be dominated by the U.S. Seeing incipient crisis, the Green, or environmental, movement promoted a technology that would depend on renewable resources and on low-energy processes that would produce biodegradable products, recycle waste, and address problems of the health and nutrition of the world.

In 1973 the German government, seeking a new and "greener" industrial policy, commissioned a report entitled *Biotechnologie* that identified ways in which biological processing was key to modern developments in technology. Even though the report was published at the time that recombinant

DNA (deoxyribonucleic acid) was becoming possible, it did not refer to this new technique and instead focused on the use and combination of existing technologies to make novel products.

Nonetheless the hitherto esoteric science of molecular biology was making considerable progress, although its practice in the early 1970s was rather distant from the world of industrial production. The phrase "genetic engineering" entered common parlance in the 1960s to describe human genetic modification. Medicine, however, put a premium on the use of proteins that were difficult to extract from people: insulin for diabetics and interferon for cancer sufferers. During the early 1970s what had been science fiction became fact as the use of DNA synthesis, restriction enzymes, and plasmids were integrated. In 1973 Stanley Cohen and Herbert Boyer successfully transferred a section of DNA from one *E. coli* bacterium to another. A few prophets such as Joshua Lederberg and Walter Gilbert argued that the new biological techniques of recombinant DNA might be ideal for making synthetic versions of expensive proteins such as insulin and interferon through their expression in bacterial cells. Small companies, such as Cetus and Genentech in California and Biogen in Cambridge, Massachusetts, were established to develop the techniques. In many cases discoveries made by small "boutique" companies were developed for the market by large, more established, pharmaceutical organizations.

Many governments were impressed by these advances in molecular genetics, which seemed to make biotechnology a potential counterpart to information technology in a third industrial revolution. These inspired hopes of industrial production of proteins identical to those produced in the human body that could be used to treat genetic diseases. There was also hope that industrially useful materials such as alcohol, plastics (biopolymers), or ready-colored fibers might be made in plants, and thus the attractions of a potentially new agricultural era might be as great as the implications for medicine. At a time of concern over low agricultural prices, such hopes were doubly welcome. Indeed the agricultural benefits sometimes overshadowed the medical implications.

The mechanism for the transfer of enthusiasm from engineering fermenters to engineering genes was the New York Stock Exchange. At the end of the 1970s, new tax laws encouraged already adventurous U.S. investors to put money into small companies whose stock value might grow faster than their profits. The brokerage firm E. F. Hutton saw the potential for the new molecular biology companies such as Biogen and Cetus. Stock market interest in companies promising to make new biological entities was spurred by the 1980 decision of the U.S. Supreme Court to permit the patenting of a new organism. The patent was awarded to the Exxon researcher Ananda Chakrabarty for an organism that metabolized hydrocarbon waste. This event signaled the commercial potential of biotechnology to business and governments around the world. By the early 1980s there were widespread hopes that the protein interferon, made with some novel organism, would provide a cure for cancer. The development of monoclonal antibody technology that grew out of the work of Georges J. F. Köhler and César Milstein in Cambridge (co-recipients with Niels K. Jerne of the Nobel Prize in medicine in 1986) seemed to offer new prospects for precise attacks on particular cells.

The fear of excessive regulatory controls encouraged business and scientific leaders to express optimistic projections about the potential of biotechnology. The early days of biotechnology were fired by hopes of medical products and high-value pharmaceuticals. Human insulin and interferon were early products, and a second generation included the anti-blood clotting agent tPA and the antianemia drug erythropoietin. Biotechnology was also used to help identify potential new drugs that might be made chemically, or synthetically.

At the same time agricultural products were also being developed. Three early products that each raised substantial problems were bacteria which inhibited the formation of frost on the leaves of strawberry plants (ice-minus bacteria), genetically modified plants including tomatoes and rapeseed, and the hormone bovine somatrotropin (BST) produced in genetically modified bacteria and administered to cattle in the U.S. to increase milk yields. By 1999 half the soy beans and one third of the corn grown in the U.S. were modified. Although the global spread of such products would arouse the best known concern at the end of the century, the use of the ice-minus bacteria—the first authorized release of a genetically engineered organism into the environment—had previously raised anxiety in the U.S. in the 1980s.

In 1997 Dolly the sheep was cloned from an adult mother in the Roslin agricultural research institute outside Edinburgh, Scotland. This work was inspired by the need to find a way of reproducing sheep engineered to express human proteins in their milk. However, the public interest was not so much in the cloning of sheep that had just been achieved as in the cloning of people, which had not. As in the Middle Ages when deformed creatures had been seen as monsters and portents of natural disasters, Dolly was similarly seen as monster and as a portent of human cloning.

The name *Frankenstein*, recalled from the story written by Mary Shelley at the beginning of the nineteenth century and from the movies of the 1930s, was once again familiar at the end of the twentieth century. Shelley had written in the shadow of Stahl's theories. The continued appeal of this book embodies the continuity of the fears of artificial life and the anxiety over hubris. To this has been linked a more mundane suspicion of the blending of commerce and the exploitation of life. Discussion of biotechnology at the end of the twentieth century was therefore colored by questions of whose assurances of good intent and reassurance of safety could be trusted.

See also **Breeding, Animal, Genetic Methods; Breeding, Plant, Genetic Methods; Cell Culturing; Cloning, Testing and Treatment Methods; Feedstocks; Genetic Engineering, Methods; Genetic Engineering, Applications; Medicine; Synthetic Foods**

ROBERT BUD

Further Reading

Beier, F.K., Crespi, R.S. and Strauss, J. *Biotechnology and Patent Protection: An International Review.* OECD, Paris, 1985. Classic overview of patenting issues.

Bud, R. *Uses of Life: A History of Biotechnology*, Cambridge University Press, Cambridge, 1994. This volume explores many of the issues dealt with in this piece.

Commission of the European Communities. FAST: Subprogramme C 'Biosociety', FAST/ACPM, 79, 1G–3E, 1979. One of the very first overall analyses of the field, its findings were included in the book published as *Eurofutures.*

FAST Group. *Eurofutures: The Challenge of Innovation.* Butterworth, London, 1987.

Gaskell, G. and Bauer, M., Eds. *Biotechnology: The years of controversy.* NMSI Trading, London, 2001.

Bauer, M.W. and Gaskell, G., Eds. *Biotechnology: The Making of a Global Controversy.* Cambridge University Press, Cambridge, 2002.

Hall, S. *Invisible Frontiers: The Race to Synthesise a Human Gene*, 2nd edn. Oxford University Press, Oxford, 2002. This tells the story of the race between Biogen and Genentech to make human insulin one of the very first products of genetic engineering.

Kevles, D. *In the Name of Eugenics: Genetics and the Uses of Human Heredity.* Alfred A. Knopf, New York, 1985.

Kloppenberg, J.R., Jr. *First the Seed: The Political Economy of Plant Biotechnology 1492–2000.* Cambridge University Press, Cambridge, 1988.

Orsenigo, L. *The Emergence of Biotechnology: Institutions and Markets in Industrial Innovation.* Pinter, London, 1989. Overview of the birth of biotechnology.

Parascandola, J., Ed. *The History of Antibiotics: A Symposium.* American Institute of the History of Pharmacy, Madison, 1980.

Rothman, H., Greenshields, R. and Callé, F.R. *The Alcohol Economy: Fuel Ethanol and the Brazilian Experience.* Pinter, London 1983.

Teitelman R. *Gene Dreams: Wall Street, Academia, and the Rise of Biotechnology.* Basic Books, New York, 1989.

Thackray A., Ed. *Private Science: Biotechnology and the Rise of the Biomolecular Sciences.* University of Pennsylvania Press, Philadelphia, 1998.

Turney J. *Frankenstein's Footsteps: Science, Genetics and Popular Culture.* Yale University Press, New Haven, 1998.

Wilmut, I., Campbell, K. and Tudge, C. *The Second Creation: The Art of Biological Control by the Scientists Who Cloned Dolly.* Headline, London, 2000.

Blood Transfusion and Blood Products

Blood has always had a cultural significance, symbolic of the essence of life; but the process of transfusion—replacing blood with blood—only became an accepted and reliable practice in the twentieth century.

William Harvey's demonstration of blood circulation in 1628 opened up the possibility of transfusion. In 1665 an English physiologist, Richard Lower, described the first successful transfusion between dogs. The first human transfusion came two years later: Frenchman Jean-Baptiste Denis transferred blood from a lamb to a sick boy, who reportedly recovered. The experiment was repeated but, following several deaths, was banned by 1678.

Interest revived in the nineteenth century when the role of blood as an oxygen transporter was understood. James Blundell at Guy's Hospital in London used transfusions to revive women who hemorrhaged after childbirth. But there were two main problems. First, outside the body, blood would quickly clot, stopping free flow. Second, many patients had severe, sometimes fatal, reactions. Karl Landsteiner solved this second problem in 1901 with his discovery of blood groups. He noticed a human serum sample "clumped" the red blood cells from some people but not others. Using new immunological theories, Landsteiner realized agglutination was due to the presence or absence of specific antigens on the red blood cell. Some individuals have antigen A, some B, some both, and some neither, leading to four blood groups or types: A, B, AB, and O. Not all groups are compatible; mixing incompatible groups causes potentially fatal clumping. Landsteiner's discovery would ultimately make blood transfusion safe (he was awarded the Nobel Prize for physiology in 1930 for this work), but clinicians initially ignored the importance. In 1908 Reuben Ottenberg introduced typing and cross-matching of donors and recipients, but compatibility testing was not immediately adopted.

Improvements in surgery drove the early twentieth century reintroduction of transfusion in America. In 1902 Alexis Carrel reported the possibility of direct transfusion, sewing a donor's artery to the recipient's vein (anastomosis). George Crile pioneered the technique, carrying out over 200 animal transfusions before progressing to humans, but Carrel received the recognition following publicity describing a transfusion between a surgeon and his 5-day-old son.

Direct transfusion avoided problems of coagulation but required delicate and painstaking surgery. It was difficult to quantify the amount of blood transferred, which could be lethal. Other surgeons experimented with semidirect methods of transfusion. W. G. Kimpton and M.S. Brown, along with many others, developed specialized equipment using canola, syringe, needles, stopcocks, and valves. Coating vessels with paraffin wax minimized clotting. Coagulation was overcome in 1914; three doctors (Agote in Argentina, Hustin in Belgium, and Lewisohn in the U.S.) independently demonstrated that sodium citrate could be used as an anticoagulant. Adding small concentrations to blood did not harm the patient but prevented clotting. Indirect transfusion was now possible.

World War I accelerated the pace of change. With increasing demand, the indirect method was perfected. Blood was collected in citrate–glucose solution, refrigerated and transported in bottles to the front lines. Transfusion spread from North America to previously skeptical Europe. By the war's end, it was a practical and relatively simple treatment that saved thousands of lives. The focus then turned to donor recruitment. The need for blood typing became clear; rapid testing procedures allowed selection of appropriate blood. In 1921, Percy Lane Oliver set up the first transfusion service with the British Red Cross. It was a "walking donor" service in which volunteers of known blood groups were available on demand, donating blood wherever it was needed. The idea spread, and donor panels were set up in Europe, the U.S., and the Far East during the 1920s and 1930s. The first blood bank was established in 1932 at Leningrad Hospital in Russia.

The outbreak of World War II prompted another dramatic expansion in blood donation services. A huge logistical operation supplied blood to the front lines and to civilian casualties; by 1944 U.K. donors provided 1200 pints a day.

Plasma, the yellow serum that carries red cells, became a common transfusion fluid, used to treat shock by restoring blood volume. Using Flosdorf and Mudd's lyophilization process, plasma was freeze-dried. Removing water under high vacuum left a dry powder, stable for months. Adding sterile water reconstituted the plasma. Other plasma warwork had long-term impact. Edwin Cohn, from Harvard Medical School, developed a process of cold ethanol fractionation to break plasma down into components. The most important product, albumin, was isolated from Fraction V. Packaged in glass ampoules, this concentrated ready-to-use liquid had vital antishock capabilities. Other products were developed from fractions: gamma globulin, fibrin foam, and blood-grouping globulins. The Plasma Fractionation Project expanded to an industrial scale, with collaboration between universities and pharmaceutical companies.

After the war, civilian blood transfusion expanded. The U.K. Blood Transfusion Service was established in 1946, recruiting voluntary donors with the promise of a cup of tea. More controversially, donors in the U.S. were paid. Developments in blood typing and screening ensured compatibility. Over twenty genetically determined blood group systems were identified, including Rhesus positive and negative.

Collection equipment improved as disposable equipment replaced glass flasks and rubber tubing. In 1950 Carl Walter introduced the plastic collecting bag, having experimented with polymers to find one suitably robust, inert, and immune to extreme temperatures. The new bags reduced contamination and allowed economical ultra-low temperature freezing of blood. Using cryoprotectants like glycerol, red blood cells were preserved for long periods, allowing stockpiling of rare blood types.

Processing developments continued through the century. Today, blood is collected into 450-ml plastic packs with anticoagulant solution. Using a closed system of satellite bags, it is centrifuged with minimal risk of contamination. The red cells, platelets, and plasma components are separated into individual bags, ready for further processing. More than 17 preparations of blood components are available, including clotting factors (such as Factor VIII for hemophiliacs), and antibodies for vaccine production. Whole blood is used only rarely, but no part of a blood donation is wasted.

The wide availability of blood components has facilitated dramatic advances in surgery. Blood

BORANES

transfusion is commonplace in hospitals and clinical blood transfusion is a specialty in its own right. In the U.K. alone, over 2.5 million units of blood are collected each year, and demand continues to rise. In the last two decades of the century however, there was concern about virus transmission through transfusion. Public scandals in France, Canada, and Japan, where patients and particularly hemophiliacs became infected with HIV as a result of transfusions, led to comprehensive monitoring at all stages of donation. Testing for HIV was introduced in 1986 and for hepatitis C in 1991. Blood labeling was internationally standardized in 1992.

Artificial blood substitutes may be the future. Blood volume expanders and hemodilutants (isotonic electrolyte solutions) are already widely used. The search for an artificial oxygen transporter is underway. Possibilities include microencapsulated hemoglobin, recombinant hemoglobin, or perfluorochemical emulsions.

See also **Hematology**

NICOLA PERRIN

Further Reading

Wintrobe, M., Ed. *Blood, Pure and Eloquent*. McGraw-Hill, New York, 1980.
Starr, D. *Blood: An Epic History of Medicine and Commerce*. Alfred A. Knopf, 1998.
Creager, A. Biotechnology and Blood: Edwin Cohn's Plasma Fractionation Project, 1940–1953, in *Private Science: Biotechnology and the Rise of the Molecular Sciences*, Thackray, A., Ed. University of Pennsylvania Press, Philadelphia, 1998.
Contreras, M., Ed. *ABC of Transfusion*. BMJ Books, London, 1998.
Keynes, G. *Blood Transfusion*. John Wright & Son, Bristol, 1949.
Schneider, W. Blood transfusion in peace and war, 1900–1918. *Soc. Hist. Med.*, 10, 01, 1997, 105–126.

Useful websites

National Blood Service (UK): http://www.blood.co.uk
Blood Transfusion Safety, World Health Organization: http://www.who.int/bct/

Bombs, *see* **Fission and Fusion Bombs**

Boranes

Boranes are chemical compounds of boron and hydrogen. During the 1950s, the U.S. government sponsored a major secretive effort to produce rocket and aircraft fuels based on boron hydrides. Much of the information initially available to the

U.S. effort was contained in a book written by the German chemist Alfred Stock in 1933. When burned in air, the energy released by various boron hydride compounds, as measured by their heat of combustion, is 20 to 55 percent greater than the energy released by petroleum-based jet fuels. It was expected that this greater energy content of the boron fuels would translate into equivalent higher payloads or ranges for rockets and aircraft.

All of the boron fuel manufacturing processes started with the production of diborane, a compound composed of two boron atoms linked to six hydrogen atoms. Initially, this was produced by reacting lithium hydride with boron trifluoride or boron trichloride in diethyl ether as a solvent. This entailed a need to recover and recycle the expensive lithium. A later process produced diborane by reacting sodium borohydride with boron trichloride in the presence of catalytic amounts of aluminum chloride, using a solvent called diglyme.

Diborane is a gas that is highly toxic and pyrophoric (can catch fire spontaneously on contact with air), and the boron hydrides burn with a brilliant green flame. Diborane was condensed to a liquid at $-80°C$ and transferred to high-pressure cylinders which were stored at $-10°C$ to minimize degradation in storage.

The diborane had to be converted to liquids suitable for storage in aircraft or rocket fuel tanks at the normal operating temperature for those tanks. By 1952, diborane and pentaborane were being produced in pilot plant quantities. There were two major fuel-producing contractors on the project. One of the two produced liquid fuels by alkylating the diborane with ethylene. The other contractor pyrolyzed the diborane to pentaborane and decaborane and then alkylated those with propyl, ethyl, or methyl groups.

Pyrolysis is a process in which diborane is diluted with hydrogen and circulated through a carefully heated tube in which 60 to 70 percent of the diborane converts to pentaborane, a lesser amount converts to decaborane, and the remainder converts to waste boron hydrides (which would have to be recycled to recover the relatively scarce boron) and hydrogen gas. Alkylation is a process in which hydrocarbon groups were attached to the boranes, for example by reacting pentaborane with propyl chloride in the presence of aluminum chloride catalyst to produce propyl pentaborane. Alkylation produced the desired physical properties at the expense of reducing the heat of combustion of the resulting fuel in direct proportion to the amount of hydrocarbon added.

Pentaborane is a toxic, colorless, pyrophoric room temperature liquid with a 60°C boiling point. There are two forms, but the stable compound consists of five boron atoms linked to nine hydrogen atoms. Decaborane is a toxic white crystalline solid at room temperature, melting and subliming at 100°C. Decaborane consists of 10 boron atoms linked to 14 hydrogen atoms. It can be handled in air without igniting, although it did oxidize in air at a rate dependant on the temperature, becoming reliably pyrophoric at 100°C.

Pentaborane was HEF-1 (high-energy fuel-one). Propyl pentaborane was HEF-2, ethyl decaborane was HEF-3, and methyl decaborane was HEF-4. HEF-3 and HEF-4 were reliably nonpyrophoric at room temperature. Each of the alkylated boranes was primarily monoalkylated, but each also contained some di-, tri-, and tetra-alkylated boranes. The fuels made by direct alkylation of diborane had somewhat different compositions, but were required to meet the same specifications as the materials made by alkylating pentaborane or decaborane.

In a period of four years, 1956 through 1959, five commercial plants were built to produce alkylated boranes:

Niagara Falls, New York—$5.5 million cost to produce 100 tons per year HEF-2

Lawrence, Kansas—$4 million cost to produce fuel equivalent to HEF-2

Model City, New York (Navy)—$4.5 million cost to produce 240 tons per year HEF-2

Model City, New York (Air Force)—$45 million cost to produce 5 tons per day HEF-3

Muskogee, Oklahoma (Navy)—$38 million cost to produce 5 tons per day HEF-3 equivalent

The first three plants were completed and went into operation before June 1958. Including the pilot plants, $100 million dollars were spent on plant construction, and even the largest plants listed above were considered small plants. If approved for use in the supersonic B-70 bomber, each plane could burn 20 tons per hour of HEF-3 without the afterburners and 80 tons per hour with the afterburners.

The toxic and pyrophoric nature of the boron hydrides made them difficult and dangerous to work with. All plant personnel were required to carry a gas mask and to put it on before performing any work with the boron hydrides. Many people did experience boron hydride poisoning, even though the strong odor of the boranes did provide ample warning of exposures. Fires were frequent and some explosions occurred. A total of eight people were killed in five of the accidents. The perseverance of the people on the project under these circumstances amply demonstrates their dedication to the success of the project.

During the period from 1955 to 1959, news of the project swelled from a trickle to a flood in technical publications and newspapers and became more and more accurate. Many contractors were vying for "a piece of the action" in what seemed to be the infancy of a new fuel industry. The launch of the Russian Sputniks in the fall of 1957, followed by an authoritative but false claim that the Russians had used boron fuels, helped to bring the enthusiasm to its high point in 1958. Although Russia was working on boron fuels, both the U.S. and Russian programs were driven in part by intelligence reports of the other side's efforts.

Why then, did the U.S. government cancel the project in August 1959, just when the two largest plants under construction were nearing completion? Basically, too much reliance had been placed on heat of combustion as a measure of how well the boron fuels would perform. The crucial fact was that boric oxide was produced in the combustion. In jet engines, the boric oxide fouled the rotating fan blades. In rocket engines, a solid combustion product cannot contribute to thrust as gaseous combustion products do. Hence, the boron fuels were unsuitable for the intended uses.

Curiously, the people doing rocket fuel tests probably knew the truth as early as 1957. This was evident in impractical suggestions that the boron fuels be burned with fluorine rather than oxygen, or be burned in combination with hydrogen. It seems that nobody wanted to be the first to blow the whistle. The project also had at least two fringe benefits: it was an excellent training ground for many chemists and engineers, and the chemistry of boron received a major boost in interest and understanding. Nobel prizes were awarded in 1976 to William Lipscomb for his work on the structure of boranes and in 1979 to Herbert C. Brown for his development of boron reagents important to organic synthesis. Bonding and structure of the boranes is quite unusual (see Figure 3). In the early 1940s, Herbert Brown had demonstrated the hydrolysis of sodium borohydride as a means of generating low-pressure hydrogen. The conclusion of World War II ended the construction of a sodium borohydride plant being built for that purpose, but the same process was considered at the close of the twentieth century as a source of low-pressure hydrogen for fuel cells to power the next generation of automobiles. Additional work is being done on carborane

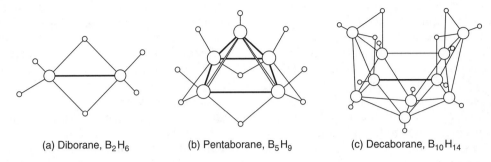

(a) Diborane, B_2H_6 (b) Pentaborane, B_5H_9 (c) Decaborane, $B_{10}H_{14}$

Figure 3. Structural diagrams of (A) diborane, (B) pentaborane, and (C) decaborane. The large circles represent boron atoms; the small ones represent hydrogen atoms. Note the unusual three-dimensional structures of the higher borane molecules and also the unusual "bridge" hydrogen atoms, which are shared by two boron atoms.

polymers, though none are yet offered commercially. The technical literature on boron chemistry has swelled immensely.

See also **Rocket Propulsion, Liquid Propellant**

ANDREW DEQUASIE

Further Reading

Dequasie, A.E. *The Green Flame.* American Chemical Society, Washington D.C., 1991.

Holtzman, R.T. *Production of the Boranes and Related Research.* Academic Press, New York, 1967.

Stock, A. *Hydrides of Boron and Silicon.* Cornell University Press, Ithaca, 1933.

Breeding, Animal: Genetic Methods

Modern animal breeding relies on scientific methods to control production of domesticated animals, both livestock and pets, which exhibit desired physical and behavioral traits. Genetic technology aids animal breeders to attain nutritional, medical, recreational, and fashion standards demanded by consumers for animal products including meat, milk, eggs, leather, wool, and pharmaceuticals. Animals are also genetically designed to meet labor and sporting requirements for speed and endurance, conformation and beauty ideals to win show competitions, and intelligence levels to perform obediently at tasks such as herding, hunting, and tracking.

Prior to the twentieth century, humans carefully chose and managed animals with desired qualities for breeding. Unaware of genetics, people relied for centuries on observation, experience, and chance to breed selectively livestock and pets that displayed valued traits such as sturdiness, gentle temperaments, and coat colors and textures with the expectation that their offspring might also have those characteristics. Breeding of specific lineages, which produced exceptionally vigorous animals, was recorded by breeder associations as ideal specimens, honored with awards, and promoted. Animals that lacked prized assets were removed from breeding herds. Males were castrated. Sometimes culled animals were slaughtered. In eighteenth century England, Robert Bakewell initiated many fundamental animal-breeding concepts such as keeping breeding records with accurate pedigrees, evaluating young male animals with progeny tests, emphasizing family attributes with careful inbreeding, and breeding only the best animals.

During the early twentieth century, geneticists recognized the potential of agricultural genetics. Applying scientific principles from Mendelian plant breeding experiments to animal breeding, scientists developed methodologies to manipulate some of the 30,000 to 40,000 genes in farm animals to improve such factors as growth rates, fur quality, and milk production. At Iowa State University in the U.S., animal husbandry professor Jay L. Lush (1896–1982) pioneered quantitative biometrics techniques in the 1930s and 1940s. He is recognized globally as the founder of modern animal breeding and genetics based on statistical analysis. Because of Lush, an international animal breeding study center was established at Iowa State. Such research initiated the transition of animal breeding from an amateur to primarily professional activity.

Line-breeding programs for cattle began in the U.S. in 1934. Breeding strategies are pursued because breeders aspire to develop high-quality animals, which earn higher market prices, produce greater yields, receive larger sporting purses, or win more prestigious competitions. These victories and performance tests, which assess yields and growth rates, validate genetic breeding programs. Breeders strive to create efficiently and consistently uniform types of animals that appeal to consumers. Purebred line breeding enables breeders to develop and rely on breeding stock from lineages known to produce certain characteristics such as meatier beef cattle, faster horses, stronger oxen, or tastier swine. Sheep breeders use genetics to achieve desired fleece pigment, weight, and fiber diameter, curvature, and durability. These traits are reinforced when genetically similar animals that share ancestors are bred. Inbreeding involves mating closely related animals such as siblings in an effort to emphasize genetic traits in offspring. However, recessive genes associated with undesired characteristics that breeders cannot visually detect in the parents are sometimes paired during fertilization, resulting in inferior offspring.

In contrast, outcrossing involves mating unrelated animals representing the same breed. Breeders select animals that display specific qualities, hoping that the offspring will demonstrate all of the targeted traits. Animals of different breeds within a species are crossbred to enhance strengths associated with each breed. Sometimes breeders cross animals representing varying species such as a goat and sheep producing a geep. Most cross-species hybrids are artificially created, although some occur in the wild. Mules, the offspring of a horse–donkey cross, are sterile and considered more reliable and stronger workers than their parents.

Dog hybridization commonly occurs and historically accounts for the creation of most modern breeds such as Yorkshire terriers, which were derived from an amalgamation of breeds. Mongrels are the most familiar canine hybrids. Some breeders intentionally cross breeds in an attempt to emphasize and improve breed qualities in offspring. Hybrids usually inherit their parents' best physical, or phenotype, characteristics and lack any genetic material that might cause deficiencies. They are considered more vigorous and resistant to genetic defects and diseases than purebred dogs and tend to be nonshedding, thus nonallergenic. Cockapoos, a cross between Cocker Spaniels and Poodles, are popular because of their gentle dispositions and attractive appearances. Some people purposefully mate dogs with wolves or coyotes in an attempt to create aggressive hybrids for protection. Some natural genetic mutations occur such as the Munchkin cat breed, which has short legs as a result of a dwarfism gene.

Breeding agendas changed during the twentieth century due to industrialized agriculture and cultural attitudes regarding nutrition. Through World War II, fatty swine were valued for the lard they produced which had many practical applications. Breeders cultivated hogs genetically prone to produce large amounts of fat until the postwar urban market shifted to prefer lean pork. From the 1950s, fatty swine were culled from breeding programs, and many pig breeds became extinct. As breeding became more selective to meet consumer demands, animals became physically and genetically more uniform. Such uniformity was also essential for animals to be processed by automated slaughter machinery. Animals' genetic diversity essential for sustainable agriculture was threatened as more breeds became extinct. Groups such as the American Livestock Breeds Conservancy formed with the aim of preserving animal breeds. Members are encouraged to register their animals and store sperm in semen banks. Heirloom animals, representing endangered breeds from past centuries, are protected on farms where breeding efforts strived to replenish stock. Some consumers protest at the use of genetics to control commercial animal reproduction and question how animal welfare and product quality are affected.

By the late twentieth century, genetics and mathematical models were appropriated to identify the potential of immature animals. DNA markers indicate how young animals will mature, saving breeders money by not investing in animals lacking genetic promise. Scientists also successfully transplanted sperm-producing stem cells with the goal of restoring fertility to barren breeding animals. At the National Animal Disease Center in Ames, Iowa, researchers created a gene-based test, which uses a cloned gene of the organism that causes Johne's disease in cattle in order to detect that disease to avert epidemics. Researchers also began mapping the dog genome and developing molecular techniques to evaluate canine chromosomes in the Quantitative Trait Loci (QTL). Bioinformatics incorporates computers to analyze genetic material. Some tests were developed to diagnose many of several hundred genetic canine diseases including hip dysplasia and progressive retinal atrophy (PRA). A few breed organizations modified standards to discourage breeding of genetically flawed animals and promote heterozygosity.

See also **Artificial Insemination and *In Vitro* Fertilization (Farming); Farming, Growth Promotion; Genetic Engineering, Applications**

ELIZABETH D. SCHAFER

Further Reading

Bixby, D.E., Christman, C.J., Ehrman, C.J. and Sponenberg, D.P. *Taking Stock: The North American Livestock Census*. American Livestock Breeds Conservancy, Pittsboro, 1994.

Bourdon, R.M. *Understanding Animal Breeding*, 2nd edn. Prentice Hall, Upper Saddle River, 2000.

Cameron, N.D. *Selection Indices and Prediction of Genetic Merit in Animal Breeding*. CAB International, Oxford, 1997.

Clark, A.J., Ed. *Animal Breeding: Technology for the 21st Century*. Harwood, Amsterdam, 1998.

Comstock, R.E. *Quantitative Genetics with Special Reference to Plant and Animal Breeding*. Iowa State University Press, Ames, 1996.

Maijala, K., Ed. *Genetic Resources of Pig, Sheep, and Goat*. Elsevier, New York, 1991.

Padgett, G.A. *Control of Canine Genetic Diseases*. Howell Book House, New York, 1998.

Scott, J.P. and J.L. Fuller. *Genetics and the Social Behavior of the Dog*, reprint edition. University of Chicago Press, Chicago, 1998.

Willis, M.B. *Practical Genetics for Dog Breeders*. Howell Book House, New York, 1992.

Breeding, Plant: Genetic Methods

The cultivation of plants is the world's oldest biotechnology. We have continually tried to produce improved varieties while increasing yield, features to aid cultivation and harvesting, disease, and pest resistance, or crop qualities such as longer postharvest storage life and improved taste or nutritional value.

Early changes resulted from random cross-pollination, rudimentary grafting, or spontaneous genetic change. For centuries, man kept the seed from the plants with improved characteristics to plant the following season's crop. The pioneering work of Gregor Mendel and his development of the basic laws of heredity showed for other first time that some of the processes of heredity could be altered by experimental means.

The genetic analysis of bacterial (prokaryote) genes and techniques for analysis of the higher (eukaryotic) organisms such as plants developed in parallel streams, but the rediscovery of Mendel's work in 1900 fueled a burst of activity on understanding the role of genes in inheritance. The knowledge that genes are linked along the chromosome thereby allowed mapping of genes (transduction analysis, conjugation analysis, and transformation analysis). The power of genetics to produce a desirable plant was established, and it was appreciated that controlled breeding (test crosses and back crosses) and careful analysis of the progeny could distinguish traits that were dominant or recessive, and establish pure breeding lines. Traditional horticultural techniques of artificial self-pollination and cross-pollination were also used to produce hybrids. In the 1930s the Russian Nikolai Vavilov recognized the value of genetic diversity in domesticated crop plants and their wild relatives to crop improvement, and collected seeds from the wild to study total genetic diversity and use these in breeding programs. The impact of scientific crop breeding was established by the "Green revolution" of the 1960s, when new wheat varieties with higher yields were developed by careful crop breeding. "Mutation breeding"—inducing mutations by exposing seeds to x-rays or chemicals such as sodium azide, accelerated after World War II. It was also discovered that plant cells and tissues grown in tissue culture would mutate rapidly. In the 1970s, haploid breeding,

Figure 4.   To increase the genetic diversity of U.S. corn, the Germplasm Enhancement for Maize (GEM) project seeks to combine exotic germplasm, such as this unusually colored and shaped maize from Latin America, with domestic corn lines. [*Photo by Keith Weller. ARS/USDA*].

which involves producing plants from two identical sets of chromosomes, was extensively used to create new cultivars. In the twenty-first century, haploid breeding could speed up plant breeding by shortening the breeding cycle.

Although the ability to make genes change by inducing mutation and selecting beneficial varieties was well known, the development of a crop or plant with desired characteristics was lengthy, haphazard, and difficult when done by these conventional breeding techniques.

In the 1970s, recombinant DNA technology made it possible to manipulate sequences of DNA and to combine DNA from two or more sources. To be useful, the recombinant molecule must be replicated many times. Gene cloning provided a new dimension to plant biotechnology by enabling intentional directed changes to the genotype of a plant using enzymes to identify, remove, invert, and splice genes and by using mobile pieces of DNA (plasmids) to carry genes from cell to cell. Along with the knowledge of genes, understanding of the biochemistry of cells has been critical. For example knowing how genes make use of the information stored in their DNA to produce proteins and how they switch genes on and off allows us to manipulate gene expression.

One of the most important food crop plants is wheat, a plant that is difficult to modify genetically since it has a huge genome (17,000,000 kilobases) or six times more than a human cell. Because the rice genome (40,000 kilobases) is structurally similar to parts of wheat, and rice is also an important staple crop of inherent commercial importance, a novel strategy has been developed. The similar genes are identified in rice, and this can be used as a model to locate the genes for wheat.

By 2000, researchers had established the complete gene sequence (genome) for the first plant (*Arabidopsis thaliana*, a small plant in the mustard family) and are beginning to understand how genes switch on and off. This information will allow studies of how cells respond to external stimuli. To do this, it is necessary to study the total protein content of a cell (the proteome) as it changes. Genetic engineers can now access gene-sequencing data readily. It is not published in conventional scientific journals but in one of three databases that includes information from around the world. These databases can be searched for the information the plant breeder needs.

Plants can be modified in a directed way by gene addition (cloning) or gene subtraction (genes are removed or inactivated). Plants are now engineered for insect resistance, fungal resistance, viral resistance, herbicide resistance, changed nutritional content, improved taste, and improved storage. A significant commercial example is the development of Bt maize. Maize crops can be devastated by the European corn borer (*Ostrinia nubiliasis*), which tunnels into the plant and thus evades insecticide sprays. The gene product that was introduced to the plant was from a bacterium producing *Bacillus thuringiensis* toxin (Bt toxin), a substance known as a δ-endotoxin. This toxin is a protein that is 80,000 times more toxic to insects than organophosphate insecticides. Its advantage as an insecticide is that the protein accumulates in the bacterium in the form of an inactive precursor and can be introduced into the plants without any effect. When the maize is eaten by the insect, its digestive enzymes attack the protein, the resulting small protein molecules bind to and damage the insect gut, and it starves to death.

Early work to produce Bt maize was carried out with the CryIIA (b) version of the protein. This contains 1115 amino acids, but the toxic activity results from the segment of amino acid from 29 to 607. Rather than try to isolate the gene from the bacterium, scientists made a length of DNA by artificial gene sequencing (putting sequences of bases together in a desired order). This allowed them to modify the way the gene would be expressed in maize. The synthetic DNA was placed into a vector (a piece of DNA used to transport it), and this was then forced into maize embryos. The embryos were grown into mature plants, and the individual plants were tested to be sure the new gene was present and to determine its activity. Other plants that have been engineered to produce δ-endotoxin include rice, cotton, potatoes, and tomatoes.

Legumes (such as cow pea and beans) have been engineered to produce proteinase inhibitors. This strategy is particularly effective against larvae that feed on weeds. Other gene addition projects have led to plants that are able to withstand fungi, bacteria, and viruses or to resist the toxic effects of herbicides used to control weeds.

Gene deletion or inactivation is also used to genetically modify a crop plant. The first genetically modified (GM) food to be approved for sale in the UK was a tomato. This was developed with antisense technology in which a gene is removed from the chromosome and reinserted in the reverse direction so that it does not function. The gene in question was the polygalacturonidase gene. Inactivation or partial inactivation of this gene delays ripening, enabling the grower to leave the fruits on the plant until they ripen and develop a full flavor while preventing softening that makes transportation difficult. Other characteristics that result from gene subtraction include the prevention of discoloration by removal or reduced expression of the polyplenol oxidase or a lowered starch content by reduction of starch synthetase. The first GM plants were engineered for such traits, but breeders are increasingly developing engineered traits that improve nutritional content by increasing the proportion of essential amino acids, reducing the starch content, increasing the vitamin content, or adding higher monosaturated, or unsaturated, fatty acid content. One of the best examples of this technology is the development of golden rice, which contains the provitamin A gene. Rice is a staple diet in many parts of the world where a major problem is childhood blindness caused by vitamin A deficiency. Consumption of golden rice will prevent this serious condition. Because the beta-carotene, or provitamin A, is expressed as a golden color, identification of the GM variety of rice is possible.

Genetic engineering of our food supply has generated much debate, strong views, and heated argument. Our ability to change the genetic makeup of cells is seen by some as interfering with life itself. There is also concern about the possible impact of GM crops on the food supply chain and farming practices around the world, while there are powerful counterarguments that without GM crops wide-scale starvation will be a fact of life for many people. Many have questioned the motives of the multinational companies that have developed herbicide resistant crops while also manufacturing the herbicide. From a scientific point of view the important issue is whether GM crops are safe. This question can be answered both from a food safety and an environmental safety perspective. In the context of food safety, all GM crops to be marketed in Europe must pass a stringent assessment as defined by the European Novel Food Regulation (258/97). The food is assessed rigorously by leading experts who look at and evaluate complex scientific evidence. Details of the genetics, intended use, nutritional aspects, toxicology, and potential allergenicity are all scrutinized. It is true to say that we know far more about the GM food crops that have been through this process than we do about the so-called conventional agricultural crops that have resulted from years of hybridization and selection. We do however have a tradition of consumption and recognize that any potential problems might cause a serious allergic response in some sensitive

people, as has been the case with peanuts. Today there is a great deal of information available relating to the development and approval of GM crops, but public reaction, particularly in Europe, has led to the withdrawal of foods containing GM ingredients from some supermarkets. Certainly the use of antibiotic-resistant markers, which were once used to identify the transformed plants during the cloning process, is no longer tolerated since there is a very small theoretical risk that the antibiotic resistance could, upon consumption of the food, be transferred to the cells in the flora of the human gut. However, it is true to say that despite consumption in the U.S. of significant amounts of GM foods, no evidence of harm has ever been reported.

Another area of concern is potential harm to the environment from GM foods. Predicting the effects that GM plants might have on the soil or the flora and fauna in the environment can be done on model systems, but ultimately there is a need for field scale evaluations to measure potential environmental effects of GM foods as compared to the non-GM crop over several seasons in a variety of locations. Until such experiments are concluded, there is a moratorium on the commercial growing of GM crops in the U.K. It is important that these field scale evaluations (which are only done after carefully controlled releases) go ahead and that the results are evaluated robustly and peer reviewed. Evidence of safe consumption and lack of environmental hazard along with public education and total transparency are required if the public is to accept GM products in agriculture.

See also **Biotechnology; Food, Processed and Fast; Genetic Engineering, Applications; Pesticides**

JANET M. BAINBRIDGE

Further Reading

Feitelson, J.S, Payne, J. and Kim ?. *Bacillus thuringiensis* and beyond. *Biotechnology*, 10, 271–275, 1992.

Smith C.J.S. *et al. Nature* 334, 724–726, 1994.

Advisory Committee on Novel Foods and Processes, Annual Report, 2000, FSA/0013/0301.

Bridges, Concrete

A complex interplay between societal change, the development of the internal combustion engine, and the impact of World War I, led to an explosion in the number of road vehicles in the immediate postwar years—and a totally inadequate nineteenth century legacy of roads to accommodate them. Following the first International Road

Congress in 1923, vast and expensive road-building programs were undertaken in the U.S. and Europe, particularly in Germany, during the 1930s. After World War II highway construction continued to grow in an attempt to keep pace with the popularity of the car for private transport. Concrete—strong in compression but weak in tension—is not particularly satisfactory as a running surface. It can easily crack, unlike tarmac, though in the 1960s its use as a surface did become widespread. Otherwise, however, concrete became omnipresent in twentieth century road construction, and in the myriad of bridges, large and small, associated with highway networks.

Because of unreinforced concrete's limitations, nineteenth century concrete bridges were built as arches, following the usual form of bridges in natural stone, which has similar strength and weakness. Paradoxically, it was the incorporation of the "rival" and far more expensive material steel—equally strong in compression but also with high tensile strength—that enabled concrete thus reinforced to become the twentieth century's most widely used material for bridges carrying road and sometimes rail traffic.

The French engineer François Hennebique notably developed techniques for embedding steel reinforcing rods in concrete for structural effectiveness, and had built around 100 reinforced concrete bridges by 1900. Already his designs were exploiting the material's potential for strength and slenderness in arch designs that were much thinner than was possible with stone or mass concrete; but the first, and arguably still the greatest, designer of reinforced concrete bridges was the Swiss, Robert Maillart. Though Maillart only built in his native country and despite the fact that none of his spans exceeded 90 meters, the elegance, economy, and inventiveness of his designs, combined with their fitness for purpose and for their often sublimely beautiful locations, made him one of the most influential twentieth century bridge designers following his death in 1940.

Even more than simple reinforcement, prestressing opened up new possibilities in concrete bridge building. As so often with technical innovation, the concept—of tightening metal rods or strands within concrete to further increase its tensile strength—had been around for a long time before it became a practical proposition. That it did so is due to the observation and tenacity of another French engineer, Eugene Freyssinet, who around 1910 first observed the tendency of concrete to "creep" (its continuing slow shrinkage after solidification) on his first reinforced concrete bridges.

In later projects he introduced the practice of jacking the arch halves apart after casting, and inserting extra concrete at the crown between them to counter the effects of the shrinkage. The scale in particular of his Plougastel Bridge, completed in Brittany in 1930 and then by far the world's largest reinforced concrete bridge, made it necessary for him to study and evaluate the effects of creep as exactly as possible. His researches led him to the view that "locking in" tensile strength by incorporating steel strands in the concrete, stretched to a precisely calculated extent, was a viable structural system—and indeed would effectively produce a new building material.

Freyssinet's first prewar attempt to mass-produce and market prestressing failed, but after World War II he successfully built six single-span prestressed concrete bridges across the River Marne in France. Prestressed concrete, either pre- or posttensioned, rapidly became the material of choice for some short-, most medium-, and more rarely, some long-span bridges. In pretensioning, the concrete is poured around tendons that have already been stressed against an anchor frame, this being released when the concrete has hardened so that the tensile strength locked into the strands is imparted to the concrete adhering to them. In posttensioning, the strands are threaded through voids cast into already-hardened concrete, and then tightened.

Reinforced concrete and prestressed concrete are used in several structural forms in modern bridge building. For most short single spans, the concrete is cast as an arch or a solid slab *in situ* (literally "on site") on formwork and around the mesh of reinforcing rods. For simple spans from around 16 to 20 meters, the reinforced slab is usually cast with voids to lighten the weight of the concrete. Prestressing is normally introduced when longer spans are required. Such bridges can be anything from a single span across a road to the literally thousands that comprise the 38-plus kilometer Lake Pontchartrain Bridges in Louisiana. Beams, of rectangular or T-shaped cross-section, are prestressed and precast offsite, craned into place on supporting piers, and topped with deck slabs. For yet longer spans, sections of prestressed concrete box girder (see the entry "Bridges, Steel") may be are joined together between supports of up to 200 meters and more. These bridges often have the appearance of a wide, shallow arch, though rarely do they act structurally as a true arch, in which forces are carried around and down the arch and into abutments. Instead, the structure acts as a beam, with gravity creating compression forces along the top of the span and tension forces along the bottom, which the prestressing withstands.

The longest-span wholly prestressed concrete bridges are, however, true arches. The archetype is the 305-meter-span Gladesville Bridge in Sydney, completed across the Parramatta River in 1964. Its design was unusual, in that it followed the same voussoir principle that the Romans used for their masonry arches, in which wedge-shaped stones were cut to form segments of a semicircle. In the case of the Gladesville Bridge, the voussoir units are hollow prestressed concrete boxes, each precast in the shape necessary to form the giant shallow arch of the main bridge structure, from the upper surface of which the even shallower curve of the precast road deck is carried on slender upright prestressed piers.

Even longer concrete spans do exist in which a concrete deck or concrete pylons may form part of a cable-stayed or a suspension bridge. However, as all bridges in these forms also incorporate steel, always in the hangers of both types and in the cables of suspension bridges, they are discussed in a separate entry, as are all-steel suspension bridges built in the early twentieth century.

See also **Bridges, Long Span and Suspension; Bridges, Steel; Concrete**

David J. Brown

Further Reading

Brown, D.J. *Bridges: Three Thousand Years of Defying Nature*. Mitchell Beazley, London, 1993. A well-illustrated popular history.

Concrete Bridge Development Group. *Durable Bonded Post-Tensioned Concrete Bridges*. Concrete Society, London, 1996.

Dekenkolb, O.H. and American Concrete Institute. *Concrete Box Girder Bridges*. State University Press, Iowa, 1977.

Gies, J. *Bridges and Men*. Cassell & Company, London, 1964. No longer up-to-date, but an excellent popular narrative of bridges and their designers.

Heins, C.P. and Lawrie, R.A. *Design of Modern Concrete Highway Bridges*. Wiley, Canada, 1984.

Leonhardt, F. *Bridges: Aesthetics and Design*. Architectural Press, London, 1982. A seminal work on bridge design by one of the greatest twentieth century bridge engineers.

Ryall, M., Peake, G. and Harding, J., Eds. *Manual of Bridge Engineering*. Thomas Telford, London, 2000. A definitive text on the design and construction of all types of modern bridge, published under the auspices of the Institution of Civil Engineers.

Wallbank, E.J., Department of Transport and G. Maunsell and Partners. *The Performance of Concrete in Bridges: A Survey of 200 Highway Bridges*. Her Majesty's Stationery Office, London, 1989.

Useful Websites

http://www.bridgeweb.com/

Bridges, Long Span and Suspension

From the beginning of the twentieth century, bridge spans in excess of 300 meters became increasingly common. Depending on considerations of location, use, and loading—not to mention aesthetic and engineering aspiration—these could be suspension, arch, or cantilever structures. When spans of 1000 meters or more began to be contemplated from around 1930 however, a suspension bridge was the only answer. The breakthrough structure was New York's George Washington Bridge; its clear span of 1067 meters almost doubled that of the previous record-holder, the 564-meter Ambassador Bridge in Detroit completed only two years earlier. Nonetheless, within a few years the leading edge of enterprise had passed to the West Coast, with the simultaneous construction of the San Francisco Bay Bridge complex (twin 704-meter suspension spans plus a tunnel and a cantilever), and the 1280-meter-span Golden Gate Bridge, opened in 1937.

The Bay Bridge was designed under the supervision of California State highway supremo, Charles H. Purcell, while the Golden Gate was the one great work of the engineer Joseph Strauss. Others however, like Othmar Ammann, Leon Moisseiff, and David Steinman, were signature designers who left a monumental legacy in steel across the length and breadth of the U.S.

In the design of the many large suspension bridges constructed in the burst of activity in the 1930s, stability of the unprecedented extended decks was a primary concern. The response of long, narrow unsupported structures to wind forces was not precisely understood, though the steel plate girder deck of the George Washington Bridge was correctly deemed by its designer, Ammann, to be wide (36.3 meters) and thick (3 meters) enough to be stiff and stable without a supporting truss. The longer and narrower Golden Gate, by contrast, was given a 7.6-meter-deep stiffening truss, whilst the need for the Bay Bridge to carry two levels of traffic necessitated a 9-meter-deep truss to carry the second deck. Other long-span bridges for which far lower volumes of traffic were anticipated, however, invited a combination of narrowness and slenderness in the pursuit of elegance which proved ultimately disastrous when Moisseiff's Tacoma Narrows Bridge ("Galloping Gertie") collapsed in a wind of only 68 kilometers per hour (km/h) in November 1940. Like that of the George Washington Bridge, its deck was a straight-sided plate girder, but only 11.9 meters wide and 2.4 meters thick over its 853-meter length, and its designed-in flexibility led to twisting under quite moderate winds that progressively increased under its own momentum until failure occurred.

For the next 25 years no long-span suspension bridge was designed without a stiffening truss, including the magnificent 1158-meter-span Mackinac Straits Bridge between Great Lakes Huron and Michigan—the crowning masterpiece of David Steinman's career—and the 1298-meter-span Verrazano Narrows Bridge in New York. This, the last great work of Othmar Ammann, took the world record span from the Golden Gate in 1964 and held it for 17 years.

The challenge of designing a slender, untrussed but aerodynamically stable deck was finally met successfully by a team of British engineers led by Ralph Freeman when the 987.5-m-span Severn Bridge in England was given an "aerofoil" with trailing edges tapered in cross-section on both sides. This minimized wind resistance and avoided the buffeting eddies of wind created by the square sides of the earlier American plate girder designs. Aerodynamic decks became the norm for a new generation of large suspension bridges in Europe, including the two Bosporus Bridges in Turkey (1074 meters and 1090 meters), the record-breaking 1410 meter span Humber Bridge in England, opened in 1981, and the even larger East Bridge across the Great Belt in Denmark, completed in 1998 with a main span of 1624 meters.

East Asia, however, saw the late-twentieth century's greatest activity in long-span bridge design and construction, the most intensive building program taking place in Japan with three major links comprising over 20 individual bridges completed between the islands of Honshu and Shikoku in the 1980s and 1990s. These included no fewer than eight large suspension bridges, among them the world-record holder at the end of the century, the Akashi Kaikyo, opened in 1998 with a main span of 1991 meters.

Suspension bridges have their decks literally suspended by hangers (usually vertical) from thick steel cables formed from thousands of steel strands spun back and forth between the tops of the pylons (towers) and then compacted and wrapped. This technique was pioneered in the nineteenth century by John Roebling, most famous as the designer of the Brooklyn Bridge in New York. Cable-stayed bridges, equally literally, have their decks "stayed" by being directly connected to supporting masts

with straight cables. Beyond this, there are many variables in design: pylons single or double or configured as a A-shape or an inverted Y; and anything from a single cable connecting pylon and deck to many, either fanning out from near the top of the pylon or parallel to each other. The first modern cable-stayed bridges were three in number, designed and built in close proximity to each other in 1952 in Düsseldorf, by the great German engineer Fritz Leonhardt.

In the remainder of the century economy and esthetics made the cable-stayed form the design of choice for the majority of spans upward of 300 meters in most parts of the world. Two landmark structures, completed respectively in 1994 and 1998, showed cable staying being used on a scale for which only suspension spans would previously have been conceived. These were the 856-meter-span Pont de Normandie in northern France and the 890-meter-span Tatara Bridge, yet another element in Japan's great Honshu–Shikoku complex. As the century closed, cable-stayed spans over 1 kilometer were being actively planned, though the suspension concept envisaged for the long-dreamt-of bridge between Italy and Sicily, at 3.3 kilometers, still far exceeds any possible extension of the cable-staying principle.

See also **Bridges, Concrete; Bridges, Steel**

DAVID J. BROWN

Further Reading

Brown, D.J. *Bridges: Three Thousand Years of Defying Nature*. Mitchell Beazley, London, 1993. A well-illustrated popular history.

Gies, J. *Bridges and Men*. Cassell & Company, London, 1964. No longer up-to-date, but an excellent popular narrative of bridges and their designers.

Gimsing, N.J. *Cable Supported Bridges: Concept and Design*, 2nd edn. Wiley, Chichester, 1997. A definitive study by one of the late twentieth century's most distinguished designers.

Leonhardt, F. *Bridges: Aesthetics and Design*. Architectural Press, London, 1982. A seminal work on bridge design, by one of the greatest twentieth century bridge engineers.

Pugsley, A. *The Theory of Suspension Bridges*, 2nd edn. Edward Arnold, London, 1968.

Ryall, M., Peake, G. and Harding, J., Eds. *Manual of Bridge Engineering*. Thomas Telford, London, 2000. A definitive text on the design and construction of all types of modern bridge, published under the auspices of the Institution of Civil Engineers.

Useful website

http://www.bridgeweb.com/

Bridges, Steel

Though techniques for smelting steel had been known in principle since antiquity, only from the mid-nineteenth century did its large-scale production as a practical structural material become a reality. Stronger than wrought iron and more ductile than cast iron, its superior qualities were exploited in three great steel bridges, each in a different structural system, built between 1870 and 1890. The triple-arch St. Louis Bridge in Missouri, with its two levels for road and rail, the suspension Brooklyn Bridge in New York, and the double-cantilever Forth Rail Bridge in Scotland neatly prefigured the resourcefulness with which twentieth century bridge engineers would continue to exploit the material in long-span structures. With growing understanding of the structural potential of steel, and improvements in its tensile strength and other properties, bridges continued to progressively increase in span.

By the end of the nineteenth century a succession of major steel arch bridges had been constructed, notably in southern France, but one of the twentieth century's first such milestone structures was completed in 1907 across the Zambezi Gorge in Zimbabwe by the British engineer, Ralph Freeman. Its span of 152 meters was virtually doubled less than ten years later, when the American engineer Gustav Lindenthal completed his huge 298-meter arch truss Hell Gate Bridge across an arm of the East River, New York. This bridge held its world record for an arch span for 15 years, until the Bayonne Bridge, designed by one of Lindenthal's assistants named Othmar Ammann, was completed across the Kill Van Kull, also in New York, in 1931. Unlike the rail Hell Gate Bridge, the Bayonne was a much lighter structure for road traffic only, and at span of 503.6 meters had been deliberately designed to just exceed a mighty rival, the Sydney Harbor Bridge (Figure 6). Begun four years earlier but completed a few months later in March 1932, the latter remains the widest of all long-span bridges, built to carry both rail and roadway side by side. As steel arch structures, these two giants on opposite sides of the world were only exceeded in 1978 when the 518-meter-span New River Gorge Bridge was completed in West Virginia. Unlike its three predecessors, whose decks are suspended from the arches, the New River Gorge Bridge—as its name indicates—spans a deep river valley, and therefore carries its deck atop the steel arch.

No steel arch bridge was ever the world's longest span *per se*. Cantilever bridges, on the other hand, in their time were world-beaters. In these, arms extend from both sides of rigid bases and either

Figure 5.   Sfalassa Bridge.

Figure 6. Sydney Harbor.

meet in the middle or support a central span. In the case of the Forth Bridge, there had been three supporting towers and two clear 521-meter spans, double world-beaters when it was completed in 1889. Superficially similar in profile, but with a single and even longer main span of 549 meters, the Quebec Bridge in Canada took over the record when it was completed in 1917 after 13 years' work and two major construction disasters involving 85 fatalities. Steel cantilever bridges continued to be a popular structural type for large bridges, particularly in the first half of the twentieth century. Two notable examples were the Queensboro Bridge in New York (the first to be constructed in high-strength nickel steel), and the Carquinez Straits Bridges in San Francisco. For the longest spans of all, however, favor shifted to the suspension bridge with a series of constructions in the U.S. whose single greatest "leap forward" was the mighty 1067-meter-span George Washington Bridge in New York, completed in 1931. Though throughout the remainder of the century even longer-span suspension bridges continued to be designed and built (discussed in the entry Bridges, Long-Span/ Suspension), the George Washington, with its truss-framed double deck and unique open 183-meter skeletal steel towers, remains perhaps the archetypal all-steel suspension bridge.

The second half of the twentieth century saw two distinct developments in bridge design: the first, cable staying, is discussed in the entry on long span and suspension bridges. The second is the box-girder. It also had nineteenth century progenitors in the iron tubular rail bridges erected by Robert Stephenson at Menai and Conway, but the post-World War II steel box girder was a far smaller and lighter affair, its design actually built on new understanding of the behavior of thin-walled boxes and tubes in torsion (twisting) derived from the design of wartime aircraft fuselages. A box girder is essentially a hollow beam formed from a series of open-ended boxes made of steel welded together, with stiffened horizontal plates or flanges at the top and bottom connected by vertical or angled side-plates or webs. Box-girder bridges were quickly found to be strong, economical in material, and relatively quick to construct, but progress in their use came to a temporary halt in 1970 when two major failures occurred within three months, one at Milford Haven in Wales and the other on the West Gate Bridge over the River Yarra at Melbourne. In both cases joints between boxes failed, at Milford Haven above a column and at the West Gate at the midpoint of a completed span. Though the collapses seemed similar, and both involved fatalities, extensive

Figure 7. Garabit Viaduct.

investigations showed rather different causes: failure of a pier support diaphragm in the former and inadequacies in erection methods and site organization in the latter. Greater understanding of the forces acting within and upon the steel components led to new and much more stringent construction standards, and in the decades since, steel box girders have continued to be a widely used form of bridge construction all over the world.

Figure 8. Victoria Falls.

Steel is also very commonly used for constructing footbridges, and particularly towards the end of the century, there was a proliferation of creative invention in their design. A pair of braced girders or a small steel box remains the simplest forms of structure, whilst for a longer span a truss form will deliver a lighter and economical bridge. If a landmark structure is needed, however, the variations and combinations of—in particular—cable-stayed and arched designs seems virtually limitless. A notable inspiration for other designers has been the numerous memorable bridge profiles created by the Spanish architect and engineer Santiago Calatrava. Pedestrian decks may be braced by virtually invisible stainless steel wires extending from vertical or angled masts, suspended from slender tubular arches, or as in the truly innovative Gateshead Millennium Bridge designed by Wilkinson Eyre, curved in plan and pivoted to raise in conjunction with its balancing steel arch to allow shipping to pass beneath, an action which recalls the form and action of a gigantic "blinking eye."

See also **Bridges, Concrete; Bridges, Long Span and Suspension**

DAVID J. BROWN

Further Reading

Brown, D.J. *Bridges: Three Thousand Years of Defying Nature*. Mitchell Beazley, London, 1993. A well-illustrated popular history.

Chatterjee, S. *The Design of Modern Steel Bridges*. BSP Professional Books, Oxford, 1991. A key textbook, particularly on plate and box-girder bridges.

Corus Construction Centre. *The Design of Steel Footbridges*. Corus U.K., Scunthorpe, 2000.

European Convention for Constructional Steelwork and British Constructional Steelwork Association. *Proceedings of the International Symposium on Steel Bridges, Institution of Civil Engineers, 25–26 February 1988*. ECCS, Brussels, 1988.

Gies, J. *Bridges and Men*. Cassell & Company, London, 1964. No longer up-to-date, but an excellent popular narrative of bridges and their designers.

Leonhardt, F. *Bridges: Aesthetics and Design*. Architectural Press, London, 1982. A seminal work on bridge design by one of the greatest twentieth century bridge engineers.

Matildi, P. and Matildi, G. *Ponti Metallici: Esperienze Vissute*. Costruzioni Cimolai Armando SPA, Pordenone, Italy, 1990. A useful Italian perspective on the subject, with many construction details and captions in English as well as Italian.

Ryall, M., Peake, G. and Harding, J., Eds. *Manual of Bridge Engineering*. Thomas Telford, London, 2000. A definitive text on the design and construction of all types of modern bridge, published under the auspices of the Institution of Civil Engineers.

Steel Construction Institute. *Integral Steel Bridges: Design Guidance*. SCI, Ascot, 1997.

Tordoff, D. and British Constructional Steelwork Association. *Steel Bridges*. BCSA, London, 1985.

Useful websites

http://www.corusgroup.com/
http://www.bridgeweb.com/

Building Acoustics

An important element in a properly functioning building is correct building acoustics. Achieving a low level of background noise in a classroom, for example, will ensure that the teacher's voice is audible; the sounds of an orchestra will be optimal in a concert hall with proper acoustics. The systematic study of room acoustics began at the end of the nineteenth century, and consequently a scientific understanding of building acoustic design is almost entirely a twentieth century phenomenon.

The means to achieve low noise levels in buildings were developed during the twentieth century. One of the greatest differences between old and new auditoriums is the low noise levels achieved in those built from the mid-twentieth century onward. Noise from external sources can enter a room through vibration paths (structure-borne transmission) or can pass directly into the building through adjacent walls (airborne transmission). Where very low noise or vibration levels

are needed in auditoriums, recording studios, and operating theaters, vibration isolation (springs and resilient materials) are used, as are physical breaks in vibration paths. Airborne noise is reduced by the use of constructions such as double partitions separated by air gaps containing absorbent materials. The failure to achieve the desired background noise levels is often due simply to poor workmanship.

Building service equipment such as boilers, etc., should be mounted on vibration isolators, and the structure and airborne paths should be considered in the design. Ventilation outlets may require silencers; low-velocity air conditioning is favored because it is quieter. Noise generated within rooms can also be a problem. For example the twentieth century saw a great increase in the use of atria, which can be very noisy, reverberant spaces due to footfall noise and the sounds of noisy items such as escalators. Perversely, a few rooms require the application of noise in order to achieve confidentiality. Masking noise from a radio is often added to hospital waiting rooms to provide privacy for consultations.

By the end of twentieth century the most critical design requirements for correct acoustics in rooms were well established. For speech, the requirement is often for intelligibility rather than fidelity. This requires the speech to be louder than the background noise and limits the room size that can be used without electronic enhancement. The direct path between the speaker and listener should not be obstructed, and the audience should be as close as possible to the speaker; placing seats on an incline is useful. Speakers should always face the audience, however, the audience can surround the speaker in small theaters. There should be hard surfaces close to the speaker and listeners to create beneficial reinforcing sound reflections. Where the stage area of a theater is very absorbent due to the scenery, reinforcing reflections from the ceiling and proscenium arch are vital. Surfaces that generate late-arriving reflections are usually treated with absorption to prevent echoes. Many of these principles were exploited in ancient amphitheaters, but it is only in the twentieth century that the scientific reasons for good or bad acoustics were understood. These requirements are easier to achieve when the speakers are located in one place, such as the stage of a theater and difficult in courts and debating chambers where there are many different speaker positions.

The room sound should not reverberate excessively, otherwise the sound of one syllable will run into the next syllable. Acoustic absorption is used to reduce reverberance. During the twentieth century, various technologies for testing speech intelligibility were developed. In the early 1970s, T. Houtgast and J.J.M. Steeneken reported on their studies showing a direct correlation between modulation reduction factors and speech intelligibility. Their studies are the basis for the Speech Transmission Index (STI) program, an objective measure used in performance specifications. Perceptual tests may also be carried out, but these tests are rather slow to do.

Developments in electronics have influenced building acoustics. Speech reinforcement and public address systems are used for emergency evacuation and day-to-day messaging. Electronic reinforcement will sound unnatural if the speech appears to come from the loudspeakers rather than the speaker. This is solved by applying delays to the sound coming from the loudspeaker. The Haas (precedence) effect states that the sound that arrives first—the first sound heard—will usually determine the perceived location of the sound. For a room with a high ceiling such as an auditorium, a single large cluster of loudspeakers may be used. For a room with a low ceiling, a series of loudspeakers placed along the length of the room may be used. Each loudspeaker covers a different area and has different delays. To prevent feedback, sound from the loudspeakers should not be directly picked up by the microphone. Loudspeakers can be sited forward of the speaker and directional microphones used. If a space is overly reverberant, only the frequency ranges important for speech intelligibility are reproduced. Operators need to be trained to use these sound systems correctly, otherwise the speech produced may be of poor quality.

Sound reproduction rooms such as control rooms in studios tend to be small. Low-frequency resonances of the room cause coloration, but the audible effects are reduced by appropriate choice of room dimensions and techniques for the application of resonant acoustic absorption. These consist of vibrating membranes over a cavity, with resistive material such as mineral wool in the cavity. At higher frequencies, the coloration caused by early-arriving loud reflections must be minimized. Absorbers (which remove sound energy) or diffusers (which spatially and temporally disperse sound energy) can be used. Absorber technology was developed throughout the twentieth century, while diffuser designs date only to the mid-1970s. In the 1970s, revolutionary new diffuser designs used phase gratings based on mathematical sequences; designs even exploited the emerging

mathematical discipline of fractals. More modern diffusers use numerical optimization algorithms to provide curved diffusers that complement contemporary architecture.

A great concert hall acts as an extension to the musical instruments, embellishing and improving the sound produced by the musicians. In 1895–97, physicist Wallace C. Sabine, in tests aimed at correcting the poor acoustics in the lecture hall at Harvard University's Fogg Art Museum, established the need to obtain the correct reverberance. Sabine has been called the father of architectural acoustics, and was the first to apply quantitative measures to the design of a concert hall (the first auditorium that was designed by Sabine was the new Boston Music Hall, opened October 15, 1900). In the last 30 years of the twentieth century, additional parameters to reverberance have been identified as important. For example, acousticians now understand how the hall shape influences sound quality. In previous centuries, trial and error had established the 'shoebox' shape (long, high, and narrow) as providing a good sound; it is now understood that this shape worked because it provided beneficial side reflections. Circular and elliptical shapes risk focusing sound-creating "hot spots." The influence of amphitheaters, theaters, and early cinema can be seen in fan-shaped halls built in the mid twentieth century, but this shape can lead to a lack of side reflections. New shapes created for halls over the last half of the twentieth century include the Vineyard Terrace, which subdivides the audience area so the dividing walls produce beneficial early reflection. In the last decades of the century, acousticians developed a better understanding of the needs of musicians. Unless musicians receive reflections from surfaces around and above the stage, they cannot hear themselves or others, and so cannot create a good tone and blend or play in time.

Multipurpose halls create problems given the different acoustic demands of different events. The general principle is to design for a primary purpose, and then adjust the acoustics for other uses. Absorption is used to deaden a concert hall ready for electronic music, and electronic means are used to make a speech theater more reverberant for music. Electronic enhancement began with the assisted resonance system in the 1960s, and there was slow and steady growth in the use of enhancement systems in theaters from that time. Sound is picked up from the stage using microphones. The signals are then delayed and played from loudspeakers to create extra reflections and so increase reverberance.

The biggest influence that electronics has had on building acoustics has been the computer. Sophisticated computer-based instrumentation has allowed accurate measurement of building acoustics. Computer-based prediction models have enabled the improved understanding and design of acoustic technologies, from building elements to the whole rooms. Much of the mathematics used by acoustic engineers was developed in the nineteenth century, but this has only been exploitable at the end of the twentieth century using computers. There was also increased interest in virtual acoustic prototypes, which would allow building acoustics to be listened to in virtual environments, allowing nonacoustic experts to more readily understand the principles of good acoustic design.

See also **Audio Recording, Electronic Methods; Loudspeakers and Earphones**

TREVOR COX

Further Reading

Barron, M. *Auditorium Acoustics and Architectural Design.* Spon Press, 1993.

Beranek, L.L. *Concert and Opera Halls: How They Sound.* Acoustical Society of America, 1996.

Houtgast, T. and Steeneken, H.J.M. The modulation transfer function in room acoustics as a predictor of speech intelligibility. *Acustica,* 28, 66–73, 1973.

Rayleigh, J.W.S. *The Theory of Sound,* vols. 1 & 2. Dover Classics of Science & Mathematics, 1976.

Sabine, W.C. *Collected Papers on Acoustics.* Harvard University Press, Cambridge, MA, 1922. Republished by Acoustical Society of America, 1993.

Sharland, I. *Woods Practical Guide to Noise Control.* Woods Acoustics, 1990.

Smith, B. J., Peters J. and Owen S. *Acoustics and Noise Control*, Longman, 1996.

Templeton, D., Ed. *Acoustics in the Built Environment.* Architectural Press, 1997.

Thompson, E. *The Soundscape of Modernity: Architectural Acoustics and the Culture of Listening in America, 1900–33.* MIT Press, Cambridge, MA, 2002.

Steeneken, H.J.M. and Houtgast, T. A physical method for measuring speech transmission quality. *J. Acoust. Soc. Am.* 67, 318, 1980.

Building Equipment, *see* **Construction Equipment**

Buildings, Designs for Energy Conservation

Obtaining our energy supplies from coal, oil or gas (the fossil fuels) results in the production of a range of pollutants that have adverse effects on people, forests, wildlife, and aquatic environments. By the last quarter of the twentieth century, increased fossil fuel consumption was believed by many to have a profound effect on Earth's climate as result of the release of so-called greenhouse gases such as carbon dioxide (CO_2).

In most countries in the twentieth century, the energy consumed in buildings represented a substantial proportion of nationwide energy consumption. In higher latitude regions, the majority of this energy demand has historically been energy for homes to provide space heating, followed by energy for hot water, for powering appliances, and for lighting. In nondomestic buildings in these regions the demand has historically been dominated by electricity for lighting, appliances, and ventilation and cooling. While space and water heating can be the largest proportion of household energy consumption, electricity consumption can be as important in terms of upstream CO_2 emissions if it is generated in a fossil fuel electricity generating station. Architects, builders, and engineers have struggled to balance the demand for energy, particularly in the industrialized countries that are heavily energy dependent, with environmental and cost concerns. The oil crisis of 1973, following an embargo of oil directed primarily against the U.S. by Middle Eastern oil-producing companies, and the OPEC oil crisis of 1979 was the end of the era of cheap energy. Energy conservation emerged as a concern for both designers and consumers, particularly in countries solely dependent on imported oil. The government in Korea, for example, asked people to "think poor," reduced the number and size of electric light bulbs in government and corporate buildings, and discouraged the use of elevators, air conditioning, and street lighting. Later policies supported use and development of energy conservation technologies. In the U.S. and also in Japan, large-scale research and development funding resulted in building guidelines and technologies for energy conservation that are discussed in this entry.

Space Heating

In residential buildings in high latitudes and other colder regions it is possible to substantially reduce or eliminate the need for space heating energy by a combination of building techniques:

- Very high standards of thermal insulation of walls, roofs, and floors
- High thermal performance glazing of windows
- Airtight construction (e.g., sealing cracks between frames and windows)
- Energy-efficient ventilation (passive ventilation or efficient mechanical ventilation)

- Efficient heat recovery in ventilation systems (a heat exchanger recovers heat from the indoor air being expelled)
- Compact building form to reduce surface area to volume ratio
- Thermostatically controlled responsive space heating delivery systems
- Use highest building density possible

In technology terms, new materials are key to thermal insulation and glazing. Cellulose insulation, in the form of sawdust and paper, has long been used in walls and ceilings. Modern blown or sprayed cellulose fiber is made from fire-retardant paper, and is popular for uninsulated wall cavities. Fiberglass has been used since the 1930s as a building insulator. Plastics developed in the 1930s were used in foam form after World War II as thermal insulation (e.g., polyurethane, polystyrene or styrofoam). Ceramic fibers, commercially available from the 1970s, are used industrially as a replacement for asbestos but are not widely used in building insulation. Low-emissivity coatings (very thin, transparent layers of metals or oxides) on glass were developed in the early 1980s to reduce heat transfer to the outside. Use of a vacuum between glass panes or filling the gap between the glass panes with low-conductivity gas such as argon or krypton at atmospheric pressure reduces conductive and convective heat transfer.

Energy demand for space heating can be further reduced if passive solar measures such as the following are included:

- Orienting the main aspect of the building and the majority of glazing to face the noon-day sun (toward the south in the northern hemisphere and toward the north in the southern hemisphere)
- Designing buildings so that the most frequently used rooms are on the side nearest the midday sun
- Including materials of high thermal capacity inside the building envelope to reduce temperature swings and absorb solar heat gains
- Locating sun spaces on building aspects facing the midday sun, provided that they are not unheated or air conditioned
- Avoiding overshading the building during the winter months

Passive solar heating through architectural design has been in use since at least the fifth century BC. Archaeological evidence of house plans and the writings of scholars such as Socrates, show that ancient Greece and Rome were well aware of passive solar design. Anasazi cliff dwellings of the American Southwest were also orientated to maximize winter sun and summer shade.

Cooling

Domestic buildings in tropical regions of the world and Middle Eastern deserts have traditionally been designed with shading and air circulation patterns in mind. The rise of industrialization and commerce places more individuals in factories and office buildings and has increased the energy needs for such settings.

In many nondomestic buildings, the energy required for lighting and equipment as well as cooling may be more significant than that for space heating. In lower latitude regions and regions subject to very high solar gains, the major energy demand has historically been for providing space cooling or air conditioning, followed by hot water provision and electricity for powering appliances and lighting. It may be possible to reduce energy demand for cooling by taking account of the following:

- Orienting the main aspect of the building to face away from the midday sun (toward the south in the northern hemisphere and toward the north in the southern hemisphere)
- Minimizing the glazing that is not solar protected on the western and eastern façades
- Maximizing the ceiling heights of the rooms
- Including north-light roof lights to maximize day lighting without incurring high solar heat gains
- Designing the building to incorporate solar shading devices to keep high temperature solar gains from entering the building
- Including materials of high thermal capacity inside the building envelope
- Incorporating lighter colors and reflective surfaces on the building to reflect solar gains from the building
- Designing the building to incorporate ventilation cooling towers or solar chimneys to achieve *passive stack cooling*
- Where appropriate considering the use of evaporative cooling techniques
- Considering the use of earth tubes to precool ventilation air
- Where feasible, considering controlling the building temperature to achieve night cooling of thermally massive components of the building to keep the building cool during the day

- Where wind speeds are sufficient, considering the use of wind energy for cooling utilizing wind ventilation towers, wind scoops, or wind catchers

Wind catchers in traditional Arabic architecture used a "chimney" with one end underground and the other set over a specific height on the roof. The air trap operated according to temperature differences, with difference in air density resulting in air flow that reversed.

Other Electrical Demands

Once the larger problems of space heating or cooling have been addressed, other electrical demands should be considered. In hot water heating, the energy required for providing domestic hot water can be minimized by switching from baths to showers and using atomizing showerheads to limit water flow instead of "power-showers." Careful selection of low water consumption and low-temperature washing machines should be used in preference to more profligate models. Hot water demand is also driven by lifestyles and the conservation awareness of householders; for example, the frequency of taking a shower or bath, doing the laundry, or using dishwashers in preference to hand-washing dishes.

All buildings used by people require lighting at certain times of day, but buildings can be designed to minimize the use of artificial lighting by maximizing the use of day lighting, providing manually controlled task lighting, and providing daylight and motion-sensing controls, which can regulate artificial lighting levels and switch off unnecessary lights when no one is in the room. Building design can employ north-facing (or south-facing in the southern hemisphere) glazed façades for tasks which require good lighting, and minimize the depth of rooms from a window to maximize the level of daylight penetration. Electricity consumption can also be minimized by avoiding the use of incandescent lamps and instead using more efficient lamps such as fluorescent tubes or compact fluorescent lamps (CFLs). Better still, light emitting diode (LED) lamps were becoming available for certain purposes by the late 1990s (see Lighting Techniques).

By the 1980s and 1990s, many countries had instituted standards for energy efficiency that applied to such household appliances as refrigerators, dishwashers, and washing machines and dryers. U.S. federal law from 1978 mandated efficiency standards for major residential energy-using equipment, including heating, cooling, and water heating equipment, but standards were not finalized until legislation in 1987. Certain states, especially California, have efficiency standards for this equipment that are more stringent than the federal standards. In 1993 the Environment Protection Agency introduced the voluntary Energy Star label program to promote energy-efficient products. Consumers were urged to select energy-efficient models for conservation and cost reasons. Another option was to use portable equipment powered by batteries, such as portable computers, which are likely designed to be more energy efficient than standard electric equipment. In 1995 Energy Star was extended to cover new homes and commercial and industrial buildings.

Obtaining Building Energy Supply from Low-CO$_2$ Sources

The carbon dioxide emissions from fossil fuel energy sources can be reduced substantially by means of cogeneration or combined heat and power (CHP) stations, which deliver both heat and electricity. This is because a higher proportion of the primary energy contained in the fuel at the power station is used compared to electricity-only power stations that do not make use of the waste heat. For example, in the U.K. the waste heat lost in power stations is comparable to the amount of heat provided by gas to heat buildings. Historically these CHP stations have been small power stations located in built-up areas that use heat mains (hot water pipes) to deliver heat to buildings within a certain radius of the CHP station. This approach has been successfully utilized in many European countries, particularly Denmark.

Small scale CHP is also employed for large buildings or small groups of buildings and can be successfully employed provided there is a good match with both the heat and the electricity demand (see Domestic Heating). There is also research underway in various locations (including Woking in the U.K.) into the use of fuel cells to provide electricity and heat from gas but with much reduced emissions. Fuel cells are electricity-generating devices that combine oxygen and hydrogen together, and the only important emission is water(see Fuel Cells). Research in the late twentieth century to develop fuel cells for vehicle propulsion could improve the economic prospects for fuel cell CHP and provide a means of utilizing emission-free hydrogen fuels derived from alternative or renewable energy sources of energy in the future. There are also field trials underway in the U.K. and in other European countries for so-called micro-CHP units that can be used instead of gas boilers in buildings to produce both electricity and heat within the building. These micro-CHP units are mainly based around high efficiency Sterling engines, although some are based around fuel cells.

Obtaining Building Energy Supply from CO_2-Free and CO_2-Neutral Sources

Alternative or renewable energy sources such as solar, wind, or small-scale hydro energy and the low temperature heat stored in the ground are considered CO_2-free. Ground-coupled energy is primarily solar energy stored in the ground as low temperature heat; there may also be locations that are able to use geothermal heat from hot aquifers, which can provide hot water, and in some cases steam to power turbines to generate electricity. Most biological sources of energy (biofuels) are considered CO_2-neutral if they are derived from sustainable sources of biomass feedstocks. Building design can incorporate these energy sources in a variety of ways.

Solar energy:

- Trapping passive solar heat gains through windows or via solar sun-spaces.
- Incorporating active solar thermal collectors on roofs or façades to heat or preheat domestic hot water.
- Incorporating solar thermal collectors on roofs or façades of the building to actively trap solar energy for space heating via an intermediate insulated heat store.
- Incorporating photovoltaic (PV) modules on roofs or façades to generate electricity from light.
- Using high temperature solar collectors to power heat engines in climates that have high levels of direct sunlight.

Wind energy:

- Purchasing electricity from an electricity distribution company supplied by a wind power plant. Various "green electricity" rates or tariffs are available in many countries to assure their consumers that the electricity sold comes from a renewable energy source, frequently wind energy.
- Utilizing a small wind turbine located on a tower near to the building.
- Utilizing one or more community medium or large-scale wind turbines for group housing or neighborhood, village, or town. The town of Swaffham in the east of England obtains 70 percent of the annual electricity needs of its householders from two large wind turbines located nearby.
- Utilizing one or more medium or large scale wind turbines for educational, hospital, public, commercial, industrial, or agricultural buildings to offset electricity demand and generate an income from the export of excess electricity.
- Incorporating building integrated wind energy devices such as the *Aeolian Roof*™ or *Aeolian Tower*™ (and similar systems patented by the author) to generate electricity from wind passing over the building or be used to assist in the ventilation of the building.

Ground-coupled energy:

- Utilizing underground "earth-pipes" to temper incoming ventilation air.
- Utilizing a ground-source heat pump in combination with vertical or horizontal underground collectors to capture low temperature heat for space heating, hot water provision or cooling.
- Utilizing a local aquifer to provide "free cooling" of the building.

Small-scale hydroelectricity:

- Purchasing electricity from an electricity distribution company supplied by small-scale hydro power plant.
- Developing a small-scale hydroturbine for a building or group of buildings if a stream or river with sufficient fall and water flow is available locally.

Biofuels:

- Locally harvested wood fuels from sustainably managed woodlands can be used as a source of heat via wood pellet, wood chip, or log boilers when buildings are located in highly wooded or forested areas. (The supply must be within a minimum energy-balance distance from wooded areas if the wood fuel is transported by fossil-fueled transportation; otherwise the fuel consumption may exceed the energy content of the wood fuels being transported.) Both heat and electricity can be provided via cogeneration or CHP plant.
- In certain locations, agriculture crop residues may be used or converted into fuels and converted into electricity—for example,

the straw-powered generation stations in various European countries and at Ely in the U.K.—or used in a CHP plant.

- In certain locations energy crops may be grown and the fuels derived from them used in buildings if they are available locally.
- Gaseous fuels can be extracted from the anaerobic digestion of sewage, animal manure, crop wastes, food wastes, and certain components of municipal waste streams. The gas produced in this way can be used in buildings directly or via community or neighborhood cogeneration or CHP plants and district heating networks.
- It may be possible to recover energy from components of the waste stream by incineration, although this is a controversial approach due to concern about pollutants contained in their emissions, and it therefore requires very careful sorting of the waste to be used for combustion. It is an approach used in a number of European countries as waste-to-energy CHP plants.
- It is also possible to process components of the waste stream via pyrolysis (thermal decomposition in the absence of oxygen) to produce liquid or solid fuels, which can be used in conventional or CHP power stations.

See also **Air Conditioning; Domestic Heating; Electricity Generation and the Environment; Lighting Techniques; Power Generation, Recycling; Solar Power Generation; Wind Power Generation**

DEREK TAYLOR

Further Reading

Cofaigh, E., Olley, J.A. and Lewis, J. *The Climatic Dwelling: An Introduction To Climate-Responsive Residential Architecture*. James & James, London, 1996.
Crosbie, M.J. *Green Architecture: A Guide to Sustainable Design*. American Institute of Architects Press, Washington D.C., 1994.
Daniels, K. *The Technology of Ecological Building, Basic Principles and Measures, Examples and Ideas*. Birkhauser/Princeton Architectural Press, 1997.
Levine M.D., Martin N., Price L., Worrell E. *Energy Efficiency Improvement Utilising High Technology: An Assessment of Energy Use in Industry and Buildings*. World Energy Council, London, 1995.
Smil, V. *Energy in World History*, Westview, Boulder, CO, 1994.

Useful website

Department of Energy, Environmental Energy Technologies Division: http://eetd.lbl.gov/BT.html

Buildings, Prefabricated

Prefabricated buildings are assembled from components manufactured in factories. They differ in several ways from "stick-built" structures which are fabricated entirely on site. Typically, prefabricated components are mass produced out of the weather on indoor assembly lines. This method insures that parts can be replicated countless times with little or no variation. Economies of volume reduce costs, and precision measuring and cutting by stationary machine tools lessens waste. As work takes place on assembly lines, it is subject to constant inspection and quality control. Component assemblies made in immovable fixtures and forms further ensure that the finished work is precise and true. Thus, the quality of buildings made from parts fabricated on assembly lines has far greater chance of being accurate and uniform than those made in the field.

There are three basic types of factory made buildings and they differ according to the extent to which their components are assembled. Modular or sectional buildings consist of one or more finished rooms that leave the factory ready to be united on site with matching components. Panelized structures are produced as finished wall panels with both exterior and interior coverings in place or semifinished with only exterior cladding. At the building site, they are erected and joined with similar members to form walls. The precut building is delivered to the work site completely disassembled and consists of all the needed lumber cut to length. The latter is probably the most widely known type of prefabricated building.

One of the earliest examples of a prefabricated building was a house built in England in 1624. Its disassembled parts were shipped to Cape Ann, Massachusetts, where it was reerected. During the 1840s, buildings fabricated in New York, England, and as far away as China were shipped to California where they were used as housing for the influx of gold seekers. Prefabricated houses were being made in limited numbers by at least two U.S. factories by the 1890s.

In 1906, one of the first companies to offer mass-produced prefabricated housing was Alladin Redi-Cut Houses of Bay City, Michigan. Alladin kit houses consisted of precut and numbered pieces of lumber and they remained in production in various forms until 1981. Their renown and sales were eclipsed by Sears, Roebuck and Co., which over the years sold as many as 100,000 homes through its mail order catalog. House kits consisted of precut lumber, directions, and all the materials and paint needed to complete the structure. As many as 450 designs were manufactured in Sears' factories between 1908 and 1940 when production ceased. Although they were neither trendsetting designs nor technologically innovative, they offered home buyers a choice of inexpensive products.

Generally, mass-produced buildings were the work of unnamed architects or designers. However, immediately after World War I, several well-known architects considered the potential of this building technique to meet Europe's pressing need for good and economical housing. A new type of architectural creativity was called for, and (Swiss-born) French architect Le Corbusier (aka: Charles Edouard Jeanneret) embraced the idea, developing several schemes for mass-produced prefabricated housing. However, despite his strong advocacy he produced little actual housing. Likewise, Walter Gropius of the Bauhaus in Germany, who had advocated industrialized worker housing as early as 1910, proposed structures composed of standardized flat roofed "Building Blocks," but little came of his proposals. During the mid-1930s, even noted America architect Frank Lloyd Wright, considered large scale, low-cost housing. While not actually advocating the mass production of housing, he realized the value of regularity or standardization of characteristics. This resulted in his design for small and relatively inexpensive Usonian Houses with standardized layouts, modular dimensions, and a modest degree of detailing. No more than 60 of these houses were built during a 15-year period beginning in 1936.

The sudden need for new housing just prior to and during World War II was met in part with prefabricated structures. One direct result of the demand was the development of the semicircular-shaped steel Quonset hut. That prefabricated structure was based on the British-built Nissen hut devised in the early 1930s. Construction methods as well as interior finish and insulation for the corrugated pressed steel Quonset were improvements over the Nissen hut. Quonsets were built in a number of sizes and served as housing for troops, warehouses, hospitals, and offices. Following the war, their use continued and broadened to include churches, shops, and emergency housing. Nonetheless, the need for domestic housing remained acute, and in the U. S. the situation was ideal for new approaches to prefabrication and mass production.

In 1947, New York developer Abraham Levitt and sons began to address the housing shortage on a scale previously unknown. In a Long Island potato field they commenced work on what would eventually become Levittown. Rather than focusing on building individual prefabricated houses, they set about building and selling an entire mass-produced community of standardized structures. The five simple types of homes they offered were built on concrete slabs. The homes were assembled of precut lumber shipped in from the Levitts's own lumbermill in California. There were great cost advantages to builder and buyer due to the tremendous volume of materials being used. Nails were needed in such quantity that the Levitts erected their own nail factory. Employing assembly line techniques, teams of workers performed specific tasks as they moved from one structure to the next and, by 1948, they were erecting 30 houses per day. When the development was finally completed in 1951, more than 17,000 homes had been constructed.

Carl Strandlund of Chicago established the Lustron Corporation and in 1948 began production of a new all-steel prefabricated house. Although the first prototype lightweight all-steel house was constructed by Albert Frey and A. Lawrence Kocher for a building products exhibition in New York in 1931, developer and consumer interest never materialized. However, with a pressing national need for housing immediately after the war, massive surplus industrial capacity along with generous federal government loans were available to address the problem. Assembly lines were set up in Ohio in a former aircraft factory manned by workers hired from the automotive industry. They stamped, punched, welded, and bolted sheet steel into house components. No wood was used in the construction, and exposed metal surfaces were preserved with a long-lasting porcelain enamel finish like that used on household appliances. Despite the suitability of the project to assembly line techniques and its acceptance by the public, production and financial difficulties put the Lustron Corporation in bankruptcy by 1950. Only 2500 prefabricated steel homes were built.

Nonetheless, in the decades following World War II, other all-metal prefabricated buildings were successfully manufactured and marketed and became a significant part of the industry. The Butler Manufacturing Company of Kansas City, Missouri was one of the first manufacturers of preengineered steel commercial structures. Devised by brothers Wilbur and Kenneth Larkin in 1939, their buildings used rigid steel frames. It was only after the war when restrictions on the sale and use of steel were lifted that Butler Buildings were first marketed. The rigid internal framework followed wall and roof lines and required no internal supports or bracing. This left greater interior open space than in other buildings of comparable size. Precision manufacturing as well as quality control resulted in perfectly fitting components. Not only were the buildings readily assembled, but the economical use of materials insured a reasonable price. Butler produced a broad range of buildings for commercial and industrial users in an international market supplied by a network of manufacturing plants throughout the world. The company expanded the technology when it offered multistory and long-span preengineered steel buildings.

During the 1950s and 1960s the industry worked to dispel the belief, which began during World War II, that prefabricated buildings were shoddy and poorly made. One approach was to abandon the term prefabricated and refer to buildings as being either factory made, preengineered, or manufactured. Although the name changed, traditional assembly line prefabrication methods continued during the second half of the century. Standards were kept high and maintained by quality control and production oversight.

Prefabrication moved in a new, but not altogether permanent direction in the early 1970s, when architect Kisho Kurakawa unveiled plans for his Nakagin Capsule Tower in Tokyo. Designed and built between 1970 and 1972, it consisted of a steel reinforced concrete tower core filled with capsule living spaces. The modest size capsules were preassembled lightweight welded steel modules built and finished offsite. After delivery to the tower, they were hoisted into place and secured by four bolts. If necessary, the units could be detached and replaced.

At the close of the twentieth century, production of prefabricated buildings was strong, and the industry continued worldwide growth. Countries as disparate as Romania and Australia supported domestic manufacture, while in North America, Canada was a major international exporter of these buildings. In 1999, in the U.S. alone there were no less than 670 manufacturers who prefabricated wooden buildings and 579 who carried out the process in metal.

See also **Construction Equipment**

WILLIAM E. WORTHINGTON, JR.

Further Reading

Arieff, A. and Burkhart, B. *Pre Fab*. Gibbs Smith, Salt Lake City, 2002.
Carter, D.G. and Hinchcliff, K.H. *Family Housing*. Wiley, New York, 1949.
Rabb, J. and Rabb, B. *Good Shelter: A Guide to Mobile, Modular, and Pre Fabricated Houses, Including Domes*. Quadrangle/ Times Books, New York, 1975.
Stevenson, K.C. and Jandl, H.W. *Houses By Mail*. Preservation Press, Washington D.C., 1986.
Watkins, A.M. *The Complete Guide to Factory Made Houses*. E.P. Dutton, New York, 1980.

C

Calculators, Electronic

An electronic calculator is a calculating "machine" that uses electronic components, such as integrated circuits, transistors, and resistors to process the numbers that have been entered through a keyboard.

The electronic calculator is usually inexpensive and pocket-sized, using solar cells for its power and having a gray liquid crystal display (LCD) to show the numbers. Depending on the sophistication, the calculator might simply perform the basic mathematical functions (addition, subtraction, multiplication, division) or might include scientific functions (square, log, trig). For a slightly higher cost, the calculator will probably include programmable scientific and business functions.

Either way, the calculator is small, self-powered, and relatively disposable. Yet it was not always that way. The room-sized Z3 developed by Konrad Zuse and the ENIAC of the mid-twentieth century are considered the world's first digital computers (see Computers, Early Digital), but they really were little more than electronic calculators designed to work through a series of numbers. The fact that these early behemoths had the computing power of today's $10 student calculator does not lessen their importance to the engineers and scientists of the mid-1940s. This was a time when people would spend days, weeks, or months calculating formulae using mechanical adding machines. Or they would use slide rules and preprinted tables to help determine certain scientific solutions, knowing that the final number was close but not perfectly accurate to the nth degree.

The need for sophisticated and speedy manipulation of numbers was of major importance. It took the birth and growth of the electronic age for this to happen. Vacuum tubes (valves) allowed the beginning of the computing age, transistors brought it down to size, and integrated circuits were the catalyst to make electronic calculators truly possible and accessible to the world. All this happened in the space of thirty years.

In 1962 British company Sumlock Comptometer designed and sold the first all-electronic calculator, a desktop model named the Anita (an acronym for A New Inspiration To Arithmetic, as the story goes). This model used small vacuum tubes as electronic switches. The reliability was not infallible, but it was much faster than the electro-mechanical calculators available at the time and very desirable to people who needed the number-crunching "computing" power (see Figure 1).

Figure 1. Anita Mk VIII, one of the world's first desktop calculators, launched in 1961.
[*Courtesy of www.vintagecalculators.com.*]

In the early 1960s, Japanese manufacturers were among the first to become interested in electronic calculators and the first to develop all-transistor devices. Sony showed a prototype at the New York World's Fair in 1964, and Sharp was the first to sell a production model soon after. Even though it cost the same as a small car ($2500) at the time, Sharp's CS-10A sold well to those who needed its speed and power. However, at 25 kg it was not portable.

It was the invention of the integrated circuit, the heart of modern-day computers, that made the most serious mark in electronic calculators. An integrated circuit can reduce the size of the circuitry in an electronic device to one hundredth (or more!) of the original size, and offer computing power far greater than was otherwise possible.

In the late 1960s, calculator companies, particularly Japanese, began to work with American semiconductor companies who were developing the ability to design and manufacture integrated circuits.

In 1967 one such company, Texas Instruments, had a breakthrough. Their engineers took an electronic calculator design with hundreds of discrete components (transistors, diodes, relays, capacitors) that covered the top of a desk and successfully reduced most of the electronics to four small integrated circuits. Interestingly enough the project, code-named Cal-Tech, was created not to start Texas Instruments' entry into calculators, but to interest calculator makers in using their integrated circuit products. The integrated circuit was then a new and unproven product in the world of electronics, and the Cal-Tech was intended to prove that they were effective. Texas Instruments' effort worked—Canon became very interested in the Cal-Tech and eventually created a similarly designed calculator (the 1970 Pocketronic) with the help of Texas Instruments' engineers. In the same period, other business machine companies began to develop and market similar products.

One small Japanese company, Busicom, began to work with a fledgling company called Intel to develop an integrated circuit to use as the brains for Busicom's 141-PF desktop calculator. Eventually, Intel bought back the rights to the circuit design. This design, the model 4004 microprocessor, is the grand predecessor to all of Intel's current Pentium products.

Companies like Bowmar and Summit in the U.S., and Sinclair and Aristo in Europe, would develop very small pocket-sized models in the early 1970s. Soon after, hundreds of large and small companies worldwide would develop and sell hand-held calculators, thanks to the availability of inexpensive, calculator-function integrated circuit chips.

A major milestone occurred in January 1972 when Hewlett Packard (HP) sold the world's first scientific pocket calculator. This model was in so much demand that even though it cost US$395 (two weeks wages for most engineers) there was a 6-month backlog to buy one. The HP-35 was so powerful that it rivaled some small computers and brought computing power directly to the hands of its users. As the years passed in the 1970s and 1980s, the production cost of integrated circuits (once costing US$100 each) dropped to less than $1. Calculators which once cost US$300 dropped to $10 or less. By that time, simple models were even distributed free as inexpensive promotional tools.

At the end of the twentieth century, the electronic calculator was as commonplace as a screwdriver and helped people deal with all types of mathematics on an everyday basis. Its birth and growth were early steps on the road to today's world of computing.

See also **Calculators, Mechanical and Electromechanical; Computers, Early Digital; Integrated Circuits, Design and Use; Transistors**

GUY D. BALL

Further Reading

Ball, G. and Flamm, B. *Collector's Guide to Pocket Calculators.* Wilson/Barnett, Tustin, CA, 1997.
Greenia, M.W. *History of Computing: An Encyclopedia of the People and Machines That Made Computer History.* Lexikon Services, Antelope, CA, 2000.
Haddock, T. *Personal Computers and Pocket Calculators.* Nostalgia Publishing, La Centre, 1993.
How the Computer Got Into Your Pocket. *Invention and Technology*, Spring 2000.
Johnstone, B. *We Were Burning: Japanese Entrepreneurs and the Forging of the Electronic Age.* Basic Books, New York, 1999.
Lukoff, H. *From Dits to Bits: A Personal History of the Electronic Computer.* Robotic Press, Ontario, 1979.
Reed, T.R. *The Chip: How Two Americans Invented the Microchip and Launched a Revolution.* Random House, New York, 2001.
Various. Special 15th Anniversary Issue. *Electronic Engineering Times*, November 16, 1987.

Calculators, Mechanical and Electromechanical

The widespread use of calculating devices in the twentieth century is intimately linked to the rise of large corporations and to the increasing role of mathematical calculation in science and engineering. In the business setting, calculators were used to efficiently process financial information. In science and engineering, calculators speeded up routine calculations.

At the beginning of the nineteenth century, mechanical calculators were already in widespread use, and by 1822 Charles Babbage was at work on his difference engine (which he never completed because of the mechanical complexity of thousands of brass cogs and gears). Based on technology developed over several centuries, early twentieth century calculators can be divided into two major types: (1) the slide rule; and (2) the adding machine and related devices.

Invented by William Oughtred in the seventeenth century, the slide rule is based on the physical relationship between two logarithmic scales. In the mid-nineteenth century, accurate methods for reproducing such scales on instruments were developed, and the slide rule became much more widely available. By the early twentieth century, slide rules were commonly used by scientists and engineers for calculations involving multiplication, division, and square roots.

During the first half of the twentieth century, slide rules underwent a gradual evolution. Plastic gradually replaced wood and aluminum in construction, and additional scales were added to more expensive slide rules for specialized computation. In addition, a variety of specialized slide rules were developed for particular applications, most notably electrical engineering.

Slide rules had several advantages over manual calculation. First, with practice, calculations could be made much more quickly than by hand. Second, slide rules were compact and light, and so could be used almost anywhere. Finally, calculations could also be carried out with a reasonable degree of accuracy. However, slide rules also had one important disadvantage—the results of their calculations were approximations rather than exact numbers. This was acceptable for most scientific and engineering calculations, but unacceptable for financial calculations, which had to be exact. As a result, the use of slide rules was limited primarily to scientific and engineering applications.

The second major type of mechanical calculator was the adding machine and related devices. While a variety of adding machines were developed beginning in the seventeenth century, a series of major innovations in the late nineteenth century led to the widespread marketing of machines that were much more reliable, compact, and easy to use. The combination of increased demand from large business and government organizations, along with

improvements in machine tools, materials, and manufacturing techniques, triggered a period of intense innovation that continued until World War I. The key mechanical innovations were all patented before 1900, but putting them into practice took some time, and it was not until the first years of the twentieth century that mechanical calculators began to sell in large numbers (see Computers, Uses and Consequences).

Two key mechanical innovations just before 1900 made this expansion possible. The first, invented almost simultaneously by the American Frank S. Baldwin and the Swede Willgodt Theophil Odhner, consisted of a round disk with moveable radial pins that could be extended beyond the edge of the disk. Input for calculation was a function of varying the number of pins extended by the action of levers, which then meshed with a register mechanism. This design, known both as Baldwin type and Odhner type, proved much more compact and reliable than previous systems.

The second was the introduction of the keyboard for data entry. The first practical machine to use a keyboard for data entry, the Compometer, was invented in 1885 by the American Dorr Eugene Felt, who formed the Felt and Tarrant Manufacturing Company in 1887 to produce his device. The keyboard entry system greatly simplified the operation of calculating machines, and was soon copied by other manufacturers, such as William Seward Burroughs, who founded the American Arithmometer Company (which became the Burroughs Adding Machine Company in 1905). These machines had nine rows of keys, one for each digit (1 to 9). The number was entered by pressing one digit in each column. There was no zero key because zero was represented by the absence of a keystroke in the corresponding column.

During the first half of the twentieth century, the market for mechanical calculators was divided into roughly three categories of machines. The first, and largest in terms of total numbers produced, was adding machines. These machines, produced in large numbers in standard designs that changed little over time, were used primarily for basic accounting by small businesses. Their manufacture was dominated by three large firms: the Burroughs Adding Machine Company and Felt and Tarrant in the U.S. and Brunsviga in Germany, though a variety of smaller firms also competed.

The second type of mechanical calculator was the four-function calculator, which were similar mechanically to adding machines, but performed multiplication and division in addition to addition and subtraction. Such calculations could be carried out on adding machines by trained operators, but four-function calculators were faster and allowed the use of less highly trained operators. Four function machines were used primarily in medium-sized businesses whose volume of calculation did not justify the use of more expensive specialized machines. They were also used by engineers, which led to the development of machines in the 1950s that included specialized functions such as calculating square roots, such as the Friden SRW model of 1952 (which weighed 19 kilograms). No one firm dominated the market for these machines, and there was considerable competition for market share.

The third type of mechanical calculator was a group of devices known as accounting machines (also called book-keeping machines during the 1920s and 1930s). These devices were used to enter data onto standard forms and then to perform accounting calculations. They could also prepare balances and print the results. Used by larger firms whose volume of calculation could justify the investment in specialized machinery, in some applications accounting machines competed directly with punched card tabulating systems, such as those developed by Herman Hollerith in 1890 and later developed by the successor to Hollerith's company, International Business Machines (IBM). Accounting machines were also used in specialized niche applications.

All three types of calculator could be either hand or motor driven in operation. Motor-driven mechanical models first appeared just after 1900. By replacing the hand crank with a small electric motor, these machines were less tiring to operate and could reliably perform repeat operations, simplifying the construction and operation of machines that performed multiplication and subtraction. As a result, motor-driven calculators were commonplace on the desks of engineers by the 1940s. Hand-driven machines continued to sell well, however, because they were quieter, lighter, smaller, and less expensive.

Regardless of the type of mechanical calculator, firms found that to be successful they had to provide a high level of service to customers. Calculators were sold, not bought, and firms maintained large sales forces to educate customers as to the capabilities of machines and to anticipate customer's needs. Calculator manufacturers also had to service their machines, which as mechanical devices need constant maintenance to function reliably, and train operators in the correct operation procedures.

The manufacture and sale of calculators was a widespread industry, with major firms in most industrialized nations. However, the manufacture of mechanical calculators declined very rapidly in the 1970s with the introduction of electronic calculators, and firms either diversified into other product lines or went out of business. By the end of the twentieth century, slide rules, adding machines, and other mechanical calculators were no longer being manufactured.

See also **Calculators, Electronic; Computers, Analog; Computers, Uses and Consequences**

MARK CLARK

Further Reading

Aspray, W., Ed. *Computing Before Computers.* Iowa State University Press, Iowa, 1990.
Cortada, J.W. *Before the Computer: IBM, NCR, Burroughs and Remington Rand and the Industry They Created, 1865–1956.* Princeton University Press, Princeton, 1993.
Hopp, P.M. *Slide Rules: Their History, Models, and Makers.* Astragal Press, 1999.
Jerierski, D.V. *Slide Rules: A Journey Through Three Centuries.* Astragal Press, 2000.
Marguin, J. *Histoire des instruments et machines à calculer: trois siècles de mécanique pensante, 1642–1942.* Harmann, Paris, 1995.
Martin, E. *The Calculating Machines (Die Rechenmaschinen): Their History and Development (1925).* Translated and edited by Kidwell, P.A. and Williams, M.R. MIT Press, Cambridge, MA, 1992.
Turck, J.A.V. *Origin of Modern Calculating Machines (1921).* Arno Press, 1972.

Cameras, 35 mm

The origins of the 35 millimeter (mm) camera lie in the increasing availability of motion picture film stock during the early years of the twentieth century. The 35 mm format was first used in Edison's Kinetoscope, a moving picture viewing device patented in 1891, and was later adopted as the standard film gauge by cinematographers after 1896. The earliest 35 mm film was very slow by the standards of the day and not ideal for still camera work. However, as the quality of the film improved, the potential virtues of small size and convenience of handling began to appeal to camera designers.

Although three Spanish inventors took out a British patent for a still camera using 35 mm film as early as 1908, the first 35 mm still camera sold to the public was probably the American Tourist Multiple camera of 1913. Patented a year later, the design was for a camera taking a 15-meter magazine of standard 35 mm cine film, allowing 750 still exposures. Another American 35 mm camera, the

Simplex, appeared in 1914. The Simplex camera allowed the photographer to switch between two picture sizes, giving 400 full-frame pictures or 800 half-frames, or any number in-between if the sizes were mixed. The Jules Richard Homeos camera, a stereoscopic 35 mm camera, was patented in France and England in 1913 but did not reach the market until a year later.

Sales of the cameras above were limited by the outbreak of World War I in 1914, as was the development and marketing of further models. Unsurprisingly perhaps, the biggest selling 35 mm camera of the decade was produced in America, where public spending was least affected by war. However, the Kodak 00 Cartridge Premo box camera of 1916 was an unusual model using unperforated 35 mm film, a concept that was to have little influence on future development. More important was the work of Oskar Barnack, an engineer working with cine cameras for the Leitz Optische Werke in Germany. In 1913 Barnack designed and built for his personal use a small still camera to make use of "short ends" of cine film. War prevented immediate development but Barnack's little camera was the prototype of the Leica, a camera that was to have a profound influence on future camera design.

The original Leica is sometimes thought of as the first 35 mm camera but it was far from that. During the early 1920s several new 35 mm models were marketed throughout Europe, some enjoying modest success. Nevertheless, the most significant camera of the period was undoubtedly a developed version of Barnack's 1913 camera, the Leica of 1925. Manufactured by Leitz, a renowned optical instrument company, its most important features were the compact integrated design and high-precision construction along with the quality of its lenses, which allowed fine prints to be produced from tiny negatives. The Leica convinced the photographic world that the 35 mm camera was worthy of consideration by the serious photographer. Improved models included provision for a screw fitting interchangeable lens system, which partly promoted another important innovation, the coupled rangefinder. The first coupled rangefinder 35 mm camera was the Leica II introduced in February 1932. A month later the giant German photographic company, Zeiss Ikon, produced its own precision constructed 35 mm camera, the Contax. From the outset, the Contax came with a coupled rangefinder and a bayonet fitted interchangeable lens system. Both cameras had their own reloadable film cassettes, a system that was to lead to universal cassette standardization.

More manufacturers began to exploit the 35 mm format. A noteworthy example was the 1934 Retina camera, the first in a series of well-made 35 mm self-erecting folding cameras marketed by the German arm of Eastman Kodak. The Retina models proved popular and provided a comparatively inexpensive alternative to the Leica and Contax for the keen amateur photographer. The first orthodox mass-produced 35 mm camera to be produced in America was the Argus model A camera of 1936. It was an immediate success as were subsequent models, one of which (the CC model) had an integral photoelectric exposure meter, the first American camera to be so equipped. The Argus C-range cameras, widely known as "bricks" in the trade, were particularly popular. Other significant cameras of the period include the German Robot camera of 1934, the first purpose-designed 35 mm camera with a spring-driven motor drive, and two 35 mm single-lens reflex cameras, the Russian manufactured Sport of 1935 and, more influentially, the German Kine Exakta of 1936. The latter camera was a particularly important pointer to the future.

Coupled rangefinder cameras continued as the classic 35 mm design throughout the 1940s. The postwar Leica remained essentially similar to its original form until a radical redesign led to the introduction of the M series cameras in 1954. The Contax also underwent little major development. Both however, continued to significantly influence camera manufacturers, particularly in Japan. In the immediate postwar years Canon produced a series of Leica inspired rangefinder cameras while the Nikon S of 1954 was firmly based on the Contax II. This period also saw the production of enormous numbers of cheaper 35 mm cameras of varying quality.

Good-quality 35 mm rangefinder cameras from both Europe and Japan remained popular for most of the 1950s. It was the end of the decade before single-lens reflex cameras from Japan began to seriously challenge their supremacy. Yet only a few more years passed before reflex cameras from the likes of Pentax, Nikon, and Canon, incorporating important innovations such as the pentaprism viewfinder, instant-return mirror and through-lens metering had almost completely displaced range-finder cameras from the quality 35 mm market. A limited number of fixed lens nonreflex 35 mm cameras with excellent specifications continued to enjoy good sales, as did many so-called "automatic" snapshot cameras. Nevertheless, there was a steady decline in the use of the 35 mm photographic system during the last third of the twentieth century. Its main representative today is the high-quality, pentaprism single-lens reflex camera.

See also **Cameras, Automatic; Cameras, Single Lens Reflex; Cameras: Lens Designs, Wide Angle, Zoom**
JOHN WARD

Further Reading

Coe, B. *Cameras: From Daguerreotypes to Instant Pictures.* Marshall Cavendish Editions, London, 1978.

Coe, B. *Kodak Cameras: The First Hundred Years.* Hove Foto Books, Hove, 1988.

Condax, P.L, Tano, M.; Takashi, H.; Fujimura, W.S. *The Evolution of the Japanese Camera.* International Museum of Photography, Eastman House, Rochester, 1984.

Dechart, P. *Canon Rangefinder Cameras 1933–68.* Hove Foto Books, Hove, 1985.

Hicks, R. *A History of the 35 mm Still Camera.* Focal Press, London, 1984.

Janda, J. *Camera Obscuras. Photographic Cameras 1840–1940.* National Museum of Technology, Prague, 1982.

Miniature Camera in *The Focal Encyclopedia of Photography*, vol. 2, fully revised edn. London, Focal Press Ltd, London, 1965.

Rogliatti, G. *Leica The First 60 Years.* Hove Foto Books, Hove, 1985.

Rotoloni, R. *Nikon Rangefinder Camera.* Hove Foto Books, Hove, 1983.

Tubbs, D.B. *Zeiss Ikon Cameras.* Hove Foto Books, Hove, 1977.

Cameras, Automatic

The quality and precision of a photograph is basically dependent on the aperture, the light sensitivity of the film, and the shutter speed of the camera. As photography became a sought-after source of record keeping for the amateur, demand increased for a camera that could automatically align this seemingly nefarious collaboration of technical details. The evolution of the automatic camera inevitably reflects the mechanical, electronic, and high-tech progress of the twentieth century, but primarily employed its own specific mechanical innovations.

Prior to the twentieth century, cameras were a fairly cumbersome collection of enormous photographic plates and large boxes, usually mounted on massive tripods in order to keep camera perfectly steady for an image to burn the film without blurring. In the 1870s faster-exposure plates were introduced and the shutters that controlled light exposure were built for the first time inside the camera, reducing some of the bulk. By the turn of the century a wide variety of cameras were available for every special purpose imaginable, including gigantic cameras that produced large prints and double-lensed "stereo" cameras that created three-dimensional and panoramic images. But the amateur photographer was left out of these innovations, too inconvenienced to master the complicated equipment and heavy, expensive photographic plates.

In 1888, American George Eastman brought "Kodak No. 1" to market, as the first hand-held camera, which used rolls of film in place of the photographic plate. While this "snapshot" camera was quite popular, quality photographs from a hand-held camera that could manipulate the necessary components—aperture, shutter speed, and film speed—were decades away. The Kodak No. 1 was set at a single speed shutter of 1/25 of a second, and it used a fixed-focus lens good for any subject more than 2.5 meters away. The camera came preloaded with a roll of 100 pictures, which the photographer had to send, along with the camera, back to Kodak to have developed and reloaded.

It wasn't until 1925, at the Leipzig Fair, that German E. Leitz introduced the Leica camera, which used a higher-quality 35 mm film, a fast focal-plane shutter, and a more precise f/3.5 lens. These adaptations created a camera that could take high-quality photographs under a variety of conditions. The focal plane shutter could give a range of exposures, from 1/1000 second to 1 second, and the lever that operated this mechanism also advanced the film. The lens, which was fast enough to allow for indoor photography, was also connected to a rangefinder that focused easily at the hand of the photographer. The Leica ushered in a new era of photographic conveniences, and soon photojournalism became a fixed—and critical—customer of these new hand-held cameras.

Soon thereafter, the growing needs of photojournalists and discriminating amateur photographers inspired the creation of similar handheld cameras with slight improvements over the Leica. In 1927, Rolleiflex introduced the twin-lens reflex camera, employing a separate focus lens through which the photographer looked, and a parallel lens that captured the image. Ten years later, a single-lens reflex (SLR) hand-held called the Exacta came to market, which allowed the photographer to easily focus through one 35 mm lens. The shutter of the first SLR cameras was dependent on a fast-moving slit that exposed different parts of the subject at different times across the film, distorting fast-moving objects. But the technology of the SLR cameras eliminated the parallax error—or the difference between the image that goes through the lens and the image seen through the viewfinder—of the smaller viewfinder cameras. Using a mirror and a diaphragm, the SLR camera allows the photographer to both compose the image and focus through the 35 mm lens.

Following these improvements, a series of cameras that appealed to the novice photographer came on the market. Most notably, American Edwin H. Land created a camera that used a film capable of producing developed shots on demand—the precursor to the Polaroid camera. After World War II, a sudden plethora of camera manufacturers produced cameras that used 110 film and electronic controls, taking Eastman's snapshot Kodak No. 1 a step further by allowing a slight range of shutter speed, but still employing the same basic fixed-focus lens.

Lens and optical companies soon designed their own versions of the Kodak and Leica amateur-friendly cameras. By 1946, Japanese companies developed 35 mm cameras. All these cameras still required imprecise threading of film, manual control over focus, aperture and shutter speed. The next step in improvements came in the form of film rolls that snapped into a fitted compartment and threaded themselves. Eventually, automatic control over exposure was also incorporated, using a built-in light meter that reads the level of illumination, then sets either the aperture, or shutter speed, or both together. In general, these two forms—the leaf shutter and the focal-plane shutter—are still used today. Originally in Leicas's historical camera, the leaf shutter consists of a series of overlapping blades in the lens, powered by a spring-loaded shutter button. When the shutter is pressed, the blades open and shut according to the settings. The focal-plane shutter, on the other hand, is not in the lens but the camera itself, directly in front of the film. It is a faster mechanism, allowing for faster exposure, utilizing two curtains that open slightly—also by a spring-driven shutter button—exposing the film to a window of light.

Automation of shutter speed and aperture accompanied the development of the built-in exposure meter. This is done by "reading" the light reflected from the object being photographed by a needle in the light-measuring device. The needle measures the light, moving into position. When the shutter button is released, the needle becomes fixed, and a scanner arm moves until it hits the needle. The end of the scanner arm then engages the devices that control the shutter and the aperture diaphragm.

By the late 1950s, camera companies included extras like timers, and by the late 1960s, Polaroid created the first color instant film and the first "instamatic" camera, which was similar to the 110-film cameras of the forties, but with color film.

Autofocus, present in the 1950s in its most basic form, utilized a motor that spun the lens's focus ring, which was usually inside the lens. Over time, manufacturers built the motor in the body of the camera, but the poor quality of these early systems earned autofocus a bad reputation (especially among expert photographers). By the 1970s, however, the American-company Honeywell earned a patent for phase-detection autofocus, which was applied to Konica's C35 AF camera. Basically, the system works by taking light from the subject that passes through the lens and the semitransparent reflex mirror. Another mirror directs the light toward the autofocus module, and an array of light-sensitive charge-coupled devices, or CCDs. The distance between the CCDs determines the focality of the image. A small circuit board then controls a motor that moves the focusing ring of the lens.

By the 1980s, compact, "pocket-sized" 35 mm cameras with autofocus came onto the market, employing systems developed mostly by Japanese manufacturers, who fine-tuned the Honeywell technology and added computer technology and microelectronics to the mechanism controlling the systems. In addition, the perils of setting aperture and shutter speed were further relieved by a new kind of light meter employing a panel of semiconductor sensors that translate the light level into electrical energy. The light meter then "reads" the film speed by special markings on the outside of the 35 mm cartridge and accounts for shutter speed to adjust for the correct aperture.

By the early 1990s, both automatic focus and light meters required nothing of the shooter. Central microprocessors, now fairly common in cameras, activate several motors that control focus, shutter speed and aperture, in a single click. Automatic cameras at the beginning of the 21st century are closer to computers, especially as digital cameras have become a standard piece of equipment for the amateur.

See also **Cameras, 35 mm; Cameras, Polaroid; Cameras, Single Lens Reflex**

LOLLY MERRELL

Further Reading

Auer, M. *The Illustrated History of the Camera from 1839 to the Present*. New York Graphic Society, Boston, 1975.

Dobbs, W.E. *Your Camera and How it Works*. Ziff-Davis, Chicago, New York, 1939.

Horne, D.F. *Lens Mechanism Technology*. Adam Hilger, London, 1975.

Jenkins, R. *Images and Enterprise: Technology and the American Photographic Industry, 1839 to 1925*. Johns Hopkins Studies in the History of Technology, Maryland.

Johnson, W.S. *Photography from 1839 to Today*. Klotz, George Eastman House, Rochester, 1999.

Ray, S.F. *Camera Systems: A Technical Guide to Cameras and Their Accessories*. Focal Press, London, 1983.

Tubbs, D.B. *Daguerre to Disc: The Evolution of the Camera*. California Museum of Photography, Riverside, 1983.

Cameras, Digital

Digital photography constitutes the most revolutionary development in photography since the early experiments of image making on chemically sensitized materials in the 1820s, just as digital computing technologies revolutionized the manner in which the written word was recorded in the latter half of the twentieth century. The invention of flexible film in the late nineteenth century profoundly changed the course of photography, but the adoption of electronic media 100 years later as a substitute for film constitutes an even more fundamental upheaval, with broad implications for the future of visual communication. At the end of the twentieth century electronic imaging seemed poised to render chemically processed photographs obsolete, and the future of film appeared to be in doubt as improvements in the quality and capacity of electronic imaging occurred at a dizzying rate.

Digital photography blends three distinct technologies: traditional photographic methodology, video, and digital computer systems. The advent of television and the cathode-ray tube provided both the technology and the conceptual framework for the production of images on an electronic screen instead of film or paper. The digital computer offered the means for rendering not only text but pictures, broken into tiny picture elements called pixels, which could approximate continuous-tone pictorial representations if the number of constituent pixels was large enough and their size small enough to resist detection by the eye.

Video and computer technologies are electronic cousins, although the first uses analog means and the latter digital. A half century's experience in viewing images on a television screen provided a conceptual basis for the easy reception and popularization of digitally produced images on a computer monitor, and the notion of transmitting televised imagery in the air has its parallel in the transmission of digital imagery via the Internet. Very quickly the concept of imprinting digital images on paper would threaten the traditional chemically based photographic print.

The computer scanner initially served as an intermediate tool between traditional photography and the digital camera. The rise of the Internet created a demand for transmitting photographic imagery, and scanners converted photographs into electronic files which could be viewed on a computer monitor and shared, either by recording them on a medium such as a floppy disk for exchange with other computer users, or by transmitting them via various file-sharing protocols, including the familiar e-mail systems in common use. Computer printers could convert electronic image files into prints, which eventually rivaled the ordinary chemically based photographic print. The rise of electronic image editing and manipulation programs, such as Adobe Photoshop, served to aid in the preparation of scanned images. At first digital photography was largely confined to this conversion of film-based photographs into electronic copies, but soon a new generation of cameras emerged to produce direct electronic images without an intermediate chemically processed photograph at all. This method, interacting directly with a computer, is especially attractive because it completely eliminates the time-consuming steps of chemical processing and the scanning.

There are two types of electronic still cameras, the still video camera and the digital. Both are similar to film cameras except that a built-in light-sensitive computer chip or "imaging array," either a charge-coupled device (CCD) or a complementary metal-oxide semiconductor (CMOS), takes the place of film to capture images. The electronic era in still photography essentially began in 1981 with the introduction of Sony's Mavica video camera, but still video was an analog system. The Dycam digital camera initiated the fully digital era in 1990.

When light enters the digital camera through the lens and strikes the chip, it emits an electrical charge which is measured electronically and is then sent to the electronic memory (buffer) of the camera. It is then compressed into a format; for example, a JPEG, and transferred to the memory card or disk. Some cameras need to wait for the completion of this process before another picture can be taken, while others have a buffer large enough to hold several pictures, thus enabling rapid "burst" shooting for a group of consecutive pictures. The CCD has millions of receptors to record the amount of light striking them. Each sensor represents a pixel or image element. The information on the chip is read one horizontal line at a time into the internal memory of the camera, combining the individual pixels into an image, before it is saved to the memory medium of the camera, from which it can be viewed or printed via a computer.

Color can be recorded with a digital camera in several ways. Some cameras use three separate sensors with a separate filter, and light is directed to the sensors by a beam splitter. Another method is to rotate a series of red, green, and blue filters in front of a sensor, but the continuous movement of the filter wheel requires stationary subjects. Another sophisticated idea places a permanent filter over each sensor and utilizes an interpolation process to approximate color patterns. Most consumer cameras use a single sensor with alternating rows of green–red and green–blue filters in what is called a Bayer filter pattern.

The quality of a digital photograph depends upon a complex combination of factors. The higher the resolution, the more closely the image resembles a continuous-tone analog image. Resolution in film-based photography relates to both the ability of the camera lens and the recording medium to reproduce or "resolve" fineness of detail; in digital photography the number and size of the pixels that constitute the image, control resolution. The larger the number and the smaller the size of the pixels, the greater the resolution. Since higher resolution results in greater file size, practical considerations usually mandate a compromise on resolution. Low resolution can be satisfactory for viewing images on a computer monitor, but images to be printed normally require a higher resolution. Most digital cameras permit the photographer to select a "capture" resolution, which can later be modified with the computer and the editing program, but the photographer must also consider the amount of camera system memory the image will consume. If a preferred resolution setting will utilize too much memory, it may be necessary to select a higher image compression to reduce the file size. The need to manipulate and balance these factors to arrive at acceptable results demonstrates the inherent limitations of digital photography and is one reason that film photography still maintained a strong presence at the end of the twentieth century. Yet the rapid pace of technological advances in computers, software, and digital cameras has led many to assume that film-based photography's days are numbered.

The convenience of digital media has led many professional and other serious photographers to abandon film-based photography altogether. As the tempo of modern life has accelerated, most newspaper photographers and other photojournalists have embraced digital photography exclusively. "Wet" darkrooms were dismantled because photographers in the field, using expensive, sophisticated digital cameras, can transmit images electronically and get them into print in a fraction of the time required with film cameras; also, chemical photography is considered environmentally unfriendly. Photographic manufacturers such as Eastman Kodak, Fujifilm, Canon, Pentax, Olympus, Leitz, and Nikon devote an increasingly higher percentage of their product lines to digital cameras, joined by such familiar electronic firms as Sony, Samsung, and Panasonic.

With the major exception of the image recording system, film and many digital cameras have similar design principles, employing a range of shutter speeds and aperture sizes to control exposure. Some digital cameras employ an entirely different exposure system in which the chips in the image-sensor array turn on and off at varying intervals in order to capture more or less light. Other digital cameras vary exposure by varying the strength of the electrical charge that a chip emits in proportion to the amount of light received. Despite the radical difference of this method, the camera's sensitivity or "speed," which may be manually variable, is typically rated according to standard film speed nomenclature. The resolution rating of a camera generally is expressed in megapixels, ranging from one to four and higher. The higher the megapixel rating of the camera, the more expensive it tends to be. The larger the pixel count, the larger an acceptable print that can be made, but the higher the camera price and the more file storage space an image requires.

Early digital cameras used internal or "on-board" memory for file storage, requiring the periodic transfer of images to a computer after the memory was filled before additional pictures could be taken. This limitation was solved by the advent of removable memory devices, such as cards or disks. Most later digital cameras employed a liquid crystal display screen to preview pictures, which functioned as a viewfinder, as well as to review images stored in the camera, as with a computer monitor; this feature represents a substantial advantage over film cameras, permitting the user to redo an unsatisfactory image before leaving the scene. A variety of innovative special-purpose and special-feature digital cameras were also marketed.

DAVID HABERSTICH

Further Reading

Aaland, M. and Burger, R. *Digital Photography*. Random House, New York, 1992.

King, J.A. *Digital Photography for Dummies*, 4th edn. Wiley, New York, 2002.

B&H. Photo-Video, Inc. *Digital Photography Source Book*, 3rd edn. B&H, New York, 2003.

Kasai, A. and Sparkman, R. *Essentials of Digital Photography*. Translated by Hurley, ENew Riders, Indianapolis, 1997.

Ritchin, F. *In Our Own Image: The Coming Revolution in Photography: How Computer Technology Is Changing Our View of the World*. Aperture, New York, 1990.

Useful Websites

Nice, K. and Gurevich, G.J. *How Digital Cameras Work*: http://www.howstuffworks.com/digital-camera.htm

Cameras, Disposable

The first disposable camera was probably "The Ready Fotografer" of 1892, which used a pinhole aperture instead of a lens. But the single-use concept lay dormant for decades. At the end of the twentieth century digital photography threatened to make photographic film obsolete, but one factor forestalling this revolution was the popularity of "disposable" 35 mm rollfilm cameras. They recapitulate several themes of technological and socioeconomic significance for the history of amateur photography from the late nineteenth century through the twentieth century.

The first theme is epitomized by the Eastman Kodak slogan, "You press the button, we do the rest," for the Kodak camera of 1888 and its successors. The Kodak was a simple box camera, preloaded with rollfilm for 100 exposures, intended for an inchoate amateur market. When the customer finished the roll, the entire camera was sent to the company for processing and printing, and the reloaded camera was returned to the owner. Thus one could take pictures without handling film at all, let alone indulging in messy darkroom work. The primary innovation that attracted consumers was the processing service, since the darkroom was the domain of professionals and specialists. Rollfilm also eliminated the need to handle individual glass plates. The amateur market for cameras and film was created virtually at a single stroke. The other advantage of the system—freedom from loading and unloading film—was forgotten when later cameras accustomed users to performing these tasks. Perhaps having a camera on hand at all times seemed preferable to waiting days for a reloaded Kodak to return.

Another advantage of the Kodak was its small size compared to the large, tripod-mounted cameras of professionals. It was hand-held in operation and light to carry. Later, cameras became even more portable; for example, by using a bellows so that the camera could be collapsed when not in use. Folding cameras were called "pocket" cameras, and Kodak appropriated the word as a brand name, although one needed large pockets to accommodate them. The trend toward smaller cameras was well established by the early twentieth century when the 1924 Leica appeared. This precision camera used 35 mm film, was eminently suitable for a pocket, and its superior optics made enlargements from its small negatives feasible. The first Leicas were difficult to load, however, and that issue plus the high price made it the choice of professionals and dedicated amateurs, not casual snapshooters. 35 mm cameras with wide-aperture lenses and adjustable shutter speeds were versatile but expensive and were considered "serious" cameras. Meanwhile Kodak and other manufacturers continually introduced new fixed-focus, inexpensive cameras for amateurs who could not afford adjustable cameras or preferred simplicity. At first, these simple cameras were limited to exposures in bright sunlight, but eventually battery-operated flashbulb guns were added to camera kits.

In the 1970s, many inexpensive cameras used 110 format film, whose frame size was about half that of 35 mm. The film came preloaded in plastic cartridges that could drop into the camera like an audio cassette. Some were essentially novelty semidisposable cameras, such as a lens assembly attached to a keychain, which snapped onto a 110 film cartridge. Frequently disdained, 110 cameras nevertheless had features that anticipated aspects of 35 mm disposable cameras. When the latter half of the twentieth century was swept by a culture of "convenient," disposable consumer products, particularly those utilizing plastics (many snapshot cameras were already being manufactured from plastics), the stage was set for a disposable 35 mm plastic camera. Fujifilm introduced one in 1986 and Kodak soon followed.

Disposable cameras are popular because they are small, pocketable (thanks to lenses of short focal length), inexpensive, and ultraconvenient. Some models incorporate electronic flashguns, ideal for photographing friends and family socializing indoors. One simply removes the packaging, advances the film to the first exposure, and begins shooting, delivering the entire camera to the processor when the roll is finished. Even owners of expensive cameras purchase disposables if they have forgotten to pack the "good" camera for a trip or want pictures in situations that might place expensive cameras at risk of damage or loss. The disposable camera uniquely fulfills the almost-forgotten premise of the 1888 Kodak, and suggests that many people found film handling a reason to avoid taking photographs, even with automatic take-up and threading. Displays of disposable cameras near supermarket cash registers attest to their popularity as impulse purchases. Makers emphasized the "fun" of using disposable cameras, and users indeed indicated fancy-free feelings. According to the Photo Marketing Association, one-time use cameras are targeted primarily to people aged under 25, and are often used by teens.

Some executives were worried that a low-end new product might have harmed established brands. When Kodak planned its "single-use" camera, staff of the company's film division vigorously opposed it, fearing that photographs using inexpensive plastic lenses would compare unfavorably with those from 35 mm cameras. However, it didn't matter. The Kodak Funsaver of 1994 was purchased for specific jobs: people wanted a camera on vacation but either didn't own one or had forgotten to bring one. The Funsaver competed, not with 35 mm cameras, but with nonconsumption, and customers were pleased with the quality.

Disposable cameras have appeared in many designs and styles. In 1989 the single-use Kodak Stretch 35 Camera produced 3.5- by 10-inch (90 by 250 millimeter) panoramic prints, a format previously available only with expensive specialized cameras. The single-use Kodak Weekend 35 camera was an all-weather camera that could take pictures underwater down to a depth of 2.5 meters, and offered low-risk beach photography. Disposable cameras with built-in electronic flash are popular for parties, weddings, and other social events. (The flash unit contains a large capacitor— 120 to 160 microfarads, rated for 330 volts.)

Early disposable cameras had poor lenses and films were less advanced, but later disposables were greatly improved. They used molded aspherical lenses, which would be expensive in a traditional glass lens. These shapes enable a simple lens to render sharp images without increased cost, since the camera is entirely molded out of plastic anyway. Disposable cameras do not have a rewind mechanism, so they operate backward compared to a conventional 35 mm camera: as the film is exposed, it is rolled into its canister. In 1998 Polaroid introduced a single-use camera which the user returned to Polaroid for recycling, and the Polaroid JoyCam appeared in 1999. Kodak added a switchable single-use camera with a choice of two print formats and faster flash recharge time in 1998.

Manufacturers, who claim to be committed to recycling, prefer the term "single-use" over the popular "disposable." Kodak was especially stung by environmentally minded consumer criticism of its disposable cameras. In response, the company began a "take-back" program in 1990 to reuse and recycle the cameras. Kodak's initiative is more extensive than that of other brands. About 86 percent of a camera's weight is recycled or reused. Most of the remaining weight of flash models is the battery, which is reused or donated to charity. The outer covers of Funsaver cameras are recycled. The chassis, basic mechanisms, and electronic flash systems are tested and reused. Components that fail inspection are ground up and added to the raw material for molding new cameras. Used lenses are ground up and sold to outside companies for other products. Once camera parts have been used ten times, they are recycled into new components. Kodak pays photofinishers for used cameras, providing a financial incentive to collect them, although it may be inadequate. Kodak eventually began to reimburse labs for the costs of sorting, storing, and shipping, but processors faced a dizzying array of types, brands, and procedures. At the end of the twentieth century, studies showed that less than half of disposable cameras were actually recycled.

Disposable cameras are excessively packaged, although the paperboard and foil containers can be recycled. In mid-1995 Kodak announced that it had recycled or reused 50 million single-use cameras; and in 1996, 70 million. In 1999 Kodak said it had saved 18 million kilograms of waste by preventing the disposal of 250 million single-use cameras.

In the twenty-first century it appears that the disposable camera, often hailed as the savior of film-based photography against the digital onslaught, may collide with the new medium. Ritz Camera, based in Beltsville, Maryland, was the first photographic chain to nationally launch a fully digital disposable camera, selling for about $11. Ritz claimed the throwaway camera had many features of more expensive digital cameras, including automatic flash and timer and the ability to delete unwanted photos. This product combined two of the fastest-growing segments in photography. According to the Photo Marketing Association, single-use cameras grew more than 15 percent annually over a five-year period, representing about 19 percent of film rolls processed. Digital camera sales grew 23 percent in

2003 and represented almost a third of consumer photography. Ritz's Dakota Digital Camera produced more expensive but inferior images when compared to disposable film camera results. Some predict that professional cameras will ensure the survival of film due to unbeatable versatility and image quality, but disposable cameras will also survive because they are so cheap. In 2002, 170 million disposable cameras were sold in the U.S. Digital camera sales were predicted to eclipse traditional film cameras in 2003, according to the Photo Marketing Association, but disposable film cameras still reigned supreme, with 214 million units sold in 2003.

See also **Cameras, 35 mm; Cameras, Automatic; Cameras, Digital; Cameras, Polaroid**

DAVID E. HABERSTICH

Further Reading

At your disposal. *Photo Technique*, London, April 1998. Five professional photographers testing single use cameras.

Blaker, A.A. To toss or not to toss? Should you toss out disposable 35 mm cameras after a single use? Not necessarily. *Darkroom Photography*, 1990, 48–64.

Coe, B. *Cameras: From Daguerreotypes to Instant Pictures.* Crown Publishers, New York, 1978.

Goldsmith, A.A. *The Camera and Its Images.* Newsweek Books, New York, 1979.

The Columbia Electronic Encyclopedia, 6th edn. Columbia University Press, Columbia, 2003.

Useful Websites

Christie, R.D. The Disposable Camera: http://www.ee.washington.edu/conselec/CE/kuhn/labs/camera/fl.htm

Eastman Kodak Co. History of Kodak: From Glass Plates to Digital Images: http://www.kodak.com/US/en/corp/aboutKodak/kodakHistory/kodakHistory.shtml

Grepstad, J. Disposable Camera Photography: http://home.online.no/~gjon/stretch.htm

History of Photography and the Camera: http://inventors.about.com/library/inventors/blphotography.htm

Modern Photography: http://inventors.about.com/gi/dynamic/offsite.htm?site=http://www.infoplease.com/ce6/ent/A0838872.html

Wells, H. Reloading and Adapting Single-Use (Disposable) Cameras: http://mywebpages.comcast.net/hmpi/Pinhole/Articles/Disposable/Disposable.htm

Cameras, Lens Designs: Wide Angle and Zoom

The function of camera lenses is to refract light rays and bend them to form an image inside the camera of a subject outside the camera, to be captured on a photosensitive surface such as film, or more recently, on photosensitive surfaces in digital cameras. Lens design strategies have always been dedicated to improving the quality and fidelity of the camera's images because basic lenses generally are afflicted by various limitations and imperfections, called aberrations.

The story of lens design has been largely a tale of the struggle to maximize lens performance—seeking to increase the effective aperture size (or speed) and improve resolving power and sharpness—while eliminating the aberrations or imperfections which degrade the images they produce. Probably the first camera lens to be used in nineteenth century photography was the singlet landscape lens suggested by William Hyde Wollaston for camera obscuras in 1812. The lens, produced in a meniscus shape, had some of its faults corrected by experimentally locating the aperture at the optimum distance from the lens, and it was used on simple cameras for many decades. The first design improvement for this basic photographic type was the Chevalier lens for daguerreotype cameras, consisting of two glass elements cemented together.

The first lens of high relative aperture that facilitated portraiture with light-sensitive materials of extremely low sensitivity, such as the daguerreotype, was designed by Josef Petzval in 1840. It contained a telescope objective at the front, widely spaced from another telescope objective at the rear. This basic design was a classic type, utilized widely throughout the history of photography for still cameras, motion picture projectors, as well as microscopes. Originally used for portraiture, the Petzval design was eventually supplanted by other designs.

The most popular camera lens of the latter nineteenth century until 1910 was the symmetrical "duplet" consisting of two identical lens groups on either side of a central diaphragm, because this design eliminated many types of lens aberrations. Other activity in nineteenth century lens design included the use of new types of glass and experimentation with both symmetrical and asymmetrical systems. By 1893 a new basic design, the "triplet," was introduced, consisting of three air-spaced singlets or three air-spaced lens groups. In the twentieth century enormous quantities of triplet lenses were manufactured for moderately priced folding cameras and other cameras. Considerable research went into methods of modifying the triplet design to produce faster, higher-aperture lenses.

The relationship between the focal length of a camera lens and image size has been somewhat arbitrarily codified over the years. A "standard" lens is usually considered to be one whose focal length approximates the diagonal dimension of the rectangular image produced by the camera, on the theory that this produces "normal" perspective. This is a simplistic notion, as the perception of perspective is more complex than many realize, but the standard remains. Thus a 35 mm camera is routinely equipped with a lens of about 45 to 55 mm. A lens of shorter focal length is considered "wide angle," although it became fashionable at the end of the twentieth century for photographers who prided themselves on selecting a wide-angle lens as their personal "standard" workhorse. A specially constructed lens of extreme wide angle, called a fisheye lens, has scientific applications, but it was popularized as an experimental tool for seemingly distorted and intentionally bizarre effects. A lens of greater than standard focal length is called "long-focus" or "telephoto." A telephoto lens denotes a particular design type, however, which is intended to produce a lens significantly shorter than its effective focal length would suggest; that is, the total length from the front lens to the image plane is actually less than the focal length. The second principal plane lies outside the positive end of the lens system, or the rear principal plane will be in front of the front component. Not every long-focus lens is a true telephoto, despite the confusing tendency of photographers to use such nomenclature uncritically. Telephoto construction requires a positive front element that is widely separated from a negative rear component. As it is more difficult to correct aberrations in a telephoto lens, a long-focus lens of conventional construction may be preferable; a telephoto is used when the need for compactness equals or outweighs the concern for optimum quality.

Perhaps the single most important contribution of the twentieth century to photographic lenses, especially since it came to be utilized almost universally, was the application of hard, permanent coatings on lenses to reduce the surface reflectivity of the glass. The fact that coatings could reduce reflections was discovered by H. Dennis Taylor in 1896 when he worked with tarnished lenses, but attempts to reproduce such tarnish with precision were unsuccessful. In 1936 John Strong suggested depositing a thin layer of a low-index material such as calcium fluoride onto lens surfaces,. The thin film reduces surface reflections through interference, in which beams of light passing through the coated lens interfere with each other and eliminate the reflected light. Commercial lens coatings were offered in December 1938, and the method was subsequently greatly improved. In addition to the very real technological advance that lens coating represented, the esthetic appeal of coated lenses evidently had an impact on the market. Advertising images of cameras with gleaming lens coatings of subtle, attractive colors—browns, blues, magentas, and purples—helped to make amateur photography one of the most important hobbies of the twentieth century. Lens coatings tended to connote precision and drew upon a ready market of gadget collectors.

A "zoom" lens has some means for continuously varying the focal length of the lens while retaining image focus at the film plane. Originally popular for motion-picture photography for the obvious advantage of both continuous and seamless zoom effects, as well as the ability to change the apparent camera-to-subject distance rapidly without the need to change lenses. They were extremely difficult to design and manufacture with adequate corrections, however, and they did not become plentiful until the mid-twentieth century. The best known zoom lens at that time, the Zoomar, had 22 separate lens elements in the 16 mm motion-picture model and an even larger number in its 35 mm counterpart. In the late twentieth century low-cost manufacture of zoom lenses made them more available to the amateur market, and many 35 mm cameras were routinely sold with variable-focus lenses. Eventually small "point-and-shoot" 35 mm cameras became ubiquitous, and most were equipped with zoom lenses in lieu of interchangeable lens capability.

All modern lenses are corrected achromatically, which means that two different colors will focus at the same distance between the lens and focal plane, but the difficulty increases with faster telephoto and zoom lenses. Apochromatic (APO) lenses are corrected to focus three different colors in the same plane. Such designs were expensive until the 1990s. Engineers rely upon computers to develop improved optical and glass manufacturing technology to provide high-quality low-dispersion glass. The availability of computerized lens design undoubtedly constitutes one of the most important advances in the history of twentieth century photographic optics.

Unusual lenses have been manufactured for special purposes. One of the most striking is the anamorphic lens developed for widescreen CinemaScope motion pictures in the 1950s. This lens could be used to compress laterally a widescreen image into the normal dimensions (1:1.85 aspect ratio) of a conventional 35 mm movie frame, then expand the image to fill the CinemaScope screen, with its extreme aspect ratio (1:2.35).

Another exciting late twentieth century innovation relating to camera lenses actually has only an indirect relationship to optics. This was the advent of automatic focus or autofocus camera systems: properly speaking, these methods involve the use of a motor to focus the lens for the user, operating via signals received through such technologies as SONAR, infrared pulses, and computer analysis.

See also **Cameras, 35 mm; Cameras, Automatic; Optical Materials**

DAVID E. HABERSTICH

Further Reading

Brandt, H.-M. *The Photographic Lens.* Focal Press, London, 1956.

Kingslake, R. *Lenses in Photography: The Practical Guide to Optics for Photographers.* Garden City, New York, 1951.

Kingslake, R. *A History of the Photographic Lens.* Academic Press, Boston, 1989.

Lockett, A. *Camera Lenses: A Handbook to Lenses and Accessories for Amateur and Professional Photographers*, 5th edn, completely revised by Lee, H.W. Sir Isaac Pitman & Sons, London, 1962.

Neblette, C.B. *Imaging Processes and Materials*, Sturge, J., Walworth, V. and Shepp, A., Eds, 8th edn. Van Nostrand Reinhold, New York, 1989.

Neblette, C.B. *Photographic Lens Manual and Directory.* Morgan & Morgan, New York, 1959.

Neblette, C.B. and Murray, A.E. *Photographic Lenses*, revised edn. Morgan & Morgan, Dobbs Ferry, 1973.

Smith, W.J. *Modern Lens Design.* McGraw-Hill, New York, 1992.

Wildi, E. *Photographic Lenses: Photographer's Guide to Characteristics, Quality, Use and Design.* Amherst Media, Buffalo, CO; and Turnaround, London, 2002.

Useful Websites

Shepperd, R. Marvels of Lens Design, in *Outdoor Photographer*, Nov. 1998.
http://www.outdoorphotographer.com/content/pastissues/1998/no/lens.html

Cameras, Polaroid

Edwin Herbert Land and Polaroid, the company that he founded, produced the first self-developing cameras by employing synthesized sheet material that could align light waves. Discovered in 1934, this plastic was dubbed Polaroid for its resemblance (-*oid*) to polarization. Initially, the company used the plastic in sunglasses and filters. Concluding that the sheeting had potential as part of an instant camera, Land assigned a team to develop such a product.

An immediate sensation, the first Polaroid Land camera went on sale in 1948 in an upscale department store. Marketed as a luxury item for amateur photographers, the camera cost more than other types yet quickly found buyers among people drawn by the appeal of instant results. Since instant photography skips the step of submitting film to developers, the Model 95 also found a market among those who wanted to keep intimate photos of lovers away from the eyes of others. Professional photographers often purchased the camera to test setup conditions. More than half a million Model 95 cameras sold in the first five years of production. The camera contained a folding bellows that connected the lens housing to the body, which was covered in imitation leather. A latch popped the camera open and, when locked into the closed position, the sensitive interior was fully protected meaning that no lens cap was necessary. The camera produced 3.25- by 4.25-inch (90 by 110 mm) sepia-toned photographs, eight to a roll. It weighed just over 2 kilograms when loaded with a film roll, making it difficult for many people to handle. An exposure value (EV) number from one to ten described settings for aperture and shutter speed. Polaroid recruited General Electric to design an inexpensive exposure meter to read the light of the scene expressed in the same EV numbers. With the meter clipped to the camera, the photographer checked the meter then set the same number by turning a wheel on the shutter board before taking a properly exposed picture. Within a few years, this EV system would be adopted, with some modifications, as the standard of the amateur photographic industry.

To produce an instant photograph, the camera exposed an image on a roll of negative photographic paper pulled down from the top of the camera into view of the lens. This paper met a set of rollers with positive paper pulled up from the bottom of the camera. Interspersed on the positive roll were small pods of chemicals: the standard developer hydroquinone and the typical fixer sodium thiosulphate. As the papers pulled through the rollers together and out of the camera body, the pressure of the rollers burst the reagent pod, spreading chemicals evenly through the middle of the positive–negative sandwich. The chemical reaction took about sixty seconds to complete, at which time the photographer peeled the positive print from the negative. The process was not foolproof and the film did not always peel easily.

The Model 95 failed to produce consistently good photographs because Polaroid had been unable to produce a high-quality, easy-to-use film. The company solved this potential customer satisfaction problem by encouraging camera users to think of photography as a creative process filled with trial and error. In 1950, Polaroid introduced black and white film but consumers soon reported a fading problem. After determining that the problem was caused by contaminants present in the air, Polaroid reconstructed the positive print and instructed consumers to add the annoying step of painting pictures with a protective coating. Famed photographer Ansel Adams, serving as a paid consultant, suggested ways to improve the tonal value of Polaroid film and persuaded the company to market 4- by 5-inch (100 by 130 mm) sheet film. The smooth surface of the film showed smudges, picked up glare, and revealed fine-detail flaws in the photograph. Color film, Polacolor, became available in 1963. Polacolor records the three primary colors in three light-sensitive layers. When the film is exposed, some dye is trapped while the remainder transfers to the receiving layer thereby reproducing the color of the subject. The film did not hold its color over the years. Wastage and quality of film remained concerns, and Polaroid was never able to match the quality and consistency of Kodak's equivalent 35 mm film.

While Polaroid worked to perfect its film, it continued to introduce instant cameras. Like other photography companies, Polaroid made most of its profit on film, giving it an incentive to reduce camera prices as much as possible. In 1954, the moderately priced Highlander reflected the company's plan of offering increasingly lower-priced models. The 1963 Automatic 100 had pack film that developed outside the body, so the photographer could take a series of shots without waiting for the film to develop. In 1965, the Swinger became Polaroid's first low-cost model. Named after a slang word for "fun person" to appeal to the teenage market, this camera had a high-impact plastic boxy body. It used roll film that produced small black and white photographs and tended to jam if the camera suffered rough treatment. The Swinger contained an innovative exposure control device. When the user gripped a small red stick that projected above the shutter, an electric bulb inside the camera illuminated a checkerboard display in the viewfinder just above the image of the scene to be photographed. By rotating the control stick, the photographer could open or close the aperture to admit enough light to balance the brightness of the bulb which was keyed to light sensitivity of the film. When the two brightnesses were balanced, the checkerboard spelled YES. Later versions of the

Swinger included a T-bar strap to make the camera easier to hold while extracting film.

In 1968, the Big Swinger replaced the Swinger before the Colorpack II succeeded it in 1969. The Colorpack accepted black and white or color film, had an electronic shutter for automatic exposure control, an electric eye, a more precise lens, and a built-in flashgun that relied upon four-shot flash-cubes. In 1971 the company added a self-timer that gave the photographer three seconds to enter the picture frame before the shutter snapped. Additionally, a beeper sounded when the picture was ready to be peeled from the negative. A third device also activated when insufficient natural light existed for a good picture. In 1972, SX-70, the first pocket-sized Polaroid camera, used cast plastic technology to fold light through internal lenses and mirrors. The 116 mm lens was slightly wide-angle to make the best compromise among all the possible situations under which camera users might be operating. Portraiture remained difficult since the lens tended to flatten and distort anything filling up the frame at a short distance. To remedy this problem, Polaroid eventually developed a telephoto lens attachment. Later Polaroid cameras featured automatic focus, a built-in flash with an automatic recharge feature, a frame indicator displaying the numbers of pictures remaining in a film pack, and a close-up adaptor.

In the 1980s, the popularity of Polaroid cameras dipped as cheap 35 mm cameras and one-hour photo shops permitted consumers to produce better-quality photos without sacrificing a great deal of developing time. In the 1990s, the advent of digital photography further eroded the market for Polaroids as photographers began to use computers to transmit images although the cameras remained in heavy use for identification purposes, particularly for licenses issued by motor vehicle bureaus.

See also **Cameras, 35 mm; Cameras, Automatic**

CARYN E. NEUMANN

Further Reading

Adams, A and Baker, R. *Polaroid Land Photography*. New York Graphic Society, Boston, 1978.

Carr, K.T. *Polaroid Transfers: A Complete Visual Guide to Creating Image and Emulsion Transfers*. Amphoto Books, New York, 1997.

Langford, M. *Instant Picture Camera Handbook: All About Your Instant Picture Camera and How to Use It*. Alfred A. Knopf, New York, 1980.

McElheny, V.K. *Insisting on the Impossible: The Life of Edwin Land*. Perseus Books, Reading, 1998.

Olshaker, M. *The Polaroid Story: Edwin Land and the Polaroid Experience*. Stein & Day, New York, 1978.

Wensberg, P.C. *Land's Polaroid: A Company and the Man Who Invented It*. Houghton Mifflin, Boston, 1987.

Cameras, Single Lens Reflex (SLR)

The principle of the single-lens reflex (SLR) camera, reflecting the light path by means of an angled mirror behind the lens so as to project an image onto a horizontal glass screen, dates back to the prephotographic portable camera obscura. However, the early photographic experimenters using this type of instrument soon found that the light loss inherent in the arrangement was unacceptable and adopted cameras with a direct light path.

Although the 1839 camera made by Giroux for Daguerre (the earliest photographic camera to be widely sold to the public) had a reflex viewing mirror, the first true SLR camera was patented by the Englishman, Thomas Sutton in 1861. Sutton's camera incorporated an internal mirror that reflected the image formed by the lens up into a horizontal glass screen. The mirror was swivelled up to cover the screen during exposure. The relative insensitivity of the wet collodion plates in use at the time and the delay imposed by the need to prepare them directly before exposure negated the great advantage of the reflex camera, continuous inspection allowing adjustment of focus. Few of Sutton's cameras were made and the reflex principle was almost completely abandoned until "fast" gelatin dry plates were introduced in the 1870s.

The first SLR camera to be at all widely sold was patented by an American, Calvin Rae Smith, in 1884. Called the Monocular Duplex camera, it was fitted with an internal mirror attached to a wedge-shaped unit pierced with an aperture. Raising the unit allowed light from the lens through to a plate at the back of the camera, thus acting as a shutter. The English company, Perken, Son and Rayment, marketed a similar camera in 1888. S. D. McKellen's Detective camera, also sold in the U.K. during the same year, was equipped with a roller-blind shutter fitted between lens and mirror. Encapsulating the advantages of the SLR design, a contemporary review noted that an "exact and full-size picture" could be seen "up to the very moment of firing."

During the last decade of the nineteenth century, several new SLR camera designs were produced. Innovations included cameras designed to use rollfilm or magazines of cut film instead of plates, and models fitted with focal plane shutters. The ability to frame and focus moving subjects up to the moment of exposure led to several models fitted with long focus lenses being advertised as particularly suitable for naturalists or for photographing wildlife. The most influential design of the period was the American Folmer and Schwing Graflex camera of 1898. By the standards of the day this was a compact model, the lens panel being fitted to a bellows extension moved by a rack and pinion mechanism. The English Soho Reflex camera introduced in 1905 and built to a similar pattern was immensely popular and, along with derivatives of the Graflex, was marketed until after World War II. Both types were much favored by press photographers. The amateur market was first widely exploited in the 1920s and 1930s as many relatively inexpensive rollfilm reflex cameras began to appear.

A great drawback of the early reflex camera was its bulk. Folding or collapsible models were produced from as early as 1903. Many of the best examples came from German manufacturers, perhaps the finest being the Miroflex models first produced in the late 1920s. There was a broad trend at this time towards compact all-metal cameras and a notable feature of the period was the success of precision-constructed 35 mm cameras such as the Leica and the Contax. A 1933 reflex-mirror box attachment for the Leica camera has been described as the first application of the reflex principle to 35 mm photography. The first production 35 mm SLR camera was probably the Russian Sport camera of 1935, followed a year later by the much more influential and widely sold Kine Exakta by Ihagee of Dresden. Although basically a 35 mm version of a 127 rollfilm camera, the Kine Exakta was a well-made instrument equipped with a fast standard lens and built-in flash synchronization.

The development of 35 mm SLR cameras to the point where they displaced rangefinder models derived from the Leica and Contax as the serious photographers choice was primarily due to the introduction of the pentaprism viewfinder. For the first time, it made possible a compact reflex camera that allowed right-way-up, right-way-round, eye-level viewing. More than one pentaprism patent was filed in the 1940s but the most influential production, SLR camera with a pentaprism viewfinder was the Contax S of 1948. The Contax was soon followed by a series of 35 mm pentaprism reflex cameras produced increasingly in Japan.

From the1950s, the evolution of the 35 mm pentaprism SLR camera continued by way of a series of innovations made possible by advances in light sensor design and electronic components and circuitry. Many important features were pioneered in European cameras but were developed and refined by Japanese companies. Asahi Pentax models of the 1950s were the first widely sold SLR cameras to incorporate modern instant return mirrors and automatic diaphragms. Similar features appeared in the modular-constructed Nikon F camera of 1959. When later equipped with through-lens metering and marketed with a wide range of lenses and accessories, the early Nikon F was arguably the most influential camera of the period. The first SLR cameras with through-lens metering (the microelectronic coupling of exposure meter to shutter and diaphragm to provide fully automatic exposure control) offered for sale to the public were again Japanese products: the Topcon RE Super of 1963, closely followed by the Pentax Spotmatic. Since 1970, the quality camera market has been dominated by the versatile, Japanese manufactured, fully automatic, 35 mm single-lens reflex model, such as the Canon A and Nikon F series cameras. Perhaps the only exception is the square picture format camera with interchangeable film magazine exemplified by the 500C Hasselblad, favored in certain branches of professional photography.

See also **Cameras, 35 mm; Cameras, Automatic; Cameras, Lens Designs, Wide Angle, Zoom**

JOHN WARD

Further Reading

Coe, B. *Cameras: From Daguerreotypes to Instant Pictures.* Marshall Cavendish, London, 1978.
Coe, B. *Kodak Cameras: The First Hundred Years.* Hove Foto Books, Hove, 1988.
Condax, P.L, Tano, M., Takashi, H. Fujimura, W.S. *The Evolution of the Japanese Camera.* International Museum of Photography, Rochester, 1984.
Gaunt, L. *Canon A Series.* Focal Press, London, 1983.
Gernsheim, H. and Gernsheim, A. *The History of Photography.* Thames & Hudson, London, 1969.
Hicks, R. *A History of the 35 mm Still Camera.* Focal Press, London, 1984.
Janda, J. *Camera Obscuras. Photographic Cameras 1840–1940.* National Museum of Technology, Prague, 1982.
Kingslake, R. Folmer and Schwing-Graflex, in *The Photographic Manufacturing Companies of Rochester.* New York. Rochester, 1997.
Lothrop, E.S. *A Century of Cameras.* New York, 1973.
Reflex Camera, in *The Focal Encyclopedia of Photography*, vol. 2, fully revised edn. Focal Press Ltd, London, 1965.

Cancer, Chemotherapy

When German bacteriologist Paul Ehrlich coined the term "chemotherapy" early in the twentieth century, he referred to drugs that had affinities for certain cells, much as nature's "magic bullets"—antibodies—targeted pathogens. The notion of bullets and targets remained a popular image in anticancer chemotherapy for about 50 years, and it fit especially well into the so-called "war on cancer" during the 1970s and 1980s in the U.S. Subsequently, however, researchers found that anticancer drugs did not kill tumor cells directly but rather induced them to die naturally in a process called apoptosis, or programmed cell death. This later concept, based mainly on nucleic acid studies of cell biology, shifted the paradigm of anticancer therapy from models of war to those of communication, and resulted in 2001 in the first signal transduction inhibitor (STI) drug with many more drugs of that type under development. The STIs act by inhibiting the transmission of a signal, in the form of messenger ribonucleic acid (RNA), from the chromosomal gene to where peptides (preproteins) are assembled, and thus blocking the gene's expression in a cancer mode.

As the twenty-first century began, about 50 anticancer drugs (grouped by the mode of action as alkylating agents, antimetabolites, antibiotics, enzymes, plant alkaloids, and biologics and usually prescribed in various combinations) were effective as primary or adjunct treatments in patients with localized or metastized (tumors spread to a distant site) disease.

Modern chemotherapy, as it developed during and after World War II, was made possible by trends in science, medicine, and economics. They included

1. An accumulation of knowledge in science (pharmaceuticals, tissue culture, inbred mice for tumor lines, microscopy, and basic science)
2. Available capital, largely from the pharmaceutical industry's financial success with antibiotics
3. An apparent need as seen in the relative decline of heart disease mortality and the rise of cancer deaths
4. The social acceptability of pursuing "cures" over "prevention"

The first drug to arise in this matrix was the alkylating agent mechlorethamine (trade name Mustargen) in 1947. Its development resulted from the studies of Alfred Gilman and colleagues at Yale University carried out in 1942 on the effects of nitrogen mustard gas on lymphoid tissue, including lymphosarcomas, or malignant tumors. As a group, alkylating agents work within cells by attaching to and breaking DNA (deoxyribonucleic acid) strands, interfering with the cell's replication and protein synthesis, and thereby inhibiting cancer cell growth. This action was further demonstrated in such drugs as busulfan (Myleran) in 1953 for chronic myeloid leukemia; chlorambucil (Leukeran) in 1953 for chronic lymphocytic leukemia; melphalan (Alkeran) in 1953 for multiple myeloma; and cyclophosphamide (Cytoxan) in 1957 for lymphomas.

Generally, anticancer drugs worked best against fast-growing "liquid" cancers such as the leukemias and lymphomas and were less effective against slow-growing "solid" tumors that metastasized early. While the overall cancer cure rate using various chemotherapies was around 15 percent, some drugs showed spectacular results. Mercaptopurine (Purinethol), the first antimetabolite chemotherapy in 1953, virtually reversed the mortality from 80 percent to 20 percent for children suffering from acute lymphocytic leukemia. Antimetabolites work at the cellular level by substituting DNA base analogs that interfere with the cell's chromosome synthesis and prevent it from replicating. Gertrude Elion and George Hitchings, who pioneered this class of drugs in the 1940s, shared the 1988 Nobel Prize in chemistry. By the time they received the prize, antimetabolite drugs were also indicated as a treatment for HIV (human immunodeficiency virus) infection.

Antibiotics constituted a third type of cancer chemotherapy, beginning with Harvard University researcher Sidney Farber's 1954 clinical studies of dactinomycin (Actinomycin-D) among children with Wilm's tumor (kidney cancer). As investigators around the world discovered more species of streptomyces, they learned that antitumor antibiotics had less toxic side effects than antibacterial antibiotics and mainly worked by inhibiting DNA and RNA (ribonucleic acid) functions and protein synthesis. The array in this class grew and included bleomycin (Japan, 1965), daunorubicin (Belgium, 1964), doxorubicin (South Africa, 1969), mitomycin (Japan, 1956), and mitoxantrone (Germany, 1982). Similarly, investigators in the early 1950s learned that enzymes, particularly asparaginase, restricted tumor-cell proliferation by blocking protein synthesis. It was marketed under the trade name L-Asparaginase in 1953.

Anticancer plant alkaloids (plants containing nitrogen and usually oxygen, especially in seed plants), first discovered during the 1960s, are actually among the oldest known medicines. The *Leech Book of Bald* from tenth century England mentioned the mandrake plant, also known as May apple, as a cathartic. It also contains a toxin that was discovered in 1970 to act as a mitotic inhibitor and preventor of cell division in resistant testicular tumor cells and small cell lung cancer. By then, however, the *Vinca* plant derivatives vincristine (Oncovin, 1961) and vinblastine (Velban, 1965) had already proved able to bind microtubes during metaphase cell division, thus restricting mitosis. Curiously, another alkaloid (paclitaxel) isolated from the bark of the Pacific yew tree was shown to have a similar effect by a completely opposite action. In 1979 Susan Horowitz of the Albert Einstein College of Medicine in New York demonstrated that paclitaxel promoted so much polymerization during metaphase microtube elongation that it paralyzed phase transition, thereby stopping growth. In 1994, after the politically contentious Pacific Yew Act of 1991 protected the natural source *Taxus brevifolia*, the drug was manufactured in a semisynthetic process under the trade name Taxol and indicated for use in treating ovarian and breast cancer. Hundreds of plants with anticancer properties were subsequently discovered—as the Chinese Institute of Medicinal Plants most recently catalogued—but most exceed acceptable toxicity for most patients.

Among the increasing number of biologic anticancer agents (i.e., those produced naturally by mammalian cells rather than synthesized) are: antiestrogenic hormones such as tamoxifen; interferon, which works by stimulating natural killer cells and macrophages against tumors; interleukin-2, a T-lymphocyte growth promoter; the cytotoxic protein called tumor necrosis factor; and monoclonal antibodies, which can detect tumors as well as deliver small molecule drugs or radioactive isotopes by targeting human tumor antigens. Also in this group of biologic agents are the STIs, the first of which was marketed as Gleevec in 2001 in the U.S. (elsewhere named Gilvec) and approved for treating chronic myeloid leukemia. Administered as a pill, Gleevec targeted tyrosine kinase, which the DNA transposition in chronic myeloid leukemia activated, causing a 10- to 25-fold increase in white blood cells. By inhibiting the protein kinase, Gleevec showed dramatic therapeutic results with minimal side effects. This led many researchers to speculate that protein kinase "cocktails" could have a similar role in making cancer a manageable disease with much reduced mortality, similar to that of the protease inhibitors used in treating AIDS. By early 2002, over 170 other signal transduction inhibitors were in various stages of testing.

During the last decade of the twentieth century, the Human Genome Project produced a "rough draft" map of all human genes. As that project is completed in the first decade of the twenty-first century, "patient discovery" may well drive chemotherapy research more than "drug discovery." As gene expression analysis creates a molecular classification of cancers and identifies both the individuality of the disease and the predictable efficacy to toxicity response in each patient, "new" anticancer drugs may come from novel discoveries, or from finding precise applications for drugs that have been known since medicine began.

See also **Cancer, Radiation Therapy; Hormone Therapy; Medicine**

G. TERRY SHARRER

Further Reading

De Vita Jr., V.T. Principles of chemotherapy, in *Cancer Principles & Practice of Oncology*, Da Vita Jr., V.T., Hellman, S. and Rosenberg, S.A., Eds., 4th edn. J.B. Lipincott, Philadelphia, 1993.

Laszlo, J. *The Cure of Childhood Leukemia: Into the Age of Miracles.* Rutgers University Press, 1995.

Porter, R. *The Greatest Benefit to Mankind: A Medical History of Humanity.* W.W. Norton, New York, 1997.

Pratt, W.B., Ruddon, R.W., Ensminger, W.D. and Maybaum, J. *The Anticancer Drugs*, Oxford University Press, New York, 1994.

Waalen, J. Gleevec's glory days. *Howard Hughes Medical Institute Bulletin*, 14, 5, 10–15, 2001.

Cancer, Radiation Therapy

Prior to the advent of the x-ray and radioactive isotopes, cancers were treated by removing them surgically. Many cancers grew undetected because imaging techniques had not yet been developed. The Roentgen ray, later referred to as the x-ray and named for the German physicist and discoverer Wilhelm Conrad Roentgen in the late nineteenth century, would change both the diagnosis and treatment of cancer. (See X-Rays in Diagnostic Medicine.)

Skin damage arising from careless use of x-rays led to early ideas for the therapeutic use of radiation on human tissue, though its ionizing effect was not understood at the time. Today scientists understand that some cancer cells are more susceptible to damage from ionizing electromagnetic radiation than are ordinary cells. The therapeutic use of x-rays began in 1896 when Emil Grubbe used x-radiation to treat a patient with breast cancer. In 1902, Guido Holtzknecht created a chromo-radiometer, which could record and measure the radiation dose administered to a patient, paving the way for systematic use of radiation therapy. One of the first applications of x-ray therapy was to treat ringworm infection. If enough radiation was applied to kill the fungus, hair on the skin fell out, but if too much was applied, the hair did not grow back and the skin was burned. Thus the nemesis of treatment was defined: the narrow margin between enough and too much. By the 1920s, however, x-ray machines were routinely used in hospitals for clinical treatment.

The French physicist, Henri Becquerel is credited with the discovery of natural radioactivity when he observed that uranium salts produced images on nearby photographic plates. Other radioactive elements, first polonium and then radium, were discovered by French chemists Marie and Pierre Curie. Perhaps the first medical application of radium occurred when Marie's husband Pierre burned his arm with it. The Curies lent radium to Paris physicians, and one early documented case in 1907 described the removal of a child's facial angioma using a crossfire technique.

The first x-ray therapy was administered using radon gas in tubes. The challenge was then and continued to be the manipulation of a ray (which is straight) into a body that has contours, many layers, and organs of different densities and sensitivities. The objective was to obtain adequate tissue depth with rays without destruction of surrounding cells.

In 1913, the first external beam machine used a cathode, or "Coolidge," tube for treating superficial tumors. It was referred to as an orthovoltage applicator and was very slow. Although these machines were extremely limited, they introduced a new genre of technology to medicine known as x-ray therapy, later called radiation therapy. In the 1930s, the use of so-called radium bombs (telecurietherapy) led to refined treatment times, shielding to avoid exposure of healthy tissue, and prescribed doses of radiation. However, results at that time were only palliative.

The Van de Graaff generator, built in 1931, was able to build up a high electrostatic charge and thus high voltages (up to 1 million volts). A medical Van de Graaff, in which electrically accelerated particles were used to bombard atoms and produce radiation, was first used in a clinical setting in 1937. Throughout the 1940s, radiation treatment with the Van de Graaff allowed a very narrow, targeted beam, higher energy (about 2 megavolts), and less treatment time than with gas tubes. But as with most other cancer treatments of the 1940s, these efforts were still palliative and only temporarily relieved pain or reduced the size of a tumor.

The first circular electron accelerator, named the betatron, was built in 1940 by Donald Kerst and Robert Seber. Originally designed for research in atomic physics in the U.S., the betatron was soon adopted for clinical use. Its first clinical application was by Konrad Gund in 1942 in Germany during World War II, and it was first used by Kerst in the U.S. in 1948. Both directly produced electrons, and x-rays produced by accelerators were an ideal source for therapy and a considerable improvement over the energy that could be achieved with gas and vacuum tubes (higher energy rays have better penetration properties). The betatron energy range of 13–45 megavolts, with 25 megavolts being optimal for therapy, made the device suitable. Linear electron accelerators were developed simultaneously by D.W. Fry in England and William Hansen in the U.S. The first patient to be treated in London with a linear accelerator was in 1953.

Until the new specialty of radiation oncology was recognized in the 1960s in the U.S., diagnostic radiologists administered radiation therapy. Subsequently, the European Society for Therapeutic Radiation and Oncology as well as many other organizations of these specialists have formed worldwide.

In the early 1950s, a group of Canadian scientists isolated a highly radioactive cobalt-60 isotope from a nuclear reactor. This provided a source of gamma rays, popularly and misleadingly referred to as a "cobalt bomb," which could be directed at patients. The Cobalt 60 machine emitted gamma rays of 1.25 megavolts at a distance of 50 to 60 centimeters, and could penetrate deep tissues. Because of the danger of exposure to these rays, buildings in which the machines were located were required to have walls of very thick lead. Many of the original cobalt gamma ray systems have been replaced with linear accelerators.

In 1975, the development of proton beam radiation allowed for higher doses of radiation to target tissues while sparing adjacent cells. Since that time much of the progress in radiation therapy has been through the application of other technologies. Refinements have included more stable machines, radiation at higher rates, modifications to the treatment table, mobility of various machines, higher energy outputs, and collimators (a device to direct the beam). Energy is now described in millions rather than thousands of electron volts. Most machines treat patients in the range of 10 to 25 megavolts or 18 to 20 megavolt photons. The new collimator takes the place of lead positioning blocks that were previously limited in shape and size. One system consists of 25 moving parts that can shape the direction of the treatment to conform to the target tumor.

Miniaturization of technology has allowed for as many as 120 motors to fit in the head of certain machines to deliver radiation to the patient. The newest system of external beam radiation is IMRT (intensity modulated radiation therapy), which links the treatment planning system to the linear accelerator and the multileaf collimator. IMRT has reduced the amount of radiation to surrounding tissues and provided high-resolution images of the patient's anatomy.

By the 1990s, highly sophisticated imaging technology with the use of computed tomography (CT) scans, magnetic resonance imaging (MRI), and ultrasound facilitated more accurate treatment planning for radiation oncology. Radionuclides delivered to bone tissue can identify malignant tissues in a bone scan. More recently, the use of positron emission tomography (PET) has enabled physicians to image metabolic processes and to track tumor metastases from lung and bone tumors.

Another form of radiation therapy, brachytherapy, uses application of a source to tissues a short distance away. Typical sites are the lung, where high doses of radiation are given over a two-week period through a catheter placed in the lung, and the cervix, where cesium-137 is placed in the vagina for a few hours. Radioactive iodine and palladium are used to treat prostate cancer by placing or implanting "seeds" (tiny titanium cylinders containing the radioactive isotope) in the gland, and radioactive palladium is also used in implants for tumors of the tongue. This is known as implant therapy.

One tumor site that posed the most difficult problems for both external beam radiation therapy and implant technology was the brain because it is covered by bone. The Gamma Knife, developed in 1968 by Lars Leskall and Borge Larsson in Sweden, is an instrument that delivers a concentrated radiation dose from Cobalt-60 sources. It fires 201 beams of radiation into the skull that intersect at the target site. No single beam is powerful enough to harm surrounding tissue, but the cumulative effect of this precision tool destroys the tumor.

See also **Cancer, Chemotherapy; Cancer, Surgical Techniques; Nuclear Magnetic Resonance (NMR, MRI); Particle Accelerators: Cyclotrons, Synchrotrons, Collider; Particle Accelerators,**

Linear; Positron Emission Tomography (PET); Tomography in Medicine

LANA THOMPSON

Further Reading

Khan, F. *The Physics of Radiation Therapy*. Williams & Wilkins, Baltimore, 1995.

Mould R.F. *A Century Of X-Rays and Radioactivity in Medicine*. Institute of Physics Publishing, Bristol, 1993.

Oldham, M. Radiation physics and applications in therapeutic medicine. *Physics Education*, 36, 6, 2001.

Rosenow, U. 50 years electron therapy with the betatron. *Zeitschrift für Medizinische Physik*. 9, 2, 73–76, 1999.

Thomas, A., Isherwood, I. and Wells, P.N.T., Eds. *The Invisible Light 100 Years of Medical Radiology*. Blackwell Science, Oxford, 1995.

Cancer, Surgical Techniques

If tuberculosis, with its treatments of rest and fresh air, was the dominant scourge of the nineteenth century; cancer, with its dramatic and often devastating surgical and pharmaceutical interventions, became the dreaded curse of the twentieth. Medical understanding of the body had shifted from a humoral framework in which disease and illness, including tumors, were believed to be caused by imbalances of bodily fluids, to an anatomical–pathological framework in which tumors understood as solid aberrations to the structure of a particular organ. This shift, coupled with breakthrough developments of anesthesia, antisepsis and asepsis, dramatically increased the use of surgical intervention as a means of treating patients with cancer.

In the early part of the twentieth century, treatment of cancer tended toward radical surgery due to the belief that complete removal of the offending tumor—along with surrounding tissue, and often the entire organ—would prevent a recurrence of the cancer. The results were often hideous disfigurement and poor quality of life—many times with only marginal gains in length of survival.

In treating breast cancer, William S. Halsted (1852–1922), who spent much of his career at the Johns Hopkins Hospital in Baltimore, Maryland, was one of the earliest proponents of the radical mastectomy. Halsted reported his surgical findings to the American Surgical Association in 1898 and 1907. At the latter meeting, Halsted reported that two fifths of his patients who were followed more than three years after their operation were considered to be three-year cures. Unfortunately, two thirds of his patients died of breast cancer despite the radical surgery. Notwithstanding these dismal statistics, Halsted's approach was adopted as standard treatment due in large part to the fact that he trained numerous surgeons at Hopkins who perpetuated his technique of radical mastectomy as the first line of intervention for breast cancer. Even early on, however, the radical mastectomy endured its share of criticism, particularly from the English surgeon Geoffrey Keynes, who decried its disfiguring effects.

Despite criticism of radical surgery, surgical removal of cancers became even more aggressive by midcentury, leading to the "super-radical" mastectomy popularized by such figures as Owen Wangensteen of the University of Minnesota and George T. Pack of the Memorial Sloan-Kettering Hospital in New York City. Similarly, total excisions of lung and stomach cancers became the norm by the 1950s. Gynecological surgery followed a similar path with super-radical surgeries of the uterus and vagina being the standard treatment at the turn of the twentieth century. Medical luminaries such as Drs. Christian Albert Theodor Billroth, Ernst Wertheim, and Alexander Brunschwig were themselves early proponents of this kind of radical surgery, despite statistics of high mortality. By midcentury, a number of radical surgeries became commonplace: hemicorporectomy (surgical removal of the lower half of the body), hemipelvectomy (amputation of a lower limb through the sacroiliac joint), super-radical mastectomy, complete pelvic exenteration (removal of internal organs and tissues), and surgical removal, or resection, of head and neck tumors. The search for a cure for cancer as well as the radical surgical interventions were replete with military metaphors—the war on cancer, a crusade for better health, battling the illness, advances in research—which characterized the aggressive assault on the disease while implicitly justifying the physical carnage that often resulted.

In the last few decades of the twentieth century, however, surgical techniques for the treatment of patients with cancer became more sophisticated and refined and also became more responsive to consumer demands. Ultraradical cancer surgery with its mutilating effects was, to a great extent, discontinued. Where radical procedures were necessary, innovative developments in plastic and reconstructive surgery, coupled with intensive physical and occupational rehabilitation programs, strove to offset and diminish disfigurement and disability as lingering postsurgical effects. Methodologies to evaluate the morbidity and mortality of surgery patients improved during the twentieth century despite the fact that the gold standard double-blind research approach is not ethically applicable to surgery patients, in contrast to medical patients.

Surgical treatment of cancer is increasingly performed in concert with other kinds of treatment modalities, such as chemotherapy, radiation, hormonal therapies, psychosocial support, and rehabilitation. This sharing of responsibility for the cancer patient among surgeons and other specialists also illustrates the much greater input that nonsurgeons have with the care of cancer patients.

Surgical treatments of cancers have proven to be most effective when the tumor is discrete and located within one segment of the body. Other surgical treatments of cancer include preventive surgery (a partial or total removal of an organ where there is a high risk that a cancer will develop), diagnostic procedures (such as biopsies), cytoreductive surgery (where most of the cancer is removed surgically and the remainder is treated with radiation or chemotherapy), and palliative surgery (where the goal is to alleviate the symptoms of the cancer, rather than curing the cancer itself).

In the waning years of the twentieth century, laser surgery evolved as the latest innovation in treating cancer. The appealing possibilities of laser surgery echo those of the early decades of the century, when x-ray and radium treatments held the promise of removing the cancer without invasive surgery, resulting scars, or the possibility of disturbing the tumor site in such a way as to prompt cells to proliferate and metastasize. However, radiation-induced injuries—including edema, sterility, flesh burns, and even death—and the difficulty in limiting radium treatments specifically to cancer cells without damaging the normal cells, posed challenges to these nonsurgical treatments. Laser surgery seems to answer these challenges by offering highly specific targeting and destruction of cancer cells, minimal invasiveness to the body, and a high degree of control by the surgical team. By the early twenty-first century, leading academic medical centers were running clinical trials to gauge the effectiveness of laser surgical intervention on the treatment of various cancers.

Conventional cancer therapy in the U.S. continues to revolve primarily around the three disciplines of surgery, oncology, and radiation therapy. However, with patient concerns shifting from mere survival of the disease to the quality of life during treatment of the disease, unconventional therapies have emerged as a complement to the standard approaches. Often referred to as alternative, complementary, or unorthodox therapies, they include nutritional therapies, acupuncture, meditation, prayer, therapeutic touch, and herbal supplements to name but a few. Although the debate continues—particularly in the U.S.—over the efficacy of such therapies, patients often seek to integrate some combination of them along with conventional means.

See also **Anesthetics; Cancer, Chemotherapy; Cancer, Radiation Therapy; Medicine; Surgery, Plastic; Surgery, Keyhole and Micro**

KAYHAN PARSI

Further Reading

Cantor, D. Cancer, in *Companion Encyclopedia to the History of Medicine*, Bynum, W.F. and Porter, R., Eds. New York, Routledge, 1993, pp. 537–561.

Lerner, B.H. *The Breast Cancer Wars : Hope, Fear, and the Pursuit of a Cure in Twentieth-Century America*. Oxford University Press, New York, 2001.

Lerner, M. *Choices in Healing: Integrating the Best of Conventional and Complementary Approaches to Cancer*. MIT Press, Cambridge, MA, 1994.

Patterson, J.T. *The Dread Disease: Cancer and Modern American Culture*, Harvard University Press, Cambridge, MA, 1987.

Rettig, R.A. *Cancer Crusade: The Story of the National Cancer Act of 1971*. Princeton University Press, Princeton, 1977.

Rutkow, I.M. *American Surgery: An Illustrated History*, Lippincott-Raven Publishers, Philadelphia, 1998.

Trohler, U. Surgery (modern), in *Companion Encyclopedia to the History of Medicine*, Bynum, W.F. and Porter, R., Eds. Routledge, New York, 1993, pp. 984–1028.

Cardiac Pacemakers and Valves, *see* **Cardiovascular Surgery and Implants: Pacemakers and Valves**

Cardiovascular Disease, Diagnostic Methods, *see* **Angiography; Electrocardiogram; Nuclear Magnetic Resonance (MNR) and Magnetic Resonance Imaging (MRI); Tomography in Medicine; Ultrasonography in Medicine**

Cardiovascular Disease, Pharmaceutical Treatment

By the end of the twentieth century, deaths from heart disease topped the list of causes of death in the U.S. with a near seven-fold increase over the previous 100 years to 268 annually per 100,000 of total population. In 1900 only two effective drugs were available for heart disease. Digitalis (which slows the heart when abnormal rhythms are present) had been discovered by William Withering in 1775 when a lady suffering from dropsy was treated successfully by folk doctors with herb tea containing foxglove leaves. Nitrates (which dilate the coronary arteries) had been used for angina (cardiac pain) since 1879. Surprisingly, both these drugs remained in use at the end of the century.

By 1946 it had become apparent that hypertension, or high blood pressure, was the prime predisposing factor in the development of arteriosclerosis and therefore myocardial infarcts (heart attacks). Although a class of drugs called the prostaglandins had been introduced in 1930 for blood pressure control, they were largely ineffective. The ganglion blockers such as guanethidine caused intolerable side effects, and it was not until the 1960s that effective therapy that was well tolerated by patients became available.

Beta-adrenergic blockers
Developed in the 1960s by Sir James Black, the so-called beta blockers such as propanolol and atenolol have an inhibitory effect on adrenaline and so slow the heart and reduce both blood pressure and cardiac contractility. These actions reduce myocardial oxygen requirements. They are an effective treatment for prevention of angina, hypertension, some disorders of cardiac rhythm, and heart disease secondary to an overactive thyroid. There is evidence of both primary and secondary prevention of cardiac infarcts (when a coronary artery is blocked off and a portion of heart muscle dies when deprived of oxygen.

Calcium channel blockers or antagonists
Introduced in 1968, these drugs induce relaxation of smooth muscles in blood vessels, including the coronary arteries which then dilate, by impeding the inward movement of calcium ions. By reducing vascular resistance, the blood pressure falls. They also have an effect on some abnormalities of cardiac rhythm. This class of drugs includes nifedipine, verapamil, and diltiazem.

Fibrinolytic drugs
The end-point of coronary artery thickening and blockage is a myocardial infarct. A series of controlled studies in the early 1980s (including the GISSI Trial of 1983 in Italy) showed the value of fibrinolytic therapy. These are drugs that break down the protein fibrin, which is the main constituent of blood clots. Streptokinase was the agent first introduced, and tissue Plasminogen Activator (tPA) is now the preferred agent as it is fibrin-specific. If given intravenously as soon as possible after an acute myocardial infarction, mortality is significantly decreased and damage to the heart muscle minimized. The drug must be given quickly, hence the emphasis in emergency departments on "door-to-needle time." Survival following an infarct is also improved by the use of the appropriate antiarrhythmic drugs early in treatment. Despite these therapeutic advances, in-patient mortality is between 10 and 20 percent.

Angiotensin-Converting Enzyme (ACE) Inhibitors
The first of these drugs was captopril, introduced in 1980 and soon followed by many others, all similar in action. They inhibit the action of the enzyme which converts angiotensin I to angiotensin II. This peptide is a potent vasoconstrictor and also stimulates the release of aldosterone and vasopressin, both of which increase the blood pressure. Initially, ACE inhibitors were introduced for the treatment of hypertension, but it soon became apparent that they were beneficial for patients in heart failure where the renin–angiotensin–aldosterone system has been activated. ACE inhibitors are now a crucial component in the treatment of this condition. The effects are related not only to lowering the blood pressure but also to the peripheral vasodilation, which has an "off-loading" effect in reducing the cardiac workload. They also have an effect on remodeling the musculature of the left ventricle of the heart, which becomes thickened in hypertension and some cardiomyopathies (primary disorders of the cardiac muscle). In diabetics, even with normal blood pressure, these drugs exert a strongly reno-protective action which has been shown to delay the onset of kidney failure. Many physicians believe that the ACE inhibitors are among the most important therapeutic advances of the twentieth century.

Diuretics
The treatment of the accumulation of fluid in the legs, abdomen, and lungs is central to the treatment of the failing heart. There was no effective therapy until 1919 when a mercurial compound used intravenously for the treatment of syphilis was found to induce diuresis (excretion of large quantities of urine). Mercurial diuretics remained the only therapeutic option until effective oral diuretics were developed from 1957 onward. The first of two main groups of drugs were the thiazides, which block the reabsorption of sodium and potassium in the renal tubules. The "loop" diuretics, such as frusemide, also inhibit reabsorp-

tion of salts, but they act in the loop of Henle in the kidney. All the diuretics can cause serious electrolyte imbalances and frequent monitoring of blood chemistry is mandatory.

Lipid-lowering agents

By 1980 the Kerala project in Finland had shown that intervention in a population with a high death rate from coronary artery disease associated with high cholesterol levels was very effective, and the Framingham study in Massachusetts had been delivering the same message (among many others) since 1945. The ion-exchange resin cholestyramine worked, but patients found it difficult to take. More effective and far better tolerated were the statins (e.g., simvastatin). This class of drugs acts to inhibit an enzyme essential for the synthesis of cholesterol.

Antiplatelet drugs and anticoagulants

Aspirin can be described as an old drug reinvented because of the observations, 100 years after its introduction as an analgesic, that it interferes with the activation of platelets, which are essential to the blood clotting process. It reduces mortality if taken at the onset of an acute coronary event and reduces the risks of further infarction if taken long term. Newer antiplatelet drugs include ticlopidine. Anticoagulants such as warfarin or heparin are more controversial and are not as widely used in the acute phase of myocardial infarction as they were in the 1950s and 1960s. Heparin, however, is of great importance in the crescendo angina syndrome and warfarin is used to prevent blood clots breaking off from the heart and moving around the circulation. This can occur in several circumstances when blood flow through the cardiac chambers is reduced.

By 2000, patients who had been spared the ravages of the tuberculosis and pneumonia that killed so many in 1900 had available a cocktail of drugs to control blood pressure, angina, cholesterol, heart failure, and hinder blood clotting. However, they were getting older, fatter, less active, and were often still smoking.

See also **Cardiovascular Disease, Diagnostic Methods; Cardiovascular Surgery**

John J. Hamblin

Further Reading

Fleming, P.A. *Short History of Cardiology*. Rodopi BV Editions, 1996. Not that short, very detailed.

Jowett, N. and Thompson, D. *Comprehensive Coronary Care*. Scutari, 1995. Comprehensive & detailed.

Laurence, D.L. *et al. Clinical Pharmacology*. Churchill Livingstone, London, 1997. Standard textbook Good historical perspective.

Opie, L. *Drugs for the Heart*. W.B. Saunders, Philadelphia, 2000. Very detailed account of current practice.

Timmis, A. and Nathan, A. *Essential Cardiology*. Blackwell, London: 1997. Short textbook, covers all fields of current cardiology.

Cardiovascular Surgery, Pacemakers and Heart Valves

The living, pulsing heart has long awed, yet intimidated, physicians and surgeons. Until recently, a disrupted heartbeat or an occluded heart valve brought certain death for patients, while various kinds of congenital heart malformations killed thousands of children every year.

At the turn of the century, the development of techniques for suturing heart wounds created initial interest in the surgical treatment of mitral stenosis, a narrowing of the mitral valve opening. In 1902, the English physician, Sir Lauder Brunton observed:

> "One is impressed by the hopelessness of ever finding a remedy [for mitral stenosis] which will enable the auricle to drive the blood in a sufficient stream through the small mitral orifice, and the wish that one could divide the constriction as easily during life as one can after death. The risk that such an operation would entail naturally makes one shrink from it."

Braunton's own attempt at a surgical cure (proposed in 1902, but not attempted until 1923) was unsuccessful as were most other early efforts. In 1925, however, surgeon Henry Souttar of the London Hospital, opened a mitral valve using only his gloved finger to separate the valve leaflets. The 19-year-old patient recovered. Souttar's achievement was both rare and precious. By 1929, published reports of 12 patients surgically treated for mitral stenosis showed only two survivors.

Likewise, physiologists in the late nineteenth century understood the electrical nature of the heart's contractions and developed rudimentary measuring devices using electrodes to create crude electrocardiograms that measured electrical activity in heart muscle. The new technology, however, required intricate interpretation before it could be successfully employed as a serious diagnostic tool. By 1907, physiologists had a basic understanding of what caused the heart to beat—the electrical origin in a bundle of tissue in the upper right atrium (the sinus node) and its conduction through the ventricle. When this signaling went awry, clinical symptoms became apparent. Irregular heartbeats constituted one set of problems. Another was evident in patients who had either a partial or complete "block" in conductivity between the upper and lower chambers of their hearts. Such patients had a clinical profile of a slow pulse accompanied by what were originally described as "fainting fits." These "Stokes–Adams" attacks indicated that the ventricle was not providing an adequate supply of oxygenated blood to the brain. It was this condition, moreover, that inspired researchers to develop electric stimulators for the heart.

Before World War II, however, improvements in blood transfusion, anesthesia, and suturing techniques for blood vessels had begun to make surgery safer and led to the expansion of efforts to operate on the living heart. Combat itself finally dispelled the concept of the heart as a fragile organ, however, when it became possible to remove shrapnel and close the wound to a beating heart without killing the patient. The stimulus for an artificial heart-pacing device as well as advances in valve surgery came after World War II and the development of open heart surgery.

In 1947, Claude Beck of Western Reserve University successfully used an electrical defibrillator internally on a patient whose heart had stopped during routine surgery. In 1952, Paul M. Zoll announced that he had kept a Stokes–Adams patient alive through a bedside defibrillation device. Zoll's most important insight was recognition that he should pace stimulate the patient's ventricle, rather than the atrium, but his treatment, employing 130 to 150 volts was too painful for extended patient use. Implanting part of the pacemaker would lower the voltage requirement, and this idea was pursued at the University of Minnesota Medical School. In operating on the hearts of small children, C. Walter Lillehei occasionally found that his sutures damaged conduction cells in the heart and produced a postsurgical heart block which killed some of his young patients. Lillehei commissioned an external pulse generator, devised by Earl Baaken, a Medtronics engineer, for use in his patients as a temporary assistance device he hoped would allow these conductive cells to heal naturally. When a myocardial pacing wire was inserted into the heart wall and connected to the external device, Lillehei discovered that "one or two volts drove the heart beautifully." The wire lead could be removed without further surgery. Active children required a portable unit, and dependence on the hospital's

electrical system inspired the creation of a battery-powered portable unit. The Medtronic 5800, invented in 1958, became one of Medtronic's earliest marketing successes.

Shortly after Beck's success using internal defibrillation, Charles Philamore Bailey, a Philadelphia surgeon, performed the first deliberative and successful intracardiac operation on a 24-year-old young woman with severe mitral stenosis in 1948. Within two years, mitral valve surgery was being performed successfully around the world. Also during the 1940s, Charles Hufnagel, began the formidable task of developing a workable prosthetic aortic valve. In September 1952, Hufnagel successfully implanted a prosthetic "ball" valve into a patient at Georgetown University Medical Center. Without a heart–lung machine, Hufnagel could not replace the faulty valve; he could only position the prosthetic as a "check valve" to correct for the effects of the diseased aortic valve. Nonetheless, the accomplishment was impressive as a first step in creating workable replacement valves.

Heart surgery had lagged technologically behind other surgical specialties in the 1940s. There were a few operations to correct specific abnormalities, but patients risked a mortality rate of 50 percent or more. Heart catheterization techniques, in which a small tube could be moved through a large vein and into the heart chambers, allowed steadily improving visions of diseased hearts, and became an ordinary hospital procedure and part of the standard workup for most patients scheduled for heart surgery during this time. In addition to better preoperative understanding of their patients' heart abnormalities, surgeons required time to fix heart problems surgically. Experiments with hypothermia in cardiac surgery proved that a patient's heart could be stopped "cold" and then warmed up to resume its normal rhythm, allowing surgeons precious time to complete more complicated surgical procedures. Hypothermia and a riskier cross-circulation technique using a live donor were the only options for oxygenating blood during surgery until technological advances brought the first heart–lung bypass machines into the operating suite beginning in 1953. Such advances were critical in the subsequent development of artificial replacement heart valves. Cardiac pacing also remained closely associated with open heart surgery throughout the 1950s.

Between 1957 and 1960, at least eight research groups designed and tested fully or partially implantable cardiac pacemakers in humans. One patient who received one of the earliest pacemakers was still alive and on his twenty-sixth pacemaker in 2000. Between 1961 and 1963, a great number of implanted pacemakers failed and were replaced with improved models. According to one author, Jeffrey:

> "these were inventions in the early stage of product development. The only justification for using them at all with human beings—a compelling one—was that the patients had little chance of survival without some electrical assist for their heartbeats."

Although some nuclear powered pacemakers were implanted in the U.S. in the 1970s, smaller, more powerful lithium batteries and advanced circuitry made newer pacemakers more attractive. Moreover, by 1972, transvenous insertion of small pacemakers under local anesthesia virtually replaced the major chest surgery that had been required for implanting pacemakers in the 1960s. The development of the heart–lung machine made it possible to remove damaged valves and replace them with mechanical prosthetic valves. The first truly successful artificial valve was developed by surgeon Albert Starr and engineer Lowell Edwards and used in a mitral valve replacement in 1960 at the University of Oregon Medical School. Many variations on this design were quickly tested both in the laboratory and in patients using improved materials and streamlined designs. Mechanical valves, however, do present ongoing challenges. The perfect valve would produce no turbulence as blood moved through the valves, would close completely, and would be constructed of a material that discouraged the formation of blood clots. The danger of clot formation required that patients with mechanical valves take lifelong anticoagulants to prevent clots from forming. Although mechanical valves are extremely durable, they can fail, and when they do they often fail catastrophically. In 1968, Viking Bjork, a Swedish professor created a new mechanical valve with a tilting disk design held in place by welded struts. The design was intended to reduce the risk of blood clots, a significant problem with previous implants. Manufactured by the Shiley Company, approximately 85,000 Bjork Shiley valves were sold worldwide between 1978 and 1986. An engineering flaw (strut failure) in hundreds of the valves, however, caused about 250 reported deaths and even more lawsuits. Nonetheless, for long-term use mechanical valves are still used in about 65 percent of the over 60,000 heart valve replacement surgeries every year.

In both the heart valve industry and the heart pacemaker industry, product failures in the late 1960s and early 1970s created concerns among consumers that led to major changes, both within each industry and within government circles. In 1976, Congress enacted the Medical Device Amendments of 1976 to clarify the scope of the Food and Drug Administration's authority over medical devices. Under the 1938 Food, Drug, and Cosmetic Act, the FDA had taken action against many quack medical devices, and had acted on occasion against medical devices under the drug provisions of the 1938 law, but regulatory actions under such conditions were cumbersome, risky, and expensive. Both industry and the FDA wanted clarification over the agency's authority over the burgeoning medical device field. Dr. Ted Cooper, Director of the NIH National Heart and Lung Institute, chaired a committee that surveyed problems in the medical device industry. Concluding that "many of the hazards seem to be related to problems of device design and manufacture," the report cited over 10,000 patient injuries in the medical literature between 1963 and 1969. Signed into law in 1976, the new law governing medical devices classified them according to their perceived risk. Thus Class 3 devices would be the most tightly regulated, while Class 1 products including such "devices" as sterile bandages, would require less stringent regulation. New Class 3 devices were also subject to pre-market approval by the FDA: manufacturers had to demonstrate that they were "safe, reliable, and effective" before they could be put on the market. Debate continues over the long-term effects of government regulation on industry innovation in the medical device field, as in the pharmaceutical industry, but it remains clear that medical device firms, including manufacturers of heart valves and of cardiac pacemakers, have resisted any tendency to make valves and pacemakers mere medical commodities. Competing through innovation has kept prices high and competition in these technology driven industries alive and well into the twenty-first century.

See also **Electrocardiogram (ECG); Hearts, Artificial**

SUZANNE WHITE JUNOD

Further Reading

Foote, S.B. *Managing the Medical Arms Race: Innovation and Public Policy in the Medical Device Industry*. University of California Press, Berkeley, 1992.

Fye, W.B. *American Cardiology: The History of a Speciality and its College*. Johns Hopkins University Press, Baltimore, 1996.

Howell, J.D. *Technology in the Hospital: Transforming Patient Care in the Early Twentieth Century.* Johns Hopkins University Press, Baltimore, 1995.

Kirk J. *Machines in Our Hearts: The Cardiac Pacemaker, The Implantable Defibrillator, and American Health Care.* Johns Hopkins University Press, Baltimore, 2001.

Johnson, S.L. *The History of Cardiac Surgery, 1896–1955.* Johns Hopkins University Press, Baltimore, 1978.

Rothman, D.J. *Beginnings Count: The Technological Imperative in American Health Care.* Oxford University Press, New York, 1997.

Cars, *see* **Automobiles**

Catamarans

The first catamarans (boats with two hulls) were derived from the traditional paddled rafts used by natives of Polynesia. These boats, which were essentially two canoes joined with logs, powered by a single sail, provided great stability and allowed them to travel vast distances on the open ocean.

In Europe, the history of catamarans begins with Sir William Petty (1623–1687). Petty studied the hydrodynamics of model hulls and found that a long, slim hull travels more easily through water than a wide one. The ratio of hull length to hull width, now called the "fineness," "slenderness," or "displacement-to-length" ratio, is the key determinant of hydrodynamic performance. Within certain constraints that we will look at below, a slender hull has less drag and creates less of a bow wave, both of which slow down the movement of boats through the water.

The major disadvantage of such a hydrodynamic hull is its instability. A heavy wind or larger sails, rather than providing more speed, would capsize the hull. The accepted way to increase stability in a monohull while keeping it relatively streamlined is with a heavy keel along the bottom of the hull. However, this decreases the power-to-weight ratio of the boat and forces it to float lower in the water, increasing drag. Petty realized that the stability of a wide boat could be combined with the speed of a thin one by connecting two fine hulls together, producing a light, stable sailing platform. Petty had discovered the key advantage of multi-hull boats—high transverse stability relative to their displacement.

Despite these early advances, catamaran development did not really take off until the late nineteenth century, when the established yacht designer Nathaniel Herreshoff, from Rhode Island, USA, built the Amaryllis, patented in 1876. Herreshoff had designed a bare-bones structure with a huge sail area and two very light, fine hulls, providing an awesome power-to-weight ratio. Despite beating the fleet of New York Yacht Club's Centennial Regatta, the boat was so unusual it was deemed unfair and banned from all future races.

As skepticism began to wane during the twentieth century, designers began to take advantage of new, cheaper, lighter construction techniques. Because catamarans can use their width for stability instead of ballast, they were better placed to benefit from the development of the new breed of light but strong composite materials. Fiber-reinforced plastics (FRPs) began to be used in boat construction in the 1940s. By the 1970s over three quarters of all boats were made from FRPs. The majority of these used glass-reinforced plastic (GRP, or fiber glass), which is light, strong, cheap, and easy to maintain. Carbon fiber (CFRP), a composite material borrowed from the aerospace industry in the 1970s, provided even greater stiffness, strength and lightness, but remains very expensive. Polyester (in the 1950s), and more recently, carbon fiber- and Kevlar-reinforced sails helped to further reduce weight and improve performance. Boat design, which previously relied heavily upon intuition and experience, benefited from the codification of fluid dynamics models and sophisticated CAD (computer-aided design) tools.

Such innovations allowed faster racing catamarans to be developed, and allowed for bigger and more robust boats. In the world of offshore sailing, large catamarans have set and broken countless speed records. These boats are so efficient that they can travel at twice the speed of the wind that fills their sails (when the wind is at right angles to the direction of travel).

The advantages of two hulls (transverse stability and lightness) are clearest with sailing boats, but engine-powered catamarans can also benefit from the increased stability and decreased drag. Increasingly, fast ferries are built as catamarans, giving a wide, flat, and stable cargo platform. These ferries have taken over much of the role once held by hovercrafts and hydrofoils.

Considering the advantages of two hulls, one might assume that the fast, stable catamaran is the ideal shape for all boats. But there are a few disadvantages that tend to limit their applications to fast, high-performance craft. First, narrow hulls and lack of displacement means that catamarans are bad at transporting large, heavy cargo. Second, catamarans are not efficient when traveling slowly, due to the complexities of fluid dynamics. Slender hulls have a greater "wetted surface area," which is a key constituent of drag at low speeds. As such, catamarans suffer from a "resistance hump," which must be overcome before increases in power lead to large increases in speed. Third, catamarans are only stable while both hulls remain in the water. Once the boat heels beyond a certain angle, it becomes very unstable and will capsize easily. By contrast, a traditional, weighted mono-hull will experience a strong righting moment right up until it capsizes. Finally, a catamaran encountering large waves undergoes greater and more complex strains on its structure because it has more points of contact with the water. This presents challenges to designers, who want to take advantage of lightweight (but expensive) materials while also building a robust boat.

Multihull boats (including trimarans with, as you might expect, three hulls) have provided designers with new ways to think about and advance sea travel and racing. The combined advantages of speed, lightness, and stability have overcome hydrodynamic constraints that were once thought insurmountable. It remains to be seen whether there is an upper limit on the speed of these vessels, and whether advances in construction techniques can keep pace with the increasing strains on these boats as they get faster. More radical designs, using hydrofoil technology and fixed-wing sails, look set to challenge speed records further into the twenty-first century.

See also **Hovercraft, Hydrofoils, and Hydroplanes**

JACK STILGOE

Further Reading

The Practical Encyclopedia of Sailing: The Complete Guide to Sailing and Racing Dinghies, Catamarans and Cruisers. Lorenz Books, 2001.

Claughton, A. Wellicome, J. and Shenoi, A. *Sailing Yacht Design: Theory*, Prentice Hall, 1999. For an in depth look at general design issues for sailing boats.

Larsson, L. and Eliasson, R. *Principles of Yacht Design*, International Marine Publications, Camden, Maine, 1995.

Smith, C.S. *Design of Marine Structures in Composite Materials*. Chapman & Hall, London, 1989.

White, R. and Wells, M. *Catamaran Racing: For the '90s.* Ram Press, 1992.

Useful Websites

http://www.therace.org. With information about the catamarans involved in this extreme, round-the-world contest.

Cell Culturing, *see* **Tissue Culturing**

Cell Phones, *see* **Mobile (Cell) Telephones**

Ceramic Materials

Ceramics are synthetic, inorganic, nonmetallic solids that generally require high temperatures in their processing, and are used in many industrial applications. Many of them are metal oxides (compounds of metallic elements and oxygen) but others are compounds of metallic elements and carbon, nitrogen, or sulfur. Typically, ceramic materials have favorable engineering properties such as high mechanical strength, chemical durability, hardness, low thermal and electrical conductivity, as well as relatively low density, which make them comparatively light and suitable for application in, for example, automotive engines. (Magnetic ceramics, used in magnetic memory, are described in the entry on Alloys, Magnetic.)

There are mechanical drawbacks that scientists and engineers have successfully reduced during the last few decades: they are brittle and lack ductility, have poor resistance to mechanical and thermal shock, are difficult to machine because of their hardness, and are sensitive to catastrophic failure because of the potential presence of microvoids.

One of the origins of manufacture of engineering ceramics is in techniques for firing clay for porcelain, tableware, and construction materials. However, twentieth century technology and the chemical, electrical, electronic, automotive, aerospace, medical, and nuclear industries have made new demands on high-performance materials. In the electrical industry, for example, ceramic insulators have been used in electric power lines since the 1900s and since the 1920s in spark plug insulators.

Naturally occurring steatites, a class of magnesium silicate minerals also known as soapstone, were known and used in the late nineteenth century for burners in stoves and gas lights, but were developed and processed as improved electrical insulators that had low loss at high frequencies and high temperatures. They were therefore used as insulators in microwave electronics for radio relays in telephone networks in the 1940s. During and after World War II the applications of ceramics in electronics extended to compact capacitors, piezoelectric transducers for use in telecommunications, resistor and semiconductor compositors as well as magnetic materials and other energy converters.

Apart from electric and electronic applications abrasives have been a field in which ceramic materials, in this case silicon carbide (SiC) and aluminum oxide (Al_2O_3), excelled. During the nineteenth century it became clear that abrasive products like natural sandstone used in grinding wheels no longer satisfied industrial demands. In 1891 Edward G. Acheson, an electrical engineer from Monongahela City in Pennsylvania, combined a mixture of clay and powdered coke in an electrical furnace. This resulted in shiny crystals (silicon carbide, SiC) which proved immensely suitable for polishing precious stone. Until the invention of boron carbides in 1928, silicon carbide had been the hardest synthetic material available. Silicon carbide's high thermal conductivity, strength at high temperatures, low thermal expansion and resistance to chemical reaction made it the material of choice for manufacturing bricks and other refractories at high temperatures, for example for industrial boilers and furnaces, and tiles covering the space shuttles.

The manufacture of aluminum oxide abrasives closely followed the development of silicon carbide. In 1897 scientists at the Ampere Electro-Chemical Company, New Jersey, made the first successful attempts at aluminum oxide manufacture with rock bauxite, of which aluminum oxide is the main ingredient. Today aluminum oxide, sometimes with the addition of zirconium oxide, is indispensable for producing highly precise and ultrasmooth surfaces in the automotive and aerospace industries.

During the twentieth century, high-performance engineering materials were increasingly required for structural applications. In particularly erosive, corrosive, or high-temperature environments, materials such as metals, polymers or composites could not fulfill the demands made on them. In 1893 Emil Capitaine, a German engineer who played a role in the development of the internal combustion engine, suggested the use of porcelain and fire clay in engines, though it is not clear what for; a decade later engineers at the Deutz Motor Company in Cologne experimented with ceramic materials for use in stationary gas turbine blades. During World War II interest in new high-performance materials grew rapidly, in Germany not so much for surpassing steel alloys but for replacing metals such as chromium, nickel, and molybdenum, which were in short supply in the armament industry. Experiments with aluminum oxide seemed promising for use in gas turbine blades but the material's susceptibility to thermal shock created insurmountable difficulties. In order to reduce these problems and to impart greater ductility and thermal shock resistance to aluminum oxide, engineers during and after World War II tried to mix ceramics with different metals to make composite materials. Although this did not yield the expected results at the time, the idea of reinforcing a ceramic matrix with metal fibers proved useful. Metal components have also been made stronger by the incorporation of ceramic fibers, whiskers, or platelets; special ceramic fibers and whiskers are also incorporated in a ceramic matrix to increase toughness and reduce the risk of catastrophic failure due to incipient cracks and microvoids. Automobiles and aircraft engines of the future may have ceramic matrix composite components such as brake disks, and turbine parts for high-temperature jet engines.

After the war, it soon became clear that for many applications a ceramic material like aluminum oxide had too many drawbacks, a conclusion based largely on research in Germany. Attention therefore shifted from oxide ceramics to ceramic materials such as silicon nitride or silicon carbide. During the 1970s the oil price shock accelerated interest in thermal efficiency of power plants, enhanced by environmental legislation in countries such as the U.S. or Germany. If the combustion temperature is raised for greater efficiency, the turbine parts must be oxidation-, impact-, and thermal-shock-resistant. From that time onward, government-sponsored research and development programs, especially in the U.S., Japan, Britain, France and Germany, have advanced research on ceramic materials such as silicon nitride and silicon carbide which, among other assets, are more oxidation resistant than super alloys. To date, however, none of these ceramic materials has been successfully adapted to the proposed high-temperature gas turbines. Advanced structural ceramics have also been employed in nuclear power as heat-resistant control rods.

Because of their good biocompatibility, ceramics are employed in medical and dental applications such as false teeth, implants, and joint replacements; in the automotive industry they are useful as catalysts, catalyst supports or sensors.

Apart from silicon nitride and silicon carbide, zirconium oxide is an excellent engineering ceramic. Zirconium oxide offers chemical and corrosion resistance at far higher temperatures than aluminum oxide. Stabilized zirconium oxide produced by addition of calcium, magnesium, or yttrium oxides, exhibits particularly high strength and toughness. This and its low thermal conductivity led to applications in oxygen sensors and high-temperature fuel cells. Over recent decades it has become possible to better cope with the intricacies of engineering ceramic materials but many questions are still unsolved. The ceramics' first-rate potential in various demanding applications make these efforts worthwhile.

See also **Alloys, Magnetic**

HANS-JOACHIM BRAUN

Further Reading

Braun, H.-J. Engineering ceramics: research, development and applications from the 1930s to the early 1980s, in *Science–Technology Relationships*, Herlea, A., Ed. San Francisco Press, San Francisco, 1993, pp. 161–167.

Braun, H.-J. and Herlea, A., Eds. *Materials: Research, Development and Applications*. Brepols, Turnhout, Belgium, 2002.

Brook, R.J., Ed. *Concise Encyclopaedia of Advanced Ceramic Materials*. Pergamon Press, Oxford, 1991.

Kingery, D.W.D., Ed. *High Technology Ceramics. Past, Present, Future: The Nature of Innovation and Change in Ceramic Technology*. American Ceramic Society, Westerville, 1986.

Mason, T.D. Industrial ceramics, in *The New Encyclopaedia Britannica, Macropaedia*, 15th edn, vol. 21. Encyclopaedia Britannica, Chicago, 2003, pp. 246–265.

Richerson, D.W. and Dunbar, B.J. *The Magic of Ceramics*. American Ceramic Society, Westerville, 2000.

Changing Nature of Work

Two men, Fredrick Winslow Taylor (1856–1915) and Henry Ford (1863–1947), are often credited with transforming the relationship between humans and machinery in the workplace and beyond, lending their names respectively to overlapping twentieth century socio-technical systems, Taylorism and Fordism. Subsequent developments in production technologies like automation, flexible specialization, and "lean production" were compelled to engage with their enduring legacy.

Taylorism

Taylor's *Principles of Scientific Management* (1911) set out a methodical engineering logic for eliminating wasteful human effort. In the famous "Schmidt" experiment Taylor turned his close observations of "inefficient" working practices and his training in engineering exactitude to determine in minute detail only those motions essential to the completion of a defined work task, in this case the handling of pig iron. Taylor thus answered his own rhetorical question: "Why can we not apply the same principles of efficiency to the hand and muscle of man that we apply to the design of machines?" External measurement and control could manipulate human activity to make it more "machine-like." Thus the technical *conception* of work design was radically separated from the laboring *execution* of work activity. Job analysis and time and motion study allowed work previously done by a single person to be broken down into more simple, repetitive tasks. Pay rates were calculated more precisely according to relative time spent handling work or merely supervising the machine at work. Large gains in productivity seemed to support Taylor's insistence that the technical organization of work be standardized and synchronized to allow human labor to perform efficiently.

Many objections have been aimed at the Taylorist rationalization of work. One is that Taylor had a very crude understanding of worker motivation as animated by economic progress. He anticipated that the elimination of waste, the removal of complicated decision making for the worker through the use of scientific procedures and piecework pay schemes would appeal greatly to a basic human desire to improve their own material well-being. A second objection, made by radical and Marxist critics like Harry Braverman (1974), is that Taylorism was far from being a scientifically disinterested exercise. Instead, Taylorism merely provided a rationalization for passing control of the production process away from labor and into the hands of technicians and supervisors who performed functions essential to the class interests of employers. A third objection raised by psychologists working in the behavioral science tradition is that Taylorism was not scientific enough. It lacked scientific rigor in human physical or psychological capacities and was based on arbitrary decisions and informed guesswork about the mind–body coordination of "average" or "experienced" workers. A further objection is that workers are not merely passive factors of production animated by the single viewpoint of a management-designed socio-technical work system. Workers can actively modify or subvert even highly detailed and closely monitored work systems like scientific management, both informally as members of a work group and more formally through trade unionism. Finally, though Taylorist ideas became widely diffused, especially under the impact of the production demands of two world wars, they were rarely adopted as an entire scientific system but were pragmatically deployed by engineers and managers moved job planning from the shop floor into the offices of production engineering departments.

Fordism

With the development in 1908 of the Model T automobile Henry Ford invented what he himself called "mass production." Others like Antonio Gramsci gave it the eponymous title "Fordism" to indicate the ways that mass production and mass consumption brought an entirely new way of life into existence. In his contribution to the thirteenth edition of the *Encyclopaedia Britannica* on "Mass Production," Ford set out his new production philosophy—simplicity.

> "Three plain principles underlie it: (a) the planned orderly and continuous progression of the commodity through the shop; (b) the delivery of work instead of leaving it to the workman's initiative to find it; (c) an analysis of operations into their constituent parts."

These principles mechanically controlled the pace at which work passed before the worker. From Ford's 1913 experiment with a conveyor belt for the assembly of magnetos at the Highland Park plant in Detroit the "three plain principles" were generalized to the assembly of other automobile components and, in the 1920s, to a broad range of consumer goods. Huge gains in productivity meant that small batch goods previously limited to the luxury market became mass produced in standard forms for relatively low prices. Inside the car plant Ford's real achievement may have been less to do with the continuous pace of the work than with the simplicity of easily assembled, inexpensive, perfectly interchangeable parts fashioned to high tolerances and amenable to the same gauging system.

On the consumption side, Ford's success was based on low, and falling, costs and durable and easily serviced car design and materials. However, Ford's initial advantages slipped in the 1920s when Alfred P Sloan's new functional management system at General Motors adopted a thorough standardization of mechanical items coupled with annual design changes to the external appearance of the car's "hang-on" features. Yet, for all Ford and Sloan's innovations, autoworkers were viewed as interchangeable parts of the production process, something rendered artistically in the controversial 1932 Detroit Industry frescoes painted by the radical Mexican muralist Diego Rivera. To offset labor discontent, in 1914 Ford offered workers "the $5 day," effectively doubling wages for those workers that qualified and created a "Sociological Department" to enforce respectable social behavior among the workforce. As turnover at car plants decreased workers became increasingly dissatisfied with their experiences of working on the assembly line and car plants became centers of industrial conflict and unionization drives.

Automation

Automated machine tools took the process of component standardization, interchangeability and technical control a stage further. Automation provided a managerial solution to the high-cost burden and inflexibility of jigs and fixtures, which left high levels of control and discretion in the hands of toolmakers. Machine tools like center lathes or milling machines are indispensable to all machine-based metal-working processes, including the capital goods production of machine tools themselves. Machine tools rotate or otherwise set in motion the tool piece or the work piece (or both) to reshape the work piece to a predetermined size, contour, and surface finish.

In the 1930s and 1940s, tracer technology increasingly deployed hydraulic or electronic sensing devices to cut contours into materials according to templates. More sophisticated sensing and measuring devices like precision servometers were only one example of industrial innovation during World War II. The challenge was how to completely automate machine tools. Unlike automated manufacturing processes, where single-purpose, fixed automation proves cost effective for high volume output, automated machine tools are more versatile, multipurpose machines geared towards one-off or small-batch volumes. Automated machine tools combine machine versatility with off-the-job control and information records. One solution was "record-playback," developed around 1946 by General Electric and Gisholt, and some smaller companies. A record was made on magnetic tape of a machine set in motion under the instruction of a machinist, allowing the automatic production of identical parts by replaying the tape and precisely reproducing the machine tool movements.

A second solution, numerical control (N/C), was developed by John Parsons, a Michigan subcontractor who manufactured rotor blades for U.S. Air Force helicopters. N/C involved what David Noble (1984) calls "an entirely different philosophy of manufacturing." N/C transferred the knowledge of the machinist to programmers who made up the tapes prior to the planned work arriving at the machine. Early programming involved manually writing unique subroutines for each particular geometric surface, a time-consuming and tedious operation. In 1956 Douglas Ross, a young engineer and mathematician at MIT, developed a systematized solution, automatically programmed tools (APT) that combined the versatility of five-axis control with an overarching skeleton computer program. The U.S. Air Force funded N/C research and enforced its diffusion among its contractors, whose cost-plus military contracts subsidized their adoption of the expensive APT system, despite the operational difficulties caused by the use of large, complex computing systems.

N/C was viewed enthusiastically by engineers, managers, and the Air Force as a hyper-Taylorist solution to human error and recalcitrant worker attitudes. As Peter Drucker stated, "automation is a logical extension of Taylor's scientific management" (in Kranzberg and Pursell, 1967). However, controlling labor through automatic machinery was far from complete since the implementation of such complex, expensive and fallible equipment was contested by the machinist union and a continuing need for highly skilled machinists came to be recognized by many manufacturers. N/C techniques were further enhanced by advances in electronics such as computer-aided design and manufacture (CAD/CAM), which seemed to threaten even the skills of design engineers. Such arguments tend to rely on technological determinism, where technology is viewed as an independent force shaping social relations rather than being shaped by them. CAD/CAM has, however, had the support of engineers, reflecting their senior social positions inside the hierarchy of command and the division of labor.

Women and Industrial Technology

Until the final quarter of the twentieth century the relationship between gender and technical change was subsumed under the simple equation of masculinity with machinery. The prevailing view tended to see technology as neutral and that men's physical strength equipped them to adopt machinery more readily than women. Such biological reasoning failed to account for the social shaping of technology by existing ideologies of gendered work. After all, machinery tends to be introduced to reduce physical effort in the production process. Technical work often fosters a struggle for control of nature in 'raw materials' through the controlled use of energy, something many feminist writers view as intrinsically male attributes.

Women are subject to systematic discrimination in pay and access to skilled occupations, even in industries such as textiles where women predominate. Women also tend to be regarded, falsely, as transient or peripheral to the core of the labor market and passive in the face of employer controls. Indeed, the labor force has been increasingly feminized, partly due to the expansion of services and part-time work. This reflects a domestic division of labor that casts women as "carers" in the family. The ideology of women as home makers has also had profound consequences for the way industrial technologies were introduced into the domestic sphere. Women's unpaid domestic labor was subject to an invisible industrial revolution in the twentieth century—from hand power to electric power, to gas and electric appliances for cooking, cleaning, waste disposal, to central heating, to prefabricated housing, and so on.

Industries that employ relatively low-cost and abundant female labor face little incentive to invest in expensive labor-saving technical innovation. Automation is subject to technical as well as social limits. For example, the skilled hand–eye coordination of female sewing machinists has proven difficult to fully substitute with technology. Industries such as clothing traditionally deploy a highly defined sexual division of labor, with men designated the high status of "tailor" and women the lower status of "seamstress/" As technology entered the industry craftsman tailors abandoned machinery to women. Men retained craft status in the cutting room until cutting and pattern process were separated and routinized by the advent of Taylorism and male craft workers were substituted by deskilled female workers.

Studies of the introduction of word processing show how office technologies such as typewriters, dictation machines, copiers, printers, and filing cabinets, over which women workers previously exerted some control and autonomy, were replaced by the "intelligent office" of networked word and data processing through advanced computer and telecommunication systems. A loss of status and control accompanies female secretaries and typists who previously typed their copy but now simply key-in data and follow screen prompts. Tasks such printing, filing, and typing become fragmented and physical movement is reduced by systems that store the data electronically.

The Rise of Services

The socio-technical system of simplifying and replacing labor with machinery set in motion by Taylor and Ford and deepened by process automation began to falter in the developed economies in the1960s. On the other hand, state policies in support of industrial imitation allowed newly industrializing countries (NICs) such as Taiwan and South Korea to successfully industrialize. A new international division of labor began to emerge as multinational corporations were able to take advantage of the division and subdivision of production and advanced transport and communication techniques to relocate labor-intensive subprocesses to low-cost Third World locations. As manufacturing was subdivided internationally, industrial employment went into absolute decline in Western Europe and the U.S. The Western adoption of neoliberal principles in the 1980s helped further stimulate what many described as "globalization." This did not mean that the entire world became fully integrated. Instead, productive investment strategies created a triad of regionally linked advanced economies, North America, Europe, and South East Asia, exacerbating the development gap between the triad and the NICs, and the low-industrialized countries making up the Third World.

Accompanying the deindustrialization of Western economies was a continuous rise in the absolute numbers employed in the service sector. However, even with fewer workers manufacturing output continued to rise, indicating manufacturing's socio-technical capacity for increasing productivity per worker. Due to its heterogeneous nature, "services" are notoriously difficult to define. *The Economist* defined services as "anything sold in trade that could not be dropped on your foot." Services are often viewed as some kind of residual category, what is left over after manufacturing, agriculture, and extractive sectors are accounted for. Yet services now appear to be central to most developed economies. Services can be broken down into two basic concepts:

1. Service *industries*, where the final output takes a nonmaterial form such as financial services
2. Service *occupations*, which includes jobs not directly involved in producing material goods such as transport, sales, clerical, administrative, professional, and cleaning services.

Postindustrialism or Industrial Services?

The rise of service industries and service occupations has been accompanied by much dispute over how to characterize changes to society and economy. For some, like Daniel Bell in the U.S. and Alain Touraine in France, "industrial society" is being superseded by "postindustrial society." Bell describes a trajectory of socio-economic development, which passes through preindustrial, industrial, to postindustrial society. Preindustrial society was technologically limited to a brute human struggle with nature based on raw, muscle

power." Industrial society was premised on energy-driven sectors that coordinated men and machines for the mass production of goods. Postindustrial society is

> "organized around knowledge, for the purpose of social control and the directing of innovation and change" [Bell, 1973].

On Bell's calculations, by 1970, 65 percent of the U.S. labor force worked in services such as information and professional services, health, education, welfare, cultural, and leisure services and research, design and development.

Touraine similarly identifies a shift from manufacture-based industrial society to a knowledge-based postindustrial society. Where Bell draws conservative conclusions about social and political development, Touraine identifies new, radical forms of protest. In industrial society the workplace became the organizing center for social movements based on economic class which appealed to the national state for redress of grievances. In postindustrial society, new social movements like ecology and antinuclear movements emerge that reject technocratic integration into the economy and the state. This reflects the more diffuse sources of political power in a knowledge-society characterized by Touraine as the "management of the data-processing apparatus."

Both Bell and Touraine make their claims about postindustrial society by placing a conceptual emphasis on knowledge." Yet the relationship between knowledge-based services and physical goods manufacturing is more complex than their image of separate, contiguous sectors allows for. Indeed much of the rise in services can be accounted for by the greater support necessary for advanced technical production systems. Additionally, an increased dependence on knowledge may still rely heavily on production in the capital goods sector of intelligent machines and knowledge infrastructures.

Gershuny and Miles identify the emergence of a "self-service" economy that defies the linear logic of the supplanting of industrial society by post-industrial society." For them *service functions* can be met by a range of organizational and technical forms. Since services tend to be more labor-intensive and less amenable to large productivity gains arising from technical innovation, households can elect to service themselves by substituting cheaper manufactured goods such as cars, washing machines, videos, microwave ovens, chilled ready-made meals, and ready-to-assemble furniture, for market-based services such as transport, laundries, cinema, restaurants, and so on. The self-service economy therefore represents a further extension of manufacturing into everyday life rather than its eclipse by "knowledge." The rise of the self-service economy also provides a good example whereby a range of technical innovations in the manufacture of consumer goods cluster in the 1950s to reinforce what Joseph Schumpeter referred to as a "wave of innovation." Schumpeter's model of discontinuous, disruptive periods of technical change contrasts with the linear, continuous model of socio-technical succession in post-industrial theories.

After Fordism

Fordism reached a high point in the 1950s during the postwar economic boom. Fordism was never an all-embracing socio-technical system and the term does not cover the wide variety and processes even in the manufacturing sector. At best, Fordism represented the leading edge of development and is not an accurate description of the entire structure of production between the 1920s and 1970s.

Two main schools of thought emerged to characterize the period after Fordist industrial leadership. First, Michael Piore and Charles Sabel in *The Second Industrial Divide* (1984) deployed an institutional approach to technological development which focused on the contingent influence of particular national political and economic differences. They discuss the crisis of Fordism in terms of external shocks such as labor discontent, and the 1973 oil crisis for triggering inflation and recession which repressed demand and investment and led to the fragmentation of mass markets. This trend was exacerbated by the internal structural tensions of mass production where the rigidities of Fordist production could not adapt to the saturation of the mass market and consumer demands for more differentiated and individualized products. Piore and Sabel advocated the adoption of "flexible specialization" as a solution to the crisis of Fordism. Flexible specialization deploys newly versatile production technologies based on computerized systems to efficiently switch production from high volume mass production to small batches of customized goods.

In the work of the French Marxists Michel Aglietta, Robert Boyer, and Alain Lipietz, a second approach focused on "neo-Fordism" as the latest stage in capitalism's development. Fordism enters crisis in this account because of the contradiction between the "regime of accumulation,"; that is, the macroeconomic principles around which capitalism stabilizes for a period of time, and "the mode of regulation,"; that is, how a particular national economy is regulated both within and between countries. After the crisis of overproduction in the 1930s, the Fordist regime of accumulation was characterized by mass production where real wages rose in line with productivity, a stable rate of profit, and full employment. This was buttressed by a new mode of regulation consisting of institutionalized collective bargaining, a developed welfare state, and the control of credit by central banks. Neo-Fordism is characterized by flexible production technologies allied to differentiated and segmented consumption, and a less comprehensive welfare state.

Where the institutionalists focus on the contingent national factors, the regulationalists focus on the structural contradictions of whatever manifestation capitalism takes. Both approaches have been criticized for empirically neglecting the resilience of mass production and mass consumption. Taylorism has deepened into other areas such as the mass production and consumption of fast foods, giving rise to the further rationalization of all social institutions in what George Ritzer called the "McDonaldization" of society. Both have also been criticized for conceptually isolating or reifying a particular production technology and erecting an elaborate theory of socio-economic and cultural change around it.

The Coming of the Information Age

Multinational companies in Europe and the North America also began to adopt "Toyotism," or "lean production," after the Japanese manufacturing philosophy of Taiichi Ohno. Toyotism challenged the top-down controls over labor of Taylorism by a system of workers' own continual improvements to the production process called *kaizen*. All the stages in the creation and circulation of the commodity were to be synthesized through the *kan-ban* system by developing a long-term relationship between the manufacturer, the dealer, and the consumer through a "build-to-order" system rather than the unpredictable "build-to-economize" system of mass production, which regularly resulted in crises of overproduction. On the other hand, the flexible worker needed to become multiskilled to cope with the new processes and technologies.

New technologies based on information processing systems are transforming certain aspects of work, though the extent of this should not be exaggerated. In the early 1980s the Fiat car factory at Cassino claimed to be the first fully robotized plant in Europe. Teleworking, for instance, was claimed to herald a new era of social organization, where the home would increasingly absorb the functions of the workplace through being networked to the Internet. In the late 1990s the available evidence indicated that only very small numbers "tele-commute" with little daily reduction in car use, although the myth of workplace transcendence and travel reduction remains strong. Information and communication technologies are also being used to stretch and intensify the working day, with the average office worker estimated to be spending three hours per day sending and receiving some 150 electronic messages. Strict control over and close monitoring of labor is even more evident in technologically intensive workplaces such as call centers, which proliferated on the edge of low-cost, metropolitan centers in the 1990s. Call centers utilize automated call distribution systems to direct incoming or outgoing calls to customer service operators, housed in large sheds that resemble advanced telecommunications factories.

See also **Automobiles; Farming, Agricultural Methods; Farming, Mechanization; Iron and Steel Manufacture; Technology and Leisure; Urban Transportation**

ALEX LAW

Further Reading

Aglietta, M. *A Theory of Capitalist Regulation: The US Experience*. New Left Books, London, 1979.

Baker, E.F. *Technology and Women's Work*. Columbia University Press, New York, 1964.

Bell, D. *The Coming of Post-Industrial Society*. Heinemann, London, 1973.

Braverman, H. *Labor and Monopoly Capital*. Monthly Review, New York, 1974.

Cavendish, R. *Women on the Line*. Routledge & Kegan Paul, London, 1982.

Cockburn, C. *Machinery of Dominance: Women, Men and Technical Know-How*. Pluto Press, London, 1985.

Edwards, R. *Contested Terrain: The Transformation of the Workplace in the Twentieth Century*. Heinemann, London, 1979.

Flink, J.J. *The Car Culture*. MIT Press, Cambridge, MA, 1979.

Ford, H. Mass production, in *Encyclopaedia Britannica*, 13th edn, suppl. vol. 2, 1926, pp. 821-823.

Gershuny, J.I. and Miles, I.D. *The New Service Economy: The Transformation of Employment in Industrial Societies*. Macmillan, London, 1983.

Gillespie, A. and Richardson, R. Teleworking and the city: myths of workplace transcendence and travel reduction, in *Cities in the Telecommunications Age: The Fracturing of Geographies*, Wheeler, J., Aoyama, Y. and Wharf, B., Eds. Routledge, London, 2000.

Gorz, A. *Reclaiming Work: Beyond the Wage-Labour Society*. Polity Press, Cambridge, 1999.

Gramsci, A. Americanism and Fordism, in *Selections from the Prison Notebooks*, Hore, Q. and Nowell-Smith, G., Eds. Lawrence & Wishart, London, 1971.

Greenbaum, J. *Windows on the Workplace: Computers, Jobs and the Organization of Office Work in the Late Twentieth Century*. Monthly Review Press, New York, 1995.

Hirst, P. and Thompson, G. *Globalisation In Question*. Polity Press, Cambridge, 1996.

Hounshell, D. *From the American System to Mass Production 1800–1932*. John Hopkins University Press, Baltimore, 1984.

Jaffe, A.J. and Froomkin, J. *Technology and Jobs: Automation in Perspective*. Fredrick A Praeger, New York, 1968.

Knights, D and Willmott, H., Eds. *Gender and the Labour Process*. Gower, Aldershot, 1986.

Kranzberg, M. and Pursell, C.W., Eds. *Technology in Western Civilization, Volume II: Technology in the Twentieth Century*. Oxford University Press, New York, 1967.

Lacey, R. *Ford: The Men and the Machine*. Heinemann, London, 1986.

Lipietz, A. *Mirages and Miracles: The Crisis of Global Fordism*. Verso, London, 1987.

Littler, C.R. *The Development of the Labour Process in Capitalist Societies*. Heinemann, London, 1982.

Noble, D. *Forces of Production: A Social History of Industrial Automation*. Alfred A Knopf, New York, 1984.

Pollard, S. *The Genesis of Modern Management*. Edward Arnold, London, 1965.

Pollert, A. *Girls, Wives, Factory Lives*. Macmillan, London, 1981.

Ritzer, G. *The McDonaldization of Society*. Sage, London, 2000.

Schumpeter, J. *The Theory of Economic Development*. Harvard University Press, Cambridge, 1934.

Smith, C. *Technical Workers: Class, Labour and Trade Unionism*. Macmillan, London, 1987.

Taylor, F.W. *Scientific Management*. Harper & Row, New York, 1947.

Taylor, P. and Bain, P. An assembly line in the head: the call centre labour process. *Ind. Relat. J.*, 30, 2, 101–117, 1998.

Thompson, P. *The Nature of Work: an Introduction to Debates on the Labour Process*. Macmillan, London, 1989.

Womack, J.P., Jones, D.T. and Roos, D. *The Machine That Changed the World*. Macmillan, New York, 1990.

Wood, S., Ed. *The Degradation of Work?* Hutchison, London, 1982.

Chemical Process Engineering

The chemical industry expanded dramatically during the twentieth century to become a highly integrated and increasingly influential contributor to the international economy. Its products seeded and fertilized the growth of other new technologies, particularly in the textiles, explosives, transport, and pharmaceutical industries. The industry also became a major supporter of industrial research, especially in the U.S. and Germany. The production of chemicals during the century can be described as a history of products, processes and professions.

At the turn of the twentieth century, the chemical industry was dominated by two classes of products. While a variety of chemicals had been produced on a small scale since antiquity, the nineteenth century had seen the rise of large-scale chemical industries. The manufacture of bulk (or heavy) inorganic chemicals such as sulfuric acid and bleaching powder became important industries by midcentury. Coal-tar derivatives were another important category of product. The production of illuminating gas from coal from the 1820s generated significant waste materials which subsequently were processed into useful products in their own right; for example, naphtha, used as lamp fuel and solvent; creosote, for the protection of timber (especially railway sleepers); and dyes. Synthetic dyes, originating with the discovery in 1856 of mauve from a coal-tar derivative by the Briton W.H. Perkin, became an industry increasingly dominated by Germany from the 1870s; by World War I, Germany was producing 80 percent of the world output of dyestuffs and other fine chemicals reliant on complex processes. By the mid-twentieth century, these products had been eclipsed by spectacular increases in the production of petroleum, petrochemicals, synthetics and pharmaceuticals. Nevertheless, the industry remained little appreciated by the public because most of its output was in the form of intermediate products for use by other industries.

The larger scale changes in chemical production can be better understood in terms of processes rather than as discrete products. Indeed, Hardie and Pratt (1966) describe the history of the chemical industry in terms of the history of its processes; that is, the succession of actions that transform raw materials into a new chemical product. Such conversion may involve chemical reactions (e.g., the production of soda alkali and sulfuric acid in the LeBlanc process); physical change (e.g., oxidation by roasting, or distillation by boiling and condensation); or physical manipulation (e.g., by grinding, mixing and extruding).

Professions, too, had an important influence on the trajectory of the industry. Two models, developing side by side, supplied the design and production labor of the chemical industries in different countries. At the turn of the century, chemists held sway over the technical side of much of the business of chemical manufacturing. But the growing economic importance of bulk chemicals over the previous few decades had been accompanied by the emergence of new specialists in the design and operation of chemical plants. From the 1880s certain groups of technical experts dissatisfied with their standing in the industrial hierarchy made repeated attempts to wrest control of certain industrial practices from chemists. By the first two decades of the twentieth century, groups of senior workers employed in American and British chemical manufacturing were seeking to distinguish themselves from analytical chemists, who had a low status, by claiming as their own the task of scaling-up chemical processes from the laboratory to the industrial level. In America and Britain, a new breed of worker—the chemical engineer—developed as embodying this special expertise. The growth of a "process" perspective developed in lock-step with this nascent profession. This culminated in the formation of the American Institute of Chemical Engineers (1908) and, in Britain, the Institution of Chemical Engineers (1922).

The leaders argued that the development of new manufacturing processes could be understood within the conceptual framework of industrial chemistry. Early chemical engineers used their knowledge of physical chemistry as a way of distinguishing themselves both from self-trained "factory hands" and from mechanical engineers who had become involved with chemical plant design.

The key to legitimating chemical engineering as a profession was the evolution of a new cognitive base—the unit operation. Rudiments of the idea can be found in the writings of George E. Davis, a Manchester chemical engineer, from the late 1880s (Davis, 1901). The unit operations evolved more explicitly at several American universities, especially the Massachusetts Institute of Technology. The unit operations were first described as such in 1915 by Arthur D. Little, who defined them as discrete physical processes employed in chemical manufacturing, such as distillation, roasting, filtering, and condensation. Each described a particular way in which material could be transformed physically—for example, by the reduction in size of solid matter, or by the mixing or separation of solids, liquids, or gases. These basic processes would be performed in sequence to obtain a final product. While the number and order might vary from chemical to chemical, any manufacturing process could be understood in terms of the same set of building blocks. The steps could either be carried out on batches of material, or performed as a continuous process in which the material is transformed at different locations in a plant. This intellectual transformation made possible a scaling up of production and greater efficiency in an industry that had blighted the landscape by serious air and water pollution from its waste products during the nineteenth century.

The particular technical requirements of petroleum refining—namely physical operations such as distillation—may explain the acceptance of the conceptual framework of the unit operations by manufacturers in America, where oil had been pumped since the late nineteenth century. Hence the American profession appears to have been shaped by at least three factors: the desire of certain groups of technical workers to differentiate themselves from the more heterogeneous and less prestigious category of "chemists"; their appropriation of a cognitive realm (first theoretical chemistry, and then unit operations) as a way of underpinning claims to technical expertise; and a gain in legitimacy in academic, engineering and industrial circles following the successful application of this intellectual apparatus to the design and operation of large chemical plants.

By contrast to the American case, a unique occupation combining mechanical and chemical expertise failed to coalesce in Germany, Austria and Switzerland, major chemical producers before and during the twentieth century. In the German occupational model, labor was divided strictly between industrial chemists and mechanical engineers working in tandem. One reason for this difference is that the petrochemical industry there was negligible, and much more complex chemical syntheses—traditionally based on low-volume batch processes and requiring the specialized scientific knowledge of chemists—dominated the dyestuffs and pharmaceuticals industries. The occupational specialty of *Verfahrenstechnik* became organized around specific industrial products and their manufacturing processes, and chemical plants through most of the twentieth century were designed and maintained by a combination of chemists and mechanical engineers (Buchholz, 1979).

In Britain, the chemical engineering perspective gained ground during the First World War. The organizer of government chemicals factories, Kenneth B. Quinan, was credited with introducing clear methods of analyzing the problems of chemical design and production. Quinan's techniques of statistical control for administration achieved impressive manufacturing efficiencies, and were important in helping to demonstrate that chemical engineers were competent enough to handle the problems of rapidly and efficiently scaling-up

facilities from laboratory demonstration to "pilot plant" to "production plant" size. British chemical engineers stressed the importance of industry-wide research, academic training, and accreditation to carry forward the expansion triggered by the wartime chemical factories. While promoting similar professional and educational aims as their American counterparts, they also emphasized government–industry–professional cooperation (so-called "corporatism") and the efficient recovery and use or disposal of byproducts.

World War I had a catalyzing effect in ramping up chemical production and in defining the specialists responsible for it. It also played an important role in the coalescence of representative bodies that came to promote the industry, discipline and profession. The supply of chemical munitions was central to the waging of the war. At the outbreak of World War I, only the German chemical industry had been prepared for the required scale of explosives production; high explosives, in particular, were produced by few factories. For other materials such as cordite (the principal propellant used by the British military, and consisting principally of a mixture of nitroglycerin and nitro-cellulose extruded in thick cords), there appeared to be no prospect of a large-scale demand after the war which would induce existing manufacturers to extend their works. Thus in Britain, the government became directly involved in the wartime management of chemical production. In the U.S., the Du Pont company was subsidized by the government to increase production. The "chemists' war" eventually mobilized thousands of technical workers with chemical and engineering backgrounds.

These wartime experiences dramatically altered postwar chemical production. Chemical firms—traditionally secretive about the processes they employed—were now larger and less isolated. While some postwar factories were sold for little more than scrap, other explosives plants were readily converted to civilian products such as rayon production, for which the production and raw materials were not dissimilar to gunpowder production. To many such transformed plants came experienced managers and designers from the ordnance factories.

This mixing of experience in the commercial and government spheres, and the heightened international competition after the war, encouraged firms to combine and expand. In the U.S., the DuPont company, which had begun as an explosives manufacturer, expanded into dyestuffs and cellulose products during and after World War I. In

Germany, IG Farben amalgamated in 1925 from dyestuff firms that had been associated since 1916. As the largest chemical company in the world, it continued to be the most prolific inventor of new synthetic materials. In Britain, Imperial Chemical Industries (ICI) was formed in 1926 by the merger of a number of companies, and the Association of British Chemical Manufacturers was formed to unite industry interests for the first time. In France, RhÔne-Poulenc was created in 1928 by the merger of two fine-chemical firms.

New technologies, too, signaled new opportunities between the wars. The most prominent of these was high-pressure plant, pioneered in Germany by the Haber–Bosch process for the synthesis of ammonia (1909). This demanding technology, employing gases at high pressure and temperature, was inherently large-scale and capital-intensive. The plant required expertise in compressors, pumps, reacting vessels, control devices and moving machinery. These demanding technologies were emblematic of a new scale and approach to process engineering.

New materials for chemical plant were also a frequent source of enthusiasm and optimistic forecasts. Novel metal alloys and corrosion-resistant materials had been sinks for speculative capital for decades, and proliferated with somewhat more success during the interwar period. Developments in tantalum, tellurium, lead, copper, and nickel alloys transformed the environment of chemical plants. Stainless steel, first used in German and Austrian nitric acid and munitions plants after 1916, became increasingly applied from the 1920s in an industry which had (prewar) relied on cast and wrought iron, mild steel, lead, copper, tin, wood, and bitumen. Chemical stoneware, vitreosil and inert resins similarly multiplied the options available to designers, and improvements in welding techniques allowed stronger, more reliable vessels than could be produced by forging or riveting. Other singular technological innovations became just as pervasive: electric motors, for example, helped promote a transformation in the practice of pumping fluids, which had frequently been based on pneumatically driven lifts and "eggs" before World War I. A postwar flurry of books, periodicals and advertisements supported this new awareness of a coherent chemical process industry.

World War II further scaled up chemical production capacity in many countries. Most factories were organized either as government-owned, company-operated (GOCO) plants in America or as "agency" factories in Britain, in which a firm constructed and operated plants for government production quotas. As with the previous war, there were long-lasting consequences for production capacity, product expertise and commercial relationships.

During the decade and a half after World War II, chemical engineering became firmly established. The chemical and process industries evinced an unprecedented expansion, and chemical engineers found new and more senior places within them. The discipline became firmly rooted in universities and colleges, increasing the number of degree-trained chemical engineers many fold.

New chemical products and technologies transformed the postwar process industries. The principal products were petroleum and petrochemicals, synthetics, and pharmaceuticals.

During World War II, the refining of petroleum on a very large scale had developed rapidly. "Cracking" processes, producing shorter-chain derivatives from the long-chain molecules of heavy petroleum by thermal dissociation via a series of furnaces, yielded a variety of economically feasible products (see Cracking). The subsequent production of "petrochemicals," or chemical products synthesized from petroleum-based raw materials or feedstocks, developed most quickly in America, with its ample oil reserves. In petrochemical processes, petroleum is separated into various components by conventional distillation and catalytic cracking to yield a series of lighter compounds, which are separated, reacted with other compounds and combined to synthesize new chemical products.

By contrast, in Germany, which had been committed to a policy of self-sufficiency in resources during World War II, fuels had been based on coal rather than petroleum reserves. Britain, too, lacked established oil reserves in the first half of the century, and until the war its chemical industry relied on fermentation alcohol, coal, and tar derivatives for organic chemical production. Wartime demand and new products altered priorities. ICI, for example, previously a general chemicals manufacturer, had become a major producer of wartime aviation fuel in Britain. ICI's oil hydrogenation facility provided the basis of a petrochemical industry; Shell began producing a detergent based on wax cracking and after the war ICI sought associations with petroleum refiners to provide its feedstocks for nylon and polyethylene.

The development of new petrochemical products transformed the post-World War II chemical market, as much by replacement as by expansion. The business in solvents from sugarcane molasses, for example, was undermined by the development of alcohol from ethylene. The alcohol market itself eroded in the early 1950s. Similarly, the dehydration of alcohol for use as motor fuel—an important process in the early years of the war—was abandoned. The coke oven industry withered with the employment of oil to replace coke in blast furnaces.

Like petrochemicals, the production of synthetic materials had prewar origins but came to have a dramatically greater economic influence after the war. Thermosetting products such as phenol-formaldehyde and epoxy resins, first produced in the 1890s, became practical products early in the twentieth century when their chemistry began to be studied. "Bakelite," for example, found a market from 1909 in the electrical and motor industries for insulating materials and compression mouldings. Similarly Rayon, based on cellulose, had been available from the 1890s and manufactured according to methods, and manpower, largely familiar to the traditional textile industry. Until World War I, most polymers were obtained from the chemical byproducts of gas works and coking ovens. Nylon, however, developed during the 1930s by DuPont in America, required production processes drawn from the chemicals industry. Its British counterpart, ICI, similarly established new processes for synthetic materials, from low-density polyethylene in 1939 (used extensively in wartime applications) to Terylene fibers (1949), Teflon (1950) and polyacrylonitrile fibers (1960). Here again, the developing discipline of chemical engineering promoted problem solving including heat transfer (because the reactions could proceed rapidly), filtration and mass transfer. The need for continued and intensive physical research was obvious: the characteristics of melt flow were distinct and complex for each new material, and the design of process equipment often a black art. Nevertheless, many such synthetic materials could be processed on equipment adapted from that developed earlier for the rubber industry for mixing, rolling and extruding.

By the mid-twentieth century there was a growing list of processes for producing chemicals using biological operations. Scientific brewing methods, for example, had flourished from the late nineteenth century. Several strands of research combined in the period before World War I to promote what has become known latterly as biochemical engineering or biotechnology. A rubber shortage between 1907 and 1910 encouraged searches for synthetic alternatives. Fermentation vied with distillation as a key unit operation in wartime production. After World War I, when

petroleum reserves appeared limited, the production of synthetic fuels again helped to promote interest in biological methods. Nevertheless, the large manufacturers paid little attention to such processes in the interwar period, when the employment of coal-tar as a raw material appeared most promising.

World War II revitalized the development of biochemical processes. The production of synthetic rubber and pharmaceutical drugs assumed significant dimensions by the end of the war. The case of penicillin, in particular, illustrates how chemical engineering methods were adopted and adapted gradually, and how technology transfer operated between countries and disciplines. Penicillin production became the last major use of British agency factories during World War II. American pharmaceutical firms such as Pfizer, Merck, Squibb and Eli Lilly became major antibiotics manufacturers in the postwar years based on more efficient mould-growing technology developed there.

With the end of World War II, the international chemical industry was transformed. In 1951, the Allies broke up the IG Farben chemical cartel in Germany into a number of firms; the three largest—Bayer, Hoechst and BASF—independently extended manufacturing of pharmaceuticals, plastics and other chemical products. Similarly, the Swiss firm CIBA (Chemische Industrie im Basel), which had focused on textile dyes in the first half of the century, diversified into pharmaceuticals. The prewar activities of seeking oil and raw chemicals from colonies or possessions was extended from the 1950s to petroleum refining and chemical plants, notably in the near East and India.

By the 1950s, chemical engineers were beginning to provide a technical competence recognized in certain spheres, particularly in the design of plant for new processes. Indeed, the employment of chemical engineers in general plant contracting was an occupational niche that developed significantly after World War II. Contracting was an outgrowth of the independent consulting work and equipment brokering in which many early chemical engineers dabbled. The post-World War II version, however, was no longer small scale and short in duration. Nor was contracting any longer dominated by firms that also fabricated equipment. The American style of financing, design and fabrication of chemical plant was largely exported to Europe after the war.

The scale of production increased dramatically after the war, the capacity of new ethylene plants, for example, rising from some 10,000 tons per annum in 1950 to 450,000 tons per annum in 1970.

In Japan and West Germany, polymer production increased ten-fold during the 1960s. In the Soviet Union, conversion of the chemical industry from coal-based to petroleum-based feedstocks did not get underway until 1958. By 1970, some 90 percent of the world's chemicals were produced by the so-called "developed" countries.

Such expansion was based on continuous flow processes. On both economic and technical grounds, the refining of petroleum and the manufacture of petrochemicals were best undertaken as continuous processes. Instead of relying on the combination of premeasured components, flow processes required careful monitoring of reactants (which in turn depended increasingly on instrumentation such as flow meters and feedback control systems). Following the crude oil shortage of the early 1970s and a consequent quadrupling of prices, combined with growing public opposition in the West over pollution, such design elements allowed chemical process plants to be made increasingly efficient in the recovery and use of their byproducts.

The chemical industry had always been cyclic, and the demand for chemical products and industry workers by the early 1960s was made even more unstable by waves of investment. This economic cycle was exacerbated by delayed feedback: new investment in the increasingly large-scale chemical plants was followed by over-capacity, a reduction in profits and cuts in investment. Employment in the industry soared as plants were designed, constructed and started up, declined as a fewer operators were required to man them, and plummeted as further plant design was deferred. The rate of expansion of the international chemical industry slowed considerably by the 1970s. So too, did the number of jobs, journals, courses and professorial appointments, the rate of students graduating, and book publishing; all were to reach saturation by the early 1970s. During the last decades of the century this characteristic of restrained growth was sustained.

See also **Chemicals; Cracking; Detergents; Dyes; Feedstocks; Green Chemistry; Materials and Industrial Processes; Nitrogen Fixation; Oil from Coal Process; Plastics, Thermoplastics**

SEAN F. JOHNSTON

Further Reading

Aftalion, F. *A History of the International Chemical Industry*. University of Pennsylvania Press, Philadelphia, 1991.

Bradbury, F.R. and Dutton, B.G. *Chemical Industry: Social and Economic Aspects*. Butterworths, London, 1972.

Buchholz, K. Verfahrenstechnik (chemical engineering)— its development, present state and structures. *Soc. Stud. Sci.*, 9, 33–62, 1979.

Davis, G.E. *A Handbook of Chemical Engineering*, 2 vols. Davis Brothers, Manchester, 1901, 1904.

Divall, C. and Johnston, S.F. *Scaling Up: The Institution of Chemical Engineers and the Rise of a New Profession*. Kluwer, Dordrecht, 2000.

Haber, L.F. *The Chemical Industry 1900–1930: International Growth and Technological Change*. Clarendon Press, Oxford, 1971.

Hardie, D.W.F. and Pratt, J.D. *A History of the Modern British Chemical Industry*, Pergamon, Oxford, 1966.

Hayes, P. *Industry and Ideology: IG Farben in the Nazi Era*. Cambridge University Press, Cambridge, 1987.

Lesch, J.E., Ed. *The German Chemical Industry in the Twentieth Century*. Kluwer, Dordrecht, 2000.

Mowery, D.C. and Rosenberg, N. *Paths of Innovation: Technological Change in 20th-Century America*. Cambridge University Press, Cambridge, 1998.

Reader, W.J. *Imperial Chemical Industries: A History*, 2 vols. Oxford University Press, Oxford, 1970, 1975.

Reuben, B.G. and Burstall, M.L. *The Chemical Economy*. Longman, London, 1974.

Shreve, R.N. *Chemical Process Industries*. McGraw-Hill, New York, 1967.

Spitz, P.H. *Petrochemicals: The Rise of an Industry*. Wiley, New York, 1988.

Travis, A.S., Schröter, Harm G., and Homberg, E., Eds. *Determinants in the Evolution of the European Chemical Industry, 1900–1939: New Technologies, Political Frameworks, Markets, and Companies*. Kluwer, Dordrecht, 1998.

Chemicals

In the twentieth century, the chemical industry became an essential contributor to technological innovation, economic growth, and military power. At midcentury, *Fortune* magazine proclaimed that it was indeed a "chemical century." For the first half of the century, the industry had consisted of a diverse set of technologies serving a broad array of markets. The chemical industry provided thousands of products that were used in everything and everywhere. After the war the industry coalesced around products made from petroleum and natural gas, with plastics and polymers accounting for over half the industry's output. Another key growth area was pesticides, following on the example the wartime miracle chemical, DDT. The dramatic success of another wartime innovation, penicillin, led to the rapid growth of the modern pharmaceutical industry, based on a combination of chemistry and biology. In addition to specific products, the chemical industry provided scientific and technological knowledge needed to develop other critical technologies, such as nuclear weapons and semi-conductors. In the latter decades of the century, major innovations declined and competition increased causing chemicals to lose their high-tech image, reducing them to commodity status. The global industry has undergone massive reorganization in the wake of these new realities. Industry leaders hope that technologies based on biotechnology, green chemistry, and nanotechnology can restore the industry to its former glory.

Although people have made and used chemicals for thousands of years, the modern industry, based on large-scale production, emerged during the Industrial Revolution of the late eighteenth and early nineteenth centuries. The first industrial chemical—sulfuric acid—dates from the mid-eighteenth century when large lead-lined chambers were used to allow the oxidation of sulfur dioxide, made by burning sulfur, to sulfur trioxide, which reacts with water to produce acid. By the mid nineteenth century sulfuric acid plants had grown very large and had reached a high degree of technical sophistication, incorporating most of the techniques of modern chemical engineering. The availability of cheap sulfuric acid allowed the development of cheap alkali by the LeBlanc process, first developed in France but commercialized in Great Britain after 1810. Sulfuric acid was converted to sodium carbonate through a series of reactions with salt, limestone, and charcoal. Large quantities of acids and bases were consumed in Great Britain principally in textile operations, such as washing, bleaching, and dyeing.

Armed with these two reagents—acid and base—chemists began to experiment with a wide variety of substances, many of them organic (carbon-containing). By midcentury chemists had discovered some useful new compounds. In 1856, a young English chemist, William H. Perkin while naively trying to convert coal-tar into the valuable antimalarial quinine produced a purple colored solution instead. At the moment of this discovery extremely expensive purple was the fashionable color among Europe's elite. Using cheap coal-tar, a waste product from coal gasification plants that supplied illuminating gas to cities, Perkin developed a process to make a purple dye, mauve, by oxidizing aniline (benzene with an ammonia group substituted for a hydrogen atom). Other chemists soon discovered that the larger class of chemicals based on benzene rings would yield a rainbow of colors when reacted with acids and bases. The systematic and highly profitable exploitation of aniline dyes shifted in the 1870s to the new nation of Germany where the government, universities,

and emerging chemical companies cooperated to develop this important industry. By World War I, three German companies, Bayer, BASF, and Hoechst controlled about 90 percent of the world's dyestuffs production. German chemists isolated the chemicals made by the madder and indigo plants that produced red and blue dyes, respectively. Chemists and engineers then learned how to manufacture these chemicals from coal-tar chemicals, replacing natural dyes which were major agricultural products of several countries, especially Turkey (madder) and India (indigo). Dyestuffs chemistry led German chemists into new fields such as pharmaceuticals with the discovery of aspirin by Felix Hoffmann and salvarsan (the first effective treatment for syphilis) by Paul Erlich. Another dyestuffs-related chemical, TNT (trinitrotoluene) would play a critical role as a shell-bursting explosive in World War I.

Explosives were revolutionized by chemists beginning in the middle of the nineteenth century. Experiments with nitric acid and organic molecules resulted in nitrate groups bonding onto the organic molecules, creating highly flammable or even explosive compounds. This characteristic resulted from the molecular proximity of a fuel (the organic compound) and oxygen (there are three oxygen atoms in each nitrate group). The most notorious of these new compounds was nitroglycerin, a liquid with tremendous explosive energy that was so unstable it often detonated prematurely. In Sweden, Alfred Nobel stabilized nitroglycerin by absorbing it into diatomaceous earth to produce a putty-like substance that could be extruded into paper casings. He called his product dynamite, and beginning in the 1870s it displaced black powder in blasting operations. Dynamite was one of the technological advances that would make projects such as the Panama Canal feasible.

Even more important than dynamite was its chemical cousin, nitrocellulose, prepared by reacting nitric acid with cotton fibers. This still cotton-like material became the basis for smokeless powder, which in the 1890s began to replace black powder as the propellant in guns and cannon. Smokeless powder burned much more cleanly than black powder and was much more powerful. The new propellant made the machine guns and heavy artillery into the terribly effective weapons that turned World War I into a bloody stalemate. Smokeless powder had a tendency to decompose causing spontaneous fires and explosions, until German chemists discovered a dye-stuffs-related compound that stabilized the powder in 1908. Another key chemical in the munitions machine was TNT, which exploded on shell impact causing huge craters and saturating the air with shrapnel.

The ingenuity of chemists added another horrific element to life in the trenches—poison gas. At the Battle of Ypres in 1915, German chemist Fritz Haber orchestrated the release of 5000 cylinders of chlorine which drifted with the wind into the Allied lines. The burning, choking gas caused panic in the Allied army but the Germans were not prepared to attack and so lost the advantage of its new weapon. Afterward both sides used poison gases such as lewisite, phosgene and mustard gas throughout the remainder of the war. All of these gases contained chlorine, which could be made in large quantities using electrochemical technology.

The development of the dynamo in the 1870s made available large quantities of electricity that could be used to make chemicals, many of which could not be economically made by other methods. Perhaps, the most important example was aluminum, which had semiprecious metal status—a small pyramid of it capped the Washington Monument which was completed in 1883. Three years later, Charles Martin Hall in the U. S. and Paul Louis Toussaint Heroult in France discovered a process to make aluminum using electricity. This method is still used today.

Another important electrochemical process was the production of chlorine and caustic soda (sodium hydroxide) from salt water. Chorine was used principally in bleaching powder and sodium hydroxide became the major base, replacing earlier compounds such as sodium carbonate. This process began to be used in 1890s; several electrolytic plants were built near Niagara Falls where hydroelectric power was available, and Herbert Dow built an early plant in Midland, Michigan where there was a rich supply of brine wells.

Other important materials were made in electric furnaces, which could generate very high temperatures, invented by Henry Moissan in 1892. One new ceramic compound was silicon carbide, which is so hard that it can be used to shape metals by grinding. Another was calcium carbide which reacts with water to produce acetylene, used in early automobile head lights and in oxyacetylene metal-cutting torches. Made from coal, acetylene became an early chemical building block used to make other chemicals.

The development of the Haber–Bosch ammonia process between 1906 and 1912 was a technological and scientific tour de force that became a prototype for future chemical processes. One of the great scientific and technological challenges of the late

nineteenth century was "fixing nitrogen." Nitrogen was an essential ingredient in explosives and fertilizer. Most of the world's useable nitrogen came from nitrate mines in the Atacama desert in northern Chile. Of course, air is 80 percent nitrogen, but it is almost chemically inert because it consists of two tightly bound atoms. Chemists sought ways to break those bonds. One way to do this was to react nitrogen and hydrogen to make ammonia. On paper it looked simple; in the laboratory it did not happen under normal conditions. A solution to this apparent impasse was suggested by theoretical considerations derived from the evolving disciplines of kinetics (the rate of chemical reactions) and chemical thermodynamics (determines the feasibility of particular reactions). The ammonia reaction was found to be feasible by German chemists Walter Nernst and Fritz Haber. Their calculations showed that the reaction would occur at very high temperatures (for kinetics) and very high pressures (for thermodynamics). The challenge then became technological: was it possible to build steel vessels that could withstand temperatures of 500°C and a pressure of 200 atmospheres? After Haber was able to make ammonia in laboratory scale apparatus, Carl Bosch of the BASF Company oversaw the development of a commercial process. Some of the early reactors were made from Krupp cannons. An essential part of the process was the development of a catalyst, a substance that causes the nitrogen and hydrogen to react with each other. At BASF, Alwin Mittasch led an exhaustive search until an efficient and durable iron-based catalyst was developed. The first large plant started up in 1913 a year before World War I would make Chilean nitrates unobtainable in Germany because of Britain's dominance of the seas. Without "synthetic" nitrogen, the Germans could not have sustained their war effort for four years.

In the 1920s BASF would expand on its high-temperature, high-pressure technological base by developing processes to make methanol from carbon monoxide and hydrogen and gasoline from coal. Before the new process, methanol was obtained by distilling it from wood (hence its name wood alcohol). The synthetic gasoline project was initiated by predictions of impending shortages of crude oil. After 1929, the discovery of the east Texas oil field increases world crude supplies and the Great Depression lowered demand for gasoline, the huge investment in synthetic gasoline technology threatened the viability of the giant IG Farben chemical combine. (The major German chemical companies had merged in 1925 primarily to sustain export markets.) The project and company would be rescued by Hitler after he came to power in 1933, since a domestic supply of gasoline—Germany has no oil—would be essential in a future war.

Hitler's policy of autarky sustained another project that would have important consequences for the chemical industry—synthetic rubber. Making synthetic versions of natural materials had been a long-standing objective of the chemists and one of the foundations of the chemical industry. Dyestuffs had been the first major success, but chemists also sought to make other substances, especially silk and rubber. Until the 1920s the basic structure of these substances was a matter of scientific uncertainty. This, however, did not stop chemists from forging ahead trying to make synthetic substitutes for exotic and expensive natural materials.

The origin of synthetic materials dates to 1870 when Albany tinkerer, John Wesley Hyatt formed a solid plastic from a mixture of nitrocellulose and camphor, which he called celluloid. According to tradition, Hyatt was looking for a substitute for expensive elephant ivory in billiard balls. When his new material failed in this use, he then made celluloid look like exotic materials—ivory, amber, and tortoiseshell—so it would be used in toilet sets, toys, and numerous other trinket-like applications. Its most enduring legacy was as the film base that made motion pictures possible beginning in the 1890s. An unsuccessful use of nitrocellulose was as an artificial silk fiber that, among other deficiencies, was highly flammable.

A much better silk-like fiber was rayon, formed by dissolving cellulose to make a syrupy viscose solution that was extruded through small holes in a plate into another chemical bath that solidified the fiber. Charles Cross and Edward Bevan in Britain discovered this process in the 1890s, while attempting to make improved light bulb filaments. After 1910 the market for rayon fibers began to expand rapidly worldwide; the fashion industry embraced it the 1920s; and during the Great Depression it replaced silk in all apparel except stockings. Rayon was the biggest new product for the chemical industry in the interwar years.

Rayon was just one a growing number of products made of large molecules (or macromolecules), in this case it was natural cellulose. Chemists were beginning to make entirely new large molecules. A pioneer is this effort was Leo Baekeland who invented a hard plastic he dubbed Bakelite in 1907. The new material was made by heating phenol and formaldehyde under pressure.

Among the many uses for Bakelite was as a substitute for ivory in billiard balls.

The growing importance of and interest in large molecules in the 1920s sparked a scientific debate, especially in Germany—still the center of chemistry—about their structure. Although many chemists argued that large molecules were held together by peculiar forces, Hermann Staudinger put forth the hypothesis that large organic molecules were just that—larger versions of common organic chemicals held together by same types of chemical bonds. Following Staudinger, Wallace H. Carothers, a researcher in the DuPont Company developed methods for making large molecules, or polymers, in the laboratory. Out of this research DuPont researchers discovered neoprene synthetic rubber (1930) and nylon (1934). By 1940 neoprene had established itself as a specialty rubber and nylon had become the preferred stocking fiber. Once the mysteries surrounding polymers had been solved, chemists everywhere began to explore this large and promising new field.

Perhaps the most significant discovery, both historically and for the future chemical industry, was made in 1929 by IG Farben chemists who made a general purpose synthetic rubber from a polymer consisting of repeating units of butadiene and styrene. At the time of this breakthrough virtually all of the world's rubber came from British controlled plantations in Malaysia. By early 1942, these were all in Japanese occupied territory. The first year of American fighting was hampered by a lack of rubber which threatened to bring the effort to a thudding halt. To resolve this crisis the U.S. government organized a cooperative venture between oil, chemical, and tire companies to rapidly build up an American synthetic rubber capability. This initiative was a marked success, production went from nothing to 800,000 tons in two years. One of the major obstacles that had to be overcome was to develop processes to make enormous quantities of styrene and butadiene. Styrene was available before the war in limited quantities but butadiene was not a commercial chemical. The supply of butadiene came primarily from oil companies, which had previously concentrated on making fuels not chemicals.

In the interwar years a few companies such as Union Carbide and Shell Oil had begun to make chemicals from petroleum and natural gas. One notable product introduced in the 1930s was ethylene glycol—automobile radiator coolant antifreeze. Until World War II organic chemicals used as feedstocks for the chemical industry were distilled from coal. For example, the type of nylon DuPont commercialized was determined mainly by the abundance of benzene, a major coal impurity. After World War II the oil and chemical industries, especially in the U.S., would soon shift to petrochemical feedstocks.

The oil industry was now generating large quantities of chemicals as a byproduct of new processes developed to produce more gasoline from a barrel of crude of oil and to produce higher octane fuels, especially necessary for aviation gasoline during World War II. Crude oil is a complex mixture of hydrocarbons with varying carbon atom chain lengths. Originally, natural gasoline, which contains five to nine carbons in each molecule was distilled out of crude oil. In 1913, E. M. Burton developed a cracking process in which he subjected the heavier fractions of crude oil to heat and pressure which broke the larger molecules into smaller ones, some of which were in the gasoline-size range. In the 1930s, French inventor and engineer, Eugene Houdry, added a catalyst to this process which significantly improved its overall performance. A decade later, an improved catalytic process called fluidized bed cracking was developed by Massachusetts Institute of Technology chemical engineers and Standard Oil of New Jersey. This process has been used ever since. Also during the late 1930s oil companies began to develop processes to combine some of the smaller cracked molecules into larger ones that could boost the octane rating of gasoline. During World War II, American 100-octane aviation fuel helped Allied pilots win the air war over Europe.

A few years after the war ended, the Universal Oil Products company, a research organization, introduced a new process which had been developed by Vladimir Haensel. Called "platforming" because of its platinum catalyst, it dramatically improved the octane rating of gasoline primarily by stripping hydrogen atoms from cyclical compounds, converting them to benzene, tolulene, and xylene. These compounds were not only important for high-octane gasoline but were in great demand by the chemical industry as raw materials, especially for the booming plastics and polymers businesses.

The dramatic post-World War II expansion of the chemical industry was led by plastics and polymers. Shortages of metals and other materials during the war had prompted the U.S. government to encourage manufacturers to use plastics for a wide variety of applications. For example vinyl resin (mostly polyvinyl chloride or PVC) production increased from 2.3 million to 100 million kilograms. Although many plastics ended up in applications such as army bugles, others served essential high-technology functions. Polyethylene, a difficult to make plastic, had unique insulating properties needed for radar. DuPont's exotic polymer Teflon, which did not melt, dissolve in solvents, or stick to anything, was used as a sealant in the Manhattan Project for the atomic bomb. Clear acrylic plastic became the material for airplane windows and bomber gunner turrets.

After the war both the uses and varieties of polymers increased to fulfill the demand by consumers for whom convenience became a hallmark of modern life. A New England inventor, Earl Tupper, introduced his line of polyethylene food storage containers—Tupperware—that preserved leftovers and kept them neatly in the refrigerator. The most sensational new products were synthetic textile fibers that made clothing more affordable, machine washable, drip dry, and wrinkle free. DuPont's nylon took over the stocking market and made major inroads in other apparel. Polyester, discovered by two British chemists in 1940, was used instead of wool in suits and blended with cotton to make permanent press garments. Acrylic fibers made sweaters, especially lightweight ones, popular with postwar women. By 1956, synthetic fibers had eclipsed wool as the number two textile fiber consumed in the U.S. By 1970, synthetic fiber consumption surpassed that of cotton in apparel.

At the same time that synthetic fibers were revolutionizing modern wardrobes, new types of plastics found myriad uses. The major breakthrough in plastics was made when German chemist, Karl Ziegler in 1953 discovered a new type of catalyst that produced new kinds of polymers, notably linear polyethylene and polypropylene. These two plastics had outstanding properties such as toughness which led to many uses, especially food packaging. As polymer science matured in the 1950s, chemists made more complex and sophisticated compounds, examples being DuPont's Kevlar polyaramid and Lycra spandex fibers.

Another significant growth sector for the post-World War II chemical industry was organic-chemical based pesticides. The archetype was DDT, whose remarkable kill-on-contact property was discovered by Paul Mueller in Switzerland in 1939. Most earlier insecticides were poisons, such as lead arsenate, that had to be ingested. During the war, George W. Merck, working on biological warfare for the government, discovered that a DuPont plant growth compound called 2,4-D was actually an effective herbicide. After the war chemical companies focused research efforts on finding new insecticides, herbicides, and fungicides.

Although Rachel Carson's *Silent Spring* (1962) publicized the toxic effects of DDT on birds and raised questions about the effect of pesticides on human health, during that decade 96 new insecticides, 110 new herbicides, and 50 new fungicides were introduced. Insecticides included organophosphorus compounds, carbamates, and synthetic pyrethrins. DuPont in 1967 introduced Benlate (benomyl), the first fungicide that was taken up internally rather than being effective only on the leaf surfaces. In the 1970s Monsanto introduced its blockbuster herbicide, Roundup (glyphosate), which was suitable for a wide variety of crops. In the 1990s chemical companies, especially Monsanto and DuPont combined biotechnology and herbicides to create crop seeds that were compatible with specific herbicides, and to incorporate insecticidal properties into plants by splicing in genes from other organisms. These so-called genetically modified foods have created controversy in Europe but have met with little resistance in the U.S.

It became evident in the 1960s that the chemical industry was maturing. During the 1970s the industry was beleaguered by spikes in the cost of energy and feedstocks, and environmental legislation that required major capital investments in pollution control and abatement. By the 1980s, the chemical industry had become very competitive worldwide with growth and profits tightly linked to larger business cycles. Since then the industry had undergone massive reorganization in response to these new economic realities. For the most part, chemicals, if not the companies that make them, have become commodities. Because it still has significant research capabilities, the chemical industry is hoping that new technologies such as nanotechnology—very small molecular structures—or green chemistry—replacing petroleum with renewable feedstocks—might restore chemicals to the essential status it enjoyed in the twentieth century.

See also **Chemical Process Engineering; Dyes; Electrochemistry; Explosives, Commercial; Feedstocks; Green Chemistry; Nanotechnology; Nitrogen Fixation; Oil from Coal Process; Pesticides; Synthetic Resins; Synthetic Rubber; Warfare, Chemical**

JOHN KENLY SMITH

Further Reading

Arora, A., Landau, R. and Rosenberg, N., Eds. *Chemicals and Long-Term Economic Growth: Insights from the Chemical Industry*. Wiley, New York, 1998.

Haber, L. *The Chemical Industry, 1900–1930: International Growth and Technological Change*. Clarendon Press, Oxford, 1971.

Haynes, W. *American Chemical Industry: A History*. Van Nostrand, New York, 1945–1954.

Hounshell, D.A. and Smith, J.K. Jr. *Science and Corporate Strategy: DuPont R&D, 1902-1980*. Cambridge University Press, Cambridge, 1988.

Idhe, A.J. *The Development of Modern Chemistry*. Harper & Row, New York, 1964.

Spitz, P.H. *Petrochemicals: The Rise of an Industry*. Wiley, New York, 1988.

Chromatography

The natural world is one of complex mixtures, often with up to a 100,000 (e.g., proteins in the human body) or 1,000,000 (e.g., petroleum) components. Separation methods necessary to cope with these, and with the simpler but still challenging mixtures encountered, for example in pharmaceutical analysis, are based on chromatography and electrophoresis, and underpin research, development, and quality control in numerous industries and in environmental, food, and forensic analysis.

In chromatography, a sample is dissolved in a mobile phase (initially this was a liquid), which is then passed through a stationary phase (which is either a liquid or a solid) held in a small diameter tube—the "column." According to the differing relative solubilities in the two phases, mixture components travel through the column at different rates and become separated before emerging and detected by the measurement of some chemical or physical property. The sample size can be as small as one picogram (10^{-12} g), but tens of grams can be handled in preparative separations.

Chromatography spans the twentieth century: the separations described by the Russian botanist Mikhail Tswett in 1903 by continuous adsorption/desorption in open packed columns, commonly applied in natural-product chemistry, was followed by the Nobel Prize-winning work (awarded 1952) of Archer John Porter Martin and Richard Laurence Millington Synge. Tswett called his new technique "chromatography" because the result of the analysis was "written in color" along the length of the adsorbent column (in this case, chalk). In 1941 Martin and Synge replaced liquid–liquid extraction by liquid partition chromatography (LC), supporting the liquid stationary phase on a solid support (using silica gel), over which they passed the sample solution. In their paper they made the famous statement:

"The mobile phase need not be a liquid but may be a vapor... very refined separations of volatile substances should therefore be possible in a column in which permanent gas is made to flow over gel impregnated with a nonvolatile solvent."

This forecast led to the reality of gas chromatography (GC), developed by Martin and his coworker Anthony T. James in 1952. Their first GC comprised of a glass tube containing the support material through which a gas mixture was blown by a carrier gas. The separated analytes in the mixture were detected using a simple titration technique.

LC is applicable to all soluble substances, whereas GC can be applied to mixture components which can be heated without decomposition to give sufficient vapor pressure (measured as a few millimeters of mercury, mmHg) for each solute to pass through the column in reasonable time. The volatility range of GC can be extended by the use of a supercritical fluid (a substance above its critical temperature and pressure) as mobile phase—supercritical fluid chromatography (SFC). This is a better solvent than a gas and more diffusive than a liquid, hence permitting more rapid analysis. Capillary electrochromatography (CEC) combines the advantages of LC and electrophoresis. The mobile phase flow is maintained by an electric field rather than by applied pressure; the separation principle is again partition, but the effect of different rates of electromigration may be superposed in applications to charged analytes.

Early partition chromatography was carried out in columns with an inert support on which the stationary phase was either coated or, better, bonded. In 1958 M.J.E. Golay showed how a tortuous path through a packed bed could be replaced by a much straighter path through a narrow open tube. Long, and hence highly efficient columns could thus be fabricated with the stationary phase on the inner wall, and remarkable separations achieved by GC and later by SFC. LC took a different course because slow diffusion in liquids means that separations in open tubes require impractically small diameters. Increased efficiencies can be more simply achieved on columns packed with small silica particles with bonded organic groups.

Initial commercial GC instrumentation was based on glass packed columns, and the most famous gas chromatograph was the Pye 104 instrument with a flame ionization detector. The direct coupling of gas chromatography and time-of-flight (TOF) mass spectrometry was achieved in the mid-1950s in collaboration with W.C. Wiley, I.H. McLaren, and Dan Harrington at the Bendix Corporation. At about the same time, GC was coupled to a magnetic sector instrument. The great utility of modern GC-MS was made possible by the advent in the 1960s of carrier gas separators that removed the GC carrier gas prior to introduction of a sample into the high-vacuum mass spectrometer. In 1979, the landmark paper by Ray Dandeneau of Hewlett Packard announced the development of fused silica columns to the world. Capillary columns have more theoretical plates (a measure of column resolving power or efficiency) per meter as compared to packed columns and since they have less resistance to flow they can be longer than packed columns. This means that the average capillary column of 30 meters has approximately 100,000 theoretical plates while the average packed column of 2 meters has only 2500 plates.

High-pressure liquid chromatography (HPLC) was developed in the mid-1970s and quickly improved with the development of spherical silica packing materials and the small volume UV detector. In the late 1970s, reverse phase liquid chromatography gave improved resolution and the technique was widely accepted into the pharmaceutical industry.

By the 1980s HPLC was commonly used for the separation of small molecules. New instrumentation techniques improved separation, identification, and quantification far above the previous techniques. Computers and automation added to the convenience of HPLC. Since 1990 there has been a continuous movement from the 4.6-millimeter internal diameter columns of the 1970s and 1980s to microbore columns ranging in internal diameter from between 1 and 3 millimeters. Today this is moving into capillary columns that range from 3 to 200 microns. These capillary columns have provided an easier interface to the mass spectrometer that is now very quickly becoming the detector of choice for LC.

Other forms of chromatography are:

- Thin-layer and paper chromatography. First developed in the early work by Martin and Synge, where a layer of absorbent is spread on a glass plate, and mixtures of components are placed on the edge of the plate. Solvent is then allowed to move up the plate by capillary action, drawing the components of the mixture along by varying degrees.
- Gel-permeation chromatography. Compounds are separated on the basis of their molecular size.
- Ion-exchange chromatography. Separation of ions based on their charge using a charged support material.

Electrochromatography, capillary electrophoresis (CE and CZE), capillary gel electrophoresis (CGE), micellar electrokinetic capillary chromatography (MECC) and capillary electro chromatography (CEC) are new separation techniques employing high voltages and narrow bore capillaries they are all grouped into the area of electrochromatography. This technique utilizes the principles and advantages of electro-osmotic flow.

See also **Electrophoresis**

KEITH D. BARTLE AND PETER MYERS

Further Reading

Giddings, J.C. *Dynamics of Chromatography Principles and Theory*. Marcel Dekker, New York, 1965.

Cazes J. and Scott, R.P.W. *Chromatography Theory*. Marcel Dekker, New York, 2002.

Grob R.L. and E.R. Barry, Eds. *Modern Practice of Gas Chromatography*. Wiley, New York, 2004.

Neue, U. *HPLC Columns*.Wiley, New York, 1997.

Lee, E.M.L. and Markides K.E., Eds. *Analytical Supercritical Fluid Chromatography and Extraction*. Chromatography Conferences, Provo, Utah, 1990.

Leslie, S.E. *Milestones in the Evolution of Chromatography*. ChromSource, Franklin, Tenn, 2002.

Bartle, K.D. and Myers, P., Eds. *Capillary Electrochromatography*. Royal Society of Chemistry, Cambridge, 2001.

Brinkman, U.A.T., Ed. *Hyphenation: Hype and Fascination*. Elsevier Science, Amsterdam, 1999. Reprinted from: *J. Chromatogr.* A, 856, 1–2, 1999.

Mondello, L., Lewis, A.C. and Bartle K.D., Eds. *Multidimensional Chromatography*. Wiley, New York, 2001.

Civil Aircraft, Jet Driven

The development of jet power shortly before World War II resulted in a series of transport projects designed during the conflict, especially in England. The jet principle seemed simple in theory. Air drawn through an intake is compressed through the engine nacelle by turbine-powered rotating blades, mixed with injected kerosene, which is then burned to create energy and rear-flowing exhaust gas. In fact, the internal engine dynamics, such as compression rates, temperature, and the fuel injection system all required considerable experimentation, with military jets becoming the first to be powered that way.

In the U.K., the Brabazon Committee, a government air transport task force formed in 1942, determined in the course of its second gathering a year later that several new types of airliners would be needed after the war. Among these, "type IV" was to be a jet-driven plane, a challenge the De Havilland Company picked up with its model 106, soon known as the Comet. The machine in question went through several design changes, yet while news of its construction for British Overseas Airways (BOAC) was common knowledge, no pictures or drawings were shown to the media or the public until the machine's ground trial runs. First flown in 1949, the Comet was not the only civilian machine of its kind (Avro Canada flew its C-102 that same year), nor the first (a Vickers Viking test aircraft was), but it became the most visible as it outpaced all other projects under development, entering service in 1952.

The Comet began to revolutionize air travel by radically cutting time spent in the air, while offering a new level of comfort associated with higher altitude cruising. However, two accidents in 1954 grounded all Comet aircraft. When the cause of the crashes was discovered (unforeseen metal fatigue), the redesign of the machine delayed the introduction of the transatlantic version of the plane by four years, thus giving the advantage to a Boeing model that would set new standards for jet travel: the 707.

Initially derived from a military air refueling project, the 707 also signaled new trends in commercial aviation. First, it signaled the beginning of the dominance of the jet market by American manufacturers, one which would not be successfully challenged until the Airbus consortium marketed several new aircraft in the 1980s. Second, it reflected a shift in airliner manufacturing practices, whereby constructors learned to offer multiple versions of the same basic model to increasingly selective airlines who might turn to the competition otherwise. Boeing, for example, had to redesign the initial 707 with a wider fuselage after United Airlines chose the Douglas DC-8 competitor aircraft. Boeing also worked on a slightly smaller version of the plane, known as the 720, to meet airline demand for a faster plane over a shorter range. Airlines eager to draw on an elite clientele quickly relegated their propeller-driven planes and introduced jets which, thanks to their speed, became new heralds of modernity. New models quickly followed.

While jet plane design was focused first on placing the engine within or attached to the wing, other solutions were also devised beginning in the 1950s to deal with the matter of thrust, noise, and aerodynamics. In the realm of short- and medium-range aircraft (1500 to 4000 kilometers), the French-built Caravelle became the first aircraft to use rear-mounted twin engines. A solution that reduced noise considerably in the passenger cabin and simplified the design of the wing, the model was highly successful for its time, and inspired such a design on several other models, such as the Douglas DC-9. Eventually, the latter together with the Boeing 737 twin (with engines under the wings) took back much of the Caravelle market, thanks in part to the greater flexibility of the American manufacturers in designing fuselage extensions in response to airline demand. However, other manufacturers, like the Dutch Fokker company and the British Aircraft Corporation maintained a European presence with the Fokker 28 and the BAC 1-11 short- to medium-range machines. These, together with updated Boeing and Douglas aircraft formed the second generation of jet airliners. By then, airports had lengthened their runways to cope with jets' higher take-off speed and greater payload, and new designs, such as mobile gangways leading directly from the waiting area to the plane were being designed.

This second generation, appearing in the late 1960s and early 1970s, also became associated with mass travel due to the introduction of wide-bodied aircraft. The development of a new type of jet engine, the turbofan, improved the thrust ratio 20 to 1, with the new engines amounting to only 4 percent of the total aircraft weight (against 9 percent for the 707). This massive increase in thrust made the completion of the such planes as the three-engined Douglas DC-10 and Lockheed Tristar, and the giant Boeing 747 possible.

The 747 was originally designed in response to a U.S. military call for a heavy (over 350 tons) long-distance jet transport. Boeing lost the bid to Lockheed's C-5A Galaxy plane, but was able, thanks to Pan American Airways' initial order, to develop a commercial wide body (characterized by twin aisles with up to ten seats per row). Able to carry some 350 passengers across the Atlantic in one fell swoop, later versions of the 747 would sport a capacity increase up to 450 passengers (with a consequent reduction in comfort and seating space). Boeing also offered two Japanese carriers a version designed for frequent take-offs and landings intended to link Japanese cities commuter-style, which could seat 550 people. Generally, however, the 747 and other wide-bodied long-range aircraft became associated with affordable travel beginning in the late 1970s, when airline deregulation allowed for greater competition.

However, early in its career the 747 was often found to be much too big for many links. The airline business thrives on filling seats, and it was not uncommon that the 747 flew well under capacity. While the DC-10 and L-1011 remedied this problem

(they sat on average 250 passengers at capacity) the added troubles of worldwide recession and the double oil shocks of the 1970s meant that airlines began requesting smaller machines that would also use less fuel. The twin-engine jets that appeared in the early 1980s filled that gap in all ranges.

The first aircraft manufacturer to recognize this niche was the European Airbus consortium, which began to offer the A-300 with up to 250 seats in 1972. Although sales were close to nil at first (analysts suggest the machine came on the market too soon), by the late 1970s a smaller, more versatile model for up to 200 passengers, the A-310, began picking up sales. This third generation jetliner faced as main competition Boeing's new 767 of comparable size. More recent versions of both planes, were used to lengthen the ETOPS time rule (Extended Twin Engine Operations) that required aircraft with fewer than three engines to always be within 120 minutes of an airfield in case of single engine failure. With the advent of other twinjets, both short range and long range, the struggle between Airbus and Boeing has continued to this day, and spread into the realm of fourth-generation airliners.

The latest generation of jetliners, though outwardly similar to earlier ones, is characterized by "glass cockpits" (where liquid crystal displays have replaced dials and buttons), fly-by-wire systems that initially existed on military aircraft alone (computers send electronic signals to the controls), and more efficient engines with more complex wings (to minimize fuel consumption and noise pollution). Such technological advances have contributed to reducing the number of cockpit occupants from four for the first generation jetliners, to three by the second (the navigator disappeared), to two by the latest (the engineer is no longer needed). At the same time, however, the increase in distances covered on some flights (16,000 kilometer stretches are now routine), have made the inclusion of a relief crew essential. At the same time, the increase in passenger numbers, the need for revenue, and the competitive market have replaced the veneer of prestige characterizing early jet travel with visions of delays, poor service, and air traffic congestion.

See also **Aircraft Design; Turbines: Gas, in Aircraft**

GUILLAUME DE SYON

Further Reading

Davies, R.E.G. *A History of the World's Airlines*. Oxford University Press, Oxford, 1964.

Davies, R.E.G. and Birtles, P. *De Havilland Comet.* Paladwr Press, McLean, 1999.

Dierikx, M. and Bouwens, B. *Building Castles of the Air: Schipol Amsterdam and the Development of Airport Infrastructure in Europe, 1916–1996.* Sdu Publishers, The Hague, 1997.

Gunston, B. *Airbus: The European Triumph.* Osprey, London, 1988.

Heppenheimer, T.A. *Turbulent Skies: The History of Commercial Aviation.* John Wiley, New York, 1995.

Jarrett, P. Ed. *Modern Air Transport. Worldwide Air Transport from 1945 to the Present.* Putnam, London, 2000.

Newhouse, J. *The Sporty Game.* Alfred A. Knopf, New York, 1982.

Sabbagh, K. *21st-Century Jet: The Making and Marketing of the Boeing 777.* Scribner, New York, 1996.

Civil Aircraft, Propeller Driven

The advent of early commercial propeller aircraft followed in the wake of World War I, when manufacturers modified bomber models in response to airline demand. Thus, the Farman Company developed the Goliath model, which was used in early Paris to London links beginning in 1919. Along similar lines, using a modified Vickers Vimy, British pilots Alcock and Brown were able to cross the Atlantic that same year. However, their exploits and those of Charles Lindbergh notwithstanding, propeller aviation in the interwar years focused primarily on short and medium range aircraft. As engine and airplane design improved the range of machines, some were used for long-distance mail transport.

With every nation, including defeated Germany, eager to invest in civil aeronautics, hundreds of civil aircraft initially derived from military machines appeared in European skies. The most famous machine built specifically for transport purposes, however, was the German Junkers F-13. As the first all-metal airliner it proved extraordinarily sturdy and, combined with the Junkers Company's heavy investment in foreign airlines, helped spread the use of that machine. Its small size, however (it carried only four passengers), meant that newer models would be required as traffic demand expanded. Other manufacturers such as Fokker supplied very successful machines, including the three-engined F-VII. Other manufacturers offered single-engine machines, like the Latécoère 28 or the Potez 25, both in service on French airmail routes.

In the interwar years, the push to expand aircraft range came primarily from lucrative airmail contracts. In both Europe and the U.S., new solutions were thus devised to solve the dual problem of speed and distance covered. In North America, a mix of indigenous and foreign machines shared the scene, while in Europe, long distance fascination was focused on the Atlantic Ocean.

Early airlinks with South America were run partly by boat, but modified planes like the Latécoère 28 "Comte de la Vaulx" successfully bridged the South Atlantic in 1930, though Germany sought to check French expansion by building an airship service. In addition, the Dornier company tried (but failed) to develop the Giant Dornier DO-X to fly across the Atlantic regularly (it once lifted 169 passengers on a demonstration flight), but had greater success supplying the German Lufthansa with Dornier Wal ("Whale") aircraft, which were refueled after a sea landing by specially positioned ships in the middle of the South Atlantic. Other midrange solutions included the British experimentation with a tandem system involving the Short S-20 "Mercury" being launched from a "Maia" mother plane. With the exception of the German system, most such solutions remained experimental.

By the 1930s, despite European progress in the realm of aircraft such as the Dewoitine 338, the Handley Page Hermes, the Junkers 52 Trimotor and the Focke-Wulf 200 Condor, the primary thrust towards the modern airliner was to be found in the U.S. There, the Ford Trimotor, derived from a freight model, was used for coast-to-coast passenger service in two days beginning in 1929. Four years later, however, the Boeing Company designed and flew what is considered the first modern airliner. Its model 247 was a twin-engine all-metal machine that could carry ten passengers, and flew on average 120 kilometers per hour faster than other airplanes in service, thanks to its improved aerodynamics. Airline and aircraft manufacturing competition, however, led the Douglas Company to take the next step when it agreed to design a twin-engine plane, the DC-1, for TWA. Improvements to this model led to the introduction of the DC-2, which had a considerable range for a serial-built land plane at the time (some 1000 kilometers). Other manufacturers entered the fray, yet in the realm of twin-engine transport Douglas remained an uncontested leader with the introduction of its next model, the DC-3, which could accommodate 21 passengers. Used heavily in World War II, a few surviving examples of the more than 10,500 built remain in service to this day.

In other categories, several manufacturers tried their hand at speed, with Lockheed offering the Orion model, and Heinkel building the He-70. In such cases, however, the design of the machines sacrificed comfort, and proved more useful for mail transport than passenger service.

In parallel to the development of land planes, giant flying boats were studied and tested over long distances. Deemed safer than land planes and faster than airships, these appeared to offer the ideal solution to the problem of long-distance passenger transport. England, France, Germany, and the U.S each witnessed the development of several types of flying boats after the failed DO-X, some with six engines, but World War II interrupted most testing and the projects became military ones. A notable exception was the Boeing 314 model, in service with both Pan American Airways and BOAC, which allowed for a wartime service between the U.K. and the U.S.

During World War II, the main developmental thrust came from the U.S., where the war effort sped up already existing projects. Douglas, which had flown its four-engine DC-4 model in 1938, completed development of the machine, and Lockheed began development of its Constellation model. Consequently after World War II, European airlines eager to catch up bought American planes, despite pressure from their respective governments to support national industries. By 1947, pressurized versions of the Constellation as well as an improved version of the DC-4, the DC-6, and the introduction of the Boeing 377 Stratocruiser signaled further dominance of the long-range market by U.S. machines (the pressurized cabins allowed flights above 3000 meters, thus avoiding most of the dangerous weather systems).

In the medium- and short-range markets, France and Great Britain were able to produce a few piston-engined propeller aircraft, but none was as successful as the DC-3, which remained in use throughout the 1950s and early 1960s.

The propeller era would have likely come to an end with the advent of the jet were it not for the development of the turboprop engine, in which instead of gasoline-powered engines with heavy cylinders and pistons, new, kerosene-fuelled turbines moved the propeller. The result was that while fuel consumption was on average slightly higher than a piston engine, the energy output could be almost double for the same weight (thus increasing speed and payload), and the safety level far greater. Several turboprop airliners were thus developed, including the British Vickers Viscount, the Soviet Tupolev 114, and the American Lockheed Electra. While longer-range turboprops gave way to jets, the short- and medium-range market held their ground, and new models appeared in the 1960s, including the Fokker 27 and the British Aerospace HS-748. In Canada, the De Havilland Twin Otter became a favored machine for its rugged performance and capacity to operate from short runways in cold weather, and gained worldwide success.

Finally, in the 1980s, several models were developed to service the commuter market, including the Swedish-built Saab 340 twin, and its main competitor, the Italo–French ATR 42. Although later developments in the small jet plane market may displace such turboprop planes, their capacity to operate from short runways will keep such machines in the market for years to come, especially in remote areas where airport development is more challenging.

See also **Aircraft Design; Civil Aircraft, Jet Driven; Internal Combustion Piston Engine; Turbines, Gas**

GUILLAUME DE SYON

Further Reading

Davies, R.E.G., *A History of the World's Airlines*. Oxford University Press, Oxford, 1964.

Dierikx, M. *Fokker, a Transatlantic Biography*. Smithsonian Institution Press, Washington D.C., 1997.

Gwynn-Jones, T. *Farther and Faster: Aviation's Adventuring Years, 1909–1939*. Smithsonian Institution Press, Washington D.C., 1988.

Heppenheimer, T.A. *Turbulent Skies: The History of Commercial Aviation*. Wiley, New York, 1995.

Jarrett, P., Ed. *Biplane to Monoplane. Aircraft Development 1919–39*. Putnam, London, 1997.

Jarrett, P. Ed. *Modern Air Transport. Worldwide Air Transport from 1945 to the Present*. Putnam, London, 2000.

Propliner Magazine quarterly, 1979–present.

Stroud, J. European post-war piston engined airliners. *Aeroplane Monthly*, a monthly series, 1992–1995.

van der Linden, F.R. *The Boeing 247: The First Modern Airliner*. University of Washington Press, Seattle, WA, 1991.

van der Linden, F. R. *Airlines and Airmail: The Post Office and the Birth of the Commercial Aviation Industry*. The University Press of Kentucky, Lexington, 2002.

Civil Aircraft, Supersonic

When the jet age began revolutionizing air travel by halving flight time and doubling passenger numbers in the 1950s, supersonic transport (SST) appeared to be the logical next stage in the progress of commercial aviation But while the dynamics and challenges of supersonic flight were clearly defined (the speed of sound, Mach 1, ranges between 1000 and 1200 kilometers per hour)), applying them to civilian transportation remained extremely difficult. The challenges included propulsion, weight distribution (on military aircraft the payload is easily predetermined), passenger comfort and, as it would soon become clear, economic efficiency. Several nations undertook studies of supersonic transport in the late 1950s including the U.S., the Soviet Union, the U.K, and France.

In the U.S., an SST project was announced by President Kennedy in 1963 and saw the involvement of major aerospace giants Boeing and Lockheed. However, while Boeing gained the upper hand, the proposals it offered for the design of a 300-seat aircraft proved too unrealistic. Public and political impatience, combined with skyrocketing costs, prompted a termination of governmental funding in 1971 and the project was abandoned, having reached a total cost of US$1 billion.

In the Soviet Union, too, several designs for SSTs were considered, with Tupolev receiving the go-ahead to design the model 144. First flown on 31 December 1968, ahead of the Anglo-French Concorde, the machine was also the first passenger airliner to exceed Mach 2, and was shown at several Paris Air Shows. However, the machine required considerable redesign between the first and the second prototypes, and several problems, such as cabin noise, were never satisfactorily solved. One prototype crashed at the 1973 Paris Air Show, but testing continued and service with the Soviet airline Aeroflot did happen, albeit briefly, in the late 1970s between Moscow and Alma Ata. The production aircraft were then withdrawn from service and stored, though one saw testing service through a National Aeronautics and Space Administration (NASA) grant in the 1990s.

In the U.K, the impulse towards SST studies began in the mid-1950s, when the Ministry of Supply formed a Supersonic Transport Aircraft Committee. The STAC suggested two types of SST projects, one medium range, one long range, with speeds ranging from Mach 1.3 to 2.0. By 1959, the long-range option was favored as a means to make British aerospace industry more competitive in the jet age, since Boeing had taken the lead over the ill-fated British Comet jetliner. British company Bristol had gained considerable experience in the

jet engine realm, however, by developing the Olympus military motor. The staggering cost projection for such an aircraft, however, meant that when the British government awarded the British Aircraft Corporation (BAC) a development contract in 1960, it recommended that international partners be found to help cushion the investment risk. Openings to American companies revealed a different design philosophy, which projected either a Mach 3 airliner, or a 300-passenger giant using expensive titanium. Across the Channel, however, French state manufacturer Sud Aviation, based in Toulouse, had done studies parallel to those of BAC. Following heavy technical and political negotiations, a deal was signed in 1962 for the joint development of an SST, whereby the engine development was 60:40 in favor of Britain, while the airframe work proportion was reversed in favor of France.

While more financial and political hurdles lay ahead (the project was almost cancelled in 1965 by the British Labour government of Harold Wilson), technical challenges became some of the toughest any aircraft constructor had to overcome. Among these were the extreme temperatures a Mach 2 aircraft would experience while cruising at an altitude high above conventional jetliners (18,000 meters instead of 9,000). This kinetic heating would extend the fuselage some 16 centimeters despite the intense cold, and different sections would heat up to temperatures between 90 and 120°C. The windows consequently had to be designed extremely small to avoid stress problems.

Similarly, the British Olympus engines had to function at a variety of speeds, and provide acceleration with afterburners. They were based on military engines, but required further refinement to carry passengers and control fuel consumption. For example, the air rushing into the engines at Mach 2 would actually be too fast and its shock wave would destroy the powerplant. Engineers came up with an ingenious system of variable size inlets and doors that changed the boundary layer and slowed the air down during cruising speed (Figure 2). The gas dynamics of the

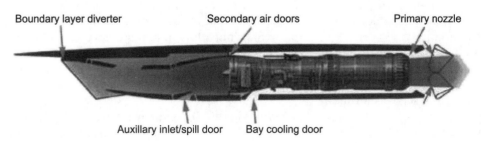

Boundary layer diverter   Secondary air doors   Primary nozzle

Auxillary inlet/spill door   Bay cooling door

Figure 2. Supersonic transport engine.

air also meant that at higher altitudes, drag would be reduced, and the engines would have to do less work. However, reaching the desired altitude involved considerable thrust, and therefore dealing with higher exhaust temperatures. On subsonic aircraft, the exhaust gas necessary to provide the thrust comes out through a narrowing nozzle. In the case of Concorde's Olympus engine, the converging nozzle was assisted by a diverging, or expanding nozzle that provided the extra thrust necessary for the plane to carry a payload.

The question of gas exhaust remained central to the development of the engines. If a Concorde cruises supersonically at about 700 meters per second, then the exhaust gases must leave the engine at least at 900 meters per second. During acceleration and climb, this could be done through afterburning, a process common on military jets, where fuel is ignited into the exhaust gases, producing the necessary energy. This would become the favored solution for take-off and acceleration. It also meant higher temperatures would affect the engines, ranging from $120°C$ at the inlet level, to $550°C$ in the compressor area, followed by even higher temperatures when fuel was burned. This required the use of titanium and nickel alloy for the compressor blades. In the case of Concorde, most of the plane is made of Hiduminium RR58, an aluminum alloy, but some areas, including the engine exhausts and areas surrounding them were made of high-strength steel and titanium; these metals would have been the primary choices had Concorde been designed to reach Mach 3, but their weight made them less likely to be selected.

The aircraft's needle shape impaired cockpit vision considerably, especially on the ground. To allow the crew to maintain visual contact with the runway during take-off and landing, a hinged nose and visor were designed to drop as low as 12 degrees in relation to the cockpit windshield, and fully retract during supersonic cruise. Other innovations included the transfer of fuel during different flight phases from one tank to another. As the plane emptied its tanks, its center of gravity would shift, which would affect its speed. Thus, in addition to the main fuel tanks situated in the wings, forward and rear trim tanks were added as a means to control the center of gravity.

The Concorde project called for the testing of two prototypes and two pre-series aircraft (designed to iron out any last minute details in assembly and flight operations, and serve as airline demonstrators). The projected production was at least 150 machines, and when the prototypes flew in 1969, sixteen major world airlines had taken options on over 70 aircraft. Yet by the time manufacturing closed down in 1979, some 16 production Concordes had been built and delivered to the only two airlines that bought it, British Airways and Air France. The causes for the commercial failure are multiple, but can be traced back to the uneconomical features of the machine, its pollution level, and the oil shocks of the 1970s. As one journalist once remarked, when compared to an early Boeing 747, which first flew the same year as Concorde did, the supersonic flew twice as fast, but consumed four times more fuel to carry four times fewer passengers (about 100). Second, as aircraft testing was underway in the 1970s, new environmental concerns began to appear in the industrialized world, including matters of atmospheric pollution and airport noise abatement. Although Concorde was not significantly noisier than other planes of its time while taxiing at an airport, its noise print at cruising altitude raised such concerns that ultimately it would only be allowed to fly at its maximum speed over oceans. Finally, the oil shocks of 1973 and 1979 provided the proverbial nails in the coffin by making fuel extremely expensive and accentuating a world economic downturn.

At the level of operations, both British Airways and Air France faced a difficult start that involved wrestling American landing rights for the machine, trouble phasing the supersonic into their respective fleets, and selecting which destinations would become money makers. Operations began in 1976, and by the mid-1980s, both airlines were making a profit on the Concorde, though this eventually required limiting their service to scheduled Paris and London to New York and Washington schedules, and luxury charter flights (Figure 3).

The crash of an Air France Concorde in July 2000 appeared to toll the end of the first passenger supersonic era, but following safety modifications to the fuel tanks, the flying certificate was reissued, and both airlines resumed operations. In 2003, however, the airline industry's economic downturn prompted both operators to announce the end of operations that year. As for a "son of Concorde," multiple projects have been announced for years, but thus far they have consistently encountered problems similar to the supersonic pioneer's, including cost and concern for the environment.

See also **Aircraft Design; Civil Aircraft, Jet Driven**
GUILLAUME DE SYON

Figure 3.   Concorde flying over Manhattan (New York City).
[*Courtesy of Adrian Meredith at www.concordephotos.com.*]

Further Reading

Davies, R.E.G. *Supersonic (Airliner) Non-Sense: A Case Study in Applied Market Research*. Paladwr Press, McLean, 1998.

Horwitch, M. *Clipped Wings: The American SST Conflict*. MIT Press, Cambridge, MA, 1982.

Knight, G. *Concorde, The Inside Story*. Stein & Day, New York, 1976.

Moon, H. *Soviet SST: The Technopolitics of the Tupolev-144*. Orion Books, New York, 1989.

Trubshaw, B. *Concorde: The Inside Story*. Sutton Publishing, Phoenix Hill, 2000.

Cleaning: Chemicals and Vacuum Cleaners

Chemicals

During the nineteenth century, rising standards of cleanliness, a new breed of domestic engineers, an increase in the number of objects people owned, and the emergence of the "germ theory" influenced the ideology and practice of cleaning. Still, throughout the twentieth century, people continued to clean with many of the same chemicals as had their predecessors in earlier centuries. The distribution of piped hot water and electrical current contributed as much to a shift in cleaning practice as any change in the nature of chemicals. Twenty-two percent of British households lacked piped hot water as late as 1961. The impact of soaps and cleaning agents that require hot water for effective use must not be overestimated.

Before the twentieth century, many cleaning agents were made at home. While manufactured pastes, liquids and polishes were available by the early nineteenth century to those who could afford them, both the novelty and the effectiveness of these have been questioned. Soda played a significant role in housecleaning in the twentieth century, in part because it was cheaply and easily acquired after Frenchman Nicholas Leblanc developed an efficient way of mass producing it from salt in the late eighteenth century. Historian Susan Strasser claims that only two cleaning agents were manufactured before 1880. Besides soda, twentieth century women continued to rely on sand, milk, salt, borax, camphor, lye, vinegar, turpentine, clay, acids, and oils.

The spread of dishwashing and laundry chemicals followed the introduction and mass manufacture of the dishwasher and washing machine, most prominently in Europe and North America after World War II. These products were often simply refinements of existing ones. Liquid soap was introduced by Minnetonka in the 1980s. In the latter three decades of the twentieth century, manufacturers responded to consumer concerns about the effect of their products on the natural environment. Chlorine, petrochemicals and phosphates were among the agents avoided in the new environmentally friendly cleaning products.

Vacuum Cleaners

The vacuum cleaner was a product of a desire for greater cleanliness and arguably one of the technologies that most abetted greater cleanliness. Along with the electric iron, the vacuum became one of most widely owned household appliances in the twentieth century. Historians agree, however, that the dissemination of the vacuum barely affected time spent in cleaning. Yearly or bi-yearly outdoor carpet and curtain shakings were replaced by the daily or weekly use of the vacuum, as standards of cleanliness rose. The history of the vacuum cleaner remains "an important example of the commercial application of the phobia against dirt" (Forty, 1982).

The portable electric vacuum did not appear for sale until the first decade of the twentieth century. By the middle of the nineteenth century, portable appliances that sucked particles by means of bellows and others with both air draft and revolving brushes had been patented. Those who could afford to, sent their carpets out to be professionally cleaned. Steam-powered industrial machines simply beat the carpet with rubber beaters. Later, machines employed steam to destroy insects and a rotary fan to blast dust out of the carpet and up a chimney. An American machine that blew compressed air through carpets inspired British civil engineer William Booths, who patented a "suction machine" in 1903. Booth's "Puffing Billy," which could be installed permanently in a building or mounted on a trolley and pulled by horses and automobiles, was exceptional among dozens of similar inventions in that it was the only one with a self-contained power source.

American David T. Kenney had developed a similar machine in the U.S., and its installation as a central vacuum in the Frick house in Manhattan set a precedent which was followed in both domestic and commercial applications. Central vacuums would not achieve widespread use, however, because their capital and installation costs were prohibitive. In addition, most American or European households had not been electrified at the time of its invention. Chapman and Skinner in San Francisco invented the first "portable" electric vacuum in 1905. It weighed 42 kilograms and used a fan 45 centimeters in diameter to produce the suction. Because of its size, it did not sell well. By 1908 consumers could purchase the heavy (18 kilograms) but portable Hoover Model O Electric Suction Sweeper, for $75.00, a considerable sum at the time. It consisted of a tin body, filled with a fan, a motor and a rotating cylindrical brush. Behind the motor, a bag attached to the handle received the refuse. This model included a flexible nozzle that could be fitted to the machine and used to vacuum upholstery. By 1926, the motor and parts were much lighter and a beating mechanism had been added to assist the brushes and suction. By this time, the machine was known simply as the "Hoover." A second type of portable electric unit made by the Swedish Electrolux Company eliminated the rotating brushes from the design and used only suction. The Model O and the Electrolux provided the functional and design paradigms for vacuum cleaners until the end of the twentieth century.

Improvements and refinements through the end of the twentieth century included more efficient and quieter motors; lighter (and later recyclable) plastic parts; disposable dust bags; flexible cord winders (electrical cords that could be "sucked" back into a neat bundle); microfilters; and streamlined designs.

In 1979, Black and Decker Company introduced the cordless vacuum, the battery-operated and soon very popular "Dustbuster." British designer and engineer James Dyson developed a bagless vacuum, which sucked and filtered dust into a plastic container that replaced the bags that Dyson claimed became clogged with dust and prevented efficient absorption. After marketing what became the fastest-selling vacuum in Britain, Dyson began working on a robotic vacuum.

See also **Detergents; Dishwashers; Laundry Machines and Chemicals**

JENNIFER COUSINEAU

Further Reading

Cowan, R.S. *More Work for Mother: The Ironies of Household Technology from the Open Hearth to the Microwave.* Basic Books, New York, 1983.

Davidson, C. *A Woman's Work is Never Done: A History of Housework in the British Isles 1650–1950.* Chatto & Windus, London, 1982.

Forty, A. *Objects of Desire: Design and Society from Wedgwood to IBM.* Pantheon Books, New York, 1986.

Gideon, S. *Mechanization Takes Command: a Contribution to Anonymous History.* W.W. Norton, London, 1948.

Hardyment, C. *From Mangle to Microwave: The Mechanization of Household Work.* Polity Press, Oxford, 1988.

Lupton, E and Miller, J.A. *The Bathroom The Kitchen and the Aesthetics of Waste: A Process of Elimination.* Princeton Architectural Press, New York, 1992.

Lupton, E and Miller, J.A. *Mechanical Brides: Women and Machines From Home to Office.* Princeton Architectural Press, Princeton, New Jersey, 1993.

Nye, D. *Electrifying America: Social Meanings of A New Technology.* MIT Press, Cambridge, MA, 1990.

Strasser, S. *Never Done: A History of American Housework.* Pantheon Books, New York, 1982.

Yarwood, D. *Five Hundred Years of Technology in the Home.* B.T. Batsford, London, 1983.

Clocks and Watches, Quartz

The wristwatch and the domestic clock were completely reinvented with all-new electronic components beginning about 1960. In the new electronic timepieces, a tiny sliver of vibrating quartz in an electrical circuit provides the time base and replaces the traditional mechanical oscillator, the swinging pendulum in the clock or the balance wheel in the watch. Instead of an unwinding spring or a falling weight, batteries power these quartz clocks and watches, and integrated circuits substitute for intricate mechanical gear trains.

When quartz timepieces first hit the market, it seemed unlikely that the expensive gadgets would sell. Instead they won over consumers and revolutionized the way timepieces are made, sold, and used. Today, quartz is the most common source of time and frequency signals, not only in clocks and watches but also in scientific instruments and in other consumer products like computers, cell phones, and television sets.

The technology that made quartz a practical time base for domestic clocks and watches developed from several independent research streams that stretched over nearly a century. Quartz, the mineral silicon dioxide, is one of Earth's most common materials. It has special properties, as the Curie brothers, Pierre and Jacques, had shown in the late nineteenth century. When subjected to electrical voltage, quartz vibrates, many thousands of times a second at a regular rate that is dependent on how it is cut and shaped (this phenomenon is known as piezoelectricity). During the 1920s, amateur radio operators, many of whom made their own crystal units, demonstrated the potential of quartz for controlling broadcast frequencies. At the same time, industrial scientists were also investigating the properties of quartz for frequency control.

The first quartz clock, constructed in 1927, was nearly as big as a room. Canadian telecommunications engineer Warren Marrison and his colleague J. W. Horton at Bell Telephone Laboratories in New York City developed the basic system that subsequent quartz timepieces would employ. To keep time, the clock counted the vibrations of a quartz crystal in an electrical circuit, subdivided those vibrations—the equivalent of a mechanical clock's "beats"—to minutes, seconds and hours that were displayed on a dial. In building his clock, Marrison had been searching for ways to monitor and maintain precise electromagnetic wave frequencies that carry radio and telephone messages. When he found that the crystal's vibrations were stable enough to hold electromagnetic waves to a particular frequency, he also realized the potential of his invention for improved timekeeping.

After Marrison, others built quartz clocks and demonstrated repeatedly that they were more accurate than the best mechanical clocks. During the 1930s and 1940s, influential research institutions throughout the world—including the Naval Observatory and the National Bureau of Standards in the U. S., the Royal Greenwich Observatory and the British Post Office in the U. K., and the Physikalisch-Technische Reichsanstalt in Germany—installed quartz clocks. These clocks remained rare, expensive and experimental scientific instruments through the 1940s, when scientists began developing atomic standards even more

accurate than quartz. At the same time, commercial quartz standards, built by such firms as General Radio Company of Cambridge, Massachusetts, served the needs of those with slightly less precise time and frequency needs than the world's elite laboratories—mainly broadcast facilities and electrical power stations.

World War II laid the groundwork for the postwar quartz revolution in the global watch industry. Because of its neutral status, Switzerland continued to make timepieces while countries engaged in the war converted their watch factories to supply military customers. As a result, Switzerland responded to high civilian demand, exported cheap mechanical movements throughout Europe and the U.S., and gained a seemingly indomitable lead in the global watch industry. American watch companies—like Hamilton, Elgin, and Bulova—survived the war years through military contracts or by assembling cheap Swiss movements into American-made cases. Meanwhile, watch production in Japan nearly ceased altogether. After the war, only a few watch firms remained worldwide, and those seeking to challenge the Swiss for a share of the market began to investigate alternative technologies potentially more accurate than the best mechanical watches.

World War II anticipated the postwar quartz revolution in another way. The enormous demand for radio communications during the war stimulated both the production of quartz oscillator units and research into the material's characteristics and behavior. This established quartz as a viable source of time and frequency signals. But quartz units and wartime electronics technology based on vacuum tubes were both still too large and power-hungry for watches and domestic clocks. The postwar advent of transistors and integrated circuits ultimately provided components small enough to be alternatives to gears, springs and tubes.

The history of the modern electronic wristwatch began in the 1960s when, in pursuit of more accurate timepieces, teams of engineers—working independently in Japan, Switzerland and the U.S.—used newly created microelectronic components to completely reinvent the wristwatch. The three teams that ultimately brought the first quartz watches to market took three completely different approaches to the task.

At K. Hattori & Company (now Seiko Corporation) in Japan, a group formed in 1959 began by building a quartz clock, and through successive products, miniaturized their quartz timekeepers to the size of a chronometer, a pocket watch and ultimately a wristwatch. Their Seiko

Astron SQ, introduced in Tokyo on Christmas Day 1969, was the first quartz watch sold anywhere in the world.

The Swiss were only a few months behind. In 1962 the industry had founded a new research laboratory, Centre Electronique Horloger (CEH) in Neuchâtel, to develop a new kind of electronic watch. By 1967 researchers there had two kinds of working quartz watch prototypes, which, along with Seiko's, made their debut at time trials at the Neuchâtel Observatory in 1967. The first Swiss quartz watches for sale—each of which contained an electronic module designed at CEH and dubbed Beta 21—were available on April 10, 1970 under the brand names of nearly 20 different watch companies.

The Americans were even further behind. The firm responsible for the first quartz watch produced in the U.S., the Hamilton Watch Company of Lancaster, Pennsylvania, launched its quartz watch project in 1967. Because the Swiss and Japanese had a clear lead in quartz analog watches, those with the traditional dial and hands, Hamilton concentrated its efforts on inventing the first solid-state digital watch. Partnering with Electro-Data, a Texas electronics company, Hamilton brought out the Pulsar in 1972. At the push of a button, the watch displayed the time of day with flashing red numerals made from light-emitting diodes (LEDs). It cost $2100—the price of a small car.

Domestic quartz clocks appeared on the market about the same time as the Pulsar, and, even though they were larger than watches, they too depended on the miniaturization of electronics. The earliest examples came from the German firms Junghans, Kienzle, and Staiger, the British firm F. W. Elliot and, in the U.S., General Time's Westclox brand. Already by the mid-1970s, quartz timepieces began to shift from expensive rarities to ubiquitous low-cost timepieces.

The introduction of the first quartz watches disrupted the entire global watch industry. Established watch firms were slow to respond, and new U.S. semiconductor companies—Texas Instruments, Fairchild, and National Semiconductor, for example—began making huge numbers of digital quartz watches, with both LED and liquid crystal displays (LCDs). Competition quickly forced prices down. In Japan, Seiko, Citizen and Casio quickly invested in new electronic watch equipment, and for a time they led in worldwide watch production. By 1978 Hong Kong, concentrating on low-end modules that cost pennies to make, exported the largest number of electronic watches worldwide. Both traditional watch manufacturers and the U.S. semiconductor companies competing in the consumer electronics market were irreparably damaged. The Swiss watch industry suffered near-fatal decline, reorganized completely, and became competitive again in the 1980s, buoyed by an entirely new product—the Swatch.

Consumers also played a role in this quartz revolution. The earliest quartz watches had, along with high prices, significant shortcomings. Watch buyers complained about short battery life, large clunky cases, and technical breakdowns caused by hard knocks and moisture. LED watches were difficult to read in daylight, and LCDs could not be seen at night. Interactions with users over these issues forced manufacturers to improve their products.

Consumers weighed in on the way the new watches displayed time too. The electronic wristwatch evolved almost overnight from a rare gadget for rich men to a cheap throwaway timepiece for men, women, and even children. Easy availability coupled with the introduction of digital watch displays, radically different in appearance from the traditional dial and revolving hands, generated passionate responses from consumers, both positive and negative. Some serious opponents of digital displays in the 1970s even predicted the day would come when the traditional dial would vanish completely and our ways of knowing time would be irretrievably diminished. The debate over the relative suitability of digital versus analog displays continues to this day. Since the mid-1980s, sales figures have shown most buyers prefer the analog dial over the digital display.

By the 1990s nearly all new watches and clocks made were electronic. As a result, nearly everyone has access, whether we need it or not, to the split-second accuracy once available to only scientists and technicians.

See also **Integrated Circuits; Transistors**

CARLENE E. STEPHENS AND MAGGIE DENNIS

Further Reading

Glasmeier, A.K. *Manufacturing Time: Global Competition in the Watch Industry, 1795–2000*. The Guilford Press, New York, 2000.

Johnstone, B. *We Were Burning: Japanese Entrepreneurs and the Forging of the Electronic Age*. Basic Books, Boulder, 1999.

Landes, D. *Revolution in Time: Clocks and the Making of the Modern World*. The Belknap Press of Harvard University Press, Cambridge, MA, 2000.

Marrison, W. The evolution of the quartz crystal clock. *Bell Sys. Tech. J.*, 27, 510–588, 1948.

Stephens, C. and Dennis, M. Engineering time: inventing the electronic wristwatch. *Br. J. Hist. Sci.*, 33, 477–497, 2000.

Weaver, J.D. *Electrical and Electronic Clocks and Watches.* Butterworth, London, 1982.

Clocks, Atomic

Atomic clocks rely on the precise timing of the oscillation frequency of certain atoms which are energized or excited to another energy state. The near immutability of the energy levels in the atomic structure of certain paramagnetic elements provides atomic clocks with the inherent reproducibility and long-term stability previously lacking in material standards such as quartz clocks or the Earth's rotation period. Some ten thousand (commercial) atomic clocks are now in operation throughout the world ensuring essential control and synchronization in a wide range of applications in science and technology, such as in the NAVSTAR satellites used in the global positioning system (GPS). The readings of a subset of more than 250 clocks maintained in about 60 national institutions are combined with those of about 12 laboratory clocks to form International Atomic Time (TAI), the basic time scale for science and technology and whose derivative, Coordinated Universal Time (UTC), which differs from TAI by an integral number of seconds, provides the atomic equivalent of mean solar time for general use worldwide.

The principle underlying atomic clocks is that when individual atoms of the same element absorb or emit energy by electrons moving from one energy level to another, the radiation produced by individual atoms has exactly the same (quantized) frequency. These resonant frequencies, stable in time and space, are which makes the atoms perfect timepieces. The oscillations are changed by environmental factors such as humidity that would decrease the accuracy of normal clocks.

The main sources of atomic timekeeping are cesium and rubidium clocks with a smaller, but significant, contribution from active hydrogen maser clocks. In the atoms of all three elements the ground state (lowest-energy state) is split into two hyperfine levels by the interaction between the spin (and consequent magnetic moment) of the single valence electron and the spin of the atomic nucleus. The energy difference, $\triangle E$, between these two closely separated states determines a resonant or characteristic frequency v given by $v = \triangle E / h$, where h is Planck's constant. Transition from the top to the lower state causes emission of a photon of that frequency. The states are nearly equally populated and the necessary unbalance; that is, alteration of the distribution of atoms in a given state, can be achieved by either magnetic state selection involving spatial separation of the atoms (selecting only those in one hyperfine state) or optical pumping using differential transfer to excited states.

The magnetic approach was first employed in the 1920s when Otto Stern and Walther Gerlach in Germany made use of high-gradient magnetic fields to deviate a beam of (silver) atoms, each hyperfine state being oppositely deflected, confirming space quantitization. The method was subsequently improved in the late 1930s by Isodor Rabi at Columbia University who added a radio-frequency field to stimulate transitions between the hyperfine states and thus brought together the elements of a magnetic resonance machine that could function as a frequency standard. Those atoms that are stimulated to change state emit light. The frequency of the radio-frequency field is varied, and eventually, a frequency is achieved that alters the states of most of the atoms and maximizes their fluorescence. A counter counts the radio-transmitter pulses, which gives a precise frequency standard that can be used to define a time standard.

In 1949 Rabi's colleague Polykarp Kusch produced a basic design concept for a cesium atomic clock, his proposals incorporating a Ramsey cavity, following the discovery in the same year by Norman Ramsey at Harvard University that coherent excitation applied at the beginning and end of an interval of beam travel was equivalent to atom-field interaction over the whole interval, thereby reducing substantially the width of the resonance line.

In the following years work began at several laboratories on cesium beam resonators based on the Kurch design. In the U.S. Harold Lyons at the (then) National Bureau of Standards achieved a linewidth of 300 hertz in 1952 with Ramsey excitation. At the Massachusetts Institute of Technology (MIT) Jerrold Zacharias developed a compact, transportable clock in 1954 which would form the basis for the first commercial production in late 1956. Meanwhile, at the U.K. National Physical Laboratory (NPL), Louis Essen and John Parry, largely following the Kurch recipe, brought a cesium beam standard into operation in June 1955. This date represents the start of atomic timekeeping for, unlike the U.S. developments the NPL resonator was linked to an existing quartz clock ensemble and the resulting atomic clock could be routinely compared with astronomical time. The calibration of the cesium frequency in terms of Ephemeris Time over a three-year period led in 1967 to the formal redefinition of the second as 9,192,631,770 periods of the cesium-133 hyperfine transition frequency.

The present uncertainty in the realization of the second by laboratory standards making use of magnetic selection is about part in 10^{14}, equivalent to nearly 1 nanosecond (10^{-9} seconds) per day while the commercial clock contributing to TAI display stabilities of the same order. The essential features of a commercial resonator are shown in Figure 4. Atoms traveling at about 200 meters per second are selected by magnet A in one of the hyperfine levels and make the transition ($F = 4$, $m = 0$)($F = 3$, $m = 0$), the Zeeman substates $m = 0$ having least dependence on the low magnetic field in the C region. Thereafter, the atoms are directed by magnet B to a surface ionization detector, evaporating as positive ions which provide the servo-output to drive the frequency of the exciting oscillator to the peak of the Ramsey pattern.

Alfred Kastler in 1950 proposed a scheme of optical pumping which would be fully exploited only many years later with the advent of diode lasers. However, the partial overlap in the spectra of the rubidium isotopes ^{85}Rb and ^{85}Rb enabled optically pumped rubidium gas cell frequency standards and clocks to be realized in the late 1950s. The main features are shown in Figure 5. The two D lines at wavelengths of 780 and 795 nanometers in the ^{87}Rb lamp in (a) are differentially absorbed in the ^{85}Rb filter cell (b), the residual light allowing transitions only from the F=1 hyperfine level (c) in the vapor cell, thus depopulating the level and making the cell more transparent. Microwave excitation at the hyperfine frequency equalizes the hyperfine populations again, resulting in increased light absorption and a consequent fall in detector output, thereby providing a control signal to the microwave source. A buffer gas in the cell shield the rubidium atoms from depolarizing contact with the cell walls and largely eliminated Doppler broadening of the resonance line.

In the 1950s Zacharias at the Massachusetts Institute of Technology had constructed a vertical cesium resonator with the intention of achieving a long Ramsey interval using slow atoms interrogated as they rose and fell under gravity through a single cavity. He was not successful using thermal atoms but the "Zacharias fountain" was finally realized by André Clairon and his colleagues at the Paris Observatory in 1993, making use of atoms cooled to a few microkelvin, state-selected by optical pumping and then periodically projected upwards at speeds of only a few meters per second to give a resonance width of about 1 hertz. Several fountain clocks are now in operation and

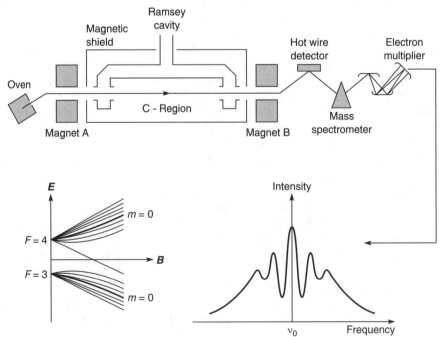

Figure 4. Features of a commercial resonator.
[*Source: Audoin, C. and Vanier, J. Atomic frequency standards and clocks, J. Phy. E: Sci. Instrum., 9, 697–720, 1976.*]

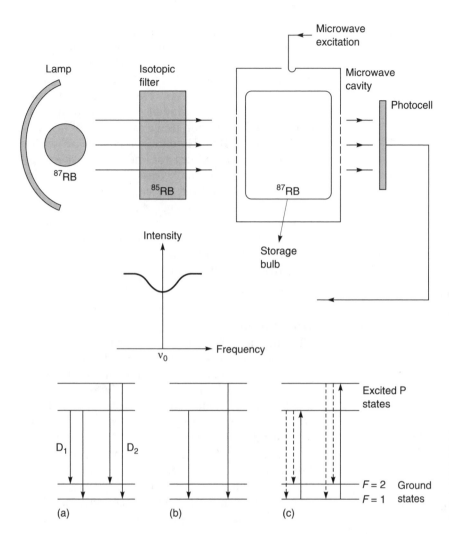

Figure 5. Features of optically pumped rubidium gas cell frequency clocks.
[*Source: Audoin, C. and Vanier, J. Atomic frequency standards and clocks, J. Phy. E: Sci. Instrum., 9, 697–720, 1976.*]

contributing to TAI with an uncertainty of about 1 part in 10^{15} equivalent to 100 picoseconds (10^{-12} seconds) per day.

The clocks so far considered have been passive devices requiring external excitation. The atomic hydrogen maser, shown in outline in Figure 6, first operated in Ramsey's laboratory at Harvard in 1960. It produces a signal at about 1420 megahertz albeit at the low level of around 10^{-12} watt.

Magnetic state selection focuses atoms in the $F = 1$, $m = 0.1$ states into the bulb immersed in a low-loss cavity resonator. A film of Teflon applied to the wall of the bulb allows atoms to make thousands of contacts with the wall while giving up energy to the cavity through the $(F = 1, m = 0) \rightarrow (F = 0, m = 0)$ transition. If this energy exceeds the cavity loss self-sustaining oscillations will result. The hydrogen maser has a short-term stability approaching 1 part in 10^{16} and about 50 hydrogen maser clocks now contribute to TAI.

See also **Clocks and Watches, Quartz; Global Positioning System (GPS)**

JAMES McASLAN STEELE

Further Reading

Audoin, C. and Vanier, J. Atomic frequency standards and clocks. *J. Phys. E. Sci. Instrum.* 9, 697–720, 1976. Still an excellent review of the subject at some remove in time and the source of Figures 4, 5, and 6.

Audoin, C. and Guinot, B. *The Measurement of Time*, Cambridge University Press, Cambridge, 2001. Original French edition, *Les Fondements de la Mesure du Temps: Comment les Fréquences Reglent le Monde*, Masson, Paris, 1998. A physicist and an astronomer combine to provide a comprehensive survey of the field.

Forman, P. Atomichron®: the atomic clock from concept to commercial product. *Proc. Inst. Electr. Electr. Eng*, 73, 7, 1181–1204, 1985. A detailed and fascinating account of the genesis of the first commercial cesium clock.

Jones, T. *Splitting the Second: The Story of Atomic Time*. Institute of Physics Publishing, Bristol, 2000. Sponsored

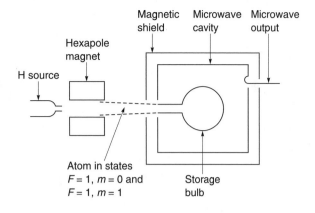

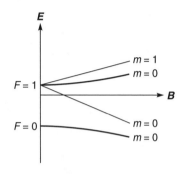

Figure 6. Atomic hydrogen maser.
[*Source: Audoin, C. and Vanier, J. Atomic frequency standards and clocks, J. Phy. E: Sci. Instrum., 9, 697–720, 1976.*]

by NPL; an excellent book for anyone interested in modern timekeeping.

Ramsey, N.F. *Molecular Beams*. Clarendon Press, Oxford, 1956. The classical reference work.

Ramsey, N.F. History of atomic clocks. *J. Res. Nat. Bureau of Stand.*, 88, 5, 301–320, 1983. An account of developments in some of which the author played a major role.

Vanier, J. and Audoin, C. *The Quantum Physics of Atomic Frequency Standards*, vols 1 and 11. Institute of Physics Publishing, Bristol, 1989. A magisterial contribution to the subject in over 2000 pages.

Cloning, Testing and Treatment Methods

Definitions

The use of the word "cloning" is fraught with confusion and inconsistency, and it is important at the outset of this discussion to offer definitional clarification. For instance, in the 1997 article by Ian Wilmut and colleagues announcing the birth of the first cloned adult vertebrate (a ewe, Dolly the sheep) from somatic cell nuclear transfer, the word clone or cloning was never used, and yet the announcement raised considerable disquiet about the prospect of cloned human beings. In a desire to avoid potentially negative forms of language, many

prefer to substitute "cell expansion techniques" or "therapeutic cloning" for cloning. Cloning has been known for centuries as a horticultural propagation method: for example plants multiplied by grafting, budding, or cuttings do not differ genetically from the original plant. The term clone entered more common usage as a result of a speech in 1963 by J.B.S. Haldane based on his paper, "Biological possibilities for the human species of the next ten-thousand years." Notwithstanding these notes of caution, we can refer to a number of processes as cloning. At the close of the twentieth century, such techniques had not yet progressed to the ability to bring a cloned human to full development; however, the ability to clone cells from an adult human has potential to treat diseases.

DNA Cloning

Cloning may also refer to the process of creating identical segments of DNA. The development of DNA sequencing and recombinant DNA techniques in the 1970s meant a gene of interest could be located and extracted from an organism, its DNA strand cut and cloned, and spliced into another "transgenic" organism. To create many copies of the desired DNA segment for gene splicing or for DNA testing, the polymerase chain reaction (PCR), invented by Kary Mullis in 1985, is used. The PCR technique mimics the natural DNA replication that takes place during cell division.

Embryo Splitting (Blastomere Separation)

Embryo splitting is a method whereby the cells of a very early embryo (two-, four-, or eight-cell stage) are separated out to continue developing individually. Since at this stage each blastomere is "totipotent," the undifferentiated cell will continue to divide, then differentiate into functionally specific cell types, and eventually develop into a complete organism. Each of the resulting organisms will be genetically identical. This process of embryo splitting is, in many respects, a technically induced means identical to that by which identical (monozygotic) twins occur naturally. The technique was first demonstrated in 1902 by the German embryologist Hans Spemann who divided a two-celled salamander embryo into two. Each cell grew to adulthood.

In the context of human reproductive fertility, this form of cloning was first used experimentally in 1992 when researchers split the cells of 17 chromosomally abnormal human embryos to determine if development would continue. While there was no intention to reimplant the embryos to develop *in vivo*, it nevertheless demonstrated the application of the technique in human reproduction.

There are, however, important differences between embryo splitting and the nuclear transfer method discussed below. The primary difference is that embryo splitting requires that an organism has already begun developing following fertilization in which the two separate gametes of parents are combined. The resulting split embryos are therefore "clones" of each other and not clones of either parent. Somatic nuclear transfer on the other hand can be used to produce a genetic replica of an already known adult being.

Nuclear Transfer

The concept of nuclear transfer was first proposed by Spemann in 1936 who suggested that it may be possible to take the nucleus of a cell, containing the full complement of genes necessary for development, and transfer it into an ovum from which the nucleus had been removed.

However, it was not until 1952 that the principle was demonstrated by Robert Briggs and Thomas J. King when they applied nuclear transfer to frogs. In their subsequent experiments throughout the 1950s and 1960s, they found that the procedure was only successful where the nuclei was taken from early embryonic undifferentiated stem cells. The use of adult and differentiated cells on the other hand resulted in very rare successes, where resulting tadpoles developed abnormally, if at all. This suggested that it was not possible to reverse the specialized differentiation of cell nuclei such that they could be induced to develop successfully into a complete adult organism.

However, the early 1960s experiments by John Gurdon seemed to suggest that adult differentiated cells taken from the intestines of South African frogs could be used to produce clones. The claim was subsequently contested on the basis that undifferentiated stem cells, which are naturally present in adult tissues, may have been mistakenly used instead.

Similar work in mammals has progressed far more slowly than in amphibians. In 1984, the Danish scientist Steen Willadsen successfully applied nuclear transfer to sheep for the first time by deriving nuclei from early embryo cells (stem cells). The next key development in the field occurred in 1996 when Ian Wilmut's team at the U.K.'s Roslin Institute announced the birth of Dolly, claiming that the sheep had been cloned

using the nuclei of an adult somatic cell rather than that of an embryonic stem cell. The nuclei had been treated to reverse the process of cell specialization. The technical shift from embryonic to somatic nuclear transfer holds the possibility of much greater flexibility in the production of cloned and transgenic mammals. By using cultured (adult) somatic cells, researchers need not be overly restricted by having to rely on a limited number of available human embryos (from aborted fetuses or unused embryos from fertility treatment).

Cloning for Therapeutic Applications

International policymaking in the late 1990s sought to distinguish between the different end uses for somatic cell nuclear transfer resulting in the widespread adoption of the distinction between "reproductive" and "therapeutic" cloning. The function of the distinction has been to permit the use (in some countries) of the technique to generate potentially beneficial therapeutic applications from embryonic stem cell technology whilst prohibiting its use in human reproduction. In therapeutic applications, nuclear transfer from a patient's cells into an enucleated ovum is used to create genetically identical embryos that would be grown *in vitro* but not be allowed to continue developing to become a human being. The resulting cloned embryos could be used as a source from which to produce stem cells that can then be induced to specialize into the specific type of tissue required by the patient (such as skin for burns victims, brain neuron cells for Parkinson's disease sufferers, or pancreatic cells for diabetics). The rationale is that because the original nuclear material is derived from a patient's adult tissue, the risks of rejection of such cells by the immune system are reduced.

See also **Fertility, Human; Genetic Engineering, Methods**

NIK BROWN

Further Reading

Kolata, G. *Clone: The Road to Dolly and the Path Ahead.* Penguin, London, 1997.

Lauritzen, P., Ed. *Cloning and the Future of Human Embryo Research.* Oxford University Press, Oxford, 2001.

Wilmut, I., Schnieke, A.E., McWhir, J. *et al.* Viable offspring derived from foetal and adult mammalian cells. *Nature*, 385, 810–813, 1997.

Wilmut, I. Cloning for medicine. *Scientific American*, 279, 6, 58–63, 1998.

Coatings, Pigments and Paints

The development and application of pigments, paints, and coatings have been an integral part of human development from Paleolithic cave paintings, the art of early civilizations, and protection of buildings from rain. During the twentieth century, understanding of chemicals and the manufacturing need for high-quality decorative and protective coatings drove rapid progression in paint technology. All paints employ the same basic ingredients: pigment to provide color; a medium to bind or suspend the pigment, including emulsions such as resins or oils; and a solvent carrier, which acts to wet the surface to ensure adhesion and thins the resin to make it easy to apply.

Early pigments such as those used in Minoan frescoes and Anasazi rock art and the use of henna as body paint originated primarily from natural sources. Clays, mineral pigments such as iron or chromium oxide, vegetable dyes, animal sources such as shells or urine, as well as precious metals and gems gave rise to a broad selection of pigments that were often unique to one region (e.g., lapis lazuli) and thus highly prized as trade items. But value wasn't limited merely to pigments. Preservation of painted surfaces required protective resins. The most successful early coating was lacquer, used in China since at least the 1300 BC. Lacquer was processed—using a highly guarded secret formula—from resin from the lacquer tree (*Rhus vernicflua*). Shellac, produced from the gum secreted by an insect native to India and southern Asia, makes a varnish when mixed with acetone or alcohol and was used from the eighteenth century. Natural plant resins dissolved in oil or solvent such as turpentine were used in the nineteenth century—evaporation of the solvent leaves a lacquer coating.

Although mineral oxides were often naturally found in soft clays or chalks, binders are needed to enable the pigment to disperse over the surface evenly. Traditional binders such as gum arabic and other natural resins, beeswax, glycerin, egg, animal glue, and linseed oil have been used for centuries. Pigment–oil mixtures enhanced the pigment, increasing brilliance, translucence, and intensity of color. By the early twentieth century, chemists understood more about the physical properties of different pigments, providing them with more strategic approaches to improving binders and carriers.

By the mid-nineteenth century, use of synthetic pigments and dyes as alternatives to natural plant or mineral pigments had increased. A wide range of cadmium yellows and reds were soon available.

By 1930 titanium dioxide, a bright white pigment with great covering power derived from titanium ore, was successfully applied to paint. Its fine particles were easily suspended in linseed oil.

Until the mid- to late 1800s paint was generally prepared onsite by painters themselves. Powders could be mixed with water (as in distemper or whitewash) or white lead paste was mixed with linseed oil. Hand-ground pigments were often coarse, and dispersal by hand resulted in unevenness of color. In 1867, D.R. Averill of Ohio patented the first ready-mixed paints in the U.S., but did not market the paints. Mass produced ready-mixed paints were developed by the Sherwin-Williams Company by 1880 and sold in tins in a wide range of colors.

By the early twentieth century, in the quest to make more durable paints and coatings for a growing manufacturing industry, cellulose became a critical ingredient in synthesized lacquer. Cellulose acetate, which is produced from a cellulose source such as paper or cotton, is mixed with acids and dispersed in solvents to form a low-viscosity lacquer. By 1928 the development of a nitrocellulose solution with a dense pigment base and low viscosity became the paint of choice for the mass production of planes and automobiles, coinciding with the start of spray application.

Research in synthetic resins and polymers led to the use of alkyd and amino resins in coatings and varnishes from the 1920s. Alkyd resins, derived from glycerol processed from animal and vegetable fats, were introduced in paints in 1927. Durable, flexible, inexpensive to produce, and with strong adhesion to most surfaces, alkyd resins were popular as industrial coatings, for example glossy finishes in the automotive industry. Urea and melamine formaldehydes were used from the 1930s in industrial and decorative laminates.

World War II stimulated discoveries in the paint industry. By the end of the war, linseed oil and the solvents that were used to cut it became scarce. Thus from the 1940s and 1950s alkyd resins began to replace linseed oil as a binder in paints. Wartime scarcity of rubber also stimulated research into synthetic latex, and a byproduct was the introduction of latex as a waterborne binder in paints. The first synthetic latex (styrene-butadiene) emulsion paints were introduced for architectural use in 1948, and were also quickly adopted, replacing oil and solvent-based paints. By the 1960s other polymers, vinyl acetate and methyl methacrylate (acrylics) gradually replaced styrene-butadiene latex as paint binders, becoming widely used in household paints. Acrylic had been used as a fast-drying automobile lacquer from the 1950s, replacing nitrocellulose lacquers, and from the 1960s was also used in automobile enamel paints. The acrylic enamel paints had drying times competitive with lacquers, improved gloss and durability, and improved resistance to ultraviolet (UV) damage. Powder coatings, developed in the 1960s as an alternative to liquid paints, are applied dry as an acrylic, silicone, or polyester resin powder and form a coating on heating by cross-linking of polymers. From the 1990s teflon (polytetrafluoroethylene) was used as a painted sealant on aircraft and automobiles.

In the 1980s polyurethane paint binders and clear varnishes evolved, using a polymer formed by reacting an isocyanate group with another group, often a hydroxyl group. An alkyd resin reacted with a polyisocyanate forms the common polyurethane varnish used in residential applications. In industrial applications, the coating reacts and cross-links in one of several ways: on application and exposure to moisture from the air, by heat curing, by mixing with a catalyst, or by heat activation of a latent catalyst. Cross-linking improves the toughness of the resin.

By the 1960s, coatings and paints were driven by chemistry, with new materials created by research groups answering the need for better, stronger, and cheaper coatings for plastic, paper, and of course, wood and metal. Formaldehyde coatings are also widely used for permanent press fabrics. Paints continue to evolve, with non-glare, UV-resistant paint, paint that deadens sound, and acrylics that absorb stains.

Environmental Concerns

Lead paint is durable, cheap to produce, and widely used as interior and exterior paint, despite its known toxicity. Industry consensus standards limiting the use of lead pigments date back to the 1950s, when titanium dioxide started to replace lead as the white pigment of choice. Use of lead in household paints in the U.S. was banned by the Consumer Product Safety Commission in 1978.

In response to the toxic properties of solvents used in varnishes, lacquers, and urethane, paint manufacturers in the 1990s shifted toward producing low VOC (volatile organic compound) paints. Today, alkyds (which can be made soluble in water) have replaced most oil-based emulsifiers, and latex has replaced solvent-cut paints. The waterborne products have disadvantages such as lower gloss, and are also more expensive to produce. Some high-solid products were devel-

oped, where some of the VOC was replaced by lower molecular-weight resins, which did not increase the viscosity. Powder coatings have near-zero VOC.

See also **Dyes; Synthetic Resins**

LOLLY MERRELL

Further Reading

Ball, P. *Bright Earth: Art and the Invention of Color.* Farrar, Straus & Giroux, New York, 2002.
Gettens, R.J. and Stout, G.L. *Painting Materials, a Short Encyclopedia.* Dover Publications, New York.
Hammond, K. A brief history of paints. *Polym. Paint Col. J.*, 193, 24–25, 2003.
Marrion, A., Ed. *The Chemistry and Physics of Coatings.* Royal Society of Chemistry, Cambridge, 1994.

Coherers and Magnetic Detectors, *see* **Radio Receivers, Coherers and Magnetic Methods**

Color Photography

Most colors in nature can be represented as a mixture of the primary colors of red, green, and blue, and almost all photographic color processes begin by recording the proportions of these colors captured from the original scene. Two different paths may be followed. One involves adding together the three primary colors to build an image—an additive process; the other starts with pure white light containing all primary colors and removing those not wanted by superimposing positive images in complementary colors: cyan, magenta, and yellow—a subtractive process.

The possibility of producing photographic images in natural colors was a dream of the pioneers of photography, but for decades the only means available was to hand color with paint or powder. As late as 1882 the English photographer, John Werge, claimed that there was "little or no probability of polychromatic pictures being obtained in the camera direct." Nevertheless, it was a series of nineteenth century discoveries and innovations that were to make color photography possible. The single most influential work was that of the Scottish physicist, James Clerk Maxwell. In 1861 he photographed a tartan ribbon through separate blue, green, and red filters to make three positive lantern slides. Using blue, green, and red light sources in a lantern, the projected images were superimposed on the screen to produce a colored picture approximating to the color of the original tartan. By the early 1890s, the American, Frederick Ives, had devised a color photography system based on Maxwell's principles. The same decade also saw the first screen plate color photographs, made by exposing a photographic emulsion through a mosaic of tiny primary color filters. These primitive processes were unsatisfactory in many respects however and color photography only became a reality in the twentieth century.

In 1904 the Lumiere brothers Auguste and Louis announced a new screen plate process prepared by a randomly scattered mixture of starch grains dyed in the three primary colors and coated with a standard panchromatic emulsion. Each grain acted as a tiny filter so that from a single exposure in a camera, three "separations" were made on the one plate. After processing, the original subject was reproduced as a glass transparency in natural colors. The process required long exposures and the final images were dense and needed bright illumination for viewing. However, it was the first fully practicable color process. When marketed in 1907 as the Lumiere autochrome process, it soon became popular with both professionals and amateurs. It received an early boost in America when Edward Steichen exhibited examples along with specimens by other notable photographers. Other screen plate processes soon followed. The Thames Plate of 1908 and the Paget process of 1914, both with regular patterned screens, required shorter exposures than the Autochrome process and enjoyed success in England. With the rising popularity of roll film cameras, the screen plate principle was transferred to celluloid film. Agfacolor and Dufaycolor roll film was widely sold throughout Europe in the 1930s.

The screen plate processes described were additive processes, but the other major means of producing color images during the first 40 years of the twentieth century produced color by subtraction. Ingeniously designed three-color cameras containing mirrors or beam splitting devices were used to produce the color records in the form of separation negatives. Cyan, magenta and yellow carbon prints were then made from those negatives. When superimposed in exact register a print in full natural color was produced. A carbon print consists of hardened gelatin containing colored pigment and is completely stable. Its great advantage at the time was that richly colored images could be produced on paper. Several color processes based on refinements of the carbon process were marketed in Europe and America. Perhaps the most successful was the Carbro (or Ozobrome) process, which derived from H. E. Farmer's 1889 discovery that bichromated gelatin in contact with

a silver image becomes insoluble in water without the action of light. In the 1920s and 1930s the Autotype Company in England successfully marketed the process worldwide as Trichrome Carbro with the slogan "no daylight required." Unlike some carbon color processes, it was comparatively easy to work and became popular with amateur photographers.

Early in the twentieth century, it seemed that color photography might be achieved by coating three different color sensitive emulsion layers onto a single film base to form a tripack. In 1912 a German, Rudolph Fischer, patented a tripack containing color-forming materials within each emulsion layer but formidable technical problems prevented commercial exploitation. Over 20 years passed before two musicians working for Eastman Kodak, Leopold Godowski and Leopold Mannes, devised a slightly different tripack system involving three thin monochrome emulsion layers, each sensitized to a different color. Unlike Fischer's technique, color couplers were not introduced until processing, which was complex and included a three-stage development followed by bleaching to remove unwanted dyes. Kodak marketed the new process on roll film, for cine cameras in 1935, and for still cameras a year later. Called Kodachrome, it was received enthusiastically by the public and marks the beginning of modern subtractive color photography.

The German Company, Agfa, was also working on tripack film and in 1936 finally introduced a commercial development of Fischer's integral tripack. It was again constructed on the three-layer principle with the important difference that the color couplers and dyes were incorporated within the emulsion layers. The Agfa system was much simpler to process and Kodak soon developed its own version, a negative–positive process introduced in the early 1940s as Kodacolor, the precursor of almost all current color negative film. A variant, Ektachrome, for color transparencies followed, although an improved form of Kodak's original Kodachrome was still made available for high-quality color slides. Also developed during the 1940s were papers coated with a three-layer emulsion incorporating dye-releasing color couplers, which allowed the production of good quality color prints (sometimes termed C-type prints). Modern improved versions are used today to produce the millions of images processed in mini labs and shops throughout the world.

See also **Cameras, 35 mm; Cameras, Automatic**

JOHN WARD

Further Reading

Coe, B. *Colour Photography*. Ash & Grant, London, 1978.

Friedman, J.S. *History of Color Photography*. American Photographic, Boston, 1944. Reprinted by Focal Press, London, 1968.

Hunt, R.W.G. *The Reproduction of Colour*. Fountain Press, Watford, 1975.

Mees, C.E.K. *From Dry Plate to Ektachrome Film*. Eastman Kodak, New York, 1961.

Sipley, L.W. *A Half Century of Color*. Macmillan, New York, 1951.

Spencer, D.A. *Colour Photography in Practice*. Pitman Greenwood, London, 1938.

Spencer, D.A., Ed. The Focal Dictionary of Photographic Technology. Focal Press, London, 1973.

Wall, E.J. *The History of Three-Color Photography*. American Photographic, Boston, 1925. Reprinted by Focal Press, London, 1970.

Combinatorial Chemistry

Combinatorial chemistry is a term created about 1990 to describe the rapid generation of multitudes of chemical structures with the main focus on discovering new drugs. In combinatorial chemistry the chemist should perform at least one step of the synthesis in combinatorial fashion. In the classical chemical synthesis, one synthetic vessel (flask, reactor) is used to perform chemical reaction designed to create one chemical entity. Combinatorial techniques utilize the fact that several operations of the synthesis can be performed simultaneously.

Historically, the first papers bringing the world's attention to combinatorial chemistry were published in 1991, but none of these papers used the term combinatorial chemistry. Interestingly, they were not the first papers describing the techniques for preparation of compound mixtures for biological evaluation. Previously, H. Mario Geysen's lab had prepared mixtures of peptides for identification of antibody ligands in 1986. Other laboratories heavily engaged in synthesizing multitudes or mixtures of peptides were Richard A. Houghten's laboratory in San Diego and Árpád Furka's laboratory in Budapest. The recollections of the authors of these historical papers were published in the journal dedicated to combinatorial chemistry, *Journal of Combinatorial Chemistry*.

Since the goal of combinatorial chemistry is the discovery of new compounds with interesting properties (biological as new pharmaceuticals or physical as new materials), the chemists want not only to make as many compounds as possible as quickly as possible, but to make these compounds as different from each other as possible in order to cover what is referred to as chemical space. This space stretches over all theoretically possible structures and conformations of all compounds within a given range of size. When the structures of a set of compounds made by combinatorial chemistry are evenly distributed over the respective chemical space, one of the compounds has a better statistical chance of being identical or at least similar to the "optimal" structure (conformation) for a desired property (e.g., biological activity), as compared to a set of compounds that cover only a fraction of the chemical space.

A large set of related synthetic compounds is typically called a library (or combinatorial library). Libraries range in complexity from a few dozen up to millions of compounds. The central feature of combinatorial libraries is that all compounds making up the library represent combinations of two or more "building blocks" which are connected by chemical reactions.

Combinatorial libraries can be classified based on their composition or synthetic history (see Figure 7). Peptide-like (oligomer) libraries are composed of repeated units of similar building blocks connected by repetition of the same (or similar) chemical reaction. Glucose-like (scaffolded) libraries are based on the multifunctional scaffold, the functional groups of which are selectively employed in attachment of various building blocks. Benzodiazepine-like (condensed) libraries are created by connecting building blocks capable of forming unique structures depending on the order of performed reactions, where original building blocks may not be readily identifiable within the resulting library structure (various strategies and building block types can be used for forming the same resulting structures). Libraries can be structurally homogeneous or heterogeneous; that is, the compounds can have identical or variable "scaffolds" or "backbone." All combinatorial libraries can also be complete (containing all theoretically possible combinations of used building blocks), or incomplete (containing only a fraction of all possible compounds). A complete combinatorial library is composed of all possible permutations of the building blocks at their respective positions. If the scaffold of a library has three attachment points (prospective diversity positions), and ten different building blocks are used for each diversity positions, then the complete combinatorial library is composed of $10^3 = 1000$ compounds.

Synthesizing a combinatorial library can be rather straightforward, as the same protocol is typically used for all compounds, so that the synthesis method has to be worked out only

Peptide-like (oligomer) library

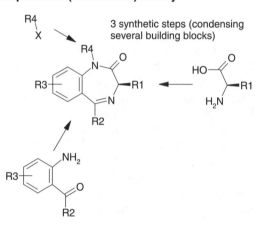

4 synthetic steps
(connecting 4 amino acids)

Glucose-like (scaffolded) library

5 synthetic steps (attachment of
5 different building blocks)

Benzodiazepine-like (condensed) library

3 synthetic steps (condensing
several building blocks)

Figure 7. Different library types—peptide-like (oligomer) libraries, glucose-like (scaffolded) libraries, and benzodiazepine-like (condensed) libraries. Oligomeric libraries are built by connecting similar building blocks by repetition of one (or several) reactions—peptides and oligonucleotides are typical examples. Scaffolded libraries are constructed by modification of individual functional groups on the template (scaffold) molecule. In condensed libraries it may be difficult to trace the character of building blocks used for their construction.

once. This, however, is not always as easy as it may sound (with issues covering the choice of synthetic strategy, in solution or on solid support, chemistry of attachment of the first building block, protection and deprotection strategies, release from solid support, etc.), as the optimal reaction conditions can vary greatly among the different building blocks used for a particular step.

An important prerequisite for combinatorial chemistry is the availability of methods for parallel synthesis. Prototypical of combinatorial chemistry techniques is Houghten's "tea bag" method. In this technique the solid support (functionalized polystyrene resin) is sealed in packets made of polypropylene mesh (Figure 8), which is permeable for solvents and reagent solutions. Up to several hundreds of such resin packets can be processed simultaneously in common reaction vessels. After each step the packets are resorted for the next synthesis step. Resorting is either based on readable alphanumeric labels, or it can be simplified by enclosing a radio-frequency tag in the tea bag. In this way up to a thousand compounds can be

synthesized using a reasonable number of reaction vessels. For example in the peptide synthesis, only 20 reactors with individual amino acids are required for the synthesis of basically unlimited number of (natural) peptides of any length in parallel.

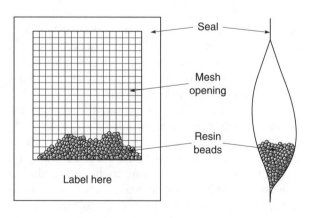

Figure 8. Schematic drawing of the "tea-bag."

A powerful, yet simple method for manual or semiautomated solid-phase synthesis of mixtures of up to millions of compounds is the "one-bead–one-compound" approach (Figure 9). It has also been referred to as "split-and-mix" or "divide–couple–recombine" approach, and is based on coupling each building block to separate portions of the solid-phase resin, followed by combining and mixing all resin portions, before dividing the resin again for the next synthesis step. By repeating this procedure three more times, and using 20 different building blocks for each synthesis step, a library of 160,000 (20^4) compounds can be readily prepared. This process yields libraries containing an individual, unique compound on each resin bead. After assembling the library on the resin, it can be either cleaved for bioassays in solution, or left on the resin for solid-phase assays. The bio-assays are typically performed on single beads, so that the screening format of one-bead–one-compound libraries is that of single compounds, rather than compound mixtures. The one-bead–one-compound library principle is based on the statistical distribution of the particles in the process. However, this statistical nature of the process can be eliminated by the use of continuously divideable solid supports, such as membranes or threads. In this case, all members of the library are guaranteed to be prepared, and none of them is prepared in more than one copy.

Michal Lebl

Further Reading

Frank, R. and Doring, R. Simultaneous multiple peptide synthesis under continuous flow conditions on cellulose paper discs as segmental solid supports. *Tetrahedron*, 44, 19, 6031–6040, 1988.

Furka, A., Sebestyen, F., Asgedom, M. and Dibo, G. General method for rapid synthesis of multicomponent peptide mixtures. *Int. J. Peptide Prot. Res.* 37, 487–493, 1991.

Geysen, H.M., Meloen, R.H. and Barteling, S.J. Use of peptide synthesis to probe viral antigens for epitopes to a resolution of a single amino acid. *Proc. Natl. Acad. Sci. USA*, 81, 3998–4002, 1984.

Houghten, R.A. General method for the rapid solid-phase synthesis of large numbers of peptides: Specificity of antigen-antibody interaction at the level of individual amino acids. *Proc. Natl. Acad. Sci. USA*, 82, 5131–1315, 1985.

Houghten, R.A., Pinilla, C., Appel, J.R., Blondelle, S.E., Dooley, C.T., Eichler, J., Nefzi, A. and Ostresh, J.M. Mixture-based synthetic combinatorial libraries. *J. Med. Chem.*, 42, 19, 3743–3778, 1999.

Lam, K.S., Salmon, S.E., Hersh, E.M., Hruby, V.J., Kazmierski, W.M. and Knapp, R.J. A new type of synthetic peptide library for identifying ligand- binding activity. *Nature*, 354, 82–84, 1991.

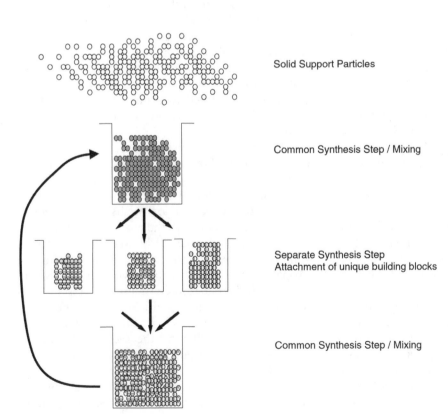

Solid Support Particles

Common Synthesis Step / Mixing

Separate Synthesis Step
Attachment of unique building blocks

Common Synthesis Step / Mixing

Figure 9. Principle of one-bead–one-compound library synthesis. Resin particles are exposed to only one reagent at a time and therefore each particle can contain only one structure. The process of separating the solid support into aliquots and mixing them is repeated as many times as there are steps of the library building using the unique building blocks. A multistep synthesis of a library, in which three of the steps use the various building blocks (10 different building blocks are used in each step), would generate 10 × 10 × 10 = 1000 different bead populations. If 1 gram of 130-micrometer polystyrene beads were used for the synthesis (1,000,000 beads), there will be in average 1,000 beads carrying the same compound. If only 1,000 beads are used for the synthesis (very unlikely), the chance of having any particular structure represented in the library would be only about 70 percent.

Lebl, M. Parallel personal comments on "classical" papers in combinatorial chemistry. *J. Combinatorial. Chem.*, 1, 1, 3–24, 1999.

Nicolaou, K.C., Xiao, X.Y., Parandoosh, Z., Senyei,A. and Nova, M.P. Radiofrequency encoded combinatorial chemistry. *Angew. Chem. Int. Ed.* 34, 2289–2291, 1995.

Stankova, M., Wade, S., Lam, K.S. and Lebl, M. Synthesis of combinatorial libraries with only one representation of each structure. *Peptide Res.* 7, 6, 292–298, 1994.

Useful Websites

Compilation of papers in molecular diversity field, 1996: http://www.5z.com.

Communications

To the many technological inventions of the nineteenth century that improved communications, such as the telegraph and the telephone, the twentieth century added motion pictures, radio, television, and the Internet. Most of the products created for improving communication in the nineteenth century improved communication between individuals. During the twentieth century, new technologies added ways for groups or organizations to communicate to other groups, marking the birth of mass communications. These new media would have profound effects on entertainment, how people received news, and politics.

Motion Pictures

The first major communication improvement to be commercialized in the twentieth century was the motion picture. Thomas Edison invented the first practical motion picture camera in the U.S., and in 1896 he showed a motion picture to the public in the New York City Music Hall. The U.S. was the pacesetter for the film industry and New York City was the early center of the motion picture business. The first narrative film was *The Great Train Robbery* (1903). During World War I, Hollywood began to replace New York as the home of the movie industry. By the 1920s, Hollywood had clearly become the movie-making capital of the world and silent-film stars such as Charlie Chaplin, Buster Keaton, and Mary Pickford established themselves there.

Technological developments in the early twentieth century made sound motion pictures possible and greatly changed the film industry. In 1926, Warner Brothers, then a relatively minor studio, released the film *Don Juan* with a synchronized orchestral accompaniment. The studio had purchased the sound-on-disk Vitaphone system from American Telephone and Telegraph (AT&T). Warner Brothers sought to make short-term profits by supplying the technology to theaters that could not afford to hire live orchestras. This first attempt was successful enough that the first talking movie—*The Jazz Singer* staring Al Jolson—was released in 1927. In addition to the orchestral accompaniment, this film also featured popular songs and dialogue. The new sound films enjoyed great success and almost all Hollywood films included sound by the late 1920s, leading to greatly increased profits for the studios.

The introduction of sound did lead to some problems and changes in the film industry. Problems included the tremendous expense now involved in the production of motion pictures and the primitive nature of microphones, which forced actors to remain almost stationary and would pick up the sound of cameras and other set noise. Changes included the replacement of many silent-era actors with new actors with stage experience due to the fact that with sound, film actors had to have pleasant sounding voices without strong foreign accents. Sound led to the production of more realistic films, including crime epics and historical biographies. Musicals also became an important film genre, including the animated musicals of Walt Disney.

Movies would become the world's leading form of entertainment until the advent of television after World War II. For many, films represented the decadence of twentieth century society, displaying modern sexual mores on the screen. Along with radio, the film industry also became big business. Furthermore, the motion picture industry centered in Hollywood served to export the culture of the U.S., as moviegoers around the world viewed U.S.-made films.

Although motion pictures were mostly used for entertainment, showings often included such current news events as wars, parades, and speeches. In the 1930s, after sound had been added, newsreels covering the week's major events were shown in most theaters along with the movies. Authoritarian governments that emerged between the two world wars became particularly adept at utilizing film as a propaganda tool. The Bolsheviks in the Soviet Union and the Nazis in Germany successfully used motion pictures for political ends. The most famous example of this political use is *Triumph of the Will*, a 1934 film by the German filmmaker Leni Riefenstahl that depicted a Nazi rally at Nuremberg.

During the war, film was used to support the war aims of governments. In the U.S., for example,

Hollywood backed the government's information campaign through the Bureau of Motion Picture Affairs, which produced commercial features with patriotic themes. Hollywood also produced documentaries such as Frank Capra's *Why We Fight*, which sought to explain the war to both soldiers and civilians.

Immediately after the war, Hollywood enjoyed a brief boom, as two-thirds of Americans went to the movies at least once a week. Soon, however, antitrust legislation, protectionist quotas abroad, and the rise of television cut into Hollywood's profits. The Cold War also greatly affected the film industry, as many suspected communists were blacklisted by the studios and film-making became more conservative. Traditional genres such as musicals and westerns continued after the war, while others grew in importance, including many lower-budget films that dealt with social problems such as racism or alcoholism. Also popular was *film noir*, which offered a dark interpretation of American society.

World War II devastated the film industries in much of Europe, the Soviet Union, and Japan. A postwar renaissance was led by Italy and its neorealist movement that attempted to show the reality of a country afflicted by warfare. Great Britain and France soon followed in reviving their film industries. Japan also was able to restore its motion picture industry after the war, as many studios were left intact. Akira Kurosawa led the Japanese revival with numerous films, including *Rashoman*.

A film industry developed in many Third World countries. India had a vibrant film industry led by director Satyajit Ray. Many of India's films provided an alternative cinema with artistic merit. At the same time, India also became the world's largest producer of low-quality films for domestic consumption, making more than 700 motion pictures in sixteen languages each year. Film was often the only access to audiovisual entertainment for the many poor and illiterate Indians.

Latin America and Africa also developed sometimes militant, alternative forms of film; for example in Cuba during the 1960s, when the country's revolution influenced world-renowned directors such as Tomás Gutíerrez Alea and Humberto Solás. The so-called *Cinema Nôvo* (New Cinema) developed in Brazil during the 1960s and spread to other Third World countries. While many Third World nations created sometimes revolutionary film genres, military dictatorships also repressed motion pictures in numerous countries such as Argentina.

Radio

By the early twentieth century, wireless communication began to appear. The first example of wireless communication was the radio. In 1895, Italian inventor Guglielmo Marconi transmitted the first wireless telegraph message. Starting in 1901, Marconi used radio telegrams to communicate with ships on the Atlantic Ocean. The usefulness of radio was seen in its use during the Russo–Japanese War in 1905. In the U.S., experimental broadcasting to a mass audience started in 1910 with a program by the famous singer Enrico Caruso at the Metropolitan Opera House in New York City. Perhaps the most dramatic example of radio's value in spreading information was its use in reporting on the sinking of the *Titanic* in 1912, which demonstrated radio's ability to allow people to experience distant events as they occurred. World War I interrupted some radio research, however the demands of military communications sped up the development of radio technology.

During the 1920s, what had been more of a hobby became a mass medium that played a central role in news reporting and entertainment. A number of experimental broadcasting stations had converted to commercial stations by broadcasting programs on a regular basis, including news such as the results of the 1920 presidential election in the U.S. In the U.S., because radio was a good way to communicate with large groups of people, broadcasting rapidly consolidated into national networks in order to attract advertising revenue to support news and entertainment programming. The Radio Corporation of America (RCA) created the first nationwide broadcast network, the National Broadcasting Company (NBC), in 1926. In Europe and some other parts of the world, governments generally controlled radio broadcasting.

Radio played an important role in twentieth century communications, as it allowed people much easier access to entertainment since many families owned radios. By the end of the 1920s, two-thirds of homes in the U.S. owned radio receivers. People no longer had to go to a concert, play, or sporting event to be entertained. Instead, they could now enjoy many forms of entertainment from the comfort of their own homes. Despite the fact that radio broadcasts could reach millions of people, the medium gave those in their homes a sense of immediacy and intimacy. Furthermore, unlike written forms of communication, no formal education was needed to enjoy radio programs. Many forms of popular entertainment shifted to the radio, allowing them to maintain and even expand their audiences. Radio offered a wide variety of entertainment genres, including dramas, comedies, sports, and music.

Besides providing entertainment, supplying news, and making money for entrepreneurs, radio also proved to be an important tool for politicians, better enabling them to mobilize the masses. Perhaps best known are Franklin Roosevelt's "Fireside Chats," which allowed the president of the U.S. to reach the public directly during the Great Depression and World War II. As was the case with film, authoritarian regimes in particular made use of radio technology. Italy's Benito Mussolini pioneered the use of radio to address the nation. In the Soviet Union, the first experimental radio broadcasts began in 1919. In 1922, a central radio station in Moscow began broadcasting. By 1924, regular broadcasts could be heard throughout most of the USSR and by 1937, there were some 90 radio transmitters in operation in Stalin's Soviet Union. Leaders in Nazi Germany also made effective use of the radio during the 1930s and 1940s. In Japan, the right-wing government utilized radio to promote its goals leading up to World War II.

Radio has also become an important means of communications in other parts of the world. In Latin America, for example, radio, along with television, is the main medium for transmitting information. In most Latin American countries, radio reaches far more people than print media, due to lower rates of literacy and lack of purchasing power. From the 1930s to the 1960s, many radio stations broadcast *radionovelas*, serial radio programs similar to soap operas. Since the 1960s, such programming has largely moved to television. As was the case elsewhere, early Latin American radio also featured variety shows, dramas, sports, talk shows, and news.

Radio also contributed to the spread of Latin American culture to other parts of the world, especially in the realm of dance and music. Argentine tango, Mexican boleros, salsa from New York's Latin community, and Brazilian samba all became popular beyond the borders of Latin America in large part because of radio airtime. Samba, for example, emerged as a musical and dance form from the poor sections of Rio de Janeiro, the capital of Brazil at the time. From its Afro-Brazilian roots, samba emerged from a locally popular form to one that had a national importance in Brazil. As samba received increased radio airplay, it seemed to unite the country and came to represent Brazilian nationalism. Performers such as Carmen Miranda, who later also became a

Hollywood film star, popularized the music on the Brazilian airwaves. Soon, listeners heard samba on their radios throughout the world, demonstrating that mass culture could spread from poorer countries to elite consumers around the globe due to communications technology such as radio.

From the 1990s, radio stations in Latin America have often become more specialized as they seek audiences. Amplitude modulation (AM) stations tend to carry news, talk, and local popular music. Also, they often cater to the interests of groups outside of the cultural and linguistic mainstream. For example, radio stations in Lima, Peru feature ethnic music and news in Quechua or Aymara languages for recent migrants from the highlands. Frequency modulation (FM) stations emphasize music, particularly national popular music or international music. International music tends to be popular among the young and affluent, while national music appeals to an older, more working-class audience.

Television

Another important twentieth century development was television, which would soon overshadow radio and motion pictures. In the U.S. during the late 1920s, many attempts were made to create an experimental telecast, and a few met with success, particularly RCA's efforts. In 1936, NBC provided 150 experimental television sets to homes in New York City and sent telecasts to them, the first show being the cartoon "Felix the Cat." By 1939, NBC was providing regular telecasts but to a limited market. When the U.S. entered World War II in 1941, however, all television projects were suspended until the war ended in 1945.

After the war, television development continued where it left off, with the invention of better television sets, creative programming, and larger markets. The first coast-to-coast program was President Harry Truman's opening speech at the Japanese Peace Treaty Conference in 1951. By the 1950s, television had become a profitable industry. Television enjoyed a "golden age" and increasingly replaced radio as the principal mass medium. Indeed, television became a key part of social life in the U.S. and other parts of the world. Following World War II, a growing number of people had more money and more leisure time, both of which were often spent on television.

While early televisions in the U.S. were largely affordable, they were often unreliable. Technological improvements soon made television much more reliable and appealing. These improvements included the replacement of vacuum tubes with the transistor and the development of color sets. In 1953, the first color telecast was made, which spread so fast that by the 1960s, most telecasts were in color. Later advancements include the spread of cable television in the 1980s, which gave viewers access to dozens of specialized channels and challenged the power of the traditional television networks. Many of the newly available cable channels, such as MTV and CNN, would have important effects on society and culture. The end of the twentieth century witnessed the rise of satellite and high-definition television, which offered viewers even more choices and improved the technical quality of television.

Television continued the process of the globalization of U.S. culture, as viewers around the world watched comedies and dramas produced in the U.S. Sporting events also helped to spread the U.S.'s cultural values. The National Basketball Association (NBA) was particularly successful in its international marketing efforts, popularizing its sport around the globe and creating stars such as Michael Jordan, who arguably became the most recognized athlete in the world. In addition, U.S.-based businesses, such as Nike, benefited from the globalization of basketball through television, as the sport helped to sell more of its athletic shoes. Yet it was not only basketball and the U.S. that dominated the use of television. During the 1986 soccer World Cup in Mexico, games were played under the midday sun in order to be broadcast during primetime in European countries.

Television grew more slowly in the Soviet Union than in the U.S. and Western Europe. As late as 1960, only five percent of the Soviet population could watch television. Television audiences grew during the 1970s and 1980s, often at the expense of film and theater audiences. By 1991, 97 percent of the population could view television, and a typical audience for the nightly news from Moscow numbered 150 million.

Television also became available in Latin American countries during the 1950s, when it was largely restricted to an upper- and middle-class urban audience. In this early phase, programming was limited to live, local productions. From the 1960s, television became much more of a mass medium. In this period, much of the programming was imported from abroad, especially the U.S. By the 1970s and 1980s, high-quality national production appeared, especially in Brazil, Colombia, Mexico, and Venezuela. The most important and successful productions were *telenovelas*, a form of the soap opera. By 2000, in some countries, such as Brazil and Mexico, perhaps 90 percent of the population had regular access to television. While the figure is lower in rural areas, even this began to change in the 1980s when satellite dishes linked to repeater transmitters allowed for increased access in remote areas.

Computers and the Internet

While film, radio, and television all had dramatic effects on communications in the twentieth century, all three were still "one-way" media that lacked any sort of interactive capabilities. The advent of the computer revolution and in particular the Internet changed this situation in the late twentieth century. While early computers had been large and slow, by the 1970s and 1980s, engineers centered in California's so-called Silicon Valley created increasingly smaller computers with greater memory capacity. After these hardware developments, improvements in software followed that allowed computer users to word process, play games, and run businesses. These technological improvements in computer hardware and software would soon have a profound effect on communications and commerce with the development of the Internet.

The creation of the Internet was the result of attempts to connect research networks in the U.S. and Europe. In the 1960s, the U.S. Department of Defense created an open network to help academic, contract, and government employees communicate unclassified information related to defense work. After crucial technological advances in the 1970s, in 1980 the Department of Defense adopted the transmission control protocol/Internet protocol (TCP/IP) standard, which allowed networks to route and assemble data packets and also send data to its ultimate destination through a global addressing mechanism.

During the 1980s, the defense functions were removed from the network, and the National Science Foundation operated the remainder, adding many new features to the network and expanding its use around the world. While government agencies were the principal early users of the Internet, by the 1980s its use had spread to the scientific and academic community. By the 1990s, the Internet had become increasingly commercialized and privatized. The rise in the use of personal computers and the development of local area networks to connect these computers contributed to the expansion of the Internet. Starting in 1988, commercial electronic mail (e-mail) services were connected to the Internet, leading to a boom in traffic. The creation of the World Wide Web and easy to use Web browsers made the Internet more accessible so that by the late 1990s, there were more than 10,000 Internet providers around the world with more than 350 million users.

In the early twenty-first century, the Internet is a critical component of the computer revolution, offering e-mail, chat rooms, access to the wealth of information on the Web, and many Internet-supported applications. The Internet has had a dramatic impact on global society. E-mail is rapidly replacing long-distance telephone calls, and chat rooms have created social groups dedicated to specific subjects, but with members living around the world. The Internet has not only changed how people communicate but also how they work, purchase, and play. Many people now work at home, using the Internet to stay in touch with the office. People have also begun to use the Internet for banking and shopping services rather than so-called "brick and mortar" locations.

The communications revolution of the twentieth century created many new social problems that will have to be addressed in the twenty-first century. While people have access to more information than ever before, that information, often unfiltered and invalidated, has created several generations of children who are seemingly immune to extreme violence. Health concerns are also an issue, as people spend less time in outdoor activities and more time sitting in front of the television or computer. The online nature of the Internet will also make privacy one of the major issues of the near future.

See also **Entertainment in the Home; Film and Cinema: Early Sound Films; Internet; Radio, Early Transmissions; Television, Cable and Satellite**

RONALD YOUNG

Further Reading

Abbate, J. *Inventing the Internet*. MIT Press, Cambridge, MA, 1999.
Barnouw, E. *A History of Broadcasting in the United States*, 3 vols. Oxford University Press, New York, 1966–1970.
Beniger, J. *The Control Revolution*. Harvard University Press, Cambridge, MA, 1986.
Baughman, J.L. *The Republic of Mass Culture: Journalism, Filmmaking, and Broadcasting in America since 1941*. Johns Hopkins University Press, Baltimore, 1992.
Burns, R.W. *Communications: An International History of the Formative Years*. Institute of Electrical Engineers, London, 2003.
Carey, J.W. *Communication as Culture: Essays on Media and Society*. Unwin Hyman, Boston, 1988.
Cook, D. *A History of Narrative Film*. W.W. Norton, New York, 1990.
Crisell, A. *Understanding Radio*. Methuen, New York, 1986.
Crowley, D. and Heyer, P. *Communication in History: Technology, Culture, Society*. Longman, New York, 1991.
Knight, A. *The Liveliest Art: A Panoramic History of the Movies*, revised edn. Macmillan, New York, 1978.
Leiner, B.M., *et al.* The past and future history of the Internet. *ACM Comm.*, 40, 2, 102–108, 1997.
McLuhan, M. *Understanding Media*. Signet, New York, 1964.
Williams, R. *Television and Society*. Fontana, London, 1979.

Composite Materials

Composite materials are defined as those that contain two or more materials that have been bonded together. Wood is an example of a natural composite material. It is made from lignin—a natural resin—which is reinforced with cellulose fibers. For thousands of years, straw has been used to reinforce mud bricks, forming a two-phase composite, and more recently, reinforced concrete has been developed. Concrete contains a cement binder with a gravel reinforcement. By adding another reinforcing material (steel rebar), concrete becomes a three-phase composite. Metal matrix and metal or ceramic composites have now also been developed.

Perhaps the most familiar class of composite materials are polymer composites. These are a class of reinforced plastics in which fibers are used to reinforce a particular polymer matrix. Phenolic laminates made from phenol formaldehyde resin and paper were developed around 1912, finding a use as an electrical insulating material. Glass is the most common fiber used to reinforce a polymer matrix, forming fiber-glass or glass-reinforced plastic. Other more expensive and higher strength materials such as carbon and, more recently, aramid fibers, are used in advanced applications such as aircraft components. Polyester, vinyl ester and epoxy resins are the most commonly used thermoset resins for the formation of the polymer matrix. PEEK (polyether ether ketone) may also be selected as the resin for applications where cost is less of a problem, such as in aerospace applications. Other resins, including phenolic, silicone, and polyurethane may be used for particular functions. The designer can alter the chemical resistance properties, service temperature capabilities, weather resistance, electrical properties, resistance to fire and adhesive properties by choosing the right type of resin.

Fibers for reinforcing can be obtained in a variety of different formats. They can be woven or multiaxial, continuous or chopped (to give a random form). Alternatively they can be variations or combinations of all these types. By selecting the right orientation of fiber lay-up, it is possible to "design-in" a variety of properties connected with the physical strength of the ultimate composite.

Composite materials are often what we call layered composites—they are made up of layers of fibers, supplied as plies or lamina. A single ply comprises fibers oriented in a single direction (unidirectional) or in two directions (bidirectional); for example, in a woven fabric. Random fiber layers are also used. These are supplied as "pre-pregs," meaning that they are prepared before being molded by using a thermoplastic binder that is applied to the reinforcement and then heating the reinforcement. This softens the binder, which can then be formed into the desired shape in a separate operation. A specialist form of composite production involves filament winding on a mandrel prior to molding, producing high-strength rods.

Polymers fall into two classes—thermoplastics and thermosetting plastics, or thermosets.

Thermoplastics include familiar plastics such as acrylic (polymethyl methacrylate), polyethylene, acrylic, polypropylene, and polystyrene. They can be heated and formed, then remelted and reformed into a new shape.

Composite materials are normally made with thermoset resins. These start as liquid polymers. Following the cross-linking process, which they undergo on heating, the liquid polymers are changed during the molding process into an irreversible solid form. Composite materials therefore have superior properties to thermoplastics as they have better heat and chemical resistance, enhanced physical properties, and more structural resilience.

The development of composite materials that possess a range of advantageous properties have inspired a range of new uses in areas ranging from global uses, such as in aerospace, transport (both by road and sea), and building applications, as well as domestic products—high-tech sporting equipment, electrical goods and office equipment.

Composites have various advantages including high strength, light weight, and some temperature resistance. Their high strength has enabled the design of composite materials that meet the taxing needs of a particular function, for example an aircraft nose cone, which needs high impact strength. A variety of resins and reinforcements (from random to woven) can be used to meet the exact physical and mechanical properties needed for a particular structure.

The light weight of composite together with their high strength properties can be designed to meet very demanding specifications, for example in Formula One racing cars or in aerospace uses. Composites can be used to produce the highest strength-to-weight ratio structures known, as for example when aramid fibers are embedded in an epoxy matrix. Due to these high strength-to-weight ratios, these composites contribute enormous weight savings to aerospace structures—a vital consideration when more weight means more fuel use. Other advantages of these composite materials are their high resistance to corrosion and fatigue.

More complicated composite materials known as composite hybrids have been developed. These are formed by adding another material such as glass or aramid fiber to the original carbon fiber and epoxy matrix. These extra materials improve mechanical properties such as impact resistance and fracture toughness. Glass-reinforced plastic can gain improved stiffness by adding carbon and epoxy to the matrix.

While most modern engineering composites are made from a thermosetting resin matrix reinforced with fibers, some advanced thermoplastic resins may also be used. Other composites, in particular phenolic composites, have filler reinforcements, which may be mineral or fibrous, or a mixture of both. Foams and honeycombs can also be utilized as cellular reinforcements to bestow stiffness together with a very light weight. An example of such a polymer composite was used in the rotor blades of the EH101 Westland helicopter. The only disadvantage of this type of reinforcement, apart from its high cost, is that if the surface is damaged and the honeycomb becomes wet, it may lose its mechanical properties.

Composite plastic materials are made using a range of processes. The original process was a hand lay-up process, which can be slow, time-consuming and expensive. New manufacturing methods include pultrusion, vacuum infusion, resin transfer molding (RTM), sheet molding compound (SMC), low-temperature curing prepregs and low-pressure molding compounds. These methods are being used in high tech areas such as aerospace, and RTM in particular looks as though it is developing into a very good value for money process.

Choosing composites for certain applications is straightforward in some cases, but in others, their selection will rely on parameters such as service life, production run, complexity of mold, cost savings in assemblage, and on the experience and skills of the designer in tapping into the ultimate potential of composites. Sometimes it is best to use composites together with more traditional materials.

The development of polymer composites has been pushed forward by the needs of the aerospace industry but has revolutionized the design of furniture, allowing the design of organic forms, as well as the design of boats, and sporting equipment where, as for example in the case of tennis rackets and racing cars, their performance has radically exceeded expectations.

See also **Ceramic Materials; Synthetic Resins; Thermoplastics**

SUSAN MOSSMAN

Further Reading

Kelly, A., Ed. *Concise Encylopaedia of Composite Materials*. Pergamon Press, Oxford, 1989.
Parkyn, B. Fibre-reinforced composites. in *The Development of Plastics*, Mossman, S.T.I. and Morris, P.J.T., Eds. Royal Society of Chemistry, Cambridge, 1994, pp. 105–114.

Computer and Video Games

Interactive computer and video games were first developed in laboratories as the late-night amusements of computer programmers or independent projects of television engineers. Their formats include computer software; networked, multiplayer games on time-shared systems or servers; arcade consoles; home consoles connected to television sets; and handheld game machines.

The first experimental projects grew out of early work in computer graphics, artificial intelligence, television technology, hardware and software interface development, computer-aided education, and microelectronics. Important examples were Willy Higinbotham's oscilloscope-based "Tennis for Two" at the Brookhaven National Laboratory (1958); "Spacewar!," by Steve Russell, Alan Kotok, J. Martin Graetz and others at the Massachusetts Institute of Technology (1962); Ralph Baer's television-based tennis game for Sanders Associates (1966); several networked games from the PLATO (Programmed Logic for Automatic Teaching Operations) Project at the University of Illinois during the early 1970s; and "Adventure," by Will Crowther of Bolt, Beranek & Newman (1972), extended by Don Woods at Stanford University's Artificial Intelligence Laboratory (1976). The main lines of development during the 1970s and early 1980s were home video consoles, coin-operated arcade games, and computer software.

Spacewar! grew out of the new "hacker" culture of the Tech Model Railroad Club (TMRC) at MIT. Intended as a demonstration program for a new PDP-1 computer donated by Digital Equipment Corporation, Spacewar! allowed players to control spaceships depicted on accurate star maps via the equally new precision cathode-ray tube (CRT) display Type-30. They maneuvered their spaceships via novel control boxes to avoid obstacles and fire torpedoes at their opponents. The result was a popular game available on PDP computers distributed to U.S. computer science laboratories in the 1960s and 1970s, such as the University of Utah's strong program in computer graphics. Nolan Bushnell, a former Utah graduate student and amusement park employee, recognized Spacewar!'s potential as a commercial product. With Ted Dabney, his co-worker at Ampex Corporation in California, he created "Computer Space" (1971) for Nutting Associates; this was a coin-operated version of Spacewar! set in attractive arcade cabinets.

Ralph Baer independently pursued the idea of creating video game consoles attached to home television sets. In 1971, he received a U.S. patent for a "television gaming apparatus," soon followed by acquisition of rights to it by Magnavox and, in 1972, by production of the first home video console, the Magnavox Odyssey. Bushnell and Dabney had by then created a new company, Atari Corporation; joined by Al Alcorn, another Ampex alumnus, Atari shipped Alcorn and Bushnell's electronic ping-pong game, "Pong," as an arcade game in November of 1972. Joining forces with the Sears department store chain, Atari released a home version of Pong in 1975. The phenomenal success of Pong stimulated competition leading to improved home and arcade consoles. The equally successful Atari 2600 VCS (video computer system), released in 1977, provided more flexibility and encouraged the separate development of game software distributed on cartridges and the hardware platforms accepting these games, at least in the home market. Activision, founded in 1979 by four former Atari game designers, was the first company exclusively focused on game software.

By the late 1970s, "home computers," single-user general-purpose computers with microprocessors, provided a new platform for electronic entertainment. Apple Computer's Apple II (1977) and the IBM Personal Computer (1981) featured color graphics, flexible storage capacity, and a variety of input devices. The Atari 800 (1979) and Commodore International's Commodore 64 (1982) retained cartridge slots for console-style games, but were also capable home computers. Games designed for computers at first resembled arcade and video console titles, but early computer games took advantage of greater flexibility, inspired by complex paper-and-pencil role-playing games such as "Dungeons and Dragons," boardgames, and Crowther's "Adventure." The original Adventure linked Crowther's experiences as an explorer of Kentucky's Mammoth and Flint Ridge cave systems to the Tolkien-inspired fantasy world of role-playing games; written in FORTRAN for the PDP-10 computer, Adventure became the prototype for "interactive fiction," games featuring scripted story lines revealed as players typed responses to textual information provided by software. The numerous text-only adventures published by Infocom during the 1980s pushed the "adventure" genre further, beginning with the wildly popular "Zork" series. Other games such as the "King's Quest" series by Sierra On-Line (1983), military simulations and role-playing games published by Strategic Simulations Incorporated (founded in 1979), Richard Garriott's "Akalabeth/Ultima" series (1979), and the sports and multimedia titles of Electronic Arts (founded in 1982) extended the simulation and storytelling capacity of computer-based games. MUD (multi user dungeon), developed by Roy Trubshaw and C. Richard Bartle at the Univesity of Essex in 1978, combined interactive fiction, role-playing, programming and dial-up modem access to a shared computer to build a virtual world on the basis of social interaction as much as structured game play; hundreds of themed multiplayer MUDs, and BBS-based games were written during the 1980s and early 1990s.

In the late 1980s, a new generation of video consoles led by the Nintendo Entertainment System (1985) and the Sega Genesis (1989) offered improved graphics and also introduced battery-powered storage cartridges that enabled players to save games in progress. Games such as Shigeru Miyamoto's "Super Mario Brothers" (1985) and "The Legend of Zelda" (1987) for Nintendo or Square's "Final Fantasy" series (1987) took advantage of these capabilities to provide deeper game experiences, flexible character building and complex, interactive environments, encouraging comparisons between video games and other narrative media such as cinema. In the 1990s, computer game designers exploited the three-dimensional graphics, faster microprocessors, networking, hand-held and wireless game devices, and the Internet to develop new genres for video consoles, personal computers, and networked environments. The most important examples included first-person "shooters"—action games in which the environment is seen from the players view—such as id Software's "Wolfenstein 3-D" (1992), "Doom" (1993) and "Quake" (1996); sports games such as Electronic Arts' "Madden Football" (1989) based on motion capture systems and artificial intelligence; and massively multi-player games such as "Ultima Online" (1997) and "Everquest" (1998), combining traits of MUDs and graphical role-playing games to allow thousands of subscribers to create avatars and explore "persistent" virtual worlds.

See also **Entertainment in the Home; Technology, Arts and Entertainment; Technology and Leisure**

HENRY LOWOOD

Further Reading

Cassell, J. and Jenkins, H., Eds. *From Barbie to Mortal Kombat: Gender and Computer Games*. MIT Press, Cambridge, MA, 1998.

Herman, L. *Phoenix: The Fall and Rise of Videogames*, 3rd edn. Rolenta Press, Springfield, NJ, 2001.

Herz, J. C. *Joystick Nation: How Videogames Ate our Quarters, Won Our Hearts, and Rewired our Minds*. Little, Brown & Company, Boston, 1997.

Kent, S.L. *The Ultimate History of Video Games: From Pong to Pokémon and Beyond: The Story behind the Craze That Touched Our Lives and Changed the World*. Prima, Roseville, CA, 2001.

Loftus, G.R. and Loftus, E.F. *Mind at Play: The Psychology of Video Games*. Basic Books, New York, 1983.

Poole, S. *Trigger Happy: Videogames and the Entertainment Revolution*. Arcade Publishing, New York, 2000.

Sheff, D. *Game Over: How Nintendo Conquered the World*, 2nd edn. Gamepress, Wilton, CT, 1999.

Sudnow, D. *Pilgrim in the Micro-World: Eye, Mind, and the Essence of Video Skill*. Warner Books, New York, 1983.

Wolf, M.J.P., Ed. *The Medium of the Video Game*. University of Texas Press, Austin, 2001.

Computer Displays

The display is an essential part of any general-purpose computer. Its function is to act as an output device to communicate data to humans using the highest bandwidth input system that humans possess—the eyes. Much of the development of computer displays has been about trying to get closer to the limits of human visual perception in terms of color and spatial resolution.

The earliest output devices for early digital computers were lamp displays (Z3 by Konrad Zuse, 1941). An early digital computer, the EDSAC (electronic delay storage automatic calculator) developed at Cambridge University in 1949,

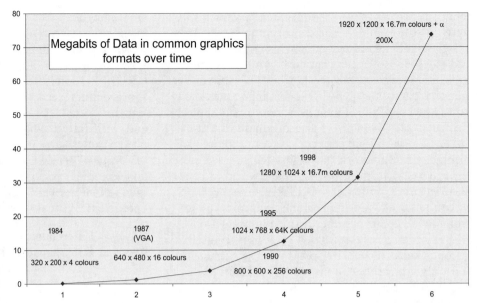

Figure 10. Increase in computer graphics bandwidth since the 1980s.

had a cathode-ray tube display monitor which could be used by programmers to view (as binary data with bright spots representing 1 and dim spots representing 0) the content of one memory store at a time. In the EDSAC there were 32 memory stores, or delay lines in total, each of which stored 32 words of 18 bits. The display monitors, with their flashing grid of dots, were only used to observe the progress of a program and monitor it. More usefully, numerical values were output to punched paper tape, or later to an attached teleprinter.

"Input systems involving modified teletype equipment for inscribing keyboarded information were often integrated with teletypes and printers for the output device. This form of input and output continued to be used on some mainframes until the 1980s, but from the 1960s VDUs (visual display units) based on the cathode-ray tube replaced teletype as the output."

The development of computer graphics technology based on cathode-ray tubes in the early 1960s by Dr David Evans and Dr Ivan Sutherland at the University of Utah, led eventually to an explosion in computer display use. By the end of the century, around 150 million computer displays were being manufactured each year. These devices had an impact on the daily life of almost everybody in the developed world.

Most computer displays of the late twentieth century are based on cathode-ray tubes (CRTs), invented by Karl Ferdinand Braun in February 1897 in Aachen, Germany. In fact the name predates the identification of the electron as "cathode rays."

In the 1970s, when interactive terminals started to be needed in high volumes, the CRT was a natural choice as it was already in high-volume production because of the existing market for television use. The high volume of CRTs and the development investment to enable color TV gave the technology an overwhelming cost advantage over competing technologies.

The first computer displays used "vector scanning"; that is, the beam of the CRT was deflected using varying voltages to paint vectors, or lines, on the screen. Thus, when the computer wanted to paint a circle, the beam moved in a circular motion. Later displays used a fixed scanning system, called raster graphics and used in televi-

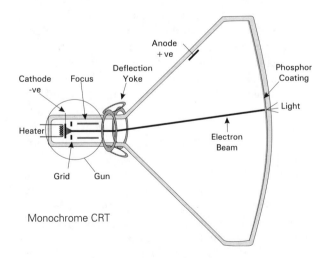

Figure 11. Line diagram showing operation of a CRT.

sions, where the image was stored in the memory of the computer as dots, or "pixels" (short for picture cells) that represent values for the color of the image at different screen locations. The computer scans through the memory and sends the values to the screen for display.

Cheaper and faster memory technologies enabled the widespread use of raster graphics in the 1970s and 1980s; CRT makers started to develop CRTs specially optimized for computer display use.

Mainframe and minicomputers used "terminals" to display the output. These were fed data from the host computer and processed the data to create screen images using a graphics processor. The display was typically integrated with a keyboard system and some communication hardware as a terminal or video display unit (VDU) following the basic model used for teletypes. Personal computers (PCs) in the late 1970s and early 1980s changed this model by integrating the graphics controller into the computer chassis itself.

Early PC displays typically displayed only monochrome text and communicated in character codes such as ASCII. Line-scanning frequencies were typically from 15 to 20 kilohertz—similar to television. CRT displays rapidly developed after the introduction of video graphics array (VGA) technology (640 by 480 pixels in 16 colors) in the mid-1980s and scan frequencies rose to 60 kilohertz or more for mainstream displays; 100 kilohertz or more for high-end displays. These displays were capable of displaying formats up to 2048 by 1536 pixels with high color depths. Because the human eye is very quick to respond to visual stimulation, developments in display technology have tended to track the development of semiconductor technology that allows the rapid manipulation of the stored image.

CRTs have major disadvantages in bulk, weight, and power consumption and from the early 1970s onwards, there were many attempts to replace them with flat panel displays.

Liquid crystal displays (LCDs) were the main competitive technology. Early LCDs, developed in the late 1960s, had difficulty matching the response speeds, viewing angles, and color performance of CRTs and were much more expensive to manufacture. LCDs using an active matrix made an initial impact in mobile computers, where the lower bulk and power advantages made the cost premium acceptable. In the late 1990s, LCD makers developed active matrix TFT LCDs that could start to get close to CRTs for performance and price and adoption was then rapid even on the

desktop. In the first decade of the twenty-first century, the CRT will be replaced by the LCD.

Most computers are used by individuals or very small groups at short viewing distances. However some are used to display to groups and two basic approaches are used—projection and direct view.

Early projection displays were based on high-brightness CRTs, but these were expensive and difficult to maintain and align. LCDs became the most popular imaging device for computer projectors, although the digital micromirror device (DMD) invented by Texas Instruments became popular for very portable projectors. In these projectors, small imaging devices (typically 23 millimeters diagonal) are used for creating an image that is illuminated by a very bright light and projected onto a screen using an optical system.

The main large-screen direct view display is the plasma display panel (PDP). Although originally conceived as a general-purpose desktop display, it proved difficult to make small pixels with efficient light output, so PDPs were developed for applications from 640 millimeters diagonal and above, designed for viewing from several meters or more. From the early development of monochrome PDPs in the 1970s and 1980s, by the 1990s full color PDPs up to 1.5 meters diagonal were in production.

For outdoor and public displays of very large size, such as those used in sports stadia, displays based on LED technology were developed. As has often been the case in display technology, blue devices proved difficult to discover and make and it was only in the 1990s that blue LEDs were first made, although red LEDs had been in production since the early 1960s and green shortly afterward. It has proved difficult to make LEDs with matched brightness, therefore LED displays are expensive, but can have a long life and high brightness.

Over many years there were attempts to develop "near to eye" personal displays, but problems in designing the high-quality low-weight and -cost optical systems and concerns about potential effects on health limited the use of this kind of display to professional and military applications.

See also **Computer-User Interface; Light Emitting Diodes; Liquid Crystals; Printers; Radar Displays**

ROBERT RAIKES

Further Reading

Foley, J.D., van Dam, A., Feiner, S.K., Hughes, J.F. *Computer Graphics, Principles & Practice.* 2nd edn. Addison-Wesley, 1995.

Keller, P.A. *The Cathode Ray Tube, Technology, History and Applications.* Palisades Press, New York, 1991.

Myers, B. *Display Interfaces, Fundamentals and Standards*. Wiley, Chichester, 2003.

Peddie, J. *High-Resolution Graphics Display Systems*. Windcrest/McGraw-Hill, 1994.

Computer Memory, Early

Mechanisms to store information were present in early mechanical calculating machines, going back to Charles Babbage's analytical engine proposed in the 1830s. It introduced the concept of the "store" and, if ever built, would have held 1000 numbers of up to 50 decimal digits. However the move toward base-2 or binary computing in the 1930s brought about a new paradigm in technology—the digital computer, whose most elementary component was an on–off switch. Information on a digital system is represented using a combination of on and off signals, stored as binary digits (shortened to bits): zeros and ones. Text characters, symbols, or numerical values can all be coded as bits, so that information stored in digital memory is just zeros and ones, regardless of the storage medium. The history of computer memory is closely linked to the history of computers but a distinction should be made between primary (or main) and secondary memory. Computers only need operate on one segment of data at a time, and with memory being a scarce resource, the rest of the data set could be stored in less expensive and more abundant secondary memory. The focus of this entry will be on primary memory technology, with attention to the early developments and the technological hallmarks.

A method of storing data on rotating "drums"—an electromechanical device—had emerged in the late 1930s. John Vincent Atanasoff successfully manufactured a device consisting of a rotating drum in which 1600 capacitors were placed in 32 rows. Later, magnetic coatings were used on drums and during World War II, this storage emerged as a reliable, rugged, inexpensive but slow memory device. Engineering Research Associated (ERA) produced a commercial device in 1947 that could store 65,000 32-bit words (over 2 million bits). Magnetic drum storage was to survive in later computers for secondary data storage, and later in disk platter form of continually shrinking proportions. It was the predecessor to the hard disk drive, the ubiquitous component of all later computers.

Early digital computers made use of electromechanical relays for binary logic. Early relay computers included Howard Aiken's automatic sequence-controlled calculator. Developed at Harvard University with the assistance of a group of International Business Machine (IBM) engineers, the Mark I was completed in 1944 and could store 72 numbers mechanically using electromagnetic decimal storage wheels, with the programming and data sequencing instructions fed into the machine via four punched tape readers. However the bulk, power requirements, and the delay time in the operation of relays were performance-limiting factors.

Thermionic valve (or vacuum tube) technology matured during World War II. Valves could operate as simple switches and a switching speed increase in the order of 1000 times over relays was realized. Although there were earlier prototypes, the completion of the ENIAC (electronic numerical integrator and computer) heralded the era of electronic computers. Built for the U.S. Army by the Moore School of Electrical Engineering at the University of Pennsylvania, it was only completed in late 1945 after the war. It consisted of 18,000 valves in which its core logic was implemented, as well as 1,500 relays which were used to store initial data loaded from a punched tape reader. Twenty "Accumulators" to store binary numbers during calculations were implemented with valve logic. Subsequently, purpose-built binary data storage valve technology emerged, although not without difficulties in production (see Selectron valve below).

The first generation of computers such as the Mark I and the ENIAC were purpose-built, and had the sequence of instructions "hard coded" or preprogrammed, for example with paper tape input. The British Colossus computer was another special purpose machine using electronic valves that implemented algorithms developed by Alan Turing and colleagues to crack encrypted German U-boat radio communications during World War II. However, mathematician John von Neumann approached the issue of storing programming code at a more fundamental logical level and initiated the concept of the stored program, whereby program code could be stored in memory much like the data was, meaning the setup for a new calculation could be expedited without rewiring. The EDVAC computer (electronic discrete variable automatic computer), first described in a draft report in June 1945 by von Neumann who was a regular visitor to the U.S. Army project at the Moore School, University of Pennsylvania, was the first design to implement the stored program concept—what came to be known as the "von Neumann" architecture. EDVAC was, however, just an idea at that stage, and not constructed by the Moore Laboratory until 1949. Von Neumann and EDVAC's designers were keen to use purely electronic memory for stored program, not drums or relays.

From around the period of the mid- to late 1940s, emerging digital computers experienced what historian Thomas Hughes describes as a "reverse salient." This was a result of the scarcity of capacious and reliable memory technology and held back the advance of digital computers. Several electronic innovations in response to this were adaptations from technology developed during the war.

The ultrasonic delay line was developed for storing analog radar information, and was easily adapted to digital storage. The delay line worked by using a piezoelectric quartz crystal to convert an electrical signal into a sonic wave pulse, which then traveled through a liquid medium at the speed of sound in a long tube, and was then converted back to an electric signal. The difference in the speed of propagation meant that a number of bits of data could be stored. The delay line was a serial device—once a pulse had entered the tube there was no way to access it until it emerged at the other end. All operations were considered as serial transfers of numbers.

Delay lines were first built at the Bell Telephone Labs by William Shockley, and later at the Moore School for the Radiation Laboratory at the Massachusetts Institute of Technology (MIT) in 1943. Mercury acoustic delay lines were adapted for use with computers by J. Presper Eckert and John Mauchly in 1946 and selected for the EDVAC, which was actually not completed until 1952. Mercury delay lines realized a hundred times storage density saving over valve technology. The first use in digital computers was in Maurice Wilkes's EDSAC (electronic delay storage automatic computer) completed in 1949 at the Cavendish Laboratory in Cambridge, U.K., where 32 "tanks" of delay lines each stored 32 words of 18 bits (totaling about 2 kilobytes).

In response to the high cost and slowness of delay lines, Frederick Williams and colleagues at Manchester University in the U.K. developed the technique of using electrostatic cathode-ray display tubes as digital stores. The cathode-ray tube was another wartime development and the persistence of the phosphor display screen once illuminated by the cathode ray provided the memory. By 1948, a storage of 1024 bits was successfully implemented. Williams' colleague Tom Kilburn made improvements that increased the capacity to 2048 bits. The Williams–Kilburn tubes (commonly known as Williams tubes) were used on several of the early stored program computers, including the Manchester "Baby" (1948) and the Manchester Mark 1 which became operational in 1949, and the Institute of Advanced Study (IAS) machine spearheaded by von Neumann at Princeton, finally completed in 1951. The Williams tube memory had a big advantage over delay line memory in that it allowed fast random access (any memory location could be addressed and read directly). The Manchester Mark I was the first to store both its programs and data in RAM (random-access memory), as modern computers do.

In 1946 Vladimir K. Zworykin, Jan Rajchman and colleagues at the Radio Corporation of America (RCA) labs built the Selectron valve (or selective storage electrostatic tube) that could store 256 binary digits (bits) of information, and was described by Herman Goldstine (who directed the outpost of the U.S. Army's Ballistics Research Laboratory at the Moore School from 1942) as "a work of great engineering virtuosity." Like the Williams tube, the Selectron was also a random-access storage device. The Selector design however took over three years to reach the market, and was scaled down from the initial specification of 4096 bits storage. The revised Selectron was used in the 1953 JOHNNIAC at the Rand Corporation. The JOHNNIAC was one of the many machines based on the Princeton Institute of Advanced Studies (IAS) report authored in 1946 by John von Neumann, Arthur Burks, and Goldstine, based on their work on digital computers at Princeton following their involvement at the Moore School.

Magnetic core memory was first suggested by Jay Forrester of the Servomechanisms Laboratory at MIT and Andre Booth of the University of London, and was an important step in miniaturization and speed of computer memory. Small ferrite rings that were laid out in a two-dimensional matrix, with the magnetic property of hysteresis "remembering" an electric pulse. Each element of data was independently accessible via a row and column addressing system. Forrester and colleagues developed a successful prototype in 1949, and core arrays storing 1024 bits were used in their Whirlwind computer installed for the Navy's SAGE anti-aircraft defense system in 1958. Core memory became the dominant means of primary storage and was widely used on large commercial and scientific computers such as the IBM 701 in 1955, which was a significant commercial success and led to IBM's establishment as the dominant player in large systems. The IBM 7094 in 1963 had a core memory with over a million bits.

Ultimately it was the solid-state semiconductor that was to provide the memory technology of the computer revolution. The transistor was invented in 1947 by William Shockley at the Bell Laboratories after eight years of research on semiconductors for radar use. It was small and had very low power requirements and the switching speed was very quick compared to valves and core memory. Although it was a functional "switch" replacement for the valve as discussed above, it did not appear as computer memory until the late 1950s. There was significant momentum with the success of magnetic core memory in the industry, and the reliability of initial mass produced transistors was a contributing negative factor. Furthermore, Bell Labs was a regulated monopoly and limited by a court decision to the telecommunications business, although they did use transistors in special purpose computers built for the military use in intercontinental ballistic missiles around 1952. However, information about transistor technology was available to other manufacturers, with Philco successfully mass producing reliable and high-performance components. These were used in the SOLO, regarded as the first general purpose transistorized computer, completed by 1958. The commercialized version was called the TRANSAC which became available in 1960. Although the IBM 7090 (first installed in 1959) had a transistorized central processing unit, it still used magnetic core memory.

The development of integrated circuit (IC) technology began in 1959 with Jack Kilby of Texas Instruments. Collections of transistors were miniaturized onto a silicon "chip." It was only in the early 1970s that semiconductor memory was commercially used in computers after earlier pioneering use in purpose-built military and Apollo-era space guidance computers in the 1960s. The Super Nova produced by Data General in 1971 was the first commercial computer to use integrated circuit transistor-transistor logic (TTL) IC chip memory, after a less-successful research computer, the Illiac-IV, broke the tradition of use of magnetic cores by using a 256-bit memory chip made by Fairchild. IBM went down a proprietary route by using monolithic semiconductor memory in the System/370 Model 145 in 1971. Intel entered the IC memory market by producing a 1024-bit memory chip in 1970.

Memory chips were rapidly embraced by the emerging minicomputers, resulting in both compact size and more importantly, low cost. Later in the 1970s, these IC memory chips together with microprocessors provided the basis of the personal computer. With the advent of large-scale integration (LSI) and very large scale integration (VLSI), the cycle of increasing density of semiconductor memory fabrication continued unabated. Although the concepts and architecture of computers remained consistent with early designs, these techniques, competitive markets, and insatiable demand for personal computers globally, resulted in a doubling of memory density almost every year to the end of the century where up to a billion transistor elements on memory chips were commonplace.

See also **Computers, Early Digital; Computer Memory, Personal Computers; Integrated Circuits, Design and Use; Processors for Computers; Transistors; Vacuum Tubes/Valves**

BRUCE GILLESPIE

Further Reading

Ceruzzi, P.E. A history of modern computing. MIT Press, Cambridge, MA, 1998.

Eames, C. and Eames, R. *A Computer Perspective, Background to the Computer Age.* Harvard University Press, Cambridge, MA, 1973.

Goldstine, H.H. *The Computer: From Pascal to Von Neumann.* Princeton University Press, Princeton, 1972.

Hodges, A. *Alan Turing: The Enigma.* Burnett Books, 1983.

Wurster, C. *Computers: An Illustrated History.* Taschen, Koln, 2002.

Computer Memory, Personal Computers

During the second half of the twentieth century, the two primary methods used for the long-term storage of digital information were magnetic and optical recording. These methods were selected primarily on the basis of cost. Compared to core or transistorized random-access memory (RAM), storage costs for magnetic and optical media were several orders of magnitude cheaper per bit of information and were not volatile; that is, the information did not vanish when electrical power was turned off. However, access to information stored on magnetic and optical recorders was much slower compared to RAM memory. As a result, computer designers used a mix of both types of memory to accomplish computational tasks. Designers of magnetic and optical storage systems have sought meanwhile to increase the speed of access to stored information to increase the overall performance of computer systems, since most digital information is stored magnetically or optically for reasons of cost.

The first systems for the magnetic storage of large amounts of digital information were developed in the early 1950s. They were designed to replace the punch card record systems then in common use in large business and government organizations. These early magnetic systems were based on reel-to-reel tape recorder technology developed in the 1930s and 1940s for sound recording. In adapting these systems for computer use, the controls for tape movement were made fully electronic so that they could be controlled by the central processing unit (CPU) in the computer. The only other significant innovation was the addition of a system whereby a loop of tape was maintained using vacuum to buffer the rapid starts, stops, and changes in direction of the tape, considerably reducing the chance of tape breakage. It is this vacuum system that accounts for the common appearance of mainframe computer tape drive systems—the tape reels mounted in the upper part of a tall cabinet and the vacuum mechanism concealed behind the panel below.

While an inexpensive form of storage, tape drives were very slow due to the one-dimensional arrangement of information. Thus, by the 1960s they were replaced by disk drives for most computer applications. However, due to their very low cost per bit for storage, tape drives continued to be used for backup storage as information was regularly copied on hard disks and other storage methods. The form of these backup tape drive systems changed over time, mirroring changes in sound, and later videotape systems. In particular, tape drive systems moved from open reel to cassette systems by the 1980s. At the end of the twentieth century, tape systems continued to be used for backup copies.

The disk drive, the most commonly used computer mass storage device of the twentieth century, was invented at International Business Machines (IBM) in the early 1950s by a team led by Reynold B. Johnson and Louis Stevens, and first marketed in 1957. An outgrowth of earlier drum memory systems, the disk drive consists of one or more flat disks that rotate at high speed. Information is magnetically recorded on and read from the surface of the disk using an electromagnetic coil, the read/write head. Information is recorded in a series of concentric tracks, and the head is moved between tracks by an actuator. Thus, in contrast to the one-dimensional storage on a tape drive, the two-dimensional form of the disk drive allows for much faster access to a particular bit of information, albeit at increased cost and complexity.

The basic mechanical operation of disk drives has not changed since their initial development. However, the capacity for data and the speed of access to it has increased by many orders of magnitude. This has resulted from a large number of gradual improvements rather than a single breakthrough. Three major areas of improvement account for most of the speed increase. First, the development of improved magnetic alloys allowed for greater recording densities without loss of accuracy. Second, improved manufacturing methods produced smoother disk surfaces. As a result, the recording head could operate closer to the disk surface, making narrower tracks and so increasing density. Finally, better recording heads and more accurate actuating mechanisms to guide them also allowed for narrower recording tracks and greater density. As a result, the average diameter of disk drives has dropped from 12 inches (300 millimeters) in the 1960s to under 3 inches (76 millimeters) at the end of the twentieth century. The smaller size and greater density of the disks means that data access becomes faster since heads do not have to move as far to read information.

The operation of removable magnetic media such as floppy disk drives, zip-drives, and the like is identical to that of hard disks. Removable media operate with lower precision than hard drives, and thus storage densities are lower. However, the advantage of removable media is its portability—data can be moved from one computer to another. In the late 1990s, a hybrid type of removable media emerged, a self-contained disk drive that could be easily plugged into computer systems without a separate power supply.

Optical storage systems operate in a fashion similar to disk drives. However, information is recorded using a laser to mark the surface of the disk, and the resulting marks are scanned using the laser and a photoelectric sensor. The primary advantage of optical storage is its resistance to bulk erasure. Unlike magnetic storage, which can be erased by exposure to a strong magnetic field, optical storage is more stable. Although in theory optical systems can have recording densities similar to magnetic media, in practice during the twentieth century magnetic systems had higher densities due to engineering differences.

The most common computer-related optical storage medium in the late twentieth century was the compact disk (CD), based on the same CD developed for recorded music in the 1970s. Information is recorded on the disk in one continuous spiral rather than in a concentric ring. The introduction of new recording media in the 1990s allowed the CD, initially a write-once, read-many system, to function much like the disk drive. Due to low cost and stability, the CD became the media of choice in the late 1990s for the distribution of software, an application previously dominated by removable floppy disks. However, magnetic recording continued to dominate hard disk applications due to lower cost and higher densities.

Both magnetic and optical recording methods have upper physical limits on recording density, but by the end of the twentieth century those limits had not yet been reached. Due to their low cost per bit, it is likely that these types of recording will be used in computer systems for some time to come.

See also **Alloys, Magnetic; Audio Recording, Compact Disk; Computer Memory, Early**

MARK CLARK

Further Reading

Bell, A.E. Next generation compact discs. *Scientific American*, 275, 1, 42–46, 1996.

Daniel, E.D. Mee, C.D. and Clark, M.H. *Magnetic Recording: The First 100 Years*. IEEE Press, 1999.

Camras, M. *Magnetic Recording Handbook*. Van Nostrand Reinhold, New York, 1988.

Computer Modeling

Computer simulation models have transformed the natural, engineering, and social sciences, becoming crucial tools for disciplines as diverse as ecology, epidemiology, economics, urban planning, aerospace engineering, meteorology, and military operations. Computer models help researchers study systems of extreme complexity, predict the behavior of natural phenomena, and examine the effects of human interventions in natural processes. Engineers use models to design everything from jets and nuclear-waste repositories to diapers and golf clubs. Models enable astrophysicists to simulate supernovas, biochemists to replicate protein folding, geologists to predict volcanic eruptions, and physiologists to identify populations at risk of lead poisoning. Clearly, computer models provide a powerful means of solving problems, both theoretical and applied.

Scientific models, or representations of nature, date to antiquity, when astronomers developed mathematical methods to predict planetary movements. However, computer models only originated in the mid-1940s. Indeed, the history of computer models is intertwined with that of computers. The second digital computer, the ENIAC (electronic numerical integrator and computer), was used to simulate an early design of the hydrogen bomb. Before the ENIAC, physicists had to use slide rules

to calculate the power of nuclear explosions. The invention of high-speed digital computers made it possible to speed up the calculation process and thereby perform faster simulation experiments by programming mathematical equations representing the laws of nature. Computers obtain solutions to differential equations by evaluating variables at specific points at given time intervals. In this way, a computer can simulate the development of a thunderstorm by evaluating pressure, temperature, wind velocity, humidity, and other variables at thousands of points in space and time. Models are thus mathematical representations of physical phenomena.

Although computing power has increased exponentially over the past six decades, modelers must still make trade-offs between resolution and complexity. A modeler can increase the accuracy and realism of a simulation by adding more spatial points and shortening the time intervals, or by adding more laws of nature to the set of equations. However, while higher-resolution models decrease the levels of uncertainty associated with computer modeling, such actions increase computing time and power, demands on personnel, and funding obligations.

Because it remains difficult to integrate all the variables of complex systems, even the most sophisticated computer models contain shortcomings. Such limitations include reliance on simplifying assumptions, spatial resolution problems, difficulties in validating models with observational data and experimental results, problems in representing small-scale processes in terms of large-scale variables, and weak agreement between different models. Such deficiencies have provoked intense debates regarding the scientific standards of model trustworthiness, especially the accuracy of modelers' assumptions. These debates are not limited to the scientific community, since models have explicit policy-making value. Computer models are therefore interesting not only for the ways in which they have changed the practice of science, but also for the political and epistemological questions raised by their use.

The significant political role of computer models is exemplified by the ongoing international debates over global climate change and the greenhouse effect. Today's global climate models evolved from weather-prediction models and general circulation models of the 1950s and 1960s. During the 1970s and 1980s, larger, faster, cheaper computers facilitated the construction of models capable of analyzing both atmospheric and oceanic general circulation. Such "coupled" models allow scientists to study the effects of increasing carbon dioxide levels on the global climate, and other phenomena that cannot be subjected to controlled laboratory tests, including acid rain, ozone depletion, and nuclear winter. Politicians have used data generated by such models to develop controversial policies, such as the 1997 Kyoto Protocol, a treaty designed to slow global warming by requiring 38 developed nations to reduce greenhouse gas emissions to an average of 5.2 percent below 1990 levels by 2012.

Another important political role played by computer modeling relates to national security. Following the end of the Cold War and the U.S. government's 1992 moratorium on nuclear testing, U.S. officials seeking to build domestic support for the Comprehensive Nuclear Test Ban Treaty while keeping the national nuclear weapons laboratories open, developed the Stockpile Stewardship Program and Accelerated Strategic Computing Initiative (ASCI). As a result, supercomputer models have been used since the mid-1990s to maintain the U.S. nuclear stockpile without detonating actual bombs. The political need to simulate the incredibly complex physics of nuclear explosions has led to tremendous advances in computational power. Over several months beginning in 2001, researchers at the Lawrence Livermore and Los Alamos national laboratories used the world's fastest supercomputer, IBM's ASCI White—capable of performing 12.3 trillion calculations per second (teraflops)—to conduct the first complete three-dimensional simulation of a thermonuclear explosion. ASCI officials anticipate the creation of 100-teraflop supercomputers, a technological advance with applications far beyond stockpile stewardship. Indeed, critics have long argued that such efforts to scale up supercomputing will facilitate the design and development of new nuclear weapons, not just the maintenance of ageing armaments.

Despite the awesome advances in supercomputer modeling techniques brought about by weaponry testing and weather forecasting, models still cannot represent complex real-world processes with absolute precision. For this reason, some groups oppose the use of model-generated data in the policymaking arena. Most notably, global warming skeptics point to the inability of global climate models to predict regional changes, and scorn modelers' assumptions and simplifications as "internal fudge factors" which generate "garbage-in, garbage-out" simulations. Such charges have led to antagonistic debates on the relationship between model trustworthiness and credible science, as well as initiatives to improve computer modeling uncertainty analysis.

The political use of computerized models has also stimulated discussions of the philosophical and epistemological implications of computer modeling. During the 1990s, some scientists and scholars in the field of science studies asserted that computer models cannot be validated, since models function as inductive arguments, which according to the philosopher Karl Popper, can only be proven false, not true. One organization that relies on computer models, the United Nations-sponsored Intergovernmental Panel on Climate Change, responded by replacing the word "validation" with "evaluation" to convey the extent of correspondence between models and the real-world processes they represent. Other analysts questioned how policymakers can expect high levels of model-based certainty, since many political decisions are made under uncertain conditions. As with any powerful new technology, debates over the trustworthiness of computer modeling techniques will persist as long as unacceptable levels of uncertainty remain. Nevertheless, computer models will likely continue to transform both the practice of science and the production of policy-relevant knowledge.

See also **Computers, Uses and Consequences**

CHRISTINE KEINER

Further Reading

Cipra, B. Revealing uncertainties in computer models. *Science*, 287, 960–961, 2000.

Edwards, P.N. A brief history of atmospheric general circulation modelling, in *General Circulation Model Development: Past, Present, and Future*, Randall, D.A., Ed. Academic Press, San Diego, 2000.

Kaufmann III, W.J. and Smarr, L.L. *Supercomputing and the Transformation of Science*. Scientific American Library, New York, 1993.

Konikow, L.F. and Bredehoeft, J.D. Ground-water models cannot be validated. *Adv. in Water Resour.*, 15, 75–83, 1992.

Kwa, C. Modelling technologies of control. *Sci. Cult.*, 4, 20, 363–391, 1994.

Oreskes, N. Evaluation (not validation) of quantitative models. *Environ. Health Perspect.*, 106, 6, 1453–1460, 1998.

Oreskes, N. Why believe a computer? Models, measures, and meaning in the natural world, in *The Earth Around Us: Maintaining a Livable Planet*, Schneiderman, J.S., Ed. W.H. Freeman, New York, 2000.

Paine, C.E. A case against virtual nuclear testing. *Scientific American*, 281, 74–79, 1999.

Shackley, S., Risbey, J., Stone, P. and Wynne, B. Adjusting to policyexpectations in climate change modelling: an interdisciplinary study of flux adjustments in coupled atmosphere-ocean general circulation models. *Climatic Change*, 43, 413–454, 1999.

Weiss, P. Computer simulates full nuclear blast. *Science News*, 161 189, 2002.

Computer Networks

Computers and computer networks have changed the way we do almost everything—the way we teach, learn, do research, access or share information, communicate with each other, and even the way we entertain ourselves. A computer network, in simple terms, consists of two or more computing devices (often called nodes) interconnected by means of some medium capable of transmitting data that allows the computers to communicate with each other in order to provide a variety of services to users.

Using the above definition, a typical computer network must include the following:

- A set of computing nodes which may range from simple personal digital assistants for retrieving electronic mail (e-mail) all the way to the largest supercomputers.
- A data transmission medium over which the nodes can communicate with each other. This medium can be optical fiber, coaxial cable, copper twisted pairs, or even air and free space for wireless transmission of data. In effect, the transmission media can be considered as the roads and highways over which data and information are transmitted.
- An interface (often called a NIC or network interface card) between the computing node and the data transmission medium to transform the digital signals used in the computer into a form suitable for the transmission medium (for example, light signals for optical fiber or electromagnetic waves for wireless transmission).
- A communication or network protocol which is the set of rules and conventions that govern the effective transmission of data over the network to ensure that communicating nodes "speak the same language," recognize, and understand each other.
- variety of devices (often called network electronics) used to perform a variety of network control tasks such as traffic switching, filtering, aggregating, routing, protocol translation, and signal conversion from one transmission medium to another (for example, from an optical fiber medium between buildings to a twisted copper pair medium within buildings).

Historically, the first communications networks were telegraphic. Communication protocols were developed to relay messages from one network station to another ("message switching"). The expansion of telephone networks in the late nineteenth century meant that each telephone could not be connected directly to every other telephone; automatic switching was needed ("circuit switching"). By the 1960s when the first computer networks were being developed, similar switching for data communications over telephone lines was needed. (The development of the first resource-sharing computer networks (at the National Physical Laboratory in England and ARPAnet) and the concurrent development of "packet switching" are described in the entries on Electronic Communications, Internet, and Packet Switching.)

A given computer network can be characterized by a number of attributes, with the most important being geographic size, protocol and type of access control used, and the way it is funded. Networks that span a local area such as one or more neighboring buildings are called local area networks, or LANs. Ethernet is by far the most widely used LAN technology today. It was developed in the mid-1970s at the Xerox Palo Alto Research Center to interconnect a number of desktop computers and printers. It was so successful that it quickly became a standard. Since those days, there have been numerous enhancements and there are standards defining Ethernets that support different network topologies and a variety of transmission media at rates of 10,100 and 1,000 million bits per second.

Networks distributed over a wide area (distances of tens to thousands of kilometers) are called wide area networks or WANs, while metropolitan area networks (MANs) span distances between those of a LAN and a WAN. The transmission media for LANs is usually provided by the organization owning the LAN, while the transmission media for WAN connectivity are usually leased from telephone or common carrier companies.

Computer networks can also be characterized by the way they are funded. Private networks, for example, are usually owned by some entity that confines use of the network and its services to its employees or customers. Public networks are owned by entities that offer network services to any individual or organization that wishes to subscribe and pay for the services provided. Networks that provide e-mail and Internet access services to any user on a fee-for-service basis are examples of public networks. Finally, community networks refer to networks that are managed and supported by a local community and provide services specially tailored to its community of users.

Client/server and peer-to-peer are phrases used to describe networks with different types of hierarchy and access control. In a client/server network, there is at least one computer that is dedicated as a server to some kind of service (for example e-mail or Web service) which controls access from other computers that are its clients. A peer-to-peer network, on the other hand, does not have any dedicated servers or hierarchy among its computers. Instead, users must make decisions regarding who gets access to what, which adversely impacts system security and consequently such networks are limited to small numbers of computers.

In the very early days, computer networks were relatively small, and computer or modem vendors tended to develop proprietary protocols that allowed their machines to communicate with each other, but not necessarily with those made by others. This created the need to interconnect two or more compatible or incompatible networks together to create an Internet, or a network of networks. In the mid 1970s, TCP/IP (transmission control protocol/internet protocol) was developed and became the main communication protocol using the Unix operating system and its various derivatives. Today, TCP/IP is the de facto standard and is available for almost all computers. When sending a file from one machine to another, TCP breaks it into a number of appropriately sized data packets, which also include other information such as the type of packet, source, and destination addresses. IP then takes over, routing those packets from the source to the destination, where

TCP takes over and reassembles them into the original file. Together, these two protocols allow data and control information to be "packetized," addressed, and routed to the destination where it is reassembled and appropriately processed.

Today, the great majority of computer networks are connected to form the global Internet, which uses TCP/IP. When a user wants to access a Web server such as www.xyz.com, the symbolic destination domain, xyz.com, is translated into a specific IP address, and the user is connected to that server. Once there, the browser on the client (user) machine allows the user to look at specific files of various kinds. Similarly, when an e-mail is sent to userid@xyz.org, the source mail server sends the message to the destination server xyz.org which then stores the file in the mail folder of userid, until the latter logs into xyz.org, reads the e-mail, and deletes it.

Figure 12 shows two LANs where computers are connected via switches and routers to the global Internet. Today, the latter interconnects millions of such LANs worldwide, with a total of tens of millions of computers providing a variety of services to hundreds of millions of users. Its total traffic has been growing exponentially, and it is expected that this growth will continue as new generations of Internet applications are developed, especially those that require the digitization and transmission of voice or video packets in addition to data.

See also **Electronic Communications; Internet; Packet Switching; Telephony, Digital; World Wide Web**

JOHN SOBOLEWSKI

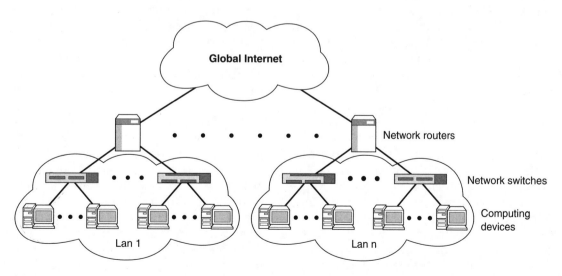

Figure 12.   Diagram of a computer network showing how network switches interconnect computers within LANs, which are then connected via routers to the Internet.

Further Reading

Comer, D. and Stevens, D. *Internetworking with TCP/IP. Vol. I: Principles, Protocols and Architecture.* Prentice Hall, Englewood Cliffs, NJ, 1994. A description of principles of communication protocols with an emphasis on TCP/IP.

Metcalf, R. and Boggs, D. Ethernet: distributed packet switching for local computer networks. *ACM Comm.*, 19, 7, 395–403, 1976. The classic paper describing the first Ethernet.

Peterson, L. and Davie, G. *Computer Networks: A Systems Approach.* Morgan Kaufman, San Francisco, 2000. A recent book describing various aspects of computer networks.

Ralston, A. *et al.*, Eds. *Encyclopedia of Computer Science.* Nature Publishing Group, Oxford, and Grove's Dictionary, New York, 2000. This contains a number of articles describing various aspects of computing and computer networking.

Sobolewski, J. Data transmission media, in *Encyclopedia of Physical Science and Technology*, 3rd edn, vol. 4, Meyers, R., Ed. Academic Press, London, 2001. Describes various data transmission media in use today.

Computer Science

Computer science occupies a unique position among the scientific and technical disciplines. It revolves around a specific artifact—the electronic digital computer—that touches upon a broad and diverse set of fields in its design, operation, and application. As a result, computer science represents a synthesis and extension of many different areas of mathematics, science, engineering, and business.

The tide that lifted science and engineering in general following World War II lifted computing as well, albeit in a less straightforward fashion. It was the need to speed up the calculation of ballistic firing tables (beyond what was achievable using mechanical differential analyzers, essentially analog calculators) after all, that led to development of the ENIAC (electronic numerical integrator and computer), completed in 1945. Science and engineering emerged from World War II and entered the Cold War with a new sense of national importance attached to them. As a result, government funding of scientific and technical research and training, especially in universities and particularly in the U.S. and the U.K., surged.

However, the interdisciplinary nature of computing complicated development of a computing discipline both politically and substantively. Establishing the jurisdiction of a profession or discipline by definition involves political issues of control and legitimacy. Since computing did not obviously and by consensus fall within the purview of any single group of practitioners or academics, many laid claim to it and to the prerogatives and resources that accompanied it. Existing academic disciplines, including applied mathematics, electrical engineering, and physics (varying on a university-by-university basis), attempted to keep computing under their respective institutional wings. Industrial practitioners, meanwhile, took issue with academic approaches in general, demanding a discipline that bore some connection with what they actually did. In the U.S., for example, this meant that even as mathematics and electrical engineering departments on some campuses engaged in a tug of war to define computer science, the Association for Computing Machinery, with its academic bent, was engaged in a similar struggle with the Data Processing Management Association, with its industrial orientation.

Substantively, no one had a clear or dominant definition of what computing practitioners did or what knowledge constituted the core of a computing discipline. As a result, efforts to define it tended to reflect the disciplinary and professional backgrounds of the people doing the defining. Computer science (the very term prompted some heated arguments) ended up adopting an eclectic mixture of knowledge from a range of existing disciplines and from industry as well. From electrical engineering came such things as switching theory and circuit analysis and design. From applied mathematics came numerical methods and algebraic analysis. From linguistics came syntactic and semantic analysis. From psychology came cognitive theory. From business came information management. All of these, and others as well, were adapted and extended to focus on computing, leading to distinctive areas such as computer architecture (the organization of computer hardware), algorithmic analysis (the determination in principle of the time and storage efficiency of algorithms), computability theory (the analysis of which classes of problems are in principle solvable by a computer), artificial intelligence (the development of computer systems that can learn and otherwise replicate human intelligence), human–computer interaction (the analysis and application of the psychological processes governing communication between humans and computers), information storage and retrieval (the organization and use of databases), software engineering (the development of software in a disciplined and systematic fashion), and programming languages (the analysis and design of programming languages using formalized principles and notations), among others.

Computing professionals worried about how to unify this range of concerns in a fashion deeper than their common focus of the computer. This search for "fundamental principles" akin to those in established disciplines generated a great deal of angst, and to some extent still does. However, identifying principles with both the requisite breadth and the desired depth has proven very problematic. Often the two qualities represent a trade-off.

Computing professionals also sought to be viewed as such by others. Industrial practitioners sought this recognition from the both the government and the general public. In the U.S., for example, programmers and system analysts were outraged when, in 1972, the Department of Labor ruled that they were not professionals. At the same time, academics openly fretted about their professional standing relative to other academic fields and the potential impact on government research funding. While some of this same anxiety was evident in Europe, it was generally milder. This was partly due to the legitimacy lent by different professional structures, which generally awarded professional societies in Europe much more authority to regulate their professions. The British Computer Society, for example, has full and explicit authority to accredit curricula and to nominate Chartered Engineers. Moreover, debates about whether computer science was in fact a science were more muted in those parts of Europe that placed this collection of knowledge under the rubric of "informatics." (This represented a certain irony as the European inclination toward mathematical formalism actually meshed better with the "science" label than some of the American formulations.) Even today, computing professionals continue to express concern over their professional status as reflected in research funding and the perceptions of public and peers alike.

In hindsight, part of the difficulty of defining a computing discipline was the tendency of the individuals and institutions involved to treat the process as a zero-sum game. From this perspective, there could only be a single computing discipline and its orientation and content therefore were subject to a great deal of dispute. More recently, however, there has been growing acceptance of the notion that computer science is only one of several related but distinct computing disciplines including computer engineering (focused on computer hardware), software engineering (focused on the software that runs on the hardware), information systems (oriented toward business), and, in Europe, various specialized forms of informatics

(focused on the application of computing within a specific domain such as medicine). Relatively specialized disciplines may possibly prove more amenable to the discovery or development of fundamental principles as well. The realization that computer science need not serve as the sole discipline for computing may ease a number of the stresses and strains that have characterized the development of computer science for most of its brief history.

See also **Artificial Intelligence; Software Engineering**

STUART S. SHAPIRO

Further Reading

Aspray, W. Was early entry a competitive advantage? US universities that entered computing in the 1940s. *IEEE Ann. Hist. Comput.*, 22, 3, 42–87, 2000. An analysis of the early institutional development of computer science at five major U.S. universities.

Croarken, M. The emergence of computing science research and teaching at Cambridge, 1936–1949. *IEEE Ann. Hist. Comput.*, 14, 4, 10–15, 1992. An analysis of the institutional origins of computer science at Cambridge University in the U.K.

Denning, P., Comer, D., Gries, D., Mulder, M., Tucker, A., Turner, A.J. and Young, P. Computing as a discipline. *ACM Comm.*, 32, 1, 9–23, 1989.

Ensmenger, N. The "question of professionalism" in the computer fields. *IEEE Ann. Hist. Comput.*, 23, 4, 56–74, 2001. An analysis of professionalization efforts in computing in the U.S. in the 1950s and 1960s.

Giordano, R. Institutional change and regeneration: a biography of the computer science department at the University of Manchester. *IEEE Ann. Hist. Comput.*, 15, 3, 55–62, 1993. An analysis of the institutional development of computer science at the University of Manchester in the U.K.

Hartmanis, J. and Lin, H., Eds. *Computing the Future: A Broader Agenda for Computer Science and Engineering.* National Academy Press, Washington D.C., 1992.

Report of the Association for Computing Machinery's Task Force on the Core of Computer Science, presenting an intellectual framework for the discipline of computing.

Report of the U.S. National Research Council's Committee to Assess the Scope and Direction of Computer Science and Engineering. Chapter 6: What is computer science and engineering?.

Shapiro, S. Boundaries and quandaries: establishing a professional context for IT. *Inf. Technol. People*, 7. 1. 48–68, 1994. An analysis of professionalization efforts in computing in the U.S. and the U.K. focusing primarily on the 1970s and 1980s.

Computer-Aided Design and Manufacture

Computer-aided design and manufacture, known by the acronym CAD/CAM, is a process for manufacturing mechanical components, wherein computers are used to link the information needed in and produced by the design process to the information needed to control the machine tools that produce the parts. However, CAD/CAM actually constitutes two separate technologies that developed along similar, but unrelated, lines until they were combined in the 1970s.

Computer-Aided Design

Today, engineers use computers in the process of designing most products, from mechanical components to chemicals. However, computer-aided design is a relatively new practice. While engineering applications have been central to the development of the computer from its World War II origins, the commonplace use of computers in engineering design only dates to the introduction and diffusion of the minicomputer in the 1970s. Computer-aided design spread rapidly with the introduction of the microcomputer, especially the personal computer, or PC, in the 1980s.

As was the case with many of the technologies of the postwar period, research into computer-aided design began under the auspices of the military. The first electronic computers were designed and built for the war effort and used for ballistics calculations, cryptography, and atomic bomb calculations. The application of these new machines to an array of engineering problems followed shortly after the war. Finite-element analysis, the most successful method of computer-aided design, originated in airframe design in the early 1950s. In the 1940s, the introduction of the jet engine on military planes created new structural problems in the design of airframes. To take advantage of the faster speeds generated by jet engines, plane designers moved to a swept-back wing shape. They also moved to minimize the weight of the plane's structure. For structural engineers working on airframe analysis, the old methods of analysis were not accurate enough to allow them to optimize the airframe's design. At Boeing, analyzing the stiffness of an airplane's wing became the impetus to develop a new method of structural analysis. Ray W. Clough, a professor of civil engineering at the University of California, Berkeley who was working at Boeing in 1952 and 1953, traced the inaccuracy of the existing methods of analysis to the fact that the wing was modeled strictly as a frame, while the metal skin covering the frame was ignored. At the time, no method existed for mathematically modeling the metal skin. M. J. Turner, the head of Boeing's Structural Dynamics Unit, suggested that the team

consider the patterns of strain within the skin as a way to take into account the stiffness of the skin. The wing's skin could be idealized as a mesh of arbitrary elements. Each element would add stiffness to the wing in a proportion drawn from the strain patterns determined experimentally. This mesh could be represented mathematically as a matrix, and the Boeing engineers knew that matrix notation lent itself to digital computation.

Coined finite-element analysis, or FEA, by Clough in 1960, the spread of the new method depended on access to high-speed digital computers. As a result, through the 1960s the use of the method was mainly confined to large corporations, usually with defense connections, though many companies in the automobile industry also came on board. As minicomputers became available and more affordable in the 1970s, FEA also became more accessible to smaller and mid-sized companies. As these computers offered greater graphical capabilities, engineers could literally draw their designs using computer-aided drafting programs, and feed those graphical models into FEA programs. With the introduction and remarkable spread of personal computers in the 1980s, FEA became accessible to every company, and its integration with software for graphical representation became complete. Commercial FEA packages were often bundled with drafting packages. FEA was the core of an information processing system for engineers. Using this type of software, engineers could perform computer-aided analysis of everything from structural analysis to heat transfer. Regardless of particular application, the key to the method remained the notion of laying a mesh over the design and dispersing it into arbitrary, finite elements that can be mathematically represented as matrices.

Computer-Aided Manufacturing

Like computer-aided design, the most successful development in computer-aided manufacture began as a collaborative project between a corporation and academe under the auspices of a military contract. While the notion of automating machine tools predated World War II, the direct development of tools to actually control the processes of machining began at the Massachusetts Institute of Technology (MIT) in the late 1940s. In 1949 John T. Parsons, of the Parsons Corporation, contacted the MIT Servo-Mechanisms Laboratory for advice on the mathematics connected with Parsons' contract to produce an automatic contour cutting mill for the U.S. Air Force. The Parsons Corporation had made its name machining helicopter rotor blades, and was looking to get into the business of machining wing sections for new Air Force prototype airplanes. Parsons believed that by using punched cards to feed information to machine tools he could increase both the accuracy and speed of production. The partnership between Parsons and MIT coined the phrase numerical control, or NC, to describe the operation at hand. NC became the heart of computer-aided manufacture. With NC, each part to be machined could be described mathematically. The machine tool's motions were controlled by servomotors, which responded to digital commands programmed into the machine to determine a path for the cutting head to follow.

Parsons dropped out of the project in 1950 when he realized that fulfilling his Air Force contract did not entirely mesh with MIT's plan to fully automate machining using digital servo control. Still, he is rightly credited as the inventor of NC. Work on the automatic milling machine continued at MIT until 1956. Under MIT's direction, research into automation of machine tools focused more on control and feedback systems than on machining. Still, MIT's system of automation became commercially available in the early 1950s, through a Cambridge, company called Ultrasonic, with very close ties to the servo lab.

Like FEA, the diffusion of NC machining initially stayed close to the defense industries from which it was developed. Only with the development of cheaper, more powerful computers could NC become common outside the rarified atmosphere of military contracts.

CAD/CAM

In the 1970s, cheaper, more powerful minicomputers meant that many engineering companies were using the same computer to design and manufacture parts. The common use of computers led to the marriage of CAD and CAM, so that the information produced in the design of a component could be directly transferred through the computer to the shop floor. The graphical representation of components on the computers played a key role in putting CAD and CAM together. Engineers laid out a part on the computer in order to perform various analyses on it; that same information could then be used to set up an operation for producing the component. Again, the Air Force played a central role as a powerful proponent of CAD/CAM with its $100 million integrated computer–integrated manufacturing, or

ICAM program. Through ICAM, the Air Force was able to foster corporate and academic, leading to faster, less risky development and dispersal of new CAD/CAM technology.

At the turn of the twenty-first century, parts as varied as airplane wings and fishing rods are designed and manufactured using desktop computers. CAD/CAM stands for a different way of making artifacts, as well as an integration of conceptual design and shop floor production. It has also ushered in a new method of oversight, whereby one individual can supervise and control both design engineering and manufacture on a single computer.

ANN JOHNSON

Further Reading

Bennett, S. *A History of Control Engineering, 1930–1955.* Institute of Electrical Engineers, London, 1993.

Diebold, J., *Automation.* American Management Association, New York, 1983.

Downey, G. CAD/CAM saves the nation?: Toward an anthropology of technology. *Knowledge Soc.,* 9, 143–168, 1992.

Johnson, A. From Berkeley to Boeing: civil engineers, the Cold War, and the origins of finite element analysis, in *Growing Explanations: Historical Perspectives on the Sciences of Complexity*, Wise, M.N., Ed. Duke University Press, Durham (in press).

Mindell, D.A. *Between Human and Machine: Feedback, Control, and Computing Before Cybernetics.* Johns Hopkins University Press, Baltimore, 2002.

Noble, D.F. *Forces of Production: A Social History of Industrial Automation.* Oxford University Press, Oxford, 1984.

Computers, Analog

Paralleling the split between analog and digital computers, in the 1950s the term analog computer was *a posteriori* projected onto pre-existing classes of mechanical, electrical, and electromechanical computing artifacts, subsuming them under the same category. The concept of analog, like the technical demarcation between analog and digital computer, was absent from the vocabulary of those classifying artifacts for the 1914 Edinburgh Exhibition, the first world's fair emphasizing computing technology, and this leaves us with an invaluable index of the impressive number of classes of computing artifacts amassed during the few centuries of capitalist modernity. True, from the debate between "smooth" and "lumpy" artificial lines of computing (1910s) to the differentiation between "continuous" and "cyclic" computers (1940s), the subsequent analog–digital split became possible by the multitudinous accumulation of attempts to decontextualize the computer from its socio-historical use alternately to define the ideal computer technically. The fact is, however, that influential classifications of computing technology from the previous decades never provided an encompassing demarcation compared to the analog–digital distinction used since the 1950s.

The list of what is currently placed beneath the heading of pre-electronic analog computers seems inexhaustible: calendars; sundials and orreries; astrolabes and planetariums; nomograms and graphs; special and general purpose slide rules; slide rule-based assemblages; computing boards and tables; material models of all kinds (including scale models, linkages, artifacts with integrators and differentiators, curve tracers and kinematic mechanisms of which the most known are the various planimeter configurations); harmonic analyzers and synthesizers in the tradition of Lord Kelvin's tide predictor of 1876; mechanical, electromechanical, and electrical analyzers (such as the differential analyzer proposed by Charles Babbage in the 1820s, but never completed); electrolytic tanks; resistive papers and elastic membranes to model two dimensional problems; and an incredible array of fire-control computers, extending from the simple computing mechanism of the anti-aircraft acoustic detectors of the 1920s to the complex anti-aircraft director of the 1940s—a machine that consisted of thousands of parts and required several men for its operation. What escapes this already large list are artifacts that defy easy classification because they were extremely unique structures and were uniquely named (such as those used for tidal calculations in the Netherlands or as the Bell Labs isograph) and they may amount into an even more lengthy listing.

The long list of analog computers corresponds to an equally long list of computer architectures. By constitution idiosyncratic, analog computers cannot be defined by reference to shared technical characteristics. This explains why all available technical definitions of the analog computer never went beyond the attempt to define the technical difference between the analog and the digital computer. All things considered, the technical difference between an analog slide rule and a digital desktop machine was through encasing the motion of the latter into a box (blackboxing), which divided the visible computing action into an open or public, and a closed or private (encased) distinction. In general, the analog computer became conceptualized by the visible mechanical and electrical motion that mediated between the input and the output of numerical data whereas by comparison, the digital computer, given that this motion was invisible, became conceptualized by the restricted attention to the input and output of the numerical data themselves.

In 1920, Vannevar Bush of the Massachusetts Institute of Technology (MIT)—a leading figure of interwar computing who later became a founder of postwar science and technology policy—while seeking to connect electric power transmission lines into networks, experimented with machines as complex and fully electrified as an artificial line and with machines as simple and mechanical as a nomogram (or alignment charts). Between 1927 and 1943, Bush supervised the development at MIT of a new spectrum of electromechanical machines to be used in electric power transmission, ranging from his special-purpose network analyzer (an electrical machine that modeled in miniature a large-scale power grid by measuring varying voltages) to the general-purpose differential analyzer (mechanical), completed in 1931 with the aid of his students. The MIT differential analyzer was widely copied in Europe and in America between 1934 and the early 1950s.

How network analyzers became used in other scientific and technical computations, including the computation of subatomic structures by Douglas Hartree's analyzer at Manchester, reinforces the subjectivity of the demarcation between a special and a general-purpose computer. Flow diagrams accumulated by the intensive use of the analog network analyzer were subsumed under the digital computer software (programs) of electric power transmission—an example typical of the continuity between the richness of analog hardware (early twentieth century) and digital software (late twentieth century). Bush's differential analyzer used mechanical integrators, as did Kelvin's tide predictor (though Bush did not know of Kelvin's ideas at the time), torque amplifiers, drive belts, shafts, and gears and was driven by electrical servomotors. Unlike the tide predictors, it could solve differential equations for any application, leading to demand for further machines.

General Electric's Edith Clarke, who started as a "computor" (i.e., a person who produced computations using a desk calculator) before becoming an electrical engineering analyst (i.e., a person who directs computors), developed several unusual electric power transmission slide rules and graphical calculators in the 1920s. The most unusual electric power transmission computers were perhaps those of Vladimir Karapetoff, an electrical engineering professor at Cornell University from

1908–1939. Named after distinguished electric power transmission specialists, the "Heavisidion" and the "Blondelion," Karapetoff's computers may be classified as combinations of kinematic computing mechanisms with linkages, a class of computers that became better known by the work of Antonin Svoboda in the 1940s.

The military counterpart of this civilian line of computer development points to the same diversity. In the military fire control of artillery, the range of what we now call analog computers extended from something as simple as the slide rule to something as complex as an aircraft computing bombsight, or its inverse, a radar-controlled anti-aircraft gun director that directed guns to fire based on target positions of aircraft and correcting for shell speed and flight time. It took Macon Fry a series of seven *Machine Design* articles (September 1945–February 1946) to describe the basic constituents of pre-electronic fire-control computing mechanisms. The Bush differential analyzer at the Moore School of Electrical Engineering at the University of Pennsylvania, built in 1942, was used in World War II ballistics calculations. Another mechanical analog computer, the Norden bombsight, which was used to compute the trajectory of the first atomic bomb in 1945, perhaps qualifies as the most important computer of the twentieth century. Interwar fire-control technology suggests that the role of the state in fostering important changes in computing technology has much deeper roots than canonically assumed. The military contracts awarded to manufacturers of digital calculating machines and punched card machinery to produce analog electromechanical fire-control computers set the stage for the crucial transfer of technology that made IBM's postwar triumph possible. Noticeably, bombsight technology was a national secret, second only to the atomic bomb. From bombsights to analyzers, the pre-1950 ideology of intelligent machines in the history of electromechanical analogs prepared for the acceptance of the electronic computer as a thinking artifact after 1950.

From the General Electric versus Westinghouse competition in renting or selling network analyzers to electric power utilities, to the legendary clash between the Sperry Gyroscope and the Norden bombsights, the business antagonisms of recent decades are prefigured in the history of the development and use of electromechanical analog computers'. Some of the firms involved; for example, Vickers (Europe) and Sperry (U.S.), maintained a presence in their field for many decades. We possess considerable information about a few items, but practically nothing about

other equally important artifacts—not to mention that our knowledge barely delves into the knowledge of how to use these artifacts. For example, the development of the aforementioned electric power transmission computers cannot be adequately understood without reconfiguring Clarke's method of "symmetrical components" or Bush's "operational calculus." A better engagement with the history of pre-electronic analogs would doubtless identify many surprising contributors. We have several studies on Bush's network analyzers but very little on the artificial lines of Bush's dissertation advisee, Arthur Kennelly—not to say that we need to learn more about the computers of Kennelly's early employer, Thomas Edison, who computed his first electric networks by relying on miniatures constructed by his Austrian employee, Dr. Herman Claudius.

The analog electronic computer emerged in one of two ways: either by reconfiguring pre-World War II electromechanical analogs, or, as was more common, by the initial configuration of novel electronic computer designs afforded by the post-Word War II cheapening of their constituent electronic components. In the face of rapidly increasing costs for special-purpose software to run the general-purpose electronic digital computer, the costs for which decreased rapidly, crucial advances in fields as important as that of aeronautical design rested almost exclusively throughout the third quarter of the twentieth century, on the development and use of the electronic analog computer.

By the early 1960s, the peak period in the development of the electronic analog computer, its subclasses and uses were so prolific that it took Stanley Fifer a series of four volumes simply to introduce them. Between the late 1940s and the early 1970s, the total production of books and journals on electronic analog computers and associated technologies (e.g., hybrid computers, analog-to-digital and digital-to-analog converters—see Computers, Hybrid) filled rooms of libraries in polytechnic institutions around the world. Now covered by a quarter century of dust, the pages of these publications tell a story that challenges the canonical explanation of the technical inevitability of the victory of the digital over the analog. It is a story that juxtaposes the specific merits of concrete utilization of analog speed to the abstract rhetoric about the customary virtues of the digital accuracy.

All evidence suggests that analog electronic computers became indispensable as controllers of industrial processes (direct analogs). Of the innumerable uses of electronic analog computers—in contexts that privileged solutions in real or faster-than-real time—the best studied are those from the military. Rockets, missiles (including intercontinental), and military aircraft developed concurrently with military analog electronic computers. This coevolution became possible by a series of pioneering military projects launched in the late 1940s and the early 1950s. In the U.S., these projects were subcontracted to private firms; in Britain, they were developed at in-house facilities. The result was the eventual development of a rich global network of prospering electronic analog computer manufacturers. The military analog computers were mathematical machines (indirect analogs) used for the solution of the differential equations corresponding to the phenomena to be computed. They were programmed, which means assembled and reassembled, by connecting units or modules of units of alternating or direct current operational amplifiers, used to perform addition, subtraction, integration, and differentiation. The replacement of mechanical amplifiers (the most famous of which was the wheel-and-disk integrator) by electronic versions resulted in the development of electronic differential analyzers (also indirect analogs). The electrical network analyzer, first cousin of the mechanical differential analyzer (a more direct analog), was also restructured by the inclusion of electronic components. The result was a developmental line that concluded in electronic computers similar to Westinghouse's Anacom.

Beginning in the late 1940s, the analog–digital debate continued and escalated into the 1950s. The value of the analog as a tool for demonstration and visualization of computing problems helped it decisively to remain in the market in the 1970s as a tool for educational and instructional purposes. Those who have studied the analog–digital debate from a perspective that is sensitive to the analog side see a social tradeoff between the analog and the digital instead of an inevitable technical evolution from the analog to the digital. Extensions of this debate included, for example, the contrast between the adaptability in use afforded by the hardware malleability of the analog computer versus the generality of use permitted by the software flexibility of digital computer.

The contemporary proponent of the analog seeks to counter the obvious respect to the precision of a solution being paid by a mental worker, mathematician, or accountant, on the one hand, with the pleasure of the "feeling" the solution enjoyed by a manual worker, craftsman or technician, on the other hand. In the first part of the twentieth century, engineers argued that "feeling," "insight," and "intuition" of the problem under computation would be lost in choosing the "invisible," "inner," and "private" workings of the digital desktop machine over the fully "visible," "external," and "public" act of computing with the analog slide rule. In his comparison of analog and digital electronic installations of the 1950s and the 1960s, James Small finds something similar.

Digital computer installations were frequently organized on a closed-shop basis, with computer operators being frequently responsible for scheduling computer usage while users often simply submitted their problem. The fate of its solution was left to numerical analysts and programmers who, first, had no physical insight, and second, need not know how the computer worked. By contrast, analog computer installations usually operated on an open-shop basis, with engineers having direct access to the computer, configuring and reconfiguring it in a process of a computation that was capable of dynamically adjusting to identifications of error sources due to the mathematical model or dysfunctional components. Thus seen, the story of the analog computer is the story of living labor (variable capital) as the other side of automatic machinery (constant capital). As such, the analog versus digital debate of the third quarter of the twentieth century is the new version—that of the electronic era—of a struggle that is as old as the capitalist mode of production.

The historiographical retrieval of the significance of (electronic) hybrid computing, which was a mode of computing that was developed to combine the advantages of the analog and the digital (see Computers, Hybrid), provides further indication of the analog–digital relationship being more historically complex than previously assumed. One may simply start by commenting upon the notoriously problematic identification of analog computer subclasses. For Fifer, the class of analog computers, as distinct, supposedly, from that of digital computers, contained the option of constructing a computer (as opposed to a "simulator-tester", a "trainer" or a "controller") that was indirect-functional (as opposed to direct); compressed-fast time (as opposed to real or extended-slow time); general (as opposed to special); and electronic (as opposed to mechanical, electrical or electromechanical). It is unclear how an electronic general-purpose compressed-time indirect computer differed from an electronic digital computer, which means that in Fifer's classification (1961) the digital electronic computer was actually a subclass of the analog. It is then a matter of interpretation whether the analog actually lost to the digital or it was simply superseded by the digital.

Historians of the digital computer find that the experience of working with software was much closer to art than science, a process that was resistant to mass production; historians of the analog computer find this to have been typical of working with the analog computer throughout all its aspects. The historiography of the progress of digital computing invites us to turn to the software crisis, which perhaps not accidentally, surfaced when the crisis caused by the analog ended. Noticeably, it was not until the process of computing with a digital electronic computer became sufficiently visual by the addition of a special interface—to substitute for the loss of visualization that was previously provided by the analog computer—that the analog computer finally disappeared.

See also **Calculators, Mechanical and Electromechanical; Computers, Early Digital; Computers, Hybrid**

ARISTOTLE TYMPAS

Further Reading

Aspray, W. Edwin L. Harder and the Anacom: analog computing at Westinghouse. *IEEE Ann. Hist. Comput.* 15, 2, 35–52, 1993.

Bennett, S. *A History of Control Engineering, 1930–1955.* Peter Peregrinus, London, 1993.

Bromley, A. Analog computing devices, in *Computing before Computers*, Aspray, W., Ed. Iowa State University Press, Ames, IA, 1990.

Bush, V. Instrumental analysis. *Trans. Am. Math. Soc.*, 42, 10, 649–669, 1936.

Fifer, S. *Analogue Computation: Theory, Techniques, and Applications, vols. I–IV.* McGraw-Hill, New York, 1961.

Fry, T.C. Industrial mathematics. *Bell Syst. Tech. J.*, 20, 3, 255–292, 1941.

Hartley, M.G. *An Introduction to Electronic Analogue Computers.* Methuen, London, 1962.

Higgins, W.C., Holbrook, B.D. and Emling, J.W. Defense research at Bell Laboratories. *Ann. Hist. Comput.*, 4, 3, 218–244, 1982.

Hopp, P.M. *Slide Rules: Their History, Models, and Makers.* Astragal Press, Mendham, NJ, 1999.

Horsburgh, E. M., Ed. *Modern Instruments and Methods of Calculation: A Handbook of the Napier Tercentenary Exhibition.* Bell, London, 1914.

Jackson, A.J. *Analog Computation.* McGraw-Hill, New York, 1960.

von Jerzerski, D. *Slide Rule: A Journey through the Centuries.* Astragal Press, Mendham, NJ, 2000.

Leffingwell, W.H. *The Office Appliance Manual.* National Association of Office Appliance Manufacturers, 1926.

Lipka, J. *Graphical and Mechanical Computation.* Wiley, New York, 1918.

Karplus, W. *Analog Simulation: Solution of Field Problems.* McGraw-Hill, New York, 1958.

McFarland, S.L. *America's Pursuit of Precision Bombing, 1910–1945.* Smithsonian Institution Press, Washington D.C., 1995.

Mindell, D. *Between Human and Machine: Feedback, Control, and Computing Before Cybernetics.* John Hopkins University Press, Baltimore, 2002.

Owens, L. Vannevar Bush and the differential analyzer: the text and the context of an early computer. *Technol. Cult.*, 27, 1, 1986, 63–95.

Persson, P.-A. Transformation of the analog: the case of the Saab BT 33 Artillery Fire Control Simulator and the introduction of the digital computer as control technology. *IEEE Ann. Hist. Comput.*, 21, 2, 52–63, 1999.

Paynter, H.M. *A Palimpsest on the Electronic Analog Art.* Philbrick Researches, Boston, 1955.

Small, J.S. General-purpose electronic analog computing, 1945–1965. *IEEE Ann. Hist. Comput.* 15, 2, 8–18, 1993.

Small, J.S. Engineering technology and design: the post-Second World War development of electronic analogue computers. *Hist. Technol.*, 11, 33–48, 1994.

Small, J.S. *The Analogue Alternative: The Electronic Analogue Computer in Britain and the USA, 1930–1975.* Routledge, London, 2001.

Tomayko, J.E. Helmut Hoelzer's fully electronic analog computer. *IEEE Ann Hist. Comput.*, 7, 3, 227–240, 1985.

Various. Twentieth century analog machines. *IEEE Ann Hist. Comput.*, Special Issue, 18, 4, 1996.

Wilden, A. *System and Structure: Essays of Communication and Exchange.* Tavistock, London, 1972.

Computers, Early Digital

Digital computers were a marked departure from the electrical and mechanical calculating and computing machines in wide use from the early twentieth century. The innovation was of information being represented using only two states (on or off), which came to be known as "digital." Binary (base 2) arithmetic and logic provided the tools for these machines to perform useful functions. George Boole's binary system of algebra allowed any mathematical equation to be represented by simply true or false logic statements. By using only two states, engineering was also greatly simplified, and universality and accuracy increased. Further developments from the early purpose-built machines, to ones that were programmable accompanied by many key technological developments, resulted in the well-known success and proliferation of the digital computer.

Early digital computers made use of electromechanical relays, adapted from telephone exchanges. Relays were essentially on–off switches, and implementation of binary logic was simply a matter of wiring, with the logical element called the flip-flop being the fundamental component of binary information storage. This was a simple combination of two basic logic "gates," themselves constructed from an arrangement of switches.

Among the first recorded digital computers was a machine built by a German, Konrad Zuse. In 1936 he constructed a binary calculating machine from electromechanical relays in his living room. This performed binary arithmetic, floating point calculations and program control was by a punched tape, widely used in telegraphs. His work was independent of projects elsewhere in the U.K. and U.S. and did not result in proliferation or market success.

The first round of digital computers in the U.S. also made use of electromechanical relay technology and was seeded by Claude Shannon's 1937 master's thesis at the Massachusetts Institute of Technology (MIT). He described a way of using symbolic logic to improve electrical switching circuits and described an electrical circuit made from relays and switches that could add binary numbers. Since relay technology was widely used in telephone exchanges, it was mature and broadly available. Bell Telephone Laboratories in the U.S. tested ideas on a small scale with their relay interpolator, which consisted of 500 relays and later the ballistic computer, which contained 1300 relays. Their work cumulated in 1944 with an all-purpose computer with 9000 relays and operated by 50 teletype consoles.

International Business Machines (IBM), which had established itself in the office adding and tabulating machine market, took up ideas put forward by Harvard University's Howard Aiken to solve existing problems of "insufficient means of mechanical computation." Under IBM's sponsorship and capitalizing on IBM's existing punched card readers and writers, Aiken's Mark 1 computer (also known as the IBM automatic sequence controlled calculator) was developed at Harvard University with the assistance of a group of IBM engineers and also used electromechanical relay technology. Completed in 1944 at 15.5 meters long and 2.4 meters high, it was considered the first automatic general purpose digital machine. It had four tape readers in which data and program code was fed, and subroutines stored. A more capacious Mark II followed. The machines were used by the U.S. Navy for tackling ballistic problems, one of the recurring needs during wartime for "number crunching."

In the U.K., pioneering efforts in digital computing were made by mathematician Alan Turing, whose idea of a "Universal Turing Machine" helped break the conceptual ground for automated computers tackling mathematical problems. His 1936 paper "On Computable Numbers" theoretically described a machine that could do any calculation that could be done by a human. During World War II, his work with developing computing systems at Bletchley Park was fundamental in cracking German U-boat "Enigma" communications and keeping the Atlantic shipping lanes open, ultimately contributing towards the Allied victory. The purpose-built 1944 "Colossus," built by Thomas Flowers of the Post Office Research Laboratories was built from relays from Post Office telephone exchanges, and operated with a fixed sequence for deciphering the Germans U-boat communications encrypted with Enigma codes.

These machines marked the transition of computers from calculating machines to automated computers. However the bulk, power requirements, and the delay time in the operation of relays were performance-limiting factors.

The development and application of the electronic vacuum tube (also known as the valve) as a switching unit in digital computers permitted vastly improved performance compared to relay technology. The first unit to employ large-scale use of valves for digital computing was the ENIAC (electronic numerical integrator and computer) which consisted of over 18,000 valves (see Figure 13). Built by John Mauchly and John Prosper Eckert at the Moore School of Electrical Engineering at the University of Pennsylvania to compute ballistic tables for the U.S. Army, it was only completed after the close of the war in late 1945. Although it could only store 20 binary words, its unprecedented calculation speed attracted attention from scientists and professionals, in particular a key member of the Los Alamos Manhattan Project for the atomic bomb, mathematician John von Neumann. The ENIAC was used for other scientific purposes such as weather predicting and other fluid dynamic calculations, including calculations for the fusion hydrogen bomb. However, programming the ENIAC meant setting arrays of function switches and a maze of patch cables. This was time consuming and meant that it could not be used for computing whilst it was being programmed. This drove the development of the stored program computer, whereby the computational instructions could be prepared on tape and loaded quickly into memory.

The EDVAC (electronic discrete variable calculator) was the first stored program electronic digital computer and was also built by the Moore School under contract to the U.S. Army. It was highly influenced by von Neumann's work in

Figure 13.   Two women wiring the right side of the ENIAC with a new program.
[*U.S. Army Photo from the archives of the ARL Technical Library.*]

formal logics since he provided critical early direction into the logical structure and design of the computer in 1945, together with Mauchly and Eckert. Von Neumann abstracted the logical design from the hardware, and introduced the concept of the "stored program" whereby programming instructions are stored as numbers in memory alongside numerical data. The EDVAC, installed at the Ballistic Research Laboratory at Aberdeen Proving Ground, MD in 1949 but not fully complete until 1952, also introduced mercury delay lines for primary memory. Developed by John Eckert for the project, these devices increased storage density by a factor of 100 over valve devices.

The ideas and work of the Moore School rapidly proliferated thereafter. A summer course on computer design based on the von Neumann concept was run by that institution in 1946, and was attended by British and American professionals. Also in 1946, von Neumann, Herman Goldstine and Arthur Burks published a comprehensive report of their work at Princeton's Institute of Advanced Study Electronic Computer Project, where they had established themselves following their involvement with the Moore School. The report detailed the operation and architecture of their work on digital computers, and has been described as the blueprint for "modern" digital computing. The authors' insistence that it be placed in the public domain, as is the tradition with scientific publishing, was certainly an important factor in the spread and adoption of this knowledge. The design proposed for the EDVAC became the basis for several machines, including in England, the electronic delay storage automatic computer (EDSAC) built by Maurice Wilkes at the Mathematical Laboratory at Cambridge University completed in 1949, and the Manchester Mark I.

The first digital computing machine to store both program and data in random-access memory (RAM) was Manchester University's Mark I, built by Geoff Toothill, Frederick Williams and Tom Kilburn in 1949, using cathode-ray tubes for fully electronic memory. The Mark I is considered by many to be the first stored program computer, since it was in use before the EDVAC was fully operational.

The Whirlwind was developed at the Servomechanism Laboratory at MIT in Boston, and was sponsored by the U.S. military for real-time simulation of the effect of engineering changes on aircraft behavior. The project began in 1946 and while it drew from the work at Princeton and Moore Schools, the designers' background in automatic control and electrical engineering gave it a certain technological uniqueness. The Whirlwind project resulted in some key technological developments such as magnetic core memory and the parallel synchronous method of handling information. Developed by Jay Forrester in 1949, magnetic core memory was an important step in miniaturization and speed of computer memory. The Whirlwind was only commissioned in 1958, finding use as the control and prediction computer for the SAGE air defense system. This relied on real-time communications to remote radar stations for which a key digital communication technology was developed—the modem. However, it has been argued that Whirlwind's time was over before it ever began: the system had become redundant by the Cold War proliferation of intercontinental ballistic missiles.

The UNIVAC is regarded as the first commercial digital computer. It was designed and built by J. Presper Eckert and John Mauchly who had worked extensively on the engineering of ENIAC project. The acronym stems from "universal automatic computer," the intention being that it would have universal appeal to scientists, engineers and business. Although the core functional "organs" were similar to the EDVAC, it was a more robust machine and required less maintenance. The UNIVAC incorporated some key improvements that vastly speeded up data processing which included a magnetic tape system for secondary memory storage and a data buffering mechanism to the delay line primary memory storage. The first commercial installation of a UNIVAC was at the U.S. Census Bureau in 1951. LEO (Lyons electronic office) however went into commercial office action a few months before UNIVAC. LEO, based on Cambridge University's EDSAC, was used by Lyons & Company Limited Bakery for data processing. With interest from other companies, Lyons later sold LEO II computers to other British firms.

IBM responded to UNIVAC in 1952 with the Model 701, which was technologically similar to the UNIVAC with the exception that it featured cathode-ray tube memory. This was a technology appropriated from the television industry, the phosphorescence of the screen providing the ability to store binary signals. Another key feature was the use of a magnetic oxide-coated drum for secondary storage (the predecessor of the hard disk) as well as a lightweight plastic tape tertiary memory. This had a lower inertia than the metal tapes of the UNIVAC and sped up tape operations considerably. Due to IBM's massive market presence and infrastructure, the 701 was a considerable market success.

Following the success and proliferation of these computers in business and scientific research organizations from the early 1950s, a plethora of startups presented the global market with a diverse range of digital computers. These companies were initiated by key staff members from the pioneering organizations, such as Seymour Cray and William Norris. Besides technological improvements in processing speed, primary and mass storage memory density, architecturally digital computers did not diverge from the classic "von Neumann architecture." Furthermore, military-sponsored research in the U.S. produced key advances such as the transistor, the integrated circuit and computer networking (from which the Internet grew) although personal computing was more of a civilian initiative. Larger demand for computers in all spheres of society spawned a global industry, which ultimately drove down prices and increased miniaturization, and resulted in the digital computer becoming ubiquitous in most industrialized societies by the turn of the century.

See also **Calculators, Mechanical and Electromechanical; Computer Memory, early; Computers, Analog; Computers, Mainframe; Encryption and Code Breaking; Vacuum Tubes/Valves**

BRUCE GILLESPIE

Further Reading

Aspray, W. *John von Neumann and the Origins of Modern Computing.* MIT Press, Cambridge, MA, 1990.

Ceruzzi, P.E. *A History of Modern Computing.* MIT Press, Cambridge, MA, 1998.

Eames, C. and Eames, R. *A Computer Perspective, Background to the Computer Age.* Harvard University Press, Cambridge, MA, 1973.

Goldstine, H.H. *The Computer: From Pascal to von Neumann.* Princeton University Press, Princeton, 1972.

Hodges, A. *Alan Turing: The Enigma.* Burnett Books, London, 1983.

Hughes, T.P. *Rescuing Prometheus: Four Monumental Projects that Changed the Modern World.* Vintage Books, 1998.

Irvine, M.M. Early digital computers at Bell Telephone Laboratories. *Ann. Hist. Comput.*, 23, 3, 22–42, 2001.

Computers, Hybrid

Following the emergence of the analog–digital demarcation in the late 1940s—and the ensuing battle between a speedy analog versus the accurate digital—the term "hybrid computer" surfaced in the early 1960s. The assumptions held by the adherents of the digital computer—regarding the dynamic mechanization of computational labor to accompany the equally dynamic increase in computational work—was becoming a universal ideology. From this perspective, the digital computer justly appeared to be technically superior. In introducing the digital computer to social realities, however, extensive interaction with the experienced analog computer adherents proved indispensable, especially given that the digital proponents' expectation of progress by employing the available and inexpensive hardware was stymied by the lack of inexpensive software. From this perspective—as historiographically unwanted it may be by those who agree with the essentialist conception of the analog–digital demarcation—the history of the hybrid computer suggests that the computer as we now know it was brought about by linking the analog and the digital, not by separating them. Placing the ideal analog and the ideal digital at the two poles, all computing techniques that combined some features of both fell beneath "hybrid computation"; the designators "balanced" or "true" were preserved for those built with appreciable amounts of both.

True hybrids fell into the middle spectrum that included: pure analog computers, analog computers using digital-type numerical analysis techniques, analog computers programmed with the aid of digital computers, analog computers using digital control and logic, analog computers using digital subunits, analog computers using digital computers as peripheral equipment, balanced hybrid computer systems, digital computers using analog subroutines, digital computers with analog arithmetic elements, digital computers designed to permit analog-type programming, digital computers with analog-oriented compilers and interpreters, and pure digital computers.

There were many types of analog-to-digital converters, including raster, chronometric, rotating drum, null detector, direct counting, incremental steps, star wheel, fixed interval, and commutator and brush, which transformed continuous analog information into discrete numerical data. Their common feature was an accurate standard of spatial or temporal measurement to set the analog information into digital form.

Given the tendency towards digitalization, the literature on analog-to-digital converters was more expansive than its opposite: digital-to-analog converters. In addition to converters, bilateral operation could rely on multiplexers and demultiplexers to permit the converters to be time shared, and to hold devices, buffers, timing, and control circuitry. The increasing demand for hybrid interface equipment brought about display devices (plotters, scopes, etc.) that accepted digital outputs and produced analog-appearing inputs, and conversely, similar artifacts (sketchpads, tablets, and light pens) that accepted analog-type inputs and converted them to digital signals suitable for computer input.

The way analog and digital computers could be used together was classified into two broad categories: unilateral operation in which information flew across the interface between the analog and the digital sections in one direction, and bilateral operation in which the flow across this interface was in both directions. Noticeably, in bilateral operation the analog or the digital computer could be regarded as playing the part of a complex and elegant input or output device. It therefore follows that claiming that the digital won the analog–digital debate is similar to arguing that the history of hybridization ended with the victory of a unilateral hybrid computer in which the information flowed only to the digital section.

Early hybridization, called analog–digital conversion techniques, suffered from the lack of adequate software, inadequacies of vacuum-tube linkage elements, and the unsuitability of available digital computers for online operation. Initially, most hybrid computer systems employed large digital computers, such as the IBM 7094 or the Univac 1103. During the 1960s, some smaller digital such as the Scientific Data Systems 930 were also used. Eventually, hybrid computers were used in sampled-data system simulation, random process simulation, optimization, simulation of distributed parameter systems, studies in guidance and control of missiles and space vehicles, simulation of man-machine systems, and numerous other application contexts. Hybrid speed was, for example, advantageous in problems with random variables where a large number of solutions were required for statistical analysis.

One notable development was the configuration of a hybrid computing Monte Carlo technique, which permitted accumulation of statistics taken from thousands of fast analog-computer runs. The hybridization of electronic analog and digital computers was also undertaken in order to overcome computational problems in the design and development of intercontinental ballistic missiles (ICBMs). Combining the analog's speed and the digital's accuracy through conversion was first tried at Convair Astronautics in 1954 and at the Ramo-Woolridge Corporation in 1955. Both combinations were undertaken to create the Atlas ICBM system. In the late 1950s and early 1960s, several experimental hybrids undertaken by Electronic Associates, Inc. (EAI) and Packard-Bell were built by linking analog and digital through interfaces. In the mid-1960s, Comcor/

Astrodata and EAI introduced fully transistorized analog computers and highly integrated hybrid computer systems began to be manufactured commercially.

By the 1960s, hybrid equipment was used by nearly every firm and government research and test facility working on missile design and other aerospace applications. Space exploration provided the impetus for the further development of the hybrid computer. James Small argues that the manned space program took over where the problems of cost, complexity, and hazards in ICBM development and simulation left off. Including the astronaut in the control loop increased computing complexity. Large, fully transistorized true-hybrids were used in the Gemini and the Apollo programs and in the Saturn-V rocket system. Hybrids were employed in a wide range of space applications, including the simulation of space vehicle docking, engine and navigation system design, and astronaut instruction.

By the late 1960s, more than ten firms were manufacturing hybrid computer systems, and while sales of analog computers that were not fully hybrid-compatible were decreasing, sales of hybrid computers were increasing. Sales of hybrids peaked at approximately $50 million per annum in the late 1960s, falling substantially when aerospace budgets were eventually reduced in the 1970s. They remained, however, between $25–$30 million through 1975. As late as 1989, there were 300 large installations of hybrids worldwide. The decline of the hybrid cannot, however, be explained by economic necessity alone. Small notes that by 1970 at least 75 percent of all simulations were performed on digital computers, even after studies showed that the hybrid computer offered speed advantages greater than the fastest digital computers by as much as 100 to 1. In army material control studies, advanced hybrid computer systems (fourth generation hybrids) showed potential savings of 30 to 1 by using hybrids, amounting to a potential market of $300 million in the U.S. alone. All specialists agreed that hybridization had resulted in increased computing hardware utilization. For some, hybridization was also a healthy response to an unfortunate specialization of interest and skills by computer engineers, which was the outcome of the fanatical separation between analog and digital computers. Hybridization helped the efficient transformation of analog schematics into digital flow charts, and the creation of utility libraries meant less duplication of programming work. Others, however, feared that hybridization would bring about a dependence on automatic programming techniques that could isolate engineers from the computer, thereby replacing the open-shop computing environments of the analog computer by the closed-shop computing environments of the digital computer.

See also **Computers, Analog; Computers, Early Digital**

ARISTOTLE TYMPAS

Further Reading

Bekey, G.A. and Karplus, W.J. *Hybrid Computation.* Wiley, New York, 1968.

Fifer, S. *Analogue Computation: Theory, Techniques, and Applications*, vol. II. McGraw-Hill, New York, 1961.

Hydman, D. E. *Analog and Hybrid Computing.* Pergamon Press, Oxford, 1970.

Korn, G.A. and T. M. Korn. *Electronic Analogue and Hybrid Computers*, 2nd edn. McGraw-Hill, New York, 1972.

McLeod, J. *Simulation: The Dynamic Modeling of Ideas and Systems with Computers.* McGraw-Hill, New York, 1968.

Small, J.S. General-purpose electronic analog computing, 1945–1965. *IEEE Ann. Hist. Comput.*, 15, 2, 8–18, 1993.

Small, J.S. Engineering technology and design: the post-second world war development of electronic analogue computers. *Hist. Technol.*, 11 33–48, 1994.

Small, J.S. *The Analogue Alternative: The Electronic Analogue Computer in Britain and the USA, 1930–1975.* Routledge, London, 2001.

Susskind, A.K., Ed. *Notes on Analog–Digital Conversion Techniques.* Wiley, New York, 1957.

Computers, Mainframe

The term "computer" currently refers to a general-purpose, digital, electronic, stored-program calculating machine. The term "mainframe" refers to a large, expensive, multiuser computer, able to handle a wide range of applications. The term was derived from the main frame or cabinet in which the central processing unit (CPU) and main memory of a computer were kept separate from those cabinets that held peripheral devices used for input and output.

Computers are generally classified as supercomputers, mainframes, minicomputers, or microcomputers. This classification is based on factors such as processing capability, cost, and applications, with supercomputers the fastest and most expensive. All computers were called mainframes until the 1960s, including the first supercomputer, the naval ordnance research calculator (NORC), offered by International Business Machines (IBM) in 1954. In 1960, Digital Equipment Corporation (DEC) shipped the PDP-1, a computer that was much smaller and cheaper than a mainframe.

These more affordable machines, called minicomputers, became increasingly popular in the commercial sector after the introduction of the PDP-8 in 1964 and the PDP-11 in 1970, and widely increased the base of consumers and applications. Microcomputers were introduced in the 1970s, including such machines as the Kenback I, the Altair, and the Apple; these further broadened the computer market (see Computers, Personal; Computers, Supercomputers).

All computers contain processors, memories, and peripherals. The CPU manipulates instructions and data that are stored in its registers and in main memory. The CPU and memory consist of electronic components, allowing current processing speeds in the billions of instructions per second. Nanotechnology is being researched for even faster speeds. All storage within the CPU and main memory is currently volatile; if power is removed, the stored values are lost. The peripheral devices typically have mechanical components that limit their speed. These include nonvolatile secondary storage devices, which hold programs and data for input and output on demand, as well as devices used to interface with humans, such as monitors, printers, and terminals. Other devices provide for computer-to-computer interactions, such as those used by networks. Special-purpose processors perform various specialized functions, such as floating-point operations or input and output of data.

Early electronic computing machines were analog, with computed values analogous to input values. Analog machines, first designed by Vannevar Bush at MIT in 1927–1935, were widely used during World War II for calculating ballistics tables for weapons systems. They were soon overshadowed by digital computers, in which all data and control information (programs) are encoded with discrete values. A great many inventors contributed to the development of the various components of modern computers. Following are some highlights of this process.

In 1940, George Stibitz demonstrated a communications cable between his complex calculator and a teleprinter, which was a forerunner of modern digital communication ports. Konrad Zuse, in association with Helmut Schreyer, is credited with building the first working general-purpose program-controlled computing machine, the Z3, in 1941. John Atanasoff and Clifford Berry developed an electronic digital computing machine between 1939 and 1942; their ABC machine utilized vacuum tubes (valves) for switches. Influenced by the work of George Babbage, between 1939 and 1943 Howard Aiken led the development of the Harvard Mark I, a program-controlled digital calculator with electromechanical components. By 1943, under the direction of Tommy Flowers, the Colossus, a large-scale, electronic, digital computing machine was developed for use in deciphering German code during World War II (see Computers, Early Digital). John Mauchly and J. Presper Eckert designed a general-purpose digital electronic machine, the ENIAC (electronic numerical integrator and computer), which was completed in 1946. Between 1946 and 1952, John von Neumann spearheaded the development at Princeton of the Institute of Advanced Study (IAS) machine, which stored its programs in memory together with data. The late 1940s saw the introduction of Maurice Wilkes's EDSAC (electronic delay storage automatic computer) at Cambridge University and the EDVAC (electronic discrete variable calculator) at the Moore School, University of Pennsylvania, both with stored programs in a memory of cathode-ray tubes and mercury delay lines. In 1948, IBM marketed the 604 Electronic Calculating Punch Card Machine, with input and output data stored on punched cards and programs stored on a plugboard. Also in 1948, Geoff Toothill, Frederick Williams and Tom Kilburn built a forerunner, called "Baby," for Manchester University's Mark I, the first operational stored-program digital computing machine. In Sydney, Australia, Maston Beard and Trevor Pearcey supervised the building of the Council for Scientific and Industrial Research Computer (CSIRAC) between 1949 and 1951. In 1951, computers became commercially available, with Ferranti's Mark I (based on the Manchester Mark I) and Eckert's and Mauchly's Computer Corporation's (EMCC) UNIVAC I. In the 1950s, the IBM 702 and 704 were developed for business as well as scientific applications. Other early computer manufacturers included Burroughs (Atlas or AN/GSQ-33 Computer, later merged with Sperry-Rand to form Unisys), Remington Rand (acquired EMCC, later merged into the Sperry-Rand Corp.), Texas Instruments (first commercial use of silicon transistors in computers), Fujitsu (Japan's first commercial computer), the Dutch PITT (ZERO and ZEBRA), and Olivetti (Italy's first commercial computer).

IBM became the leader of the mainframe computer market, first with its series 7000 line, and then, overwhelmingly, when it introduced the System/360 family of computers in 1964. The System/360 was a line of computers of increasingly greater size, speed, and number of input and output ports, with a corresponding increase in cost. All the System/360 computers had the same instruction set, so that software systems developed for machines at the low end of the market could also execute on system upgrades. They featured integrated circuitry, a main memory that was unusually large for the time, and microprogramming. Other innovations included the introduction of different instruction sets for different data types, support for multiprogramming, and the pipelining of instructions. These features were widely influential on the manufacture of successive mainframes, and IBM has continued to dominate the mainframe market to this day (see Computers, Uses and Consequences). Other current mainframe manufacturers include Fujitsu-Siemens, Hitachi, and Unisys.

Mainframes once each filled a large room, cost millions of dollars, and needed a full maintenance staff, partly in order to repair the damage caused by the heat generated by their vacuum tubes. These machines were characterized by proprietary operating systems and connections through dumb terminals that had no local processing capabilities. As personal computers developed and began to approach mainframes in speed and processing power, however, mainframes have evolved to support a client/server relationship, and to interconnect with open standard-based systems. They have become particularly useful for systems that require reliability, security, and centralized control. Their ability to process large amounts of data quickly make them particularly valuable for storage area networks (SANs). Mainframes today contain multiple CPUs, providing additional speed through multiprocessing operations. They support many hundreds of simultaneously executing programs, as well as numerous input and output processors for multiplexing devices, such as video display terminals and disk drives. Many legacy systems, large applications that have been developed, tested, and used over time, are still running on mainframes.

See also **Computer Memory, Early; Computer Networks; Computers, Uses and Consequences; Processors for Computers; Systems Programs**

Trudy Levine

Further Reading

Ceruzzi, P. *A History of Modern Computing*. MIT Press, Cambridge, MA, 1998.

Cortada, J.W. *Historical Dictionary of Data Processing Technology*. Greenwood Press, New York, 1988.

Gschwind, H.W. and McCluskey, E.J. *Design of Digital Computers*. Springer Verlag, New York, 1975.

Push, E.W. *Building IBM: Shaping an Industry and its Technology*. MIT Press, Cambridge, MA, 1995.

Stallings, S. *Computer Organization and Architecture*, 5th edn. Prentice Hall, 2000.

Tanenbaum, A. *Structured Computer Organization*, 4th edn. Prentice Hall, 1999.

Computers, Personal

A personal computer, or PC, is designed for personal use. Its central processing unit (CPU) runs single-user systems and application software, processes input from the user, sending output to a variety of peripheral devices. Programs and data are stored in memory and attached storage devices. Personal computers are generally single-user desktop machines, but the term has been applied to any computer that "stands alone" for a single user, including portable computers.

The technology that enabled the construction of personal computers was the microprocessor, a programmable integrated circuit (or "chip") that acts as the CPU. Intel introduced the first microprocessor in 1971, the 4-bit 4004, which it called a "microprogrammable computer on a chip." The 4004 was originally developed as a general-purpose chip for a programmable calculator, but Intel introduced it as part of Intel's Microcomputer System 4-bit, or MCS-4, which also included read-only memory (ROM) and random-access memory (RAM) memory chips and a shift register chip. In August 1972, Intel followed with the 8-bit 8008, then the more powerful 8080 in June 1974. Following Intel's lead, computers based on the 8080 were usually called microcomputers.

The success of the minicomputer during the 1960s prepared computer engineers and users for "single person, single CPU" computers. Digital Equipment Corporation's (DEC) widely used PDP-10, for example, was smaller, cheaper, and more accessible than large mainframe computers. Timeshared computers operating under operating systems such as TOPS-10 on the PDP-10—co-developed by the Massachusetts Institute of Technology (MIT) and DEC in 1972—created the illusion of individual control of computing power by providing rapid access to personal programs and files. By the early 1970s, the accessibility of minicomputers, advances in microelectronics, and component miniaturization created expectations of affordable personal computers.

Innovation in software design during the 1960s and 1970s also shifted attention from computers as large calculating machines to their potential use as technologies of personal productivity. Work at Douglas Engelbart's Augmentation Research Center at the Stanford Research Institute (SRI), beginning in 1961; David Evans's and Ivan Sutherland's Computer Science Laboratory at the University of Utah and their own company, Evans & Sutherland, founded in 1968; and Xerox's Palo Alto Research Center (PARC), founded in 1969, resulted in prototypes of graphical user interfaces such as Engelbart's NLS software and PARC's Alto, a desktop workstation. The Xerox Star (1981) showed that personal workstations equipped with powerful graphical user interfaces, networking, and office productivity software could be manufactured, but due primarily to its high cost, it was a commercial failure.

The first microcomputers of the mid-1970s, many little more than component kits, were inexpensive and featured simple designs that invited experimentation. The Altair 8800, manufactured by Micro Instrumentation Telemetry Systems (MITS), was the first, making its debut as the cover story of the January 1975 issue of *Popular Electronics*. Encouraged by the article to mail-order kits and build their own computers, hobbyists gathered to share information about microcomputers in clubs such as the Homebrew Computer Club in Palo Alto, California, founded in March 1975. The "computer liberation" philosophy of writers such as Ted Nelson, author of *Computer Lib/Dream Machines* (1974), inspired them to learn about microcomputers as a way of steering computing power "to the people."

Despite the enthusiasm it unleashed, the Altair was not a powerful computer. Ed Roberts, the founder of MITS, was poorly prepared for its success. He realized the need for supporting the claims in *Popular Electronics*, especially the need for software. He hired Bill Gates and Paul Allen—then college students—to write a version of the BASIC (beginner's all-purpose symbolic instruction code) programming system for the Altair; they completed the work in February 1975 after establishing a company, Microsoft, to license the software to him. Dozens of new companies emerged between 1975 and 1977 to manufacture microcomputers with many differences in hardware and software. Many of them were built on the 100-pin bus known as S-100, which had been used by the Altair and the IMSAI 8080 (1976) to provide a channel for the microprocessor to communicate with hardware devices, including third-party add-ons such as Cromemco's Dazzler graphics board (1976). However, it was not a reliable standard. In 1977, Gary Kildall completed a version of the CP/

M operating system with a basic input/output system (BIOS) that could be written for each new machine, while the rest of the operating system did not need to be changed. CP/M became an informal software standard for computers that did not rely on proprietary graphics displays, and "CP/M computer" became the generic term for these microcomputers.

Apple Computer was founded in 1976 by Steve Jobs and Stephen Wozniak in order to sell Wozniak's elegantly designed Apple I microcomputer. They launched the Apple II computer at the first West Coast Computer Fair in 1977. Like CP/M computers, the Apple II integrated features such as disk storage with an expandable bus architecture that created an after-market for cards fitting into expansion slots. Sales of the Apple II benefited from new software available for it, such as Daniel Bricklin's and Robert Frankston's VisiCalc, a spreadsheet program in great demand. Jobs worked with engineers, retailers and customers to define personal and business markets for the Apple II, establishing Apple Computer as a leader of the new microcomputer industry. Competitors, such as Radio Shack and Commodore Business Machines followed similar strategies in seeking a customer base beyond the Homebrew hobbyists.

The microcomputer industry rapidly matured in the early 1980s. In August 1981, IBM launched the IBM PC after more than a year of development under the codename Acorn. Its designers utilized Intel's new 8088 microprocessor, beginning the transition to a new generation of more powerful, 16-bit microcomputers. Despite this, the PC was technologically conservative. The MS-DOS operating system written by Microsoft for the IBM PC differed little from CP/M, and the machine itself offered a familiar array of hardware. Yet, the IBM PC, merely by putting IBM's stamp of approval on the "personal computer," significantly expanded the business market for microcomputers.

Graphical user interfaces (GUI) began to appear on personal computers in the early 1980s. At Apple, Jef Raskin proposed a networked personal computer in 1979 utilizing the style of GUI introduced at PARC, just as the company was formulating strategy for a new generation of computers to succeed the Apple II/III line. Raskin's proposal influenced development of the Lisa (1983), a failure in the marketplace. By then, Jobs had taken control of Raskin's Macintosh project, which produced a more economical and compactly designed version of this technology. The Macintosh was introduced in January 1984 with a splashy marketing campaign that depicted it as a personal computer "for the rest of us," contrasting its innovative design to IBM's staid business-oriented computer.

Despite the fact that GUI interfaces such as VisiCorps VisiOn (1983) and Microsoft Windows (1985) were available for the IBM PC, IBM and Apple by the mid-1980s established alternatives to personal computing. On the one hand, The Macintosh was a closed, tightly controlled architecture emphasizing usability. On the other, the open architecture of the PC using MS-DOS gave manufacturers flexibility in hardware design, with the BIOS linking their machines to the de facto operating system "standard." The evolution of these PC "clones" tracked progress in Intel microprocessors, beginning with the fully 16-bit 8086 chip and continuing through the Pentium processor line, or followed versions of Microsoft's MS-DOS and Windows. The companies that produced them (Compaq, Dell, etc.) paced the phenomenal growth of the PC industry during the 1980s and 1990s.

See also **Computers, Mainframe; Computers, Uses and Consequences; Electronic Communications; Software Application Programs**

Henry Lowood

Further Reading

Bardini, T. *Bootstrapping: Douglas Engelbart, Coevolution, and the Origins of Personal Computing.* Stanford University Press, Stanford, 2000.

Ceruzzi, P.E. *A History of Modern Computing.* MIT Press, Cambridge, MA, 1998.

Friedewald, M. *Der Computer als Werkzeug und Medium: Die geistigen und technischen Wurzeln des Persona Computers.* Verlag für Geschichte der Naturwissenschaften und der Technik, Berlin, 1999.

Linzmayer, O.W. *Apple Confidential: The Real Story of Apple Computer, Inc.* No Starch Press, San Francisco, 1999.

Freiberger, P. and Swaine, M. *Fire in the Valley: The Making of the Personal Computer.* Collector's Edition. McGraw-Hill, New York, 2000.

Goldberg, A., Ed. *History of Personal Workstations.* ACM Press, New York, 1988.

Levy, S. *Hackers: Heroes of the Computer Revolution.* Dell, New York, 1984.

Levy, S. *Insanely Great: The Life and Times of Macintosh, the Computer that Changed Everything.* Penguin, New York, 1994.

Computers, Supercomputers

Supercomputers are high-performance computing devices that are generally used for numerical calculation, for the study of physical systems either through numerical simulation or the processing of scientific data. Initially, they were large, expensive, mainframe computers, which were usually owned by government research labs. By the end of the twentieth century, they were more often networks of inexpensive small computers. The common element of all of these machines was their ability to perform high-speed floating-point arithmetic—binary arithmetic that approximates decimal numbers with a fixed number of bits—the basis of numerical computation.

Almost all the early electronic computers were called supercomputers or supercalculators. Two early machines, the Massachusetts Institute of Technology (MIT) Whirlwind (1951) and the International Business Machine (IBM) Stretch (1955) achieved their performance using innovative circuitry. The Whirlwind pioneered the use of magnetic core memory, while the Stretch used high-speed transistors. Though subsequent supercomputers used even more advanced electronics technology, most achieved their speed through parallelism, by performing multiple calculations simultaneously.

Numerical weather prediction, one of the standard problems for supercomputers, illustrates the parallel nature of many scientific calculations. The basic analysis of the earth's weather system, done in the 1920s, imposed a grid upon the surface the earth and developed a series of differential equations to describe how the weather changed at each point of the grid. The equations described such things as the absorption of solar energy, the dissipation of the wind, the movement of humidity, and the fluctuations in temperature. Though some values are passed from point to point, many of the calculations can be done in parallel. This can be best appreciated by comparing two points on opposite sides of the globe. As the weather in London has little connection to the weather in Tahiti, the calculations for the two locations can be done at the same time. Other kinds of problems that shared this kind of geometric parallelism include stress analyses of structures, the processing of graphical images, the simulation of fluid or airflow, and the simulation of nuclear explosions. This last problem was crucial to the evolution of supercomputers. Many supercomputers were developed for the U.S. Defense Department in the 1960s and 1970s. The U. S. nuclear research laboratories helped to test these machines, debug them, and create software for them.

Early attempts to exploit parallelism produced supercomputers called vector machines. Vector processors can perform the same instruction repeatedly on a large amount of data. These machines are often associated with the designer Seymour Cray. First employed by Sperry-Rand, he helped to found two supercomputer manufacturers: Control Data Corporation (CDC) and Cray Research. His first parallel machine was the CDC 6400, which was announced in 1964. It was a "look-ahead" machine—the elements of the machine could work autonomously and its control unit could execute instructions out of sequence. If it encountered an instruction that needed circuits that were currently in use, it would jump to the next instruction and try to execute it. He expanded this idea in his next two machines, the CDC 7600 and the Cray I.

The vector processor of the Cray I exploited parallelism within floating-point arithmetic. As floating-point numbers are recorded in scientific notation, the fundamental operations, such as addition, subtraction, multiplication, and division are more complicated than the arithmetic of integers. Floating-point addition has four separate steps and requires three separate additions or subtractions. Cray devised logic circuits to perform each step of a floating-point operation. He then arranged these circuits in logical assembly lines, similar in concept to the assembly lines of an automobile factory. The assembly line for addition had four stages, each corresponding to one of the steps. As data passed down these lines, they stopped at each stage to have part of the floating-point operation performed. Once the pipe was filled with data and was processing it, it would produce additions four times faster than a processor that did one complete floating-point addition at a time.

Vector processors were capable of great speeds. If all parts of the Cray I were operating, it would produce 133 million floating-point operations per second, or mega-FLOPS. Yet, the problems of achieving this speed illustrated the software problems for supercomputer designers. Unless a program made heavy use of the vector processor, the machine ran at a fraction of its top speed, an observation known as "Amdahl's Law." The vector hardware could be a thousand times faster than the main processor or even a million times faster, but if the program used it only 5 percent of the time, the supercomputer would run only 5 percent faster than a conventional machine because the supercomputer would always be waiting for the main processor to complete its work. As a consequence, programmers radically learned to rewrite their code in order to keep the vector processor fully occupied, a process known as "vectorization." They commonly discovered that vectorized programs altered the order of basic

operations and required substantially more memory to hold intermediate results. Vectorization could be a time-consuming process and difficulty to justify unless the program was run repeatedly. However, by the early 1980s, software developers had created optimizing FORTRAN (from "formula translation," a computer language) compilers that automatically rewrote programs for a vector machine.

During the late 1980s, supercomputer research began to shift from large vector machines to networks of smaller, less expensive machines. These machines promised to be more cost effective, though computer scientists often discovered that these clusters or constellations or processors could be more difficult to program than vector machines. They experimented with different configurations of processors, the extent to which they shared common memory, and the way in which they were controlled. Some machines, such as the Burroughs Scientific Processor, used a single stream of instructions to control an array of processors that worked from a single memory. At the other extreme, the Cray X-MP had independent processors, each of which had large dedicated memories. Between the two, fell machines such as the Connection Machine. It consisted of a large number of processors, each of which had a small amount of independent memory. An important step in this work came in 1994 when a group of NASA Beowulf project. Beowulf machines could be assembled from commercial processors using standard network hardware. Few programmers could utilize the full power of these machines, but these were so inexpensive that many were able to use them. The original NASA Beowulf cost $40,000 and had the potential of computing one billion floating-point operations per second.

With the advent of inexpensive supercomputers, these machines moved beyond the large government labs and into smaller research and engineering facilities. Some were used for the study of social science. A few were employed by business concerns, such as stock brokerages or graphic designers.

See also **Computer Modeling; Computers, Mainframe; Processors for Computers**

DAVID GRIER

Further Reading

August, M.C., Brost, G.M., Hsiung, C.C. and Schiffleger, A.J. Cray X-MP: the birth of a supercomputer. *Computer*, 22, 1 45–52, 1989.

Bell, G. and Gray, J. What's next in high-performance computing? *ACM Comm.*, 45, 2, 91–95, 2002.

Cybenko, G. and Kuck, D. Supercomputers-revolution or evolution? *IEEE Spect.*, 29, 9, 39–41, 1992.

Frenkel, K.A. Special issue on parallelism. *ACM Comm.*, 29, 12, 1986.

Hord, M. *Understanding Parallel Supercomputing*. Institute of Electrical and Electronics Engineers, New York, 1998.

Murray, C. *The Supermen: The Story of Seymour Cray and the Technical Wizards Behind the Supercomputer*. Wiley, New York, 1997.

Padmanabhan, K. Cube structures for multiprocessors. *ACM Comm.*, 33,1, 1990.

Computers, Uses and Consequences

Towards the close of the last century the computer was claimed to be the most revolutionary artifact of twentieth century technology. It transformed business and industrial production, engineering and the sciences, as well as everyday life, and brought about a "computer revolution," a "computer age," an "information age," and an "Internet age," as many observers noted. It would therefore appear that the economy and the society of the industrial nations are more and more based on computer technology. Added to this is the ongoing debate over the risks and negative consequences of the increasing ubiquity of computer technology in our society; for example, the effects of computer technology on employment and data privacy protection, and the case of the risks of using software for the control of technological systems such as nuclear weapons systems. The overall picture shows that on the one hand, the computer seems to be a machine that has repeatedly stirred up illusions and visions in society since its invention (e.g., the vision of computers as intelligent machines), and on the other hand, people often fear the consequences of its uses.

The modern computer—the (electronic) digital computer in which the stored program concept is realized and hence self-modifying programs are possible—was only invented in the 1940s. Nevertheless, the history of computing (interpreted as the usage of modern computers) is only understandable against the background of the many forms of information processing as well as mechanical computing devices that solved mathematical problems in the first half of the twentieth century. The part these several predecessors played in the invention and early history of the computer may be interpreted from two different perspectives: on the one hand it can be argued that these machines prepared the way for the modern digital computer, on the other hand it can be argued that the computer, which was invented as a mathematical instrument, was reconstructed to be a data-processing machine, a control mechanism, and a communication tool.

The invention and early history of the digital computer has its roots in two different kinds of developments: first, information processing in business and government bureaucracies (see Calculators, Mechanical and Electromechanical); and second, the use and the search for mathematical instruments and methods that could solve mathematical problems arising in the sciences and in engineering (see Computers, Early Digital).

Origins in Mechanical Office Equipment

The development of information processing in business and government bureaucracies had its origins in the late nineteenth century, which was not just an era of industrialization and mass production but also a time of continuous growth in administrative work. The economic precondition for this development was the creation of a global economy, which caused growth in production of goods and trade. This brought with it an immense increase in correspondence, as well as monitoring and accounting activities—corporate bureaucracies began to collect and process data in increasing quantities. Almost at the same time, government organizations became more and more interested in collating data on population and demographic changes (e.g., expanding tax revenues, social security, and wide-ranging planning and monitoring functions) and analyzing this data statistically.

Bureaucracies in the U.S. and in Europe reacted in a different way to these changes. While in Europe for the most part neither office machines nor telephones entered offices until 1900, in the U.S. in the last quarter of the nineteenth century the information-handling techniques in bureaucracies were radically changed because of the introduction of mechanical devices for writing, copying, and counting data. The rise of big business in the U.S. had caused a growing demand for management control tools, which was fulfilled by a new ideology of systematic management together with the products of the rising office machines industry. Because of a later start in industrialization, the government and businesses in the U.S. were not forced to reorganize their bureaucracies when they introduced office machines. This, together with an ideological preference for modern office equipment, was the cause of a market for office machines and of a far-reaching mechanization of office work in the

U.S. In the 1880s typewriters and cash registers became very widespread, followed by adding machines and book-keeping machines in the 1890s. From 1880 onward, the makers of office machines in the U.S. underwent a period of enormous growth, and in 1920 the office machine industry annually generated about $200 million in revenue. In Europe, by comparison, mechanization of office work emerged about two decades later than in the U.S.—both Germany and Britain adopted the American system of office organization and extensive use of office machines for the most part no earlier than the 1920s.

During the same period the rise of a new office machine technology began. Punched card systems, initially invented by Herman Hollerith to analyze the U.S. census in 1890, were introduced. By 1911 Hollerith's company had only about 100 customers, but after it had been merged in the same year with two other companies to become the Computing-Tabulating-Recording Company (CTR), it began a tremendous ascent to become the world leader in the office machine industry. CTR's general manager, Thomas J. Watson, understood the extraordinary potential of these punched-card accounting devices, which enabled their users to process enormous amounts of data largely automatically, in a rapid way and at an adequate level of cost and effort. Due to Watson's insights and his extraordinary management abilities, the company (which had since been renamed to International Business Machines (IBM)) became the fourth largest office machine supplier in the world by 1928—topped only by Remington Rand, National Cash Register (NCR), and the Burroughs Adding Machine Company.

Origin of Calculating Devices and Analog Instruments

Compared with the fundamental changes in the world of corporate and government bureaucracies caused by office machinery during the late nineteenth and early twentieth century, calculating machines and instruments seemed to have only a minor influence in the world of science and engineering. Scientists and engineers had always been confronted with mathematical problems and had over the centuries developed techniques such as mathematical tables. However, many new mathematical instruments emerged in the nineteenth century and increasingly began to change the world of science and engineering. Apart from the slide rule, which came into popular use in Europe from the early nineteenth century onwards

(and became the symbol of the engineer for decades), calculating machines and instruments were only produced on a large scale in the middle of the nineteenth century.

In the 1850s the production of calculating machines as well as that of planimeters (used to measure the area of closed curves, a typical problem in land surveying) started on different scales. Worldwide, less than 2,000 calculating machines were produced before 1880, but more than 10,000 planimeters were produced by the early 1880s. Also, various types of specialized mathematical analog instruments were produced on a very small scale in the late nineteenth century; among them were integraphs for the graphical solution of special types of differential equations, harmonic analyzers for the determination of Fourier coefficients of a periodic function, and tide predictors that could calculate the time and height of the ebb and flood tides (see Computers, Analog).

Nonetheless, in 1900 only geodesists and astronomers (as well as part of the engineering community) made extensive use of mathematical instruments. In addition, the establishment of applied mathematics as a new discipline took place at German universities on a small scale and the use of apparatus and machines as well as graphical and numerical methods began to flourish during this time. After World War I, the development of engineering sciences and of technical physics gave a tremendous boost to applied mathematics in Germany and Britain. In general, scientists and engineers became more aware of the capabilities of calculating machines and a change of the calculating culture—from the use of tables to the use of calculating machines—took place.

One particular problem that was increasingly encountered by mechanical and electrical engineers in the 1920s was the solution of several types of differential equations, which were not solvable by analytic solutions. As one important result of this development, a new type of analog instrument—the so called "differential analyzer"—was invented in 1931 by the engineer Vannevar Bush at the Massachusetts Institute of Technology (MIT). In contrast to its predecessors—several types of integraphs—this machine (which was later called an analog computer) could be used not only to solve a special class of differential equation, but a more general class of differential equations associated with engineering problems. Before the digital computer was invented in the 1940s there was an intensive use of analog instruments (similar to Bush's differential analyzer) and a number of machines were constructed in the U.S. and in Europe after the model of Bush's machine before and during World War II. Analog instruments also became increasingly important in several fields such as the firing control of artillery on warships or the control of rockets (see Computers, Analog). It is worth mentioning here that only for a limited class of scientific and engineering problems was it possible to construct an analog computer—weather forecasting and the problem of shock waves produced by an atomic bomb, for example, required the solution of partial differential equations, for which a digital computer was needed.

The Invention of the Computer

The invention of the electronic digital stored-program computer is directly connected with the development of numerical calculation tools for the solution of mathematical problems in the sciences and in engineering. The ideas that led to the invention of the computer were developed simultaneously by scientists and engineers in Germany, Britain, and the U.S. in the 1930s and 1940s. The first freely programmable program-controlled automatic calculator was developed by the civil engineering student Konrad Zuse in Germany. Zuse started development work on program-controlled computing machines in the 1930s, when he had to deal with extensive calculations in static, and in 1941 his Z3, which was based on electromechanical relay technology, became operational.

Several similar developments in the U.S. were in progress at the same time. In 1937 Howard Aiken, a physics student at Harvard University, approached IBM to build a program-controlled calculator—later called the "Harvard Mark I." On the basis of a concept Aiken had developed because of his experiences with the numerical solution of partial differential equations, the machine was built and became operational in 1944. At almost the same time a series of important relay computers was built at the Bell Laboratories in New York following a suggestion by George R. Stibitz. All these developments in the U.S. were spurred by the outbreak of World War II. The first large-scale programmable electronic computer called the Colossus was built in complete secrecy in 1943 to 1944 at Bletchley Park in Britain in order to help break the German Enigma machine ciphers.

However, it was neither these relay calculators nor the Colossus that were decisive for the development of the universal computer, but the ENIAC (electronic numerical integrator and computer), which was developed at the Moore School of Engineering at the University of Pennsylvania. Extensive ballistic calculations were carried out there for the U.S. Army during World War II with the aid of the Bush "differential analyzer" and more than a hundred women ("computors") working on mechanical desk calculators. Observing that capacity was barely sufficient to compute the artillery firing tables, the physicist John W. Mauchly and the electronic engineer John Presper Eckert started developing the ENIAC, a digital version of the differential analyzer, in 1943 with funding from the U.S. Army.

In 1944 the mathematician John von Neumann turned his attention to the ENIAC because of his mathematical work on the Manhattan Project (on the implosion of the hydrogen bomb). While the ENIAC was being built, Neumann and the ENIAC team drew up plans for a successor to the ENIAC in order to improve the shortcomings of the ENIAC concept, such as the very small memory and the time-consuming reprogramming (actually rewiring) required to change the setup for a new calculation. In these meetings the idea of a stored-program, universal machine evolved. Memory was to be used to store the program in addition to data. This would enable the machine to execute conditional branches and change the flow of the program. The concept of a computer in the modern sense of the word was born and in 1945 von Neumann wrote the important "First Draft of a Report on the EDVAC," which described the stored-program, universal computer. The logical structure that was presented in this draft report is now referred to as the "von Neumann architecture." This EDVAC report was originally intended for internal use but once made freely available it became the "bible" for computer pioneers throughout the world in the 1940s and 1950s. The first computer featuring the von Neumann architecture operated at Cambridge University in the U.K.; in June 1949 the EDSAC (electronic delay storage automatic computer) computer built by Maurice Wilkes—designed according to the EDVAC principles—became operational (see Computers, Early Digital).

The Computer as a Scientific Instrument

As soon as the computer was invented, a growing demand for computers by scientists and engineers evolved, and numerous American and European universities started their own computer projects in the 1940s and 1950s. After the technical difficulties of building an electronic computer were solved, scientists grasped the opportunity to use the new scientific instrument for their research. For example, at the University of Göttingen in Germany, the early computers were used for the initial value problems of partial differential equations associated with hydrodynamic problems from atomic physics and aerodynamics. Another striking example was the application of von Neumann's computer at the Institute for Advanced Study (IAS) in Princeton to numerical weather forecasts in 1950. As a result, numerical weather forecasts could be made on a regular basis from the mid-1950s onwards.

Mathematical methods have always been of a certain importance for science and engineering sciences, but only the use of the electronic digital computer (as an enabling technology) made it possible to broaden the application of mathematical methods to such a degree that research in science, medicine, and engineering without computer-based mathematical methods has become virtually inconceivable at the end of the twentieth century. A number of additional computer-based techniques, such as scientific visualization, medical imaging, computerized tomography, pattern recognition, image processing, and statistical applications, have become of the utmost significance for science, medicine, engineering, and social sciences. In addition, the computer changed the way engineers construct technical artifacts fundamentally because of the use of computer-based methods such as computer-aided design (CAD), computer-aided manufacture (CAM), computer-aided engineering, control applications, and finite-element methods (see Computer-Aided Design and Manufacture). However, the most striking example seems to be the development of scientific computing and computer modeling, which became accepted as a third mode of scientific research that complements experimentation and theoretical analysis (see Computer Modeling). Scientific computing and computer modeling are based on supercomputers as the enabling technology, which became important tools for modern science routinely used to simulate physical and chemical phenomena. These high-speed computers became equated with the machines developed by Seymour Cray, who built the fastest computers in the world for many years. The supercomputers he launched such as the legendary CRAY I from 1976 were the basis for computer modeling of real world systems, and helped, for example, the defense industry in the U.S. to build weapons systems and the oil industry to create geological models that show potential oil deposits (see Computers, Supercomputers).

Growth of Digital Computers in Business and Information Processing

When the digital computer was invented as a mathematical instrument in the 1940s, it could not have been foreseen that this new artifact would ever be of a certain importance in the business world. About 50 firms entered the computer business worldwide in the late 1940s and the early 1950s, and the computer was reconstructed to be a type of electronic data-processing machine that took the place of punched-card technology as well as other office machine technology. It is interesting to consider that there were mainly three types of companies building computers in the 1950s and 1960s: newly created computer firms (such as the company founded by the ENIAC inventors Eckert and Mauchly), electronics and control equipments firms (such as RCA and General Electric), and office appliance companies (such as Burroughs and NCR). Despite the fact that the first digital computers were put on the market by a German and a British company, U.S. firms dominated the world market from the 1950s onward, as these firms had the biggest market as well as financial support from the government.

Generally speaking, the Cold War exerted an enormous influence on the development of computer technology. Until the early 1960s the U.S. military and the defense industry were the central drivers of the digital computer expansion, serving as the main market for computer technology and shaping and speeding up the formation of the rising computer industry. Because of the U.S. military's role as the "tester" for prototype hard- and software, it had a direct and lasting influence on technological developments; in addition, it has to be noted that the spread of computer technology was partly hindered by military secrecy. Even after the emergence of a large civilian computer market in the 1960s, the U.S. military maintained its influence by investing a great deal in computer in hard- and software and in computer research projects.

From the middle of the 1950s onwards the world computer market was dominated by IBM, which accounted for more than 70 percent of the computer industry revenues until the mid-1970s. The reasons for IBM's overwhelming success were diverse, but the company had a unique combination of technical and organizational capabilities at its disposal that prepared it perfectly for the mainframe computer market. In addition, IBM benefited from enormous government contracts, which helped to develop excellence in computer technology and design. However, the greatest advantage of

IBM was by no doubt its marketing organization and its reputation as a service-oriented firm, which was used to working closely with customers to adapt machinery to address specific problems, and this key difference between IBM and its competitors persisted right into the computer age.

During the late 1950s and early 1960s, the computer market—consisting of IBM and seven other companies called the "seven dwarves"—was dominated by IBM, with its 650 and 1401 computers. By 1960 the market for computers was still small. Only about 7,000 computers had been delivered by the computer industry, and at this time even IBM was primarily a punched-card machine supplier, which was still the major source of its income. Only in 1960 did a boom in demand for computers start, and by 1970 the number of computers installed worldwide had increased to more than 100,000. The computer industry was on the track to become one of the world's major industries, and was totally dominated by IBM.

The outstanding computer system of this period was IBM's System/360. It was announced in 1964 as a compatible family of the same computer architecture, and employed interchangeable peripheral devices in order to solve IBM's problems with a hotchpotch of incompatible product lines (which had evoked large problems in the development and maintenance of a great deal of different hardware and software products). Despite the fact that neither the technology used nor the systems programming were of a high-tech technology at the time, the System/360 established a new standard for mainframe computers for decades. Various computer firms in the U.S., Europe, Japan and even Russia, concentrated on copying components, peripherals for System/360 or tried to build System/360-compatible computers.

The growth of the computer market during the 1960s was accompanied by market shakeouts: two of the "seven dwarves" left the computer business after the first computer recession in the early 1970s, and afterwards the computer market was controlled by IBM and BUNCH (Burroughs, UNIVAC, NCR, Control Data, and Honeywell). At the same time, an internationalization of the computer market took place—U.S. companies controlled the world market for computers—which caused considerable fears over loss of national independence in European and Japanese national governments, and these subsequently stirred up national computing programs. While the European attempts to create national champions as well as the more general attempt to create a European-wide market for mainframe computers failed in the end, Japan's attempt to found a national computer industry has been successful: Until today Japan is the only nation able to compete with the U.S. in a wide array of high-tech computer-related products (see Computers, Mainframe).

Real-Time and Time-Sharing

Until the 1960s almost all computers in government and business were running batch-processing applications (i.e., the computers were only used in the same way as the punched-card accounting machines they had replaced). In the early 1950s, however, the computer industry introduced a new mode of computing named "real-time" in the business sector for the first time, which was originally developed for military purposes in MIT's Whirlwind project. This project was initially started in World War II with the aim of designing an aircraft simulator by analog methods, and later became a part of a research and development program for the gigantic, computerized anti-aircraft defense system SAGE (semi-automatic ground environment) built up by IBM in the 1950s.

The demand for this new mode of computing was created by cultural and structural changes in economy. The increasing number of financial transactions in banks and insurance companies as well as increasing airline traveling activities made necessary new computer-based information systems that led finally to new forms of business evolution through information technology.

The case of the first computerized airline reservation system SABRE, developed for American Airlines by IBM in the 1950s and finally implemented in the early 1960s, serves to thoroughly illustrate these structural and structural changes in economy. Until the early 1950s, airline reservations had been made manually without any problems, but by 1953 this system was in crisis because increased air traffic and growing flight plan complexity had made reservation costs insupportable. SABRE became a complete success, demonstrating the potential of centralized real-time computing systems connected via a network. The system enabled flight agents throughout the U.S., who were equipped with desktop terminals, to gain a direct, real-time access to the central reservation system based on central IBM mainframe computers, while the airline was able to assign appropriate resources in response. Therefore an effective combination of advantages was offered by SABRE—a better utilization of resources and a much higher customer convenience.

Very soon this new mode of computing spread around the business and government world and became commonplace throughout the service and distribution sectors of the economy; for example, bank tellers and insurance account representatives increasingly worked at terminals. On the one hand structural information problems led managers to go this way, and on the other hand the increasing use of computers as information handling machines in government and business had brought about the idea of computer-based accessible data retrieval. In the end, more and more IBM customers wanted to link dozens of operators directly to central computers by using terminal keyboards and display screens.

In the late 1950s and early 1960s—at the same time that IBM and American Airlines had begun the development of the SABRE airline reservation system—a group of brilliant computer scientists had a new idea for computer usage named "time sharing." Instead of dedicating a multiterminal system solely to a single application, they had the computer utility vision of organizing a mainframe computer so that several users could interact with it simultaneously. This vision was to change the nature of computing profoundly, because computing was no longer provided to naïve users by programmers and systems analysts, and by the late 1960s time-sharing computers became widespread in the U.S.

Particularly important for this development had been the work of J.C.R. Licklider of the Advanced Research Project Agency (ARPA) of the U.S. Department of Defense. In 1960 Licklider had published a now-classic paper "Man–Computer Symbiosis" proposing the use of computers to augment human intellect and creating the vision of interactive computing. Licklider was very successful in translating his idea of a network allowing people on different computers to communicate into action, and convinced ARPA to start an enormous research program in 1962. Its budget surpassed that of all other sources of U.S. public research funding for computers combined. The ARPA research programs resulted in a series of fundamental moves forward in computer technology in areas such as computer graphics, artificial intelligence (see Artificial Intelligence), and operating systems. For example, even the most influential current operating system, the general-purpose time-sharing system Unix, developed in the early 1970s at the Bell Laboratories, was a spin-off of an ambitious operating system project, Multics, funded by ARPA. The designers of Unix successfully attempted to keep away from complexity by using a clear, minimalist design approach to software design, and created a multitasking, multiuser operating system, which became the standard operating system in the 1980s (see Computers, Systems Programs).

Electronic Component Revolution

While the nature of business computing was changed by the new paradigms such as real time and time sharing, advances in solid-state components increasingly became a driving force for fundamental changes in the computer industry, and led to a dynamic interplay between new computer designs and new programming techniques that resulted in a remarkable series of technical developments. The technical progress of the mainframe computer had always run parallel to conversions in the electronics components (see Computer Memory, Main). During the period from 1945 to 1965, two fundamental transformations in the electronics industry took place that were marked by the invention of the transistor in 1947 and the integrated circuit in 1957 to 1958. While the first generation of computers—lasting until about 1960—was characterized by vacuum tubes (valves) for switching elements, the second generation used the much smaller and more reliable transistors, which could be produced at a lower price. A new phase was inaugurated when an entire integrated circuit on a chip of silicon was produced in 1961, and when the first integrated circuits were produced for the military in 1962. A remarkable pace of progress in semiconductor innovations, known as the "revolution in miniature," began to speed up the computer industry. The third generation of computers characterized by the use of integrated circuits began with the announcement of the IBM System/360 in 1964 (although this computer system did not use true integrated circuits). The most important effect of the introduction of integrated circuits was not to strengthen the leading mainframe computer systems, but to destroy Grosch's Law, which stated that computing power increases as the square of its costs. In fact, the cost of computer power dramatically reduced during the next ten years.

This became clear with the introduction of the first computer to use integrated circuits on a full scale in 1965: the Digital Equipment Corporation (DEC) offered its PDP-8 computer for just $18,000, creating a new class of computers called minicomputers—small in size and low in cost—as well as opening up the market to new customers. Minicomputers were mainly used in areas other than general-purpose computing such as industrial applications and interactive graphics systems. The PDP-8 became the first widely successful minicomputer with over 50,000 items sold, demonstrating that there was a market for smaller computers. This success of DEC (by 1970 it had become the world's third largest computer manufacturer) was supported by dramatic advances in solid-state technology. During the 1960s the number of transistors on a chip doubled every two years, and as a result minicomputers became continuously more powerful and more inexpensive at an inconceivable speed.

Personal Computing

The most striking aspect of the consequences of the exponential increase of the number of transistors on a chip during the 1960s—as stated by "Moore's Law": the number of transistors on a chip doubled every two years—was not the lowering of the costs of mainframe computer and minicomputer processing and storage, but the introduction of the first consumer products based on chip technology such as hand-held calculators and digital watches in about 1970 (see Calculators, Electronic). More specifically, the market acts in these industries were changed overnight by the shift from mechanical to chip technology, which led to an enormous deterioration in prices as well as a dramatic industry shakeout. These episodes only marked the beginning of wide-ranging changes in economy and society during the last quarter of the twentieth century leading to a new situation where chips played an essential role in almost every part of business and modern life.

The case of the invention of the personal computer serves to illustrate that it was not sufficient to develop the microprocessor as the enabling technology in order to create a new invention, but how much new technologies can be socially constructed by cultural factors and commercial interests. When the microprocessor, a single-chip integrated circuit implementation of a CPU, was launched by the semiconductor company Intel in 1971, there was no hindrance to producing a reasonably priced microcomputer, but it took six years until the consumer product PC emerged. None of the traditional mainframe and minicomputer companies were involved in creating the early personal computer. Instead, a group of computer hobbyists as well as the "computer liberation" movement in the U.S. became the driving force behind the invention of the PC. These two groups were desperately keen on a

low-priced type of minicomputer for use at home for leisure activities such as computer games; or rather they had the counterculture vision of an unreservedly available and personal access to an inexpensive computer utility provided with rich information. When in 1975 the Altair 8800, an Intel 8080 microprocessor-based computer, was offered as an electronic hobbyist kit for less than $400, these two groups began to realize their vision of a "personal computer." Very soon dozens of computer clubs and computer magazines were founded around the U.S., and these computer enthusiasts created the personal computer by combining the Altair with keyboards, disk drives, and monitors as well as by developing standard software for it. Consequently in only two years, a more or less useless hobbyist kit had been changed into a computer that could easily be transformed in a consumer product.

The computer hobbyist period ended in 1977, when the first standard machines for an emerging consumer product mass market were sold. These included products such as the Commodore Pet and the Apple II, which included its own monitor, disk drive, and keyboard, and was provided with several basic software packages. Over next three years, spreadsheet, word processing, and database software were developed, and an immense market for games software evolved (see Computers, Software, Application Programs). As a result, personal computers became more and more a consumer product for ordinary people, and Apple's revenues shot to more than $500 million in 1982. By 1980, the personal computer had transformed into a business machine, and IBM decided to develop its own personal computer, which was introduced as the IBM PC in 1981. It became an overwhelming success and set a new industry standard.

Apple tried to compete by launching their new Macintosh computer in 1984 provided with a revolutionary graphical user interface (GUI), which set a new standard for a user-friendly human–computer interaction. It was based on technology created by computer scientists at the Xerox Palo Alto Research Center in California, who had picked up on ideas about human–computer interaction developed at the Stanford Research Institute and at the University of Utah. Despite the fact that the Macintosh's GUI was far superior to the MS-DOS operating system of the IBM-compatible PCs, Apple failed to win the business market and remained a niche player with a market share of about 10 percent. The PC main branch was determined by the companies IBM had chosen as its original suppliers in 1981 for the design of the microprocessor (Intel) and the operating system (Microsoft). While IBM failed to seize power in the operating system software market for PCs in a software war with Microsoft, Microsoft achieved dominance not only of the key market for PC operating systems, but also the key market of office applications during the first half of the 1990s (see Computers, Personal).

Networking

In the early 1990s computing again underwent further fundamental changes with the appearance of the Internet, and for the most computer users, networking became an integral part of what it means to have a computer (see Computer Networks). Furthermore, the rise of the Internet indicated the impending arrival of a new "information infrastructure" as well as of a "digital convergence," as the coupling of computers and communications networks was often called.

In addition the 1990s were a period of an information technology boom, which was mainly based on the Internet hype. For many years previously, it seemed to a great deal of managers and journalists that the Internet would become not just an indispensable business tool, but also a miracle cure for economic growth and prosperity. In addition, computer scientists and sociologists started a discussion predicting the beginning of a new "information age" based on the Internet as a "technological revolution" and reshaping the "material basis" of industrial societies (see Internet).

The Internet was the outcome of an unusual collaboration of a military–industrial–academic complex that promoted the development of this extraordinary innovation. It grew out of a military network called the ARPAnet, a project established and funded by ARPA in the 1960s. The ARPAnet was initially devoted to support of data communications for defense research projects and was only used by a small number of researchers in the 1970s. Its further development was primarily promoted by unintentional forms of network usage. The users of the ARPAnet became very much attracted by the opportunity for communicating through electronic mail, which rapidly surpassed all other forms of network activities. Another unplanned spin-off of the ARPAnet was the Usenet (Unix User Network), which started in 1979 as a link between two universities and enabled its users to subscribe to newsgroups. Electronic mail became a driving force for the creation of a

large number of new proprietary networks funded by the existing computer services industry or by organizations such as the NSF (NSFnet). Because networks users' desire for email to be able to cross network boundaries, an ARPA project on "inter-networking" became the origin for the "Internet"—a network of networks linked by several layers of protocols such as TCP/IP (transmission control protocol/internet protocol), which quickly developed into the actual standard.

Only after the government funding had solved many of the most essential technical issues and had shaped a number of the most characteristic features of the Internet, did private sector entrepreneurs start Internet-related ventures and quickly developed user-oriented enhancements. Nevertheless the Internet did not make a promising start and it took more than ten years before significant numbers of networks were connected. In 1980, the Internet had less than two hundred hosts, and during the next four years the number of hosts went up only to 1000. Only when the Internet reached the educational and business community of PC users in the late 1980s, did it start to become an important economic and social phenomenon. The number of hosts began an explosive growth in the late 1980s—by 1988 there were over 50,000 hosts. An important and unforeseen side effect of this development became the creation of the Internet into a new electronic publishing medium. The electronic publishing development that excited most interest in the Internet was the World Wide Web, originally developed at the CERN High Energy Physics Laboratory in Geneva in 1989 (see World Wide Web). Soon there were millions of documents on the Internet, and private PC users became excited by the joys of surfing the Internet. A number of firms such as AOL soon provided low-cost network access and a range of consumer-oriented information services. The Internet boom was also helped by the Clinton–Gore presidential election campaign on the "information superhighway" and by the amazing news reporting on the national information infrastructure in the early 1990s. Nevertheless for many observers it was astounding how fast the number of hosts on the Internet increased during the next few years—from more than 1 million in 1992 to 72 million in 1999.

The overwhelming success of the PC and of the Internet tends to hide the fact that its arrival marked only a branching in computer history and not a sequence. (Take, for example, the case of mainframe computers, which still continue to run, being of great importance to government facilities and the private sector (such as banks and insurance companies), or the case of supercomputers, being of the utmost significance for modern science and engineering.) Furthermore it should be noted that only a small part of the computer applications performed today is easily observable—98 percent of programmable CPUs are used in embedded systems such as automobiles, medical devices, washing machines and mobile telephones.

See also **Computer and Video Games; Computer Displays; Computer Memory; Computer Science; Computers, Hybrid; Computer–User Interface; Software Engineering**

ULF HASHAGEN

Further Reading

Abbate, J. *Inventing the Internet*. Cambridge, MA, 1999.

Aspray, W. *John von Neumann and the Origins of Modern Computing*. Cambridge, MA, 1990.

Campbell-Kelly, M. and Aspray, W. *Computer: A History of the Information Machine*. New York, 1996.

Campbell-Kelly, M. *From Airline Reservations to Sonic the Hedgehog: A History of the Software Industry*. Cambridge, MA, 2003.

Ceruzzi, P. *A History of Modern Computing*. Cambridge, MA, 1998.

Cortada, J. *Before the Computer: IBM, NCR, Burroughs, and Remington Rand and the Industry They Created, 1856–1956*. Princeton, NJ, 1993.

Edwards, P. *The Closed World: Computers and the Politics of Discourse in Cold War America*. Cambridge, MA, 1996.

Hashagen, U., Keil-Slawik and Norberg, A., Eds. *History of Computing: Software Issues*. Berlin, 2002.

MacKenzie, D. *Mechanizing Proof: Computing, Risk, and Trust*. Cambridge, MA, 2001.

Norberg, A. and O'Neill, J. *Transforming Computer Technology: Information Processing for the Pentagon, 1962-1986*. Baltimore, 1996.

Petzold, H. *Rechnende Maschinen: eine historische Untersuchung ihrer Herstellung und Anwendung vom Kaiserreich bis zur Bundesrepublik*. Düsseldorf, 1985.

Roland, A. *DARPA and the Quest for Machine Intelligence, 1983–1993*. Cambridge, MA, 2002.

Computer–User Interface

A computer interface is the point of contact between a person and an electronic computer. Today's interfaces include a keyboard, mouse, and display screen. Computer user interfaces developed through three distinct stages, which can be identified as batch processing, interactive computing, and the graphical user interface.

Early computing machines used punch cards as a method to input instructions to be calculated. Punch-card technology was used during the first half of the twentieth century for business machines and it was applied to early computers. The ENIAC (electronic numerical integrator and computer), developed between 1943 and 1945 at the Moore School of Electrical Engineering at the University of Pennsylvania, used punch card technology and was the first electronic computer. Human operators supplied instructions to the ENIAC on punched cards and then waited for the results. For the following two decades, punch cards were the main input devices for computer–user interfaces.

In the postwar era, new types of interactive computer user interfaces were invented. The technologies required to enable people to interactively operate a computer were first developed in American government-funded projects, such as the SAGE (semi-automatic ground environment) air defense system, based on MIT's Whirlwind Computer. Project SAGE introduced magnetic core memory as a method of information storage to replace punched cards. Additionally, SAGE designers wrote software to run the computer's CRT (cathode-ray tube) that displayed information to human operators. The system converted radar information into pictures generated on the CRT and operators could select information by pointing to a computer-generated target with a light pen. This type of interactive interface was further advanced by Ivan Sutherland's Sketchpad project.

In 1962, Sketchpad introduced the concept of computer graphics by enabling an operator to create sophisticated visual models on a display screen that resembled a television set. Operators created models by putting information into the computer with a light pen and keyboard. Computer-generated drawings were immediately updated on the screen as different commands were selected.

Sketchpad was part of a new movement that examined how users interacted with computers. "Human–computer interaction" (HCI) is a term used to describe the psychology of how people interact and use computers. HCI combines computer science, psychology and ergonomics to create software to be used as part of the computer–user interface. The study of HCI has led to the development of computer systems that are easier to use. Ergonomics, or human factors, is the study of applying psychology to interface design to make computers easier to operate and understand. As computers became more interactive, the ergonomic features of interface design advanced to the point at which average people could easily operate them.

A pioneer in developing interactive computing was Douglas C. Engelbart. While working at the Stanford Research Laboratory in the 1960s, Engelbart created many of the tools that are commonly used with today's personal computers, including the mouse, graphical interfaces, hypertext, multiple windows, and teleconferencing. Engelbart's system, the "augment system," replaced the light pen with a mouse. The original mouse was a small input device in the shape of a box with hidden wheels. As the mouse rolled on a flat surface the wheels signaled a cursor or symbol displayed on the computer screen.

Engelbart's augment system interactively displayed text, graphics, and video images. Computer commands were input through a keyboard, mouse, and five-finger keyset. In December 1968, Engelbart and his team demonstrated his vision of interactive computing at the Association for Computer Machinery/Institute of Electrical Engineers Fall Joint Computer Conference in San Francisco. Sitting in the audience was a young computer scientist named Alan Kay.

In the early 1970s, Kay (along with members of Engelbart's team) joined the newly established Xerox Palo Alto Research Center (PARC). The PARC researchers developed the hardware and software necessary for the next generation of computer–user interfaces, which included pictures, windows, and menus.

At Xerox PARC, researchers designed and built a new computer system called the Alto, which had enough power to drive a full-screen graphical image. The Alto used bitmap technology that created a one-to-one correspondence between the picture elements on the screen and the bits in the computer's memory. Bitmapping enables users to scale letters and mix text and graphics together on a display screen. The Alto's text-editing software added a new feature called "what you see is what you get" (WYSIWYG). Images displayed on the screen visually resembled the computer-generated information output on a printed page. These new technological developments enabled PARC researchers to invent the WIMP interface. WIMP interfaces incorporate windows, icons, a mouse, and pull-down menus into a visual interface design.

Kay and his team further enhanced the WIMP design by transforming it into the graphical user interface (GUI). GUIs enable computer users to execute commands by pointing and clicking on icons. The icons displayed on the screen resemble familiar office objects such as file folders, file cabinets, and in-mail and out-mail baskets. Objects displayed on the screen are manipulated with a mouse and the screen's work area metaphorically represents a desktop. PARC's GUI added the desktop metaphor to computer–user interfaces.

In 1979, after seeing a demonstration of the Xerox PARC technology, Steve Jobs saw the possibility of using graphical interfaces to make computers more "user-friendly." He applied PARC technology to the Macintosh and it became the first commercially successful GUI. In 1984, Apple advertised the Macintosh as a "user-friendly" system that easily enabled average people to operate a computer. The Macintosh also popularized the desktop metaphor developed at PARC. Following the success of the Macintosh interface, a number of different companies introduced graphical and WIMP-styled interfaces, including International Business Machines (IBM), Digital Research, and Commodore. However, the most successful GUI was Microsoft Windows, which still dominates the interface marketplace. Currently, GUIs have become the most popular method of computer–user interaction.

Today's graphical interfaces support additional multimedia features, such as streaming audio and video. In GUI design, every new software feature introduces more icons into the process of computer–user interaction. Presently, the large vocabulary of icons used in GUI design is difficult for users to remember, which creates a complexity problem. As GUIs become more complex, interface designers are adding voice recognition and intelligent agent technologies to make computer user interfaces even easier to operate.

See also **Computer Displays**

SUSAN B. BARNES

Further Reading

Barnes, S.B. Engelbart, D.C. Developing the underlying concepts for contemporary computing. *IEEE Ann. Hist. Comput.*, 19, 3, 16–26, 1997.

Campbell-Kelly, M. and Aspray, W. *Computer: A History of the Information Machine*. Harper Collins, New York, 1996.

Freiberger, P. and Swaine, M. *Fire in the Valley: The Making of the Personal Computer*, 2nd edn. McGraw-Hill, New York, 2000.

Hiltzik, M.A. *Dealers of Lightning: Xerox PARC and the Dawn of the Computer Age*. Harper Collins, New York, 1999.

Kay, A.C. *User Interface: A Personal View, in The Art of Human-Computer Interface Design*, Laurel, B., Ed. Addison-Wesley, Reading, MA, 1990.

Kay, A. *The Early History of Smalltalk in Programming Languages*. Bergin, T.J. and Gibson, R.G. ACM Press, New York, 1996.

Rheingold, H. *Tools for Thought*. Simon & Schuster, New York, 1985.

Rose, F. *West of Eden: The End of Innocence at Apple Computer*. Penguin, New York, 1989.

Shneiderman, B. *Designing the User Interface: Strategies for Effective Human-Computer Interaction*, 3rd edn. Addison-Wesley, Boston, 1997.

Smith, D.K. and Alexander, R.C. *Fumbling the Future: How Xerox Invented, then Ignored, the First Personal Computer.* William Morrow, New York, 1988.

Concrete, Reinforced

Reinforced concrete was in its infancy at the opening of the twentieth century, but it was very quickly adopted worldwide as an economic and versatile construction material. Employing fairly basic materials—sand, crushed stone or gravel, cement, and steel—it found use in all the existing aspects of construction, including buildings, roads, bridges, dams, reservoirs, and docks. It also served the century's new applications, such as air raid shelters and the pressure vessels of nuclear reactors. By the end of the twentieth century, concrete in its various forms—plain, reinforced, and prestressed—was probably the most widely used construction material in the world.

Although concrete had been used from at least Roman times, it was not until the last decades of the nineteenth century that the idea of reinforcing it was applied to construction. Until then, concrete was used as a cheaper and more versatile substitute for stone and brick masonry, with which it shared the properties of being strong in compression but weak and brittle when subject to tension or bending. Not until rods or bars of wrought iron, and later steel, were embedded in concrete was it considered worthwhile to use the resulting reinforced concrete with confidence for floor and roof slabs, beams, trusses, cantilevers, and other elements that work by bending.

Proprietary reinforced concrete systems were patented and used almost simultaneously in several countries from the 1880s, notably the U.S., France, and Germany, followed quickly by Britain, where the Frenchman François Hennebique had taken out a patent in 1892. His system used the plain round mild steel bar, but other systems adopted alternative profiles (Figure 14) both to satisfy the need for originality if a patent were to be granted, and to ensure good grip or bond with the concrete. This was essential if the concrete and its reinforcement were to work together. The other key component of modern concrete—Portland cement, patented in 1824 by Joseph Aspdin—had by this time been developed as a strong and reliable product.

Guidance on the use, design, and construction of reinforced concrete quickly became available. The first textbook in the U.K., by Marsh, appeared

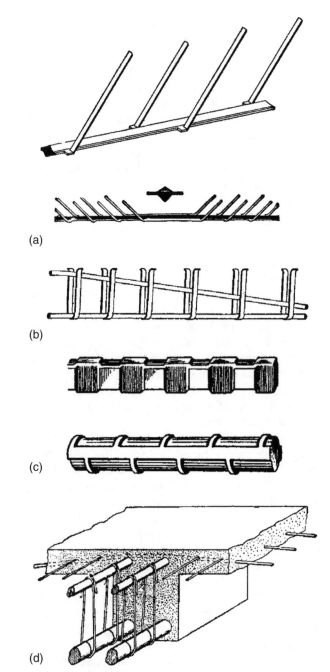

Figure 14. Some common early twentieth century reinforcing systems: (a) Coignet; (b) Hennebique; (c) Indented bar; and (d) Kahn.
[*Source: Jones, B.E., Ed. Cassell's Reinforced Concrete. Waverley Book Company, London, 1920.*]

in 1904. Subsequently, codes of practice for design and standards for materials were introduced. Subsequently, as later, research and development made major contributions to the understanding of concrete behavior through materials testing and load tests that studied the behavior of concrete structures. Many committees, learned societies, and trade associations were established during the century, providing the means of exchanging experience and understanding, both nationally and internationally.

As patents expired during the earlier decades of the century, reinforced concrete construction became freely available to contractors. This was a mixed blessing, as some lacked the skills and experience of the system providers, who took pains to ensure that their products were soundly designed and built.

A notable development was discovered by Eugène Freyssinet in 1928. He pioneered prestressing, using tensioned steel to precompress the whole of the concrete section, so that when loaded in service it did not develop tension stresses. The result was a more efficient member with improved durability.

The major reconstruction program undertaken in many countries in the decades after World War II made very extensive use of concrete. Inexperience, and pressures of time and money, resulted in many cases of inadequate cover of reinforcement and other defects so that subsequently, major repairs were needed to remedy the consequences of corroding reinforcement. By the end of the century, however, the issues of durability and how to achieve it were widely understood, so that any competent engineer or contractor could now be expected to build soundly in concrete.

Much research, development, and innovation was applied during the century to the essential materials used in concrete—cement, aggregates, and reinforcement.

Cement

Cements were developed for particular applications, such as rapid setting or resisting aggressive environments such as seawater. Others made use of industrial waste products such as ground granulated blastfurnace slag and fly ash (the ground-up clinker from coal-burning power stations). Inevitably, some materials proved to have unwelcome side effects, such as the very quick-setting high alumina cement, which was found in the 1970s to lose strength over time, even at room temperature. This was originally thought to occur only in warm damp environments. Similarly, calcium chloride, commonly added to the concrete mix to accelerate setting, especially in cold weather, was found subsequently to increase the risk of reinforcement corrosion. Use of both materials is now restricted by codes of practice.

Aggregates

The most common aggregates remain sand and crushed rock or gravel. However, commercial incentives and, later in the twentieth century, environmental issues such as opposition to new quarrying for natural aggregates, led to alternatives being sought and developed. Lightweight aggregates were made cheaply from waste materials—initially clinker from coal-burning and broken bricks, and later (once again) fly ash. Fly ash, when heated, forms pellets suitable for use as aggregate. Lightweight concrete offers savings in the supporting structure, and also offers better thermal insulation than normal-weight concrete. Concern over energy conservation issues in buildings from the 1970s meant that lightweight concrete blocks have become very widely used for the inner leaf of cavity wall construction, meeting onerous building regulation requirements for thermal insulation.

Reinforcement

With improved steel-making techniques and the need to ensure good bond to the concrete, the ribbed hot-rolled high-yield strength steel bar had displaced the plain mild steel bar to become the norm in the U.K., Europe, and elsewhere by the end of the century. Higher strength steel was needed for prestressed concrete. Forms developed and still in use included rods that could be pretensioned against the molds for precast unit manufacture, and cables or threaded bars that could be post-tensioned inside sheaths cast into the concrete, the latter approach often being used for larger beams, especially in bridges, cast in situ.

Concerns over durability led to the use of galvanized and epoxy-coated or stainless steel reinforcement, particularly on bridges and in car parks where deicing salt carried by vehicles can soak into concrete and accelerate the corrosion process. Although more expensive than plain steel, the greater initial cost may be outweighed by the potential savings from reduced—and necessarily disruptive—future maintenance and repair costs. Similar arguments apply to carbon fiber polymer-based reinforcement, which was still in its infancy at the turn of the twenty-first century.

Long-established and widely used throughout the twentieth century were asbestos sheet, corrugated asbestos, and woodwool, whose names belie the fact that all three are early forms of fiber-reinforced cement. Flat asbestos-cement sheet originated in Austria around 1900, while the corrugated form—which could span longer distances, and so was ideally suited for pitched roofs on factories and sheds—was being made in Britain by 1914. Woodwool is also believed to have originated at about this time in Austria, making use of waste timber shavings bound together with cement. Pressed into slabs, the woodwool provides lightweight roofing and walling panels with good thermal insulation properties.

Steel and polypropylene fibers have been used to reinforce concrete, particularly ground-bearing slabs. Their advocates argue that they reduce the incidence of cracking, although care is needed to obtain an even distribution of the fibers throughout the concrete mix. Glass fiber, of very light weight and capable of being molded into esthetically pleasing curves, has found use in thin glass-reinforced cement (GRC) cladding panels.

See also **Bridges, Concrete; Concrete Shells; Dams**
MICHAEL BUSSELL

Further Reading

Hamilton, S.B. *A Note on the History of Reinforced Concrete in Buildings: National Building Studies Special Report No. 24*. Her Majesty's Stationery Office, London, 1956.

Jones, B.E., Ed. *Cassell's Reinforced Concrete*. Waverley Book Company, London, 1920.

Mainstone, R.J. *Developments in Structural Form*. Architectural Press, Oxford, 1998.

Marsh, C.F. *Reinforced Concrete*. Constable, London, 1904.

Neville, A.M. *Properties of Concrete*. Longman, Harlow; and Wiley, New York, 1981.

Sutherland, J., Humm, D. and Chrimes, M., Eds. *Historic Concrete: Background to Appraisal*. Thomas Telford, London, 2001.

Concrete Shells

Of all the developments in the structural engineering of buildings in the last century, the concrete shell was surely the most spectacular. It provided the means of covering vast areas with a shell of reinforced concrete just a few centimeters thick. Like most developments in building engineering, the origins of shell structures have many strands. Roman engineers constructed domes and barrel vault roofs made of brick or concrete spanning of up to 40 meters, but these were relatively thick—over a meter at their thinnest part. In Gothic cathedrals, at up to 20 meters, spans were more modest but they were often much thinner—as little as 200 millimeters. There were also vernacular precedents, most prominently the thin tile vaults widely used in Catalonia from the seventeenth century which, made using quick-setting gypsum mortar, had the advantage that they could be built without the need for a supporting structure during construction. The idea was exported to the U.S. and patented in the late nineteenth century by Guastavino who used them in many hundreds of buildings, including a spectacular roof at the Pennsylvania Railway station.

The characteristic of a shell structure is the very high ratio of span to thickness. While a concrete beam or floor might achieve a ratio of 20 or perhaps thirty to one, a shell structure can achieve ratios of many hundreds to one. Eduardo Torroja's market hall (1933) at Algeciras in Spain is some 48 meters in diameter, yet is just 90 millimeters thick; and his roof at Madrid race course (1935) cantilevers 13 meters with a shell just 50 millimeters thick. (Figure 15).

The "ideal" concrete shell is a curved sheet of solid material carrying its loads in the plane of the shell since it is too thin to resist bending. Since a roof must be able to carry a variety of different loads caused by wind or snow, real concrete shell roofs do need to have some ability to resist bending. A shell curved in just one direction, like a sheet of paper curved to form a barrel vault, is far too flexible to be useful and most concrete shells gain their rigidity by being curved in two directions, for instance like an egg shell. Such a shell can carry normal loads by arching, which develops reactive tension forces in the plane of the shell. It thus effectively carries its loads as in-plane shear forces. The greater the curvature of a doubly curved shell, the greater its ability to resist bending. However, it is also a requirement of most roofs that they be not too curved; and there is also the need to ensure the thin shell surface does not buckle in compression. There is, then, a need to stiffen the shell itself and this has generally been achieved in three ways—thickening and reinforcing the shell, folding or corrugating the shell, or providing stiffening ribs. In fact, nature had developed all these approaches in the structures of plants, seashells and, of course, the egg shell (which has a span-to-thickness ratio of about 130). Finally, of course, the reinforced concrete shell needs to have sufficient thickness for two layers of reinforcement, roughly orthogonal, and sufficient concrete cover to keep water out.

The earliest concrete shells were developed by reinforced concrete engineers and contractors who, especially from the 1910s to the 1930s, were eager to develop new applications for the new material. They did this by careful trials and experimentation with the use of little more than simple statics to help justify their designs. For the most part, the

Figure 15.   Stand at Madrid race course (1935) by Eduardo Torroja.
[*Photo: Cement and Concrete Association.*]

drivers for these developments were economic rather than architectural and most concrete shells were used for industrial buildings, railway stations, airports and the like. The French engineer Eugene Freyssinet built the first of his many shell roofs at a glass factory in 1915 and by 1921 he had constructed the remarkable airship hangers at Orly, some 60 meters high and spanning over 85 meters. The corrugated reinforced concrete shells were just 90 millimeters thick. Sadly they did not survive American bombs in 1944.

Different concrete engineers developed or exploited shells in different ways and, unusually among engineering structures, individual styles can be recognized among the great exponents.

Luigi Nervi's domed roof over the large sports hall for the Rome Olympics in 1958 spans some 100 meters and the corrugated concrete shell is just 25 millimeters thick; it is fabricated from precast units made from Nervi's own "ferro-cement" in which the reinforcing "bar," or rather wire, is just a few millimeters in diameter (Figure 16).

Figure 16.   Sports hall for Olympic Games, Rome (1958) by Luigi Nervi.
[*Photo: Cement and Concrete Association.*]

Figure 17. Restaurant Los Manantiales at Xochimilco in Mexico (1958) by Felix Candela.
[*Photo: Cement and Concrete Association.*]

Figure 18. Market hall roof, Plymouth, by British Reinforced Concrete.
[*Photo: Bill Addis.*]

Felix Candela, on the other hand, was master of the hyperbolic paraboloid. This family of anticlastic curved surfaces have the great advantage that they can be generated using straight lines, making the timber formwork especially easy to construct. His remarkable restaurant Los Manantiales at Xochimilco in Mexico (1958) is made from his version of ferro-cement. It spans 42.5 meters and yet is just 42 millimeters thick, except at the free edges which are slightly thicker (Figure 17).

The majority of concrete shells built throughout the century were, of course, neither large nor architecturally memorable, and were often a standard product offered by contractors. However, many of these had an almost vernacular charm of their own, and were significant engineering feats. The roof over Plymouth market, built by British Reinforced Concrete in 1960, is one good example (Figure 18).

As shell sizes grew, it became necessary to understand better the stresses and deflections in shells. In the absence of useful engineering theory, engineers developed, from the 1930s to 1960s, increasingly sophisticated techniques for testing small models of shell roofs and scaling up the predictions a hundred or more times for the real thing (Figure 19). As in other areas of structural engineering, techniques of model analysis have

largely (but not entirely) been replaced by computer-based methods, and the complex shell roof of the Sydney Opera House (1957–1962) was probably the first example of a building structure that could not have been constructed without the analytical work done using what were then the largest computers in the world (with the power of today's pocket calculator). Dozens of alternatives were analyzed before choosing one that satisfactorily met all the various constraints.

Just as computing power was unleashed in the 1960s, however, shell roofs began their demise for commercial reasons—their demand for enormous quantities of temporary supporting structures made them time consuming and labor intensive. Generally speaking they lost out to long-span steel structures and, from the 1970s, a variety of tension structures using cables and taut membranes. Nevertheless, shells still have their place when the circumstances are right. The Swiss engineer Heinz Isler, for instance, has enabled his shells to remain competitive since the 1970s in the face of strong competition from steel alternatives by developing a particularly ingenious system of reusable formwork (Figures 20 and 21).

See also **Buildings, Prefabricated; Concrete, Reinforced; Dams**

BILL ADDIS

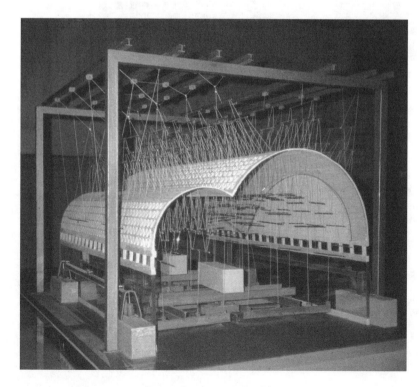

Figure 19. Model used in design development of the roof over the 60-meter-long pelota court at Recoletos, Spain, by Eduardo Torroja. The 1:200 scale model helped confirm the structural behavior of the 80-millimeter-thick shell spanning some 32.6 meters.
[*Photo: Bill Addis.*]

Figures 20 & 21.   Reusable formwork for a shell roof over an indoor tennis court, near Norwich, England by Heinz Isler.
[*Photos: Heinz Isler, Tony Copeland.*]

Further Reading

Anchor, R.D. Concrete shell roofs, 1945–1965. *Proc. Instn. Civ. Engrs. Structs. Bldgs.* 116 381–389, 1996.

Chilton, J. *Heinz Isler*. Thomas Telford, London, 2000.

Faber, C. *Candela: The Shell Builder*. Architectural Press, London, 1963.

Nervi, P.L. *Aesthetics and Technology in Building*. Harvard University Press, Harvard, 1966.

Mainstone, R.J. *Developments in Structural Form*, 2nd edn. Architectural Press, Oxford, 1998.

Morice, P.B. and Tottenham, H. The early development of reinforced concrete shells. *Proc. Instn. Civ. Engrs. Structs. Bldgs.*, 116, 373–380, 1996.

Smith, A.C. Concrete shells before 1930: technique and architecture. *International Association for Bridge and Structural Engineering (British Group)—Colloquium on History of Structures: Cambridge 1982*. Institution of Structural Engineers, London, 1982, pp. 137–142.

Torroja, E. *The Structures of Eduardo Torroja*. F.W. Dodge Corporation, New York, 1958.

Constructed World

The term "constructed world" has shallow and deep significations. In the shallow sense, it refers to a contingent assemblage of those artifacts that have in fact been fabricated by human beings. In the deep sense, the term suggests that the world itself as a unity may be taken to be a human construction. That the proportion of human experience engaged with artifacts has, especially since the Industrial Revolution, been dramatically increasing, itself tends to promote a shift from the shallow to the deep meaning. Related terms include "built," "engineered," and "technological world" or "environment."

That what might be constructed is not just products, processes, or systems, but a whole world, is an idea of unique twentieth century provenance. Although its most prominent manifestations are undoubtedly in relation to technology, during the 1900s the concept of construction increasingly became the basis for interpretations of art, architecture, psychology, education, economics, politics, ethics, knowledge, and even mathematics. From the vantage point of such a comprehensive if eclectic constructivism, all of human history is prefatory to an ethos of world fabrication that has been influenced by and in turn influences contemporary technology.

Despite, or perhaps because of, the overwhelming prominence of human construction in the twentieth century—from consumer goods through buildings to cities, from macroscale projects such as the U.S. Interstate Highway system and the European Channel Tunnel to genetic engineering and nanoscale mechanics, also including the unintended anthropogenic impacts on global biodiversity and climate—there exists no systematic overview of the world as an artifact. Instead, the (intentional and unintentional) complexity of the constructed world has thus far been conceived only piecemeal through a plurality of analytic and reflective approaches, among them history, architecture, urban planning, product design, and a diversity of related issues.

History

The history of such humanoid constructions courses over a million-year trajectory in which artifice remained subordinate initially to direct relations with nature (in hunting and gathering cultures) and then to social organization (in the rise of those axial civilizations characterized by farming and literacy in Mesopotamia, Egypt, and India). This broad distinction between artifice subordinate into natural and social milieux remains defensible even when qualified by the evidence for large-scale human terraforming, perhaps unintentional, prior to the development of literacy.

Mythological assessments of human construction include the stories of Abel and Cain (Genesis 4), the Tower of Babel (Genesis 11), Prometheus, Icarus, and more. Philosophical efforts to assess

the relationship began with Plato's critique of *techne* practiced independently of wisdom (*Gorgias*) and Aristotle's implicit distinction between cultivation and construction. For Aristotle, the primary *technai* are those that cultivate nature, thereby helping her bring forth more fruitfully that which she is in principle able to bring forth on her own: the arts of agriculture, medicine, and education. Of real but subordinate interest are the constructive arts that produce artifacts such as structures, roads, and ships. Indeed, one way to frame the trajectory of human history over the last 5000 years is from cultivation to construction.

Certainly modernity arose in the fifteenth century in part as a conscious attempt to privilege constructive invention over cultivation. Francis Bacon, among others, called not just for the cultivation of nature but its systematic transformation, and cited as paradigmatic inventions to be imitated the printing press, gunpowder, and the compass. Galileo Galilei and others likewise proposed an augmentation of the human senses by means of the telescope, microscope, and related scientific instruments. It is the new commitment to inventive reconstruction in both the laboratory and the world that formed the basis for an historical emergence two centuries later of the Industrial Revolution. Indeed, the twentieth century in particular has witnessed the instrumentalization of the human sensorium that began in the laboratory and went public to alter the means of communication in commerce, politics, and entertainment (telephone, motion pictures, radio, television, and the Internet).

This historically unprecedented degree of technical mediation by means of tools, machines, and information technologies undermines all efforts to apply to the twentieth century the characterization of previous epochs by reference to the distinctive material substrates (Stone Age, Bronze Age, Iron Age, etc.). Although proposals have been made to describe the 1900s as the age of electricity, the atomic age, or the computer age, in truth it is more accurate to define the century not in terms of some specific technology but simply as the technological age—with diverse and ever-diversifying technologies serving as multiple means of world construction.

Even more reflective of the distinctive twentieth century consciousness of the world as construction is the effort to complement retrospect with prospect to forecast what will happen next: technological change, if not progress. Futurology, with roots in prophetic sociology and science fiction, has nevertheless proved largely ineffectual. Relying more on trend analysis and imagination, it fails to engage the constructors themselves or to bring under effective economic or political directives the operative means operative for shaping the future.

Architecture

Efforts to go beyond futurology to develop a systematic analysis of the constructive elements in human affairs that might engage political and economic power grew out of the tradition of reflective building that finds classic expression in *De Architectura* by Vitruvius (circa 90–20 BC). Originally architecture designated the art of the master builder of the primary structures of the city (temples, palaces, monuments) and the layout of urban spaces in a manner that would reflect cultural ideals about the cosmic place and relations of humans. According to architectural historian Vincent Scully (1991) human builders have two basic options: to imitate natural forms or to oppose them. Compare, for instance, the architecture of indigenous peoples of the southwestern U.S., whose horizontal and earth-toned pueblos blend into a landscape defined by geological sedimentation and erosion, with the vertical thrust of those archetypical twentieth century buildings known as skyscrapers that dominate the cityscapes of Chicago or New York. On the ground, likewise, the modern city is laid out not to conform with a typology and the variegated paths of animal ambulance but as a block grid that extends into an instrumentally surveyed countryside, imposing simplification and legibility over the complex and intimate contours of rivers and mountains. Indeed, as Mumford (1961) states, as the constructed world became more and more extensive, the "city that was, symbolically, a world" was superceded by "a world that has become, in many practical aspects, a city."

Twentieth century transformations in the architecture of the constructed world have been driven by changes in materials, energy, transport and communication, and the commodities of peace and war. The first three achieved during the mid-1900s the apotheosis of developments with roots in the Industrial Revolution. Traditional construction materials such as wood and brick first became standardized and mass produced (e.g., dimensioned lumber), and then superceded as structural elements by iron, steel, and reinforced concrete; coal as an industrial energy source was complemented by oil, gas, and then nuclear power, with energy distribution and end-use itself accumulating

from the mechanical and chemical to the electrical and electronic; alongside pre-twentieth century boats and railroads there moved with increasing speed and numbers the inventions of automobiles and airplanes, while communication networks competed with those of transportation to make human world construction a dynamic planet-covering web. The 1960s images of the earth from space, with lighted continents and pollution plumes, visually defined the paradox of multiple-scale human dominance and its responsibilities—even, some argued, its limits.

Focusing first on the static aspects of this dominance, structural engineer David Billington (1983) has analyzed the influence of the new materials of steel and reinforced concrete on structures. For Billington, twentieth century structures are defined by the intersection of three factors: efficiency, (i.e., the scientifically guided pursuit of minimal materials use); economy, the market-monitored effort to reduce monetary cost; and the understated achievement of elegance through maximum symbolic expression (given the least amount of materials and cost). In structures of spare democratic utility such as bridges, tall buildings, and free-spanning roofs over industrial workplaces and warehouses, aircraft hangers, and sports complexes, architectural engineers came into their own.

"Structural designers give form to objects that are of relatively large scale and of single use, and ... see forms as the means of controlling the forces of nature to be resisted. Architectural designers ... give form to objects that are of relatively small scale and of complex human use, and ... see forms as the means of controlling the spaces to be used by people" [D. Billington, 1983, p. 14].

Bridges can be designed by engineers without architects; houses by architects without engineers. The engineered integration of efficiency and economy is realized in an esthetic of structural simplicity and thinness, as illustrated by the prestressed concrete bridges of Robert Maillart in Switzerland, the exposed steel tube x-bracing of Fazlur Kahn's John Hancock Center in Chicago, and the ribbed-concrete dome of the Palazzetto dello Sport by Pier Luigi Nervi in Rome.

Unlike structural engineering, early twentieth century architecture was less able to achieve an esthetic integration of science and democratic commerce, in part because it had to contend with well-established traditions of symbolic expression of the built world: the political iconography of Greek and Roman columns, the religious expression of the church spire, the solid façade of the bank, the decoration of Victorian domesticity. As the world-city emerged, architecture found itself caught in a cross-fire between scientific rationalism, industrial commercialism, and poetic romanticism—unclear which way to turn. The fundamental choice appeared to be between acceptance of technology or opposition to it. The winning synthesis was to take the scientifically rationalized artifact, that is, the machine, as an ideal for commercial exploitation and esthetic adaptation. In the architectural profession—itself now internally split into engineer, architect, and construction worker—this synthesis became a search for ways to design buildings that organized space in such a way as to parse human interactions into appropriate routines and to reduce resistance to their rapid interactions while minimizing the labor of construction of buildings for assembly lines, business offices, and large urban populations. The uniquely twentieth century architecture of these ubiquitous constructions, so named by a 1932 exhibition at the New York City Museum of Modern Art, was an "International Style" whose principles were an emphasis on "volume rather than mass," "regularity rather than axial symmetry," and the proscription of all "arbitrary applied decoration." This style, also known as modernism, was the first truly original building form since the rise of twelfth century Gothic.

The international style rejects the building patterns of premodern cultures (Greek, Roman, Gothic) in favor of shapes grounded in the efficient use of new materials and energies. Although steel and concrete were used initially to imitate Roman columns and Gothic arches, just as electric lights were first made to look like candles or gas lamps, in short order both became a flexible means for the design of indeterminate space and openness instead of determinate mass and enclosure. Geometric simplicity stripped of all ornamentation and standardized in modular forms at all levels, from structural members to external façade and finishing elements, contributed both to ease of construction and functional utilization.

Two leaders of this international modernism were Walter Gropius and Le Corbusier. Gropius, as the director of the Bauhaus in Germany, an engineering and product design school of great influence, eagerly embraced the machine esthetic in both buildings and their furnishings. Le Corbusier likewise condemned traditional building, redefined the house as "a machine for living in," and promoted the construction of whole cities of high-rise concrete apartment houses in repeating

blocks connected by open roadways. The high-rise building made possible by the steel frame and electric elevator became a progressively simplified form, as illustrated by the now destroyed World Trade Center towers in New York and the Sears Tower in Chicago, emblematic of that modernist international architecture that dominated the first half of the twentieth century.

Without wholly rejecting the international style, the second half of the century nevertheless witnessed a rising attraction of more complex and interesting architectural spaces—an attraction most visually manifest in a postmodern ironic complexity that playfully revived traditional forms layered over the retained modernist structural elements. The popularity of postmodernism had, however, a counterpoint in the discovery and defense of vernacular architecture.

Urban Planning

As indicated, the constructed world consists not just of structures designed by architects but of cities, including urban and suburban systems, linked with transportation and communication networks across landscapes constructed for farming, recreation, and preservation. Although architecture classically included issues of city design, urban planning has in the twentieth century become an independent profession, due to the manner of engineering and construction work.

At the beginning of the century, urban planner Ebenezer Howard proposed a vision of the garden city at odds with what would emerge as the international style. For Howard the problem of increased urban population was not to be solved simply by efficient modular housing inspired by the standardization and interchangeability of parts and machine construction, but by recognizing what he called the "twin magnets" of the town and the country. The benefits of towns are high wages, sociability, and culture, yet at the cost of high prices and congestion. The countryside is the source of natural beauty and quiet, at the risk of boredom and lack of aspirations.

> "But neither the Town magnet nor the Country magnet represents the full plan and purpose of nature .... Town and country *must be married*, and out of this joyous union will spring a new hope, a new life, and new civilization" [E. Howard 1965 [1902], p. 48].

This utopian vision became a major basis for criticism of the rationalist esthetic of high modernist architecture. Whole new small, mixed-use towns exhibiting an interweave of superblocks with narrower loop streets and cul-de-sacs instead of the repeating box grid were actually constructed in, for instance, Letchworth and Welwyn, England, and Radburn, New Jersey. Such experiments failed to live up to their promises of creating truly self-sustaining communities, as they became enclosed by larger suburban sprawl. Other influences of the garden city ideal can nevertheless be found in landscape architecture and the design of major urban parks, not to mention the construction of state and national parks and forests in both the U.S. and Europe, and eventually throughout the world.

The most practical innovation of early twentieth century urban planning was, however, the establishment of zoning laws that allowed for the political regulation of building practices. By the middle of the century architects and city planners were increasingly working together, with efforts also being made to enhance democratic participation in urban planning. The more grandiose schemes of Le Corbusier (who proposed a rebuilding of Paris) or Robert Moses (the New York state and city official who controlled its park and transportation development for more than 30 years), were moderated by local interests. Between them, social critics such as Jane Jacobs (1961) and urban planners such as Constantive Doxiadis (1963) brought realism and a more inclusive or interdisciplinary holism to thinking about the constructed world on the larger scale. The last half of the century also witnessed a new awakening of efforts among planners to take the natural environment into account in urban planning. Here the work of Ian McHarg (1969) exercised formative influence.

Product Design

Parallel to the architectural development of a machine esthetic at the level of structures, in tension with the organic ideals of urban planners, the commodities of peace and war were undergoing their own constructive transformations. Tools (dependent on human energy and guidance) were increasingly complemented if not replaced by machines (driven by nonhuman energy but still directed by human agents) and eventually semi-autonomous machines (requiring only indirect human guidance via feedback systems or programs), with the tools to machines transition continuing from the nineteenth century and dominating during the first half of the twentieth, and the rise of automation highlighting the second half. Distinctive of the century as a whole was the construction of a new type of household commodity—the electrical appliance—and then the electronic tool-machine represented most popularly by radios, televisions, and computers.

Prior to the rise of modern technology, the design of artifacts serving daily life was embedded in the craft of making—a virtually universal activity. Almost everyone was an artisan in the home, workshop, or field, and thus at one and the same time a person who conceived, fabricated, and used the indigenous basics of material culture. People "designed" things in the course of constructing them, so that making seldom involved any substantial moment of thinking through or planning beforehand, but proceeded as intuitive cut-and-try fabrication, guided by indigenous materials, traditions, and community. What has come to be called consumer product testing took place right in the making and immediate using by the maker, with the result that the commodities from regimes of craft production typically exhibit a certain practical artistic quality and honesty.

The Industrial Revolution's replacement of human power with coal- and steam-driven prime movers, its gearing of power into repetitive motion, and the required divisions of labor in manufacture, brought forth two needs: (1) the need for the designer as standard pattern maker so that artifacts could be mass produced; and (2) a need for the designer as style giver so that they could be mass marketed. Such a separation of design from construction and use could not help but open the door to a qualitative decline in the commodities produced, in reaction to which there emerged diverse efforts to reintroduce "art" into the new regime of industrial production; that is, to reunite what had been separated.

In the early stages, various arts and crafts movements sought to revive aspects of preindustrial modes of production, but at the beginning of the twentieth century the industrial design movement took a different approach, applying to quotidian commodities the principles being pursued in modernist architecture. Indeed, Gropius at the Bauhaus promoted modernist, technological simplification both in buildings and in streamlined furniture (see the famous Marcel Breuer chair). As one leading historian of product design has summarized the movement:

"By the end of the Second World War, the practice of styling mechanical and electrical goods to make them appear clean, crisp, geometrical and, above all, modern, had become commonplace. Cars, electric razors, radios, food-mixers, typewriters, cameras, washing-machines, and so on, were all given body-shells reflecting the machine esthetic of efficiency and functionalism'' [*P. Sparke 1986, pp. 49–50*].

In the last half of the century, however, in product design as in architecture, questions arose about notions of rational objectivity and universality, especially in a market dependent on advertising. The psychological requirements of the mass consumer were granted increasing legitimacy, so that expendability and playful symbolism began to replace stricter rationalisms. In counterpoint to a culture of waste and simulacra however, designers such as Victor Papanek called first for a new applied realism (1971) and then respect for the ecological imperative (1995) in product design. The question of sustainability emerged in relation to both human markets and the natural environment.

In summary, the constructed world is a historical phenomenon that has during the twentieth century emerged on three levels: the intermediate level of buildings or structures (architecture), the large-scale level of cities and landscapes (urban planning), and the small-scale level of consumer goods (product design). There are nevertheless other levels of and perspectives on construction that have been passed over here: the microlevel construction in biotechnology and genetic engineering and nanoscale engineering design, politics and warfare (construction through destruction), the economics of globalization, information technology and the construction of the networked world, and the multiple media-based transformation of life and leisure. There also remains the need for a broadly based, general understanding of construction that would unite such levels and approaches.

See also **Bridges; Construction Equipment; Concrete Shells; Dams; Highways; Skyscrapers; Tunnels and Tunnelling**

CARL MITCHAM

Further Reading

Alexander, C. *The Timeless Way of Building*. Oxford University Press, New York, 1979. Attempts to think the unconscious principles of vernacular design.
Billington, D.P. *The Tower and the Bridge: The New Art of Structural Engineering*. Basic Books, New York, 1983.
Coates, J.F., Mahaffie, J.B. and Hines, A. *2025: Scenarios of US and Global Society Shaped by Science and Technology*. Oakhill Press, Greensboro, NC, 1997.
Doxiadis, C.A. *Architecture in Transition*. Oxford University Press, New York, 1963.
Gropius, W. *The New Architecture and the Bauhaus*. MIT Press, Cambridge, MA, 1965. Original German edition 1935.
Hitchcock, H.-R. and Johnson, P. *The International Style*. W.W. Norton, New York, 1966. First published as *The International Style: Architecture Since 1922*. Museum of Modern Art, New York, 1932.
Howard, E. *Garden Cities of To-Morrow*. MIT Press, Cambridge, MA, 1965. First published 1902. Proposes satellite cities separated by green belts as a solution to dehumanizing mass urbanization.
Jacobs, J. *The Death and Life of Great American Cities*. Random House, New York, 1961. Argues that modernist architectural rationalism destroys natural urban growth and meaning.
Jencks, C. and Kropf, K., Eds. *Theories and Manifestoes of Contemporary Architecture*. Academy, Chichester, 1997. A collection of texts reflecting the explosion of architectural theory that characterized the last half of the twentieth century.
Le Corbusier [pseudonym of Jeanneret-Gris, C.E.] *Towards a New Architecture*. Brewer & Warren, New York, 1927. Original French edition 1923). The classic statement of the machine esthetic of modernist architecture.
Margolis, J. *Historied Thought, Constructed World: A Conceptual Primer for the Turn of the Millennium*. University of California Press, Berkeley, CA, 1995.
Mumford, L. *The City in History: Its Origins, its Transformations, and its Prospects*. Harcourt Brace Jovanovich, New York, 1961.
Papanek, V. *Design for the Real World: Human Ecology and Social Change*, 2nd edn. Van Nostrand Reinhold, New York, 1984.
Papanek, V. *The Green Imperative: Natural Design for the Real World*. Thames & Hudson, New York, 1995.
Rudofsky, B. *Architecture without Architects: A Short Introduction to Non-Pedigreed Architecture*. Doubleday, Garden City, NY, 1964. The catalog of a photography exhibit at the Museum of Modern Art in the mid-1960s.
Scott, J.C. *Seeing Like a State: How Certain Schemes to Improve the Human Condition Have Failed*. Yale University Press, New Haven, CT, 1998. Describes how modernist construction has manifested itself in politics and social planning.
Scully, V. *Architecture: The Natural and the Manmade*. St. Martin's Press, New York, 1991. An interpretive overview of Western architecture.
Sparke, P. *An Introduction to Design and Culture in the Twentieth Century*. Allen & Unwin, London, 1986.
Thomas Jr., W.L. Sauer, C.O., Bates, M. and Mumford, L. Eds. *Man's Role in Changing the Face of the Earth*, 2 vols. University of Chicago Press, Chicago, 1956. Collects over 50 papers by major scholars from an extended collaboration. Vol. 1 provides historical background; vol. 2 analyzes the processes by which humans are modifying the water, climate, soils, biota, as well as the ecology of wastes and urbanization. Includes extensive discussion summaries.
Venturi, R., Brown, D.S. and Izenour, S. *Learning from Las Vegas*. MIT Press, Cambridge, MA, 1972. Architecture can move beyond modernist formalism by appreciating the human pleasures of a playful commercialism.

Construction Equipment

The twentieth century had barely begun when 102 American-made steam shovels began excavating the Panama Canal. Those coal-fired giants devised in the nineteenth century for building railroads were one of the few large-scale earthmoving tools available in 1904. They were massive in size and weight and required a team of workers to operate. While heavy chains transmitted power to the excavator's bucket, the rail-mounted unit itself was not self-propelled, but had to be towed from work site to work site. Subsequent earthmoving machines evolved quite differently.

Although the focus of much twentieth century construction work was on road building, there was foundation work for buildings of all sizes, as well as civil engineering projects such as dams. The horse-drawn graders and scrapers used for leveling work on these undertakings during the first decades of the century had changed little from their nineteenth century origins. The first entirely new machine to appear was a tractor, which moved on crawler tracks and was used for towing earthmovers. It evolved from a wheeled, gasoline-powered agricultural tractor designed by Benjamin Holt in 1908 for use on the soft farmland of California.

During the 1920s the versatility of the crawler tractor was enhanced by the addition of a front-mounted scraper blade. Thus, the tractor itself became an earthmover. The technique of moving earth by a blade pushed from behind came directly from an earlier horse-powered device known as a bulldozer. It was from this that the name for the tractor and blade combination was derived. Blades were still accessories in the 1930s and their manufacture was, for the most part, by independent companies that supplied the tractor builders. The first hydraulically controlled blade appeared in 1925. But despite the compactness and precision offered by that system, the industry was slow to adopt the improvement and cable control remained the norm until the late 1930s. During the late 1920s, the efficiency of blades was improved by the addition of teeth to their leading edge. It was almost another decade before they were adopted industry wide. The decade closed with the introduction of the self-propelled grader.

The 1930s was an active decade in the earth-moving industry and there were several watershed events. None had greater impact than the introduction of diesel power. Although high-speed diesel engines were mass-produced in Germany during the late 1920s for tractor and lorry propulsion, an American-made diesel engine changed the industry. It was the first mass-produced diesel engine developed, manufactured, and applied as power in a mass-produced vehicle—the Caterpillar tractor. Diesel engines were superior to gasoline and oil power plants for use in heavy machinery for several reasons. Not least of all was the concen-

trated weight of the diesel engine, which was an asset in earthmovers where traction was a factor of weight. The engines provided power and lugging capacity not found in any other power source.

Despite their practicality, bulldozer tractors were most efficient when operated over relatively short distances and they were capable of moving only relatively small volumes. When a number of cubic meters of earth were to be moved over a distance of more than a few meters, it was best done by a scraper or bottom-dump trailers or rear-dump trucks. The other significant event of that era occurred when the metal tires on these vehicles were replaced with pneumatic heavy-duty rubber tires. Not only was the equipment more manageable on rough terrain, but its potential top speed increased. Nowhere was this more significant than on scrapers, which in 1938 were revolutionized with the introduction of the diesel-powered self-propelled LeTourneau Tournapull wheel tractor and scraper. This followed a decade-long trend in earthmoving machinery toward increased capacity and horsepower.

During World War II earthmoving machinery proved especially useful in the speedy preparation of supply depots, aircraft landing strips, and fortifications. The end of the war marked a turning point in the industry. In the immediate aftermath of the war, recovering Europe provided a ready market for American machinery exports. To meet the growing demand, some American companies invested in plants there. As conditions improved during the 1950s, a vital and creative domestic earthmoving equipment-manufacturing industry developed. Companies such JCB and Priestman in Britain and Atlas and Demag in Germany were among the many that arose throughout Europe.

The hydraulically operated excavator—a descendant of the steam shovel and the succeeding power shovel—was introduced in Germany in 1954. Up to that time, the control functions of power shovels were through cables. The industry's embrace of the excavator with components roughly analogous to the human arm and hand and a fluidity of movement to match was so thorough that power shovels were no longer used as a construction tool.

Of the many versatile machines developed during the early 1950s, the wheeled loader—also known as a front-end loader, bucket loader, or tractor shovel—was an immediate and widespread success. The nimble and highly maneuverable rubber-tired tractor with front-mounted hydraulically controlled bucket could be used to dig, lift, and quickly fill waiting dump trucks. The versatility and value of these machines increased tremendously in the mid-1950s when JCB in Britain and Case in the U.S. marketed factory-made units in which tractor loaders were joined with the boom, dipperstick, and bucket of the backhoe. The loader or backhoe became the most widely used tool on small-scale building projects.

American companies, which dominated the industry until the 1960s, faced a steady increase in competition from both European and Japanese machinery builders during that decade. By the late 1960s, machines were reaching the practical limit in their size and during the 1970s builders devoted their efforts toward improved equipment productivity. This was accomplished through increased horsepower, refinements in hydraulic controls, easier serviceability, and operator comfort.

Developments in earthmoving technology were not always embraced simultaneously or universally. Hydrostatic drive, in which fluids transmitted power to the wheels or tracks, was first used in Europe during the 1960s and by the mid-1980s was almost standard on all smaller machines built there. Large American manufacturers preferred mechanical drive, which had long proven its reliability and cost effectiveness. During the 1980s there was a general adoption of high-drive sprockets for crawler track tractors. Through modifications to the drive system and the addition of a third sprocket, the drive for crawlers was moved from ground level where it was subject to damage, to a safer and more easily serviced elevated position. The operation of hydraulic systems was optimized by the incorporation of electronics. Finally, in the 1990s a variety of diminutive-sized machines such as the mini-excavator were developed which all but eliminated the need for handwork at some job sites.

See also **Power Tools and Hand-Held Tools**

WILLIAM E. WORTHINGTON, JR.

Further Reading

Grimshaw, P.N. *Excavators*. Blandford, Poole, 1985.
Grosser, M. *Diesel: The Man and the Engine*. Atheneum, New York, 1978.
Harris, F. *Construction Plant: Excavating and Materials Handling, Equipment and Methods*. Garland, New York, 1981.
Haycraft, W.R. *Yellow Steel: The Story of the Earthmoving Equipment Industry*. University of Illinois, Urbana, 2000.
Leffingwell, R. *Caterpillar*. Motorbooks International, Osceola, 1994.
Oberlender, G.D., Ed. *Earthmoving and Heavy Equipment*. American Society of Civil Engineers, New York, 1986.
Young, J.A. and Budy, J. *Endless Tracks in the Woods*. Crestline, Sarasota, 1980.

Contraception, Hormonal Methods and Surgery

The discovery of hormones in 1898 and subsequent developments in biological science during the twentieth century were necessary before hormonal methods of contraception could be made available. Physical barrier methods of birth control and douching after intercourse had been tried with varying degrees of success for centuries, but the idea of preventing pregnancy with a pill was a dream of such women's health advocates as Margaret Sanger, founder of the Planned Parenthood clinic

The history of the development of oral contraceptives must include the work of scientists in Japan in 1924 and Austria in 1927 who devised what became known as the "rhythm method" of birth control. In independent research, both groups realized that a woman's fertile period is approximately midway in the menstrual cycle (that is, counting from the first day of her period until the beginning of the next period) and that pregnancy could be avoided by abstaining from sex at that time. The connection between hormone levels and ovulation was clearer when scientists at the University of Rochester in New York identified the ovarian hormone progesterone in 1928 and recognized its importance in preparing the uterus for the implantation and sustaining of pregnancy.

The next steps included the isolation of estrogen by Edward Doisy of Washington University, St. Louis, Missouri in the 1930s and the discovery of a way to make synthetic progesterone by chemical professor Russell Marker in 1941. This would become the basis for hormonal birth control. Following the Food and Drug Administration (FDA) approval of Enovid in 1957 as a treatment for severe menstrual disorders, it was only three years until the manufacturer Searle received FDA approval to market Enovid as an oral contraceptive, and it was quickly named "the pill." Within five years over 6.5 million women were taking it, and oral contraceptives had become the most popular form of birth control in the U.S., resulting in a revolution in birth control technology that expanded women's reproductive choices.

There are a number of hormonal contraceptives, including the combined pill; the progestin-only pill (POP), or minipill; hormonal injections such as Depo-Provera; contraceptive implants such as

Norplant; patches such as Ortho-Evra; or the Nuvaring, a combination of the hormones estrogen and progestin. Each can be highly effective, if used according to instructions. Similarly, surgical methods such as tubal ligation for women or vasectomy for men can be equally effective, and both procedures became more widespread in the last three decades of the twentieth century.

Oral contraceptives (OCs) refer to pills containing both estrogen and progestin, although there are "minipills," which contain only progestin. When taken consistently, pills can prevent ovulation, thereby eliminating the midcycle pain that some women experience at the time of ovulation. Pills can also decrease menstrual bleeding, thus decreasing the likelihood of iron deficiency anemia. Birth control pills come in packs of either 21 "active" pills, or 28 pills, with 21 "active" pills and 7 placebo pills, designed to keep the user in the habit of taking a daily pill even during her period. It is also possible to prevent a cycle by taking extra pills from a separate package. Some forms of the pill have been proven to improve acne conditions.

Oral contraceptives have been determined to protect against some forms of cancer, including ovarian and endometrial cancer. In addition, the pill reduces anemia due to iron deficiency since it lessens the amount of blood lost during a woman's menstrual cycle. Despite these positive effects, the pill does not protect users from sexually transmitted diseases, and the pill's effectiveness rests on the user remembering to take to the pill every day.

Minipills, the progestin-only pills, are less effective than the combined OCs, since they completely change a woman's menstrual cycle, which can lead to a bloated feeling or increased weight gain. These pills contain a lower dose of progestin, and no estrogen; for these pills to be effective, users must take them at the same time every day. The progestin in these pills thickens the cervix mucus, making it difficult for sperm to enter the uterus or fallopian tubes. The risk for pregnancy is, however,greater for the minipill. For users of the combined pill, the risk of pregnancy is 3 percent, while for users of the minipill, the risk increases to 5 percent.

For those who have difficulty remembering to take the daily pill or for whom OCs are not recommended, there are other options, including Norplant, in which six 34-millimeter-long Silastic rods that release levonorgestrel are inserted into the woman's upper arm in a fanlike shape just under the surface of the skin. This method, although effective, can be expensive. The erratic bleeding patterns that result also may not appeal to many women; approximately 60 percent of Norplant users report irregular bleeding patterns within the first year of use.

Depo Provera injections, which are administered in the arm or buttocks every 90 days, are highly effective and fairly inexpensive, with an average cost of $40 per injection. Women receive Depo injections every 13 weeks, with each injection containing progestin, which thickens the cervix mucus. However, Depo shots can cause the same side effects as the pill and Norplant, often resulting in headaches, weight gain, nervousness, and dizziness. Depo shots have been proven to cause the most significant weight gain, with an average of 7.5 kilograms after six years of use. It is important that women receive these shots every 13 weeks for the shots to be effective; when used effectively, about 3 in 1000 will experience an unexpected pregnancy in the first year of use.

The most common surgical method of birth control for men is vasectomy. The procedure involves the removal of a portion of the vas deferens, thus resulting in male sterility; this procedure does not affect the male sex drive or ability to ejaculate. Less than 1 in 1000 couples will experience an unexpected pregnancy in the first year of sterilization; the sterility, however, is not immediate, and the male is still fertile for three months, or 20 ejaculations, after the completion of the vasectomy.

In the tubal ligation procedure for women, the fallopian tubes are tied off, and a section of each tube is removed, thus resulting in sterility. Eggs released from the ovaries each month are blocked from reaching the uterus, thereby preventing fertilization by sperm. Less than 1 in 100 couples will experience an unwanted pregnancy in the year following a tubal ligation.

See also **Contraception, Physical and Chemical Methods**

JENNIFER HARRISON

Further Reading

Davis, A.J. Advances in contraception. *Curr. Reproduct. Endocrinol.* 27, 3 597–610, 2000.

Hatcher, R. *et al. Contraceptive Technology*, 16th revised edn. Irvington Publishers, New York, 1994.

Useful Websites

Health and Wellness Resource Center. Contraception: http://galenet.galegroup.com/

Statement on Hormonal Methods of Contraception. International Planned Parenthood Federation. IMAP

Statement on Barrier Methods of Contraception: http://www.ippf.org

For history of the development of contraceptive methods, especially hormonal methods, see PBS online, American Experience: http://www.pbs.org/wgbh/amex/pill/timeline

Contraception, Physical and Chemical Methods

Contraception refers to the deliberate means of preventing pregnancy by interfering with the normal processes of ovulation, fertilization, and implantation of the sperm. The process that could eventually lead to pregnancy begins with the maturation of the ovum (egg), thereby preparing the lining of the uterus for the fertilized egg. Birth control methods can prevent this process; however, no type of birth control is 100 percent effective. Failure rates for contraceptive methods can be divided into "perfect use" and "typical use," the former derived from clinical trial data of highly motivated patients who have received support and reminders from research personnel, and the latter from reports from the average user.

The use of birth control methods has a varied history. The 1873 Comstock laws in the U.S. prevented the distribution of sexually explicit materials, which would include the purchase of condoms or spermicides from Europe. However, it was not until the efforts of Margaret Sanger and her sister Ethel Byrne in opening a clinic to dispense contraceptive devices in 1916 that the revolution in birth control methods began. Although the New York City Vice Squad shut down the clinic, Sanger's efforts resulted in a victory for birth control by giving physicians the right to provide contraceptive measures to their patients. The so-called sexual revolution of the 1960s and the women's rights movements, particularly in the U.S. and the developed nations gave further impetus to the attempts by women to better understand and control pregnancy and family size. Controversy surrounding these issues continued into the twenty-first century.

Physical and chemical methods of birth control can be divided into three categories:

1. Physical barrier methods, including the condom, diaphragm, and cervical cap, which work to prevent the sperm from getting to and fertilizing the egg (only the condom protects against sexually transmitted diseases).
2. Chemical barrier methods, such as spermicides, which kill sperm on contact; most contain nonoxynl-9, are placed in the vagina, work best in combination with a barrier method such as a condom, and come in the form of jelly, foam, tablets, or transparent film.
3. Intrauterine devices (IUDs), which are inserted into the uterus and can remain for up to ten years; they work to prevent a fertilized egg from implanting in the lining of the uterus but may also have negative effects.

Significant developments in effective physical methods of contraception in the nineteenth century date back to 1832 when Massachusetts physician Charles Knowlton promulgated douching after intercourse using a syringe with water-based solutions to which salt, vinegar, chloride, or other elements would be added. In 1838 the German physician Friedrich Wilde offered his patients the "Wilde cap," a small cervical cap that would be placed over the cervix between menstrual periods. Although of limited success at the time, his device is considered the precursor of the diaphragm. Charles Goodyear, who invented the technique to vulcanize rubber in 1839, also provided the means to manufacture rubber condoms, douching syringes, and diaphragms.

Perhaps the most common contraceptive methods in the twentieth century are the physical and chemical barrier methods, and when used in combination, these methods are highly effective. Physical barrier methods, such as condoms, diaphragms, and cervical caps, are the only methods that protect against sexually transmitted infections, although the protection is the greatest with condoms. Chemical methods of contraception, such as spermicides, are available in foam, cream, jelly, film, suppository, or tablet form. Inserted into the vagina, spermicides contain a chemical that destroys sperm. Some types of spermicide require a ten-minute waiting period before intercourse, and must remain in the vagina for at least six hours after intercourse.

Condoms, which come in both male and female forms, work to keep the sperm and egg apart. If used correctly, male condoms only result in a pregnancy rate between 3 and 14 per 100 women per year. The male condom, a sheath that covers the penis during sex, is usually made of latex, but for those with latex allergies, condoms made with synthetic materials such as polyurethane are also available. However, only latex condoms have been proven to be highly effective in preventing sexually transmitted infections. Some condoms contain spermicide, which may provide additional contraceptive protection. The female condom, a polyurethane pouch with two flexible rings, one of which is inserted into the vagina, while the outer ring rests on the labia during intercourse, is less effective than the male condom. The Reality female condom was approved for use in the U.S. in 1993, and although it may provide some protection against sexually transmitted diseases, it is nowhere near as effective as the male latex condom. The estimated yearly failure rate for the female condom ranges from 21 to 26 percent, compared to a failure of 15 percent failure rate for the male condom.

Diaphragms, which must be fitted by physicians and women's health providers, should not be inserted longer than six hours before coitus and must be left in the vagina for at least six hours after coitus, but no longer than 24 hours. The diaphragm, a flexible rubber disk with a rigid rim, is designed to cover the cervix during and after intercourse so that sperm cannot reach the uterus. Spermicidal jelly or cream must be inserted inside the diaphragm for it to be effective. When used with spermicides, the diaphragm has a failure rate of 6 to 18 percent. There are some drawbacks, including the fact that diaphragms need to be kept clean and free of holes, and they should be refitted every other year. There is a chance that the diaphragm could become dislodged if the woman is on top during intercourse. Use of the diaphragm can also result in increased bladder infections.

Similarly, the cervical cap, which must be manually inserted by the woman, must also be used with a spermicide. Approved for contraceptive use in 1988, the cervical cap is a dome-shaped rubber cap that fits snugly over the cervix and must be fitted by a physician or women's health provider. Although it is more difficult to use, it may be left in place for up to 48 hours. There could be an increased occurrence of irregular Pap smear tests in the first six months of usage. The cap has a failure rate of 18 percent, and it may not be useful for all women since it is only available in four sizes and may be difficult to fit some women.

A less commonly used form of contraception, the intrauterine device (IUD), was first available in 1965 and has tended to result in the most satisfaction. Two forms of IUDs are available, a copper-containing IUD, the ParaGard T380a, marketed by Ortho-McNeil Pharmaceutical, Inc., and a progesterone-releasing IUD, Progestasert, produced by the Alza Corporation. The IUD interferes with sperm mobility and fertilization, although the timing of the IUD insertion is controversial. Printed literature states that an IUD should be placed in the vagina within five days of a menstrual cycle; however, an IUD can be inserted at any time a clinician is certain the patient is not pregnant. Ninety percent of IUD users are over the age of 35.

See also **Contraception, Hormonal Methods and Surgery; Fertility, Human**

JENNIFER HARRISON

Further Reading

Davis, A.J. Advances in contraception. *Curr. Reproduct. Endocrinol.*, 27, 3, 597–610, 2000.
Hatcher, R., *et al. Contraceptive Technology:* 16th revised edn. Irvington Publishers, New York, 1994.

Useful Websites

Health and Wellness Resource Center. Contraception: http://galenet.galegroup.com/
Statement on Barrier Methods of Contraception. International Planned Parenthood Federation, IMAP Statement on Barrier Methods of Contraception: http://www.ippf.org
For history of the development of contraceptive methods, see PBS online, American Experience: http://www.pbs.org/wgbh/amex/pill/timeline

Control Technology, Computer-Aided

The story of computer-aided control technology is inextricably entwined with the modern history of automation. Automation in the first half of the twentieth century involved (often analog) processes for continuous automatic measurement and control of hardware by hydraulic, mechanical, or electromechanical means. These processes facilitated the development and refinement of battlefield fire-control systems, feedback amplifiers for use in telephony, electrical grid simulators, numerically controlled milling machines, and dozens of other innovations.

Computational control in the decades before the 1950s usually meant cybernetic control involving automatic, closed-loop mechanisms. Massachusetts Institute of Technology (MIT) mathematician Norbert Wiener, who derived the word cybernetics from the Greek word for "steersman," considered the discipline—a direct forebear of the modern discipline of computer science—to be a unifying force created by binding together the theory of games, operations research, logic, and the study of information and automata. Wiener declared human and machine fundamentally interchangeable:

"[C]ybernetics attempts to find the common elements in the functioning of automatic machines and of the human nervous system, and to develop a theory which will cover the entire field of control and communication in machines and in living organisms." [*Wiener, 1948*].

Industrial, corporate, and government interest in fully electronic computer-aided control technology grew dramatically in the 1950s as access to electronic mainframe computers and transistor technology became more readily available. Indeed, information processing and computer programming emerged from the decade as virtual synonyms for control. In 1956, for instance, American Edmund Berkeley—editor of the journal *Computers and Automation* and founder of famed New York robotics company Berkeley Enterprises—defined control as the method by which one might "direct the sequence of execution of the instructions to a computer." Increasingly, computer control technology exploited open-loop systems, where a set of hard-wired commands or programmed instructions are executed in service to a predefined goal rather than uninterrupted governance.

American information technology management pioneer John Diebold explained in 1952 that realizing a truly automatic factory required "a machine that, once set up, performs a series of individual computations or steps in the solution of a problem, without further human intervention." Engineers touted the many advantages of computer hardware and increasingly *soft*-ware solutions in batch and continuous control operations, including the separation of function from material extension, mass fabrication, programmability, and flexibility. In the 1950s, computer controls found their way onto automobile assembly lines, railroad freight-sorting yards, and foundry grounds and into applications as various as packaging machines, furnace dampers, and iron-lung regulators. Early in this decade the Arma Corporation, an American Armed Forces contractor, introduced programmable machine tool automation—often referred to as numerical control or N/C—with its Arma-Matic lathe system. Later in the decade the Unimation ("Universal Automation") Company designed a programmable robot used to pull hot die-cast automobile parts out of their molds. In 1959 TRW and Texaco in Port Arthur, Texas, achieved one of the earliest installations of digital online process control in industry for the purpose of petrochemical catalytic polymerization.

In the 1960s computer control grew in importance in contexts beyond industrial production and logistics: office management, banking, air traffic control, passenger reservations, and biomedical diagnostics. The movement of computer control technology into previously sacrosanct workplace domains precipitated an outpouring of both positive and negative emotion. Among the ambiguous sets of social and cultural dilemmas weighed carefully during the control revolution of the late twentieth century were prospects for deskilling versus the pursuit of higher intellectual achievements; potential for mass unemployment versus information technology manpower shortages; increased managerial power versus decentralization of control; and inflexible working conditions versus profound achievements in workplace safety.

In the late 1960s and throughout the 1970s computer and cognitive scientists began experimenting with expert control systems, including neural networks, fuzzy logic, and autonomous and semiautonomous robots. Artificial neural networks incorporate an array of parallel-distributed processors and sets of algorithms and data operating simultaneously—presumably emulating human thought processes. The history of neural networks extends back to the cybernetic gaze of Wiener and the construction of the so-called McCulloch–Pitts neuron. This neural model, laced together with axons and dendrites for communication, represented a total cybernated system akin to the control processes of the human brain. Its namesake was derived from its American developers, the neurophysiologist Warren McCulloch and the mathematical prodigy Walter Pitts. Pitts in particular saw in the electrical discharges meted out by these devices a rough binary equivalency with the electrochemical neural impulses released in the brain. Thus it was but a short step from artificial to animal control and communication.

Fuzzy logic emerged at roughly the same time as neural network theory. The fuzzy concept—which allowed for intermediate degrees of truth rather than simple binary control—was first proposed by University of California, Berkeley computer scientist Lotfi Zadeh in 1965. Zadeh derived the idea from his study of imprecise human linguistics. Fuzzy logic is now usually subsumed under the artificial intelligence niche of subsymbolic artificial intelligence, and better known as "soft computing." Soft computing seeks control of haphazard, imprecise, and uncertain operations to form consensus opinions and make "semiunsupervised" decisions. Soft computing, a discipline still in its infancy, is currently focused on several areas of application, including computer-integrated manufacturing, computer-aided design, image processing, handwriting recognition, power system stabilization, and decision support.

In the 1980s a "long winter" of greatly reduced federal funding descended on artificial intelligence research—including neural networks and fuzzy logic. Despite this, the 1980s and 1990s witnessed a general renaissance in robotic computer-aided control. The twentieth-century pursuit of robotic incarnations of the control process extends back to Karel Čapek's fictional play *R.U.R.* (Rossum's Universal Robots). In the 1950s, Berkeley began creating hundreds of "Robot Show-Stoppers," playful mechanical creations with electronic brains. Among them were Squee, a robot squirrel that gathered tennis ball "nuts," and James, an android department store greeter. Control of Berkeley's robots was achieved using mechanical relays, phototubes, contact switches, drive motors, and punched paper tape. Robots developed in the next decade included sophisticated tactile sensors and embedded heuristic programming simulating the rule-of-thumb reasoning of human beings.

Throughout the twentieth century enthusiasts and critics hotly debated the merits of intelligent robot control in human societies. By 1970 professional roboticists began turning away from robot control and toward robot assistance. Robots—including those acting at a distance (telerobotics)—remained firmly tethered to their human masters. Examples of semiautonomous robots of the last two decades of the twentieth century include the Martian rovers developed by the National Aeronautics and Space Administration's (NASA's) Jet Propulsion Laboratory and TOMCAT (teleoperator for operations, maintenance, and construction using advanced technology), which repairs high-voltage power lines without service interruptions.

See also **Artificial Intelligence; Computer Science; Software Engineering**

Philip L. Frana

Further Reading

Anderson, J.A. and Rosenfeld, E. Eds. *Talking Nets: An Oral History of Neural Networks.* MIT Press, Cambridge, MA, 1998.

Berkeley, E.C. and Wainwright, L. *Computers: Their Operation and Applications.* Reinhold, New York, 1956.

The Computer Control Pioneers: A History of the Innovators and Their Work. Instrument Society of America, Research Triangle Park, NC, 1992.

De Wit, D. *The Shaping of Automation: A Historical Analysis of the Interaction Between Technology and Organization, 1950–1985.* Verloren, Hilversum, 1994.

Diebold, J. *Automation: The Advent of the Automatic Factory.* Van Nostrand, New York, 1952.

Goldberg, K., Ed. *The Robot in the Garden: Telerobotics and Telepistemology in the Age of the Internet.* MIT Press, Cambridge, MA, 2000.

Grabbe, E.M., *et al.* Eds. *Handbook of Automation, Computation, and Control,* 3 vols. Ramo-Wooldridge, Los Angeles, CA, 1958.

Mindell, D.A. *Between Human and Machine: Feedback, Control, and Computing Before Cybernetics.* Johns Hopkins University Press, Baltimore, 2002.

Noble, D.F. *Forces of Production: A Social History of Industrial Automation.* Oxford University Press, Oxford, 1986.

Wiener, N. Cybernetics. *Scientific American,* 179, 14–19, 1948.

Zadeh, L.A. *Fuzzy Sets and Applications: Selected Papers.* Wiley, New York, 1987.

Control Technology, Electronic Signals

The advancement of electrical engineering in the twentieth century made a fundamental change in control technology. New electronic devices including vacuum tubes (valves) and transistors were used to replace electromechanical elements in conventional controllers and to develop new types of controllers. In these practices, engineers discovered basic principles of control theory that could be further applied to design electronic control systems.

The voltage and current regulation technology of electrical power networks was among the fields that witnessed an early transition from electromechanical to electronic control. In the early twentieth century, the most prevalent voltage–current regulator was the "Tirrill" regulator made by the General Electric Company. A Tirrill regulator had a constantly vibrating contact that moved a resistor in and out of a generator's excitation circuit. It used the presence or absence of the resistor to reduce or raise the output voltage. Electromechanical arrangements like the Tirrill regulator were gradually replaced in the 1930s by a fully electronic design using gas-filled thyratron tubes.

The thyratron was developed by General Electric in the 1920s. Known as the "grid-controlled rectifier," a thyratron had two modes of operation. When the grid voltage was less than a threshold, no current flowed between the anode and the cathode (off state). Once the grid voltage exceeded the threshold, the thyratron operated as a rectifier (on state). As long as the thyratron was triggered to the on state, it remained on until the voltage across the anode and cathode vanished. Thus unlike a vacuum tube whose conductivity continuously varied with the grid voltage, a thyratron kept the high-conductivity state after being triggered, no matter whether the grid voltage dropped below the threshold later. The average output voltage of a thyratron varied with the duration of the on state, which was determined by timing of the rising edges of the voltage at the grid. A thyratron-based voltage regulator therefore adjusted the timing of the grid voltage according to the fluctuation of the output voltage (Figure 22).

A similar control device to thyratron was discovered in the age of semiconductors. Developed by M. Sparks, L.W. Hussey, and William Shockley at the Bell Telephone Laboratories in the early 1950s, a thyristor was a solid-state device constituted of a *p–n–p–n* structure. It had an anode connected to the first *p*-type block *p*1, a cathode to the second *n*-type block n2, and a gate to the second *p*-type block *p*2. Like the thyratron, the thyristor was a gate-controlled rectifier. It remained off when the gate current was inadequate to trigger its conduction. Once the gate current exceeded a threshold, the device suddenly transformed into a rectifier and remained in the on state. Therefore by adjusting the timing of the gate current's rising edges, a thyristor could also be used to construct a voltage regulator (Figure 23). With the same functional characteristics as thyratron tubes and the additional advantages of solid-state devices, thyristors eventually replaced thyratrons in power electronics.

The control technology of telephone networks was another early electronic system. A telephone network's typical control problem was how to

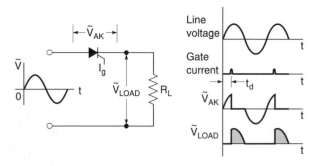

Figure 23. A thyristor regulator.
[*Source: Sze, S.M. Semiconductor Devices: Physics and Technology. Wiley, New York, 1985, p.153. © Wiley 1985. This material is used by permission of John Wiley & Sons, Inc.*]

make the output follow the time-variant input faithfully. The attenuation and dispersion of telephone signals were particularly serious when the communication distance was long. In the 1910s, the solution was to employ vacuum-tube amplifiers along a communication path to repeat signals, but these amplifiers had the problems of oscillation and nonlinear distortion. In 1923, Harold Black at the Bell Lab developed the concept of the negative feedback amplifier. In his design, the amplifier amplified the difference between the input and a feedback signal coupled from the amplifier's output (if the amplifier amplified the sum of the input and the feedback, then it would be a positive feedback amplifier). The negative feedback design reduced an amplifier's unstable resonance and extended the linear range of operation (Figure 24). Black was not the first person to come up with a negative feedback design, but he was the first to point out the fundamental difference between negative and positive feedback. Following Black, the Bell Lab engineers in the 1920s and 1930s conducted research on a mathematical theory of feedback amplifiers. Hendrik Bode and Harry Nyquist, for instance, developed representational tools, known as Bode plot and Nyquist diagram, for the design and analysis of feedback circuits.

The control in process industries (petroleum, chemicals, etc.) in the early twentieth century might not have been saliently affected by electronic technologies, but its improvement paved the way toward fundamental understanding of a control theory. An early process controller used either an electrical relay with a solenoid-operated valve or a motor-operated valve. In the first case, the electrical control signal turned the relay on or off, moving the valve in a two-stage fashion. In the second case, the control signal set the motor and

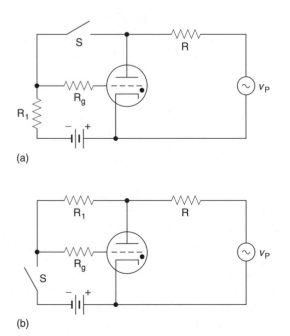

(a)

(b)

Figure 22. A thyratron control circuit.
[*Source: Happell, G.E. and Hesselberth, W.M. Engineering Electronics. McGraw-Hill, New York, 1953, p. 438. Reproduced with permission of The McGraw-Hill Companies.*]

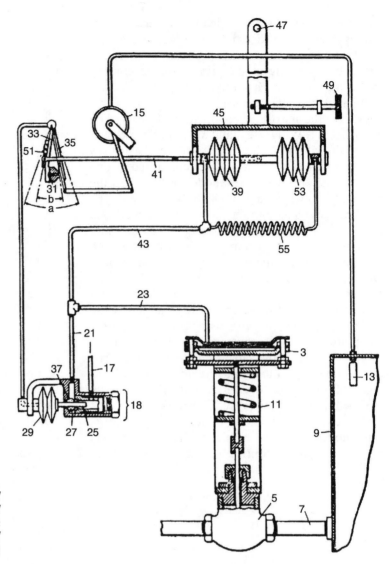

Figure 24. Basic mechanism of the Stabilog.
[*Source: Bennett, S. A History of Control Engineering: 1930–1955. Institution of Electrical Engineers, London, 1993, p. 42. Reproduced with the permission of the Institution of Electrical Engineers.*]

drove the valve in a continuous fashion: the valve motion followed the time integral of the control signal. Both methods had shortcomings. The two-stage controller introduced harmful surges at the on–off transitions. The integral controller's controlled variable could not follow promptly the variation of its controlling variable—it had a time lag. In 1931, the Foxboro Company introduced a combined proportional–integral "Stabilog" controller, and the time lag was significantly reduced with the incorporation of this proportional control. In the early 1930s, Ralph Clarridge at the Taylor Instruments Company discovered that when using the time derivative of the measured variable as the control signal, the speed of reaction could further increase. In 1935, Taylor Instruments built a small experimental factory using a controller combing the proportional, integral, and derivative (PID) actions. The PID controller retained the advan-

tages of the three individual controllers but compensated for their respective shortcomings. In fact, the PID control discovered and promoted by instrumental engineers in the 1930s was anticipated in the early 1920s by the engineer Nicholas Minorsky when he worked on theoretical problems in automatic steering. The PID scheme dominated industrial control until the 1950s.

Theoretical understanding grew with electrical control technologies. The study of telephonic repeaters produced a theory of feedback amplifiers. The development of process controllers confirmed and elaborated the theoretical work on the PID control. A unified control theory began to take shape in the late 1930s when Massachusetts Institute of Technology (MIT) researchers such as Harold Hazen, working on gunfire control systems, combined the work on feedback amplifiers, the analytic techniques involved in manufacturing control, and the stability analysis of power networks to form a theory of servomechanism.

See also **Fly-By-Wire Systems; Rectifiers; Smart and Biomimetic Materials**

CHEN-PANG YEANG

Further Reading

Bennett, S. *A History of Control Engineering: 1800–1930.* Institution of Electrical Engineers, London, 1979.

Bennett, S. *A History of Control Engineering: 1930–1955.* Institution of Electrical Engineers, London, 1993.

Fuller, A.T. The early development of control theory. *Trans. Am. Soc. Mech. Eng.*, 98, 109–118, 1976.

Happell, G.E. and Hesselberth, W.M. *Engineering Electronics.* McGraw-Hill, New York, 1953.

Hughes, T. *Networks of Power: Electrification in Western Society 1880–1930.* Johns Hopkins University Press, Baltimore, 1983.

Mindell, D. Opening Black's box: rethinking feedback's myth of origin. *Technol. Cult.*, 41, 405–434, 2000.

Stevens, R.A. Control engineering: historical background. *IEE Proc.*, 121, 396, 1974.

Sze, S.M. Semiconductor Devices: Physics and Technology. Wiley, New York, 1985.

Cracking

Three major innovations emerged to meet the dramatically higher needs for quantity and better quality motor fuel in the twentieth century: thermal cracking, tetra ethyl lead, and catalytic cracking (including the fluid method). William M. Burton of Standard Oil, Indiana, in 1911–1913 raised gasoline fractions from petroleum distillation from 15 to 40 percent by increasing both temperatures and pressures. Benjamin Stillman Jr. of Yale University had discovered in 1855 that high temperatures could transform or "crack" heavy petroleum fractions into lighter or volatile components, but at atmospheric pressure over half of the raw material was lost to vaporization before the cracking temperature (about 360°C) was reached. English scientists James Dewar and Boverton Redwood discovered and patented in 1889 a process using higher pressure to restrain more of the heavier fractions and increase the volatile components. Burton and his team, not aware of this patent at the start of their work, similarly used higher pressure to improve his gasoline yields. Their process by 1913 employed a drum 9 by 2.5 meters diameter over a furnace and a long runback pipe of 300 millimeter diameter that carried vapors to condensing coils and simultaneously allowed heavier fractions to drip back into the drum for more cracking. A tank connected to the condensing coils separated gasoline from the uncondensed

hydrocarbon vapors. They used a comparatively low (5.1 atmospheres) pressure because they relied on riveted plates. Burton's team improved the batch process in the next three years with false bottom plates in the still to improve cleaning time of coke deposits, and a bubble tower to improve fractionation of cracked vapors and increase gasoline yield. Refinery manager Edgar M. Clark took cracking to the next step by using tubes of about 100 millimeters diameter as the primary contact with the furnace, which increased cracking time and allowed pressures of about 6.8 atmospheres, and decreased maintenance time because of reduced coke deposition and fuel costs.

The dramatic increase in gasoline yield and the royalties that Indiana charged its licensees encouraged many to experiment with cracking. The U.S. Patent Office granted several patents for general principles or for overlapping techniques. Universal Oil Products Company (UOP) received two such patents in 1915 and 1921 incorporating the principles of continuous processing and clean circulation. Their Dubbs process (after Jesse Dubbs and his son, Carbon Petroleum Dubbs) allowed uncracked oils or reflux to be returned continuously and while still hot to the cracking zone to increase production of gasoline, and to reduce coke formation and cleaning time. The process could also charge lower value and heavier oils (such as fuel oil).

Although UOP had substantially revised the 1915 Dubbs patent from its original filing date in 1909 primarily to pursue infringement claims against the Burton process, the 1921 patent and considerable inventive work by talented personnel helped create a successful process development firm. Smaller as well as larger firms (e.g., Shell) could license the process because of its uniform royalty rate, UOP's continued technical improvements, and a solid defense against litigation from competing processes. UOP licensees and scientists further refined the Dubbs process with a hot oil pump to increase the capacity from 500 to 700 barrels per day, and contributed to the many advances in structural development, such as seamless steel tubing, electric welding, alloy steel tubing and parts, and high-temperature fans and insulation. Finally, UOP scientists like Gustav Egloff successfully urged UOP's leadership to focus on the next step of cracking—catalysis. Egloff recruited the Russian expert in catalytic reactions, Vladimir Ipatieff, who in turn brought German scientists such as Hans Tropsch to the firm.

Other cracking processes were based on broad patents having much in common with each other, which encouraged a patent war, and then a patent pool in 1923 between four major processes. Among the major innovations were the Cross method's pressures of 600 psi (40.8 atmospheres) in 1921, achieved with a forged reaction chamber having walls 3 inches (75 millimeters) thick. This and a new centrifugal oil pump from the design of a German concern, Weiss, allowed the Cross process and Standard Oil of New Jersey to increase the cracking temperature, the rate of cracking, the throughput, and the quality of the gasoline.

Charles F. Kettering, head of DELCO and then General Motors' research division, also focused on the quality of gasoline in addressing engine knock. He and a team under his assistant, Thomas Midgley Jr., tested chemical compounds more or less systematically, and later worked with Robert Wilson of Massachusetts Institute of Technology (MIT) to compose a periodic table of more promising antiknock elements based on the spaces in their outer electron shells. In December 1921 after five months of research on organometallic compounds of the most promising elements, the team tested an obscure compound—tetraethyl lead, or TEL. Its antiknock abilities were obvious, and General Motors soon allied with its major stockholder, DuPont, to develop and manufacture the fuel additive. Standard Oil of New Jersey quickly developed a superior method to manufacture the additive, and bargained with General Motors to form Ethyl Gasoline Corporation to make and sell TEL. Poor Jersey Standard quality control led to a U.S. government investigation in 1925 of tens of lead poisoning cases in its factory, but scientific methods and equipment were not sophisticated enough to yield sufficient data to ban the material from gasoline. In attempts to sell their product to the public, Ethyl Gasoline led a drive to construct an octane rating for antiknock characteristics.

A major roadblock to increasing TEL sales was the Sun Oil Company, which already marketed a high-octane gasoline in 1927, and invested heavily in the early 1930s in developing the first catalytic cracking (Houdry) process. French industrialist and inventor Eugene Houdry had tried to interest major European oil firms in his process in the mid-1920s, but his design was not advanced enough and IG Farben's hydrogenation process appeared to be the petroleum process of the future. Houdry first worked with the Vacuum Oil Company (later Socony-Vacuum), and then interested Sun in the high octane rating (91) of his process. Sun tenaciously avoided buying the TEL additive, and plunged into developing the fixed-bed Houdry

process, making significant improvements, from the design of the apparatus to the development of a synthetic catalyst. The process was semicontinuous, since oil feed would be switched from one reaction or regeneration chamber to another while the first's catalyst pellets would be cleaned of carbon. Sun's improvements increased the percentage of catalytic gasoline per charging stock from 23 to 43 percent, and by 1939 Sun and Socony had constructed or were constructing plants with a capacity of 221,675 barrels a day of charge.

The Houdry fixed-bed plants produced 90 percent of all catalytically cracked aviation gasoline in the U.S.'s first two years of World War II. Socony and the Houdry Process Corporation created a moving bed catalytic cracking (MBCC) process that continuously circulated pellets from cracking to regeneration chambers, decreased the use of steel and turbines, which were needed elsewhere in the war, reduced losses in costs of heating catalysts, and enabled the use of heavier feedstocks. Later versions dramatically increased the ratio of catalysts to charging stock, and used airlift instead of mechanical means to move the catalyst, which simplified the equipment and reduced investment costs.

Meanwhile other major firms had set about to circumvent the Houdry process and develop one of their own, to be named the fluid catalytic cracking (FCC), or fluid bed process. Jersey Standard led this endeavor with its early work and ventures with IG Farben, which had the best experience in catalytic hydrogenation for coal and petroleum by the late 1920s. Building up its catalytic research and development capabilities in the 1930s, Jersey refused to pay the steep royalty payments demanded by Eugene Houdry in 1938, and soon allied with Indiana Standard, M.W. Kellogg Company, IG Farben, British Petroleum, Shell, Texaco, and UOP in the Catalytic Research Associates to construct their own catalytic cracking process. Jersey's huge monetary and research and development resources, and its ties to top academic scientists, helped it to dominate this tremendous project.

As early as the mid-1930s Jersey had researched a process with a powdered catalyst combined with oil vapors at 400 to 540°C. Varying vapor mixes and pressures moved the mixture through a reaction chamber much like a fluid. The catalysts were separated from the vapor before it condensed and regenerated in a separate chamber. Advantages over a fixed-bed process included reduced use of steel with the separate reaction and regeneration units, a simplified continuous process with a uniform flow and constant physical conditions, more contact between the catalyst and the oil, and the possibility of increased size to provide more economies of scale.

Later innovations, mostly by major petroleum or process design firms, include microspherical catalyst particles in 1948 to improve flow and reduce catalyst splitting and loss, high percentage alumina catalysts in 1955 to boost yield, and zeolitic catalysts in 1964 to increase yields, decrease deactivation by coke deposition, and enable shorter reaction time with lower catalyst-to-oil ratios. Zeolitic catalysts encouraged much more cracking in the transfer line (riser) before the reactor bed, while residual oil cracking became popular in the mid-1970s with increased crude oil prices and decreased demand for heavy fuel oil.

See also **Cryogenics, Applications; Feedstocks; Oil from Coal Process**

KENNETH S. MERNITZ

Further Reading

Enos, J.L. *Petroleum Progress and Profits: A History of Process Innovation.* MIT Press, Cambridge, MA, 1962.

Gibb, G.S. and Knowlton, E.H. *History of the Standard Oil Company (New Jersey): 1912–1927, The Resurgent Years.* Harper & Brothers, New York, 1956.

Giebelhaus, A.W. *Business and Government in the Oil Industry: A Case Study of Sun Oil, 1876–1945.* JAI Press, Greenwich, CT, 1980.

Larson, H., Knowlton, E.H. and Popple, C.S. *History of the Standard Oil Company (New Jersey): 1927–1950, New Horizons.* Harper & Row, New York, 1971.

Leslie, S.W. *Boss Kettering.* Columbia University Press, New York, 1983.

Mernitz, K.S. Governmental research and the corporate state: the Rittman refining process. *Technol. Cult.,* 31, 83–113, 1990.

Robert, J.C. *Ethyl: A History of the Corporation and the People Who Made It.* University Press of Virginia, Charlottesville, VA, 1983.

Rosner, D. and Markowitz, G. Deadly fuel: autos, leaded gas, and the politics of science, in *True Stories from the American Past,* Graebner, W., Ed. McGraw Hill, New York, 1993, pp. 126–141.

Williamson, H.F. *et al. The American Petroleum Industry: The Age of Energy, 1899–1959.* Northwestern University Press, Evanston, IL, 1963.

Wilson, J.W. *Fluid Catalytic Cracking Technology and Operations.* Pennwell, Tulsa, OK, 1997.

Crop Protection, Spraying

Humans have controlled agricultural pests, both plants and insects, that infest crops with a variety of biological and technological methods. Modern humans developed spraying pest management techniques that were based on practical solutions to combat fungi, weeds, and insects. Ancient peoples introduced ants to orchards and fields so they could consume caterpillars preying on plants. Chinese, Sumerian, and other early farmers used chemicals such as sulfur, arsenic, and mercury as rudimentary herbicides and insecticides. These chemicals were usually applied to or dusted over roots, stems, or leaves. Seeds were often treated before being sowed.

As early as 200 BC, Cato the Censor promoted application of antipest oil sprays to protect plants in the Roman Empire. The nineteenth century potato famine and other catastrophic destruction of economically significant crops including vineyard grapes emphasized the need to improve crop protection measures. People gradually combined technological advances with biological control methods to initiate modern agricultural spraying in the late nineteenth century. Such crop protection technology was crucial in the twentieth century when large-scale commercial agriculture dominated farming to meet global demands for food. Individual farms consisted of hundreds to thousands of acres cultivated in only one or two crop types. As a result, spraying was considered essential to prevent devastating economic losses from pest damage associated with specific crops or locales.

People recognized the benefits of spraying to cover agricultural plots efficiently and evenly with protective substances and fertilizers. In the late nineteenth century, researchers developed effective inorganic chemical pesticides. Farmers initially used hand sprayers while walking through fields. By 1880, a commercial spraying machine was introduced. Four years later, the state of Utah passed a law requiring fruit trees to be sprayed.

Agricultural mechanization enabled spraying technology to advance. Machinery was used to apply spray uniformly. Automated timers regulated spraying at specified intervals. The type of machinery used depended on the type of chemical being applied. Sprays such as Monsanto's Roundup were made to be compatible with varying spraying equipment and situations. Technological advances included the design of tractors and the spray tanks that they carried or towed.

Typical sprayers distribute liquid drops forced through nozzles by pressure. The diameter of drops determines how many can be sprayed and how far they travel. A greater quantity of smaller drops can be sprayed at a time than larger drops. The smaller drops do not travel as far as the large drops because they are slowed more quickly by air resistance, but smaller drops more effectively cover plants. Sprayers are adaptable for a variety of tasks ranging from protecting fields to individual trees.

Specialized sprayers discharge liquids with technological processes associated with atomizers, centrifuges, and nebulizers. Compared to atomizers, sprayers require less energy and are inexpensive. Sprayers' drop size, however, necessitates more liquid to spray than atomizers and are not as comprehensive in their coverage. Fans in atomizers create an air jet, which shatters liquid drops into smaller droplets that spread over a larger surface than drops from sprayers. Atomizers require less time than sprayers to treat an area and their increased plant coverage improves chemical performance. The lightweight droplets are sometimes diverted by wind but are more likely to be absorbed by soil than to runoff, as is the case with spray droplets. They are, however, prone to rapid evaporation. Atomizers cost more and need more energy than sprayers. They are often worn like backpacks and equipped with wands to spray.

In contrast, nebulizers deliver drops with a fog of steam or gas that envelops plants. Nebulizer fogs are described as coarse or wet, or fine or dry depending on drop size. These devices are usually found in greenhouses because the tiny particles are vulnerable to weather conditions. Nebulizers are more effective than other forms of spraying but are expensive to buy and maintain. Most centrifugal sprayers are run manually and have a small electric motor powered by batteries. Gravity pulls liquids in centrifugal sprayers through rotary disks to create small particles.

Small airplanes were first utilized for crop spraying in the 1920s. Many pioneering crop dusters were World War I veteran pilots. Aviation enables large areas to be treated in a brief time compared to hand or machinery-pulled spraying tools, and isolated places could be easily reached by air. In the southern U.S., airplanes were used to spray calcium arsenate over cotton fields in an attempt to destroy boll weevils which threatened to ruin the important cash crop. Although more expensive, helicopters can reach places that planes cannot go and their rotating blades push chemical sprays toward their targets. Wind limits all agricultural aircraft spraying applications.

DDT (dichloro-diphenyl-trichloroethane) was identified as an effective insecticide in 1939. During the early 1940s, DDT transformed crop-spraying protection. Researchers addressed the increase of pest resistance to chemicals, especially DDT, in the 1950s and 1960s. Genetic engineering efforts to produce transgenic plants resistant to

pests and diseases were expanded as a supplement or alternative to chemical sprays.

Some people recognized risks associated with spraying. In 1892, a Canadian law prohibited the chemical spraying of blooming trees in order to protect bees collecting pollen. Twentieth century critics of spraying stressed the expensive costs of sprays, possible health risks to agriculturists and consumers, and the threat of environmental damage and pollution. They warned of agriculturists becoming too dependent on sprays. When spray schedules were disrupted or cancelled, pests became more problematic. Scientists sought alternative and less hazardous methods. The British Royal Commission on Arsenical Poisoning established residue limits in 1903. Twenty years later, the British government threatened to refuse importation of U.S. apples that had been exposed to chemical sprays. The U.S. Food and Drug Administration devised spraying standards for exported fruit.

Rachel Carson outlined her concerns regarding pesticides applied by spraying in *Silent Spring* (1962). Some scientists also voiced ecological concerns related to sprayed chemicals that protected crops. R.F. Smith and Robert van den Bosch devised the concept of the more environmentally sound integrated pest management (IPM) in 1967. As a result of the expanding environmental movement, DDT was banned in many places worldwide in the 1970s. Several laws regulating spraying, including licensing users, were enacted. Such official recognition continued into the 1990s when more pathogenic species of insects, weeds, and animals became resistant to chemical sprays and new spraying formulas and techniques were tested.

See also **Farming, Agricultural Methods; Genetic Engineering, Applications; Pesticides**

ELIZABETH D. SCHAFER

Further Reading

Anderson, M.I. *Low & Slow: An Insider's History of Agricultural Aviation.* California Farmer Publishing, San Francisco, 1986.
Cate, J.R. and Hinkle, M.K. *Integrated Pest Management: The Path of a Paradigm.* National Audubon Society, Washington D.C., 1994.
McWhorter, C.G. and Gebhardt, M.R., Eds. *Methods of Applying Herbicides.* Weed Science Society of America, Champaign, Illinois, 1987.
Rasmussen, H. *Crop Dusters: "Props in the Crops."* Osceola, WI, Motorbooks International, 1986.
Van den Bosch, R. *The Pesticide Conspiracy.* University of California Press, Berkeley, 1978.

Wilkins, R.M., Ed. *Controlled Delivery of Crop-Protection Agents.* Taylor & Francis, London, 1990.

Yuste, M.-P. and Gostincar, J. *Handbook of Agriculture.* Marcel Dekker, New York, 1999.

Cryogenics, Applications

The field of cryogenics, and its applications, has evolved from intertwined interests of the nineteenth century—the scientific interest to liquefy the remaining "permanent" gases, the residential interest for indoor environmental control, and the commercial interest in food preservation. A rich diversity of applications has resulted, ranging in daily impact and technical sophistication from frozen foods to rocket science. A survey of this variety is summarized in the following paragraphs.

Biomedical Applications

Although the medical field has benefited indirectly from cryogenics through the superconducting imaging technologies of magnetic resonance imaging (MRI) and SQUID detectors, direct benefits arise through various forms of cryosurgery and transport of biologicals. Significant use of a liquid nitrogen (LN_2)-based cryosurgical probe was initiated in the mid-1960s for treating brain tissue to alleviate the effects of Parkinson's disease. Throughout the century dermatologists have used a spray or swab application of LN_2 to destroy undesirable tissue. In the 1980s physicians were able to precisely treat tumors in the liver and prostate with cryosurgical devices and the the emerging imaging technologies of ultrasound, computer tomography, and MRI. In the 1990s cooling for the cryoprobes became increasingly convenient as the integration of miniature and micro cryocooler technologies replaced the use of LN_2. A significant benefit of the cryosurgical devices is the ability through partial, reversible cooling to investigate the efficacy of freezing tissue areas before permanent treatment. Additional treatments are emerging for breast cancer, benign tissue in gynecology, heart arrhythmia, and others.

Improvements to cryogenic storage vessels through the twentieth century have enabled a wide variety of biological samples to be preserved. Beginning in 1956, cryopreservation of bull semen spread the practice of scientific breeding to developing nations, resulting in a significant addition to world supplies of milk and meat. Transport dewars have also been used for storing blood, corneas, virus samples, biopsy specimens, and heart valves.

Cryocoolers

In 1956 J.W.L. Kölher at the Philips Company in Holland developed the first commercial Stirling refrigerator, by reversing the Stirling cycle previously used for power generation. Beginning in the 1960s this compact, closed-cycle cryogenic refrigerator, or cryocooler, has been utilized extensively for infrared detection in security systems, astronomy, and in military applications for night vision goggles and heat-seeking devices. A family of commercial cryocoolers, differing in their thermodynamic cycles, have been developed through the latter half of the twentieth century, including the Gifford McMahon, Vuilleumier, and miniature versions of Joule–Thomson and Reverse-Brayton cycles in the 1960s and 1970s, and the orifice pulse tube in the 1980s and 1990s. The advantages of compactness and ability to create cryogenic temperatures without liquid cryogens have resulted in continually expanding applications of cryocoolers. Examples include the over $200 million per year cryopump industry, cold electronics, fetal heart monitors, and remote-site liquefaction of air, nitrogen and oxygen.

Food Processing

Beginning in 1923 with Birdseye's processing plant for frozen fish, the food industry has capitalized on the benefits of rapid freezing afforded through the use of cryogenics. Additional benefits such as retained texture, color, and flavor have maintained the growth of this industry throughout the twentieth century.

Materials

During the latter half of the twentieth century a cryogenic process was developed to improve the strength and wear-resistance of certain metals. Tests in 1944 in which alloy steel was cold-cycled in LN_2 resulted in minor improvements to wear resistance, but also increased brittleness. In the early 1970s a different approach tested by the Advanced Engineering Metal Parts Division of the Sperry Rand Corporation revealed 150 to 300 percent increased life for machining tools. R.F. Barron at Louisiana Tech University extended the investigation, determining an optimum process for high-speed lathe tools and demonstrating improvements in wear resistance by factors of 2 to 5 with five different tool steels. The optimized process involves very slow cooling (2°C per hour) to 80°Kelvin, followed by a 1- to 2-hour soak in LN_2. Although the warm-up rate to ambient conditions is not crucial, a mild temperature at

150°C for 1 hour is important for eliminating brittleness. The process aids the conversion in steels from a large-grain austenitic microstructure to the denser, finer grain-sized martensitic microstructure with increased hardness and wear resistance. Other applications of this technology emerged in the 1990s with the establishment of more than 50 small-scale companies treating gun barrels, musical instruments, guitar strings, and engine parts for motorcycles and racecar engines.

Space

Space exploration has been distinctively enabled by the cryogenics industry. The availability of large quantities of liquid hydrogen and liquid oxygen facilitated the development, from the 1950s onward, of rockets and other space vehicles. Space technology now utilizes most of the cryogens for propulsion, sensor cooling, and life support. The space applications of helium, beyond pressurizing other cryogen tanks, are especially unique. The unusual properties of superfluid helium (liquid helium below 2.17°Kelvin) such as zero viscosity, unusually high heat transport, and a thermo-mechanical pressure, first explored during the 1930s, have been ingeniously utilized where zero gravity necessitates special requirements for coolant transfer and containment. Helium refrigeration technologies have enabled space-based infrared detectors to operate at 60 millikelvin.

Sub-Kelvin Technology

Liquid helium has enabled research at very low temperatures primarily through the use of two different refrigeration techniques. The adiabatic demagnetization technique, developed by Giauque and MacDougall in 1933, utilizes cooling provided by the magnetocaloric effect in certain paramagnetic salts to produce temperatures down to 1 millikelvin. The dilution refrigerator, proposed by H. London in 1951, and first developed at the University of Leiden by Das, Ouboter, and Taconis in 1965, evaporates pure liquid He^3 into a dilute mixture with superfluid He^4 to produce temperatures down to 2 millikelvin. He^3 is the lighter isotope (2 protons, 1 neutron) of helium. From the mid-1960s to the mid-1990s dilution refrigerators became the preferred method of reaching sub-Kelvin temperatures. However, with the growing convenience of superconducting magnets, compact adiabatic demagnetization devices are again becoming popular. The introduction and growth of superconducting magnets, enabled by

ready supply of liquid helium, is described elsewhere.

See also **Cryogenics, Liquifaction of Gases; Food Preservation: Cooling and Freezing; Superconductivity, Applications; Tomography in Medicine**

JOHN M. PFOTENHAUER

Further Reading

Barron, R.F. *Cryogenic Systems*. Oxford University Press, New York, 1985.
Baust, J.G., Gage, A.A., Ma, H. and Zhang, C.M. Minimally invasive cryosurgery-technological advances. *Cryobiology*, 34, 373–84, 1997.
DeGaspari, J. The big chill. *Mech. Eng.*, 122, 11, 94–97, 2000.
McGrath *et al.* Cryobiology update. *Cold Facts*, 15, 3, 1–26, 1999.
Miller, J.P. The use of liquid nitrogen in food freezing, in *Food Freezing: Today and Tomorrow*. Springer Verlag, Berlin, 1991.
Mendelssohn, K. *The Quest for Absolute Zero: the Meaning of Low Temperature Physics*. Taylor & Francis, London, 1977.
Shachtman, T. *Absolute Zero and the Conquest of Cold*. Houghton Mifflin, Boston, 1999.
Timmerhaus, K.D., Fast, R.W. and Kittel, P., Eds. *Advances in Cryogenic Engineering*. Plenum Press, New York, 1960.
Walker, G. *Cryocoolers: Part 1 Fundamentals; Part 2 Applications*. Plenum Press, New York, 1983.
Walker, G. *Miniature Refrigerators for Cryogenic Sensors and Cold Electronics*. Oxford University Press, New York, 1989.

Cryogenics, Liquefaction of Gases

The modern air liquefaction industry generates annual sales in the tens of billions of dollars through production of liquefied oxygen, nitrogen, argon, and specialty gases. Following liquefaction, air is fractionally distilled to separate the various components. The cryogenic liquids are then stored and transported in specially insulated vessels and shipped to thousands of customers for a variety of applications.

Oxygen is utilized in the manufacture of steel, other metals, and glass, in the pulp and paper industry, in chemical processes, and for life support in medical applications. Nitrogen finds application in chemical processes such as the production of ammonia, in rapid-freeze food processing, in biomedical systems that preserve blood, tissue or semen, and in ecological recycling of car tires. Argon provides an inert atmosphere for the manufacturing of steel and aluminum, for welding, and for fabricating electronics components. It is also used for lighting, as are neon, krypton, and xenon. The two coldest cryogenic liquids, hydrogen and helium, are utilized in various industries, hydrogen finding uses in chemical and food processing, pharmaceuticals, fuel propellant for spacecraft, and in emerging fuel cell technologies. Helium is used extensively in the aerospace industry, as an inert atmosphere in metal processing, for leak detection in HVAC and vacuum systems, and as a coolant for superconducting magnets and low temperature research.

Producing and using the liquid cryogens has been a leapfrog process in which the possibility of obtaining the pure gases has spawned new applications, with the resulting applications motivating new possibilities in the cryogenics industry. With the exception of utilizing oxygen for making steel, and nitrogen for agricultural fertilizer, none of the applications cited above existed in 1900. It was only in 1895 that the patents enabling commercial production of liquid air, oxygen, and nitrogen, were issued to Carl von Linde in Germany and William Hampson in Great Britain. Figure 25 schematically diagrams the mechanical components and flow of gas through the Linde–Hampson air liquefier. As the air is compressed it rejects heat to its surroundings. Additional cooling of the pressurized air occurs in the counter-flow heat exchanger from the colder return-gas. Exiting the heat exchanger, the pressurized air is expanded

through a valve resulting in a drop in pressure and temperature, to the extent that some of the air exits the valve in liquid form. The remaining cold gas returns up through the counterflow heat exchanger. Being warmed by the incoming pressurized air, it returns to the suction side of the compressor near room temperature. Adding the same mass in gaseous form at the inlet of the compressor then compensates for the amount of liquid that is extracted from the machine.

In 1902 Georges Claude improved the design of air liquefiers by replacing the expansion valve with an expansion piston, producing liquid more efficiently because the gas is cooled by "doing work" as it pushes against the piston during the pressure drop. In the same year, and because of the increased availability of oxygen, oxyacetylene torches replaced less effective air-acetylene torches used in welding. Although this advance was enjoyed only in Europe up to 1906, it spread along with the air liquefaction industry to other continents and is still in widespread use today.

The British Oxygen Company introduced the air liquefaction process to the U.S. at the world's fair in St. Louis in 1904. By January of 1907, the Linde Air Products Company was established in Buffalo, New York. Within 14 years the worldwide production of liquid oxygen doubled. To reduce distribution costs, cryogenic vessels holding liquefied gases increasingly replaced gas cylinders. However, inadequate insulation in the cryogenic vessels prohibited distribution over the large distances encountered in the U.S. Here the applications necessitated improvements to the liquid cryogen industry. In 1900, the best thermally insulating containers were those invented in 1892 by Sir James Dewar. These laboratory-based double-walled glass vessels used a high degree of vacuum in the space between the walls to achieve their insulating properties, but were impractical for industrial purposes. The "powder in vacuum" design, developed and commercialized by Leo Dana at the Linde Division of Union Carbide in the 1930s and 1940s, provided adequate insulation while reducing vacuum requirements and improving constructability. In this approach, the double-wall space of metal containers is filled with carefully selected powders. Containers utilizing this design today transport liquid oxygen, nitrogen, and argon worldwide for periods of several weeks without significant evaporation.

The 1940s brought the first commercial liquefaction of natural gas to improve the distribution of the popular fuel to homes and industries. Natural gas, used from the mid-1800s, is obtained

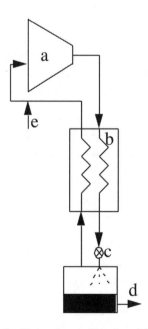

Figure 25. Linde–Hampson air liquefier: (a) compressor; (b) counterflow heat exchanger; (c) expansion valve; (d) liquid air outlet; and (e) make-up air inlet.

from underground cavities and is primarily composed of methane. The 600-fold increase in the density of liquid natural gas (LNG) as compared to the gas at ambient conditions, motivated the development of the LNG production facilities. The first commercial LNG plant was installed in Cleveland, Ohio in 1941 and liquefied 200 cubic meters per day. A storage tank failure in 1944 and subsequent fire caused serious damage and loss of lives, delaying growth of the industry for over ten years. Subsequent installations, with improved safety, were established first as a floating barge in 1956, and later as permanent land-sites worldwide. Today, the multibillion dollar LNG industry transports LNG via 41,000-liter-tank trucks and 125 000-cubic meter-capacity ships.

While the liquid air industry was beginning in the U.S., the scientific race to condense the remaining unliquefied gas approached its finish line. On 10 July 1908. Kamerlingh Onnes at the University of Leiden succeeded in liquefying helium at 4.2°Kelvin. Liquid helium remained a laboratory curiosity until the invention of a convenient helium liquefier by Sam Collins of the Massachusetts Institute of Technology (MIT) in 1947. In 1952 the National Bureau of Standards in Boulder, Colorado was established to develop a large-scale hydrogen liquefier. By the 1960s liquid helium and hydrogen were being used extensively for emerging programs in space exploration and superconductivity. Union Carbide's development of multilayer insulation (MLI) or superinsulation enabled the 100-fold improvement to the cryogenic storage vessels that allowed large-scale transport of liquid helium and liquid hydrogen.

See also **Cryogenics, Applications**

JOHN M. PFOTENHAUER

Further Reading

Dana, L.I. with Donnelly, R.J. and Francis, A.W., Eds. *Cryogenic Science and Technology: Contributions by Leo I. Dana*. Union Carbide, 1985.

Isalski, W.H. *Separation of Gases*. Oxford Science, Oxford, 1989.

Mendelssohn, K. *The Quest for Absolute Zero: the Meaning of Low Temperature Physics*. Taylor & Francis, London, 1977.

Shachtman, T. *Absolute Zero and the Conquest of Cold*. Houghton Mifflin, Boston, 1999.

Scott, R.B. *Cryogenic Engineering*. Met-Chem Research, Boulder, CO, 1988.

Scurlock, R.G. A brief history of cryogenics, in *Advances in Cryogenic Engineering*, 37A, Fast, R.W., Ed. Plenum Press, New York, 1991.

Timmerhaus, K.D. and Flynn, T.M., *Cryogenic Process Engineering*. Plenum Press, New York, 1989.

Williamson Jr., K.D. and Edeskuty, F.J. *Liquid Cryogens: Volume I Theory and Equipment, Volume II Properties and Applications*. CRC Press, Boca Raton, FL, 1983.

Crystal Detectors, *see* **Radio Receivers, Crystal Detectors and Receivers**

Crystals, Synthetic

A crystal is an immobilized arrangement of atoms, ions or molecules packed in a regular array. Although naturally occurring crystals can be things of beauty, it is rare to find a fault-free specimen of any considerable size. Crystals can now be synthesized as gems, but more often they are manufactured for utility. This discussion on synthetic crystals will outline some of the techniques that are used in the manufacture of technologically important crystalline materials other than semiconductors, which are treated elsewhere in this encyclopedia.

One of the first applications of crystalline materials was in specialist optics. In 1828 William Nicol of Edinburgh developed a prism based on calcite ($CaCO_3$) crystals which could polarize light. Two such prisms were used in a polaroscope (an instrument for measuring the concentration of optically active chemicals) and the design remained unchanged until the 1980s. At first, piezoelectricity (a voltage causing mechanical deformation and vice versa in a crystal) was a curiosity but it assumed critical importance in 1917 when Paul Langevin assembled a mosaic of thin quartz crystals as a passive sonar array for submarine detection. Quartz crystals became very important as radio engineers struggled to "lock" the frequencies of transmitters so that more users could be accommodated in the limited radio-frequency spectrum.

We can assume that by the end of World War I, the demand for certain crystals outstripped the supply of suitable naturally occurring materials. Fortunately, techniques for manufacture were already to hand and others have been developed throughout the twentieth century. These have been driven by technological advances, such as the requirements of the semiconductor industry. The demand for synthetic ruby grew enormously after it was used in the first lasers in 1960.

For the purposes of classification the plethora of techniques can be listed as: growth from a melt, growth from solution, growth involving high temperature, growth involving high pressure, and growth involving both high temperature and pressure. However, before discussing these we should remember that nature does not like a state of order and that is exactly what a crystal is. Therefore, crystals will only grow if there is a net benefit, such as a significant decrease in energy. The physical chemistry of the crystal formation is concerned with phase diagrams and with equilibria. The latter dictates that growth must normally be a slow process if defect-free crystals are to be obtained.

Although the French chemist, Edmond Fremy, had been producing commercial quality gemstones in 1877, the first major advance was the production of ruby using the Verneuil process (developed by French chemist Auguste Victor Louis Verneuil) in 1902. In this technique a seed crystal is held under an oxy-hydrogen flame and the base material (alumina, Al_2O_3 powder with an addition of chromia, Cr_2O_3 as a contaminant to give color) is fed through the flame. Other materials such as sapphire, rutile (titanium dioxide, a gem that is also used as a white pigment in paints, paper, plastics, sunscreens, and cosmetics), spinel (a mineral that can be used to imitate blue sapphire or aquamarine; also has good optical qualities for laser applications), and strontium titanate (has no natural counterpart, but can imitate diamond; and as a ceramic has been widely used for various electronic applications) were later made using this method. Synthetic gems have the same chemical composition and crystal structure as natural gems, but lack irregularities or tiny inclusion imperfections that give real gems flaws that produce their unique appeal.

The Czochralski method published by Jan Czochralski in 1918 involves dipping a seed crystal into a container of molten base material. The seed is rotated as it is slowly withdrawn and the material adhering to it assumes the same crystal orientation. This process has been used to make ruby, sapphire, spinel, and yttrium-aluminum-garnet (YAG), an important material for lasers. The Czochralski technique was later to become the cornerstone of the silicon industry and bars of single-crystal silicon up to 250 millimeters in diameter are now in regular production. Czochralski growth of III–V semiconductors such as gallium phosphide presents considerable problems because of the volatility of phosphorous. This has been overcome by adding boron oxide to the melt; which is immiscible and floats on top, rather like oil on water, inhibiting evaporation.

Growth from aqueous solution is historically much older than any of the other techniques, but it remains very important in the manufacture of certain crystals; for example, Rochelle Salt (potassium-sodium tartrate) is a much more effective piezoelectric material than quartz and in 1917 Alexander Nicolson at Bell Laboratories demonstrated that it could be used for sonar applications. In single-crystal form, potassium dihydrogen phosphate (KDP) is a nonlinear optical material used for doubling, tripling and quadrupling the frequency of the output from high-power lasers. In 1998 the world's largest KDP crystal (250 kilograms) was grown at the Lawrence Livermore National Laboratory for use in the Megajoule Laser Fusion Experiment.

The same growth techniques can be applied in nonaqueous systems. In the particular case of growth of crystals of materials that have an extremely high melting point the solvent is chosen for its flux properties; that is, its ability to induce growth at significantly lower temperatures. Thus aluminum oxide crystals have been grown in a flux of lead fluoride at 840°C. Once the system has cooled down the flux is dissolved in nitric acid to reveal alumina crystals.

Aqueous growth under high pressure (hydrothermal growth) is a useful means of growing otherwise difficult crystals. It is currently the preferred technique for the production of synthetic quartz.

There have been many attempts to produce synthetic diamond but the crystallization of carbon requires both high temperature and high pressure. It was not until materials such as tungsten carbide became available in 1930 that work could commence on high-pressure containment systems. The use of self-sealing techniques, such as that developed by Percy W. Bridgman, were also essential. Diamonds were produced by Baltzar von Platen and colleagues at the ASEA Laboratory in Stockholm in 1953, but the work was kept secret. Thus in 1955 Francis Bundy, Tracy Hall, Herbert Strong and Robert Wentorf at General Electric were able to claim the first commercial production of synthetic diamond. Today, films of diamond can be produced using less severe conditions. This has developed as an offshoot of a technique called vapor phase epitaxy, which is very important in silicon device manufacture and involves the growth of crystalline layers by chemical vapor deposition (CVD). For example gaseous silane (SiH_4) or silicon trichloride ($SiHCl_3$) is carried through a reaction chamber in a stream of hydrogen. At the center of the reactor is a graphite block held at approximately 1200°C with one or more silicon substrates resting on top. The silane breaks down on the hot surface to yield silicon and hydrogen and the deposited material has the same the crystal orientation as the substrate,

This summary lists some of the methods for growing technologically important crystals. Improvements continue, but total innovations are rare. One of the few exceptions has been molecular beam epitaxy (MBE), where crystals are grown under extreme high vacuum conditions. It is costly, but yields tailor-made materials called "super-lattice crystals" which have considerable promise for the future. Perhaps the next step is routine crystal fabrication under zero gravity conditions in space.

See also **Semiconductors, Crystal Growing and Purification; Thin Film Materials and Technology**

DONARD DE COGAN

Further Reading

Gilman, J.J. *The Art and Science of Growing Crystals*. Wiley, New York, 1963.

Hazen, R.M. *The New Alchemists: Breaking Through the Barriers of High Pressure*. Times Books, Random House, New York, 1993.

Hazen, R.M. *The Diamond Makers*. Cambridge University Press, Cambridge, 1999.

Hughes, R. Synthetic corundum, in *Rubies and sapphire*. RWH Publishing, Boulder, CO, 1998.

Scheel, H. J. Historical introduction, in *Handbook of Crystal Growth*, Hurle, D.T.J., Ed., vol. 1. Elsevier, Amsterdam, 1993.

Scheel, H.J. and Fukuda, T., Eds. *Crystal Growth Technology*. Wiley, New York, 2003.

D

Dairy Farming

Throughout the world, especially in the Northern Hemisphere, milk, cheese, butter, ice cream, and other dairy products, have been central elements of food production. Over the centuries improvements in cattle breeding and nutrition, as well as new dairy techniques, led to the increased production of dairy goods. Hand-operated churns and separators were used to make butter and cream, and those close to a barnyard had access to fresh milk.

By the late nineteenth century, new science and technology had begun to transform dairy production, particularly in the U.S. and Europe. Rail transportation and iced and refrigerated boxcars made it easier to transport milk to more distant markets. Successful machinery for separating milk from cream came from the DeLaval Corporation in 1879, and the Babcock butterfat tester appeared in 1890. The first practical automated milking machines and commercial pasteurization machines were in use in the decades before 1900. Louis Pasteur's contribution to the dairy industry—discovering the sterilization process for milk—was substantial. By heating milk, pasteurization destroys bacteria that may be harmful to humans. The pasteurization process also increases the shelf life of the product by eliminating enzymes and bacteria that cause milk to spoil. Milk is pasteurized via the "batch" method, in which a jacketed vat is surrounded by heated coils. The vat is agitated while heated, which adds qualities to the product that also make it useful for making ice cream. With the "continuous" method of pasteurization, time and energy are conserved by continuously processing milk as a high temperature using a steel-plated heat exchanger, heated by steam or hot water. Ultra-high temperature pasteurization was first used in 1948.

Key to the mechanization of the dairy industry were improvements in the mechanical milking machine. The first vacuum pump machines were patented in the 1860s, and the Mehring hand- and foot-powered model was popular in the 1890s. This machine was powered by a person sitting down, pushing pedals with their feet, and milking two cows at one time. Widespread adoption of the milking machine did not come until the introduction of the DeLaval milker in 1918. Using these early machines, one person could milk 30 cows twice per day; with improvements in milking machines by the end of the twentieth century, over 200 cows could be milked by one person. What had been seen as "women's work," the tending of dairy herds, had become a business process in which farm factories produced raw milk for shipment into central dairies for processing. Elaborate dairy barns gave way to high-tech milking parlors where cows could be milked by machine three times per day rather than once or twice, and a premium was placed on sanitary conditions. By the late twentieth century, it was not uncommon in California to see dairy operations numbering up to 50,000 animals. Numerous devices were developed to improve mechanical milking, and the intermittent "pulsator"-type milker was one of the more important innovations. Finding a device that could be kept sanitary, that would maximize production per cow and that would not harm the animal's teats, were all obstacles in developing a practical milking machine. The "thistle"-type machine, so called because of the appearance of the equipment that fit on the cow's udders, was developed in the 1910s and was the basis for the design of all subsequent milking machines. In the late twentieth century machines, the cow's teats are inserted into rubber suction cups, which are surrounded by stainless steel cups. The vacuum of the automatic machine inflates and deflates the rubber cup, with the milk flowing onto a central collection unit. Modern dairy operations are designed to be easily cleaned, and the emphasis on health and safety with the homogenization and pasteurization processes helped to increase consumer confidence in the nutritional value of dairy products.

In addition to the mass production of flavor-controlled, vitamin-fortified homogenized milk by 1919, new delivery technologies made dairy products more accessible to the public. By 1938 bulk tank trucks with coolers transported milk from dairies as dairy operations grew larger and on-farm storage demands increased. Milk bottle fillers in the 1940s could fill 4500 liters per hour, compared with earlier models that bottled 1100 to 1500 liters per hour. Plastic-coated paper containers were introduced in the 1930s, and the first all-plastic milk containers appeared in stores in 1964. By then machines could fill 23,000 liters per hour.

Increased production of high quality dairy products and better distribution of them resulted in greater consumption of dairy products in the twentieth century. The United States Department of Agriculture (USDA) created the Department of Agrostology in 1895 to study how different grasses and feedstuffs affected milk production and milk quality. The USDA's Bureau of Dairy Industry fostered better herd management and breeding techniques, more efficient and sanitary production, and new and better types of dairy products for the consumer. Agricultural researchers contributed to the dairy industry with herd improvements through artificial insemination of cattle (1938) and embryo transfer (1980). Scientists also provided a variety of antibiotics including penicillin to control mastitis and other frequent dairy diseases. Genetic modification to enhance production also came into play with the introduction of recombinant bovine somatotrophin (rBST) in 1994. That same year the U.S. Nutrition Labeling and Education Act became law, requiring producers to list rBST on the label if it is in their milk.

At the end of the twentieth century northern Europe and North America produced three-fourths of the world's raw milk, with India and South America the major secondary producers. While low-fat dairy products became increasingly popular by the 1990s, per capita ice cream consumption in the U.S. increased to 7.5 kilograms per year. In the state of California, not traditionally associated with dairy production, 20 percent of U.S. milk production came from 2,200 dairies that produced some 16 billion kilograms of raw milk.

See also **Farming, Agricultural Methods; Farming, Growth Promotion; Farming, Mechanization**

RANDAL BEEMAN

Further Reading

El-Hai, J. *Celebrating Tradition, Building the Future: Seventy-Five Years of Land O' Lakes.* Land O' Lakes, Minneapolis, 1996.

Hurt, R.D. *American Agriculture: A Brief History.* Iowa State University Press, Ames, 1994.

Moore, E.G. *The Agricultural Research Service.* Praeger, Washington D.C., 1967.

Porter, A.R. *et al. Dairy Cattle in American Agriculture.* Iowa State University Press, Ames, 1965.

Weimar, M.D. and Blayney, D.P. *Landmarks in the US Dairy Industry.* U.S. Department of Agriculture, Washington, D.C., 1984.

Dams

For thousands of years, dam and water storage technologies have allowed civilizations to flourish in parts of the world where dry climates would otherwise limit human settlement. As early as 3000 BC, civilizations along the Tigris, Euphrates, Ganges, and Nile Rivers constructed earth and stone dams across these large rivers. These structures allowed them to store water for agriculture and create complex societies on that basis.

A dam consists of a mass of earth, timber, rock, concrete, or any combination of these materials that obstructs the flow of water. A dam can either divert water or store it in a reservoir, the artificial body of water that a dam creates. Diversion dams (weirs) raise the elevation of a river and divert water into a canal for transport to a mill, power plant, or irrigated field. Storage dams impound water in a reservoir.

There are three major types of dams—gravity, arch, and buttress. Gravity dams rely for stability on their weight to resist the hydrostatic, or water, pressure exerted by the reservoir. Arch dams, built along arcs that curve upstream into reservoirs, are most commonly found in narrow canyons with hard rock foundations. The arch dam transmits the horizontal water thrust to the abutments. Multiple-arch dams consist of a number of single arches supported by buttresses. Like gravity dams, buttress dams rely on gravity for stability, but require less material than standard gravity structures. They resist hydrostatic loads by using the same engineer-

Figure 1.   Construction at Glen Canyon Dam, Colorado River Basin Storage Project, 1963.
[*Photograph courtesy of the Department of the Interior, U.S. Bureau of Reclamation*].

ing principles of the flying buttresses that braced the high walls of Gothic cathedrals.

Engineers of the late Roman Empire built the first known arch dams and several buttress dams. In medieval times, Moorish engineers harnessed water from the Sierra Nevada and constructed massive irrigation systems in southern Spain. Persia's Ilkhanid Mongols built the world's tallest arch dams at that time as well. In the sixteenth century, Spanish engineers initiated a new era of dam building by applying mathematical principles to dam design. From the late eighteenth to early nineteenth centuries, French and English engineers, including Bernard Belidor, Charles Bossut, Charles Augustin Coulomb, and Henry Moseley, designed dams further based on mathematical principles. By the end of the 1800s, Europeans had initiated large dam projects around the world, particularly in areas of British control. British engineers completed Egypt's Aswan Dam in 1902, enlarged it in 1934, and rebuilt it upstream in 1960. The dam resulted in increased agricultural and industrial capacity for the region's growing population.

The twentieth century heralded the development of large, multipurpose water projects that could control floods, store water for irrigation, municipal, and industrial use, improve the navigability of waterways, and provide water for the generation of hydroelectric power. Indeed, one of the century's great advances came with the development of technologies that could generate electricity from the motion of falling water. The introduction of concrete as a construction material allowed dam designers from the 1910s to the 1930s to consider thin shells and complex curved shapes to minimize the volume of concrete required and the overall cost of arch dams. The face of a double-curvature (dome) arch dam is curved in two directions: from side to side, transferring force into the canyon walls, and from top to bottom, transferring force into the canyon floor. Posttensioned steel rods or cables commonly reinforce concrete arch dams and gravity dams, reducing the cross-section (and volume of materials) for a conventional gravity dam of the same height. Vertical steel rods, stressed by jacks and securely anchored into the rock foundation, resist the tendency of the thin shells to overturn.

The U.S. Bureau of Reclamation's Hoover Dam (originally called Boulder Dam) (a curved gravity dam, see Figure 2) served as an early prototype for the century's multipurpose water projects. Built along the Colorado River in 1935, Hoover Dam stands nearly 222 meters high—the tallest dam in the world at that time. The dam created Lake Mead, the largest man-made lake in the U.S. Hoover Dam Power Plant's 17 large turbines generate a total capacity of 2,074,000 kilowatts of electrical power for Nevada, Arizona, and California.

Hoover Dam served as a model for multipurpose water projects around the world, including government projects in Pakistan and India. Other examples of major late twentieth century dam projects include Syncrude Tailings Dam in Alberta, Canada; Rogun Dam in Tajikistan; Tehri Dam in the Indian Himalayas; and the Itaipú Dam on the Paraná River between Brazil and Paraguay. By harnessing large rivers' water and energy, these dams allow for massive agricultural and industrial development in areas with natural limitations.

Perhaps the twentieth century's most ambitious dam project was Three Gorges Dam in China, the largest construction scheme in China since the Great Wall. Started in 1992 and completed through the turn of the twentieth century, the dam was built in order to rid the fabled Yangtze River of its deadly floods, provide an important source of electrical energy for China, and open up

Figure 2.   Hoover Dam.

the inland Chongdong region to commerce. Plans were first developed in the 1920s. They were revived in the mid-1950s, when devastating floods along the Yangtze raised the recurring need for new water policy measures. The project was finally started in 1993. The dam, almost 2.5 kilometers wide and 180 meters high, created a reservoir 66 kilometers long and hundreds of meters deep, at a cost of approximately $24 billion. On its completion in 2009, the first set of generators will produce an amount of power expected to create as much electricity as 18 nuclear power plants.

Controversy over Three Gorges Dam reflects the larger social, environmental, and economic questions of dam building. In the late twentieth century, the desire to protect rivers, riparian habitat, and human health hindered the construction of large dams and even mandated the removal of some. In China, critics of Three Gorges Dam claim that slowing the flow of the Yangtze River creates an environmental hazard by allowing sources of pollution to collect. Indeed, the inevitable buildup of silt behind large dams poses a technological problem that has not yet been solved by engineers. Cultural and social concerns also raise questions about the viability of large dams. The reservoir created by Three Gorges Dam drowned valuable antiquities sites, including evidence of the ancient

Ba people. It also flooded more than 100 towns, forcing the resettlement of 1.13 million people—the largest resettlement in the history of dam building. Three Gorges Dam, like all major dam projects, reflects the need to balance industrial growth with environmental and social concerns.

See also **Concrete Shells; Irrigation Systems; Hydroelectric Power Generation**

JESSICA B. TEISCH

Further Reading

Downing, T. and McGuire G., Eds. *Irrigation's Impact on Society*. University of Arizona Press, Tucson, 1974.

Greer, C. *Water Management in the Yellow River Basin of China*. University of Texas Press, Austin, 1979.

Jackson, D.C., Ed. *Dams*. Ashgate, Brookfield, 1997.

Justin, J., William, C. and Hinds, J. *Engineering for Dams*, vol. 3. Wiley, New York, 1945.

Schnitter, N. J. *A History of Dams: The Useful Pyramids*. Balkema, Rotterdam, 1994.

Smith, N. A. F. *A History of Dams*. Davies, London, 1971.

Wittfogel, K. *Oriental Despotism: A Comparative Study of Total Power*. Yale University Press, New Haven, 1957.

Wolf, D.E., Wolf, D.D. and Lowitt, R. *Big Dams and Other Dreams: The Six Companies Story*. University of Oklahoma Press, Norman, 1996.

Worster, D. *Rivers of Empire: Water, Aridity, and the Growth of the American West*. Oxford University Press, New York, 1985.

Dentistry

In the early twentieth century a visit to the dentist was usually for a toothache, cured by extraction or replacement of missing teeth that had already been extracted. Preventive and restorative technologies existed but were not widespread, and materials were limited. Fluoride was known to have prophylactic qualities against caries, a vaccine was being developed for periodontal disease, and a zinc phosphate cement was used to seal fissures in the occlusal surface of molars. Diagnostic procedures included x-rays but x-rays were not generally applied in dentistry until about 1916. Anesthesia, in the form of nitrous oxide gas or ether had been in use in dentistry since 1863 (earlier than in surgery). In 1904, Alfred Einhorn introduced procaine (trade name Novocain), an injectable local anesthetic, to the practice of dentistry. It was superior to cocaine, which had been used previously, in that it did not cause tissue sloughing and was not addictive. The combination of anesthesia and electrical drills prepared the way for increased extraction and restorative work.

Public health leaders had an interest in dentistry, and throughout the world clinics were opened by philanthropists following the example of American industrialist George Eastman. By 1925 the inverse relationship between caries, or tooth decay, and tooth mottling as a result of natural fluoride in drinking water impelled further epidemiological studies, and by 1962 many U.S. cities and up to 30 other countries were adding sodium fluoride to their water supply. Fluoride treatments in the form of topical applications were provided for children, and dental hygienists visited schools showing students how to brush and floss their teeth. As a result of better dental education, the prevention of caries and dental restorations helped improve overall public health. Dental specialties such as orthodontics, endodontics, oral surgery, prosthodontics, pedodontics, and periodontics rapidly developed, each with their specialized technologies. Along with local anesthesia, topical numbing, use of a suction machine, and a reclining chair with leg rests, psychological amenities such as music or television available to distract and ease the patient's anxiety and ability to undergo lengthy procedures if necessary.

Technologies and Materials

Technologies used in diagnostic work include x-rays, study models, and an examination using explorers and probes to find cavities and measure periodontal pocket depth. The first x-ray equipment specially designed for dental radiography was introduced around 1923, but early machines were not shockproof. The modern dental unit in which x-ray tube and high-voltage transformer are both contained in a single comparatively small housing was introduced in 1933. Early x-ray films were made from celluloid but had to be cut and wrapped, and were dangerous because of their flammability. The x-ray machine evolved with film technology. As films became faster, exposure to radiation for both dental personnel and patient could decrease. A long tube replaced the short cone in order to increase the distance between target (patient) and film.

Periapical and bite-wing views require 16 films and show individual teeth. Oral surgeons, orthodontists, and prosthodontists require additional information about oral pathology, impacted teeth, developmental dentitions, jaw relationships, and the temporomandibular joint as part of their diagnostic work. In the 1950s, a panoramic machine was built that traveled around the patient's head. It only required one exposure to radiation. The image produced is on a large film and shows the complete dentition, each root and below, along with occlusal relationships. By the end of the twentieth century, digital intraoral photography that instantly magnifies the teeth and underlying structures on a computer screen had been invented.

Early formulas for impression materials contained shellac, talc, glycerin, and fatty acids. A superior product, hydrocolloid (reversible agar–agar material) was heated, poured into trays, then cooled in the mouth and used to make the initial impression for inlays and crowns. After the impression was made, it was filled with a type of dental stone more accurate than plaster of Paris for the model. Today, a study model impression is taken with alginate, an irreversible colloid material, mixed with water. This is placed in a tray in the patient's mouth, removed when set. Plaster of Paris, a gypsum hemihydrate compound, is poured into this shell to create the model.

Restoration of teeth, whether with an amalgam filling, inlay, or crown, requires preparation with the use of an abrasive dental drill. The clockwork motor-driven drill was first replaced by electric plug-in drills in 1908. These early drills operated at 600 to 800 revolutions per minute (rpm). By mid-century a high-speed hydraulic drill, which operated at 60,000 rpm, was developed by the National

Institute of Standards and Technology (NIST) and the American Dental Association. Modern water-cooled, turbine-powered drills rotate at speeds between 300,000 and 400,000 rpm. An assortment of different shapes of burrs that fit the head can be interchanged for ease of manipulating the shape of the preparation.

The first modern materials for restorations of small cavities were made in the early nineteenth century from gold foil, and from the 1900s, gold alloys. The process required special pliers, an alcohol lamp, a gold foil annealer, and a mallet. The process was lengthy and very uncomfortable for the patient. Silver–mercury amalgams were first used from 1832, but the poor quality limited use until a standard manufacture was developed in 1895. Modern amalgam is made from mercury mixed with silver, copper, zinc, and tin. As it hardens, the shape of the occlusal surface is carved and then burnished and polished after it assumes its final hardness. Baked porcelain inlays had been in use from 1862, once an effective dental cement had been developed. By mid-twentieth century, tooth-colored silicates or synthetic porcelain replaced gold and provided an alternative to amalgam. The powder and liquid was quickly mixed on a glass slab to a thick consistency with a spatula, and then transferred to the tooth. The material was smoothed and after setting, polished with a special attachment to the dental handpiece. Bonding, a process using composite resin materials, has replaced older techniques and materials. The revolution in laser technology has allowed for gentle application of a polymer/monomer that creates a tooth shade indistinguishable from the natural tooth that neither discolors nor decomposes.

When a large segment of tooth is destroyed, crowns and bridges are fabricated. Early crowns and bridges employed gold alloys joined to gold. In mid-century, crowns made of porcelain fused to gold were developed. This was esthetically superior to the gold crowns and stronger with respect to the chewing surface.

The filling of a tooth root canal is necessary when decay has invaded the internal tooth structure. Root canals were first filled with silver points, which were cut and measured to fit the walls of the excavated tooth. An instrument called a reamer was used to clean out the canal, followed by filing and irrigation to clean and shape the interior portion. Gutta percha, an inert material made from the latex of a tree from Malaysia, gradually replaced silver as the material of choice for root canal filling.

Permanent and Removable Restorations

A denture is a prosthesis, an artificial device, made to take the place of missing teeth. It is either partial or full, and the appliance is either permanently cemented in the mouth or a clasp is made to hook the partial restoration onto existing teeth. In 1930, Vitallium, a chrome–cobalt alloy was introduced for making these clasps. Replacements for decayed or lost teeth have been produced for millennia. Charles Goodyear's invention of vulcanized rubber, which could be molded against a model of a patient's mouth, allowed cheap and self-retaining denture bases from 1864 (earlier false teeth on gold bases had been held in place by springs). In 1919, gingival-colored base material helped to make a more realistic prosthesis. Although implants were being fabricated from pyrolytic carbon in 1919, they did not gain widespread application until the 1990s. Despite high success rates, they are still considered experimental. Dentures today (teeth and base) are made from an acrylic material (methylmethacrylate) that is colored to simulate gum tissue and the tooth. A veneer or laminate is made from porcelain to fit the facial surface of a tooth that has been discolored or chipped or is otherwise esthetically unacceptable.

Dentures were improved through the invention of the articulator (a dental machine that works as closely as possible to the way the mouth works) by the Swiss Alfred Cysi in 1909. Esthetics developed to match the patient's face with tooth shape.

Impressions for dentures were taken with rubber mercaptan-based materials that came in two tubes, one with a paste and the other with an accelerator made from lead peroxide and sulfur. These were mixed, placed in trays, applied to the arches, and removed after setting. Silicone, and later acrylic, compounds replaced rubber base impression material as well as hydrocolloid. Today, impressions for dentures are made in the same way as study models.

Fixed restorations require retention by cement, usually composed of zinc oxide–eugenol or zinc phosphate compounds. An improved cement in the 1970s added polyacrylic acid, which made it less irritating to the dentine and stronger. Glass ionomers (zinc oxide and polycarboxylate) had the advantage of better retention and were also used as a filling material for anterior teeth. A silicate cement (fluoroaluminosilicate glass and polycarboxylic acid) was introduced in the late 1970s in England. Cements made from calcium silicate and organic acid are also used to line the floor of a cavity before filling. Bone cements, used for attaching implants, or prostheses that require incorporation with osseous structures are now made from methacrylate and apatite compounds.

Orthodontia

Edward Hartly Angle established a school of orthodontia in 1900 and a year later established the American Society of Orthodontists. Orthodontics is the specialty that moves teeth to their optimum place in the oral cavity for both esthetics and function. The first bands used in orthodontia, made from stainless steel, were applied to each tooth and attached to wires that were tightened at timely intervals to move the teeth. Because these took up space in the mouth, it was routine to remove teeth. In the 1990s, braces made from very thin nickel titanium could preclude the need for extractions. To overcome resistance to treatment with metal braces, a new clear sapphire bracket was designed. A totally invisible lingual brace was designed for amenable cases. Cements used in orthodontia must adhere closely enough not to trap bacteria but also be removable at the appropriate time. Acrylic adhesives have made this possible.

After the teeth have been moved to their desired positions and the braces removed, it is necessary for the bone to grow back into the areas that have been altered. Removable appliances made from methymethacrylate are then prescribed until the teeth are solidly held in the jaws. Headgear, an external apparatus that caused many teenagers embarrassment, has been replaced with the use of tiny magnets.

For more complex problems, orthognatic surgery is sometimes undertaken to correct severe jaw problems, after which banding is performed to complete the esthetic result.

The AIDS Epidemic

The most revolutionary change in technology resulted after the mysterious case of dentally transmitted acquired immune deficiency syndrome (AIDS) in 1987 in Florida. After exhaustive research by the Centers for Disease Control, the young woman's case was traced to her HIV-infected dentist. The overhaul in protective gear for dental professionals now includes masks, eye protectors, gloves, and polyethylene disposable covers for the head, lamp, and everything that comes into contact with the patient. A rubber dam once used only for root canal procedures can provide a barrier between the tooth and saliva. Instruments are either sterilized or disposable. Although the mouth is not a sterile field and cannot be kept sterile, mandatory laws are enforced regarding the maintenance of dental offices and disposal of fluids, tissues, and supplies.

LANA THOMPSON

Further Reading

Asbell, M.B. *Dentistry: A Historical Perspective.* Dorrence, Bryn Mawr, Pennsylvania, 1988.

Barton, R.E., Matteson, S.R. and Richardson R.E., Eds. *The Dental Assistant,* 6th edn. Lea & Febiger, Philadelphia, 1988.

Foley, G.P.H. *Foley's Footnotes: A Treasury of Dentistry.* Washington Square East, Wallingford, PA, 1972.

Glenner, R.A. *The Dental Office: A Pictorial History.* Pictorial Histories, Missoula, 1984.

Guerini, V. *A History of Dentistry, from the Most Ancient Times Until the End of the Eighteenth Century.* Milford House, New York, 1909. Reprinted by Liberac, Amsterdam, 1967.

Langland, O.E. and Sippy, F.H. *Textbook of Dental Radiography,* 2nd edn. Thomas, Springfield, 1984.

Lyons, A. and Petrucelli, R.J. *Medicine: An Illustrated History.* Abrams, 1987.

Ring, M.E. *Dentistry: An Illustrated History.* Abrams, New York, 1985.

Seide, L.J. *A Dynamic Approach to Restorative Dentistry.* W.B. Saunders, Philadelphia, 1980.

Skinner, E.W. *The Science of Dental Materials,* 6th edn. W.B. Saunders, Philadelphia, 1967.

Stecher, P.G. Ed. *New Dental Materials.* Noyes, Park Ridge, 1980.

Weinberger, B.W. *An Introduction to the History of Dentistry With Medical and Dental Chronology and Bibliographic Data,* 2 vols. Mosby, St Louis, 1948.

Detergents

Detergents are cleaning agents used to remove foreign matter in suspension from soiled surfaces including human skin, textiles, hard surfaces in the home and metals in engineering. Technically called surfactants, detergents form a surface layer between immiscible phases, facilitating wetting by reducing surface tension. Detergent molecules have two parts, one of which is water-soluble (lyophobic) and the other oil or fat-soluble (lyophilic). They are adsorbed onto surfaces where they remove dirt by suspending it in foam. Some also act as biocides, but no single product does these things equally well.

The oldest detergent, soap, has been known since antiquity. Made by boiling animal fats and vegetable oils with caustic alkalis, soap is sodium or potassium stearate:

$$CH_3(CH_2)_{16} - COO^- \, Na^+$$

Lyophilic————Lyophobic

Soap combines with calcium and magnesium ions to form an insoluble scum in 'hard' water. For this reason synthetic detergents, less affected by hardness in the water, have replaced soap for many purposes.

Synthetic detergents were first manufactured in Germany in 1917 where they replaced soap, releasing animal fats for food and other uses during World War I. These early products were propyl- and butyl-naphthalene sulfonates, but in the late 1920s and early 1930s sulfonated fatty alcohols such as sodium lauryl sulfonate were found to be better detergents.

Sodium lauryl sulphate

$$CH_3(CH_2)_{10}CH_2 — OSO_3^- \ Na^+$$

Lyophilic   Lyophobic

In America, where petroleum products were available, alkylbenzene sulfonates were used, and by the end of World War II these had displaced the alcohol sulfates as general cleaning agents, though the latter are still used in shampoos and fine detergents. Today alkylbenzene sulfonates account for about half of all synthetic detergents produced in the U.S. and Western Europe. Until the 1960s the alkyl group commonly used was tetrapropylene ($C_{12}H_{24}$). These detergents were marketed in both powder and liquid form; they have very good detergent properties and produce copious lather.

Sodium tetrapropylenebenzene sulfonate

$$C_{12}H_{24}.C_6H_4—SO_3^- \ Na^+$$

Lyophilic ————————————Lyophobic

As the large lyophilic part of all these molecules is negatively charged, they are called anionic detergents, by far the most common type. But there are also cationic, nonionic, and amphoteric detergents. Cationic detergents are widely used as textile fabric softeners. Some are germicides or fungicides and are used as sanitizers in hospitals. The nonionic alcohol ethoxylates are used in liquid detergents for the laundry, but they are not produced as powders. The amphoterics are used as foam stabilizers.

The apparently simple process of cleaning soiled surfaces proceeds in several stages. First, in reducing the surface tension, detergents allow the aqueous bath to wet the surfaces thoroughly. In the case of textiles they enable the water to penetrate the fibers, and in dyeing processes this helps the dye to spread evenly. Next, detergent molecules form a layer of electrically charged ions at the interface between the water and the surface to be washed. When the soil is bound to the surface by an oily layer, the detergent breaks this up into individual droplets, forming a colloid that is dispersed in the foam. The soil then passes into the washing water, usually facilitated by mechanical action and high temperature. Each of these stages has been scientifically investigated, with the results used to invent new detergents or to modify old ones to improve efficiency.

The largest quantities of synthetic detergents consumed in the household are spray-dried powders used in laundering fabrics. They are expected to produce a variety of effects in addition to their detergent action. Compounds that remove hardness from water are added to enhance the detergent power together with anciliary compounds such as bleaches, perfumes, and fabric brighteners. Protein stains such as egg, milk, or blood are difficult to remove by detergents alone and must be broken down by proteolytic enzymes to make them soluble or at least permeable to water. Hence in "biological" washing powders, the detergent is fortified with such enzymes to make it more effective. Proteolytic enzymes were first tried in washing powders in Germany in the 1920s with only moderate success, but improved strains of faster-acting enzymes were widely added to powders in Europe by the late 1960s, and later in the U.S. lipases and amylases are now added to break down fats and starches. Unfortunately these enzymes have a toxic effect on some people.

Useful as they are, the tetrapropylenebenzene sulfonates have caused serious environmental problems. The foam they produce remains on the surface of wastewater in drains and sewers. It retards biological degradation in sewage treatment plants and is difficult to remove in water regeneration. In rivers they reduce the solubility of oxygen in the water, killing plant life and fish. After intensive research in the 1960s the problems were reduced when simpler alkyl groups more easily broken down by bacteria replaced tetraproplyene.

Hard surface cleaners include products containing detergents for domestic, institutional, and industrial use. Some have added germicides, fungicides, or insecticides. They differ in strength according to their purpose and the way they are designed to operate, but all rely on the use of a mechanical scrubbing action, either manual or by machines. Domestic hard surface cleaners are used to remove a variety of soils from dried and baked-on food deposits (carbohydrates, fatty acids, proteins, etc.) to petroleum greases and oils. They are designed to act efficiently on many different surfaces and are used in homes and offices to clean mirrors and windows, sinks and baths, bathroom and kitchen fixtures and appliances, table and counter tops, floor coverings, and patios. Cleaners for these purposes should act without damage to the surface and without leaving stains. Most are liquid and contain mild abrasives and solvents as well as detergents; those intended for use where food is prepared must be nontoxic.

Although detergents are usually thought of as water-soluble, there are also oil-soluble substances that hold foreign matter in suspension. Called dispersing agents (dispersants) rather than detergents, these substances employ the same physical principles but are strongly lyophilic and are used widely in engineering operations. They are added to lubricating oils in automotive engines to prevent the formation of hard deposits on cylinder walls and to inhibit corrosion. They are also added to petrol to prevent the build-up of gummy residues in the carburetor and to dry-cleaning fluids to facilitate the removal of oily soils from fabrics.

Special detergents also have many uses in industry and engineering. In the oil industry for example, surfactants are sometimes used to promote oil flow in porous rocks, or to flush out oil left behind by a water flood. In this case a band of detergent is put down before the water to create a low surface tension and thus allow the oil-bearing rock to be scrubbed clean. The water is often made viscous by adding a polymer to prevent it breaking through the surfactant layer. Detergents are also widely used in industrial flotation processes to separate lighter particles from a mixture with heavier materials.

See also **Laundry Machines and Chemicals**

N. G. COLEY

Further Reading

Davidsohn, A.S. and Midwidsky, B. *Synthetic Detergents,* 7th edn. Longman, Harlow, 1987.

Kirk-Othmer. *Encyclopaedia of Chemical Technology,* 4th edn. Wiley, 1997, vol. 23, pp. 478–541.

Lange, K. R., Ed. *Detergents and Cleaners.* Hanser, New York, 1994.

Schramm, L. L. *Surfactants: Fundamentals and Applications in the Petroleum Industry.* Cambridge University Press, Cambridge, 2000.

Spitz, L., Ed. *Soaps and Detergents.* American Oil Chemists Society, Champaign, 1996.

Tadros, T.F. *Surfactants.* Academic Press, London, 1984.

Schwuger, M.J. *Detergents in the Environment.* Vol. 65 of the Surfactant Science Series. Marcel Dekker, New York, 1997.

Lai, K.-Y. *Liquid Detergents.* Vol. 67 of the Surfactant Science Series. Marcel Dekker, New York, 1997.

Showell, M.S. *Powdered Detergents.* Vol. 71 of the Surfactant Science Series. Marcel Dekker, New York, 1997.

Diabetes Mellitus

The disorder called diabetes mellitus, translated from the Greek, means "flow of urine like honey." In the second century AD Aretaeus the Cappadocian (or more likely one of his assistants) made the diagnosis by dipping a finger in urine and tasting the sweetness. By 1889 Josef von Mering had shown that diabetes could be produced in animals by removing the pancreas. It was known that the pancreas secreted a substance into the duodenum that was essential for absorption of food. Additionally, it was postulated that the pancreas had a function related to glucose control. By 1916 the substance that the pancreas released into the blood stream and that was central to control of blood glucose levels was named insulin although it had never been isolated. Frederick Banting and Charles Best were the first to produce a pancreatic extract that lowered blood sugar in 1922, and they received a well-deserved Nobel Prize for their discovery. The first patient to be treated was a 14-year-old girl in Toronto. Prior to this, what we now call type I diabetes had a mortality rate of 100 percent within two years. Few would dispute that the discovery of insulin was one of the outstanding achievements of the twentieth century.

Insulin is a pancreatic hormone that facilitates the uptake and metabolism of glucose from the blood into muscle where it is essential for energy requirements. Without insulin, blood sugar levels rise, and fat has to be used to supply the body's energy requirements. The end products of fat metabolism are highly acidic, and when the blood becomes acidic, coma and death ensue.

The original pancreatic extracts were crude and short acting, and they contained many impurities. Insulin was crystallized in 1926, and in the 1930s it was combined with zinc, protamine, or globin in various proportions to achieve different lengths of action. Protein purification procedures were developed in the late 1960s to produce monocomponent insulins. Between 1951 and 1971 Frederick Sanger and Dorothy C. Hodgkin worked on determining the structure of insulin, and their work laid the foundation for the next major development. A report in 1978 of an anticipated shortage of animal insulin led to the development of synthesized human insulin. The two ways of doing this were either by enzymatic conversion of porcine insulin or by recombinant DNA technology. Human genes are synthesized for the insulin molecules A or B chains then joined by recombinant methods with sequences coding for β-galactosidase. Plasmids containing the modified genes are inserted into a special strain of the bacteria *E. Coli*, which after fermentation produce the insulin that is cleaved from the bacterial proteins and purified. Both these techniques produced a peptide indistinguishable from pancreatic human insulin. By the late 1980s, completely synthetic insulins were available that were characterized by manipulation of the insulin molecule to modify its pharmacokinetics (speed of onset and duration of action). These products were termed insulin analogs (1992) and could equally well be described as "designer" insulins.

A series of studies in the 1980s showed that tight control reduced the microvascular complications of diabetes. Thus, physicians were at last able to achieve good 24-hour control of blood sugar levels by using different types of insulin and multiple injection techniques to mimic the normal pattern of insulin secretion according to the patient's eating, sleeping, and exercise habits.

The treatment of type II diabetes (which afflicts older, overweight people) was difficult because of insulin resistance, a key feature of this type of diabetes, which made injected insulin less satisfactory. In 1942, during a typhoid epidemic in occupied Europe, researchers in a trial for a new compound of the sulfonamides (antimicrobial agents introduced in the 1930s) noted severe hypoglycemia (low blood sugar) in the subjects. The sulfonylurea group of drugs, which stimulate the pancreas to produce more insulin, were the result of further research; examples include tolbutamide, glibenclamide, and gliclazide. The diguanides were rather more physiological; their mode of action was to increase glucose uptake, but they were not free of side effects. The thiazolidinediones, introduced in 1996 (rosiglitazone and pioglitazone), also reduce insulin resistance and increase the uptake of insulin from the blood into the cells. As with the diguanides, this is a more physiological approach. By 2000 they had not been in use long enough to be assessed.

The first successful pancreas transplant was performed in 1967 by Dr. Richard C. Lillehei at the University of Minnesota, and is now usually performed at the same time as a kidney transplant for Type I diabetes patients with severe complications. Islet cell transplants are still experimental, although pig islet cells have been transplanted into immunodeficient mice and dogs, with rejection prevented by immune-suppressing drugs.

At the end of the twentieth century, diabetes was the prime example of self-management of a serious disease. Patients inject themselves with insulin, using disposable "dial-a-dose" syringes. They determine the dose by measuring their own blood glucose levels. This can be done by using a spring-loaded lancet to prick the finger and produce a drop of blood that is applied to a strip impregnated with a reagent area containing glucose oxidase. The resulting color change is measured by a small reflectance meter. This gives a blood sugar reading in millimoles per liter within 15 seconds. When these systems were introduced in 1978, there was some skepticism as to whether patients could cope, but they have proved highly adaptable with even five-year-olds rapidly becoming competent in measuring their sugar levels and taking the appropriate dose of insulin. Later developments included digital devices that were even easier to use.

The decrease in the complications of diabetes during the century has been due to several therapeutic advances. Renal damage and hypertension (high blood pressure) have been reduced by the use of the angiotensin-converting enzyme (ACE) inhibitors, which are effective in protecting the kidney and controlling blood pressure and thereby further reducing the risk of heart disease. Until the 1980s renal dialysis was not considered an appropriate treatment for the kidney failure of diabetics, but, fortunately, that rigid and inflexible attitude no longer applies. Erectile failure is a problem for diabetics, but even the U.K. government accepted that diabetics were allowed to have sildenafil (Viagra), which enhances penile blood flow, on National Health Service prescriptions. Laser photocoagulation of diabetic retinopathy was introduced in 1984. As a result of retinal screening and laser treatment, blindness is now rare in diabetic clinics. Finally, the pioneering work of D. Hockaday and K.G. Alberti in 1972 pointed to a physiological approach to diabetic coma using frequent very low doses of insulin, and mortality was reduced drastically.

See also **Diagnostic Screening; Hormonal Therapies**

JOHN J. HAMBLIN

Further Reading

Bliss, M. *The Discovery of Insulin*. Paul Harris, Edinburgh, 1983.

Day, J. *Living with Diabetes Treated with Insulin*. Wiley, New York, 1998. Written for patients. Excellent insight into modern management.

Hakim N., Ed. *Pancreas and Islet Cell Transplantation.* Oxford University Press, Oxford, 2002.

Jovanoviv-Peterson, L. and Peterson, N. *A Touch of Diabetes.* Wiley, London, 1998. Written for patients. Excellent introduction for nonmedical readers.

Pickup, J. & Williams, G., Eds. *Textbook of Diabetes.* Blackwell, London, 1991.

Wass, J. and Shalet, S., Eds. *Oxford Textbook of Endocrinology and Diabetes.* Oxford University Press, Oxford, 2002.

Diagnostic Screening

The early twentieth century physician used sight, smell, palpation, auscultation, and taste to aid in diagnosis of disease. Microscopy, chemistry, and x-rays were in their infancy, thus it was experience and the senses, not technology, that dominated the physical examination and the practice of medicine. There were few commercial laboratories until after World War I. Doctors returning from the battlefield benefited from their experience with wartime government diagnostic labs and promoted the value of having such facilities.

The technologies of diagnostic screening (imaging and biochemical technologies) developed concomitantly with discoveries in human physiology and pathology. The more subcellular information about a particular disease, the more specific a screening tool could be developed. For example, the complement fixation phenomenon was described in 1901 by Jules Bordet and Octave Gengou, and was later the basis of a number of diagnostic serological tests. Such tests determine whether or not a sample of serum has antibodies to a particular bacteria or virus. In 1905 Erich Hoffmann and Fritz Schaudinn discovered that the bacterium *Spirochaeta pallida* (now known as *Treponema pallidum*) was the causative agent in syphilis. Synthesizing this information, August Paul von Wassermann, Albert Neisser, and Carl Bruck used the complement fixation test to find *Treponema pallidum* in human blood before microscopy had allowed for visualization of extremely small organisms.

Although mammography was first used in 1927 to visualize palpable breast tumors, it was not until the late 1950s that Jacob Gershon-Cohen and Robert Egan demonstrated that cancers that did not manifest visibly could be detected with mammography. This led to the hypothesis that early detection of breast cancer by screening in asymptomatic women would or could prevent the cancer from spreading or increasing in size. In 1967 the machine consisted of an x-ray spectrum and tube mounted on a three-legged stand and a French company, CGR, produced the Senographe. By 1980, it was possible to reduce exposure time, and improve resolution and accuracy because of the chemical makeup of new films. A motorized compression device was added in that decade as well as an add-on component to assist in a breast biopsy should the procedure be necessary. In 1990, rhodium was added to the x-ray tube for better penetration of tissue by the ray. The last decade of the twentieth century heralded digitization in technology, and digital technology was incorporated in mammography to image spot views of suspicious breast tissue. When screening identified an area that required investigation, this additional view was taken. The digital cassette allowed for processing time to be reduced. Although most of the technical challenges of mammography have been solved, the procedure is best utilized in patients over the age of 50 because their breast tissue is generally less glandular and dense than that of younger women. The technological challenge for the next century will be to discriminate between dense normal and opaque pathologic tissue.

The first pregnancy tests in the 1920s screened for the presence of chorionic hormone in a woman's urine. The urine was injected into a rabbit, and after three days the animal would be killed and its ovaries examined for changes caused by human chorionic gonadotrophin (HCG). If changes were present, the woman was pregnant. As information about inherited diseases was developed, genetic screening tests were performed prior to conceiving. Sickle cell anemia testing, and a test for Tay–Sachs disease were available by 1971. In 1985, Unipath introduced a home pregnancy test that took 30 minutes and used litmus paper embedded with antibodies to test urine for the presence of chorionic hormone. Screening for genetic defects such as Huntington's disease (Huntington's chorea) has been the work of Elena Cattaneo in Milan and Erick Wanker in Berlin whose work screens for chemicals that prevent agglutination or clumping of neurons, a common characteristic of Huntington's chorea.

Screening for cardiovascular disease is based on the knowledge of certain risk factors associated with hypertension and heart disease. If these risks can be identified, patients can modify their lifestyles to improve their chances of a healthy life. Screening tests consist of the identification of chemical predictors in the blood such as cholesterol and by physical recordings such as electrocardiograms (ECG) and imaging technologies including echocardiograms and Doppler studies. The forerunner of the ECG was developed in 1901 by Willem Einthoven, a Dutch physiologist who created a rudimentary galvanometer using magnetic poles and a silver-coated string. The acceptance and further development of the technology were encouraged by the English physician, Sir Thomas Lewis (c. 1911), who stated that an examination of the heart was incomplete if this new method was neglected.

In 1928, S.A. Ernstine and A.C. Levine used vacuum tubes instead of a string galvanometer for amplification of the electrocardiogram. Until this time, the ECG machine was a table. One of the first portable models was made by Hewlett Packard. It weighed 27 kilograms and was powered by a 6-volt automobile battery. The next improvement was the incorporation of chest leads and then a central terminal (Wilson's Central Terminal) which could be placed anywhere on the body and connected to the other leads. In 1938, placement of the leads, wiring, and positions was defined in the U.S. and U.K. so that readings could be standardized. A few years later, Emanuel Goldberger increased the voltage on the central terminal and added three more leads, bringing the total to twelve. An expensive technology that did not use skin electrodes was developed in 1963 by Gerhard M Baule and Richard McFee, but because of its prohibitive cost, it was not adopted. By the early 1990s three additional electrocardiogram leads were used by researchers in Detroit, Michigan to increase sensitivity in the detection of myocardial infarction.

Combined technologies of electrocardiography, and Doppler ultrasound have allowed for the simultaneous imaging and recording of heart waves and anatomic information. An echocardiogram displays a cross-section of the heart and its major vessels as it beats. The Doppler records the direction and velocity of blood flow through the vessels. Most recently, researchers at Johns Hopkins University in Baltimore, have developed a technology for the nonspecialist that uses computer software from the Microsoft Access relational database to screen for heart and lung disease. The physician feeds five areas of information (stethoscopic, physical exam, electrocardiogram tracings, demographic, and health information) into the program and compares it to known diagnoses stored in the database.

One of the most common uses of diagnostic screening is for diabetes. Before technologies for diabetes screening, physicians tasted urine to detect sugar. When a patient with symptoms of frequent urination, weight loss, and persistent thirst presented, the test was to add Benedict's solution to urine to detect the presence of glucose. It required two test tubes, a medicine dropper, a heat source, and the reagent. In the 1950s, Ames developed Clinitest, a product that only required one test tube and a tablet. By 1960, a paper strip (Labstix) had been developed which could be dipped in urine to test for the amount of glucose. Ten years later, blood glucose monitors were available for home use. Called a reflectance meter, it weighed 1 kilogram and was 17 cm high and 10 cm wide. In the 1980s the Glucocheck device added memory to its product. At the beginning of the twenty-first century, there are 30 different types of diabetes meters, most developed in the 1990s with computer technology.

While there are no screens available for many types of cancer, for example lung cancer, diagnostic screening for cervical cancer was developed in the 1960s and where screening has been introduced in developed countries, mortality has been significantly reduced. The Pap smear (named for George Papanicolaou) is a simple test by microscope to examine cells for changes that signify disease. Screening for prostrate cancer remains poor, with cancer most often found by digital rectal examination of men who have no symptoms of the disease. A prostrate-specific antigen (PSA) test can detect a protein released into serum as prostate cancer grows, but only one third of men with prostrate cancer are positive.

See also **Diabetes Mellitus; Electrocardiogram; Genetic Screening and Testing; Histology; Ultrasonography; X-rays in Diagnostic Medicine**

LANA THOMPSON

Further Reading

Fye W.B. A history of the origin, evolution, and impact of electrocardiography. *Am. J. Cardiol.* 73, 937–949, 1994.
Reiser, S.J. *Medicine and the Reign of Technology.* Cambridge University Press, Cambridge, 1978.
Cartwright, L. *Screening the Body: Tracing Medicine's Visual Culture.* University of Minnesota Press, Minneapolis, 1995.
Russell, L. *Educated Guesses: Making Policy about Medical Screening Tests.* University of California Press, Berkeley, 1994.

Dialysis

Dialysis, or hemodialysis, is a process of separating substances from the blood. Three quarters of a million people worldwide suffering irreversible kidney failure are maintained today using what is commonly called the "artificial kidney," or hemodialyzer, the first machine to substitute for a failing organ. Around 1960 and by coincidence two separate methods of treating irreversible renal failure developed almost simultaneously as practical treatments: dialysis and renal transplantation (see Organ transplantation).

The idea that death in uremia following kidney failure results from retention of solutes normally excreted by the kidneys emerged during the nineteenth century. Diffusion in gases or liquids was worked out in parallel, principally by Thomas Graham, with mathematical development by J. H. vant'Hoff. These two ideas underlie the possibility of treating kidney failure by dialysis of potentially toxic solutes from the blood, and hence the tissues and body water.

The first attempt to carry out *in vivo*, extracorporeal dialysis of blood was made in 1913 by John Jacob Abel (1857–1938) in Baltimore, Maryland, at the Johns Hopkins Hospital, together with Canadian physician Lawrence Rowntree and English biochemist Benjamin Turner. Extracorporeal circuits had been developed in the 1880s for perfusion of whole organs *in situ* or *ex vivo*. Dialysis needed a semipermeable membrane that would retain protein and cells but allow solutes to diffuse from the blood. A number of substances had been tried in the laboratory, but the best seemed to be collodion, developed for photography in the 1850s as sheets of cellulose dinitrate. Blood was difficult to dialyze since it clotted on contact with foreign surfaces, but about the turn of the century hirudin, the anticoagulant in leech saliva, became available in crude form, and Abel and colleagues used collodion and hirudin. Their design consisted of a number of collodion tubes immersed in a cylindrical glass bath of dialysis fluid; blood drawn from an artery was pumped through rubber tubes into the apparatus and back into a vein. They dialyzed dogs and showed urea could be removed, but they were more interested in separating physiologically active substances from blood. At a demonstration of their dialyzer in London in 1913, a staff writer at the London *Times* first used the term "artificial kidney." The team broke up after 1915 and Abel did no more work on dialysis.

The first specific attempts to treat uremia by dialysis were by Georg Haas (1886–1971) of Giessen, Germany, initially in ignorance of Abel's work. He built a machine very like Abel's in 1923 and treated the first human patients by dialyzing their blood *ex vivo* and reinfusing it intermittently. He also used hirudin as an anticoagulant, but this substance was still crude and often toxic. His patients had irreversible chronic kidney failure

and all died. Haas abandoned the work until 1927 when the new anticoagulant heparin was discovered by Henry Howell, again at Johns Hopkins, together with medical student Jay Maclean and later, retired pediatrician Emmett Holt. But Haas' patients still died, and in the face of opposition from the medical establishment, he gave up for good in 1928.

In 1923 Heinrich Necheles (1897–1979) of Hamburg was dialyzing dogs deliberately made uremic, but with a machine incorporating a new design: a flat sandwich with multiple layers of another dialysis membrane, the peritoneum from calves. However, despite his interest in uremia, he used this technique only to prepare physiological extracts from blood in dogs.

Despite the use of heparin, dialysis stalled because collodion was too complicated to produce, too fragile, and too difficult to use. A new membrane, cellulose acetate, was introduced in 1908, marketed as "cellophane" in France in 1910. By 1930 the material was formed into tubes in Chicago by the Visking Company to make "skinless" sausages and was rapidly shown to be useful in the laboratory for dialysis. William Thalhimer (1884–1961) a New York pathologist and hematologist realized in 1937 from his experience of anticoagulating and storing blood that heparin together with cellophane tubing made a practical dialyzer a possibility. He built such a machine and dialyzed dogs, but as he did no clinical work himself he encouraged others to try it in humans. In the early 1940s four individuals tried and succeeded, each without any contact with others.

In 1944 Jonathon Rhoads (1907–2002) of Philadelphia built and used a dialyzer with cellophane tubing wound in spiral on a frame within a bath of dialysis fluid and used it on a single patient. The dialysis worked, but the patient died, and he did no more. Toronto surgeon Gordon Murray (1894–1976), an expert on heparin and a colleague of Thalhimer, worked on dogs from 1940 to 1946 to perfect a similar design. In 1946 his first patient was treated three times and survived her acute kidney failure. He did only a few dialyses over the next five years, also designing and using a flat-plate sandwich type of dialyzer. Nils Alwall (1906–1986) in Lund, Sweden, worked on rabbits for four years or more before achieving success in humans in 1946. His design was similar to Murray's but was notable in that it had an outer casing permitting the pressure around the spiral to be controlled so that ultrafiltration of excess fluid could be limited. Alwall continued working in dialysis until retirement and trained many people to use his first machines, mainly in Europe but also in Australia, Israel, and even the U.S.

The best known of these researchers is Willem Kolff (1911–). In September 1945 he performed the first successful dialysis with recovery of the patient in Kampen, Netherlands. Beginning in 1943 he dialyzed 16 other patients, all of whom died. Many of them, however, had irreversible kidney failure, and some improved briefly. Although Kolff never did any animal work, for the first time he established the size parameters necessary for human use. As a result, his machine had a much larger surface area. He employed a spiral of tubing wound around a large horizontal drum, laying halfway in an open bath of dialysate, which rotated with a coupling from a Ford car engine to allow blood to flow into the cellophane tubing. Kolff believed passionately in dialysis, and during the late 1940s he built and gave away a number of machines, sent plans everywhere so people could build their own, and toured the world to advertise the new technique.

Initially facing worldwide skepticism, he went to Cleveland, Ohio in 1950. His machine was improved in Boston by physician John Merrill, surgeon Carl Walter, and engineer William Olson. Despite its many obvious disadvantages (e.g., large priming volume and uncontrolled ultrafiltration), it continued in use until the early 1960s. One major factor in acceptance of dialysis treatment during the 1950s was that, unlike the milder forms of acute renal failure where conservative management was possible, dialysis made a major difference to survival in injured soldiers with severe but temporary renal failure during the Korean War (1950–1952). New dialyzers that were much easier to use were introduced. These included the twin-coil kidney designed and built by Kolff and Bruno Watshinger of Vienna in only a couple of months in Cleveland in 1956, and the flat-plate sandwich design invented first by the chemists Leonard Skeggs (1918–) (who later invented automated clinical chemistry) and Jack Leonards, and, independently, Arthur McNeil in Buffalo, New York.

By the end of the 1950s physicians everywhere, having inadvertently started dialysis in patients with chronic irreversible disease, found a slow, miserable "second death" inevitable as the blood vessels vital for access became unusable. A solution came from the polymer industry in the form of plastic polyvinylchloride (PVC) electrical insulation tubing, which was already used for connecting patients to dialysis. Although it was first discovered in 1937, PTFE or teflon, which was not wettable and did not stimulate thrombosis in

blood, became available in the 1950s. Seattle, surgeon Warren Wintershide pointed this out to the renal physician there, Belding Scribner (1925–). He then got an engineering colleague Wayne Quinton to bend the tubes using heat and make a shunt joining an artery and vein externally so that it could be closed off from the machine between dialyses but allow blood to flow through it continuously: Suddenly, long term dialysis for irreversible disease was possible. One or two patients who started on dialysis only a few years later in 1966 remained alive at the turn of the twenty-first century. The shunts were improved by the use of another new material, silicone rubber. Then in 1965, from experience with venipuncture while working in a blood bank, James Cimino (1927–) and his colleagues in New York introduced access to veins enlarged by a surgically created arteriovenous fistula. By the end of the decade, these had become almost universal, and external shunts were already obsolete.

Initially, the reusable and then (from 1968) disposable flat-plate and coil dialyzers already in use for acute renal failure were also used for long-term dialysis. However in the early 1960s engineers at the Dow company in Michigan learned to make hollow fibers less than 100 micrometers in diameter from numerous different polymers for proposed uses in water purification, and in medicine as membrane oxygenators. Dick Stewart (1917–), a physician and chemist working with Dow, suggested their use for hemodialysis, and the first capillary hollow-fiber dialyzers were used in humans in 1967. By 1980 they were the predominant type of dialyzer in use, and from 1990 onward, used almost universally. Their great advantage, apart from efficiency, is their small size: a 30 by 15 centimeter hollow-fiber dialyzer has the same capacity as Kolff's original rotating drum dialyzer, which was more than one meter long and half a meter in diameter.

Commercial cellulose acetate dialyzing membranes continued in use until the 1990s when replaced by more permeable and more biocompatible synthetic compounds such as polysulfone. In addition, these more permeable membranes (of which the polyacrylonitrile AN-69 of 1969 is the prototype) are increasingly used to exploit convective rather than diffusive removal of solutes, as pioneered by Lee Henderson (1931–) in the U.S. and Eduard Quellhorst in Germany in the 1970s. This technology has been particularly effective in treatment of reversible acute renal failure in very ill patients, using continuous hemofiltration throughout the day.

The other major change in dialysis machinery was the introduction of continuous blending of dialysate from concentrated salt solutions and reasonably pure (but not sterile) water, pioneered by Arthur Babb for Scribner in Seattle in the early 1960s, and now universal. In addition, various online safety measures, starting in the 1950s with a bubble trap to prevent air entering the circulation and the measurement of dialysate temperature and ionic strength, have continued to be introduced through the years.

See also **Intensive Care and Life Support**

J. STEWART CAMERON

Further Reading

Cameron, J.S. *A History of the Treatment of Renal Failure by Dialysis*. Oxford University Press, Oxford, 2002.
McBride, P.T. *The Genesis of the Artificial Kidney,* 2nd edn. Baxter Laboratories, Chicago, 1987.

Dirigibles

From the time of the first balloon ascents in France in 1783, the problem of steering airships became obvious because of dependence on the wind for direction and on gas quantity and pressure inside the envelope for altitude control. It would take over a century of experimentation with gas inflation, propulsion, and altitude and directional control for airships to become practical.

The wildest solutions were proposed to solve the problem of dirigibility, but serious, scientific-based proposals also appeared. One of the first was that of Jean-Baptiste Meusnier's 1783 ellipsoid-shaped balloon, which was never built. His design included the concept of a ballonet, a gas container within the envelope (with the space in between filled with air) to keep the hydrogen and of a mechanical valve atop the balloon intended to release hydrogen on command, thus allowing better control of ascent and descent phases. While both concepts would eventually be incorporated into both balloons and dirigibles, the matter of propulsion remained difficult to solve.

The first steerable airship was flown in September 1852 by Henri Giffard but, like machines other pioneers flew over the next half century, it was handicapped by the heavy weight of the engine and accompanying fuel, to the point where there was scarcely enough space on board for a crew. With the invention of the first gasoline engine in 1896, the situation began to change, and several projects took shape by 1900, thereby defining three main types of dirigible: the nonrigid, the semirigid, and the rigid.

The nonrigid machines were the first ones designed in the nineteenth century and the most commonly in use. The envelope's shape was maintained through internal gas pressure (with the hydrogen itself placed within ballonets inside the hull). The crew section and engine(s) were attached by ropes to the envelope. Carrying capacity was limited, however, and there was a risk of hull buckling if pressure dropped, a fairly common incident. Cheap to build, such models were the most popular among inventors and armed forces eager to develop new observation platforms to supplement ballooning units.

The semirigid system attempted to solve the buckling problem by attaching the passenger and engine nacelles to a bottom keel on top of which the hull was stuck and inflated. Capable of carrying heavier loads than a nonrigid type, a semirigid airship could still be deflated and carried about the field. This system formed the basis for the "blimps" used for advertising through the twentieth century.

The rigid airship solution consisted of either a solid hull (as in the case of a couple of aluminum-clad airships) or an internal structure of girders made out of wood or aluminum. Due to the substantial size required to balance the weight of the airship with the volume of hydrogen needed, the rigid airship could carry the heaviest loads at the greatest speeds over the longest distances without buckling. The main proponent of this formula was a German nobleman, Count Ferdinand von Zeppelin (1838–1917), who argued for an armament program that would include rigid airships beginning in the late nineteenth century. In 1900, he flew his first machine, LZ-1, and through a process of trial and error, he and his engineers improved the design. By July 1908, the fourth zeppelin (as rigid airships soon became known) had successfully accomplished a 12-hour flight with 11 people on board, at a time when airplanes carried at best two people for half an hour. This and the popularity of Count Zeppelin in Germany helps explain why the rigid airship seemed assured a solid place in the development of aviation.

When World War I broke out, several nations built dirigibles for maritime patrol and artillery assistance, but rigid airships soon became the first strategic bombers, as they attacked Paris, then left the French capital to artillery and used planes to focus on London. Though failing in their intended use as destroyers of industrial infrastructure, these machines had a tremendous psychological impact on the civilian population, and in the long run convinced advocates of strategic bombing of the advantages of some form of terror warfare. Yet the rigid airship (whether zeppelin or Schütte–Lanz, the latter made of wood) had also proven far too weak in the face of weather, limited speed, and the ever-improving performance of the airplane.

Although Germany was initially banned from building even transport airships after the war, an exception was made for the LZ-126, a model delivered to the U.S. Navy at Lakehurst, New Jersey, in 1924. In the meantime, The Royal Navy's R-34, a model extrapolated from the zeppelin formula, successfully flew across the Atlantic in 1919. These events forecast the advent of the dirigible as a long-distance transport machine. Its golden age came with the first flight of the LZ-127 *Graf Zeppelin* in 1928 and ended with the fiery crash of the LZ-129 *Hindenburg* in 1937. During that time, these airships contributed to the establishment of links across the North and South Atlantic and made possible comfortable passenger transportation at a time when long-range airplanes were useful for mail, but ill-suited for an elite clientele. With the outbreak of World War II, advocates of the airship, who had relied on a combination of technical arguments and public support, had to give in, whether in Germany or in other countries.

Other nations, too, sought to take advantage of the dirigible. Great Britain planned the opening of a link to India, but the airship chosen, R-101, was rushed into service and crashed on her inaugural colonial flight in 1930. The U.S., too, faced disaster, as it lost three of its airships in storms in the 1920s and 1930s. The U.S. Navy turned instead to using blimps for submarine spotting and convoy escort and kept them in service until the early 1960s.

At the beginning of the twenty-first century, the airship experienced a rebirth of sorts. After being limited to advertising endeavors for many years, the discovery of new construction materials and the development of new missions (such as permanent coastal surveillance) have prompted the development of new projects. The Zeppelin company, which became an industrial concern after World War II, sponsored the development of a small passenger machine for pleasure cruises, while the Cargolifter corporation began to design of a new kind of sky crane intended to carry very heavy loads. While the success of such endeavors is hard to gauge, its popularity in media reports reflects the public fascination with great technology projects, as dirigibles once were.

GUILLAUME DE SYON

Further Reading

Dick, H.G., and Robinson, D. *The Golden Age of the Great Passenger Airships, Graf Zeppelin and Hindenburg.* Smithsonian, Washington D.C., 1985.

Higham, R. *The British Rigid Airship 1908–1931: A Study in Weapons Policy.* G.T. Foulis, London, 1961.

Khoury, G.A. and Gillett, J.D., Eds. *Airship Technology.* Cambridge University Press, Cambridge, 1999.

Meyer, H.C. *Airship Men, Businessmen, and Politics, 1890–1940.* Smithsonian, Washington D.C., 1991.

Robinson, D.H. *The Zeppelin in Combat: A History of the German Naval Airship Division, 1912–1918.* G.T. Foulis, London, 1971.

Robinson, D.H. *Giants in the Sky: A History of the Rigid Airship.* University of Washington Press, Seattle, 1973.

Smith, R.K. *The Airships Akron and Macon.* U.S. Naval Institute Press, Annapolis, 1965.

de Syon, G. *Zeppelin! Germany and the Airship, 1900–1939.* Johns Hopkins University Press, Baltimore, 2002.

Topping, D. and Brothers, E., Ed. *When Giants Roamed the Sky: Karl Arnstein and the Rise of Airships from Zeppelin to Goodyear.* University of Akron Press, Akron, 2000.

Dishwashers

A dishwasher is an appliance that automatically cleans, rinses and dries dishes and utensils. As is the case for many other major appliances, the basic principles and elementary technologies for the dishwasher existed in the nineteenth century. At the close of the twentieth century, household dishwashers were still considered a luxury item, owned by a minority of the population even in countries with full electrification, indoor plumbing, and near universal ownership of refrigerators, washing machines, and vacuum cleaners. Even the most sophisticated models at the end of the twentieth century could not remove accretions of baked-on food, thus ensuring the continuity of some form of hand washing. It remains common even in developed countries to wash, rinse, and dry dishes by hand.

The first dishwasher patents were filed by Americans Joel Houghton in 1850 and L.A. Alexander in 1865 for machines intended for commercial use. Houghton's device was a wooden tub that used a hand-turned wheel to splash water over the dishes, ultimately with little cleaning effect. Alexander, improving upon earlier collaborative efforts, patented a tub mounted on feet at the bottom of which was a set of blades that would be rotated by a hand crank located outside the tub. The blades would splash water against and between dishes set tangentially above them in stationary wire racks. Both machines still required a considerable amount of human labor. They needed to be filled with and drained of water and the crank turned by hand, arguably more work than the task required without a machine.

In the 1880s, Josephine Garvis Cochran of Shelbyville, Illinois designed a hand-operated mechanical dishwasher for use in her own domestic environment. The design consisted of a wooden tub containing a wire basket to hold dishes. The dishes would be rotated during the cleaning process. Hot water was sprayed into the tub and driven over dishes by plungers when a handle attached to the tub was turned. This model was later fitted with a motor. A motor was also added to the first commercially available hand-cranked model, displayed in New York in 1910 by the Walker Company. In Britain, similar models called the Polliwashup and the Blick were filled with water from a boiling kettle. The latter could be turned into a table by lowering its lid.

From the 1850s through the close of the twentieth century, dishwashers have followed the principle of propelling hot soapy water against dishes to clean them. The gradual introduction of plumbing and electricity into middle- and lower-class homes and the automation and mass production of the dishwasher by the 1950s made it available and potentially desirable to many consumers. The first freestanding dishwasher with permanent plumbing was introduced in 1920. Portable dishwashers were introduced in the 1930s that could be hooked up to the kitchen faucet or tap. Accessories such as the "aerated faucet-flo" (General Electric's term) provided an alternative faucet so that owners could access water without stopping and uncoupling a running dishwasher in midcycle.

By the middle of the twentieth century, fully electrified, large, industrial dishwashers were being marketed by companies like Toledo and Blakeslee. These galvanized iron or stainless steel machines could wash up to 11,400 dishes per hour, often using a conveyor belt. They had tank capacities of 160 liters for a wash, could pump up to 530 liters of water per minute, and weighed around 1000 kilograms.

During the latter decades of the twentieth century, manufacturers focused on refinement rather than invention. Dishwashers continued to spray hot water at high pressure onto dishes, most often by two rotating spray arms. Dishwasher interiors were made of either plastic or, in more expensive models, stainless steel. In the 1960s, exterior colors such as Avocado and Harvest Gold were introduced for dishwashers and other major kitchen appliances. By the 1960s manufacturers had introduced the built-in dishwasher, increasingly found in new homes. Built-ins came with a dwelling and stayed when it changed owners. They were situated near the sink so that the dishwasher drain hooked up to the sink drainpipe, and the entire machine connected to the household plumbing and electricity. With the advent of the built-in, front-loading dishwasher, the popularity of top-loading machines fell.

Late twentieth century affluent dishwasher owners could hide their dishwashers inside drawers paneled to match the kitchen cabinetry, with controls hidden inside. Electronic dishwashers offered timers that allowed washing to be delayed for up to nine hours and digital indicators to monitor the cycle's progress and to highlight problems. Noise reduction, introduced in the 1960s, reached its height in the 1990s and was achieved by wrapping the wash cabinet in layers of "insulation" consisting of natural and synthetic materials that would muffle the sound of the motor and water propellers.

With concern rising over the impact of dishwashers on the environment, manufacturers began to emphasize environmentally friendly features such as low water consumption, "economical" detergent dispensers, and the air-dry option. The trend toward air drying was particularly noteworthy as a reversal of priorities expressed in material culture and technology. Manufacturers in earlier decades sold heat-drying options by playing on housewives' fear of water spots caused by air drying. Fuzzy logic-based, or "smart" appliances emerged from Japan in the 1980s. A fuzzy logic chip in the dishwasher abetted the conservation of water and energy by estimating the level of turbidity (soil level) of the load and controlling the cycles accordingly.

See also **Detergents; Electric Motors, Small**

JENNIFER COUSINEAU

Further Reading

Cowan, R.S. *More Work for Mother: The Ironies of Household Technology from the Open Hearth to the Microwave.* Basic Books, New York, 1983.
Davidson, C. *A Woman's Work is Never Done: A History of Housework in the British Isles 1650–1950.* Chatto & Windus, London, 1982.
Forty, A. *Objects of Desire: Design and Society from Wedgwood to IBM.* Pantheon, New York, 1986.
Gideon, S. *Mechanization Takes Command: A Contribution to Anonymous History.* W.W. Norton, London, 1948.
Hardyment, C. *From Mangle to Microwave: The Mechanization of Household Work.* Polity Press, Oxford, 1988.

Lupton, E. and Miller, J.A. *The Bathroom, The Kitchen, and the Aesthetics of Waste: A Process of Elimination.* Princeton Architectural Press, New York, 1992.

Strasser, S. *Never Done: A History of American Housework.* Pantheon, New York, 1982.

Webb, P. and Suggitt, M. *Gadgets and Necessities: An Encyclopedia of Household Innovations.* ABC-CLIO, Santa Barbara, 2000.

Yarwood, D. *Five Hundred Years of Technology in the Home.* B.T. Batsford, London, 1983.

Useful Websites

Badami, V.V. and Chbat, N.W. *Home Appliances Get Smart*: http://www.hjtu.edu.cn/depart/xydzzxx/ec/spectrum/homeapp/app.html

Domestic Heating

Most domestic space heating systems used in the twentieth century in Britain, continental Europe and the U.S. were developments of technologies introduced much earlier. At the beginning of the twentieth century, many homes, especially in Britain, were still heated by solid fuel open fires set in fireplaces with chimneys. Closed solid fuel stoves of clay, brick or tile or metal stoves developed from the seventeenth century onward were also common methods of domestic space heating across continental northern Europe and North America, especially during the early parts of the twentieth century. Such coal- or wood-fueled stoves are typically four times more efficient than open fires. In addition to more efficient designs of metal stoves, new types of domestic room heater spread into use in the twentieth century as new fuels became available.

Gas heaters were invented in the 1850s; Pettit and Smith adapted Bunsen's gas burner for domestic heating in 1856. Electric heaters became practical after the development in the early twentieth century (patented in 1906) of a nickel–chrome heating element that did not oxidize when heated. This American invention led to Ferranti's parabolic reflector heaters in 1910 (in the U.K.) and Belling's electric heater in 1912, in which nichrome elements wound on fireclay strips would heat the strips to incandescence. Such units provided radiant heat, in which the heat travels in straight lines warming objects in its path. On the other hand, electric convector heaters, first introduced around 1910, provide warmth by convection through the movement of heated air. Fan-assisted electric convector heaters, first commercialized in the U.S. by Belling in 1936, only became popular after 1953 when a silent tangential fan heater was produced in Germany. The storage heater is another type of electric convector heater also invented at the beginning of the twentieth century. A storage heater comprises a heat-conductive casing containing ceramic heat-retaining blocks that are gradually warmed by electric elements, usually overnight. The stored heat is then released, sometimes with fan assistance, the next day. However storage heaters did not spread into use until after World War II. In Britain they were not much used until after 1962 when low-cost night electricity rates were introduced. Storage heaters provided a relative low-capital cost alternative to the central heating system.

The widespread use in Britain of individual room heaters into the late twentieth century was due to the much slower adoption of central heating than in most other developed countries. In 1970 only 30 percent of British homes had central heating, rising to 90 percent by the end of the century. In continental Europe and the U.S, where living in apartments and heating with closed stoves was more common than in Britain, domestic central heating spread much more rapidly.

Central heating is a system that heats a building via a fluid heated in central furnace or boiler and conveyed via pipes or ducts to all, or parts of, a building. There are many types of central heating system depending on the fuel used to heat the fluid—usually coal, oil, or gas (although wood fired and electric systems exist)—and the fluid itself, which may be steam, water, or air.

William Cook was the first to propose steam heating in 1745 in England. At first steam heating was mainly used to heat English mills and factories, but its development for domestic use took place mainly in America, notably by Stephen Gold who was granted a U.S. patent in 1854 for "improvement in warming houses by steam." Many different designs of steam heating system were in use in America by World War I. Steam, however, never really became popular for domestic central heating due to its relative complexity and problems such as noise and fear of explosions. Steam piped under the streets from a central boiler was used instead for "district heating" of groups of residential and commercial buildings in large cities, especially in the U.S. New York's famous steam district heating systems were introduced in the 1880s. Many district heating systems were operated as "combined heat and power" (CHP) systems in which high-pressure steam was used first to drive electricity generating turbines and then distributed at low pressure, or as condensed hot water, to heat buildings. Utilizing the heat that otherwise would have been wasted increases the overall fuel efficiency from about 30% in a conventional power station to over 70% for a CHP system. District heating and CHP systems declined in the U.S. after the 1950s, and by the early twenty-first century, were most common in Russia and Scandinavian countries, especially Denmark.

Given the drawbacks of steam for "wet" domestic central heating systems, by the early twentieth century the preferred fluid became water. This is heated in a boiler and conveyed via pipes to radiators in individual rooms (really these should be called "convectors" as they heat a space mainly by convection rather than radiation). Warm air central heating systems were also popular in the U.S., Canada, and Germany. In such systems air is heated via a furnace and distributed through sheet metal ducts to room vents. Initially hot water and air central heating relied on the density difference between the hot and cold fluid for circulation, so-called "gravity systems." Subsequently electric pumps and fans were introduced to assist circulation, thus allowing smaller pipe work and more compact designs. Whereas early systems could only be controlled by adjusting the rate of fuel burning, the introduction of central heating controls, starting with Honeywell's clock thermostat in 1905, allowed more efficient and convenient utilization of heat.

The fuel most commonly used by central heating furnaces and boilers also shifted from coal at the beginning of the twentieth century to oil and gas by the 1920s. By the 1930s most of the components of the modern domestic central heating system were in place, for example, the typical British thermostatically controlled, pumped radiator system heated by a gas boiler, which also provides hot water for bathing and so on (see Figure 3).

Following the oil price increases of the 1970s and growing concerns about the environmental impact of burning fossil fuels, attention has turned to the development and adoption of more fuel-efficient domestic heating systems that reduce the heat loss from buildings and utilize renewable energy. The first approach includes the installation of "condensing" central heating boilers in which the flue gases pass over a heat exchanger before exiting, thus increasing energy efficiency from approximately 65 percent to 90 percent. Another example of fuel-conserving heating technology is the heat pump, which extracts heat from a low temperature source, such as the air or the ground, and boosts it to a temperature high enough to heat a room. Most heat pumps work on the principle of the vapor compression cycle. The main components in such a heat pump are a compressor (usually driven by an electric motor), an expansion

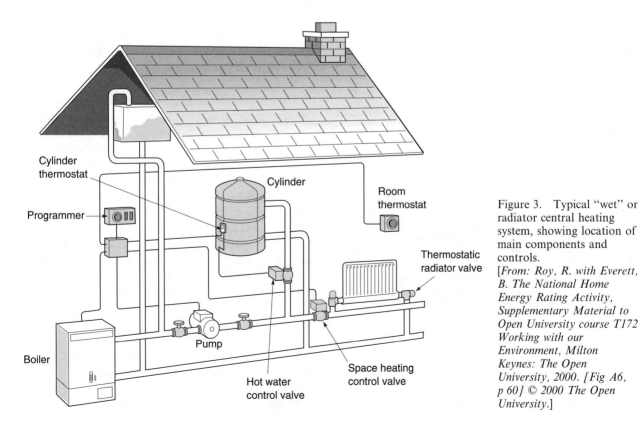

Figure 3. Typical "wet" or radiator central heating system, showing location of main components and controls.
[*From: Roy, R. with Everett, B. The National Home Energy Rating Activity, Supplementary Material to Open University course T172 Working with our Environment, Milton Keynes: The Open University, 2000. [Fig A6, p 60] © 2000 The Open University.*]

valve, and two heat exchangers, the evaporator and condenser, through which circulates a volatile working fluid. In the evaporator, heat is exchanged from the air or from pipes in the ground to the liquid working fluid, which evaporates. The compressor boosts the vapor to a higher pressure and temperature which then enters the condenser, where it condenses and gives off useful heat to the building. Finally, the working fluid is turned back into a liquid in the expansion valve and re-enters the evaporator. Heat pumps were first demonstrated for domestic use in 1927 by Haldane in England and widely adopted in parts of the U.S. in the 1950s.

Solar energy for domestic heating, after experiments in the 1930s and 1940s, was given new impetus in the 1970s by the oil crisis and again in the 1990s by environmental concerns. Solar heating technology includes both "passive" and "active" systems. In passive systems, buildings are designed and oriented to maximize useful heat gain from the sun. In active solar thermal systems, the sun is used to heat a fluid (such as water) or air that can be used immediately or stored for space heating. Some solar homes provide space heating and hot water all year round without the need for any fossil fuel energy.

See also **Air Conditioning; Buildings, Designs for Energy Conservation; Solar Power Generation**

ROBIN ROY

Further Reading

Donaldson, B. and Nagengast, B. *Heat and Cold: Mastering the Great Indoors: A Selective History of Heating, Ventilation, Air Conditioning and Refrigeration from the Ancients to the 1930s*. American Society of Heating, Refrigeration and Air Conditioning Engineers, Atlanta, 1994.

Du Vall, N. Heating and housing, in *Domestic Technology: A Chronology of Developments*. G.K. Hall, Boston, 1988.

Harris, E. *Keeping Warm (The "Arts and Living" Series, Victoria and Albert Museum, London*, HMSO, London, 1982. Fully illustrated history of domestic heating equipment focusing more on visual and functional design than on technology.

Wright, L. *Home Fires Burning. The History of Domestic Heating and Cooking*. Routledge and Kegan Paul, London, 1964. A comprehensive, mainly nontechnical guide to the history of domestic heating and cooking from prehistory to around 1960.

Useful Websites

Nagengast, B. *An Early History Of Comfort Heating*. Other articles on the development of domestic heating systems may be found on this site: http://www.achrnews.com.

Dreadnoughts, *see* **Battleships**

Dyes

The decade commencing in 1900 marked the end of half a century of remarkable inventiveness in synthetic dyestuffs that had started with William Perkin's 1856 discovery of the aniline dye known as mauve. The products, derived mainly from coal tar hydrocarbons, included azo dyes, those containing the atomic grouping $-N=N-$, and artificial alizarin (1869–1870) and indigo (1897). By 1900, through intensive research and development, control of patents, and aggressive marketing, the industry was dominated by German manufacturers, such as BASF of Ludwigshafen, and Bayer, of Leverkusen. A new range of dyes based on anthraquinone (from which alizarin and congeners were made), and generally known as vat dyes, were the first major innovations in the twentieth century. Anthraquinone was obtained by oxidation of the three-ring aromatic hydrocarbon anthracene. Vat dyes are generally applied in reduced, soluble form; they then reoxidize to the original pigment and are extremely stable.

In 1901, René Bohn, head of the BASF Alizarin Laboratory, applied the indigo reaction conditions to a derivative of anthraquinone and discovered a blue colorant that he named indanthrone, from "indigo" and "anthraquinone." He then obtained the same product directly from 2-aminoanthraquinone. Later known as Indanthrene blue RS, it was the first of the anthraquinone vat dyes, noted for remarkable fastness.

Chemists at the rival Bayer company established the structure: indanthrone consists of two anthraquinone units joined through a ring of atoms containing two nitrogens. This enabled industrial research laboratories to discover anthraquinone-based intermediates for other vat dyes. In 1904, an assistant of Bohn synthesized benzanthrone, which afforded dibenzanthrone. Other important intermediates were anthrimides, readily converted into cyclic carbazole derivatives. The majority of anthraquinone derivatives were obtained by sulfonation of anthraquinone in the 1-, or alpha, position, and conversion to 1-aminoanthraquinone. The vat dyes, though expensive, in part because they required multistep syntheses, quickly became popular because of their resistance to light and washing, and were widely used in curtains, shirting fabrics, toweling, and beachwear.

It is worth mentioning that in 1915, stable complexes between azo dyes and a number of metals, in particular chromium, were introduced for dyeing wool. Though generally duller than those of the unmetallized dyes, they were found ideal for suiting materials.

With the outbreak of World War I, the supply of German dyes to the major users in Britain and the U.S. ceased. This led to expansion of the then tiny U.S. synthetic dye industry, from which emerged that nation's organic chemicals industry. An important development in vat dye production was the building up of anthraquinone from naphthalene-derived phthalic anhydride. This was pioneered industrially in the U.S. during World War I, and removed dependence on the imported anthracene.

The absence of German dyes also spurred developments in Britain, where in 1915 Morton Sundour Fabrics of Carlisle produced the first anthraquinone vat dyes. From this endeavor emerged Scottish Dyes Ltd. of Grangemouth. In 1920, its chemists invented Caledon Jade Green, made from dihydroxydibenzanthrone (see Figure 4). It became the most widely produced vat dye in world. By 1928, when Scottish Dyes, through its acquisition by British Dyestuff Corporation, became part of ICI, almost 20 percent of dyes made in the U.K. were vat dyes.

The second major innovation in dye making took place at Scottish Dyes around 1932–1933. Chemist Arthur Dandridge and colleagues discovered a blue impurity when the intermediate

phthalimide was prepared from phthalic anhydride and ammonia. The stable product was found to contain iron, and was studied at Imperial College, London, by Reginald P. Linstead and co-workers, who proposed a constitution similar to that of the pigment in chlorophyll. It was a phthalocyanine compound, of the type first prepared in 1907. ICI manufactured the copper analog, known as Monastral fast blue. Introduced in 1934, it represented the first member of the only new structural class of synthetic dyes in the twentieth century.

In 1925, the main German firms merged their interests to create the behemoth I.G. Farben. By this time the name reflected its historical roots rather than the range of activities. In the 1930s, the Bayer division developed a red azo dye into the first of the sulfonamide drugs.

The third important discovery in synthetic colorants satisfied the quest for dyes that attached to fibers by covalent bonds rather than weak intermolecular forces. This followed from research into wool dyes at the ICI General Dyestuffs Research Laboratory at Blackley, Manchester. William E. Stephen modified azo dyes by incorporation of reactive moities, particularly cyanuric chloride (trichlorotriazine). Failing with wool, Stephen suggested to Ian Durham Rattee in October 1953 that dyeing should be undertaken with cotton, which was successful. Thus was discovered the first fiber-reactive dye, of unprecedented fastness and introduced commercially as the

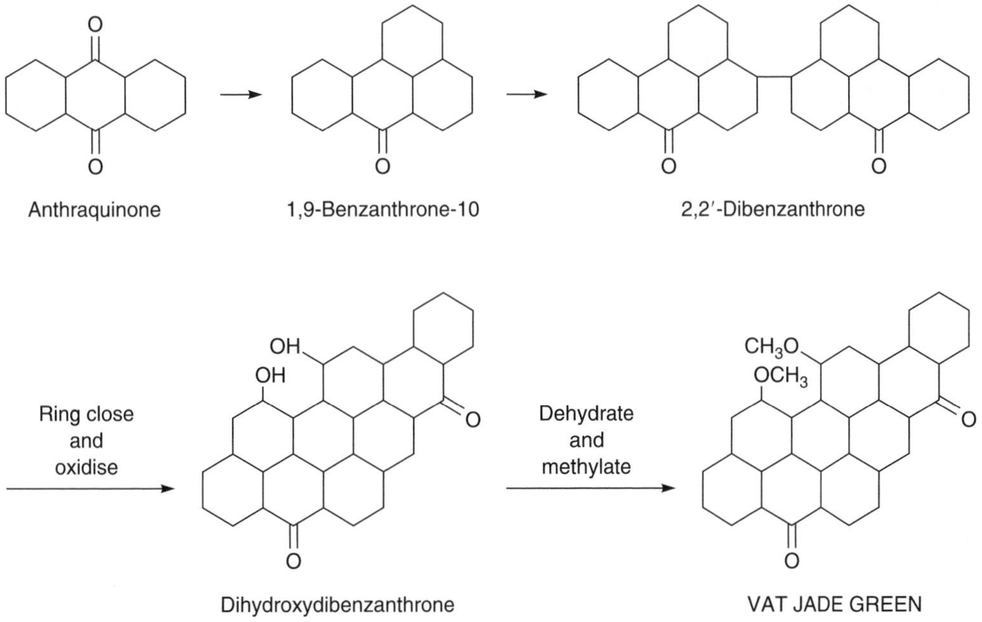

Figure 4. Flow chart for vat jade green.

ICI Procion range in 1956, on the hundredth anniversary of Perkin's discovery of mauve. The Swiss CIBA had already used the triazine grouping in dye synthesis, and the two firms came to an agreement over its application to reactive dyes. Fiber-reactive dyes displaced a number of vat dyes and reduced the incentive for research into new members of the latter class.

Semisynthetic and synthetic fibers introduced from the 1920s made new demands on the ingenuity of dye makers. Normally, dyeing is accomplished in aqueous solution, often in the presence of a fixing agent, or mordant. This is ideal for cotton, silk, and wool but not for certain synthetic fibers such as nylon and polyester that are plastic in nature; these require disperse dyes. The fiber is heated in an aqueous dispersion of a water-insoluble dye. Basic dyes are employed in the dyeing of polyacrylonitrile fibers.

While the two most important classes of dyes were anthraquinone vats and azo dyes, synthetic indigo became popular from the late 1960s with the swing towards fashionable denim and the faded look. Dyes are mainly used in textile printing and dyeing, but they have other uses, including food coloring and, when modified as pigments, for coloring plastics and synthetic fibers and for printing on paper. At the end of the twentieth century they found new and growing uses in the electronics industries, such as in inkjet printers.

From the 1970s dye making in the U.S. and Europe went into decline, due in part to reductions on tariff restrictions (U.S. and Britain) and environmental concerns (since many intermediate products were toxic). New centers for manufacturers were Asia, including Japan, India, China, and Eastern Europe.

See also **Coatings, Pigments and Paints; Fibers, Synthetic and Semi-Synthetic**

ANTHONY S. TRAVIS

Further Reading

Duff, D. A colourful tale. *Chemistry in Britain,* 37, 36–37, 2001. Celebrates a century of anthraquinonoid vat dyestuffs.

Forrester, S.D.A history of the Grangemouth dyestuff industry. *Chemistry in Britain*, 21, 1086–1088, 1985. The introduction of vat dyes into Britain.

Fox, M.R. *Dye-makers of Great Britain, 1856–1976: A History of Chemists, Companies, Products and Changes.* ICI, Manchester, 1987.

Haddock, N.H. Phthalocyanine colouring matters—their chemistry and uses. *J. Soc. Dyers Colourists*, 61, 68–73, 1945.

Lubs, H. A., Ed. *The Chemistry of Synthetic Dyes and Pigments.* Reinhold, New York, 1955. A comprehensive survey, with historical background.

Morris, P. J. T., and Travis, A. S. A history of the international dyestuff industry. *American Dyestuff Reporter*, 81, 59–100, 192–195, 1992.

Stead, C. V., Colour chemistry, in *Basic Organic Chemistry Part 5: Industrial Products*, Tedder, J. M., Nechvatal, A. and Jubb A. H. Wiley, London, 1975, pp. 315–50.

Steen, K. *Wartime Catalyst and Postwar Reaction: The Making of the United States Synthetic Organic Chemicals Industry, 1910–1930.* PhD thesis, University of Delaware, 1995. UMI Microfilm no. 9610494, 1996.

Travis, A.S. Synthetic dyestuffs: modern colours for the modern world, in *Milestones in 150 Years of the Chemical Industry*, Morris, P.J.T., Campbell, W.A. and Roberts, H.L., Eds. Royal Society of Chemistry, London, 1991, pp. 144–157.

Zollinger, H. *Color Chemistry*, VCH, Weinheim, 1987.

Zollinger, H. Some thoughts on innovative impulses in dyeing research and development. *Text. Chem. Colorist*, 23, 19–24, 1991.

E

Electric Motors

The main types of electric motors that drove twentieth century technology were developed toward the end of the nineteenth century, with direct current (DC) motors being introduced before alternating current (AC) ones. Most important initially was the "series" DC motor, used in electric trolleys and trains from the 1880s onward. The series motor exerts maximum torque on starting and then accelerates to its full running speed, the ideal characteristic for traction work. Where speed control independent of the load is required in such applications as crane and lift drives, the "shunt" DC motor is more suitable.

The electricity supply industry favored AC systems. DC motors could be adapted to operate on AC supplies, though not so well. The reversal of the current at each half cycle of the supply did not affect the direction in which the motor ran since both the field and the armature were reversed, but it induced "eddy" currents in the iron which wasted energy and heated the machine. Eddy currents could be limited by laminating all the iron, producing the "universal" motor that can be used on either AC or DC. This has been the motor used in most domestic electrical appliances, the exception being the washing machine, which requires greater power. The problem of eddy currents was also alleviated by working at a lower frequency. Some European railways employed AC series motors operating at 25 or 16 Hz rather the 50 Hz usually preferred for public electricity supply, but the railways were large enough undertakings to have their own generating stations.

One AC machine that could run from the public electricity supply was the induction motor. This was a new kind of machine, not a generator in reverse. Simple and robust, most of the world's electric motor power comes from induction motors. For most of the twentieth century, however, it had the serious limitation of being a fixed-speed machine.

DC Motors

Most DC motors have a fixed-field winding and a rotating armature with coils to which connections are made through brushes and a commutator. The windings are all on iron cores to enhance the magnetic field. As the armature and commutator rotate, the brushes make contact with successive commutator segments. The currents through those segments are switched as the commutator turns. In the series motor, the field and armature windings are in series, and the current flows through both. In the shunt motor, the windings are in parallel, and the current flowing through the field winding (which is of lighter construction than in the series motor) is controlled by an external regulator to adjust the speed of the machine.

The fact that DC motors and generators are essentially the same machines was recognized in the nineteenth century. Practical machines were introduced by Zénobe Gramme who, by 1874, had electrically driven machines in his Paris factory, although he used a single motor to turn a line shaft and not individual motors for each machine.

Large-scale use of electric motors began with electric railways and tramways in Germany, France, Britain and the U.S. in the 1880s. The deep "tube" underground railways and subways were only possible with electric traction. The main pioneer of electric railways in Europe was the Siemens Company. In the U.S., Frank Julian Sprague (1857–1934) built a railway system in Richmond, Virginia, in 1888. Sprague's most important contribution was his development in 1895 of multiple-unit control, whereby separate motors distributed along a train were all controlled from a single operating position in the driver's cab.

London's first tube was the City and South London Railway, which ran initially from Stockwell to the City about 20 meters below the surface and opened to the public in December 1890. It had locomotives pulling separate carriages. Multiple-unit control with motors distributed along the train was introduced in London on what is now the Central Line in 1902.

The first mainline railway to adopt electric traction was the Baltimore and Ohio Railroad. In 1892 an extension to this steam railway included a long tunnel where steam traction would not be practical. Electric traction was used through the tunnel and was so satisfactory that contemporary observers thought it would be extended rapidly. In practice electric traction grew quite slowly.

There were different opinions about the ideal supply voltages for electric trains, and whether the supply should be AC or DC. In 1920 a Railway Advisory Committee appointed by the British government recommended the general adoption of 1500-volt DC, which remained the standard for some years.

The introduction of the mercury-arc rectifier in 1928 was a landmark in the development of railway electrification. It permitted AC supply to the train with rectifiers on the train feeding DC motors. Most British railways now use this system with 25 kilovolt overhead wires supplied from the National Grid. Since 1960, germanium and silicon power rectifiers have replaced the mercury-arc valves, resulting in much simpler, lighter, and more reliable rectifying equipment on the trains. The railways in southern England use third-rail DC at about 700 volts and have rectifying substations at intervals alongside the track. Other countries have used a variety of systems including AC at 16b and 25 Hz and also three-phase AC. The three-phase systems in Switzerland and Italy had one phase built into the track and two overhead wires for the other phases.

AC Motors

The universal motors mentioned above have been the usual choice of motor in small power applica-tions such as sewing machines, food mixers, and vacuum cleaners. They can run at high speed, leading to a better power-to-weight ratio than other motors, and have been used in hand-held appliances such as electric drills since the 1920s. Almost every domestic motorized appliance except the washing machine uses a universal motor built into the appliance.

The first practical AC motors were the induction and synchronous motors developed by Nicola Tesla (1856–1943) in 1888; although the Italian, Galileo Ferraris (1847–1897), and others were working along similar lines. Tesla emigrated to the U.S. from eastern Europe, worked for a few years with Thomas Edison, and then joined the Westinghouse Company. Edison was a firm advocate of DC, and was firmly opposed to AC systems; George Westinghouse (1846–1914) was the leading American exponent of AC, and in changing employers, Tesla was stating his own views on the future direction of electrical engineering.

It had long been known that a pivoted perma-nent magnet or a pivoted piece of magnetic material would follow a rotating magnetic field. Tesla's great achievement was to produce a rotat-ing magnetic field from alternating currents in two or more fixed coils. His first motor had two field coils energized by alternating currents whose waveforms were 90 degrees out of step. Tesla showed in 1888 that the resultant magnetic field was constant in strength but rotating in direction. He obtained American patents covering two-phase and three-phase induction and synchronous motors. (In the synchronous motor, the rotor is a permanent magnet and keeps in step with the rotating field. In the induction motor, the rotor lags slightly behind the rotating field, but the speed of rotation is still almost constant.)

Tesla also showed that an induction or synchro-nous motor could be run from a single-phase supply if part of the field winding were connected through a capacitor or inductor to produce a second phase. Once started, the motor would run satisfactorily on the single-phase supply; in many motors, the starting capacitor or inductor is switched out of the circuit automatically, usually by a centrifugally operated switch, once the motor is running.

The Westinghouse Company bought Tesla's patents, and from 1892 they began to promote polyphase AC distribution systems and the use of AC motors in industry. They adopted the three-phase 60-Hz electricity supply which remains the U.S. standard, although 50 Hz is preferred in Europe. For induction motors up to a few horsepower, the simple and so-called "squirrel cage" rotor construction is normally used. For higher-rated motors, a wound rotor is generally preferred so that a resistance can be connected in series to reduce the starting current. The starting resistance is then cut out automatically when the motor has run up to speed. Induction motors are ideal where a constant speed is required. Because it needs no brush gear, the induction motor gives reliable service over long periods of time with little or no maintenance.

A motor that is easily confused with the induction motor is the repulsion motor. The confusion is compounded by the existence of mixed action motors that start as repulsion motors but run as induction motors. The repulsion motor is largely due to J. A. Fleming (1849–1945) and Elihu Thomson (1853–1937). Fleming, who was professor of electrical engineering at University College London and is best known for his work on the thermionic valve, made a study of the forces between conductors carrying alternating currents. In 1884 he showed that a copper ring suspended within a coil carrying an alternating current tends to twist so as to be edge on to the magnetic field. This is the basis of the repulsion motor.

Control of Motors

In many applications, motors are controlled simply by switching them on or off. If the starting current is heavy, then they may first be connected in series with a resistance, which is then cut out. Where several DC motors are working together, as in an electric train, it is usual to adopt series–parallel operation. If there are two motors and a resistance, then the controller will connect the machines to the supply in the following sequence:

1. Motors in series plus resistance
2. Motors in series only
3. Motors in parallel plus resistance
4. Motors in parallel

The principle can be extended if three or more motors are involved. When two motors are connected in series to the supply, each is effectively on half voltage, and the starting current is reduced accordingly.

In applications where precise control is needed over a wide range of speeds, then a DC motor and the Ward Leonard control system is widely used. H. Ward Leonard (1861–1915) was an American electrical engineer with a special interest in lifts. He appreciated that the ideal way of controlling a DC motor was to control its armature current. In the Ward Leonard system, developed about 1890, a first motor is used to drive a generator, and the output of that generator supplies the armature of the motor being controlled. The generator output is regulated by controlling its field current, which is of course very much smaller than the motor current. Numerous feedback systems have been devised in which the generator field current is controlled by a device that monitors any deviation of the motor's actual speed from its desired speed. Ward Leonard systems have often been used for winding motors in mines and for rolling-mill drives, but they have also been used in low-power applications where fine speed control is vital.

A two-speed variation of the induction motor is the pole amplitude modulated, or PAM, motor devised by Professor G.H. Rawcliffe in 1957. This is an induction or synchronous motor in which the field windings are so arranged that the effective number of poles can be changed by changing the connections of a few coils. The PAM motor therefore retains the reliability and robustness of the conventional induction motor while being able to work at either of two alternative speeds.

The modern approach to the ideal of an induction or synchronous motor whose speed can be varied is to provide a variable frequency supply, which was not feasible before the advent of power semiconductor devices. Transistors capable of controlling a few amperes became available in the mid-1950s, and by about 1960, thyristors (then known as silicon-controlled rectifiers) were available which could control some tens of amperes. Thyristors and other semiconductor devices are now available which can carry very large currents and switch to high voltages. The devices can also be connected together in series and parallel combinations so that there is no limit to the current and voltage they can control. They are used in inverter circuits to generate variable-frequency alternating current for supplying induction motors, and they are also used in "chopper" circuits, which switch a DC supply on and off rapidly to vary its effective voltage. Such control systems are easily regulated by reliable and compact electronic circuits that are responsive to speed or any other chosen parameter. At the end of the twentieth century, electric drives are commonly supplied as a unit incorporating motor and control equipment.

Electric Road Vehicles

Although only a small proportion of vehicles are driven by electric motors today, the electric vehicle has an important history and is likely to become more important in the future because the electric motor offers a clean and quiet power source.

At the beginning of the twentieth century, electric road vehicles were more common than gasoline-driven ones. There were two distinct types: the self-contained machine supplied by batteries and the trolley vehicle that drew current from overhead wires. By 1900 there were several hundred electric taxicabs in New York and other U.S. cities and a smaller number in London. To encourage the use of electric cars, supplier fees for battery charging were kept low. The service provided by electric vehicles was considered quite adequate until World War I, when gasoline-powered vehicles were able to travel further without refueling. The first trolley bus service (a bus also powered through overhead wires but not limited to rails) began in 1901 in Bielethal in Germany, and trolley buses appeared in Britain in 1911. Just after World War II, the number of trolley buses in the world reached a peak of about 6000.

In some countries there are still many electric vehicles, usually small delivery vehicles, in daily use. Most electric vehicles use heavy lead–acid or nickel–iron batteries. Much research effort is currently going into the quest for a better battery. A possible contender is the sodium–sulfur battery, which offers a much lighter weight battery than conventional types. It suffers from the fundamental disadvantage that it operates at about 300°C, and hot sodium would be dangerous in an accident. Gasoline, of course, is also dangerous. It may well be that further research will find a battery, using the sodium-sulfur system or some other chemical combination, that will again give electric vehicles an advantage over gasoline-driven ones. Other research is looking at the fuel cell, producing electricity directly from either hydrogen or a hydrocarbon fuel but more efficiently and with less pollution. A further variant is the hybrid vehicle in which a small internal combustion engine drives a generator that charges a battery, and the battery supplies the motor driving the vehicle. The advantage of such an arrangement is that the internal combustion engine runs at constant speed and has only to supply the average power requirement of the vehicle, which is a small fraction of its peak power requirement.

Linear Motors

Linear motors are often described as ordinary rotary motors that have been split along their length and unrolled. It follows that there are as many kinds of linear motors as there are rotary ones. Linear motors may be AC or DC commutator machines, or they may be induction or synchronous machines, to name just a few possibilities.

The idea of a linear motor was described by several experimenters in the nineteenth century, but nothing was achieved before the twentieth. The Norwegian Kristian Birkeland obtained a series of patents between 1901 and 1903 for a DC linear motor used as a silent gun, and the idea of a linear motor gun has recurred periodically. The Russian engineer N. Japolsky worked on linear motors in Russia around 1930, and during World War II the Westinghouse Company in the U.S. built a linear motor aircraft launcher called the Electropult. The aircraft sat on a trolley with windings underneath, and the fixed part also had windings. The Electropult produced 10,000 horsepower and could accelerate a 4.5-ton aircraft to 180 kilometers per hour in 4.2 seconds.

The two potential applications of linear motors that have attracted most attention throughout the twentieth century are to drive shuttles in looms and in railways. Emile Bachelet set up a company to work on both applications in 1914. The main twentieth century figure on linear motors was Eric Laithwaite (1921–1997). He was first interested in the linear motor for use in a loom and turned later to its use in transport. The loom requires a means of projecting the shuttle at high speed across the width of the cloth. Conventionally, this is done by striking the shuttle very hard and catching it on the other side, a process that wastes energy and requires considerable mechanical strength in the shuttle. Despite many attempts by a number of inventors, linear motors have still not taken over in this application.

Laithwaite made considerable progress with linear motors for transport, especially with his concept of the "magnetic river," or "maglev" (for magnetic levitation), in which a magnetic field would lift, guide, and propel an object such as a train along a track without physical contact. A few small-scale linear motor driven transport systems are in operation at airports, but despite much research the large-scale linear motor driven mass-transport system remains a dream.

One application where linear motors have achieved success is in pumping and stirring liquid metals. In the 1930s pumped liquid metal was used for heat transfer in special circumstances, where the high thermal capacity of the metal was useful, and where direct contact with the liquid metal was to be avoided. The liquid metal itself acts as the moving part of the motor, and all that is necessary to pump the metal is to fix a wound stator onto the wall of the container holding the metal. Such devices are now commonly used for stirring aluminum and other metals in furnaces and for assisting the transfer of metal to molds for casting, A specialized application is for pumping the liquid sodium coolant in nuclear reactors. Reliable operation and complete isolation of the metal being pumped are vital requirements, and the linear motor provides these.

See also **Electrical Power Distribution; Rail, Electric Locomotives; Urban Transportation**

BRIAN BOWERS

Further Reading

Hughes, A. *Electric Motors and Drives*, 2nd edn. Newnes/Butterworth-Heinemann, Oxford, 1993.

Gottlieb, I. *Electric Motors and Control Techniques*, 2nd edn. McGraw-Hill/Tab Electronics, New York, 1994.

Gottlieb, I. *Practical Electric Motor Handbook*. Newnes/Butterworth-Heinemann, Oxford: 1997.

Keljik, J.J. *Electric Motors and Motor Controls*. Delmar Learning, Albany, NY, 1995.

Valentine, R., Ed. *Motor Control Electronics Handbook*. McGraw-Hill Professional, 1998.

Yeadon, W.H. and Yeadon, A.W. *Handbook of Small Electric Motors*, McGraw-Hill Professional, New York, 2001.

Electrical Energy Generation and Supply, Large Scale

Public supply of electricity at the close of the nineteenth century was typically confined to the larger towns and cities where either a local entrepreneur, or a far-sighted municipality, established relatively small generating stations to supply local lighting loads. Many of these local power stations employed reciprocating engines to drive direct current (DC) dynamos. Overhead circuits generally carried the power no more than a kilometer or two to local businesses or the larger households in the district. Sometimes, where water-powered mills had existed previously, hydroelectric generators were established to supply the electricity consumers. As more and more people began to appreciate the convenience of electrical power and, moreover, could afford to pay for it, demand on local supplies increased and larger power stations began to be established. The invention of the electrical transformer to step-up the voltage at the generating station and step it down again to a safe level for use by the consumers, meant that higher speed alternators, often driven by steam turbines, could be employed to produce the power. High-voltage distribution reduced the losses in the circuits between the generating stations and the loads.

Development of the Grid

Large-scale electricity supply began in the second and third decades of the twentieth century with the establishment of so-called "grid" systems which provided for high-voltage interconnection of power stations. Such interconnection allowed spare generating plant on the system to be shared amongst a number of power stations. This avoided the previous need for individual power stations to carry a spare generating set to allow the consumer loads to be supplied when one of the other generating sets was out of service for maintenance or repair. The cost of providing spare capacity at each power station was becoming increasingly uneconomic as set sizes grew larger due to the strive for increased efficiency and economies of scale in their construction and operation.

Coal by Wire

By the middle of the century, the concept of a power station built in a town to supply only the load of that town had virtually disappeared, certainly in the U.K. At that time, more than 90 percent of electricity was generated in coal-fired power stations which were supplied either by coal trains or seagoing vessels. As high-voltage transmission technology developed at 220 kilovolts and above, it became more economical to locate the power stations on the coal fields themselves and transmit the electricity by high-capacity extra-high voltage circuits to the load centers. As the century progressed into the 1960s and the 1970s, individual power station sizes grew from the few hundreds of megawatts, to several thousand megawatts. Typical of the largest of the European coal-fired power stations is that at Drax in Yorkshire, which has a capacity of 4000 megawatts.

Generating Plant Technologies

Separate entries in this Encyclopedia describe the technologies of hydroelectric power generation and other renewable energy resources, turbines, nuclear reactors, and fossil fuel power stations. While coal was a dominant fuel in many countries for power generation over the first half of the century, some countries had other resources such as water, which is employed extensively in Norway and Switzerland. The great rivers of the U.S. and

Canada were also exploited to provide hydro-electric power. The generating plant at Niagara Falls operated throughout the century, but it was the 1930s and 1940s that saw the installation of many of the large hydroelectric dams such as Hoover (Boulder) on the Colorado River, Bonneville, and Grand Coulee on the Columbia River, Washington. Canada's large-scale hydro-electric plants, such as Churchill Falls in Labrador, were built somewhat later in the century. Hydroelectric power still provides for more than 50 percent of Canada's needs.

France in particular has developed nuclear power stations over the last 40 or so years of the century until they now supply over 90 percent of their electricity. The U.K. has developed many gas-fired power stations in the last decade of the century employing gas turbines, fuelled from North Sea gas, to drive the electrical generators.

Competition in the Industry

Most electricity supply systems in the developed world were originally state-owned, and many still are, except in the U.S. In the last few years of the century, however, notably in the U.K., there has been a move to break up the vast generation and distribution monopolies. The generation assets were divided up into several companies that would compete with one another on electricity price in a wholesale market, the cheapest generators being called upon to generate first. The grid system transmitted this wholesale power to distribution networks that supply individual consumers in a particular area. Both the transmission grid and the distribution network are owned by public companies with shareholders but are regulated as a monopoly service provider in terms of service and quality standards. Purchase of wholesale power and its resale to individual consumers, whether individual households or large industrial concerns, are undertaken by licensed supply companies that pay charges for the use of the transmission grid and the distribution networks.

Generating Efficiency and Combined Heat and Power

Because of limitations in the thermodynamic efficiency of steam turbines and gas turbines, the majority of the energy in the fuel supplied to the boiler in a coal-fired power station, or to the combuster in a gas turbine, is rejected as waste heat. Parasitic losses; that is, energy consumed by the auxiliary plant such as pumps and fans, lower the overall power station efficiency still further. The best coal-fired stations rarely exceed overall efficiencies of 40 percent and a gas turbine operating in open-cycle mode will have an even lower value. The situation can be improved in the former case by recovering some of the heat rejected to the condenser as the steam from the turbine is condensed back to boiler-feed water. In the latter case, the exhaust gas from the gas turbine can be used to raise steam, which can be put to use in a process such as paper making or for community heating purposes. Efficiencies in this mode of operation can exceed 70 percent. The gas turbine can also be operated in so-called "combined-cycle" mode where the steam raised from the exhaust gas can be used to drive a steam turbo-alternator set to produce more electricity. Such an arrangement can have a fuel conversion efficiency of around 60 percent.

Intercontinental Transmission

As power supply systems in individual countries have become larger, high-capacity transmission circuits have been installed to interconnect national systems and allow the export and import of power. Typical of such systems is the cross-channel interconnector between England and France which has a capacity of 2000 megawatts and which allows the U.K. to take advantage of relatively cheap French nuclear power. Other recent examples of such interconnectors include those between the Channel Islands and France and between Scotland and Northern Ireland. Many of these are based on high-voltage direct current transmission via submarine cables, rather than AC transmission, which is typically employed on land-based overhead lines. Similar high-power interconnectors between the U.K. and Scandinavia were being planned as the century closed.

Environmental Regulation

In the latter decades of the century, there was increasing concern over the environmental impacts of electricity generation. These included acid rain from the sulfur dioxide emitted by coal combustion, and the safety and waste issues with nuclear power generation. Governments increasingly began to apply regulatory limits, requiring the installation of technologies such as gas scrubbing on power station exhaust streams. The desire to protect rivers in their natural state has also limited further growth in large hydroelectric power schemes in many countries. As the century closed, the link between carbon dioxide from fossil fuel combustion and global climate change was giving rise to regulation affecting all fossil fuel generation and encouraging the development of renewable fuels for electricity generation.

See also **Electrical Power Distribution; Electricity Generation and the Environment; Energy and Power; Hydroelectric Power Generation; Turbines**

IAN BURDON

Further Reading

Edgerton, D. *Science, Technology and the British Industrial 'Decline' ca. 1870–1970.* Cambridge University Press/ Economic History Society, Cambridge, 1996.

Elliot, D. *Energy, Society, and Environment*, Routledge, London, 1997.

Hughes, T.P. *Networks of Power: Electrification in Western Society, 1880–1930.* Johns Hopkins University Press, Baltimore, 1893.

Leggett, J. *The Carbon War: Global Warming at the End of the Oil Era.* Penguin, London, 1999.

MacNeill, J.R., Ed. *Something New under the Sun: An Environmental History of the Twentieth Century.* Penguin, London, 2001.

Reddish, A. and Rand, M. The environmental effects of present energy policies, in *Energy, Resources and Environment,* Blunden, J. and Reddish, A., Eds. Hodder & Stoughton (in association with The Open University), London, 1991.

Wolf, D.E., Wolf, D.D. and Lowitt, R. *Big Dams and Other Dreams: The Six Companies Story.* University of Oklahoma Press, Norman, 1996.

Electrical Power Distribution

While the first commercial power station in San Francisco in 1879 was used for arc lighting (using a spark jumping a gap as the source of light) for street lamps, these had limited application. Edison's carbon filament lamp was the stimulus for the spread of electric lighting. A few of Edison's buildings and some private residences had their own generators, but Edison also recognized there was a need for a generating and distribution system. Edison's distribution system was first demonstrated in London, with a temporary installation running cables under the Holburn Viaduct in early 1882 that provided power for the surrounding district. The first permanent central electric generating station was Edison's Pearl Street Station in New York that went into operation in September 1882 and provided electricity (with a meter) to 85 customers in a 1 square mile (2.6 square kilometers) area. The Pearl Street Station used direct current (DC). In DC systems, the current flows in one direction, with a constant voltage. The dissipation of energy limits the size of DC systems and requires the source of electric generation to be close to the customer. Alternating current (AC) systems, in which the current changes direction (in today's public electricity supply, 50 or 60 times per second), overcame this limitation.

Transformers, invented by William Stanley Jr. in 1883, allowed voltage from an AC generator to be stepped up to high voltages for transmission and then back down again to lower voltage for utilization. At higher transmission voltages (and thus lower current), less electrical energy is lost. Power could thus be delivered efficiently over long distances and the source of generation could be far from the customers (in Edison's DC system, each generating system could only supply a few square kilometers or so). This would be particularly important once electrical energy was generated by hydropower stations, which were often many kilometers from the population or industry they served. In 1885 George Westinghouse, head of Westinghouse Electric Company, bought the patent rights to Nikola Tesla's AC polyphase system and placed its first power station in operation in 1886. The introduction of the AC induction motor in 1888 gave the AC system a great advantage in providing industry with electrical power. The AC system has been the dominant form of power transmission and distribution ever since.

Power transmission deals with the systems that move the power over very long distances from sources of generation such as hydro stations, fossil fuel plants, or nuclear power plants to central points of distribution called substations located in or near areas of large power demand such as cities and towns. In 1890, Westinghouse installed a 19-kilometer, 4000-volt transmission line from Willamette Falls to Portland. In 1897, the British Empire's first long-distance high-voltage transmission (11,000 volts or 11 kilovolts) of electric power traveled 27 kilometers from St-Narcisse on Batiscan River to Trois-Rivières, Québec, Canada. In 1903, the Shawinigan Water & Power Company installed the world's largest generator (5000 watts) and the world's largest and highest voltage line—136 kilometers and 50 kilovolts (to Montreal).

Today, the aspects of the electric power transmission system that are most apparent are the overhead transmission lines strung on the large steel towers visible in the countryside. These lines, called tower lines, transmit power over long distances. Each tower line has three power wires and two ground wires. The power wires carry the electric power and hang from insulators connected to the tower arms. The ground wires are connected at the top of the towers and protect the power carrying conductors from lightning strokes. Power transmission occurs at high voltages; 500 or 750 kilovolts are not uncommon.

Multiple tower lines are connected together at a switching station. The connection occurs at a structure called a "bus." Between the tower line and the bus is a switch called a circuit breaker. The purpose of the circuit breaker is to disconnect the tower line from the bus in times of trouble, of which there are many types. Lightning hitting the tower line and causing a short circuit, power flow through a transmission line exceeding the power delivery capacity of the line, the wind blowing the transmission line into a tree—are all examples of transmission line troubles. If such an occurrence is not quickly isolated and fixed, it may damage the line or spread to other facilities. Devices called relays continually monitor the transmission line for signs of trouble. When the relay detects trouble on a transmission line, it signals circuit breakers to open. The opening of the circuit breaker then de-energizes the transmission line and clears the trouble. The power transmission system is made up of a complex arrangement of transmission lines and switching stations that provide many redundant paths from sites of generation to locations of customer demand.

This redundancy provides reliability. Since there are multiple redundant paths, if one transmission line should come out of service because of trouble, the power will flow over the remaining lines. Power system analysis is the discipline that insures that power transmission system will deliver power to the customer with sections of the transmission system out of service. This analysis places great demands on power system design engineers.

Prior to the advent of digital computers, power system analysis was performed on large analog computers called AC boards. The AC board was composed of miniature components that could be connected together to represent the power system. Engineers would spend many hours plugging in components to represent the power system to be studied and manually recording the power flows on data sheets. Digital computer programs were introduced in the early 1960s to solve the power flow problem. Early programs could only solve systems composed of a few hundred switching stations. Improvements in analytical techniques and computer speeds today permit the solution of 30,000 switching stations. The improvement in analytical techniques permitted the design of larger and more complicated interconnections. Today, most of the North America east of the Rocky Mountains is one large interconnected electric system. As the interconnection became larger and more complicated, it

also became more difficult to operate. To insure reliable operation, computer systems called system control and data acquisition (SCADA) were introduced in the late 1960s to monitor and control the power transmission system.

A SCADA system is housed in a power company's control center, where it receives information about the operation of the transmission lines and circuit breakers. When the SCADA system detects an overloaded transmission line or a circuit breaker operation, it alerts the control room operator with an audible alarm. The alarms and system status are displayed on mimic boards, large electronic displays of the transmission power. Sophisticated computer programs provide the operator additional information including what operating moves will relieve overload facilities and what transmission flows will result if facilities trip out of service. These SCADA systems assist in providing high reliability of the transmission system in delivering power to the distribution system. Reliability of the transmission system is very important because transmission system breakdowns can cause widespread outages or blackouts.

The first widespread power outage in the U.S. was the Northeast Blackout that occurred in 1965—a relay failure blacked out 30 million people with over 20,000 megawatts of demand in the northeast U.S. and Canada. On January 3, 2001, India blacked out 220 million people in a large part of seven states for 13 hours. The most recent U.S. blackout occurred on August 14, 2003, when a huge failure blacked out New York City, Cleveland, Detroit, and Toronto affecting customers with a demand of 61,800 megawatts. On September 28, 2003, a huge blackout affected most of Italy. All of these blackouts were breakdowns in the power transmission system. Failures in the power distribution system are usually much more limited in scope.

The job of the power distribution system is to get the power from the substations to the customer. The aspects of electric power distribution that are most apparent are the wooden pole lines adjacent to streets and roads in towns and villages. The wires that carry the power from the distribution substation to the customer are connected to these poles with insulators. A pole line may contain primary circuits and secondary circuits. The primary circuit delivers the power at 4, 13, or 26 kilovolts from the substation to the pole top distribution transformer. A pole top distribution transformer resembles a large coffee can attached to the pole. The distribution transformer transforms the voltage down to the utilization or outlet voltage, which in the U.S. is 120 volts and in the UK is 240 volts, well known to international travelers whose electric appliances may require adapters for use. The secondary circuit delivers the power from the distribution transformer to the customer. Unlike the redundant network grid of the transmission systems, the distribution system is usually radial—there is just one path for the power to flow from the distribution substation to the customer. If the path gets interrupted, the customer loses power. Since the path is comprised of overhead wires on a pole line, it is subject to damage by snow, ice, tree limbs, lightning, animals, and wind. Storms can cause significant damage to a distribution system. To improve the reliability of the distribution system, the overhead lines are often equipped with sectionalizing and tie switches. These switches permit damaged parts of the system to be isolated and the power supplied by alternate paths. In more heavily populated areas, the distribution system may be placed underground. Sometimes the underground insulated wires are placed in concrete ducts and sometime they are directly buried. The duct system is more reliable, but it is also more expensive.

Re-Emergence of DC Power Distribution

With the development in the 1970s of high voltage thyristor valves that functioned as AC/DC converters it became possible to transmit DC power at high voltages (high voltage direct current or HVDC) over large distances. HVDC is particularly suitable for linking interconnected systems and for submarine links. The cross-Channel link installed between Britain and France in 1986 is one example. In the modern era of deregulated electricity supplies, cross-border energy trading between nations whose grids are synchronously interconnected can exploit the time difference of peak load periods on national systems.

See also **Electric Motors; Electrical Energy Generation and Supply, Large Scale; Electricity Generation and the Environment; Energy and Power**

PETER VAN OLINDA

Further Reading

Bowers, B., Ed. *An Early History of Electricity Supply: The Story of the Electric Light in Victorian Leeds*. Peregrinus, Stevenage, 1986.

Davis, L.J. *Fleet Fire: Thomas Edison and the Pioneers of the Electric Revolution*. Arcade, New York, 2003.

Electrical Transmission and Distribution: Reference Book. Westinghouse Electric Corporation, East Pittsburgh, 1964.

Grainger, J. *Power System Analysis*. McGraw-Hill, New York, 1994.

Harrison, K. *Currents of the Brazos: An Illustrated History of Brazos Electric Power Cooperative, Inc 1941–1991*. Nortex Press, Austin, 1991.

Jonnes, J. *Empires of Light: Edison, Tesla, Westinghouse, and the Race to Electrify the World*. Random House, 2003.

Miller, R. *Power System Operation*. McGraw-Hill Professional, New York, 1994.

Platt, H. *The Electric City: Energy and the Growth of the Chicago Area 1880–1930*. University of Chicago Press, Chicago, 1991.

Warkentin, D. *Electric Power Industry: In Nontechnical Language*. Pennwell, Tulsa, OK, 1998.

Useful Websites

Electric Power from Niagara Falls: http://ublib.buffalo.edu/libraries/projects/cases/niagara.htm

History of the U.S. Power Industry, 1882–1991: http://www.eia.doe.gov/cneaf/electricity/page/electric_kid/append_a.html

U.S. Department of Energy: http://tonto.eia.doe.gov/FTPROOT/electricity/056296.pdf

A Brief History of Public Power in the Pacific Northwest: http://www.ppcpdx.org/Side_Bar/history.htm

Mimic Board: http://www.ee.washington.edu/energy/apt/nsf/previous/MimicBoard.htm

Power system: http://www.howstuffworks.com/blackout.htm

High Voltage Direct Current Transmission Systems Technology Review Paper : http://www.worldbank.org/html/fpd/em/transmission/technology_abb.pdf

Electricity Generation and the Environment

Fossil fuel thermal generating technologies were a mainstay of both twentieth century electricity generation and environmental attention. While concern with declining urban air quality, initially at the center of this attention, dated back to the nineteenth century, it was the substantial post-World War II rise in electricity consumption that resulted in the later prominence of these concerns. The impacts of fossil fuel extraction and transportation were also a source of significant twentieth century environmental attention, but concern over atmospheric emissions dominated. Although initial concern focused on particulate emissions, attention shifted to acidic emissions from the 1970s onward, and the final decade of the century was dominated by concern with the impact of fossil fuel emissions on climate. This later concern with fossil fuel greenhouse gas emissions, primarily thermally produced carbon dioxide (CO_2) but also fugitive emissions (i.e., not caught by a capture system) such as methane from coal seams and from gas extraction and distribution systems, reinforced an increasing emphasis on alternative generating technologies. Some of these, notably macro-hydro and nuclear fission, were significant twentieth century technologies in their own right, and their environmental impacts are briefly discussed below. However, as the twentieth century closed this emphasis was increasingly turning to renewable energy technologies and the potential for significant further efficiencies in both electricity generation and consumption, including the drive to "decarbonize" electricity generation by turning away from fossil fuel technologies.

Coal-fired thermal generation was pivotal to the rising twentieth century concern with poor urban air quality. Coal contains many impurities, and the resulting particulate, and typically acidic, emissions were a major source of urban air pollution. Heavily populated and industrialized countries such as Britain were particularly affected, and London's "Great Smog" of December 5–9, 1952, is regarded as a turning point in pollution control. This event, leading to an estimated 4000 deaths, provided impetus for the U.K.'s Clean Air Acts of 1956 and 1968, which focused on curbing particulate emissions. Power station emissions were controlled at the source by both particulate removal and the use of taller emission stacks to promote dispersal over wider areas. The electrostatic precipitators used to remove particulates soon became a standard power station design feature, while the use of taller emission stacks inadvertently resulted in the issue of invisible pollutants as a significant regional concern. This concern centered on the environmental acidification resulting from the sulfur dioxide (SO_2) and to a lesser degree nitrous oxide (NO_x) emissions from coal as well as oil-fired thermal generation. Particularly marked in Europe, where British emissions were deposited on Norway and Sweden, and in North America, where U.S. emissions were similarly deposited in Canada, these concerns resulted in a milestone in pollution control, the UN Economic Commission for Europe's 1979 Geneva Convention on Long-Range Transboundary Air Pollution. While these concerns remained the focus of legislative attention throughout the rest of the twentieth century and resulted in the widespread installation of flue gas desulfurization technology, it was CO_2 emissions that soon came to dominate attention.

As the twentieth century progressed and the scale of hydroelectric installations increased they became the focus of considerable concern. Massive projects such as the "New Deal" Tennessee Valley project in the U.S. and the postwar Australian Snowy Mountains project were on a scale previously unseen, and many of their environmental impacts (such as changed hydrological regimes) only came to light many decades later. Such macro-hydro projects, particularly in the developing world, came to be seen as symbols of environmental devastation.

While the first civil nuclear power plants of the 1950s were greeted with an optimism encapsulated by the phrase "too cheap to meter," this optimism subsided as the impact and implications of the entire nuclear fuel cycle became clearer. Public disquiet was dominated by concerns with operating safety, the implications of which were graphically demonstrated by the Ukrainian Chernobyl accident of 1987, but it was the broader considerations of nuclear waste storage, decommissioning, and their economic costs that placed a political question mark over the future of nuclear technology.

Rising environmental concerns from the 1960s and the oil price shocks of the 1970s resulted in greater interest in the development of renewable energy technologies. Concern with climate change, institutionalized in the UN Framework Convention on Climate Change (1992) and the subsequent Kyoto Protocol mandating greenhouse gas emission reductions, significantly reinforced this concern in the 1990s. Photovoltaic cells, used for spacecraft from the 1950s, became a focus of terrestrial attention from the 1970s. By the end of the century both cell efficiency increases and cost decreases were dramatic, but generating costs still exceeded those of fossil fuels by a factor of around 4. The great success story of this period was wind power. Although the first 1 megawatt-plus demonstration wind turbines the 1970s overextended contemporary materials technology, turbines of more than 400 kilowatts were in widespread commercial use by the 1990s. By the close of the century, wind power was not only cost competitive with fossil fuels but was also the fastest growing of all electricity generating technologies. In addition, with increasingly sophisticated materials technologies, 1 megawatt-plus turbines were entering the commercial market. Electricity generation from waste and biomass was also increasingly common, while other renewable options such as wave and ocean power, micro-hydro, and geothermal technologies were gaining increasing attention.

The push to decarbonize electricity generation in the 1990s also resulted in proposals for fuel switching, particularly to cleaner and less carbon-intensive natural gas, which offered vastly improved generating efficiencies. Other fossil options such as cleaner coal technologies and sequestering of greenhouse gases were also pursued. Combined heat and power, in which the waste heat from thermal generation is productively captured, also became more attractive in applications that required heat inputs such as industry and domestic heating. The potential for significant energy savings from both the supply and demand sides of technological and behavioral change was widely discussed in commercial/industrial and political circles, although by the close of the century little progress had been made on this front.

See also **Biomass Power Generation; Hydroelectric Power Generation; Power Generation, Recycling; Solar Power Generation; Technology, Society, and the Environment; Wind Power Generation**

STEPHEN HEALY

Further Reading

Boehmer-Christiansen, S. and Skea, J. *Acid Politics: Environmental and Energy Policies in Britain and Germany*. Belhaven Press, London, 1991.

Elliot, D. *Energy, Society and Environment*. Routledge, London, 1997.

Goldemberg, J. *Energy, Environment and Development*. Earthscan, London, 1996.

McNeill, J. Air pollution since 1900, in *Something New Under the Sun: An Environmental History of the Twentieth Century World*, McNeill, J., Ed. Penguin, London, 2000.

McNeill, J. Energy regimes and the environment, in *Something New Under the Sun: An Environmental History of the Twentieth Century World*, McNeill, J., Ed. Penguin, London, 2000.

Reddish, A. and Rand, M. The environmental effects of present energy policies, in *Energy, Resources and Environment*, Blunden, J. and Reddish, A., Eds. Hodder & Stoughton (in association with The Open University), London, 1991.

Electrocardiogram (ECG)

The electrocardiogram (ECG or EKG) is a graphic measurement of the electrical activity of the heart produced by an electrocardiograph, or ECG machine. ECGs are used to assess heart function and are especially important in the detection of electromechanical abnormalities such as arrhythmias and myocardial infarctions (blockages in the small vessels of the heart, commonly termed a heart attack). The development of the electrocardiograph has been closely linked with advances in the science of electrophysiology, the study of the role of electricity in the functioning of living organisms. As early as 1842, the Italian physicist Carlo Matteucci demonstrated that each heart beat is accompanied by an electrical discharge. The

following year, the German physicist Emil Dubois-Raymond confirmed Matteucci's findings, coining the term "action potential" to denote the electrical charge accompanying muscular contraction. In 1856, Rudolph von Koelliker and Heinrich Muller noted a two-fold electrical charge accompanying heart systole, which is the muscular contraction of the heart to pump blood through the blood vessels of the body. Identification of this electrical charge presaged the later division of the heart's electrical signature into a series of so-called waves and complexes. Practical application of these findings to the medical field of cardiology (study of the heart) depended on the creation of a reliable means of (1) measuring the electrical activity of the heart and (2) relating the measurements to the functional anatomy of the heart (i.e., establishing the electrical signatures denoting structure, size, blood supply, electrical pathways, and rhythm of the heart). A first significant step in this direction was the creation of the capillary electrometer by the French physicist Gabriel Lippmann in 1872. This was a thin glass tube of mercury and sulfuric acid that registered variations in electrical potential that could be observed under a microscope. The capillary electrometer was used in 1878 by the British physiologists John Burden Sanderson and Frederick Page to confirm the two phases of the heart's electrical activity, later designated the QRS complex and T wave. The first human electrocardiogram produced using the capillary electrometer was published by the British physiologist Augustus D. Waller in 1887.

However, the tracing or graph produced by capillary electrometry was still crude. This was due in part to the sluggishness of the measuring needle that required a mathematical correction and in part because a crucial component of modern electrocardiography—the use of multiple leads to produce a more sophisticated picture of the heart's electrical activity—had yet to be invented. The increase in the number of measuring leads used in electrocardiography along with the corresponding degree of accuracy came in the three stages. In 1891, British physiologists William Bayliss and Edward Starling improved the capillary electrometer by adding a second lead to the right hand to the original lead over the heart itself. Using this technique, they discovered a third phase of the heart's electrical activity, which was later designated the P wave. A second stage in this development was the Dutchman Willem Einthoven's creation of a triangular arrangement of leads. His arrangement, termed Einthoven's triangle, established what have come to be known as the standard leads in electrocardiography, designated I, II, and III. First used in 1912, this system has continued to be the basis of the ECG lead arrangement.

Einthoven, a major figure in the development of modern electrocardiography, distinguished five phases of the heart's electrical action, noting that the middle pulse was in fact a complex with three separate components. Einthoven dubbed the five separate deflections of the electrocardiogram the P, Q, R, S, and T waves, terminology that is still used. Einthoven's second major contribution to the development of electrocardiography was his use of a new technology, the string galvonometer. Einthoven first employed the string galvonometer in 1901, and the following year he published the first human ECG using it. Einthoven played a key role in the production of the string galvonometer for commercial use, and the first was sold in 1908 to Edward Schafer at the University of Edinburgh. Just as important as his technical and commercial innovation was Einthoven's theoretical contribution to understanding the electrophysiology of the heart. In 1906, he published the first systematic presentation of normal and abnormal ECG tracings, thus describing a number of important electrophysiological abnormalities of the heart. Einthoven was awarded the Nobel Prize in medicine in 1924 for the invention of the electrocardiograph.

The modern arrangement of 12 ECG leads was pioneered in 1942 by Emanuel Goldberger. Although recent experiments with computerized ECGs employ 15 or more leads, the 12-lead arrangement remains standard. Other important developments involved improvements in power source and portability. The successful use of vacuum tubes was reported in 1928, marking a shift from the mechanical to the electric electrocardiograph. The first practical direct-writing ECG "cart" was developed in the 1950s. Advances in computerization, beginning in 1959 with the creation of the first analog-to-digital converter designed specifically for the electrocardiograph, resulted in the standardization by the mid-1980s of ECG carts capable of generating a computerized interpretation of ECG results. Despite the rise of new technologies such as echocardiography, which uses ultrasound to examine the heart, the ECG remains a vital tool of clinical cardiology and is likely to remain so for some time because of its simplicity, cost-effectiveness, and accuracy.

See also **Cardiovascular Disease, Diagnostic Methods; Diagnostic Screening**

TIMOTHY S. BROWN

Further Reading

Burch, G.E. and DePasquale, N.P. *A History of Electrocardiography*. Year Book Medical Publishers, Chicago, 1964.

Bynum, W.F., Lawrence, C. and Nutton, V., Eds. *The Emergence of Modern Cardiology*. Wellcome Institute for the History of Medicine, 1985.

Coumel, P. *Electrocardiography, Past and Future*. New York Academy of Sciences, New York, 1990.

Fye, W.B. A history of the origin, evolution, and impact of electrocardiography. *Am. J. Cardiol.*, 73, 937–949, 1994.

Horan, L.G. The quest for optimal electrocardiography. *Am. J. Cardiol.*, 41, 126–129, 1978.

Rautaharju, P.M. A hundred years of progress in electrocardiography 1: early contributions from Waller to Wilson. *Can. J. Cardiol.*, 3, 362–374, 1987.

Snellen, H.A. *Willem Einthoven (1860–1927) Father of Electrocardiography*. Kluwer, Dordrecht, 1995.

Useful Websites

Jenkins, D. 12-lead ECG library homepage. Electrocardiogram (ECG, EKG) Library, Llandough Hospital, Cardiff, Wales, 1999: http://homepages.enterprise.net/djenkins/ecghome.html.

Electrochemistry

Electrochemistry deals with the relationship between chemical change and electricity. Under normal conditions, a chemical reaction is accompanied by the liberation or absorption of heat and not of any other form of energy. However, there are many so-called electrochemical reactions that when allowed to proceed in contact with two electronic conductors joined by conducting wires, will generate electrical energy in this external circuit. Current between the electrodes (usually metallic plates or rods) is carried by electrons, while in the electrolyte, a nonmetallic ionic compound either in the molten condition or in solution in water or other solvents, ions carry the current.

Conversely, the energy of an electric current can be used to bring about many chemical reactions that do not occur spontaneously. The process in which electrical energy is directly converted into chemical energy is called electrolysis. The products of an electrolytic process have a tendency to react spontaneously with one another, reproducing the substances that were reactants and were therefore consumed during the electrolysis. If this reverse reaction is allowed to occur under proper conditions, a large proportion of the electrical energy used in the electrolysis can be regenerated. This possibility is used in accumulators or storage cells, sets of which are known as storage batteries.

Electrochemistry owes its rapid developments in the nineteenth century to the invention of the battery by Alessandro Volta in 1800, which provided the first source of continuous current. Within six weeks of Volta's report, two English scientists, William Nicholson and Anthony Carlisle, used a chemical battery to discover electrolysis, producing hydrogen and oxygen by passing an electric current through water. This is how the science of electrochemistry began. By 1809 the English chemist Humphry Davy had used a more powerful battery to isolate several very active metals for the first time—sodium, potassium, calcium, strontium, barium, and magnesium—by electrolyzing their liquid compounds. Michael Faraday, who was Davy's assistant at the time, studied electrolysis quantitatively and showed that the amount of energy needed to separate a gram of a substance from its compound is closely related to the atomic weight of the substance. However, the relationship between electrical charge and energy was not understood until the electron was discovered in 1896. The generation of an electrical current from gaseous hydrogen and oxygen in contact in an acidic electrolyte was first demonstrated by William Grove in 1839.

During the 1820s, Davy and Faraday attached pieces of zinc or iron to the copper sheeting on the bottom of ships, in order to prevent saltwater corrosion. Galvanic corrosion of metals occurs when two dissimilar metals are in contact in a conducting solution (such as seawater). An electrochemical cell forms, whereby the more active metal becomes the anode and corrodes (by losing metal ions) and the cathode is protected. This technique was not seriously used until 1956 when the U.S. Navy began to experiment with platinum-clad titanium anodes for the protection of their ships and submarines. Today pipelines and oilrigs are also electrochemically protected by placing zinc or magnesium "sacrificial anodes" near the steel structure.

In the twentieth century many new applications of electrochemical reactions were developed. Technologies of generation of electrical power from batteries and fuel cells are described in separate entries. Here we describe the use of electrolysis in metallurgy, electroplating and surface finishing, and analytical chemistry.

Most technologically important metals, except iron and steel, are either obtained or refined by electrolytic processes. In 1886, Charles Martin Hall in the U.S. and Paul-Louis-Toussaint Heroult in France independently and simultaneously discovered the modern method of commercially producing aluminum by electrolysis of purified alumina (Al_2O_3) dissolved in molten cryolite (Na_3AlF_6) ore. Copper is refined by electrolysis in aqueous copper sulfate solutions. Unrefined copper, with its impurities is made the anode, and thin sheets of pure copper form the cathode. As the anodes corrode away, pure metal is deposited on the cathodes. The widespread use of electrolysis for depositing protective coats on electronic conductors dates back to the middle of the twentieth century. In electroplating, the object to be coated is made the cathode in an electrolytic cell. Titanium, alkaline earth, and alkali metals are obtained by electrodeposition from molten salts, and automobile parts are chrome plated to protect them from corrosion. Surface finishing is the reverse of electroplating. Instead of coating or plating, metal is removed from the surface, leaving a smooth, clean finish.

In the chemical industry, electrolysis of seawater to obtain caustic soda (sodium hydroxide) and chlorine as a byproduct, used since 1900, has become one of the largest volume productions. Since the 1980s electrochemical process have been developed for the synthesis of a variety of inorganic compounds to the production of such synthetic fibers as nylon. Novel structures such as multilayers, nanowires, and granular composites can be formed or modified.

Electroanalysis employs electrochemical phenomena for quantitative analysis, for example detection and determination of various aqueous and gaseous ions. In electrochemical sensors the gas to be detected diffuses through a semipermeable membrane to the working electrode. At the electrode the gas is either oxidized or reduced, and the resulting current provides the signal. Electroanalytical techniques have been used for environmental monitoring and industrial quality control since the late 1960s, and more recently for biomedical and pharmaceutical analysis.

Certain technical advances in the 1980s and 1990s—such as the development of ultramicroelectrodes, the design of tailored interfaces and molecular monolayers, the coupling of biological components and electrochemical transducers, the synthesis of ionophores and receptors containing cavities of molecular size, and the development of high-resolution scanning probe microscopes—led to a substantial increase in the utility of electrochemical reactions. By the close of the century, electrochemical reactions were used in a vast range of processes and applications in multiple scientific fields and industries.

See also **Batteries, Primary and Secondary; Fuel cells**

MUZAFFAR IQBAL

Further Reading

Bard, A.J. *Electrochemical Methods: Fundamentals and Applications.* Wiley, New York, 2001.

Bockris, J.O. *Modern Electrochemistry.* Plenum, New York, 2001.

Brett, C.M.A. *Electroanalysis.* Oxford University Press, Oxford, 1998.

Bruce, Peter G., Ed. *Solid State Electrochemistry.* Cambridge University Press, Cambridge, 1997.

Dubpernell, G. and Westbrook, J.H., Eds. *Selected Topics in the History Of Electrochemistry.* Electrochemical Society, Princeton, 1978.

Ostwald, W. *Electrochemistry: History and Theory.* Verlag von veit, Leipzig, 1895. Translated by Date, N. P., for the Smithsonian Institution and the National Science Foundation, Amerind, New Delhi, 1980.

Samuel, R. *The Founders of Electrochemistry.* Dorrance, Philadelphia, 1975.

Stock, J.T., Ed. *Electrochemistry, Past and Present.* American Chemical Society, Washington D.C., 1989.

Taillefert, M. and Rozan, T.F., Eds. *Environmental Electrochemistry: Analysis of Trace Element Biogeochemistry.* American Chemical Society, Washington D.C. Distributed by Oxford University Press, 2002.

Wang, J. *Analytical Electrochemistry.* Wiley, New York, 2000.

Useful Websites

Electrochemistry Encyclopedia: http://electrochem.cwru.edu/ed/encycl/

Electroencephalogram (EEG)

An electroencephalogram (EEG) is the graphic measurement and analysis of electrical activity in the brain. The technique allows diagnostic information about brain activity to be determined noninvasively; that is, without recourse to surgery. Electrical activity in the body was studied from as early as the 1780s as seen in Luigi Galvani's work on "animal electricity." A professor at the university in Bologna, Galvani demonstrated that contact with an electric circuit would cause frogs' legs to contract. Richard Caton, an English physician, was the first scientist to report that he had recorded electrical activity in the brains of rabbits and monkeys. Caton used a galvanometer to detect the electrical signals. Galvanometers work on the principle that a current-carrying conductor in a magnetic field experiences a force. The conductor is mounted on pivots in a constant magnetic field. Once current flows through the needle, the needle deflects, with the deflection proportional to the electric current. In his experiments, Caton detected small currents of varying direction when two points were placed on the external surface of the skull.

The first human EEG was recorded by a German physician, Dr. Hans Berger (1873–1941), who published his findings in 1929. Berger used two sheets of aluminum foil as electrodes, placing one at the back and one at the front of the head. The signals were detected with a Siemen's double coil galvanometer. Electrical activity in the brain can be measured from the scalp surface due to volume conductor effects. The signal obtained will depend on the distance from the electrodes to the source of the signal and on the intervening material. The electrodes are separated from the brain by the skull and meninges (membranes) lining the brain, and large areas of the brain are comparatively remote from the electrodes. It is estimated that recordings of electrical activity through scalp electrodes can sample activity from only 0.1 to 1 percent of the neuronal population.

Although the basic measurement principles in electroencephalography are similar to those of electrocardiography, the amplitudes of the EEG potentials are approximately 1000 times smaller. Electroencephalography technology was slower to develop because of the difficulty in measuring and amplifying small signals. In the 1930s valve amplifiers were used to boost the signals detected because amplifiers increased the amplitude of a signal while maintaining the signal integrity. Pen writers were introduced in the 1940s as the recording device to graph the EEG potentials. The pen writers were driven by the signal output, and the pen moved in response to the signal changes. The pens were placed over a moving paper chart in order to record amplitude variations along a time axis.

Electroencephalography enabled clinicians to obtain diagnostic information about brain activity and brain abnormalities because EEG signals represent collective neural activity in the brain. In the 1930s EEGs found immediate application in exploring epileptic seizure disorders. Epilepsy is characterized by pronounced EEG abnormalities. American scientists Alfred L. Loomis, E.N. Harvey, and G.A. Horbart were among the first to study human sleep EEGs and EEG patterns for different stages of sleep. Walter (1936) moved EEG research into the study of other disorders with the discovery of patterns associated with brain tumors. Brain tumors can be located by a careful examination of EEG potentials along the whole contour of the scalp. A number of common EEG waveforms were subsequently characterized. Alpha waves are characteristic of the awake closed eyes state. They have amplitudes of 20–50 microvolts and a frequency of 8.0–13.5 Hz. They are not recorded if the eyes are opened. Theta waves have a frequency of 7.5–4 Hz and occur in drowsiness. Delta waves have a frequency of less than 4Hz and occur in subjects who are in deep sleep or coma. Mu waves are associated with motor activity, and lambda waves with viewing patterned visual displays

The basic measurement method has seen little change since the 1950s. Electronic amplifiers replaced valve amplifiers, which increased sensitivity and provided better signal to noise ratios. Electrodes were made smaller and placed at many points over the scalp. Electrode positioning, or montage, was standardized to ensure that each electrode was placed at the same position on the head with each subject. The 10/20 montage system defines positions on the scalp in relation to particular points on the head such as the "nosion" at the front of the head and the "inion" at the back of the head. Electrodes are placed at specified fractions of distance between these positions. The electrodes detect local differences in electrical activity between electrodes placed on the scalp. The responses are amplified by a type of electronic amplifier designed to amplify the difference between two signals. The responses are displayed as numerous graphs, each representing potential differences between electrodes in the 10/20 montage. Systems generally have 8 or 16 channels, each recording EEG data. The standard presentation is a graph of voltage as it varies with time. The EEG is displayed on monitors or printed by chart recorders.

As the frequency is also important in characterizing signals, researchers began looking at the frequency spectrum. In 1965 J. W. Cooley and J. W. Tookey introduced the use of the "fast Fourier transform" to EEG as the basis of power spectral analysis. Frequency analysis also introduced new ways of displaying frequency information as color-coded patterns. The colored patterns depict the frequency distribution of electrical activity in different areas of the brain.

The 1970s saw the development of an associated technology called evoked potential measurement. Evoked potential studies examined sensory function by recording brain responses to a specific sensory stimulus. For example visual responses may be stimulated in the subject who is viewing changing patterns on a monitor. Detection of the

evoked response among all the other brain activity that occurs simultaneously is achieved through a signal-processing technique called signal averaging. The stimulus is repeated, and, over a short poststimulus period, responses are monitored. In the poststimulus period, activity not related to the stimulus will be randomly distributed. The visual evoked responses, which are time-locked to the stimulus, will occur at the same point in relation to the stimulus. If sufficient responses are recorded, the evoked response can be distinguished from other electrical brain activity once the responses are averaged.

By the end of the twentieth century, interest in EEG measurements had waned. Computed tomography (CT) and magnetic resonance imaging (MRI) provided high-quality brain images, and functional MRI applications were developed which offered new diagnostic and research possibilities. In the 1980s, EEG brain topography was developed. This technique exploits the availability of a great number of digitized channels providing information simultaneously from an array of points across the scalp. Developments with evoked potentials led to research into their use for establishing depth of anesthesia for patients undergoing surgery. EEGs have continued to be used, however, in conjunction with positron emission tomography (PET) and MRI in studies that merge information from numerous modalities to provide more comprehensive information about brain activity.

See also **Neurology; Nuclear Magnetic Resonance (NMR, MRI); Positron Emission Tomography (PET); Tomography in Medicine**

COLIN WALSH

Further Reading

Brown, B.H., Smallwood, R.H., Barber, D.C., Lawford, P.V. and Hose, D.R. *Medical Physics and Biomedical Engineering*. Institute of Physics Publishing, U.K., 1999.

Cooper, R., Ossetton, J.W., Shaw, JC. EEG Technology, 3rd edn. Butterworth–Heinemann, 1980.

Davidovits, P. *Physics in Biology and Medicine*. 2nd edn. Harcourt/Academic Press, U.S.

Dyro, F.P *The EEG Handbook*. Little, Brown & Company, U.S., 1989.

Neidermeyer, E. *Historical Aspects (of EEG)*, in *EEG: Basic Principles, Clinical Applications and Related Fields*, 3rd edn., Neidermeyer, E. and Da Silva, F.L., Eds. Williams & Wilkins, Baltimore, 1987.

Useful Websites

http://www.neuro.com/megeeg/
http://www.epub.org.br/cm/n03/tecnologia/historia.htm

Electronic Communications

The development of digital computing and communication technology in the 1940s and 1950s was largely driven by Cold War military needs in the midst of closed-world politics. Extensive funding was provided for large-scale research and development projects during this period by the U.S. military. The origins of the communication of digital information can be attributed to the Whirlwind computer, which was developed under these conditions at the Massachusetts Institute of Technology (MIT) in Boston. This was a powerful general-purpose digital computer orientated toward real-time control and flight simulation. However, it eventually found use as the control computer in the semiautomatic ground environment (SAGE) air defense system. SAGE connected remote early-warning radar stations in the far reaches of the Arctic with control centers in the heartland to automatically direct fighter aircraft to intercept the perceived onslaught of Soviet bombers carrying doomsday nuclear arsenals. A means had to be invented to communicate digital information over long distances between the radar stations and the SAGE control centers, and this resulted in the techniques for the long-distance communication of digital data being developed. This is the origin of the modem (as it later became known) and the system became operational in 1952. The name is derived from the process of the modulation and demodulation, whereby a waveform of digital data (ones and zeros) are superimposed onto a sinusoidal carrier wave, since the square waveform of digital data cannot be sent over distances. This information is then extracted with the process of demodulation on the receiver side. It should also be noted that many other key computing technological developments resulted from the SAGE project, such as video displays, magnetic core memory, and networking among others. This technology diffused into civilian use with the SABRE airlines reservation system built by IBM for American Airlines in 1964, which inherited a lot of technological developments from SAGE. The system used modems to transmit data signals over ordinary analog telephone channels and was an early example of the general trend of the diffusion of military-sponsored computing technology into broader society.

Among other commercial data communications systems was the remote teletype terminal, built in the late 1950s. Modems are measured by the rate of data transmission in bit per second (bps or baud) and these first commercial implementations provided 110 bps. Bell Laboratories later developed the first commercial component modems, which were largely used for remote terminals and for the long-distance interconnection of computers for data transfer. By 1962 the first commercial modems were on the market. The Bell 103 was the first modem to provide full duplex transmission and had a data transmission speed of 300 bps. Further theoretical signal processing developments in optimization of encoding and compression techniques increased the speed of data transmission, another theme which would continue until the end of the century. The Bell 212 modem later provided 1200 bps. However, an important development was the setting of internationally agreed standards, with the International Telephone and Telegraph Consultative Committee (CCITT) establishing the V.21 standard in 1964, which defined 300 bps communication protocols. Handshaking—which begins before data transmission, establishes an electrical path and synchronization, and enables the two devices to send messages agreeing on a communications protocol. Communications protocols define message format and rate of transmission. Having the data communication industry produce modems to mutually agreed standards was a key factor in later proliferation, although the Bell standards continued to be used and supported.

Computing technology grew at a phenomenal pace during the 1960s with the mainframe and later the minicomputer proliferating into universities, research centers, and large corporations globally. The need for these machines and user communities to communicate and share resources led to concept of a "net," which was first discussed and proposed in 1961 by Leonard Kleinrock at MIT. His work covered much of the essential theory of data traffic behavior. Packet switching data networks as a way of achieving this was a significant shift in thinking from traditional data communications. This network would interconnect geographically dispersed computing resources, utilizing modem communications over long-distance telephone lines. The stream of digital data would be broken into packets, which were sent one at a time with no central control. Although there were research networks in the U.K. and elsewhere, the U.S. Advanced Research Projects Agency packet switching network (ARPANET) was conceived by Paul Baran at the Rand Corporation in early 1960s, and was the first to be implemented in 1969 in the U.S. Computers were connected to the network via switching nodes which were called interface message processors (IMPs) which were themselves computers and connected to each other via permanent (leased) telephone lines and adaptive equalizer modems supporting up to 56,000 bps (or 56 kbps). Communication tasks were conceptually dealt with in logical layers of functionality and utility, abstracting higher levels of application communication for lower levels of physical communication of data. This meant that exactly how the data was transported became irrelevant to an application.

The Advanced Research Projects Agency Network (ARPANET) was designed primarily for resource sharing. However, it came to be used in ways that were unintended, which arguably drove its success. Electronic messaging (e-mail, net notes, or mail) was the most significant activity on the early ARPANET. Although messaging had been present on host computers from as early as 1965, in 1971 programmer Ray Tomlinson from the company Bolt, Beranek and Newman (BBN), which designed and built the IMPs, adapted earlier code to automatically relay messages between different host computers interconnected on the net. This text message file was transmitted using the underlying file transport mechanisms. He also introduced the "username@hostname" format of messaging addressing which was later to become the standard addressing scheme for e-mail. His system was later refined and incorporated into the standard ARPANET interconnection package, and e-mail quickly became the network's most used feature, generating more traffic over the network than any other application. E-mail also opened up ways of communication that enhanced the collaboration of remotely located research groups, and was used socially between nontechnical users in ways totally unanticipated and unforeseen by the creators of the ARPANET.

However, access to ARPANET was restricted to an elite at the top research universities and institutions in the U.S. and elsewhere to a limited extent. In response to this, a separate network emerged in the later 1970s: Usenet. This was an electronic messaging system built in 1979 at Duke University, North Carolina, with no formal funding. Graduate students programmed their Unix computer to automatically link up with other Unix systems. They used homemade modems and standard Unix programs (UUCP) to build a system that automatically dialed up over ordinary telephone lines and replicated files between a series of computers. Usenet had no formal structure, and was available to all who were interested as long as they had access to the Unix operating system, which was available at low cost to the academic

and computer research community. Connections later became available via personal computers.

Usenet did not only distribute e-mail; it also facilitated a structured group discussion system called "netnews," later known as "newsgroups." These emulated an interactive newspaper, but allowed the reader to be selective. More importantly, readers could participate in discussion threads, interactively contributing to the flow of thoughts and information. The Unix developers at AT&T's Bell Labs provided a lot of support for Usenet, since they saved millions of dollars on development by using the Usenet community for help with debugging their software. This in turn enabled the network to further proliferate, since Bell Labs provided a lot of resources that helped with the distributional logistics of the network. Digital Equipment Corporation (DEC) also supported Usenet, and ultimately the spread of Usenet and Unix encouraged the sale of computers that ran the Unix operating system. The Usenet newsgroups communities provided considerable technical help for peers in an ever-proliferating range of topics that extended well beyond that of technical concerns. Usenet grew rapidly, with the number of sites and articles per day doubling yearly until 1988. Once again the scale of this growth took the original Usenet community by surprise since it was not at all anticipated. Although there was hesitation about this growth from the original developers, seemingly insurmountable problems were investigated and solved by the Usenet community. This community referred to Usenet as 'the Net'. Many communities in the U.S. and around the world established "Free Nets," based on free Usenet software. Later in the 1980s the popularity of the mass-produced personal computers and modems meant that Usenet became more accessible to individuals and small organizations via dial-up UUCP connections. Some of these Free Nets had access to the worldwide newsgroups of Usenet. A gateway through to the ARPANET was established in 1981 at The University of California at Berkeley.

Bitnet was another early network that provided a form of electronic messaging. Most notably it was the origins of the list server, a facility that provided one-to-many messaging. Once again it was entirely independent of the ARPANET developments and was primarily linked to IBM computers at academic installations on the East Coast of the U.S. It was initiated in 1981. The list server was also a very important factor in the discussion of new ideas for the improvement of the network as well as resolution and acceptance of new standards. It could thus be considered self-perpetuating, a characteristic generally thought to be unique to the (later) Internet. The Listserv program was moved over to the Unix operating system in 1991, freeing it from the proprietary hold of the IBM mainframe. The Listserv communities ultimately migrated to the Internet once it emerged, where proliferation took place on a much grander scale, for the same reasons as they did on the original Bitnet network.

The development of the personal computer (PC) in the mid-1970s by hobbyists was to have profound impact on electronic communications in broader society. PCs were made possible by integrated circuit technology, which was a spin-off of microelectronics used in 1960s aerospace projects such as the Apollo Lunar Program. By the 1980s PCs and modems had become consumer items. The Hayes modem, invented in 1977 by Dennis Hayes, provided autonomous functionality since the modem itself contained a computer in the form of a microprocessor. Thus it was able to perform "intelligent" functions such autodial and autoanswer automatically and could be programmed to respond in certain ways to different line conditions and communication protocols. Many low-level communication tasks were "off-loaded" to the modem. A consequence of this was that it allowed personal computers to be set up as central communication hosts, running bulletin board software (BBS), which provided discussion forums and electronic messaging among other more utilitarian functionality such as file transfer. Electronic messaging communities developed outside of the corporations and academic institutions and reflected diverse social interest groups, increasing the diffusion of the concept of electronic messaging and communication by attracting like-minded people to join these communities and participate. Gateways to other BBSs as well as other established networks such as Bitnet further fueled growth and acceptance of PC-based messaging. This included many independent (and in most cases noncommercial) networks such as FidoNet, DASNet and others.

Large commercial messaging systems soon followed which also became economically feasible with the popularity of the PC in the early to mid 1980s. Most of these were located in the U.S. and made extensive use of long-distance commercial data services to act as a "backbone" between distributed hosts (large computers) that were coupled together to appear logically as one system. Users connected directly to local access points via the standard modems and telephone lines, and generally were charged for the amount of time they were connected. The systems were all proprietary based on the functionality of the BBS as described above, and included CompuServe, The Source, BIX, WELL, Dialcom and many others. Some only offered e-mail, like MCI Mail and Telemail. Gateways between these systems and the Internet (as the ARPANET became in the mid1980s) and other public networks were introduced in time. A limitation of these systems was the technical barriers in establishing a connection with them. Each system required connection with unique modem communication parameters such as number of bits, parity, and checksum with each system having its own unique configuration. This required a technical literacy from users, and acted as a barrier to broader proliferation; however, many users were successful and brought together millions of users who became familiar with communicating via electronic messaging and accessing information "online" from their homes and workplaces.

It is worth noting similar developments of electronic communities in Europe and elsewhere. The Prestel system originated from work done in the 1960s by the U.K. Post Office on a standard they called Viewdata. It was an interactive data service that initially required dedicated terminals with all the electronics built and was unique in that it provided limited graphics. It was initially designed to coexist with a television set as the video display unit. The service was based on host computers on which subscribers could publish information. There were gateways to other information service providers. It was later possible to access the Prestel service from a personal computer and modem with 75 bps upstream and 300 bps downstream communication. Although it was popular, it was not commercially successful. This is in contrast to the French Minitel system (initially known as Télétel), which was similar in concept except that the French telephone network gave the Minitel Terminals away to subscribers to reduce the cost of printing telephone directories. The service was highly successful. In 1987 it was the world's largest e-mail system and 6.5 million terminals were in use by 1995.

These disparate systems and electronic communities discussed above ultimately converged and migrated onto the Internet in the 1990s, once that emerged as the dominant network. Most of the technology was software and could be "reprogrammed," drawing these communities of users together onto one system that nobody but everybody owned. For this and other reasons, the global proliferation of the Internet acted as the great unifying force for the growth of global electronic messaging or "cyber" communities, with e-mail, newsgroups and forums as the basis of communications. Many of the communities discussed above rapidly migrated to similar means of communication of the Internet, propelling its phenomenal growth.

Not all digital communication technology was successful. The telecommunication carrier and service provider community attempted to convert the subscriber analog telephone system in a deterministic way to a digital circuit with the Integrated Services Digital Network (ISDN) system. The I.120 ISDN recommendation was published in 1984 by the Comité Consultatif International Téléphonique et Télégraphique (CCITT), but it was unspecific in areas and open to interpretation. Despite extensive promulgation and marketing, ISDN did not find widespread use or success. Reasons included incompatibility between manufacturers' implementations, as well as higher bandwidth coming from analog modem developments by the time the ISDN services were working. Thus there was little incentive for users to convert to ISDN, although it did find some success later as a data link backup for permanent digital network lines.

The series of CCITT modem standards cumulated in the 56 kbps V.90 standard, finalized in 1998, although successful manufacturer-specific implementations were being used years before that. The introduction of digital subscriber lines (DSL) late in the 1990s provided a significant increase bandwidth (512 kbps). However having such "broadband" bandwidth available did not change the way users used the net, rather it merely speeded up the process, and opened up further avenues of use that users embraced, such as the transfer of digital images, sound and video files. It should also be noted that the advantages of high bandwidth were not always supported further up the network than the point of access. DSL also had line distance restrictions, and thus was restricted to urban areas.

The broad use of digital electronic message communications in most societies by the end of the century can be attributed to a myriad of reasons. Diffusion was incremental and evolutionary. Digital communication technology was seeded by large-scale funding for military projects that broke technological ground, however social needs and use drove systems in unexpected ways and made it popular because these needs were embraced. Key technological developments happened long before diffusion into society, and it was only after popularity of the personal computer that global and widespread use became commonplace. The Internet was an important medium in this regard, however the popular uses of it were well established long before its success. Collaborative developments with open, mutually agreed standards were key factors in broader diffusion of the low-level transmission of digital data, and provided resistance to technological lock-in by any commercial player. By the twenty-first century, the concept of interpersonal electronic messaging was accepted as normal and taken for granted by millions around the world, where infrastructural and political freedoms permitted. As a result, traditional lines of information control and mass broadcasting were challenged, although it remains to be seen what, if any, long-term impact this will have on society.

See also **Computer Networks; Computers, Personal; Internet; Packet Switching; World Wide Web**

BRUCE GILLESPIE

Further Reading

Abbate, J. *Inventing the Internet*. MIT Press, Cambridge, MA, 1999.

Bijker, W., Hughes, T., Pinch, T., Eds. *The Social Construction of Technological Systems*. MIT Press, Cambridge, MA, 1989.

Ceruzzi, P.E. *A History of Modern Computing*. MIT Press, Cambridge, MA, 1998.

Edwards, P.N. *The Closed World*. MIT Press, Cambridge, MA, 1996.

Grier, D. and Campbell, M. A social history of Bitnet and Listserv, 1985–1991. *Ann. Hist. Comp.*, 22, 2, 32–41, 2000.

Hauben, M. *Netizens, On the History and Impact of the Usenet and the Internet*. Institute of Electrical and Electronics Engineers Computer Society, Los Alamitos, 1997.

Hughes, T.P. *Rescuing Prometheus*. Vintage Books, New York, 2000.

Quarterman, J.S. *The Matrix, Computer Networks and Conferencing Systems Worldwide*. Digital Press, Boston, 1990.

Winston, B. *Media Technology and Society, A History: From the Telegraph to the Internet*. Routledge, London, 1998.

Electronics

Electronic systems in use today perform a remarkably broad range of functions, but they share the technical characteristic of employing electron devices such as vacuum tubes, transistors, or integrated circuits. Most electron devices in use today function as electric switches or valves, controlling a flow of electrons in order to perform useful tasks. Electron devices differ from ordinary electromechanical switches or current- or voltage-control devices in that an applied electric current or field controls electron flow rather than a mechanical device. Electronic devices are "active," like machines, but have no moving parts, so engineers distinguish them both from electromechanical devices and from other "passive" electrical components such as wires, capacitors, transformers, and resistors. When the word electronics was coined around 1930, it usually referred to so-called vacuum tubes (valves), which utilize electrons flowing through a vacuum. With the advent of the transistor in the late 1940s, a second term emerged to describe this new category of "solid-state" electron devices, which performed some of the same functions as vacuum tubes but consisted of solid blocks of metal.

Few of the basic tasks that electronic technologies perform, such as communication, computation, amplification, or automatic control, are unique to electronics. Most were anticipated by the designers of mechanical or electromechanical technologies in earlier years. What distinguishes electronic communication, computation, and control is often linked to the instantaneous action of the devices, the delicacy of their actions compared to mechanical systems, their high reliability, or their tiny size.

The electronics systems introduced between the late nineteenth century and the end of the twentieth century can be roughly divided into the applications related to communications (including telegraphy, telephony, broadcasting, and remote detection) and the more recently developed fields involving digital information and computation. In recent years these two fields have tended to converge, but it is still useful to consider them separately for a discussion of their history.

The origins of electronics as distinguished from other electrical technologies can be traced to 1880 and the work of Thomas Edison. While investigating the phenomenon of the blackening of the inside surface of electric light bulbs, Edison built an experimental bulb that included a third, unused wire in addition to the two wires supporting the filament. When the lamp was operating, Edison detected a flow of electricity from the filament to the third wire, through the evacuated space in the bulb. He was unable to explain the phenomenon, and although he thought it would be useful in telegraphy, he failed to commercialize it. It went unexplained for about 20 years, until the advent of wireless telegraphic transmission by radio waves. John Ambrose Fleming, an experimenter in radio,

not only explained the Edison effect but used it to detect radio waves. Fleming's "valve" as he called it, acted like a one-way valve for electric waves, and could be used in a circuit to convert radio waves to electric pulses so that that incoming Morse code signals could be heard through a sounder or earphone.

As in the case of the Fleming valve, many early electronic devices were used first in the field of communications, mainly to enhance existing forms of technology. Initially, for example, telephony (1870s) and radio (1890s) were accomplished using ordinary electrical and electromechanical circuits, but eventually both were transformed through the use of electronic devices. Many inventors in the late nineteenth century sought a functional telephone "relay"; that is, something to refresh a degraded telephone signal to allow long distance telephony. Several people simultaneously recognized the possibility of developing a relay based on the Fleming valve. The American inventor Lee de Forest was one of the first to announce an electronic amplifier using a modified Fleming valve, which he called the Audion. While he initially saw it as a detector and amplifier of radio waves, its successful commercialization occurred first in the telephone industry. The sound quality and long-distance capability of telephony was enhanced and extended after the introduction of the first electronic amplifier circuits in 1907. In the U.S., where vast geographic distances separated the population, the American Telephone and Telegraph Company (AT&T) introduced improved vacuum tube amplifiers in 1913, which were later used to establish the first coast-to-coast telephone service in 1915 (an overland distance of nearly 5000 kilometers).

These vacuum tubes soon saw many other uses, such as a public-address systems constructed as early as 1920, and radio transmitters and receivers. The convergence of telephony and radio in the form of voice broadcasting was technically possible before the advent of electronics, but its application was greatly enhanced through the use of electronics both in the radio transmitter and in the receiver.

World War I saw the applications of electronics diversify somewhat to include military applications. Mostly, these were modifications of existing telegraph, telephone, and radio systems, but applications such as ground-to-air radio telephony were novel. The pressing need for large numbers of electronic components, especially vacuum tubes suitable for military use, stimulated changes in their design and manufacture and contributed to improving quality and falling prices. After the war, the expanded capacity of the vacuum tube industry contributed to a boom in low-cost consumer radio receivers. Yet because of the withdrawal of the military stimulus and the onset of the Great Depression, the pace of change slowed in the 1930s. One notable exception was in the field of television. Radio broadcasting became such a phenomenal commercial success that engineers and businessmen were envisioning how "pictures with sound" would replace ordinary broadcasting, even in the early 1930s. Germany, Great Britain, and the U.S. all had rudimentary television systems in place by 1939, although World War II would bring nearly a complete halt to these early TV broadcasts.

World War II saw another period of rapid change, this one much more dramatic than that of World War I. Not only were radio communications systems again greatly improved, but for the first time the field of electronics engineering came to encompass much more than communication. While it was the atomic bomb that is most commonly cited as the major technological outcome of World War II, radar should probably be called the weapon that won the war. To describe radar as a weapon is somewhat inaccurate, but there is no doubt that it had profound effects upon the way that naval, aerial, and ground combat was conducted. Using radio waves as a sort of searchlight, radar could act as an artificial eye capable of seeing through clouds or fog, over the horizon, or in the dark. Furthermore, it substituted for existing methods of calculating the distance and speed of targets. Radar's success hinged on the development of new electronic components, particularly new kinds of vacuum tubes such as the klystron and magnetron, which were oriented toward the generation of microwaves. Subsidized by military agencies on both sides of the Atlantic (as well as Japan) during World War II, radar sets were eventually installed in aircraft and ships, used in ground stations, and even built into artillery shells. The remarkable engineering effort that was launched to make radar systems smaller, more energy efficient, and more reliable would mark the beginning of an international research program in electronics miniaturization that continues today. Radar technology also had many unexpected applications elsewhere, such as the use of microwave beams as a substitute for long-distance telephone cables. Microwave communication is also used extensively today for satellite-to-earth communication.

The second major outcome of electronics research during World War II was the effort to build an electronic computer. Mechanical adders and calculators were widely used in science, business, and government by the early twentieth century, and had reached an advanced state of design. Yet the problems peculiar to wartime, especially the rapid calculation of mountains of ballistics data, drove engineers to look for ways to speed up the machines. At the same time, some sought a calculator that could be reprogrammed as computational needs changed. While computers played a role in the war, it was not until the postwar period that they came into their own. In addition, computer research during World War II contributed little to the development of vacuum tubes, although in later years computer research would drive certain areas of semiconductor electron device research.

While the forces of the free market are not to be discounted, the role of the military in electronics development during World War II was of paramount importance. More-or-less continuous military support for research in electronic devices and systems persisted during the second half of the twentieth century too, and many more new technologies emerged from this effort. The sustained effort to develop more compact, rugged devices such as those demanded by military systems would converge with computer development during the 1950s, especially after the invention of the transistor in late 1947.

The transistor was not a product of the war, and in fact its development started in the 1930s and was delayed by the war effort. A transistor is simply a very small substitute for a vacuum tube, but beyond that it is an almost entirely new sort of device. At the time of its invention, its energy efficiency, reliability, and diminutive size suggested new possibilities for electronic systems. The most famous of these possibilities was related to computers and systems derived from or related to computers, such as robotics or industrial automation. The impetus for the transistor was a desire within the telephone industry to create an energy-efficient, reliable substitute for the vacuum tube. Once introduced, the military pressed hard to accelerate its development, as the need emerged for improved electronic navigational devices for aircraft and missiles.

There were many unanticipated results of the substitution of transistors for vacuum tubes. Because they were so energy efficient, transistors made it much more practical to design battery powered systems. The small transistor radio (known in some countries simply as "the transistor"), introduced in the 1950s, is credited with helping to popularize rock and roll music. It is also worth noting that many developing countries could not easily provide broadcasting services until the diffusion of battery operated transistor receivers because of the lack of central station electric power. The use of the transistor also allowed designers to enhance existing automotive radios and tape players, contributing eventually to a greatly expanded culture of in-car listening. There were other important outcomes as well; transistor manufacture provided access to the global electronics market for Asian radio manufacturers, who improved manufacturing methods to undercut their U.S. competitors during the 1950s and 1960s. Further, the transistor's high reliability nearly eliminated the profession of television and radio repair, which had supported tens of thousands of technicians in the U.S. alone before about 1980.

However, for all its remarkable features, the transistor also had its limitations; while it was an essential part of nearly every cutting-edge technology of the postwar period, it was easily outperformed by the older technology of vacuum tubes in some areas. The high-power microwave transmitting devices in communications satellites and spacecraft, for example, nearly all relied on special vacuum tubes through the end of the twentieth century, because of the physical limitations of semiconductor devices. For the most part, however, the transistor made the vacuum tube obsolete by about 1960.

The attention paid to the transistor in the 1950s and 1960s made the phrase "solid-state" familiar to the general public, and the new device spawned many new companies. However, its overall impact pales in comparison to its successor—the integrated circuit. Integrated circuits emerged in the late 1950s, were immediately adopted by the military for small computer and communications systems, and were then used in civilian computers and related applications from the 1960s. Integrated circuits consist of multiple transistors fabricated simultaneously from layers of semiconductor and other materials. The transistors, interconnecting "wires," and many of the necessary circuit elements such as capacitors and resistors are fabricated on the "chip." Such a circuit eliminates much of the laborious process of assembling an electronic system such as a computer by hand, and results in a much smaller product. The ability to miniaturize components through integrated circuit fabrication techniques would lead to circuits so vanishingly small that it became difficult to connect them to the systems of which they were a part. The plastic housings or "packages" containing today's micro-

processor chips measure just a few centimeters on a side, and yet the actual circuits inside are much smaller. Some of the most complex chips made today contain many millions of transistors, plus millions more solid-state resistors and other passive components.

While used extensively in military and aerospace applications, the integrated circuit became famous as a component in computer systems. The logic and memory circuits of digital computers, which have been the focus of much research, consist mainly of switching devices. Computers were first constructed in the 1930s with electromechanical relays as switching devices, then with vacuum tubes, transistors, and finally integrated circuits. Most early computers used off-the-shelf tubes and transistors, but with the advent of the integrated circuit, designers began to call for components designed especially for computers. It was clear to engineers at the time that all the circuits necessary to build a computer could be placed on one chip (or a small set of chips), and in fact, the desire to create a "computer on a chip" led to the microprocessor, introduced around 1970. The commercial impetus underlying later generations of computer chip design was not simply miniaturization (although there are important exceptions) or energy efficiency, but also the speed of operation, reliability, and lower cost. However the inherent energy efficiency and small size of the resulting systems did enable the construction of smaller computers, and the incorporation of programmable controllers (special purpose computers) into a wide variety of other technologies. The recent merging of the computer (or computer-like systems) with so many other technologies makes it difficult to summarize the current status of digital electronic systems. As the twentieth century drew to a close, computer chips were widely in use in communications and entertainment devices, in industrial robots, in automobiles, in household appliances, in telephone calling cards, in traffic signals, and in a myriad other places. The rapid evolution of the computer during the last 50 years of the twentieth century was reflected by the near-meaninglessness of its name, which no longer adequately described its functions.

From an engineering perspective, not only did electronics begin to inhabit, in an almost symbiotic fashion, other technological systems after about 1950, but these electronics systems were increasingly dominated by the use of semiconductor technology. After virtually supplanting the vacuum tube in the 1950s, the semiconductor-based transistor became the technology of choice for most subsequent electronics development projects. Yet semiconducting alloys and compounds proved remarkably versatile in applications at first unrelated to transistors and chips. The laser, for example, was originally operated in a large vacuum chamber and depended on ionized gas for its operation. By the 1960s, laser research was focused on the remarkable ability of certain semiconducting materials to accomplish the same task as the ion chamber version. Today semiconductor devices are used not only as the basis of amplifiers and switches, but also for sensing light, heat, and pressure, for emitting light (as in lasers or video displays), for generating electricity (as in solar cells), and even for mechanical motion (as in micromechanical systems or MEMS).

However, semiconductor devices in "discrete" forms such as transistors, would probably not have had the remarkable impact of the integrated circuit. By the 1970s, when the manufacturing techniques for integrated circuits allowed high volume production, low cost, tiny size, relatively small energy needs, and enormous complexity; electronics entered a new phase of its history, having a chief characteristic of allowing electronic systems to be retrofitted into existing technologies. Low-cost microprocessors, for example, which were available from the late 1970s onward, were used to sense data from their environment, measure it, and use it to control various technological systems from coffee machines to video tape recorders. Even the human body is increasingly invaded by electronics; at the end of the twentieth century, several researchers announced the first microchips for implantation directly in the body. They were to be used to store information for retrieval by external sensors or to help deliver subcutaneous drugs. The integrated circuit has thus become part of innumerable technological and biological systems.

It is this remarkable flexibility of application that enabled designers of electronic systems to make electronics the defining technology of the late twentieth century, eclipsing both the mechanical technologies associated with the industrial revolution and the electrical and information technologies of the so-called second industrial revolution. While many in the post-World War II era once referred to an "atomic age," it was in fact an era in which daily life was increasingly dominated by electronics.

See also **Audio Recording; Computers, Uses and Consequences; Control Technology—Electronic Signals; Electronic Communications; Integrated Circuits; Lasers; Lighting Techniques; Radio Receivers; Radio Transmitters; Rectifiers; Transistors; Valves/Vacuum Tubes**

DAVID MORTON

Further Reading

Dummer, G.W.A. *Electronic Inventions and Discoveries*, 4th edn. Institute of Physics Publishing, Bristol, 1997.
Fagen, M.D. et al., Eds. A History of Engineering and Science in the Bell System, 5 vols. 1975–84.
Riordan M. and Hoddeson, L. *Crystal Fire: The Birth of the Information Age*. W.W. Norton, New York, 1998.

Electrophoresis

Electrophoresis is a separation technique that involves the migration of charged colloidal particles in a liquid under the influence of an applied electric field. The word is derived from *electro*, referring to the energy of electricity, and *phoresis*, from the Greek verb *phoros*, meaning "to carry across." Electrophoresis has many applications in analytical chemistry, particularly biochemistry. It is one of the staple tools in molecular biology and it is of critical value in many aspects of genetic manipulation, including DNA studies, and in forensic chemistry.

Swedish biochemist Arne Tiselius carried out studies on proteins and colloids in the 1920s, and in 1930 introduced electrophoresis as a new technique for separating proteins in solution on the basis of their electrical charge. Tiselius was awarded the 1948 Nobel Prize in chemistry for this work, and the technique became a common tool in the 1940s and 1950s. Biological molecules such as amino acids, peptides, proteins, nucleotides, and nucleic acids, possess ionizable groups. At any given pH (concentration of hydrogen ions), these molecules exist in solution as electrically charged species either as cations (positive, or +) or anions (negative, or −). Depending on the nature of the net charge, the charged particles will migrate either to the cathode or to the anode. For example, proteins in an electric field separate according to size, shape, and charge with charges contributed by the side chains of the amino acids composing the proteins. The charge of the protein depends on the hydrogen ion content of the surrounding buffer with a high ionic strength resulting in a greater charge.

The information derived from electrophoretic methods has long been considered particularly valuable because of the gentleness of the method. The application of other separation methods, involving precipitation or aggressive chemicals, may easily cause damage to substances as unstable as biocolloids. In electrophoretic separation, the migrating substances remain in the same medium during the entire process, and the observation method is likely to give warning of eventual irreversible changes accompanying the separation.

If the material under investigation is in the form of a reasonably stable, dilute suspension or emulsion containing microscopically visible particles or droplets, then electrophoretic behavior can be observed directly. Information relevant to soluble material can also be obtained if the substance is adsorbed on to the surface of a carrier. The term zone electrophoresis refers to electrophoresis that is carried out in a supporting medium, whereas moving boundary electrophoresis is carried out entirely in a liquid phase. Most electrophoretic methods use a supporting media, such as starch powder, paper, polyacrylamide gel, or agar gel. Paper was used in the late 1940s and in 1955, Oliver Smithies used starch gel as a medium to minimize convection; unfortunately however, starch has its own small charged groups, which produce an electroendosmotic flow. In 1959 Leonard Ornstein and Baruch Joel Davis, and independently, Samuel Raymond and L. Weintraub introduced the use of polyacrylamide gel. Stallan Hjertén showed that polyacrylamide gels could act as "molecular sieves," whose pore size allowed separation of proteins by size even if their charge was the same. Polyacrylamide gels are still among the most used for electrophoretic separation of proteins, agarose gels are used for nucleic acids.

Moving boundary electrophoresis, the type first perfected in 1937 by Tiselius in separating the similar components of blood serum, has largely been superseded by simpler and less expensive methods. The original method involves dialyzing a buffered (i.e., constant pH) solution of the material under investigation in a large U-shaped tube with electrodes at each end. Separation of the proteins could be observed by observing light that was deflected by refractive index gradients at a boundary. A weakness of the method is that only partial separations can be achieved; full resolution of individual components is not possible due to diffusion and heat-driven convection.

Zone electrophoresis, developed by Stellan Hjertén in Sweden in 1967, involves particles that are supported on a relatively inert and homogenous solid or gel framework in order to minimize diffusion and convectional disturbances, and thus improve separation. Zone offers many advantages over moving boundary electrophoresis, achieving complete separation of all electrophoretically different components while the introduction of a stabilizing medium permits the use of much simpler and less expensive equipment. Perhaps most importantly, much smaller samples can be studied than in moving boundary electrophoresis, which has significant implications for DNA analysis as well as criminal detection. When filter paper is used as a medium for low voltage electrophoresis, paper strips are dipped in buffer solution and clamped between electrodes. When the separation is complete (up to 20 hours), the paper strips are removed from the electrophoresis tank and dried in an oven. The separated components are then typically located by staining with a dye that binds to the proteins.

Zone electrophoresis is the preferred form of analysis among scientists. Developments in zone electrophoresis have closely paralleled chromatography with both techniques sharing a number of supporting media and methods for estimating the separated components. However, it remains easier to separate proteins with the zone method.

Gel electrophoresis, the zone form employed with DNA, uses the frictional resistance of a gel to separate nucleic acids and proteins. Higher current densities can be used than with paper. A hot gel mixture is poured into a casting tray to assume a desired shape as it polymerizes. DNA is loaded on to the gel and electricity is applied for about 20 minutes. After staining, the separated macromolecules in each lane can be seen in a series of bands spread from one end of the gel to the other. The successive application of two separation steps in perpendicular directions (two dimensions) is known as 2-D gel electrophoresis. The first step separates proteins by charge. The second step separates the proteins by via their molecular weight. Since the mid-1980s, gel electrophoresis has been used to create genetic fingerprints for forensic or courtroom use by identifying particular DNA molecules. The number and position of bands formed on each lane of gel is the actual genetic "fingerprint" of that DNA sample. Viral DNA, plasmid DNA, and particular segments of chromosomal DNA can all be identified in this way. Another use is the isolation and purification of individual fragments containing interesting genes, which can be recovered from the gel with full biological activity. Using this technology, it is possible to separate and identify protein molecules that differ by as little as a single amino acid.

In the 1970s Rauno Virtanen was the first to use buffer-filled stationary narrow bore tubes (capillaries) to stabilize against convection. In the early 1980s James Jorgenson and Krynn Lukacs used much narrower capillaries, and developed analytic packages to simplify handling of data. Detection is usually by ultraviolet (UV) absorbance or coupling to mass spectrometers.

Electrophoresis on Earth is limited to very lightweight materials. For separation of minerals, for example, Earth's gravity causes convection currents, as well as gravitational settling. The space shuttle has been used to perform experiments in zero gravity in a free fluid, for example to produce small amounts of an electrophoretically purified protein. No commercial applications are yet developed.

See also **Chromatography; X-Ray Crystallography**

CARYN E. NEUMANN

Further Reading

Audubert, R. and de Mende, S. *The Principles of Electrophoresis.* Macmillan, New York, 1960.

Bruno, T.J. *Chromatographic and Electrophoretic Methods.* Prentice Hall, Englewood Cliffs, NJ, 1991.

Issaq, H.J. *Century of Separation Science.* Marcel Dekker, New York, 2001.

Laitinen, H.A. and Ewing, G.W., Eds. *A History of Analytical Chemistry.* American Chemical Society, New York, 1977, pp. 324–325.

Shaw, D.J. *Electrophoresis.* Academic Press, London, 1969.

Wieme, R. J. *Agar Gel Electrophoresis.* Elsevier, Amsterdam, 1965.

Encryption and Code Breaking

The word cryptography comes from the Greek words for "hidden" (*kryptos*) and "to write" (*graphein*)—literally, the science of "hidden writing." In the twentieth century, cryptography became fundamental to information technology (IT) security generally. Before the invention of the digital computer at mid-century, national governments across the world relied on mechanical and electromechanical cryptanalytic devices to protect their own national secrets and communications, as well as to expose enemy secrets. Code breaking played an important role in both World Wars I and II, and the successful exploits of Polish and British cryptographers and signals intelligence experts in breaking the code of the German Enigma ciphering machine (which had a range of possible transformations between a message and its code of approximately 150 trillion (or 150 million million million) are well documented.

In many respects the construction of the Enigma was like other rotor cipher devices in use in Europe and America in the 1930s, such as the Hagelin machine invented in the 1920s. At the heart of Enigma was a set of three or more electromechanical rotors and a movable alphabet ring inscribed with letters indicating the rotor positions. However, the Enigma distinguished itself with its reflecting rotor, which caused plaintext (the original message), once scrambled, to pass through the machine a second time in the opposite direction, resulting in double encipherment. British mathematician Alan Turing and other cryptanalysts working at the super-secret Bletchley Park broke the encipherment with several tools, including mistakes made by German users of the machine (procedural errors such as sending the same message twice or using standard message formats too often), and with the fact that a particular letter could never encipher as itself. By 1942, Allied code breakers deciphered almost 4000 German communiqués each day. By some estimates the Bletchley Park effort to break Enigma shortened World War II by two years.

Figure 1.   Marine version of an Enigma ciphering machine invented by Germany during World War II.
[*Courtesy of the photographer, Morton Swimmer, and the Bundesamt für Sicherheit in der Informationstechnik.*]

American military planners in the twentieth century—and especially in the Cold War—regarded superior information as a powerful "force multiplier" making up for the perennial problem of numerically inferior manpower. For most of the century, however, America's information advantage could be measured mainly in terms of secrecy rather than espionage. This was because spy agencies like the Army Cipher Bureau (established in 1917), the Navy Cryptanalytic Group (in 1924), and the National Security Agency (in 1952) virtually ignored in their massive number-crunching activities the issue of integrity—the corruption of data by mistakes, inflation, or enemy deception.

The key document pointing out this problem was American Bell Labs researcher Claude Shannon's "*Communication Theory of Secrecy Systems*," published in 1949. Shannon put the study of cryptography on firmer scientific ground by linking it to a formal information theory for reliable communications under noisy, error-prone conditions. His key insight was in showing that secrecy systems are "almost identical with a noisy communication system." Noise in the form of radio static, television snow, or background chatter, he argued, is very much like the process of encipherment. Cryptanalysis, conversely, involved the isolation of the signal (plaintext) from the noise (cipher key).

Public citizens in the Western world became much more engaged by the personal privacy implications of cryptographic national security in the 1950s and 1960s. Counterintelligence, they discovered, could be used not only against criminals, fascist dictators, and communists, but also against the counterculture youth movement and other political dissenters. Counterintelligence projects initiated against civil rights organizations and Vietnam War protesters for instance—which included authorized wiretapping of phones and electronic eavesdropping in homes and offices—outraged large numbers of freethinking individuals who felt that privacy is something everyone is entitled to as a right. Until about 1990, individuals had no access to the high levels of cryptography that were enjoyed by governments.

Both the problem and solution to cryptographic control were seemingly apparent in the multiply redundant and commercial interconnection of computers over telephone networks in the 1960s and 1970s. The need to address the issue of encryption to ensure privacy and security in electronic communications was highlighted by the unique enticements of networked computing as well as the extraordinary potential for malfeasance. Data security in information systems rapidly grew into a formal area of study within hardware and software engineering. Security experts realized that indirect losses—image problems as well as civil and criminal liabilities—were also a possible consequence of breaches in security. New network protective measures and access controls were added. Costs and benefits of particular security measures were calculated and recalculated.

However, public networks and computer time-sharing also allowed "every man at the console." The extraordinary movement of computational power into public hands contributed to a decentralized view of authority that competed with the surviving impetus for cryptographic national security. As Kenneth Dam and Herbert Lin have argued, "The broadening use of computers and computer networks [and cell phones] has generalized the demand for technologies to secure communications down to the level of individual citizens and assure the privacy and security of their electronic records and transmissions." Since the 1970s a growing global community of programmers ascribing to the vague hacker ethic that free information-sharing is a powerful positive good have competed with other interests motivated instead by profits of large corporations, security in electronic commerce, or the individual's demand for privacy.

Since the late 1970s, the popularization and commercialization of public-key, or "asymmetric," encryption technologies such as Pretty Good Privacy (PGP, available as freeware or low-cost commercial versions) and RSA (from the initials of Ron Rivest, Adi Shamir and Len Adleman, professors at Massachusetts Institute of Technology who developed a similar public-key approach) have weakened the hold of national governments on ownership of cutting-edge cryptographic technology. Martin Hellman, a professor at Stanford University, and two graduate students Whitfield Diffie and Ralph Merkle, discovered public-key encryption in 1976. (In 1997 it was disclosed that three employees of the British government had discovered the same approach several years earlier but had kept it a secret for reasons of national security.) Unlike classic private-key, or "symmetrical," encryption—where the same key is used to both encrypt and decrypt a message—Diffie and Hellman proposed a scheme to split the key. A *public key* would permit others to send the owner encrypted messages and verify the authenticity of messages received; a *private key* would be held only by the owner and used to decrypt messages encrypted with the public key or to sign new messages with their personal "digital signature." Because the public key could not unlock messages encrypted with the public key, that key could be distributed widely without danger of interception by a third party, a danger implicit in symmetrical encryption.

A private key (whether for encryption or signature) consists of just two very large prime numbers, and the matching public key is those two numbers multiplied together. The security arises from the fact that whereas multiplying two large numbers together is quite easy (if tedious), given just the public key there is no practical way to find out its two factors, and without doing that one cannot break the code. For example, the old Data Encryption Standard, or DES, adopted by the National Institute of Standards and Technology (NIST) in 1977 used keys that were 56 digits long, which was secure at the time. By the 1990s it was recognized that those numbers were not sufficiently large enough to be impervious to those with special hardware or distributed computing (a few days number crunching). Encryption with a security key of 128 bits length (commonly used in electronic commerce in the 2000s) increased the "strength" of the encryption by a factor of 2^{72}.

Almost immediately after Diffie and Hellman's proposal, computer, networking, and telephone companies began developing public-key technology for secure voice encryption in digital telephony. The specter of widespread use of public-key cryptography with keyspaces (the collection of all possible keys for a given cryptosystem) so large that they could never be completely checked by brute force and its possible use by organized crime or foreign governments convinced the American government to propose "key escrow." Key escrow would allow legal wiretaps and the easy decryption of intercepted digital messages with a "back door" key held in trust by the NIST and the Department of the Treasury. The government dubbed its key-escrow technology the Clipper chip (announced in 1993) and advised manufacturers of soon-to-be-released telephone security devices to replace their own security schemes with Clipper. Though Clipper was never implemented for various technical and political reasons, the key escrow idea ignited a debate by civil libertarians, so-called "cryptoactivists," government authorities and lawmakers, and public interests that has yet to subside at the dawn of the twenty-first century.

PHILIP L. FRANA

Further Reading

Blanchette, J.-F. and Johnson, D.G. Cryptography, data retention, and the panopticon society. *Comp. Soc.*, 28, 1–2, 1998.

Dam, K.W. and Lin, H.S., Eds. *Cryptography's Role in Securing the Information Society*. National Academy Press, Washington D.C. 1996.

Denning, D.E. To tap or not to tap, *Commn. Asso. Comput. Mach.*, 36, 26–33, 1993.

Diffie, W. and Landau, S. Privacy on the Line: The Politics of Wiretapping and Encryption. MIT Press, Cambridge, MA, 1998.

Hinsley, F.H. and Stripp, A. *Codebreakers: The Inside Story of Bletchley Park*. Oxford University Press, Oxford, 1994.

Kahn, D. The Codebreakers: The Story of Secret Writing. Macmillan, New York, 1967.

King, H.R. Big Brother, the holding company: a review of key-escrow encryption technology. *Rutgers Comp. Tech. Law J.*, 21, 224–62, 1995.

Levy, S. *Crypto: How the Code Rebels Beat the Government: Saving Privacy in the Digital Age*. Viking, New York, 2001.

Rosenheim, S.J. *The Cryptographic Imagination: Secret Writing from Edgar Poe to the Internet*. Johns Hopkins University Press, Baltimore, 1997.

Shannon, C.E. Communication theory of secrecy systems. *Bell Syst. Tech. J.*, 28, 656–715, 1949.

Simmons, G.J., Ed. *Secure Communications and Asymmetric Cryptosystems*. Westview Press, Boulder, CO, 1982.

Useful Websites

Electronic Privacy Information Center: http://www.epic.org

Energy and Power

At the close of the twentieth century, electricity was so commonplace that it would be difficult to imagine an existence without light, heat, and music bowing to our command at the flick of a switch. Children who could barely stretch high enough to toggle a light switch now have dominion over phenomena that less than a century ago would have been considered inconceivable. The temptations of such power have proved hard to resist.

The repercussions of the rapacious appetite for control of energy among Western industrial nations have not been confined to the lot of the individual, however. As in previous eras, when the control of mechanical or biological power carried financial, geographical, and social significance, the use and abuse of electrical energy now additionally carries environmental, political, and moral implications. Developments in energy and power in the twentieth century must therefore be considered within these broader thematic areas as the generation and consumption of energy are inextricably linked with practically the whole spectrum of human existence.

At the beginning of the twentieth century, despite the fact that many components of modern electronics such as the battery had already been invented 100 years earlier, body power was still the norm, especially in rural areas. Horses, carriages, tow paths, water mills, and the like were the standard means of transport and power for a large proportion of the population, despite the growth of electricity and the 130 supply companies that were operating by 1896 in Britain. Even in urban settings, only lighting and telegraphy were advanced to the stage where the benefits were generally enjoyed as a result of Thomas Edison's invention of the light bulb in 1879 and Alexander Graham Bell's first telephone transmission in 1878.

By 1900 in Britain the main features of an electricity supply industry had been established. The system was based on the generation of high-voltage alternating current (AC), with transformers stepping down voltages for local use. However, one obstacle that the industry had to overcome was the lack of standardization across local areas. In some parts, direct current (DC) equipment was still installed, and local voltage levels and frequencies varied considerably. Despite problems posed by these variations, at the start of the century most of the appliances that are now taken for granted had appeared. Space heaters, cookers, and lighting equipment were not yet in every home, but the very speed at which their use was adopted was testament to the flexibility and popularity of electricity. In 1918 electric washing machines became available, and in 1919 the first refrigerator appeared in Britain. They had already been introduced for domestic use in the U.S. in 1913. Electricity had been firmly accepted as the energy of the future. Demand from the residential sector started to boom and spurred further research. Most importantly, perhaps, by the 1920s in Britain the domestic immersion heater began to take over the duties of coal. The use of electric trolleys and trains, which had been running since the end of the nineteenth century, also continued to expand, and underground travel developed swiftly. Electricity also made advances in communications possible, from the telegraph and the telephone, to the broadcasting boom of the 1920s. In 1928 the construction of a British national grid system began, and it took less than ten years before the system was in operation. This alacrity is partly to be explained by the influence of World War I. The war's heavy demands on

manufacturing acted as a great incentive for the rapidly evolving electricity industry, particularly with regard to improving the efficiency of supply. Thereafter, the rebuilding and expansion of industry across the industrialized world began. In Russia, Lenin was moved to state, "Communism equals Soviet power plus electrification," as part of the propaganda for industrialization. Electricity took over the driving of fans, elevators, and cranes, driving coal-mining equipment, for example, and rolling mills in steel factories. The use of individual electric motors allowed astonishing advances in speed control, precision, and productivity of machine tools.

With World War II came devastation. Power stations and fuel supplies were inevitably considered as strategic targets for the bombers during the destructive aerial attacks by both the Axis powers and the Allies. By 1946, the estimated deficiency of generating capacity in Europe was 10,000 megawatts. According to anecdotal evidence, the victory bells in Paris were only able to ring out in 1945 because of electricity transmitted from Germany, where more industrial capacity of all kinds, including power stations, had survived. Whatever the truth of this may be, the security of electricity supply quickly became an issue of undisputed importance throughout Europe, and the fuels used in electrical generation were valuable resources indeed.

Coal

At the start of the twentieth century, there was a new worldwide optimism about coal as a resource that seemed to be available in almost unlimited amounts. Coal consumption levels rose steeply both in the U.S. and Europe, to reach a peak around 1914 and the outbreak of World War I. Between the world wars, consumption quantities remained almost static, particularly in the U.S., as other fuel types started to dominate the market. Reasons for this slow-down include the rising popularity of the four-stroke "Otto" cycle engine that is widely used in transportation even today as well as the commercialization of the diesel engine. These two technologies pushed fuel sources swiftly from solid to liquid fuels.

Nuclear Power

Nuclear fission was discovered in the 1930s. Considerable research occurred in those early years, particularly in the U.S., the U.K., France, Canada, and the former Soviet Union, in the design and construction of commercial nuclear power stations. In the early 1940s, U.S. intelligence regarding Germany's promising nuclear research activities dramatically hastened the U.S. resolve to build a nuclear weapon. The Manhattan Project was established for this purpose in August 1942. In July 1945, Manhattan Project scientists tested the first nuclear device in Alamagordo, New Mexico, using plutonium produced from a uranium and graphite-pile reactor in Richland, Washington. A month later a highly enriched uranium nuclear bomb was dropped on the Japanese city of Hiroshima, and a plutonium nuclear bomb was dropped on Nagasaki, effectively ending World War II.

The nuclear power industry suffered some notable disasters during its years of technological development. In 1979, the Three Mile Island Unit 2 (TMI-2) nuclear power plant in Pennsylvania suffered damage due to mechanical or electrical failure of parts of the cooling system. Just seven years later, on the opposite side of the Iron Curtain near an obscure city on the Pripiat River in north-central Ukraine, another disaster occurred. This accident became a metaphor not only for the horror of uncontrolled nuclear power but also for the collapsing Soviet system and its disregard for the safety and welfare of workers. On April 26, 1986, the No. 4 reactor at Chernobyl exploded and released 30 to 40 times the radioactivity of the atomic bombs dropped on Hiroshima and Nagasaki. The Western world first learned of history's worst nuclear accident from Sweden where abnormal radiation levels, the result of deposits carried by prevailing winds, were registered.

Ranking as one of the greatest industrial accidents of all time, the Chernobyl disaster and its impact on the course of Soviet events can scarcely be exaggerated. No-one can predict what will finally be the exact number of human victims. Thirty-one lives were lost immediately. Hundreds of thousands of Ukrainians, Russians, and Belo Russians had to abandon entire cities and settlements within the 30 kilometer zone of extreme contamination. Estimates vary, but it is likely that over 15 years after the event, some 3 million people, more than 2 million in Belarus alone, continued to live in contaminated areas.

Oil

Often accused of being one of the two great evils in the energy sector along with nuclear power, the oil industry grew over the course of the twentieth century to acquire significance and influence previously unimagined for any industrial sector. As the century opened, the U.S. was the largest oil producer in the world, but the discovery and exploitation of reserves in the Middle East, South America, and Mexico soon shifted the balance of the market away from the U.S., which by 1950 produced less than half the world's oil. This trend continued and by the year 2000, oil production was almost equally divided between OPEC (Organization of Petroleum-Exporting Countries) and non-OPEC countries. Even in the early years of the century, the geographical spread of supply and demand quickly created the need for a system of distribution of unprecedented scale. The distances and quantities involved led to the construction of pipelines and huge ocean-going ships and tanker trucks. The capital intensive nature of these infrastructure projects, as well as the costs of exploration and exploitation of oil fields, concentrated control of resources in the hands of a few companies with vast coffers. As the reserves from easily exploitable sites dwindled, the pockets even of governments were insufficiently deep to invest in new drilling projects, and Royal Dutch Shell, Standard Oil, British Petroleum, and others were born.

Concern about fossil fuel depletion began to be voiced around the world in the 1960s, but the issue created headlines on the international political circuit in 1970 following the publication of the Club of Rome's report *Limits to Growth*." This document warned of the impending exhaustion of the world's 550 billion barrels of oil reserves. "We could use up all of the proven reserves of oil in the entire world by the end of the next decade," said U.S. President Jimmy Carter. And although between 1970 and 1990 the world did indeed use 600 billion barrels of oil, and according to the Club of Rome reserves should have dwindled to less than zero by then, in fact, the unexploited reserves in 1990 amounted to 900 billion barrels not including tar shale.

Hydroelectric Power

Not a recent development by any stretch of the imagination, hydroelectric power was used extensively at the start of the twentieth century for mechanical work in mills and has a pedigree stretching back to ancient Egyptian times. Indeed, water power produces 24 percent of the world's electricity and supplies more than 1 billion people with power. At the end of the twentieth century, hydroelectric power plants generally ranged in size from several hundred kilowatts to many hundreds of megawatts, but a few mammoth plants supplied up to 10,000 megawatts and electricity to millions of people. These leviathans, or "temples of modern India," as India's first prime minister Jawaharlal Nehru declared, were also the cause of massive discontent from social and environmental standpoints. The displacement of local indigenous populations and failure to deliver promised benefits were just two of the many complaints. By comparison, and despite hydroelectric power's renewable credentials, the use of conventional fossil fuel technologies such as natural gas remained relatively uncontroversial.

Coal-Gas Technology

A derivative of coal as its name implies, coal-gas is produced through the carbonization of coal and has played a not insignificant role in the development of power and energy in the twentieth century. It was an important and well-established industry product as the century opened, although electricity had already started to make inroads into some of the markets that coal-gas served. Coal-gas enjoyed widespread use in domestic heating and cooking and some industrial facilities, but despite the invention of the Welsbach Mantle in 1885, electricity soon started to dominate the lighting market. The Ruhrgebeit in Germany was the most active coal-gas producing area in the world. It was here that the Lurgi process, in which low-grade brown coal is gasified by a mixture of superheated steam and oxygen at high pressure, flourished for many years. However, as the coal supplies necessary for the process became increasingly expensive, and as oil fractions with similar properties became available, the coal-gas industry swiftly declined. In fact, when the coal industry seemed to have reached a pinnacle, another rival industry—natural gas—was being born.

Natural Gas

The American gas industry developed along different lines from the European market. Each started from a different basis at the dawn of the twentieth century. The U.S. had been quick to adopt the production of coal-gas, which was used for lighting as early as 1816. After the discovery of fields of largely compatible natural gas in relatively shallow sites when searching for oil reserves, the natural gas industry expanded swiftly. Large-scale transmission mechanisms were developed with alacrity, and one noteworthy example of this came from the Trans-Continental Gas Pipeline Corporation, which completed a link from fields in Texas and Louisiana to the demand-intensive area around New York in 1951. By contrast, in Europe the exploitation of natural gas began in earnest in the years following World War II. In the Soviet Union, for example, the rich fields around Baku in Azerbaijan were connected to both their Eastern Bloc allies by 1971 and also to West Germany and Italy by over 680,000 kilometers of pipelines.

In Western Europe developments on the geopolitical level benefited Britain, which officially acquired the mineral rights for the western section of the North Sea in 1964. Just one year later, the West Sole field was discovered. Britain had already imported some natural gas from the U.S., and within 12 years had switched almost entirely from manufactured coal-gas to natural gas. This conversion was no simple operation. The differing properties of manufactured and natural gas meant that domestic and industrial appliances numbering in the tens of millions had to be altered. The British conversion scheme, which lasted ten years, is estimated to have cost £1000 million. Other similar conversion programs were carried out in Holland, Hungary, and even in the Far East.

In October 1973, panic gripped the U.S. The crude-oil rich Middle Eastern countries had cut off exports of petroleum to Western nations as punishment for their involvement in recent Arab–Israeli conflicts. Although the oil embargo would not ordinarily have made a tremendous impact on the U.S., panicking investors and oil companies caused a gigantic surge in oil prices.

There were more oil scares throughout the next two decades. When the Shah of Iran was deposed during a revolution, petroleum exports were diminished to virtually negligible levels, causing crude oil prices to soar once again. Iraq's invasion of Kuwait in the 1990s also inflated oil prices, albeit for only a short time. These events highlighted the world's dependence on Middle Eastern oil and raised political awareness about the security of oil supplies.

The "dash for gas" in the U.K.—the rapid switch from coal to gas as the dominant source of power generation fuel—was no doubt partly instigated by the discovery of home reserves there. Worldwide the new application of an old technology, combined cycle gas turbines, or CCGTs, played a significant role. During the last decades of the twentieth century, the gas turbine emerged as the world's dominant technology for electricity generation. Gas turbine power plants thrived in countries as diverse as the U.S., Thailand, Spain, and Argentina. In the U.K., the changeover began in the late 1980s and resulted in the closure of many coal mines and coal-fired power stations. As electricity industries were privatized and liberalized, the CCGT in particular became more and more attractive because of its low capital cost, high thermal efficiency, and relatively low environmental impact. Indeed, this technology contributed to the trend identified by Cesare Marchetti, which depicts the chronological shift of the world's sources of primary power from wood to coal to oil to gas during the last century and a half. Each of these fuels is successively richer in hydrogen and poorer in carbon than its predecessor, supporting the hypothesis that we are progressing toward a pure hydrogen economy.

Distributed Generation

Embedded or distributed generation refers to power plants that feed electricity into a local distribution network. By saving transmission and distribution losses, it is generally considered to be an environmentally and socially beneficial option compared with centralized generation. Technologies that contributed to the expansion of this mode of generation include wind turbines, which developed to the point where their cost of generation rivaled that of central power stations, photovoltaic cells, and combined heat and power units. These industries expanded massively in the latter years of the century, particularly in Europe where regulatory measures gave impetus and a degree of commercial security to the fledgling industries.

Hydrogen

Many industries worldwide began producing hydrogen, hydrogen-powered vehicles, hydrogen fuel cells, and other hydrogen products toward the end of the twentieth century. Hydrogen is intrinsically "cleaner" than any other fuel used to date because combustion of hydrogen with oxygen produces energy with only water, no greenhouse gases or particulate exhaust fumes, as a byproduct. At the close of the twentieth century, however, although prototypes and demonstration projects abounded, commercial competitiveness with conventional fuels was still only a distant prospect.

From almost wholly somatic sources of power in 1900, energy and power developed at an astonishing pace through the century. As the century closed, despite support for "green" power, particularly in developed nations, the worldwide generation of energy was still dominated by fossil fuels. Nevertheless, unprecedented changes seemed possible, driven for the first time by environmental and social concerns rather than technological possibilities or purely commercial considerations. Awareness of energy-related carbon emissions issues addressed by the Kyoto protocol raised questions concerning the institutional arrangements on both national and international levels, and their capacity for action in responding to public demand. After a century of development, a wide variety of institutional and regulatory regimes evolved around electricity supply. These most often took the form of a franchised, regulated monopoly within clearly defined administrative boundaries, in a functional symbiosis with government. However, each has the same basic technical model at its heart; Large, central generators produce AC electricity, and deliver it to consumers over a network. The continuing stable operation of this system on which many millions of people rely, once considered the responsibility of central governments, is changing. The increasing shift toward liberalization and internationalization is moving responsibility for energy supplies away from state-owned organizations, a trend compounded by the environmental and institutional implications of renewable energy technologies.

See also **Electrical Power Distribution; Electrical Energy Generation and Supply, Large Scale; Electricity Generation and the Environment; Fuel Cells; Lighting**

IAN BURDON

Further Reading

Bowers, B. *A History of Electric Light and Power*. Peregrinus, New York and Peregrinus/Science Museum, Stevenage, 1982.

Edgerton, D. *Science, Technology and the British Industrial 'Decline' ca. 1870–1970*. Cambridge University Press, Cambridge, 1996.

Marchetti, C. When will hydrogen come? *Int. J. Hydro. Energ*. 10, 4, 215–219, 1985.

Leggett, J. *The Carbon War: Global Warming at the End of the Oil Era*. Penguin, London, 1999.

MacNeill, J.R., Ed. *Something New Under the Sun: An Environmental History of the Twentieth Century*. Penguin, London, 2001.

Nye, D.E. *Electrifying America: Social Meaning of a New Technology, 1880–1940*. MIT Press, Cambridge, MA.

Engineering: Cultural, Methodological, and Definitional Issues

At the beginning of the twentieth century, engineering was engaged in transformations revolving around the nature and application of engineering knowledge and the education and social status of its practitioners. These transformations continued well into the century, and while the pace of change had certainly abated by the end of that period, change nevertheless remained a constant throughout.

The advent of the new "science-based" fields of electrical and chemical engineering in the late nineteenth and early twentieth centuries lent weight to the emerging view that the relationship between science and engineering was that science discovered and explained fundamental truths while engineering applied the theories produced by science to the production of technical artifacts. This led many to divide the history of engineering into two principal periods: prescientific engineering followed by science-based engineering. The latter, of course, was seen as far more effective owing to its reliance on science.

This neat conceptualization, however, is not supported by either the historical record or actual engineering practice. Engineers throughout history up to and including the present have utilized a wide variety of knowledge ranging from rules of thumb (heuristics) to tacit understanding to graphical methods of analysis to sophisticated analytical models to scientifically derived principles of material (in its broadest sense) behavior. Various attempts at categorizing these different types of knowledge have been made (Vincenti, 1993; Addis, 1990), but they are less important for the particular organizing schemes they suggest than for their explicit observation that modern engineering is not simply or even primarily a matter of applied science (although the scientific component is certainly important). One indicator of this was Project Hindsight, a study conducted by the U.S. Department of Defense in the 1960s that painstakingly analyzed the key intellectual contributions directly underlying the development of 20 core weapons systems and revealed that the vast majority constituted technological rather than scientific knowledge.

The process by which engineering knowledge grows and advances, however, does appear to echo the process by which scientific knowledge grows and advances. In broad terms, most models of the latter still invoke the notion of paradigm shifts attributed to Thomas Kuhn (1996). This model divides scientific practice into periods of normalcy and disruption. The former is characterized by shared fundamental assumptions about the world (or at least the part of it that is relevant to the area of investigation) and the way it operates. These assumptions are so basic and ingrained that they are seldom explicitly examined. Eventually, compelling empirical evidence arises that calls into question the validity of those assumptions. A new

paradigm forms to account for the new evidence (typically encountering substantial resistance) and, if it contains sufficient explanatory power, ultimately replaces the old paradigm with its new set of fundamental assumptions. Normal science then resumes within the new paradigm.

Fundamentally similar models have been proposed to describe the development of engineering knowledge. These models differentiate between normal and disruptive practice. They also tend to invoke a variation–selection–retention mechanism to explain how new solutions—embodying fundamentally new assumptions—to a problem are generated and the most effective one selected and retained.

Precisely because engineering is more than simply applied science, there is substantial room for variation in practice, including variation rooted in cultural differences. Differing professional and academic commitments to theoretical analysis on the one hand and empirical experimentation on the other, for example, will influence design goals and priorities (given that tradeoffs are inescapable) such as economy, simplicity, and esthetics. Those goals and priorities then shape both the variation and selection of technical alternatives.

While these processes for science and engineering certainly exhibit substantial similarities (in fact they represent variations on what some believe is a universal model for the generation and growth of knowledge) the values and objectives they embody are in many ways the reverse of each other. Edwin Layton Jr. (1971) has dubbed science and engineering "mirror-image twins." This is most immediately obvious in their goals. Whereas science aims to understand a phenomenon, engineering aims to solve a practical problem. In the physical sciences, the more abstract and general the work the better. Specific, concrete applications tend to garner less prestige. In engineering, on the other hand, the successful design and creation of artifacts or processes usually draws the most applause while purely theoretical work is less revered. For scientists, publication of results is viewed as an integral activity while, for engineers, publication is of decidedly less importance than actual practice.

While these observations remain largely accurate, they have by no means been immutable. The professionalization of engineering and attempts to firmly ground it in science produced an impetus to publish that began in the nineteenth century and increased throughout the twentieth. This trend was reinforced by the rise of academic engineering with its inherent need for publication outlets. The latter decades of the twentieth century witnessed a partial breakdown of the divide that had developed between practicing and academic engineers in the U.S. and the U.K. as universities were increasingly seen as a key source of new products and processes for industry. In a sense, this represented a return to the close ties between industry and engineering academics in the early decades of the century, but in this case the focal point was the production of intellectual property rather than the production of employees. This pattern was less substantial in other countries such as France and Germany, where scientific formalism in engineering had long been embraced and close cooperation between academic engineering and industry had always been prevalent.

Engineering education both motivated and reflected the move toward science-based engineering. This orientation was exactly the opposite of what characterized American engineering education through most of the nineteenth century. At the turn of the century, most practicing engineers in the U.S. had been trained through apprenticeship. Mechanical engineers were the product of a (machine) "shop culture" while civil engineers came out of a "field culture." The idea of engineers being trained in a school setting was still considered a bit odd by many, and the new scientifically oriented engineers emerging from American colleges and universities were viewed with some suspicion.

French technical education at the end of the nineteenth century, in contrast, was at its highest level the epitome of science-based engineering. French technical education was just as stratified as French society (the Revolution having dampened but not eliminated class distinctions). Its three tiers catered to very different populations and in very different ways. At the top of the hierarchy, as is the case today, were the prestigious École Polytechnique and its affiliated Écoles d'Application. The former provided education in engineering fundamentals (what today would be considered an engineering core curriculum), after which the latter would provide specialized training in a particular technical field. The vast majority of graduates ended up working for either the state or the military. The second tier consisted of the École Centrale des Arts et Manufactures, which aimed to train engineers for industry rather than government or military service. Making up the third tier were the Écoles d'Arts et Métiers, which concentrated on workshop training such as forging and machine fitting. This contrasted with the top tier, which emphasized mathematical theory. The educational thrust of the second tier was somewhere in the middle, revolving around such things as industrial chemistry and metallurgy.

Even as academia and industry throughout the industrialized world began to embrace (if they had not already) the notion of engineering as applied science, evidence of its limitations occasionally presented itself. Research at the U.S. Bureau of Public Roads between the world wars, for example, reflected the shift from empirically based research to research focused on theory and mathematical models (albeit informed by data gleaned from small-scale isolated experiments). Full-scale field studies were replaced by a search for fundamental principles that could serve as a basis for "rational" road design. This more scientific approach for road design, however, proved far less effective than the earlier efforts. Such cases reveal the complexity of the role science plays in engineering and that wholesale adoption of a scientific sensibility does not necessarily serve the ends of engineering.

Nevertheless, if science was the heir apparent to experience as a basis for engineering at the turn of the century, it was undeniably king in the aftermath of World War II. While the atomic bomb is usually seen as the most prominent example of scientific contributions to the war effort, there were plenty of others as well, including radar and the digital computer. That these were at least as much technological as scientific achievements was an unappreciated distinction. As a result, universities in the U.S. and the U.K. hastened to rid themselves of the last vestiges of practical training.

A review of university engineering education in the early 1950s sponsored by the American Society for Engineering Education fully reflected the ethos of engineering as applied science. In recognition of the increased reliance on science, the report recommended that new engineering faculty have an appropriate doctorate degree (PhD). It also called for the elimination of courses having a "high vocational and skill content" or attempting to convey "engineering art and practice" in favor of courses in engineering science, effectively sounding an official death knell for shop and field culture. The emphasis on theory and analysis was further reinforced by the general expansion of American higher education after the war, driven in part by an influx of World War II veterans. Swelling enrollments meant large class sizes, and engineering classes were no exception. Theory and analysis lent themselves to large lecture classes more readily than did design and other less scientific types of engineering knowledge.

By the end of the decade, however, employers and practicing engineers in both the U.S. and

Europe were beginning to complain of the declining ability of engineering graduates to engage in design. Accompanying these complaints was increasing criticism of engineering education that imbued students with a "blind faith" in the results of theoretical calculations and left them unable to relate mathematical engineering models to the requirements and behavior of actual artifacts. (Similar concerns have been voiced more recently regarding the results of computer-aided design tools.) Increasingly, engineering graduates, while displaying formidable analytical skills, exhibited a much-reduced ability to actually design technical artifacts. Moreover, this tendency became more pronounced with each higher academic degree. As a result, those recruited to engineering faculties were by definition those with the least inclination toward design. A 1980 international survey of engineering education found that while U.S. engineering curricula had to some extent reintroduced design as a topic of instruction, it was generally held in low esteem by the academy. In contrast, engineering education in Germany and the Netherlands incorporated a strong practical component with no diminution of status. Japan fell somewhere in between.

The twentieth century brought with it an acceleration of changes to the professional status of engineers that had been sparked by the development of large-scale industrial corporations in sectors heavily dependent on science and technology. Prior to this time, engineers, or at least those who were not part of their nation's military or government, practiced as independent professionals, typically on a contractual basis. As such, they enjoyed a degree of autonomy comparable to that of other independent professionals such as doctors and lawyers. The rise of large science and technology-based corporations changed this as, over time, increasing numbers of engineers became salaried employees rather than autonomous practitioners.

These large corporations were epitomized by firms such as General Electric (GE), American Telephone & Telegraph (AT&T) and DuPont. In addition to their need for engineers to carry out their day-to-day operations, these firms and those like them also required large pools of scientific and technical expertise to conduct research and development (R&D). These three companies, in fact, became as well known for their industrial R&D laboratories as for their other activities. That companies like GE, AT&T, and DuPont were at the forefront of this trend was not surprising. The electrical and chemical industries were considered deeply rooted in scientific knowledge right from the beginning and so were pioneers of industrial research. Other industries quickly followed suit. The years between the turn of the century and World War II saw a sweeping surge in corporate scientific and technical R&D.

This need for highly trained researchers as well as operating personnel was a key force driving the shift in engineering education from shop and field culture to a school culture. Industry and higher education worked together quite closely to shape engineering curricula that would produce employees with the requisite knowledge and skill sets. Upon entering the industrial work force, engineering graduates would often be put through internal corporate training programs designed to make them effective and loyal employees whose interests were appropriately aligned with those of their employer. Socialization was just as much an objective as technical proficiency and an understanding of company operations. The GE "test course" was one of the earliest and best known of these programs.

This shift in circumstances produced consternation on the part of engineers who worried a great deal about their status in society (and still do). This was especially true in the U.S. and the U.K., where there was a distinct absence of class-oriented mechanisms supportive of their status goals or class-based stratification that was almost wholly independent of those goals, respectively. This was unlike the situation in France, where the three-tiered system of technical education at least promised those in the top tier a modicum of professional status.

Engineers as employees confronted a fundamental tension. On the one hand, as employees they were expected to put the interests of their employers first and foremost, especially those who rose to management positions. On the other hand, as professionals they were expected to concern themselves with the interests of society as a whole. Among U.S. engineers in the early years of the twentieth century, this latter imperative crystallized under the rubric of social responsibility. Social responsibility in the sense of disinterested public service implied a measure of professional autonomy while at the same time not overly offending corporate managers.

Nowhere was this notion more firmly embraced than in the U.S. The progressive movement of the late nineteenth and early twentieth centuries had created a deep and abiding faith in the power of scientific and technical expertise in the public service. While engineers had always been involved in the development of important infrastructure—roads, bridges, dams, and so on—they began to be perceived by many, including themselves, as essential instruments of material progress and improved quality of life. Many U.S. engineers took this perspective even further, viewing themselves as the shepherds of societal progress by virtue of their commitment to rational and impartial thought and analysis. This attitude found its fullest, albeit most futile, expression in the technocracy movement between the World Wars. In seeking to apply the methods of scientific rationalism to governance, however, technocracy seriously discredited the notion of social responsibility rather than acting as its ultimate expression.

In practical terms, this tension between professionalism and corporate capitalism frequently played itself out within the engineering professional societies. (As a result of the importance of its stratified system of technical education, sector-based professional engineering societies in France have not developed in the same way or played the same role as those in the U.S. and the U.K.) Issues of membership requirements (technical versus business), ethical codes, and disciplinary mechanisms were all areas in which the clashing priorities of engineers and managers could not be entirely avoided. These tensions were verbally reconciled by equating societal progress with technological progress, thereby making society by definition the beneficiary of corporate technical activities. When it came to actions, however, there was no escaping the fact that an insistence on professional autonomy and independent thinking at some point had to come into active conflict with the corporate ethos.

Engineering professional societies in other countries often carried with them a degree of regulatory authority for their fields, and this provided a strategic avenue that U.S. engineering professional societies lacked. In the U.K., for example, the engineering professional societies accredit curricula and nominate Chartered Engineers, a mark of technical competence and achievement. Moreover, these societies set rigorous entry requirements such that even an engineering degree from an accredited curriculum is often insufficient to fully exempt a graduate from society entrance exams. This is not to say that engineers in the U.K. or elsewhere do not worry about their status in the eyes of the public or with respect to other professions. However, certain professional structures can offer a means of at least partially addressing those concerns while others are less effective in that regard.

Engineering in the twentieth century then, is not a story of the straightforward and triumphal application of science to the creation of technical artifacts. Rather, it is a very human story of myriad motivations, perceptions, and conflicts. The engineering achievements of the century are not at all diminished by recognizing that the epistemological, educational, and professional development of engineering has been as much a social process as anything else. On the contrary, the achievements become all the more impressive, and the failures all the more understandable, with an appreciation of the nonphysical forces that have been pivotal in shaping engineering from the end of the nineteenth century to the beginning of the twenty-first.

STUART S. SHAPIRO

Further Reading

Addis, W. *Structural Engineering: The Nature of Theory and Design*. Ellis Horwood, New York, 1990.

Billington, D. *The Tower and the Bridge: The New Art of Structural Engineering*. Princeton University Press, Princeton, 1985.

Bucciarelli, L. *Designing Engineers*. MIT Press, Cambridge, MA, 1996.

Buchanan, R.A., *The Engineers: A History of the Engineering Profession in Britain, 1750–1914*. Jessica Kingsley, London, 1989.

Constant, E. *The Origins of the Turbojet Revolution*. Johns Hopkins University Press, Baltimore, 1980.

Ferguson, E. *Engineering and the Mind's Eye*. MIT Press, Cambridge, MA, 1994.

Kranakis, E. *Constructing a Bridge: An Exploration of Engineering, Culture, Design, and Research in Nineteenth-Century France and America*. MIT Press, Cambridge, MA, 1997.

Kuhn, T. *The Structure of Scientific Revolutions*, 3rd edn. University of Chicago Press, Chicago, 1996.

Layton, E. Jr. Mirror-image twins: the communities of science and technology in 19th century America. *Technol. Cult.*, 12, 1971.

Layton, E. Jr. *The Revolt of the Engineers: Social Responsibility and the American Engineering Profession*. Johns Hopkins University Press, Baltimore, 1986.

Noble, D. *America By Design: Science, Technology, and the Rise of Corporate Capitalism*. Oxford University Press, Oxford, 1979.

Petroski, H. *To Engineer is Human: The Role of Failure in Successful Design*. Vintage Books, New York, 1992.

Petroski, H. *Design Paradigms: Case Histories of Error and Judgment in Engineering*. Cambridge University Press, Cambridge, 1994.

Reynolds, T., Ed. *The Engineer in America: A Historical Anthology from Technology and Culture*. University of Chicago Press, Chicago, 1991.

Seely, B. The scientific mystique in engineering: highway research at the Bureau of Public Roads, 1918–1940. *Technol. Cult.*, 25, 1984.

Vincenti, Walter. *What Engineers Know and How They Know It: Analytical Studies from Aeronautical History*. Johns Hopkins University Press, Baltimore, 1993.

Engineering: Production and Economic Growth

Production engineering as a philosophy, a theory, and a practical technique for organizing industrial processes to optimize output and minimize waste resulted from two sources. One was the rational spirit of the eighteenth century Enlightenment, which encouraged the organization of all activities according to scientific principles. This was reinforced by a metaphysical belief that replicating in industry the rational order observed in the physical universe would reduce waste and lessen injustice to the work force. This faith encouraged several of the pioneers of management theory, including Frederick W. Taylor, F.B. Gilbreth, and the "technocrats" (see below). The other chief source of production engineering was the recurrent need from the 1850s to the present to reduce waste in the primary energy industries, and to ensure that raw materials were not being squandered. The several periods of great anxiety over fuel resources (1860s, 1890s, 1920s, 1930s, and 1970s) together with the need to organize entire nations to meet the demands of global warfare stimulated the growth of comprehensive theories of how best to manage resources, industrial production, and product distribution. The British engineer Armstrong, and the philosopher–scientist W.S. Jevons were among the first to attempt large scale reviews of energy resources, fuel consumption, and the economic consequences of production, with some attempt to foresee the consequences for future ages. Jevons great work *The Coal Question* contained most of the concepts employed by much later reviews of energy and materials use on a nationwide scale. Many pioneers of rational methods in industry were closely connected with energy use, productivity, and the physical sciences, including Louis Le Chatelier (railway mechanics; physical chemistry) and Wilhelm Ostwald (chemistry, thermodynamics). In 1887, Ostwald introduced a general philosophy in which energy was a basic concept underlying natural and industrial phenomena. Called "energism" or "energetics," this philosophy anticipated some of the later ideas of Taylor, Frederick Soddy, and the technocracy movement and proposed reforming orthodox economics to make energy rather than monetary value the basic unit for measuring production. Ostwald organized an international movement called "The Bridge" to encourage energy efficiency in all activities, with the rational spirit of science as the guiding ideal. Efficient energy use, management, and organization were advocated throughout industry, but the movement was destroyed by the onset of World War I (1914–1918).

The waste of human and material resources by nineteenth century capitalism and the destructiveness of the Great War gave widespread publicity to the writings of Thorstein Veblen who contrasted the rational, creative methods and progressive values of the engineer with the irrational, destructive methods, and commercial values of the parasitic financier, lawyer, and politician. Disciples of Veblen seized on the work being done by Taylor, Gilbreth, and H.L. Gantt as providing models not only for factories producing material goods but also for a just society that would eliminate the waste of human skills and potential. The most eager disciples of applying rational methods and scientific management on a national or global scale were the advocates of technocracy ("technocrats") in the U.S. who exerted considerable influence between 1920 and 1940. Technocracy became a political movement, with a radical agenda for social reform and reorganization of the nation along production engineering lines. The movement even had a uniform code of dress and attracted more nonengineers than engineers, although its chief exponents included the engineers H. Scott, W. Rautenstrauch, and H. Gantt. American technocrats were inclined to isolationist nationalism, but others were more internationally minded and socialist. In Great Britain, Soddy and H. G. Wells were outstanding in publicizing technocratic values. The increased use of rational methods throughout industry and business, coupled with the mass production techniques and standardization of the American motor car industry, publicized the advantages of American methods between 1890 and 1920. The terms "Taylorism" and "Fordism" (for the systems developed by Henry Ford) were loosely used to identify a philosophy for organizing engineering-based industries that drew on scientific management, work study, time and motion analysis, standardization, increased mechanization, and mass production. The techniques were condemned by conservatives as inhuman and destructive of the craft tradition in industry, though they were recognized as essential for introducing an era of high consumption and increasing productivity that would trigger an endless age of continuous economic expansion. This was done in the U.S., the first nation to make the transition from a mature industrial economy to a high-productivity, high-consumption economy circa 1920. In the U.S., the transition was made relatively rapidly, but in Great Britain, which reached maturity in the 1850s, the change to a high-consumption economy was delayed until the 1950s due to technological conservatism. W. W. Rostow has analyzed this progression. British engineers were slow to recognize the import of Taylor's work in production engineering, and after a visit to London he described them as fixated on the form of equipment to the detriment of understanding general production theory. The Americans were quick to recognize rational management of industry as an essential part of engineering education, as were industrialized nations elsewhere such as Germany, France, and Japan; but Great Britain proved backward in this respect.

There has always been a conflict between engineers' values and economists' values. The values of the economist were generally taken to include those of the financier, lawyer, politician, and entrepreneur. The engineer visionaries, who stressed the rational, scientific nature of engineering, were a minority who exercised considerable influence through production engineering and the Technocrat movement. They received support and sympathy from a greater number of engineers who argued that short-sighted, selfish, financial considerations were delaying or even halting technological progress. Rationally planned, scientific mechanisms (which might be an industry or a new system of transport) had to be fitted into a much broader socioeconomic "receiving system" whose structure and activities were controlled by irrational pursuers of personal wealth. Exploitation of resources, including human, stopped the engineers from constructing a creative and liberating society, which would be better in the moral sense. Building on a philosophical foundation laid down in the Enlightenment, the general body of technocrats, as distinct from the political movement of that name, argued that engineering values promised better solutions to national and global problems than orthodox economics. They argued that conventional ways of assessing wealth and economic progress were irrational and inaccurate. The established engineering professions distanced themselves from this stance, which received stronger support from journalists, writers, teachers, and academics. In Britain, H.G. Wells advocated values similar to those of the technocrats, and leading engineers did support the movement, but as individuals rather than representatives of the profession. Engineers such as J.B. Henderson and A. Ewing reminded the profession that the engineer was no mere servant of money and politics but had a responsibility to a higher enlightenment.

Engineers should accept responsibility for their work and the uses that others made of it. The present-day use of the term technocrat is the opposite of what was originally intended.

Production engineering was much more than production of manufactured articles with minimum waste of time, material, and energy so that better use could be made of existing plant and workforce. The need to include a widespread system of production in the analysis led to improved methods of management of personnel, training, transport of materials, and organization of subsidiary activities. The analysis passed from assessing the contribution made by a particular process and measuring the efficiency of this process within a workshop or industrial site to a general review of industrial performance and efficiency within the national or global economy. This meant finding some means of quantifying the contribution made by engineering to the economy, and relating this to a more general index of economic performance. Technocracy in its various guises stressed the scientific nature of its activities, hence measurement was essential. For many centuries, the French Wheat Price Index served as a rough guide to the fortunes of the economy of France. When the French economy achieved global significance, this index provided a general indicator. The French Wheat Price Index was chosen because there were records going back to the twelfth century. Later, the Coal Price Index served the same purpose. Attempts were made to analyze these records, using Fourier analysis, to see if the component waves could be correlated with events and so reveal what caused the fluctuations in the economy—weather, warfare, political upheaval, innovation, or discoveries. It was argued that if trend curves persist, such analysis might suggest what should be done to meet future requirements. This positivist approach, with its dangers of determinism and historicism, was used by C.O. Liljegren in 1920 to help decide future policy in marine propulsion. It gave rise to an increasingly ambitious attempt to make technological forecasting a reliable enough aid to design and planning. Two indices emerged as useful to both engineers and economists: Gross National (and Domestic) Product (GNP and GDP) and its per capita expression; and electricity consumption, usually expressed as total kilowatt hours (kWh) per year and kWh per capita. The ratio between these two indices was also judged important. After 1920, the electricity generated by an industrialized nation was used to assess its rank as a modern state. The quantity of primary fuels used to generate this power was a measure of the efficiency of national industry. The amount of primary energy and the quantity of electricity required to generate the GNP of a nation per year was seen as an indicator of national technological and economic standing. For example, F. Quigley's study published in 1920 suggested that Britain was underelectrified and might suffer in consequence.

This analysis, much developed and refined, enjoyed widespread use during periods of energy crisis, and it remains in general use.

After 1920, Taylorism, Fordism, and technocracy came together to create an engineering-based approach to global production, resource use, and economics. This intensified the clash with orthodox economics, politics, and finance, and widened the gap between the Technocrats and the majority of conservative, professional engineers who feared involvement in radical politics. The Depression that began in the U.S. in 1929 and spread worldwide provided the technocrats with an opportunity and, in North America, gave the Technocrat movement its most influential period. In the USSR, Germany, Italy, Japan, and other industrial countries dominated by totalitarian regimes, technocracy came to mean the use of engineering to serve a military dictatorship with maximum efficiency and minimum considerations of conscience. As a result, the word technocrat came to mean an obedient expert who discharged assigned duties with technical competence in the service of the powerful; it has never recovered its original meaning. In the 1930s the original ideals of technocracy were pursued largely in North America and Great Britain. The technocrats in the U.S., despite being isolationist and nationalist, were the most influential inside and outside North America. The technocrats argued that the failure of the financial system in a country full of skilled workers, competent engineers, and up-to-date factories was proof that the old economics should be scrapped. The U.S. workforce wanted to work. The equipment was there. The energy was there. The engineering intelligence was there. But the financial system could not facilitate turning these resources into productive activity. Leading technocrats such as H. Scott were inspired by H.G. Wells and F. Soddy and advocated making energy units the basic currency in a new economics. Echoing Ostwald, Soddy and others, Scott said that all goods and services were converted energy, and a scientific review of a nation's activities meant quantifying all human, natural, and machine activities in energy units, which could then be used as the price of goods and services. The matter

of thermodynamic energy grade seems to have been sidestepped. Between 1932 and 1933, Scott, Rautenstrauch, and Hubbert compiled an energy survey of North America which gave widespread currency to many of the techniques, concepts, and terminology still found in energy analysis. Hubbert's contribution was outstanding. The survey charted the growth of 3000 industrial and agricultural products between 1830 and 1930 and measured production in terms of energy expended, volume of production, rate of growth, manpower per unit of production, power per unit of production, total power, total number of employees, and production man-hours. It was a standards-setting exercise, taken up by industrial nations and now a regular technique whose findings can be found in the annual volumes of statistics issued by governments all over the world. Scott's theory of making energy into a currency was successfully resisted by orthodox economists who used errors in the energy survey to discredit the ideology and political program of the technocrats. Their political program was outflanked by Franklin Delano Roosevelt's New Deal, but their analysis of energy and material use was greatly developed and widely applied during World War II and afterward during the Cold War and the energy crisis of the 1970s. The major protagonists in the World War and the Cold War embraced Taylorism and Fordism to various degrees, helped or hindered by political and ideological factors. Exploring the link between energy flows and money flows in the economy was continued and enjoyed considerable vogue in the late 1970s following the 1973 energy crisis. Despite quantification of energy investment in most goods and services, no national economy was placed on an energy-value basis. Orthodox economics and financial methods continued to dominate worldwide.

The rise of technocracy coincided with attempts to develop econometric analysis of engineering change and its consequences for the economy. Many technocrats employed econometrics, but not all econometricians supported technocracy. Christiaan Huygens, Christopher Polhem, Napoleon, Armstrong, and Jevons were a few of those who recognized the importance of engineering to a nation's economy between 1600 and 1900, but they lacked a comprehensive model of economic growth related to history. Between the world wars several comprehensive models were put forward. These models assumed that the global economy was dominated by a relatively small number of leader nations such as France and Britain in the eighteenth and nineteenth centuries and the U.S. and Japan in the late twentieth century. In 1925 N.D. Kondratieff argued that analysis of economic performance during the industrial period showed that it could be divided up into successive cycles of growth, prosperity, stagnation, recession, recovery, and so on. He further claimed that these phases repeated themselves at regular intervals of about 53 years. As long as the structure of the model lacked a rational explanation and the precise periodicity was claimed, Kondratieff's cycles were regarded as belonging to speculative metaphysics. During and after World War II, however, a school of econometrics developed that accepted the cycles as established by reliable data and looked for an explanation. S. Kuznets and Joseph A. Schumpeter argued that global economic history was associated with distinct phases that were caused by technological innovation. Innovation created new industries, and a relatively small number of industries cross-fertilizing each other launched a new era of economic development. The industries that dominated the succeeding phase of economic growth were strategic industries, created by strategic innovations. They fostered new standards of workers' skills and new management techniques, raised standards of required engineering science, and generated a fresh understanding of what the contemporary age meant by modern technology. In their periods of rapid growth, these industries were very profitable and attracted much investment. G. Mensch argued that as they became established and less modern, these industries became less profitable and attractive to investors, although they could still be important in the economy. They might be profitable enough to make the investments in them worth maintaining, but eventually the diminishing returns on investment would encourage the creation of new, dynamic industries made possible by the most recent generation of technological innovations.

Supporters of this theory claimed evidence from history. The mechanized-industrial age was launched in Great Britain by industries based on coal, iron, and textiles with transport by canal. These began in relatively few centers (Ironbridge, Cromford, and Manchester). The skills cultivated at all levels in the "first industrial nation" could not be learned easily or quickly in other nations, and Britain enjoyed a practically unassailable lead. The strategic innovations that created the industries that dominated the next phase were steam power, application of steam power to older industries, and railways. These grew out of the older technologies created in England, and so

Britain enjoyed prolonged leadership in the global economy. Later, periods were dominated by new technologies created relatively rapidly and did not evolve out of older industries. During such change points, leadership in economic growth passed to nations that developed the new skills and cultural values crucial during the next phase. Examples quoted were Germany and the U.S. in the period after 1890 when electrical engineering, industrial chemicals, and the automotive industry were strategic. Later still, electronics, aviation, rocketry, nuclear engineering, computing, and the technologies of the post-1945 age ushered in new phases of development and witnessed the rise of Japan. Though some econometricians accepted the precise periodicity of the Kondratieff model (as did Mensch), a larger number accepted that the interpretation of industrial growth was roughly correct and could be used as a guide. Many were skeptical and regarded the lessons derived from the "long-wave analysis" as due to hindsight, although the classifications might be beneficial to historians. Philosophers were suspicious of implicit determinism and historicism in the models. Much criticism was directed against the creation of long-wave trend curves, or continuous traces obtained by plotting an index of economic performance against a measure of input to industry, or against time (date). How could a major innovation be associated with a particular date? Was it legitimate to create a model by treating a succession of innovations as if each were equal in economic importance to the rest? Mensch's work attempted to deal with this issue but continued to attract adverse criticism.

The use of trend curves played a major role in large-scale econometrics. Trend curves were also used on a smaller scale to judge the extent to which a particular technology, design, or product was worthy of further investment or was approaching obsolescence. If recognized as near obsolescence, it could be replaced by a successor introduced in an orderly manner, which reduced the waste of unused potential in the old technology. The works of B. Twiss and R. Foster illustrate the use of S-curve analysis in management and business circles. Attempts to integrate small- and medium-scale S-curve analysis with the large-scale, long-term models of Kuznets, Schumpeter, and Mensch have not yet succeeded. These theories are often used as primarily qualitative guides based on history. As such, they provide valuable lessons. They suggest that it is destructive of a nation's standing, or an industry's profitability, if the nation or industry maintains investment (including intellectual skills) in declining activities that were once strategic and fails to reorganize to take control of completely new strategic industries based on recent innovations. Industrial and economic leadership may be associated in future with global networks rather than individual companies located in one nation. The changing nature of engineering is leading to industries based on nanotechnology, artificial intelligence, cybernetics, and biotechnical hybrids. The meaning of "industry" and "product" is being redefined. Contemporary and future industries may increasingly produce knowledge, patent rights, and licenses to manufacture as earlier industries produced steel and heavy equipment. The manufacture of older technologies is shifting to industrial cultures outside the first rank. This shift has caused widespread reorganization of engineering education and the profession in older industrial countries such as the U.K.

The 1973 energy crisis and the ongoing discussion concerning sustained growth, limits to growth, and environmental damage due to industrial activity gave a great impetus to neotechnocracy, long-wave econometrics, and engineer values. A few of the original technocrats still lived and carried on their original campaign, though technocracy, which survived as a movement, enjoyed little influence. The growth of innovation analysis and the need to assess the worth of expensive military projects revived interest in technological forecasting, which received funds from military sources and other government departments. The anticipated shortage of fossil fuels in the 1970s and 1980s focused attention on using trend curves to assess nearness to exhaustion of resources. Use was made of similar trend curves to link industrial production to damage to the environment. The link between energy consumption per head and GNP per capita was calculated for different countries at various stages of industrial development and used to calculate how much fossil fuel and raw materials would be needed to raise the poorer nations to the standard of living in the U.S. or Germany. Though the link between GNP per capita and energy consumption per capita was condemned as misleading, the analysis indicated the probable impossibility of abolishing world poverty in this sense, taking into account population growth, lengthening life spans, annual expansion of the economy, and expectations of a regular increase in standard of living. During the 1970s, the Florida School of analysts, associated with H. T. Odum and E.C. Odum produced studies of energy flow correlated with money flow in society and linked money value to energy value, along lines similar to those pursued by the various

technocrats in earlier years. The Odums concluded that whereas lack of money in circulation was the problem in 1929, the cause of postwar economic crises was more likely to be limited access to cheap energy and raw materials. In the 1970s the aggressive "production engineer values" of the period from 1910 to 1940 were less in evidence. The "engineer values" and the technocracy were still there, but they were presented in a more circumspect manner and in closer association with a liberal, enlightened economics that accepted limits to growth on environmental grounds.

Development of these theories and philosophies continues, as does the clash between engineers' values and economists' values. Some states require that any energy-consuming scheme be subjected to an analysis beforehand to calculate the total energy and resources investment in the project compared with the anticipated benefits. Many nations now calculate, as part of the GNP assessment, the energy investment in the goods produced and services provided.

See also **Engineering: Cultural, Methodological, and Definitional Issues.**

MICHAEL C. DUFFY

Further Reading

Akin, W.E. *Technocracy and the American Dream: The Technocrat Movement 1900–1941*. University of California Press, 1977.

Ewing, A. An Engineer's Outlook. Annual Report. British Association for Advancement of Science, 1931, pp 1–19.

Foster, R. Innovation: *The Attacker's Advantage*. Macmillan, London, 1986.

Fox, R., Ed. *Technological Change*. Harwood, 1996.

Gilbreth, F.B. *Motion Study*. Van Nostrand, New York, 1911.

Gilbreth, F.B. *Primer of Scientific Management*. Van Nostrand, New York, 1912.

Grubler, A. *Technology and Global Change*. Cambridge University Press, Cambridge, 1998.

Henderson, J.B. Invention as a Link in Scientific and Economic Progress. Annual Report. British Association for Advancement of Science, Leeds, 1927, pp 120–137.

Jevons, W.S. *The Coal Question: An Inquiry into the Progress of the Nation, and the Probable Exhaustion of Our Coal Mines*, reprint of 3rd edn. Kelley, New York, 1965.

Kondratieff, N.D. The Major Economic Cycles. Translation of Russian 1925 edn., *Review*, 4, pp 519–562, 1979.

Kuznets, S. *Economic Change*. W.W. Norton, New York, 1953.

Landes, D.S. *The Unbound Prometheus: Technological Change and Industrial Development in Western Europe from 1750 to the Present*. Cambridge University Press, Cambridge, 1969.

Liljegren, C.O. Coal, Oil or Wind? *Trans. Inst. Eng. Shipbuilders Scot.*, 64, 242–302, 1920–21.

Mensch, G. *Stalemate in Technology: Innovations Overcome the Depression.* Ballinger, Cambridge, MA, 1979.

Odum, H.T. and Odum, E.C. *Energy Basis for Man and Nature.* McGraw-Hill, New York, 1976.

Pavitt, K., Ed. *Technical Innovation and British Economic Performance.* Macmillan, 1980.

Quigley, H. *Electrical Power & National Progress.* George Allen & Unwin, 1920.

Rostow, W.W. *The Stages of Economic Growth,* 2nd edn. Cambridge University Press, Cambridge, 1971.

Schumpeter, J. *Business Cycles: A Theoretical, Historical & Statistical Analysis of the Capitalist Process.* McGraw-Hill, New York, 1939.

Schumpeter, J. *Konjunkturzyklen.* Gottingen, 1961.

Soddy, F. *Wealth, Virtual Wealth and Debt: The Solution of the Economic Paradox.* E.P. Dutton, New York, 1926.

Taylor, F.W. *The Principles of Scientific Management.* Harper Bros., New York, 1919.

Twiss, B. *Managing Technological Innovation,* 2nd edn. Longman, London, 1980.

Urwick, L. and Breck, E.F. *The Making of Scientific Management,* 2 vols. Pitman, 1951, 1953.

Walker, M. The Call to the Engineer and Scientist. Annual Report: British Association for Advancement of Science (York Meeting), 1932, pp 131–146.

Veblen, T. *Theory of the Leisure Class.* Macmillan, New York, 1899.

Veblen, T. *The Engineers and the Price System.* Huebsch, 1921.

Veblen, T. *The Instinct of Workmanship,* Macmillan, 1914. Reprinted by W.W. Norton, New York, 1964.

Entertainment in the Home

When Buckminster Fuller's futuristic Dymaxion House was displayed in Chicago in 1929, the television set, phonograph, and radio it featured were far from the everyday reality of most of the people who saw it. Less than 25 years later, these devices had become commonplace in middle-class homes across North America and Western Europe. The most significant forms of home entertainment technology of the twentieth century can broadly be divided into the audio and the visual, screen and sound, though as the century progressed the two categories increasingly overlapped. Electrification was a critical factor in the development and diffusion of these new entertainment technologies, as it was for other domestic technologies that became popular in the postwar period. Although some electronic devices could be operated by means of batteries, the consistent, popular use of most entertainment technologies in the home would not have occurred without mass domestic electrification. In the U.S. this occurred (unevenly) in the second decade of the century, and in Britain by 1926. Before World War II, live and recorded music and screen entertainment was more easily accessible to most people in public venues such as music and dance halls, cafes, and movie theaters. To some extent, the new technologies encouraged the return of public forms of entertainment to the domestic sphere.

In America, the phonograph displaced other forms of home music making, the most prominent of which was the piano in middle-class homes. By 1900, the Victor Talking Machine Company was marketing a domestic phonograph, and a few years later the Victrola became the first mass-marketed, enclosed phonograph. Its name was appropriated for the generic phonograph designed as household furniture. Futuristic designs eventually supplanted traditional disguises, and the phonograph was combined with radio and with audio cassette functions (in the early 1970s) in home stereo systems. By 1988, compact disk (CD) sales had surpassed the sales of phonographic long-playing albums, or LPs. Digital technology, which surpassed the phonograph record in quality of sound reproduction, durability, and convenience, was rapidly displacing the phonograph as the primary form of home sound technology.

The wireless technology that supported radio was a product of the late nineteenth century, but it did not become popular in homes until the 1920s. Prior to this, radio was the domain of industry, national security, and private enthusiasts. The simplification of technical controls to two knobs, the replacement of headphones with a loudspeaker, and the physical displacement of the radio from the basement and garage to the living room, transformed a primarily male activity into a genteel, feminized, domestic amusement for the whole family. By 1950, approximately 95 percent of American households had radios. In the 1990s, Canada pioneered digital radio, and in 1995 the first digital radio receivers were marketed to consumers. Researchers have predicted the full transition from AM–FM to digital radio by the first decades of the twenty-first century.

Television (TV), a more expensive and complex technology, was nevertheless much more quickly adopted by consumers than radio. It was introduced to the public in 1924 at Selfridges department store in London, England, by Scotsman John Logie Baird, but only became a household commodity after World War II. Much as inside and outside was merged in modern houses by means of the picture window, the introduction of television brought public culture into private homes. RCA and DuMont offered the first sets (all with black and white pictures) to the public in 1946. Between 1948 and 1955, nearly two thirds of

American households installed television sets, and by 1960 almost 90 percent of households had at least one television. The way people watched television during the last half of the twentieth century changed considerably due to the development of television-related technologies such as coaxial cable, communications satellites, the video cassette recorder (VCR), and the remote control for all of it. Cable, which spread slowly from the 1950s through the 1970s, and communications satellites, which gained popularity in the last two decades of the century, broadened the range of programming available to viewers. The digitization of television in the form of high-definition television (HDTV) began at the end of the twentieth century but was not expected to have a large-scale impact in homes until well into the first decade of the next century.

The VCR was the first peripheral device to television to gain widespread consumer acceptance. It had two distinct technological functions that together performed the important cultural task of giving consumers the freedom to choose programming outside of regular broadcasting. One function was time shifting, the ability to record a program for playback at a later time. The second closely related function allowed the replay of prerecorded cassettes, either commercially produced or made personally with home video recorders, introduced by Sony in 1980 with the Camcorder. In 1956, the Ampex Corporation of California introduced the industrial precursor to the consumer VCR: the VTR, or video tape recorder. By 1966, Ampex had sold over 500 VTRs for home use, but with open-reel video tape these were both unwieldy and expensive. In 1972, Sony introduced the U-matic, which sold for $1600 and was the first video cassette recorder specifically intended for home use. Although Sony's Betamax format video cassette was widely popular among consumers, the Video Home System (VHS) from JVC became the standard format until the last years of the twentieth century when laser disk technology began to make significant inroads. By 1988, more than half of television-owning households also had VCRs. In 1996, the first digital video/versatile technology (DVD) hit the consumer market in Japan and reached the U.S. market the next year. By the end of the century, DVD technology had become mainstream, and home video providers began to stock video disks alongside video tapes. Unlike video tapes, DVDs could be played independently of TV on home computers.

Home video games are considered television peripherals since they rely almost exclusively on the television screen to express their video component. The first generation of home video games was ushered in with the Magnavox Odyssey in 1972. Early versions were limited to preprogrammed games, whereas the popular Sony, Sega, and Nintendo systems developed in Japan offered the possibility of programmability and an endless number of games along with advanced graphics and fuller sound. Products such as Nintendo's Game Boy and the Sony Playstation were the objects of a home entertainment craze; over 10 million Playstations were sold between 1995 and the first years of the twenty-first century.

The practices of watching television, listening to music, and operating computers influenced the way people thought about, used, and designed their homes: from living rooms, to the recreation room/family room/games room, the kitchen, the home office, and even the bedroom and bathroom. Many postwar television watchers designated a TV room that set apart television watching architecturally from other domestic activities. Others had a TV area in the living room, or swivel stands that allowed for flexible viewing arrangements. Home entertainment technology also inspired the design of new types of furniture that reflected both the utopian and nonutopian expectations with which modernity and its products were received. Designs for cabinets and stands for televisions and stereos highlighted or hid their conspicuous futuristic designs. Alternatively, designers attempted to integrate new technologies into more traditional domestic environments by designing decorative cabinetry in fine woods to match preexisting traditional furniture.

See also **Audio Recording (all); Radio: AM, FM, Analogue, Digital; Radio, Early Transmissions; Television, Cable and Satellite; Television, Digital and High Definition; Television Recording, Tape**

JENNIFER COUSINEAU

Further Reading

Braden, D.R. *Leisure and Entertainment in America.* The Museum, Dearborn, MI, 1988.

Dale, R. and Weaver, R. *Home Entertainment: Discoveries and Inventions.* Oxford University Press, Oxford, 1993.

Ellis, D. *Split Screen: Home Entertainment and the New Technologies.* Friends of Canadian Broadcasting, Toronto, 1992.

Forty, A. *Objects of Desire: Design and Society from Wedgwood to IBM.* Pantheon, New York, 1986.

Herz, J.C. *Joystick Nation.* Abacus, London, 1997.

Peiss, C. *Cheap Amusements: Working Women and Leisure at Turn of the Century New York.* Temple University Press, Philadelphia, 1986.

Read, O. and Welch, W.L. *From Tin Foil to Stereo: Evolution of the Phonograph,* 2nd edn. SAMS, Indianapolis, 1976.

Spiegel, L. *Make Room for TV: Television and the Family Ideal in Postwar America.* University of Chicago Press, Chicago, 1992.

Spiegel, L. *Welcome to the Dream House: Popular Media and Postwar Suburbs.* Duke University Press, Durham, NC, 2001.

Wright, G. *Building the Dream: A Social History of Housing in America.* The MIT Press, Cambridge, MA, 1992.

Useful Websites

Schoenherr, S. Recording Technology History: http://ac.acusd.edu/history/recording/notes.html

Environment and Electricity Generation, *see* **Electricity Generation and the Environment**

Environmental Monitoring

As a dynamic system, the environment is changing continually, with feedback from both natural (climatic or biogeochemical) and anthropogenic (human activities) sources. Assessing the rate and magnitude of environmental processes is difficult, especially as data collection over time is limited to the last century, or even the last few decades. Since the 1980s, environmental monitoring programs have been developed as a response to concerns that environmental impact or sustainability of policy initiatives could not be evaluated adequately. So-called *State of the Environment Reports* date from this period. Many are concerned with the state of national, regional, or local environments (land, rivers, or seas); others focus on environments at particular risk, mostly due to human impact and pollution (e.g., environmental contaminants in marine or terrestrial wildlife), or those where environmental quality is significant in the context of human health (e.g., urban air quality; water resources, fisheries). Many monitoring programs include information collected remotely by satellites (see Satellites, Environmental Sensing), but this entry focuses on technologies and policies for *in situ* monitoring.

Adequate and sustained monitoring and its evaluation provides early warning of possible environmental degradation. Such information is important for the prediction of change that may be generated in the wake of a development project, such as dam construction or deforestation programs. In this context—as an element of an environmental impact assessment—monitored data are valuable for an evaluation of sustainability.

Monitoring involves the regular or continuous collection of specific parameters that depends on the type of environment being monitored, and can contribute to the identification of cause and effect relationships. Physical, chemical, and biological data may be collected in subsystems within specific environments. For example, certain organisms may be sensitive to changes in pH or nutrient status in aquatic environments, and so alterations in their population numbers and population composition can be used to detect change in water quality. These are called indicator species. Thus biological components are monitored to detect change in the chemistry of rivers and lakes. A range of socio-economic data may also be monitored; factors such as population change and a drive to increase agricultural production because of the emergence of new markets are driving forces in environmental change and are likely to stimulate alterations in the physical environment.

There are three basic types of monitoring—baseline, impact, and compliance—that reflect different stages of development. Baseline monitoring involves collection of basic environmental data (e.g., soil and water pH) prior to the start of a development project. Impact monitoring focuses on the collection of both environmental and, where relevant, socioeconomic data, during the construction phase of the project. Compliance monitoring requires regular, often periodic, data collection to determine environmental quality and change after project completion and is a means of assessing if standards of environmental protection are adequate and in accord with environmental law. Through the provision of environmental data and its evaluation, monitoring programs also contribute to the enhancement of public understanding of local and regional environments and the operation of people–environment relationships.

No monitoring program can be truly comprehensive because the physical and human environments comprise a vast range of factors that interact in complex ways. Thus an initial and vital prerequisite for any monitoring program is the choice of appropriate variables that, in turn, are dependent on the objective of the project and the goal of the monitoring program. Broadly, such variables can be classified as state, pressure, and response indicators. State indicators reflect the condition of the environmental system; examples include the composition and density of the vegetation cover, soil pH and nutrient status, and water pH and chemistry. Pressure indicators may be direct measures; for example, rate of land clearance, length of fallow, agricultural productivity, rate of wood removal from forests, rate of waste production; or indirect measures; for example, calorie consumption per individual, volume of crops produced per unit area, volumes of crops, meat or wood exported, rate of water consumption, income *per capita*, and rate of population growth. Response indicators may include soil and water conservation measures, land abandonment, and emigration rates. All may be qualitative or quantitative, or both. Each indicator and its measurement must be scientifically sound and sensitive to change. They should not cause problems for local people but should involve them where possible, as local knowledge is all-important. Each indicator and measurement must be reproducible and feasible in terms of scale, time, and finance.

Monitoring may be undertaken by various organizations, both government-related and nongovernmental. For example, the European Environment Agency of the European Union produces yearly environmental indicator reports entitled *Environmental Signals;* a special report for the millennium was also produced (OOPEC, 2002). In the US, the Environmental Protection Agency is one of several government bodies involved with environmental monitoring; between 1996 and 2001 it operated a program entitled Environmental Monitoring for Public Access and Community Tracking (EMPACT), which involved a wide range of projects on air, water, and soil monitoring in rural and urban environments. Nongovernmental agencies that produce reports on the state of the global environment based on national statistics include the Worldwatch Institute in Washington D.C.; its annual publication for the last 12 years is entitled *Vital Signs*. This publication collates statistics on "key indicators," including atmospheric, energy, agricultural, food, trade, population, disease, communication, transport, and military trends. The Worldwide Fund for Nature (WWF) produced two reports entitled *Living Planet*, in 1998 and 2002, which detail the state of the world's ecosystems and the impact of humanity on them; the data are presented on a country-by-country basis and are used to derive a Living Planet Index and Ecological Footprint which can be used for intercountry and regional comparisons. The World Conservation and Monitoring Centre, initially established in 1988 and formally linked with the United Nations Environment Programme (UNEP) since 2000, has a more specialized role insofar as it collects, collates, and evaluates worldwide data on plants, animals, and microorganisms and the ecosystems to which they belong.

Detection and characterization of environmental pollutants may be by continuous source monitoring, for example of stack emissions, industrial wastewater, or leachate sampling from landfills, or by sampling at fixed points within an areal network. The latter includes sampling of atmospheric aerosols (airborne solid or liquid particles, such as nitrous oxides and diesel particulates) at urban sites, or groundwater monitoring for nutrients and pollutants such as volatile organic compounds (benzene), heavy metals (lead, mercury, arsenic), pesticides (DDT, atrazine), polychlorinated biphenyls (PCBs), and pathogens from animal or human waste.

Indoor and outdoor ambient air quality is often measured by drawing air into an evacuated, contaminant-free container or pumping a sample through an impregnated filter paper or gas sample tube. Some optical instruments measure particle size and the amount of airborne dust particles. Colorimetric indicator methods, in which a color change of the sorbent in a gas tube gives presence or absence information, are specific to a single chemical such as arsenic.

The development of analytical chemistry techniques for chemical process control provided ready-made solutions for environmental monitoring. Infrared spectrometers, developed in the 1920s as a laboratory analytical tool, became less bulky from the 1950s and with the advent of powerful microprocessors became routinely used for measuring gaseous air pollutants, from ambient air, combustion sources, and toxic waste incinerators. Gas chromatography, a versatile analytical technique developed in the 1950s, is now available as portable units that can be used on-site for real-time direct and continuous measurement, with data logged or transmitted via radio telemetry if in a remote area. Mass spectrometry can give very fast positive identification of a broad range of compounds, whether gases, liquids, or solids. Gas-sensitive solid-state sensors, first researched in the late 1950s and commercially available from the 1970s, are now in widespread use.

Neutron activation analysis, a laboratory technique in which a sample is irradiated with high-energy neutrons, permits measurements of trace elements in atmospheric, soil and water pollution studies. Detection limits are in the parts per million to parts per billion range.

See also **Satellites, Environmental Sensing; Technology, Society and Environment**

A. M. MANNION

316

Further Reading

Mannion, A.M. *Dynamic World: Land-Cover and Land-Use Change*. Arnold, London, 2002.

OOPEC. *Environmental Signals 2002: Benchmarking the Millennium*. European Environment Agency Regular Indicator Report, OOPEC, Luxembourg, 2002.

Worldwatch Institute. *Vital Signs 200*. Worldwatch Institute, Washington D.C., 2003.

Worldwide Fund for Nature. *Living Planet Report 1998*. Worldwide Fund for Nature, Gland, Switzerland; New Economics Foundation, London; World Conservation Monitoring Centre, Cambridge, 1998.

Worldwide Fund for Nature. *Living Planet Report 2002*. Worldwide Fund for Nature, Gland, Switzerland, 2002.

Useful Websites

Environmental Protection Agency (EPA), Environmental Monitoring for Public Access and Community Tracking (EMPACT): http://www.epa.gov/empact/

Environmental Protection Agency (EPA), Environmental Technology Verification Program: http://www.epa.gov/etv/

World Conservation and Monitoring Centre (WCMC): http://www.unep-wcmc.org

Error Checking and Correction

In telecommunications, whether transmission of data or voice signals is over copper, fiber-optic, or wireless links, information coded in the signal transmitted must be decoded by the receiver from a background of noise. Signal errors can be introduced, for example from physical defects in the transmission medium (semiconductor crystal defects, dust or scratches on magnetic memory, bubbles in optical fibers), from electromagnetic interference (natural or manmade) or cosmic rays, or from cross-talk (unwanted coupling) between channels. In digital signal transmission, data is transmitted as "bits" (ones or zeros, corresponding to on or off in electronic circuits). Random bit errors occur singly and in no relation to each other. Burst error is a large, sustained error or loss of data, perhaps caused by transmission problems in the connecting cables, or sudden noise. Analog to digital conversion can also introduce sampling errors.

Early Error Detection

Claude Chappe's semaphore visual telegraph system in late-eighteenth century France incorporated rudimentary error checking: an operator had to check that the next station was correctly reproducing the signal. Error correction also existed in the form of a single control signal that signified "erase the last signal sent." The correct signal was then repeated.

In the early twentieth century, various techniques were applied to telegraph signals in order to prevent or reduce errors from transmitted signals. For cabled telephony, low transmission rates, shielding (to reduce cross-talk), and repeaters (to overcome signal loss due to attenuation) all helped to prevent errors. Errors could also be reduced by sending each message multiple times.

Telex over radio (or TOR), developed in the 1960s for sending telex to ships, converted binary code (usually the 5-bit Baudot code with 5 bits representing each text character) received by radio to text. In wireless transmissions, noise and fading cause missed or incorrect characters. Two similar and simple error detection and correction systems were developed: sending each character twice (the receiving station would request a resend if it did not see each character twice in sequence), and a read–confirm system, where each character was confirmed as received.

From the 1960s ASCII (American Standard Code for Information Interchange) replaced the Baudot code in teletype transmission: ASCII used seven bits to represent each text character, but an eighth bit called a parity bit could be added to each character to do an error check. Developed for teletype, ASCII also provided a common computer code used in most computer operating systems, e-mails, and HTML documents.

Parity checking is the oldest and simplest example of error checking code, used in teletype and computer data storage. A single bit, a parity check bit, is added to the transmitted data. The sum of a fixed number of bits can be made even (or odd) by properly setting the parity bit to a one or zero. Errors are detected on the receiving end simply by checking whether each received word is even (or odd). As with TOR, the receiver could request a retransmission of the message if the message and its error checking code are in conflict (or if the code is not received.) Check digits are similarly used embedded in credit card numbers and ISBN numbers to detect mistakes.

Parity checking has a low error detection rate, only about 50 percent. In block checking (also called longitudinal redundancy checking), a block check character (BCC) is added to the end of each block of data (a group of bits transmitted as a unit) before transmittal: the block check character is computed using binary addition to maintain parity counting longitudinally through the block. The checksum and cyclic redundancy check (CRC) algorithms similarly compute block check characters for each message block, but the character (for checksum) or series of characters (for CRC) added is a mathematical function of all the data in the message. As with parity checking, the receiver computes a second block check character to compare with the received character to determine whether the transmission is error free. Redundancy checking, which can be easily implemented and decoded in hardware, can detect up to 99 percent of errors, and is useful where it is easy to retransmit data. In TCP (the Internet's Transmission Control Protocol) checksums are computed for each message packet to detect transmission errors, and the receiver acknowledges only packets where the checksum is correct.

Retransmission is the oldest form of error correction—simply, the message is retransmitted until it is received without error. This is often called Automatic Repeat Request (ARQ) and is used in modems and increasingly in mobile Internet via wireless cellular networks. In stop-and-wait ARQ the sender waits for a reply in a fixed time frame after sending a signal. If there is no reply, the block is resent. In continuous ARQ, the receiver automatically starts a request for retransmission when an error is detected.

ARQ and parity checks are backward error correction (BEC) techniques used widely in computer communications; "backwards" because the sender is responsible for retransmitting any data found to be in error. In addition to error-detecting codes there are error-correcting codes—referred to as forward error correction (FEC), since the receiver corrects the error without requiring further input from the sender.

Error Correction Coding

Late-twentieth century error coding techniques are based on information coding theory, from work by Claude Shannon in 1948 (see Information Theory). "Information" is mathematically quantifiable as binary codes, as described above. Error correction coding adds extra bits (the "code word") to the data before transmission, which the receiver can decode to check if errors occurred during transmission. Simply repeating each bit a set number of times and assuming the most frequent value is correct can automatically correct for errors, but high repetition rates are not desirable. Shannon's coding theory showed that it was possible to encode messages in such a way that the number of extra bits transmitted was as small as possible (the Shannon limit), though his theory did not produce any actual ideal coding algorithms.

The first practical error correction codes were developed in 1948–1950 by Richard W. Hamming, a colleague of Claude Shannon's at Bell Labs, assisted by Marcel J.E. Golay. Hamming was working with an early relay computer, and was frustrated by having to restart his computations when the computer detected errors. Hamming and Golay "block codes" are linear algebraic rules for converting an information sequence of bits, of length k, say, into a transmitted sequence of length n bits, and had information transmission rates more efficient than simple repetition.

The Hamming and Golay correcting block codes are easy to encode and decode. In 1960 Irving S. Reed and Gustave Solomon (researchers at the Massachusetts Iinstitute of Technology) devised further error-correction block codes. More recently, convolution codes—in which each encoded block depends not only on the corresponding k-bit message block at the same time unit, but also on m previous blocks—have been used.

Error coding has great economic value and is used in many digital applications such as computer memory, magnetic and optical data storage media, radio-frequency links for space and satellite communications systems, computer networks, and cellular telephone networks (called channel coding).

See also **Electronic Communications; Information Theory; Telephony, Digital**

GILLIAN LINDSEY

Further Reading

Costello, D.J. Jr.; Hagenauer, J., Imai, H., Wicker, S.B. Applications of error-control coding. *IEEE Trans. of Inf. Theor.*, 44, 6, 2531–2560, 1998.

Berlekamp, R. *Key Papers in the Development of Coding Theory*. Institute of Electrical and Electronics Engineers, New York, 1974.

Hill, R. *A First Course in Coding Theory*. Oxford University Press, Oxford, 1986.

Nebeker, F. *Signal Processing: The Emergence of a Discipline, 1948 to 1998*. Institute of Electrical and Electronics Engineers History Center, 1998.

Experimental Stress Analysis

This branch of technology deals with the means of measuring strains in materials under load and, from these strains, inferring the stresses actually endured by the material. The fundamental idea underlying the design of all components of structures and machines that must carry loads is that the stress in the material should be less than, or equal to, a certain prescribed level.

Unfortunately, it is not possible to measure stress directly. Stress values inside a material must be calculated using mathematical models of both the structure and properties of the material of which it is made. When fundamental material properties such as strength and stiffness (Young's modulus) are experimentally tested, the structure is kept very simple—a wire for tests in tension or a supported beam for tests in bending. Measurements of load and the extension or deflection of these structures are then used to calculate internal stresses for simple tension or compression and for simple bending theory in the case of beams. Modern high-speed computers have enabled more complex mathematical models and have rendered complicated structures amenable to theory.

While it was common from the early nineteenth century to test full-size and scale-model prototypes of load-bearing structures, the tests provided only indirect information about actual stresses in materials. This was of limited use to engineers wanting to achieve minimum sizes and weights of geometrically complex components used; for instance, in aircraft and other high-performance machines.

While it can be important to know stresses deep within a component under load, the highest stresses are usually found at or near the surface, which is also where cracks begin when materials fail by brittle fracture. Since the early nineteenth century, surface stresses were calculated by measuring the surface strain, for instance the distance between two gauge marks, and inferring the stress knowing the Young's modulus of the material. However, this technique was of little use for components under dynamic loads or with inaccessible surfaces. To meet these challenges, two main strands of technologies for experimental stress analysis were developed during the nineteenth century.

The first made use of the phenomenon of photoelasticity. Certain transparent materials display an anisotropy, called birefringence, to light. When the incident light is polarized, the birefringence has the effect of rotating the plane of polarization of the light. The degree of anisotropy and, hence, rotation, depends on the stresses in the material or, rather, the differences between the stresses in the three principal directions; that is, $(\sigma_1-\sigma_2)$, $(\sigma_1-\sigma_3)$, and $(\sigma_2-\sigma_3)$. If a two-dimensional test piece is used, one principal stress is zero, and the magnitude and direction of the remaining two principal stresses can be established. The only point at which the photoelastic fringes give a direct indication of stress is at the model boundary where a second principal stress (perpendicular to the surface) is also zero. The varying degree of rotation of the incident polarized light is viewed in a polariscope, which displays the highly colorful interference fringes. For quantitative work, monochromatic light (usually sodium) is used to produce sharper fringes (Figure 2).

The early development of this technique was undertaken by E.G. Coker and L.N.G. Filon at University College, London in the early twentieth century using test models made of glass. Although glass is easily available, it is not highly birefringent, is difficult to cut to shape and is vulnerable to fracture. The technique became more accessible with the development of transparent epoxy resins such as Araldite in the 1940s.

The technique reached its peak utility in the 1960s in the aerospace industry, for example at

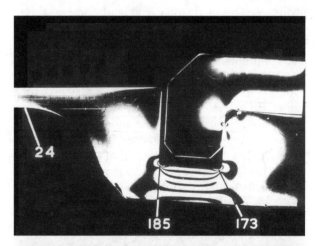

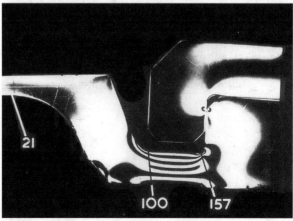

Figure 2. Photoelastic fringes photographed for stress analysis purposes. The comparison of two alternative designs for a clamp in an airengine illustrate a substantial reduction in the peak stresses in Design B.
[*Photos: Rolls Royce Photoelastic Laboratory*]

319

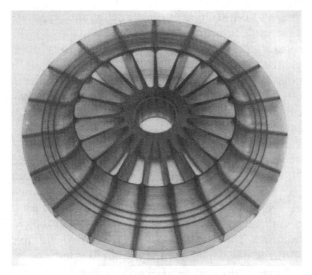

Figure 3.   Impellor machined from a solid block of Araldite prior to loading for stress analysis.
[*Photo: CIBA*]

developed. The most widely used depended on the fact that the cross-section of an electrical conductor will reduce when stretched, and this in turn increases its electrical resistance. The change in resistance is usually measured with a Wheatstone bridge. Electrical resistance strain gauges were first used by Charles Kearns in the early 1930s for studying the stresses in aircraft propellers. He ground flat a conventional carbon composite electrical resistor and mounted it on an insulating strip, which he cemented to the blade. He made the electrical connection to the static Wheatstone bridge by means of brushes and rings similar to those used in motors.

In the late 1930s, both Arthur Ruge and Edward Simmons in the U.S. used arrays of fine wires to achieve the same effect. In 1952 Peter Jackson, working with the Saunders–Roe Company on the Isle of Wight, used the new printed circuit board technology to make foil strain gauges that were much smaller, easier to fit, and more reliable. These are still in widespread use, both singly and in the form of a rosette with three gauges at 120 degrees to enable the principal stress directions to be established (Figures 4 and 5).

Despite their simplicity, electric resistance strain gauges had many practical problems, especially the low signal-to-noise ratio of the output. Mechanical brushes were replaced by mercury slip rings, direct current (DC) systems gave way to alternating current (AC) systems, and radio telemetry finally avoided the need for electrical connections. More recently, semiconductor strain gauges provide up to 50 times greater sensitivity. Two remaining problems with all strain gauges is their instability over time and their sensitivity to temperature changes. For the most careful measurement, an electrical thermometer is essential as well as constant recalibration to compensate for creep in the mounting of the gauge on the component under test.

Rolls Royce in Derby, England, where three-dimensional models of impellor or turbine disks were analyzed (Figure 3). A full-size disk, perhaps 600 millimeters in diameter, was intricately machined out of a solid block of Araldite. Dummy turbine blades were fitted. These had a scaled-down weight to take account of the difference in stiffness between the model material and actual material. The entire assembly was rotated at a scaled-down speed, in an oven of about 300°C to allow the loads to cause exaggerated deflections. The oven was then cooled to room temperature with the model still spinning so that the stresses would be frozen or locked into the Araldite. Finally, a number of slices of Araldite, just 1 or 2 millimeters thick, were cut from the model and analyzed using a conventional polariscope.

The second means for measuring surface stresses was the strain gauge, several forms of which were

Figure 4.   A foil strain gauge and a rosette formation of three such strain gauges.
[*Photo: University of Cambridge*]

Figure 5. Single strain gauges and rosettes fixed to the surface of the component under test, and wired to measuring device.
[*Photo: Fermilab*]

Another strain gauge technology developed in the 1950s was based on the long-known relationship between the pitch at which a string vibrates and the tension in the string. Vibrating-wire strain gauges were developed mainly for use in large objects such as buildings and bridges, which need to be monitored over long periods of time. A wire, perhaps 200 to 400 millimeters long is fixed at both ends, and its pitch is measured electronically as the length, and hence tension, changes due to relative movement of the two ends. Such gauges are also used in geotechnical investigation of rock movements in mines and bridge foundations.

Two other techniques for studying surface strains were developed in the 1960s. A crude, though very simple technique, was to coat a surface with a brittle lacquer. When strained, the material would crack in a direction perpendicular to the line of maximum principal stress, and the degree of cracking would indicate its magnitude. An advantage of this passive technique was that it could freeze, so to speak, the loading in a moving engine component after the engine had come to rest.

A more sophisticated technique utilized surface photoelasticity. Improvements in epoxy resins in the 1960s allowed them to be fixed directly to the surfaces of components, and the strains induced could be made visible using a reflection polariscope. This technique allowed rapid measurement of strains on complex surfaces and, with the development of holography, also allowed detailed assessment of strains and movement associated with vibration in machinery.

The arrival of optical fibers and lasers during the 1970s led to the development of optical strain gauges. Their advantage over wire strain gauges is the ability to measure the precise point along the length of an optical fiber at which it is stretched. This technology can be used to detect strains in the decks of long bridges.

While photoelasticity and strain gauge technologies have been widely replaced by finite-element stress analysis, they were of vital importance in helping to validate and calibrate such computer models in their early days. Strain gauge and photoelastic coating technologies are both still used when the high costs of setting up computer models are prohibitive.

See also **Bridges, Long Span and Suspension; Bridges, Steel; Buildings, Prefabricated; Concrete; Concrete shells**

W. A. ADDIS

Further Reading

Coker, E.G. and Filon, L.N.G., *Treatise on Photoelasticity*, 2nd revised edn. Cambridge University Press, Cambridge, 1957.

Dally, J.W. and Riley, W.F. *Experimental Stress Analysis*, 3rd edn. McGraw–Hill Education, New York, 1991.

Frocht, M.M. *Photoelasticity*. Chapman & Hall, London, 1941 (vol. 1) and 1948 (vol. 2).

Hendry, A.W. *Elements of Experimental Stress Analysis*. Pergamon Press, Oxford, 1977.

Hetenyi, M. The fundamentals of three dimensional photoelasticity. *J. Appl. Mech.*, 5, 4, 149–155, 1938.

Hetenyi, M. Handbook of Experimental Stress Analysis. Wiley, New York, 1950.

Jessop, H.T. and Harris, F.C. *Photoelasticity: Principles and Methods*. Dover, New York, 1950.

Explosives, Commercial

All chemical explosives obtain their energy from the almost instantaneous transformation from an inherently unstable chemical compound into more stable molecules. The breakthrough from the 2000-year old "black powder" to the high explosive of today was achieved with the discovery of the molecular explosive nitroglycerine, produced by nitrating glycerin with a mixture of strong nitric and sulfuric acids. Nitroglycerin, because of its extreme sensitivity and instability, remained a laboratory curiosity until Alfred Nobel solved the problem of how to safely and reliably initiate it with the discovery of the detonator in 1863, a discovery that has been hailed as key to both the principle and practice of explosives. Apart from the detonator, Nobel's major contribution was the invention of dynamite in 1865. This invention tamed nitroglycerine by simply mixing it with an absorbent material called kieselguhr (diatomous earth) as 75 percent nitroglycerin and 25 percent kieselguhr. These two inventions were the basis for the twentieth century explosives industry.

In the years from 1873 to 1920 there was unprecedented development of commercial explosives. The original dynamite was replaced with a range of nitroglycerin-based explosives prepared by mixing varying proportions of nitroglycerin with ammonium nitrate (today well known as a fertilizer) and other nitrates such as sodium and potassium nitrates. Also during this period explosives with the almost household names of blasting gelatin, gelignite, opencast gelignite, and dynamite were developed each designed specifically to achieve maximum efficiency in mining, quarrying, and civil engineering operations.

The introduction of a specialist range of explosives specifically designed to be safe (permitted) when used in dangerous conditions in gassy coal mines, where there is the risk of methane air (firedamp) explosions, occurred from 1922 to 1932. By 1970 compositions had achieved a very high degree of safety even in the most broken strata with the highest danger.

Early detonators were initiated with a length of black powder safety fuse. By 1930 a full range of electric instantaneous and delay detonators had been developed to enable sequential firing, a major advance in blasting efficiency. Today, high-precision blasting is made possible by a range of sophisticated electronic delay detonators.

There are over 200 explosive chemicals in current use, and TNT (trinitrotoluene) is perhaps the best known. TNT was first produced in 1863 by "nitrating" toluene using a mixture of strong nitric and sulfuric acids. The principal use of TNT is in military explosives, and it has been used in shells and bombs since 1902. Shortly after World War I, TNT was used in quarrying and opencast explosives in admixture with potassium nitrate and aluminum. Many other important specialist explosives include: PETN (pentaerithyritol tetranitrate), a crystalline explosive used in detonating transmission lines (Cordtex and Primacord); and RDX (cyclonite trimethylen-trinitramine) used in military applications and the manufacture of boosters for high performance and shaped charges capable of penetrating 3 centimeters of steel plate. These high-performance explosives have helped develop other technologies such as:

1. Metal forming, in which exotic metals such as titanium are precisely shaped by explosively produced hydraulic shock
2. Welding of metal pipe work by shock waves into seamless welds
3. Producing shaped charges that can cut through metal structures such as girders and caissons in demolition, engineering, and oil well perforation

The major twentieth century revolution in commercial explosives technology, however, came in 1956 with the introduction of ammonium nitrate water slurry explosive, a simple mixture of ammonium nitrate and fuel oil (ANFO) containing no explosives, which could be safely mixed on site. ANFO came into prominence following the development in the 1940s of porous ammonium nitrate prills (pellets), which absorbed the oil. The Iron Ore Company of Canada made the first serious use of ANFO in 1956. In the early 1960s the first mobile on-site bulk loading ANFO truck for opencast and quarry blasting was introduced.

The disadvantage of ANFO compositions, however, is that they are not waterproof. This disadvantage was overcome by the development of slurries in the late 1960s. Slurry is a saturated aqueous solution of ammonium nitrate in which porous ammonium nitrate prills and a sensitizing fuel are incorporated and made water resistant by the addition of thickening agents. In the early days the sensitizer was TNT, but it was soon superseded by nonexplosive aluminum. These new generation slurries were almost immediately put to use in sophisticated on-site mixing trucks that blended and loaded the slurries directly into shot holes. All the components were nonexplosive, and an explosive was only formed when mixed and delivered into shot holes. In 1990 these mobile truck operations were brought into the computer age by incorporating microprocessor control systems to provide programmable loading of shot holes.

In 1968 further development took place with the introduction of emulsion explosives by the Atlas Powder Company in the U.S. Emulsion explosives go one step further in that they do not require any chemical sensitizers whatsoever. These explosives are essentially entrapped air within a water-in-oil emulsion of ammonium nitrate, suspended as microscopically fine droplets and surrounded by a continuous fuel phase of oil and waxes stabilized by an emulsifying agent. The entrapped air acts as a sensitizer and may be in the form of ultrafine air bubbles or air entrapped in glass microballoons. The development of high-efficiency gelled cartridge emulsions in the 1970s made enormous changes, and they are now successfully used in a wide range of underground mining and civil engineering applications. The result of these major develop-

ments has been that nitroglycerin-based explosives have been replaced by ANFO and the emulsion explosives in all but the most demanding of hard rock and precision-blasting operations.

Explosives are ideally suited to provide high energy in airless conditions. For that reason explosives have played and will continue to play a vital role in the exploration of space.

See also **Warfare, High-Explosive Shells and Bombs**

JOHN DOLAN

Further Reading

Bergengren, E. *Alfred Nobel: The Man and his Work.* Nelson, London, 1960.

Brown, G.I. *The Big Bang: A History of Explosives.* Sutton Publishing, Stroud, Gloucestershire, 1998.

Cook, M.A. *The Science of High Explosives.* Krieger Publishing, London, 1958.

Croft, W.D. *Dangerous Energy: The Archaeology of Gunpowder.* English Heritage, Swindon, 2000.

Dolan, J. and Langer S.S., Eds. *Explosives in the Service of Man: The Nobel Heritage.* Royal Society of Chemistry, Cambridge, 1997.

Fordham, S. *High Explosives and Propellants,* 2nd edn. Pergamon Press, London, 1980.

Hogg, O.F.G. *Clubs to Cannon: Warfare and Weapons before the Introduction of Gunpowder.* Duckworth, London, 1968.

Meyer, R. *Explosives,* 3rd edn. VCH, Weinhein, 1987.

Urbanski, T. *Chemistry and Technology of Explosives,* 4 vols. Pergamon Press, Oxford, 1984.

F

Farming, Agricultural Methods

Agriculture experienced a transformation in the twentieth century that was vital in increasing food and fiber production for a rising global population. This expansion of production was due to mechanization, the application of science and technology, and the expansion of irrigation. Yet these changes also resulted in the decimation of traditional agricultural systems and an increased reliance on capital, chemicals, water, exploitative labor conditions, and the tides of global marketing.

A sign of the transformation of agriculture in the twentieth century was the shift from China and India as countries often devastated by famine to societies that became exporters of food toward the end of the century. As the world's technological leader, the U.S. was at the vanguard of agricultural change, and Americans in the twentieth century experienced the cheapest food in the history of modern civilization, as witnessed by the epidemic of obesity that emerged in the 1990s. Unfortunately, this abundance sometimes led to overproduction, surplus, and economic crisis on the American farm, which one historian has labeled "the dread of plenty."

Mechanization of agriculture became essential in increasing production of foodstuffs and fiber in the twentieth century. Tractors, powered by internal combustion engines were introduced in the U.S. in the early 1900s. The tractor eventually led to the near extinction of horse- and mule-powered agriculture throughout most of the developed world by the 1950s. The power take-off (PTO) attachment, hydraulics, and three-point hitches were developments in labor-saving equipment that helped prepare the soil and plant and harvest

crops. Other mechanical technologies that helped expand acreage and reduce the need for expensive animal and human labor included combine harvesters, refrigerated transportation and cold storage, chemical applicators, cotton, vegetable, and fruit-picking machines, new irrigation technology, as well as machines to test the soil, check the moisture of the crop, and move the crops from field to market.

Science also dramatically transformed agriculture in the twentieth century with the introduction of hybrid crops and genetic engineering of crops. Employing research from the system of land grant colleges such as his alma mater, Iowa State University, Henry A. Wallace (U.S. Vice-President from 1941 to 1945), Roswell Garst, and others commercialized hybrid corn in the 1920s and 1930s. In the 1940s and 1950s hybrid crops combined with petroleum-based fertilizers, insecticides, herbicides, and fungicides led to a "green revolution" in agriculture. From 1970 to 1986, new hybrid rice and wheat strains, fed by artificial fertilizers, led to threefold production in grain yields worldwide.

Western agricultural scientists were often successful in spreading green revolution technology throughout the globe, with mixed results for production, the environment, and indigenous agricultural societies. In the U.S. and around the world, researchers fought for governmental research and development money, and conducted research that could then be applied by farmers able to afford the new technologies. Often the industrialized nations, through the World Bank, the United Nations, or private foundation projects, introduced new technologies in areas where they were not suited or where they severely disrupted traditional societies or environments. For example, hybrid wheat was introduced into Mexico on a large scale with grants from the Rockefeller Foundation. This wheat was primarily grown to feed livestock, and the hybrid corn introduced in Mexico ended up as fodder because it was not good for making tortillas.

In the field of livestock husbandry for meat production, economies of scale also emerged in the industrialized nations. With fewer farmers, livestock production operations increased in scope. Many cattle, dairy cattle, chickens, and hogs were kept in giant containment facilities that housed thousands of animals that were fed grain, antibiotics, and synthetic growth hormones. Smaller farmers often provided the grain or feeder animals for these large feedlots, dairies, and poultry plants, but more often than not large corporations owned these operations as well. By increasing the size of livestock confinement facilities in many areas, producers had to cope with prodigious amount of manure that they captured in giant sewage lagoons, which often contaminated groundwater supplies and befouled the air for surrounding communities. The meat industry also had to deal with public backlash over the potential health issues of hormone-enhanced meat.

Even with these new technologies, many farmers continued to practice traditional, subsistence type agriculture or to use ancient methods such as flood irrigation in the rice regions of Southeast Asia, planting on terraces in China or Latin America, or slash and burn (swidden) agriculture, in the world's rainforests. In developed countries such as the U.S., new technologies changed agricultural systems, which resulted in the development of single-crop economies, or monocultures, dependent upon large amounts of capital and available land and major inputs of science and technology. In the Great Plains region of the U.S., a traditionally arid region, agriculturists in the 1910s and 1920s practiced "dryland" agriculture. Farmers would intensively plow their new crops, especially right before and after rain. This fell in line with the old adage that "rain would follow the plow." During the years of World War I and for a decade thereafter, wet years resulted in the expansion of planted acreage on the Great Plains, including the phenomenon of "suitcase farmers" who simply planted the crop and returned at harvest time. When the inevitable cycle of dry years returned in the 1930s, the expanded wheat acreage fell victim to drought and the legendary Dust Bowl. After World War II, farmers in the Great Plains region extensively tapped the giant Ogallala aquifer to grow corn and alfalfa in the heart of the former Dust Bowl. In areas where erosion had damaged the soil, farmers also began to practice contour plowing and terrace building as ways of preserving the soil.

Farmers throughout the world were clearly witnessing an end of traditional rural life and agriculture and the rise of global agribusiness. With the introduction of expensive machines, chemicals, and irrigation technology, less efficient, marginal, or smaller-scale farmers and sharecroppers were often forced out of business and off of the land. Governmental policies in many countries supported the removal of smaller farmers to pave the way for larger scale operations. Large multinational corporations and global marketing and transportation industries often stimulated the replacement of traditional agriculture, or made farmers more subordinate as the supplier to large industrial conglomerates. Some agricultural nations in places like Africa and Latin America complained of the unfairness of tariffs and crop subsidies that supported agriculture in developed nations, while wealthier agricultural powers worried about competition from developing countries with their lower labor costs and fewer environmental standards.

In response to the increasing economic, ecological, and human costs of large-scale agriculture, the alternative farming movement emerged in the 1950s and 1960s as a challenge to traditional agricultural systems. While noting that machinery and chemicals had created production increases, organic farmers and their supporters advocated a new "nature-based" model for agriculture rooted in the spirit of E.F. Schumacher's "small is beautiful" concept of technology.

Organic farming and sustainable agriculture systems have several central ideas, including the elimination of chemicals and the use of "natural" fertilizers, such as manure or green manure, or integrated pest management techniques that target bugs with predatory insect releases instead of poisonous pesticides. Organic sustainable farmers also believe in reducing tillage, the use of composts, and even in the mixing of plants (polyculture) to create beneficial ecological interactions between crops, or crops and the soil, pests, or fungi.

Organic farmers also tended to espouse a philosophy that smaller farms and farm communities were of great benefit to society and would be preserved and enhanced by an organic regime. By claiming to be free of harmful poisons and to be more "earth friendly," proponents of alternative agriculture received increased support from consumers and more interest from agricultural researchers who had formerly championed green revolution techniques.

At the end of the twentieth century, trends in agriculture included the rise of the precision farming movement and the growth of aquaculture on a global scale. Precision farming involves age-old farming techniques, in which the farmer is aware of which plots produced well or which tree bore the most fruit and combines this knowledge with a modern systems approach of nearly global knowledge of plant, soil, and weather conditions. Precision farming uses yield monitors (to measure moisture, weight, and other factors), field scouts, sensors, GPS (global positioning systems), and DGPS (differential global positioning systems), and statistical analysis. By having a precise knowledge of growing conditions, precision agriculture allows for the precise application of herbicides, fertilizer, and water, allowing for higher yield, lower long-term costs, and lessened environmental damage. As technology costs declined and more research was invested in the idea of precision agriculture, the movement had an increasing number of adherents, primarily in the U.S.

The revival of aquaculture was another significant trend at the end of the twentieth century. Raising fish or shellfish in confined areas has been practiced for over 4000 years. The Chinese, in particular, have excelled in aquaculture production. Aquaculture can be practiced in saltwater, brackish water, or in freshwater ponds and containers. A minimal amount of grain is required in fish farming as compared to raising beef or pork, and the most popular aquaculture species include Asian carp, tilapia, mussels, catfish, and rainbow trout. At the beginning of the twenty-first century aquaculture was the fastest growing segment of the U.S. agricultural economy, representing over one billion dollars per year. Still, the U.S. lagged behind China, India, and Japan in aquaculture production.

Agricultural systems were greatly influenced by biotechnology, especially with the advent of genetic engineering and the arrival of the first genetically modified crops in the 1980s. The practitioners of animal and plant husbandry have altered nature over the millennia to increase production and profit. But since the advent of Mendelian genetics and the successes of hybridization, the promise and the perils of biotechnology have been a central concern of the agricultural research establishment. Karl Ereky, a Hungarian, coined the term biotechnology in 1919. Many ethicists and environ-mentalists feared unforeseen consequences of literally "playing God" by splicing genes from one plant's DNA into another plant's DNA, thereby creating a new species. Scientists and farmers noted the great potential to create new strains that increased yield, were pest or drought resistant, or would grow in a particular shape or color. Still, at the end of the twentieth century, a formidable movement, particularly strong in Europe, held steadfast in opposition to genetically modified crops.

With the transportation revolution of the twentieth century, farmers became an increasing component of a global agricultural economy. Globalization of the world economy led to a more rapid adoption of technology and the elimination of traditional agriculture in the developing world. Although many nations such as Japan, the countries of Western Europe, and the U.S. continued to heavily subsidize their agricultural system, global competition increased as tariffs began to fall in the 1980s and 1990s. For example, by the end of the century, even with a 400 percent tariff on Chinese garlic, California growers stopped planting garlic and instead marketed Chinese garlic. A similar process occurred with the apple industry in Washington as lower Chinese labor costs and capital outlays wreaked havoc on competitors. It seems likely that success in feeding the world will continue to require an emphasis on sustainable agriculture and biotechnology.

See also **Breeding, Plant, Genetic; Biotechnology; Farming, Mechanization; Fertilizers; Genetic Engineering, Applications; Irrigation Systems; Pesticides**

RANDAL BEEMAN

Further Reading

Beeman, R. and Pritchard, J. *A Green and Permanent Land: Ecology and Agriculture in the 20th Century*. University Press of Kansas, Lawrence, 2001.

Bud, R. Janus-faced biotechnology: a historical perspective. *Trends Biotech.*, 7, 1989.

Emmot, A., Hall, J. and Matthews R. The potential for precision farming applied to plantation agriculture. *Proceedings of the 1st European Conference on Precision Agriculture*. Warwick, 1997.

Hurt, R.D. *American Agriculture: A Brief History*. Iowa State University Press, Ames, 1994.

Kloppenberg, J.R. *First the Seed: The Political Economy of Plant Biotechnology*. Cambridge University Press, New York, 1988.

Marcus, A.I. *Cancer from Beef: DES, Federal Food Regulation, and Consumer Confidence*. Johns Hopkins University Press, Baltimore, 1994.

Vasey, D.E. An Ecological History of Agriculture 10,000 B.C.–A.D.10,000. Iowa State University Press, Ames, 1992.

Farming, Growth Promotion

Early in the twentieth century, most farmers fed livestock simple mixtures of grains, perhaps supplemented with various plant or animal byproducts and salt. A smaller group of scientific agriculturalists fed relatively balanced rations that included proteins, carbohydrates, minerals, and fats. Questions remained, however, concerning the ideal ratio of these components, the digestibility of various feeds, the relationship between protein and energy, and more.

The discoveries of various vitamins in the early twentieth century offered clear evidence that proteins, carbohydrates, and fats did not supply all the needs of a growing animal. Additional research demonstrated that trace minerals like iron, copper, calcium, zinc, and manganese are essential tools that build hemoglobin, limit disease, and speed animal growth. Industrially produced nonprotein nitrogenous compounds, especially urea, have also become important feed additives. The rapid expansion of soybean production, especially after 1930, brought additional sources of proteins and amino acids within the reach of many farmers. Meanwhile, wartime and postwar food demands, as well as a substantial interest in the finding industrial uses for farm byproducts, led to the use of wide variety of supplements—oyster shells, molasses, fish parts, alfalfa, cod liver oil, ground phosphates, and more.

By mid-century, researchers had concluded that an unknown "animal protein factor" (APF) was a requirement for maximum animal nutrition and growth promotion. It became increasingly apparent that proteins from animal byproducts better aided livestock growth as opposed to vegetable proteins. University, industry, and government scientists concluded that a nutrient found in liver, labeled Vitamin B12, was the mysterious APF factor. Searching for a method to produce the new vitamin on an industrial scale, manufacturers found a ready solution in residues from the fermentation processes used in the production of the antibiotic "wonder drugs." Meanwhile, researchers began to notice connections between common antibiotics and animal growth. In 1949, E. L. R. Stokstad and Thomas H. Jukes of the American Cyanamid Company found that the Vitamin B12 produced from the antibiotic organism yielded more rapid growth in chicks and hogs than a pure vitamin extracted from liver. Tests using the antibiotic alone, even in subtherapeutic dosages, proved to accelerate animal growth; most initial reports indicated that both chicks and hogs responded to antibiotic feeds, while the impact on ruminants was less. More recently, antibiotics known as ionophores have become commonly used as finishing rations for ruminants. These drugs improve the fermentation efficiency of the rumen, and also reduce bloat, acidosis and other feedlot diseases that are common among animals fattened on grains rather than their natural food of forage grasses.

Hormone supplements also revolutionized animal feeding. In 1954, Wise Burroughs of Iowa State College announced the synthetic estrogen hormone diethylstilbestrol (DES) caused improved feed efficiency on sheep and cattle at very low cost and without reducing carcass quality. Other hormones also became part of animal growth strategies, such as progestins that regulate the estrus cycle and to enhance feed efficiency, and androgens that aid muscle building on the feedlot.

Meanwhile, French researchers F. Stricker and F. Greuter opened up new avenues for research with their 1928 discovery of connections between animal hormones and the milk production of dairy animals. Over the next several decades, scientists gained increasing understanding of the effect of hormones like somatotropin, but isolating natural hormones from pituitary glands of slaughtered animals remained financially infeasible. In the mid-1980s, scientists learned to isolate the cow gene responsible for the growth hormone, remove it from animals, splice it onto bacteria, use bacterial growth to synthesize additional hormones, and inject a purified culture into cows in order to increase both milk production and feed efficiency. More than an incremental step in this history, the use of recombinant DNA techniques represented a revolutionary new development.

Feed supplements are partly responsible for dramatic increases in global meat production and consumption, especially since World War II. Antibiotics and growth hormones can help speed animal growth, improve feed efficiency, increase the number and viability of eggs and animal offspring, and reduce animal diseases. In the competitive marketplace of western agriculture, most farmers quickly embraced them. Manufactured and medicated feeds spurred farmers to increase the size and capital investment of livestock operations, to manipulate the natural rhythms of animals' breeding, birth, weaning, rebreeding, and slaughter, and to move livestock into ever more confined, streamlined, and centralized operations.

These growth-promoting agents have generated controversy. Concerns about antibiotic resistance have been voiced since the early 1950s, and various studies have indicated plausible links between animal antibiotics and human disease. Although repeated efforts to regulate the use of antibiotics in subtherapuetic dosages in the U.S. have stalled repeatedly, other nations have restricted their usage. In 1954, officials in the Netherlands announced a ban on all antibiotic-enhanced feeds. In the U.K. in 1969, a special commission recommended that any antibiotics used for growth should be different from those used in human or animal medical care. Pharmaceutical firms quickly replaced these drugs with alternative antibiotics, yet several of these in turn have been banned throughout the European Union, effective in 1999.

The debate over growth hormones has been even more contentious, and agricultural feed additives have been at the center point of social, political, and ethical debates over bioengineering. News of the "DES daughters" broke in 1970. These seven young women had developed a rare form of vaginal cancer that could be traced to their mothers' use of DES as a drug intended to prevent miscarriages. The U.S. finally banned the drug in 1979, after more than two dozen other nations had already done so. Although other livestock hormones remained legal (or otherwise available) in the U.S., European policymakers and consumers both have been increasingly stringent in enforcing regulations. In recent years, even the name of the hormone technology has become controversial; opponents prefer the term recombinant bovine growth hormones, (rBGH), while supporters prefer to use the term recombinant bovine somatotropins (rbST). The BGH/rbST debate has moved beyond questions about medical or environmental objections to genetic engineering. For many, the issue is the social cost of allowing enhanced milk production technologies to eliminate many family dairy farms.

In response to some of these controversies, niche markets have emerged for livestock that are free from confined environments and for animals that are antibiotic and growth hormone free. Although such trends were evident at the end of the twentieth century, particularly in western and central Europe, it remains unlikely that such protests will slow the penetration of manufactured and medicated animals feeds into the developing world.

See also **Antibiotics, Use after 1945**

MARK FINLAY

Further Reading

Bud, R. *The Uses of Life: A History of Biotechnology.* Cambridge University Press, Cambridge, 1993.

Feed Additives: The Added Value to Feed. Nefato, Aalsmeer, Netherlands, 1998.

Johnson, A. *Factory Farming.* Blackwell, Oxford, 1991.

Jukes, T.H. *Antibiotics in Nutrition.* Antibiotics Monographs, 4, Medical Encyclopedia Inc., New York, 1955.

Krimsky, S. and Wrubel, R. *Agricultural Biotechnology and the Environment: Science, Policy, and Social Issues.* University of Illinois Press, Chicago, 1996.

McCollum, E. *A History of Nutrition: The Sequence of Ideas in Nutrition Investigations.* Houghton Mifflin, Boston, 1957.

Marcus, A.I. *Cancer from Beef: DES, Federal Food Regulation, and Consumer Confidence.* Johns Hopkins University Press, Baltimore, 1993.

Sammons, T.G. Animal feed additives, 1940–1966. *Agricult. Hist.*, 42, 305–313, 1968.

Farming, Mechanization

Mechanization of agriculture in the twentieth century helped to dramatically increase global production of food and fiber to feed and clothe a burgeoning world population. Among the significant developments in agricultural mechanization in the twentieth century were the introduction of the tractor, various mechanical harvesters and pickers, and labor-saving technologies associated with internal combustion engines, electric motors, and hydraulics. While mechanization increased output and relieved some of the drudgery and hard work of rural life, it also created unintended consequences for rural societies and the natural environment.

By decreasing the need for labor, mechanization helped accelerate the population migration from rural to urban areas. For example, in 1790, 90 percent of Americans worked in agriculture, yet by 2000 only about 3 percent of the American workforce was rural. Blessed with great expanses of land and limited labor, technologically inclined Americans dominated the mechanization of agriculture during the twentieth century. Due to mechanization, irrigation, and science, the average American farmer in 1940 fed an estimated ten people, and by 2000 the number was over 100 people. Yet even as mechanization increased the speed of planting and harvesting, reduced labor costs, and increased profits, mechanization also created widespread technological unemployment in the countryside and resulted in huge losses in the rural population.

After Nicholas Otto patented the internal combustion engine in 1876, the days of horse and steam power in agriculture were eventually eclipsed on the farm by gasoline and diesel engine power. Early steam tractors weighed as much as 5 tons or more and required substantial labor to operate and fuel. The operators were continually at risk from possible explosion of the boiler that powered these iron-wheeled beasts. A number of American manufacturers, such as the Hart-Parr Company, the John Deere Plow Company, and International Harvester, were engaged in building gasoline-powered tractors at the beginning of the twentieth century.

As tractors became more powerful, practical, and affordable, their presence increased on the farmstead. With a tricycle design for enhanced maneuverability, a PTO (power take-off) to run equipment, hydraulic lifts, a three point hitch, and pneumatic tires, tractors could pull a plow, mow and bale hay, disk weeds, or dig furrows among other tasks. In 1907 some 600 gasoline-fueled tractors were in operation in the U.S.; by 1950 nearly 3.5 million tractors were in use in the U.S. alone.

Tractors, unlike horses and mules, did not require five acres to feed every year, and they did not need rest. Henry Ford's all-purpose Fordson in 1917, and the International Harvester Farmall in 1924 enjoyed high sales in the 1920s. Other tractor improvements included roll bars, improved seating, variable hydrostatic transmission for increased speeds, and Harry Ferguson's hydromechanical servo, which allowed the tractor operator to control the depth of implements in the soil by keeping a constant load on the tractor. The first diesel designed to power movable machinery in the U.S. was built in 1930, and by the mid-1930s the first diesel tractor was sold.

From the World War I campaign to increase wheat production in the U.S. to the Soviet Union's "virgin lands" projects of the 1950s and 1960s, tractors have increased the amount of land under cultivation throughout the globe. While yields increased, so did potential environmental catastrophes, such as the American Dust Bowl in the 1930s, and the failure of much of Soviet's virgin lands program. By the end of the twentieth century the U.S. had only two major tractor manufacturers, and India led the world in tractor production, building nearly 300,000 units per year. In Africa most of the farm labor continued to be done by people or animals.

Along with the gasoline-powered tractor came other devices to increase efficiency and profits by reducing labor and expanding the scope of production. In the large wheat area of the American West,

the first combine harvesters appeared in the 1890s. After Cyrus McCormick's famous reaper appeared in the 1830s, inventors tried to find ways to combine the harvesting and threshing process. Hiram Moore and others made several attempts at manufacturing combines in the mid-nineteenth century, but it took until around 1900 for large-scale farmers on the American West Coast to employ the first commercially and mechanically viable combines.

Tractors pulled the first combines, and they used a rotary header to gather the plant, a cutting bar, and an internal thresher, but by the 1940s the machines were self-propelled. Gasoline powered engines or the PTO of a tractor operated the cutter, and the various mechanical fingers and pulleys that drew the crop into the thresher and shook the grain loose from the plant, with the grain being augered into a grain tank, and the crop residue sprayed back onto the field. Early combines required a four-man crew and could cut up to a 16-foot (5-meter) swath. Over the years combines became larger, with rubber tires, self-enclosed air conditioned cabs, headlights to allow night harvesting, and wider swathes.

Combines represented a tremendous capital investment for farmers, and they contributed to the growth in the size of individual farms and the population decline in rural communities. But whether a farmer purchased a combine, or hired a custom crew, combines were less expensive and more dependable than hiring a crew of laborers to harvest large fields. While the first combines primarily harvested wheat, mechanical engineers would later build machines to combine harvest rice, oats, soybeans, and corn by mid-century. Early models of these machines often left up to 50 percent of the grain in the field and were prone to problems when operated in muddy fields as well as mechanical difficulties. Yet even with these difficulties, combine harvesters became increasingly efficient, and they remain a symbol of the mechanization of agriculture in the twentieth century.

Inventors, corporations, mechanical engineers, and professors perfected machines to harvest grains, but cotton remained primarily a hand-picked crop until the 1950s. Cotton bolls ripened at different times, and developing a machine that could pick trash-free fiber without damaging the cotton took over 100 years. In the 1920s and 1930s, early versions of the cotton picker often used vacuum pumps and blowers to either suction or blow the cotton into large bags. Engineers John and Mack Rust experimented with combines that

Figure 1.   Mechanical corn picker, 1939, Grundy County, Iowa.
[*Courtesy U.S. Department of Agriculture.*]

used smooth wet spindles to pick the crop and barbed and serrated spindles to twist the lint from the boll. In the mid-1930s the Rust machine could pick almost 180 kilograms of cotton per hour. In 1948 the International Harvester Corporation was the first company to successfully market a mechanical cotton picker. It used rotating metal drums with wetted spindles that pulled the lint from the bolls, rubber doffers to remove the fibers from the spindles, and an air conveyor to blow the cotton into a container.

Cotton harvesters became more efficient, less expensive, and more economically viable for farmers, especially those with larger holdings. Additional improvements reduced the amount of plant material, or trash, in the cotton, and by the late 1960s most of the cotton crop in the U.S. was machine picked. An early two-row cotton picker displaced 80 workers, and in the U.S. alone millions of workers were technologically unemployed by the mechanization of the cotton harvest.

Like cotton, sugar beet harvesting was difficult to mechanize. Sugar beets have deep roots and heavy foliage. Growers relied on a large number of laborers to grow, harvest, and transport the crop. Beets had to be pulled from the ground, topped with a knife, and then loaded onto trucks or wagons. In the 1930s it took about 30 hours of

labor to harvest an acre of sugar beets. A successful harvesting device would require the machine to top the beets and to remove the excess dirt and clods from the crop. Early machines topped the beets at ground level and lifted the roots with a spiked wheel onto a conveyor belt and into a truck. By the 1970s sugar beet harvesters had been enhanced to the point where they could harvest 24 acres in a ten-hour day.

Mechanization of agriculture most often occurred in developed nations willing to invest in labor-saving technology. In the U.S., farmers in the Central Valley of California have been at the forefront of agricultural mechanization. The early combines and cotton pickers were first utilized in California, where growers were often dependent upon migratory labor. Tomatoes were another major California crop that required vast supplies of labor. Creating a mechanical tomato picker involved a number of obstacles. Tomatoes on the plant ripened at different times, the vines snarled machines, and the fruit itself was easily bruised and not suited to machines. Consequently, to develop the machine-harvested tomato required the cooperation of engineers and crop scientists. New varieties of tomatoes that ripened simultaneously and were tough skinned emerged from university and corporate laboratories and test plots, such as the work done by G. C. Hanna at the University of California, Davis.

The tomato picker developed by the 1960s worked by lifting and cutting the vines and shaking the plant on a shaking bed, where the harvested tomatoes would then fall onto a conveyor belt which transported the fruit to a bin while the vines fell to the ground. Though the early pickers had many technical problems and often malfunctioned, they could harvest 10 tons, or one third of an acre of tomatoes, per hour. Most of the tomatoes harvested mechanically ended up as catsup, juice, or tomato paste, and the process of mechanically harvesting market-fresh tomatoes remained problematic in the 1990s. Tens of thousands of farm laborers in California alone lost their jobs with the advent of tomato harvesters. The mechanical tomato harvester also allowed for a sweeping increase in tomato production worldwide.

When contemplating the mechanization of agriculture, we associate large and complicated machines roaring through a field as the symbol of technological change in the countryside. But mechanization of the farm had many smaller, less extravagant elements, such as the influence of electrification. Many rural communities in the developing world still lack electricity. In the U.S. in the early 1930s, the federal Rural Electrification Administration helped bring electricity to the farms of America. Electricity powered milking machines, cream separators, and irrigation pumps. Electric power also allowed for environmentally controlled barns and greenhouses, cold storage units, and grain dryers that facilitated the mechanization of the grain harvest. Small electric motors could make a number of tasks on the farm easier, whether it be an auger to move grain, or a grinder or arc welder in the workshop.

Many other devices helped to mechanize agriculture in the twentieth century. Automobiles, trucks, and other transportation improvements transformed rural society and created new market opportunities and new competitors for farmers. Giant feed lots confined thousands of animals for meat or dairy production using highly mechanized facilities for feeding the animals and removing the waste.

Citrus growers used wheeled machines with revolving blades to trim the trees in their orchards; in almond orchards, machines grab and shake the fruit from the tree and another machine sweeps the crop from the ground. Pesticide sprayers with tall wheels and giant booms glide over rows of vegetables. On grape vineyards plastic-fingered pickers shake grapes from the vines on their way

to be pressed into wine; and elaborate systems for drip irrigation allow for the precise application of water, fertilizers, and other chemicals.

Every year brings new changes, technological improvements and mechanical innovations, but with each new step toward mechanization there is a corresponding increase in capital costs, technological unemployment, and environmental disruption. Clearly the move from human- and horse-powered agriculture reduces the number of people required to grow food and fiber, and by the last half of the twentieth century developing countries were embracing the mechanization of their indigenous agriculture on the Western model.

See also **Dairy Farming; Farming, Agricultural Methods; Irrigation Systems**

RANDAL BEEMAN

Further Reading

Hurt, R.D. *Agricultural Technology in the Twentieth Century.* Sunflower University Press, Manhattan, 1991.
Marcus, A.I. *Technology in America: A Brief History.* Harcourt-Brace, San Diego, 1989.
Rasmussen, W. The mechanization of agriculture. *Scientific American*, September, 1982.

Fax Machine

Fax, or facsimile, technology refers to the concept of replicating printed documents across long distances and dates back to the nineteenth century, along with the advent of the telegraph. Embracing the emerging electromechanical technology of the time, several devices were developed; however diffusion was limited due to the elaborate and intricate mechanism required as well as interoperability between disparate devices. Among the first on record was Alexander Bain's "chemical telegraph," patented in America in 1843. The device used a metallic contact that sensed the raised text, which triggered the flow of electric current. In 1847 Frederick Bakewell invented the "copying telegraph," which introduced the concept of scanning the source document line by line. Both systems required pendulums and electromagnets for synchronization. However, fierce competition with Samuel Morse over long-distance telegraph lines led to legal disputes and Bain's patent was declared invalid. Although the facsimile technology did not proliferate along with the telegraph, the concept of facsimile transmission did, and isolated systems continued to emerge wherever telegraphic networks were set up. A successful system was demonstrated in France in 1862 by Abbé Casselli, and a network of commercial stations was established. In America, Elisha Gray developed a system comprised of rheostats and electromagnets. Despite some successes, the largely mechanical technology was cumbersome, and lack of international interoperability and slowness of the system compared to Morse's telegraph prevented commercial success and widespread proliferation.

At the beginning of the twentieth century, the science of reducing images to scanned lines of information which could be electrically transmitted for distant reproduction was maturing but the technology was slow to follow. In 1902 Otto von Bronk obtained a patent for the principle that an image could be constructed out a series of lines. Einstein's mathematicalization of the photoelectric phenomena and other developments in quantum physics in the early twentieth century provided the necessary scientific understanding for the emergence of photoelectric technology, which was primarily based on the element selenium. Dr. Arthur Korn invented an improved and practical fax in 1902 based on this photoelectric technology. It should be noted the development of the cathode ray tube television shared the same technological roots as the fax machine; that is, image transfer by sequentially scanned lines.

Early attempts to distribute printed images to domestic consumers in America did not find much success. However, the need to rapidly transmit photographic images for publication in newspapers and elsewhere drove the technology of the fax machine into elaborate commercial products. Along with telegraphic and telephone networks, long-distance faxing started to make use of short-wave radio communications as well. By 1910 Korn had established a network with images stations in Berlin, Paris and London, and in 1922 successfully transmitted a picture from Rome to New York by radio. In 1926 a commercial radio link for facsimile image transmission was operational between London and New York. The limitation of this system was that original image had to be a negative. Subsequently Edouard Belin developed the "Belinograph" in France in 1925. The image to be transmitted was placed on a cylinder and scanned with a powerful light beam and a photoelectric cell converted the reflected light into transmittable electrical impulses.

The American Telephone & Telegraph Company (AT&T) produced a "telephotography" machine in 1924 which was successfully used to send photos over long distances for newspaper publications. In 1934, the Associated Press news agency introduced a similar system for transmitting "wire photos" between centers. These devices perfected the spinning drum and moving photoelectric cell concept, but required mechanical precision and specialist upkeep. As their predecessors did, a continuously varying electrical signal was used to transmit the information. The limitation was that this was subject to decay and electromagnetic interference and noise along the path of transmission, although various signal modulation techniques were used to increase fidelity and resilience. This technology continued to be used in press rooms until the arrival of personal computers and the means of digitalization of images later in the century.

Other implementations of facsimile technology in this period were for public use; the "photo-telegram" service as Western Union called it. Facsimile machines were made available in public places in the 1930s for the transmission of messages although this was short lived. Facsimile technology also proved successful in augmenting international telegraph and later, telex facilities. International public fax services grew from the initial New York–London link in 1926 to 24 countries by 1950 and 65 countries by 1976. Largely these facilities were part of the public services offered by the postal services.

The American company Xerox had successfully developed the commercial photocopier in 1959 and made a global commercial success out of it. The Long Distance Xerograph (LDX), which was announced in 1964, stemmed from photocopier technology but was still a cumbersome, expensive and difficult-to-operate device. In 1966, the smaller Xerox Magnafax Telecopier operated on ordinary telephone lines, and was easier to use but still took 6 minutes to transmit a single page. At the Xerox Corporation research institute, the Xerox Palo Alto Research Center (PARC) some fundamental developments and innovations were taking place in digital technology in an atmosphere that was described as "pure invention." Several factors were driving the digitization of the fax, one being that digital information could be compressed using mathematical algorithms (by up factors of up to ten times) and once digitized, the information was not subject to decay or electrical interference during transmission.

The first digital fax machine was built by Lynn Conway at the PARC in 1972 using discrete logical components. The Sierra was a large power-hungry unit that was more of a proof-of-concept than any sort of marketable device. However digital consumer devices were being made possible by the miniaturization of digital electronics (the integrated circuit led to the large-scale integration of digital logic components into the microprocessor and the RAM memory chip, key components of the microcomputer).

International standardization of facsimile communication protocols played a vital role in the widespread adoption of the fax machine since disparate standards limited diffusion. The International Telegraph and Telephone Consultative Committee (CCITT) agreed on the Group 1 standard in 1968. A page took 6 minutes to transmit and was reduced to a series of dots which were either "on" or "off." This sequence was used to encode a signal using frequency shift keying (FSK) modulation technique. Subsequent signal processing enhancements led to the Group 2 standard in 1976, which halved the transmission time. Embracing the burgeoning digital technology of the late 1970s, the Group 3 CCITT standard was agreed to 1980, utilizing the modified Huffman coding for compression of data, which led to a transmission rate of less than 1 minute per page with an improved resolution. This standard also incorporated well-established computer data modem communication protocols, including the facility to "step-down" or scale the data communication rate according to the quality of the analog telephone line connection. This made the Group 3 fax versatile and fault-tolerant. The Group 4 standard came out in 1984 and was orientated around digital ISDN telephone services, however ISDN did not proliferate as expected and Group 3 became the most widely used standard internationally.

The impetus fell to the large Japanese consumer electronic manufacturers such as NEC Corporation and Matsushita Communications in the early 1980s to mass-produce digital fax machines. Electronic developments had enabled an easy-to-use compact solid-state device to be mass produced, and with large volumes and international compatibility, the fax machine became affordable and globally ubiquitous within a decade. Parallel to this was the development of the personal computer, and by 1985 fax peripherals were available to personal computer users which enabled faxing direct from a software application. This made "broadcast" faxing possible, whereby one message could be sent to many recipients with relative ease. Thus the fax became a mass one-too-many marketing and communication tool.

Fax technology was especially useful for international commercial communication, which was traditionally the realm of the Telex machine, which only relayed Western alpha-numeric content. A fax machine could transmit a page of information regardless of what information it contained, and this led to rapid and widespread adoption in developing Asian countries during the 1980s. With the proliferation of the Internet and electronic e-mail in the last decade of the twentieth century, fax technology became less used for correspondence. At the close of the century, the fax machine was still widely used internationally for the transmission of documents of all forms, with the "hard copy" aspect giving many a sense of permanence that other electronic communication lacked.

See also **Electronic Communications; Photocopiers; Telephony, Long Distance; Television, Electro-Mechanical Systems; Printers**

BRUCE GILLESPIE

Further Reading

Coopersmith, J. The failure of fax: when a vision is not enough. *Econ. Bus. Hist.*, 23, 272–282, 1994.
Hiltzik, M. *Dealers of Lightning. Xerox PARC and the Dawn of the Computer Age*. Orion Business, London, 2000.
Petersen, M.J. The emergence of a mass market for fax machines. *Technol. Soc.* 17, 4, 469–82, 1995.
Winston, B. *Media, Technology and Society, A History: From the Telegraph to the Internet*. Routledge, London, 1998.

Feedstocks

The word feedstock refers to the raw material consumed by the organic chemical industry. Sometimes, feedstock is given a more restricted meaning than raw material and thus applied to naphtha or ethylene, but not petroleum. The inorganic chemical industry also consumes raw materials, but the feedstock tends to be specific to the process in question, such as sulfur in the case of sulfuric acid. The development and growth of new feedstocks has driven the evolution of the organic chemical industry over the last two centuries. To a large extent, the history of this industry is the history of its feedstocks. Until the nineteenth century, the only significant raw material for the nascent organic chemical industry was fermentation-based ethanol (ethyl alcohol). Gradually, the products of wood distillation also became important, only to be overshadowed after 1860 by the coal-tar industry. As the organic chemical industry expanded in both size and scope between 1880 and 1930, the need for new feedstocks became urgent. The competition between coal and petroleum was resolved in favor of the latter in the late 1950s. The petrochemical industry has been phenomenally successful, underwriting the postwar boom in organic chemicals and plastics and weathering the oil crises of the 1970s with minimal damage. Its long-term sustainability remains an issue, and increasing attention is being paid to renewable feedstocks.

Fermentation-Based Chemicals

In dilute aqueous form (beer, wine), fermentation-based ethanol has been known for thousands of years. Pure ethanol was well known by the thirteenth century, but there was only a tiny demand, mainly for pharmaceuticals. Around the beginning of the twentieth century, it became possible to produce lactic acid by fermentation of potato starch. A few years later, Chaim Weizmann at the University of Manchester working in collaboration with the consulting firm of Strange and Graham developed a fermentation process for the production of butanol and acetone from carbohydrates obtained from grain or potatoes. Initially created to produce butanol for synthetic rubber production, it became more important as a source of acetone during World War I. In the early 1920s, the strong demand by the automobile industry for solvents, favored its co-product, butanol. To meet the American demand, the Commercial Solvents Corporation was established at Terre Haute, Indiana, to operate the Weizmann process using maize as the starting material. By 1924, the plant was producing 6500 tons of butanol. In the late 1920s, Friedrich Bergius—better known for his research on the conversion of coal to oil—developed a method of hydrolyzing wood with acids to produce sugars. The process was used in England for a while, and Bergius set up a plant in Germany, but it was not a commercial success. The German firm I.G. Farben did consider using the process to make synthetic rubber (via fermentation-based ethanol) but stayed with coal-based chemistry. The idea of using ethanol as a feedstock—from wine surpluses rather than wood—was revisited by the West German synthetic rubber producer Chemische Werke Hüls in the early 1950s, but again this route was not pursued.

Wood Distillation

Although it had been known since the seventeenth century that heating wood strongly produced wood spirit and pyroligneous acid, the industry concentrated on the production of wood tar and charcoal up to 1800 or so when pyroligneous acid was identified as acetic acid. By 1820, the industry existed in Britain (mainly in Scotland), Germany and Austria. Wood distillation in America started about 1830. During this period, acetic acid, used to make mordant salts and pigments, was the key product. The commercial production of methanol and acetone only began in the 1850s. The importance of wood distillation should not be underestimated. HIAG (Holzverkohlungs-Industrie AG) was the 77th largest German company in 1913, but it declined in the 1920s and was taken over by Degussa in 1931. This was partly a result of the introduction of new methods of making its core products, acetone and methanol.

Coal-Tar

The coal-tar industry was a byproduct of the expanding coal-gas industry, and the tar that was originally wasted became a valuable commodity. When the dry distillation of coal to produce coal-gas for lighting and heating was developed in the early nineteenth century, the coal-tar waste product was largely seen as a noxious nuisance to be burned in open pans or dumped in nearby rivers. The value of coal-tar naphtha as a solvent was soon recognized, and by the 1840s chemists had also isolated the so-called aromatic chemicals from coal-tar including benzene and phenol. Although benzene was quickly taken up as a powerful solvent, it was the development of the synthetic dye industry following William Henry Perkin's discovery of mauve in 1856 that drove the growth of the coal-tar industry. Within a decade, coal-tar was being converted into numerous chemicals by distilling chemicals from the crude tar and subsequent chemical treatment such as nitration. By the 1880s, coal-tar was also used for pharmaceuticals as well as dyes. It is an indication of the former importance of coal-tar that the first major synthetic fiber—nylon—was made from coal-tar chemicals until the 1950s in the U.S. DuPont introduced a process that used furfural, a chemical obtained from maize cobs, in 1948 and only adopted a petrochemical route based on butadiene in 1961.

Coal

Coal itself was not used as a raw material for organic chemicals until the twentieth century. Its exploitation was a combination of supply and demand. In the late 1890s, entrepreneurs had hoped to develop acetylene obtained from calcium carbide as a new method of lighting. Although acetylene gas lighting was used for many years, the entrepreneurs' hopes were largely misplaced, and there was soon a surplus of calcium carbide. This surplus was partly consumed by the manufacture of a novel fertilizer calcium cyanamide, but acetylene was now available as a versatile and relatively cheap feedstock for organic chemicals. The German chemical industry, chiefly Hoechst and Wacker, began with the manufacture of chlorinated hydrocarbons and acetic acid, which in turn was converted into acetone. At this point, at the outset of World War I, the Haber–Bosch process for nitrogen fixation energized the production of carbon monoxide from coke. (In the Haber–Bosch process, this carbon monoxide was an intermediate in the production of synthesis gas, a mixture of hydrogen and nitrogen.) In 1923, Badische Anilin and Soda-Fabrik (BASF) started the production of synthetic methanol by reacting a mixture of carbon monoxide and hydrogen under pressure. The two lines of approach—acetylene chemistry and high-pressure chemistry—were brought together by BASF's Walter Reppe, who developed a number of interesting coal-based processes.

Acetone

Acetone was one of the first major organic chemicals. It was produced by wood distillation, but the yield was low. In the mid-nineteenth century it was discovered that acetone could be made by heating calcium acetate (from acetic acid, a more abundant product of wood distillation). This "grey acetate" was exported from the U.S. to Germany. By the end of the century, acetone had become an important component of cordite smokeless powder manufacture (as a solvent for gun cotton). During World War I, the Germans developed the production of acetone from acetylene, via acetaldehyde. This process gave a major impetus to the development of acetylene chemistry. By contrast, the British relied on the Weizmann fermentation process. The petrochemical production of acetone from propylene began at Union Carbide's South Charleston, West Virginia, plant in 1928, but the production of acetone from coal-based acetylene continued in West Germany until 1963.

The Automobile

In the early 1920s there was an unprecedented demand for organic chemicals from the booming automobile industry. Car production used nitrocellulose-based lacquers, which required organic solvents. A key compound was butanol, which was initially made by the Weizmann fermentation process. BASF then developed a method of making butanol from acetylene. In the winter, motorists needed antifreeze. Antifreeze was initially based on methanol or the more expensive glycerol, but ethylene glycol (Prestone) was introduced by Union Carbide in 1927. This was one of the first organic chemical products to be made from natural gas.

Synthetic Rubber

The first synthetic rubbers were made in the first decade of the twentieth century, and industrial production was carried on in Germany during World War I. Large-scale production did not take place, however, until the mid-1930s in Soviet Russia and somewhat later in Nazi Germany. The U.S. developed a massive synthetic rubber industry during World War II. Most synthetic rubbers are based on butadiene, but it can be produced from a variety of feedstocks. The early "methyl rubber" was made from acetone, which in turn could be wood- or coal-based. Nazi Germany (and subsequently East Germany) used a lengthy synthesis based on acetylene obtained from coal via calcium carbide or by cracking natural gas. The huge quantities of acetylene needed to make synthetic rubber effectively created an "acetylene chemical industry" in Germany in the 1940s and 1950s. The Soviet industry used the Lebedev process, which converted ethanol into butadiene in one step. The German industry admired the Lebedev process and on two occasions considered switching to it. With the support of Chaim Weizmann, the midwestern farm lobby in the U.S. also promoted this process, and 82 percent of U.S. butadiene was produced from ethanol in 1943. Of far greater importance, both at the time and in terms of its eventual impact, was the development of petroleum-based routes to butadiene. Not only did petroleum account for 59 percent of the U.S.'s butadiene in 1945, it also accelerated the entry of the petroleum refiners into the chemical industry.

Petrochemicals

The petrochemical industry was a result of the massive increase in petroleum refining in the twentieth century. There were two reasons for this. As petroleum refining became more sophisticated, refiners used chemical processes to produce better gasoline (petrol). At the same time, the refineries were producing large quantities of hydrocarbon gases, which were originally burned off. In the 1930s, the American industry developed methods of making various chemicals from these gases including propanol (Union Carbide), isobutanol (Shell) and styrene (Dow). Other companies such as Jersey Standard (now Exxon) and Anglo–Iranian (British Petroleum, or BP) moved into petrochemicals in the 1930s, partly because of the contemporary interest in the hydrogenation of oil and coal. Petrochemicals is a convenient term because it combines chemicals from petroleum with chemicals from natural gas, a distinction which is all too rarely made. One of the earliest uses of natural gas in the chemical industry was the production of acetylene from Dutch natural gas in the German synthetic rubber industry by passing it through an electric arc. Subsequently it was used to replace coal in the Haber–Bosch and synthetic methanol processes. Natural gas is more important in the American chemical industry than in Europe.

Polyethylene

Imperial Chemical Industries (ICI) introduced polyethylene in 1938, and by 1962 British production had reached 155,000 tons. Just as synthetic rubber had generated a vast demand for butadiene, the expansion of polyethylene manufacture after World War II created an unprecedented demand for ethylene. When ICI first made polyethylene, it used fermentation ethanol as its raw material, thanks to its links with the Distillers Company. It soon became clear however, that this source of ethylene was both too expensive and too limited to meet the soaring demand. In Germany, the main source of ethylene was coke-oven gas, which was also in relatively short supply. During World War II, the Germans even made ethylene by adding hydrogen to acetylene. After the war, it was clear that the petrochemical route was the only one that could provide ethylene cheaply and in the quantities needed. Just as synthetic rubber had boosted the use of acetylene in the prewar German chemical industry, the production of petroleum-based ethylene for polyethylene acted as the vanguard for the petrochemical industry, especially outside the U.S.

Replacing Coal

The new petrochemical industry did not spring up overnight. The established chemical factories were based on older feedstocks, and their chemists had developed a deep understanding of the various processes they operated. Switching to petrochemicals meant building new plants, developing new processes, learning a different kind of chemistry, and even changing the location of production.

Vinyl chloride monomer for the manufacture of polyvinyl chloride (PVC)—hitherto made by adding hydrogen chloride to acetylene—was now produced by heating ethylene dichloride. The older factories housed in brick or concrete buildings were replaced by plants with shiny piping and tall column reactors sited in the open air. It is therefore not surprising that the changeover to petrochemicals took two decades in Western Europe and even longer in the Communist bloc. Petrochemical processes worked better for many industrial chemicals such as ethylene, butadiene, or acetone, but other important chemicals such as terephthalic acid (for polyesters) or aniline (for dyes) were more naturally made from coal. In order for petrochemicals to replace coal altogether, it was necessary to produce the so-called aromatic chemicals, notably benzene and phenol, from petroleum. Some types of petroleum contain aromatic compounds anyway, but this source was inadequate to meet the growing demand. New ways of making aromatic chemicals from the aliphatic hydrocarbons, such as "platforming," had to be developed. This need to introduce new processes and entirely new types of chemical plant led to the eclipse of the industrial organic chemist by the chemical engineer.

Most historians of the chemical industry have assumed that the changeover to petrochemicals was more or less inevitable, if only because petroleum was so cheap in the late 1950s and the 1960s. This consensus has been challenged by Raymond Stokes, who has argued that "the fate of coal chemistry was far from evident; even in the latter half of the 1950s it continued to dominate worldwide production methods and to provide apparently viable alternatives to petrochemicals." While his analysis of the German industry has been a useful counterbalance to the technological and economic determinism implicit in earlier studies, it is indeed hard to see how it could have turned out otherwise, given the economic and political power of the U.S. and its largest corporations. Other countries and other firms could have followed a different path, as technologically alternative routes were available—notably fermentation-based ethanol and coal-based acetylene—but these options could only survive in countries where politics rather than industrial economics were paramount, such as East Germany or South Africa.

The Oil Crisis and its Impact

The oil crisis of 1973, when the Arab-dominated Organization of Petroleum Exporting Countries

(OPEC) quadrupled oil prices was a major shock to the organic chemical industry. Higher oil prices destroyed the boom in petrochemicals, which had seen increasing profits during the 1960s, and the established players were fearful that the oil-producing countries would develop their own petrochemical plants. For the remainder of the 1970s—with another "oil shock" in 1979—there were fears that petrochemicals might become uneconomical. The industry took steps to reduce its energy consumption and to increase production yields, factors that were low priorities while oil was cheap. Perhaps the greatest effect of the oil crisis was the acquisition of the Conoco Oil Company by the chemical giant DuPont in 1981. DuPont came to regret its move into the petroleum industry, and it divested the oil business in 1998.

Having weathered the oil crises, the petrochemical industry saw profit levels decline to historically low levels amid increasing public concerns about the industry's impact on the environment. Some companies have continued their historical links with petrochemicals, including Dow and BASF. Others have turned to low-tonnage but high-margin specialty chemicals and biotechnology. In the 1990s, even oil companies have moved to divest their petrochemical interests. To a certain extent, private entrepreneurs—notably Jon Huntsman—filled the gap by acquiring petrochemical units sold off by major concerns.

The Future

Although petroleum and natural gas supplies are fairly secure at present, governments are increasingly imposing statutory requirements on industry to boost the share of renewable energy and, almost in passing, promote renewable chemicals. At present, this sector is very small and has focused on fermentation (mainly ethanol), cellulose modification, and bioplastics. Perhaps surprisingly, wood distillation has not been prominent, but there has been a revival of interest in Bergius's process for the production of sugars from wood. Nonetheless, petrochemicals will continue to dominate the organic chemical industry for the first two or three decades of the twenty-first century.

Importance of Feedstocks

There are two ways of looking at feedstocks. The first is to have an end product in mind and then choose the appropriate raw material. Alternatively, a raw material may be cheap and abundant, even a waste product, and new uses can be developed for it. The chemical industry has been brilliant at

mediating between the two approaches. It has taken waste materials and converted them into valuable products. It has perceived needs and created new ways of meeting demands. It has also mediated in another important way, namely between abundant (and hence cheap) raw materials, perhaps even the byproduct of another process, and the desired products. The result of this complex mediation is often a compromise. The raw material may be more scarce or more expensive than is desirable, and the industrial chemist may look for a more suitable raw material. On the other hand, the chemist may retain the original starting material but search for a reaction with a higher yield or one with a better end result. Although a particular raw material may have been taken up because it was originally a waste product or in surplus, its success can result in the need to manufacture it especially for that process, thereby losing the rationale for its initial selection. In this way feedstocks promote innovation. Any attempt to modernize or expand the industry usually means changing feedstocks. At the same time, however, there is usually a strong element of continuity in the processes used. The history of the organic chemical industry is one of evolution rather than revolution. Modern petrochemistry owes a large debt to the development of acetylene chemistry and the development of coal-to-oil processes in the first half of the twentieth century. A renewable chemical industry will be at least partly based on technologies that were introduced in the same period.

See also **Biotechnology; Chemicals; Dyes; Fertilizers; Oil from Coal Process; Nitrogen Fixing; Plastic, Thermoplastics; Reppe Chemistry; Synthetic Rubbers; Solvents**

PETER MORRIS

Further Reading

Aftalion, F. *A History of the International Chemical Industry From the "Early Days" to 2000*, 2nd edn. Translated by Benfey, O.T., Chemical Heritage Foundation, Philadelphia, 2001.

Greenaway, F. *et al.* The chemical industry, in *A History of Technology*, vol. VI, *The Twentieth Century, c.1900–1950*, Part I, Williams, T.I., Ed. Clarendon Press, Oxford, 1978, pp. 514–569.

Haber, L.F. The chemical industry, a general survey, in *A History of Technology*, vol. VI, *The Twentieth Century, c.1900–1950*, Part I, Williams, T.I., Ed. Clarendon Press, Oxford, 1978, pp. 499–513.

Herbert, V. and Bisio, A. *Synthetic Rubber: A Project That Had to Succeed*. Greenwood Press, Westport, CT, 1985.

Marcinowsky, A. and Keller, G. Ethylene and its derivatives: Their chemical engineering and genesis at Union

Carbide Corporation, in *A Century of Chemical Engineering*, Furter, W., Ed. Plenum Press, New York, 293–352.

Morris, P.J.T. Ambros, Reppe and the emergence of heavy organic chemicals in Germany, in *Determinants in the Evolution of the European Chemical Industry, 1900–1939: New Technologies, Political Frameworks, Markets and Companies*, Travis, A.S. et al., Eds. Kluwer, Dordrecht, 1998, pp. 89–122.

Reuben, B. G. and Burstall, M. L. *The Chemical Economy: A Guide to Technology and Economics of the Chemical Industry*. Longman, London, 1973.

Russell, C.A., Ed. *Chemistry, Society and the Environment: A New History of the British Chemical Industry*. Royal Society of Chemistry, Cambridge, 2000.

Spitz, P.H. *Petrochemicals: The Rise of an Industry*. Wiley, New York, 1988.

Stokes, R.G. *Opting for Oil: The Political Economy of Technological Change in the West German Chemical Industry, 1945–1961*. Cambridge University Press, Cambridge, 1994.

Fertility, Human

Assisted human reproduction technologies (ARTs) have flourished in the mid- to late-twentieth century. Greater understanding of human biology and reproduction has led to technological developments to assist individuals or couples experiencing infertility due to a wide range of indications. The field has evolved from a combination of technological advances (such as laparoscopy and transvaginal ultrasonography) together with pharmaceutical developments (notably purified extracts of human menopausal gonadotropic hormones) and theoretical knowledge and practical techniques taken from gynecology, genetics, urology, and associated medical specialties. However, the origins of many current ART techniques can be traced to early practices in animal husbandry.

Humans have used some less technological methods to promote their reproduction without the involvement of medical professionals, notably artificial insemination by donor (AID) or by husband/partner (AIH). The introduction of semen or concentrated specimens of spermatozoa (sperm) into a woman's reproductive tract by noncoital means can be successfully performed with instruments as simple as a turkey baster. However in recent years, fears about donor health status, risk of infection (HIV and otherwise), and legal issues (such as establishing paternity) have caused most AI to be performed in medical clinics under a physician's supervision. Some doctors avoided paternity issues by mixing sperm from several donors including the male partner, but recent advances in genetic technologies allowing paternity testing using DNA have resulted in

clarification in many jurisdictions of the legal standing of children born from AI (though issues remain for instance with custody and adoption of AID children born to lesbian couples). AI is sometimes coupled with use of gonadotrophic hormones to stimulate ovulation at the time of insemination to maximize the chances of fertilization occurring, although these drugs are associated with some risks to the women to whom they are given.

In vitro fertilization and transcervical embryo transfer (IVF-ET) had its first successful birth in 1978, in the clinic of Patrick Steptoe and Robert Edwards, who drew on embryological studies done for over 20 years in mice, rabbits, and other animals. The procedure involved laparoscopic aspiration of an oocyte (egg cell or more generally "egg") during a natural cycle (thus circumventing damaged fallopian tubes), followed by IVF using ejaculated sperm and transfer of the cleaving embryo into the woman's uterus. More generally in IVF-ET, eggs are harvested and mixed in Petri dishes either with donor sperm (AID), or with sperm from the male partner (AIH, if primary male infertility is not thought to be at issue), typically using the healthiest sperm to facilitate fertilization. Eggs may be obtained from the female or donated by another woman (e.g., in cases of premature ovarian failure, genetic abnormalities, or reduced egg production due to advanced maternal age). Most women undergo controlled ovarian hyperstimulation (as described above) prior to aspiration of eggs to increase the number of eggs that are viable. Surrogacy (the establishment of pregnancy in another woman who either donates an egg to combine with the male partner's sperm or carries a fetus produced through combination of the couple's own gametes) has occurred in cases where the female partner in a couple wishing to have a genetically-related child cannot carry a pregnancy (e.g., due to lack of uterus for congenital reasons or following hysterectomy), but the practice has been curtailed in recent years due to legal restrictions following custody disputes between surrogates and couples.

The number of fertilized embryos created and transferred differs according to anticipated success, typically related to the putative cause of infertility in the couple as well as the clinic's experience. But in recent years, improved methods have created higher success rates both in terms of creation of viable embryos as well as implantation of the embryo posttransfer (the latter had been and remains the major technological barrier to successful pregnancies via IVF). The result has been multiple gestations (often resulting in subsequent "selective reduction"; that is, termination of one or more fetuses to avoid the increased risks associated with multiple births), as well as emerging social, ethical, and legal issues associated with the status and disposition of supernumerary embryos. Consequently, many clinics have adopted more conservative approaches to the number of embryos created and transferred at any one time, and there is legislation or guidelines in some places to limit the number of embryos that can be transferred during one IVF cycle. Supernumerary embryos can be cryopreserved (typically at the four- to eight-cell stage) for later IVF cycles, donation to other infertile couples, or under certain circumstances for research. By the mid-1980s, techniques for cryopreservation were sufficiently developed to allow successful pregnancies using ET with frozen embryos, which permitted women to avoid multiple cycles of ovarian stimulation. More recently, IVF-ET has been combined with preimplantation genetic diagnosis (PGD) techniques to allow testing of embryos for genetic diseases (the technique was originally developed as an alternative to prenatal diagnosis for fertile couples with known genetic risks) and chromosomal abnormalities to allow selection and transfer only of unaffected embryos.

A number of additional ARTs have been developed in the last 20 years. Gametic intrafallopian transfer (GIFT), which involves placement of eggs (which have been removed from the follicles) together with sperm directly into the oviducts for fertilization, was first described in 1985, and is used with women with fallopian tube problems. This technique quickly became very popular because it did not require sophisticated IVF culture systems and could be done in clinics with less ART expertise and without a full IVF laboratory, and because it produced better results than IVF, perhaps because fertilization occurs in a natural environment (greater success rates are also due to patient selection, as was later recognized). Zygote intrafallopian transfer (ZIFT) involves transfer of the zygote (a fertilized egg that has not yet divided) into the oviduct after IVF, but is less frequently used. Intracytoplasmic sperm injection (ICSI) is a popular micromanipulation technique used to enhance fertilization rates, particularly for men with a reduced sperm count or with impaired sperm motility, banked sperm (obtained prior to chemotherapy or radiation), or sperm obtained through electroejaculation (e.g., in those with spinal cord injuries or recently after death, the latter being ethically and legally problematic).

Pregnancy can be achieved with only a single spermatozoon injected directly into the cytoplasm of the oocyte. The technique also can be combined with those allowing separation of male and female sperm to avoid birth of children of a particular sex (for reasons of sex selection, which is considered ethically controversial by many, or avoidance of sex-linked diseases).

See also **Artificial Insemination and** *In Vitro* Fertilization; Cloning, Testing and Treatment Methods; Genetic Screening and Testing; Genetic Engineering, Methods

RACHEL A. ANKENY

Further Reading

Chen, S.H. and Wallach, E.E. Five decades of progress in management of the infertile couple. *Fertil. Steril.*, 62, 665–685, 1994.

Cohen, C.B., Ed. *New Ways of Making Babies: The Case of Egg Donation*. Indiana University Press, Bloomington, 1996. Commissioned by the National Advisory Board on Ethics in Reproduction.

Edwards, R.G. and Steptoe, P. *A Matter of Life: The Story of a Medical Breakthrough*. Hutchinson, London, 1980.

Hildt, E. and Mieth, D., Ed. *In Vitro Fertilisation in the 1990s*. Ashgate, Aldershot, 1998.

U.S. National Research Council. *Medically Assisted Conception: An Agenda for Research*. Report of a Study by a Committee of the Institute of Medicine. National Academy Press, Washington D.C., 1989.

U.S. Congress Office of Technology Assessment. *Artificial Insemination: Practice in the United States: Summary of a 1987 Survey—Background Paper*. U.S. Government Printing Office, Washington D.C., 1988.

Royal Commission on New Reproductive Technologies. *Proceed with Care*. Minister of Government Services, Ottawa, 1993.

Seibel, M.M. and Crockin, S.L., Eds. *Family Building Through Egg and Sperm Donation: Medical, Legal, and Ethical Issues*. Jones & Bartlett, Boston, 1996.

Seibel, M.M., Bernstein, J., Kiessling, A.A. and Levin, S.R., Eds. *Technology and Infertility: Clinical, Psychosocial, Legal and Ethical Aspects*. Springer Verlag, New York, 1993.

Talbert, L.M. The assisted reproductive technologies: an historical overview. *Arch. Pathol. Lab. Med.*, 116, 320–322, 1992.

Warnock, M. *A Question of Life: The Warnock Report on Human Fertilisation and Embryology*. Blackwell, Oxford, 1985.

Wikler, D. and Wikler, N.J. Turkey-baster babies: The demedicalization of artificial insemination. *Milbank Quarterly*, 695–640, 1991.

Law and Ethics of A.I.D. and Embryo Transfer. Proceedings of a Symposium on Legal and Other Aspects of Artificial Insemination by Donor (A.I.D.) and Embryo Transfer, London, 1972. Ciba Foundation Symposium, vol. 17, Elsevier, Amsterdam, 1973.

Fertilizers

As the twentieth century opened, fertilizers were a prominent concern for farmers, industrialists, scientists, and political leaders. In 1898, British scientist William Crookes delivered a powerful and widely reported speech that warned of a looming "famine" of nitrogenous fertilizers. According to Crookes, rising populations, increased demand for soil-depleting grain products, and the looming exhaustion of sodium nitrate beds in Chile threatened Britain and "all civilized nations" with imminent mass starvation and collapse. Yet Crookes also predicted that chemists would manage to discover new artificial fertilizers to replace natural and organic supplies, a prophecy that turned out to encapsulate the actual history of fertilizers in the twentieth century.

Three basic nutrients—nitrogen, phosphorus and potassium (typically identified as NPK)—as well as several minor and trace minerals are essential requirements for productive agriculture. But nitrogen lay at the crux of fertilizer problem. The obvious solution lay in the atmosphere, where nitrogen forms 78 percent of the atmosphere's volume. In this context, German investigators led by Carl Bosch and Fritz Haber developed the process that became standard in industry—the synthetic production of ammonia by subjecting atmospheric nitrogen and hydrogen to high temperatures and pressures. German industrial giant BASF opened a nitrogen fixation plant in 1913, although the real impact of nitrogen fixation upon agricultural practice remained roughly two decades away. In the meantime, significant geopolitical battles continued over the price and supply of natural nitrogenous fertilizers, particularly the sodium nitrates of Chile. Chile's share of world nitrate fertilizer production fell from about 70 percent at the beginning of the century to under 2 percent at the end. Also, near the turn of the century, scientists demonstrated the connection between soil microbes and the ability of legumes to fix atmospheric nitrogen. This discovery created a burgeoning commercial market for legume inoculation bacteria, products that promised a new industry for "germ fertilizers" and "green manures."

Important changes affected other branches of the fertilizer industry as well. Since the early industrial era, farmers applied phosphate fertilizers derived from natural sources such as bones and phosphate rock. Since the middle of the nineteenth century, farmers typically applied phosphates in the form of superphosphates, produced by treating phosphoric minerals with sulfuric acid. Production of superphosphates peaked in 1952. Since World War II, however, ammonium phosphates have emerged as the more predominant form. Discoveries of rock phosphates in Florida and the Rocky Mountain states of the U.S. in the 1870s, in Algeria and Tunisia in the late nineteenth century, Morocco, the Kola Peninsula of the USSR, South Africa, and elsewhere have ensured an adequate supply for well-capitalized agriculturists up to the present.

In the preindustrial era, farmers applied potassium fertilizers in the form of wood ash and other natural byproducts. The potash industry was transformed in the late nineteenth century, particularly due to discovery of the Stassfurt salts of central Germany. Potash became an important political issue in World War I, as the U.S. abruptly cancelled its contracts with Germany without an adequate means of replacing them. An intense search for domestic supplies continued until potash deposits were discovered in New Mexico in the 1920s. The U.S. was a leading potash producer until of the middle of the twentieth century. Since then, the principal producers have been Canada and Russia, and additional reserves have been found in China, Thailand, Brazil, and elsewhere.

The fertilizer industry experienced further expansion after World War II due to improved techniques of producing granular, homogenized, and well-mixed fertilizers. These replaced the powder form, since they were easier to store and handle, and apply. Blended granular fertilizers were introduced in the U.S. in the 1950s. Manufacturers produced a wide range of NPK combinations appropriate for specific local soil and cropping conditions. In the 1990s, "precision farming" techniques have integrated technologies of soil nutrient sensors, global positioning satellites, and the machinery that applies fertilizer. As a result, well-equipped farmers continually and automatically adjust fertilizer delivery formulas as they drive down their fields.

Government policies also facilitated the rapid expansion of fertilizer consumption in the late twentieth century. Britain's 1947 Agricultural Act was a typical example, offering incentives for fertilizer use as part of a strategy to intensify agricultural production. In the U.S., the quasi-governmental Tennessee Valley Authority has been at the center of many technical innovations in fertilizer processing. In the former German Democratic Republic, state support for fertilizer manufactories exemplified the technocratic assumption that agricultural chemicals are an inherent good. Yet the most rapid expansion has occurred in the People's Republic of China, where officials linked shortages of nitrogenous fertilizers with the famines of the late 1950s and early 1960s. Thus China invested heavily in nitrogen fixation technology, and emerged rapidly as the largest consumer of manufactured fertilizers, both in terms of total consumption and rates of application. This in turn has spurred the most rapid rise of human population of all time. Other Asian nations, including the Republic of China, the Republic of Korea, and India also have become major producers and consumers of artificial fertilizers.

In addition to obvious links to increased agricultural production, the modern fertilizer industry has been linked with a number of concerns beyond the farm. For example, the short-lived phosphate boom on the Pacific island of Nauru offers a telling case study of the social consequences and environmental devastation than can accompany extractive industries. Further, much of the nitrogen applied to soils does not reach farm plants; nitrates can infiltrate water supplies in ways that directly threaten human health, or indirectly do so by fostering the growth of bacteria that can choke off natural nutrient cycles. To combat such threats, the European Union Common Agricultural Policy includes restrictions on nitrogen applications, and several nations now offer tax incentives to farmers who employ alternative agricultural schemes. Nevertheless, the rapidly growing global population and its demand for inexpensive food means that artificial fertilizer inputs are likely to continue to increase.

See also **Nitrogen Fixation**

MARK R. FINLAY

Further Reading

Haber, L.F. *The Chemical Industry, 1900–1930.* Clarendon Press, Oxford, 1971.

Monteón, M. *Chile in the Nitrate Era: The Evolution of Economic Dependence, 1880–1930.* University of Wisconsin Press, Madison, 1982.

Nelson, L.B., *History of the U.S. Fertilizer Industry.* Tennessee Valley Authority, Muscle Shoals, Alabama, 1990.

Smil, V. *Enriching the Earth: Fritz Haber, Carl Bosch, and the Transformation of World Food Production.* MIT Press, Cambridge, MA, 2001.

Welte, E. Die Bedeutung der Mineralischen Düngung und die Düngemittelindustrie in den letzten 100 Jahren. *Technikgeschichte,* 35, 37–55, 1968.

Wines, R.A. *Fertilizer in America: From Waste Recycling to Resource Exploitation.* Temple University Press, Philadelphia, 1985.

Fibers, Synthetic and Semi-Synthetic

Silk has long been prized as a fiber for fashion, and Robert Hooke proposed artificial silk at least as early as 1665. The invention of the electric light bulb in 1879 created a new incentive for an improved fiber: a very uniform carbon fiber was required for lamp filaments. In 1883, Joseph W. Swan invented a process of squirting a nitrocellulose solution through a dye into a coagulating bath of water and alcohol. A denitration technique was used to make the fiber less flammable. In 1885, he exhibited the material at the Inventions Exhibition as "artificial silk." Although Swan was the first to spin fibers from a nitrocellulose solution, the first commercial production of artificial silk took place in France in 1884 by Louis de Chardonnet. De Chardonnet's product, however, failed commercially as he did not denitrate the fibers to reduce their flammability, with disastrous results. In the first quarter of the twentieth century, the Chardonnet process gradually disappeared.

In Germany, Max Fremery and Johan Urban produced good quality fibers by using solutions of cellulose in a copper oxide and ammonia mixture. In 1899, production started in the Vereinigte Glanzstoff Fabriken AG. In England in 1884, Charles Cross and Edward Bevan discovered that alkali–cellulose could be converted into cellulose–xanthogenate, which could be dissolved in sodium hydroxide. This solution was called "viscose," from which fibers could be pulled. Together with Clayton Beadle, they established the Viscose Syndicate Ltd. at Kew. Viscose fibers were of very good quality, and gradually the viscose process became the main process for producing artificial silk.

After World War I there was no use for vast amounts of cellulose acetate lacquers that had been used for doping fabric-covered aircraft. Henry and Camille Dreyfuss used these lacquers to produce fibers. In 1921, they introduced acetate fibers to the U.S. "Acetate" made great inroads on the market.

After World War I it was very important to show that cellulose fibers were not merely a "surrogate" for silk, and the name "rayon"' was introduced in 1924 and gradually accepted worldwide. Though rising hemlines created a demand for cheaper alternatives to silk stockings, rayon "bagged" at the ankles and in the 1930s and 1940s uses turned more to industrial products. One of the main improvements was a "hot stretch" process to increase the strength of rayon. In this way rayon could replace cotton in tire cords. Although the basic features of the three processes to produce rayon fibers (viscose—by far the most significant process by volume—acetate, and cuprammonium) did not change much, efficiency of production continued to increase. Between 1930 and 1940, DuPont's manufacturing costs of a pound of viscose fiber fell nearly 60 percent. Until the recession of 1929, the rayon industry was a golden business.

However, the basic science of rayon was not understood. Chemists explained the characteristics of these materials by postulating forces between circular molecular structures. Most important in challenging these established concepts was Hermann Staudinger. He claimed in 1920 that normal molecular bonds could explain the products of polymerization reactions. Compounds, which precipitated during these reactions, could be explained by assuming that hundreds of molecules merged into "macromolecules." The work of Wallace H. Carothers, at E.I. DuPont de Nemours, was crucial for the acceptance of Staudinger's concepts. From 1928 onward, Carothers proved the existence of macromolecules by synthesizing new, long-chain molecules. Julian Hill, one of his research group members, discovered that he could draw fibers from the melt of these materials. In 1934, Carothers's group succeeded in making a good-quality polyamide 6,6 fiber. Technological problems were enormous: the process for spinning the fibers differed greatly from any conventional process. At the end of 1938, DuPont launched the fiber as Nylon. Nylon stockings were an overwhelming commercial success until it was rationed to replace silk in parachutes and to reinforce airplane tires. In Germany, Paul Schlack had succeeded in producing a Nylon-like fiber by the end of the 1930s. The fiber was often called Nylon 6 as opposed to DuPont's Nylon 6,6. After the war, the nylon 6 patents of IG Farben were confiscated. Many corporations started Nylon 6 production, as the technology was freely available.

During the war, research on new fibers continued. The most important of the new fibers were acrylic fibers and polyester fibers, including terylene discovered in Britain by John R. Whinfield and James T. Dickson in 1941. ICI and DuPont started producing polyester fiber (terylene and dacron) in the early 1950s. Polyester fiber had very good mechanical properties and a better light stability than nylon. Because of its very good resilience, the main application of this fiber was initially in "wash and wear" textiles.

In 1942, IG Farben found a suitable solvent to spin acrylic fibers, and after World War II, they were commercialized by Bayer, DuPont and Monsanto. Acrylic fibers had comparable properties to wool, and the chemical, thermal and light stability of the fiber was rather good. Acrylic fibers were spun by a new spinning process: after spinning, the solvent evaporated when the fibers were led through a gas stream in a spin tower. This process is called dry spinning. Nylon and polyester were spun by a melt spinning process, which implies that the polymers are molten before spinning, and not dissolved. Rayon was produced by wet spinning, which implies that the polymer is dissolved. However, the solvent is later removed from the fiber in a bath.

Polypropylene fiber was developed at the end of the 1950s, based on a breakthrough in polymer science at the beginning of the 1950s in which ordered polyethylene and polypropylene polymers

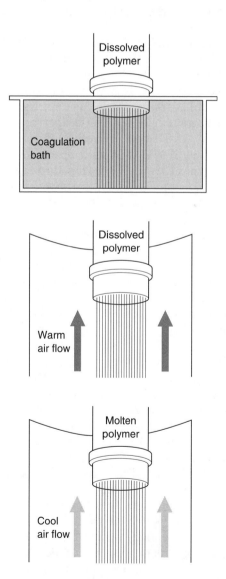

Figure 2. Wet spinning, dry spinning, melt spinning.

could be made catalytically. Polypropylene fiber was excellent for ship ropes, as its strength was high and it floated. Polyethylene was also used for fibers after DuPont invented the "flash spinning" technique by which very cheap, but low-quality polyethylene fibers could be spun. Polyethylene fibers were used to form a sheet-like wrapping material.

Particularly in the 1950s and 1960s, synthetic fibers conquered large shares of the market. Production efficiency and fiber properties increased considerably. For example polyester-spinning speeds increased from 500 meters per minute in 1961 to 4000 meters per minute in 1975, allowing fiber prices to drop dramatically in these years. Improved quality opened new markets. For example, nylon was good enough to be used in car tires after World War II, although tires reinforced with nylon were subject to so-called "flat spotting"; they had a flat spot when the car had been parked for a while.

At the beginning of the 1950s, scientists at DuPont achieved a breakthrough in polymerization. Their new method did not involve heating of intermediates. A lot of new polymers were made which could not be made before because the intermediates could not withstand heating. By this method, DuPont scientists developed the spandex fiber Lycra®. A high temperature-resistant fiber, Nomex®, was made based on a wholly aromatic polyamide. Similar fibers were introduced in Japan as Conex® and in the USSR as Fenilon®.

Other companies also started researching DuPont's low temperature polycondensation method. Celanese developed very heat-resistant and comfortable polybenzimidazole fibers. Protective clothing became more important when three of the Apollo I astronauts burned in their spacecraft in an accident during launching in January 1967.

The 1950s and 1960s were the glorious days of the fiber industry. Growth rates and profits were high. However, the market was sensitive to cyclical fluctuations, which sometimes caused financial problems at the fiber corporations. In 1967, the U.S. market collapsed. In 1970, the European fiber market was in crisis too. The recession that followed the oil crisis of 1973 put the fiber industry in the doldrums. The situation, especially in Western Europe, was dramatic. Many plants were closed down and some traditional fiber producers completely terminated their business.

In the 1960s, high-strength/high-modulus (HS/HM) carbon fibers were under development at various laboratories. These fibers were aimed at making composites. Carbon fibers were very good but brittle and expensive. They were therefore only used for military applications and in the aerospace industry. At several laboratories, it was found that HS/HM fibers could be made using aromatic or heterocyclic polyamides. In 1964, DuPont made a fiber that had an unusually high strength and modulus. Soon it was recognized that these higher strengths could possibly be used to make improved tire cords and a cheaper alternative for carbon fiber. In February 1970, DuPont announced its HS/HM fiber in public. The fiber was called Kevlar®. Other corporations had started researching HS/HM fibers too. Monsanto, Bayer and ICI made several contributions to the technology. AKZO, which improved DuPont's production process, got into an enormous patent litigation case with DuPont in the 1980s when it introduced Twaron®, a similar fiber to Kevlar. The Japanese company Teijin also developed an HS/HM fiber Technora®. In the Soviet Union, an HS/HM fiber was developed further for military use.

Kevlar was not as successful as DuPont had hoped. It was only used as a tire cord in very specialized tires. The U.S. tire cord market was gradually conquered by the steel wire reinforced radial tire, which had previously conquered Europe. New applications of Kevlar were ballistic protection, composites and replacement of asbestos.

In the 1980s, some corporations carried out research and development to develop fibers of even higher modulus and strength than Kevlar. Stanford Research Institute synthesized two polymers, PBO (poly-para-phenylene-benzo-bis-oxazole) and PBT (poly-para-phenylene-benzo-bis-thiazole), which were further developed by Dow Chemical and Toyobo. Akzo Nobel developed a similar fiber called M5®.

The polymers that were used for HS/HM fibers had a rigid rod structure. However, one also succeeded in making HS/HM fibers from flexible polymers: In 1979, scientists of the Dutch corporation DSM succeeded in spinning HS/HM fibers from a polyethylene gel and stretching it. By this process, they created an ordered structure in the polymers. This polyethylene fiber was in some applications a competitor for HS/HM aramid fibers because it was even stronger. Due to its low melting point it could not compete with aramid fibers in applications in which higher temperatures played a role.

See also **Composite Materials; Plastics, Thermoplastics; Synthetic Resins**

KAREL MULDER

Further Reading

Beer, E.J. *The Beginning of Rayon*. Beer, Paignton, Devon, 1962.

Coleman, D.C. *Courtaulds: An Economic and Social History II: Rayon*. Clarendon Press, Oxford, 1969.

Farber, E. *Great Chemists*. Interscience Publishers, New York, 1961.

Hermes, M.E. *Enough for One Lifetime: Wallace Carothers, Inventor of Nylon*. American Chemical Society/Chemical Heritage Foundation, Washington, 1996.

Hollander, S. *The Sources of Increased Efficiency: A Study of DuPont Rayon Plants*. MIT Press, Cambridge, MA, 1965.

Hongu, T., Phillips, G.O. *New Fibers*. Horwood, New York, 1990.

Klare, H. *Geschichte der Chemiefaserforschung*, Akademie Verlag, Berlin, 1985.

Morawetz, H. *Polymers: The Origins and Growth of a Science*. Wiley, New York, 1985.

Morris, P.J.T. *Polymer Pioneers*. Beckman Center for History of Chemistry, Philadelphia, 1986.

Mulder, K.F. Choosing the corporate future: Technology networks and choice concerning the creation of high performance fiber technology. University of Groningen, dissertation, 1992.

Mulder, K.F. A battle of giants: the multiplicity of industrial R&D that produced high-strength aramid fibers. *Technol. Soc.*, 21, 37–61, 1999.

Seymour, R.B. and Porter, R.S., Ed. *Manmade Fibers: Their Origin and Development*. Elsevier, London, 1993.

Staudinger, H. *Arbeitserinnerungen*. Dr. Alfred Huthig Verlag GmbH, Heidelberg, 1961.

Film and Cinema, Early Sound Films

Silent films were rarely silent. Scarcely had the gasps and screams at the Lumière brothers' train arriving at La Ciotat station died down before its successors were being screened to music provided by anything from a piano to a 100-piece orchestra. The technology needed to provide films with a "tailored" sound accompaniment had existed for almost 20 years, and in September 1896 Oskar Messter showed "sound films" in Berlin using synchronized Berliner disks. Many others followed, including talking films, included in a program at the Paris Exposition by Clement Maurice on 8 June 1900.

These were able demonstrations that pushed existing film and sound technology boundaries, but producing a reliable sound film system required three major advances:

- Providing a foolproof method of picture/sound synchronization
- Recording a wider sound frequency range at greater levels than existing acoustic methods
- Reproducing these loud and clear enough to fill a cinema auditorium

Ingenious methods of synchronizing sound recordings with films were devised, but all suffered from the fragility of records over repeated reproduction and during transport, and the risk of picture and sound going "out of sync." The solution lay in combining the two together on film—a process known as "sound-on-film." Frenchman Eugene Lauste patented the first such system in London on 11 August 1906, but it took seven years to realize this, when war intervened. Lauste's "sound gate" used two slotted iron grids through which light passed to the film. One grid was fixed, while the other slid up and down over it in response to a signal from a microphone relayed through an electromagnet, resulting in a variable density soundtrack.

With this impetus lost it fell to independent researchers to develop sound-on-film systems. Josef Engl, Hans Vogt, and Josef Engl devised Germany's "Tri-Ergon" process. Another variable density process, it used a photoelectric cell to turn sound into electric waves and then into light recorded photographically on to the film's edge. In the projector a photoelectric cell reconverted these into an electrical signal. Tri-Ergon was first demonstrated in Berlin on 17 September 1922.

The American "Phonofilm" process was developed by electronics and radio pioneer Lee de Forest in 1920. A valve produced a fluctuating light pattern in response to an electrical sound signal, exposing a series of light and dark areas on the side of a film. On projection this was read by a photocell and converted back to a sound signal. Phonofilm debuted in New York on 15 April 1923. De Forest's system enjoyed greater commercial success, and over 30 cinemas had been equipped to show its mainly musical short films within a year, but it failed, lacking major studio and distributor support.

With sound films possible technically, ways of improving sound recording and reproduction were needed. Stimulus to develop and improve sound recording came from U.S. telephone companies, notably American Bell, who formed American Telephone & Telegraph (AT&T) on 3 March 1885. A Bell subsidiary—Western Electric—became AT&T's research arm. In 1907 this was centered in New York, under director Theodore Vail, and between 1913 and 1926, breakthroughs were made there in electrical recording, playback, and amplification.

E.C. Wente patented the condenser microphone on 20 December 1916. It translated sound waves into electrical ones that could be transmitted by the vacuum tube amplifier. Improved over the years, in

1926 it became the Western Electric 394-W microphone, used to produce the first true sound films.

By 1920 electrical recording had been developed by Henry C. Harrison. It used Wente's microphone, a tube amplifier, a balanced-armature speaker, and a rubber-line recorder, recording sound in the range 50 to 6,000 Hz, an improvement over the 250 to 2,500 Hz range of acoustic recordings.

Harold Arnold developed the tube amplifier in April 1913. Based on Lee de Forest's Audion tube, Arnold found that electron flow across its electrodes improved if they were in a vacuum. Arnold built his first vacuum tube on 18 October 1913.

C.W. Rice and E.W. Kellogg, and others developed the moving coil loudspeaker. Henry Egerton patented the first balanced-armature loudspeaker driver on 8 January 1918, and E.C. Wente developed the moving coil speaker (patent filed 4 August 1926). He used a moving coil or diaphragm mechanism in a strong magnetic field. Designed to drive a theater speaker, it was installed at the Warner Theater, New York for the premiere of Warner's "Don Juan" in August 1926.

Despite sound film technology being ready by 1925, there was little interest from major studios, and it fell to an aspiring one—Warner Bros.—to seize the initiative. Sam Warner learned about Western Electric's achievements when Warner Bros. built a radio station. He saw its potential for providing a full orchestral accompaniment to Warner films wherever they were screened. On 20 April 1926 they formed the Vitaphone Corporation with Western Electric to develop this, but opted for a synchronous sound-on-disk system. A 406-mm (16-inch) disk, revolving at 33 1/3 rpm, played simultaneously with the film, synchronized by two motors held at the same speed by an electric gear. Interconnected by slip rings, the interchange of power between the armatures ensured correct synchronization when the film started, with the power source's frequency then maintaining this.

Many Vitaphone short films were made, but Warner gave its greatest showcase with the production and premiere of "Don Juan" starring John Barrymore, featuring a full-length orchestral soundtrack with sound effects. Its premiere, on 6 August 1926, left a bemused audience but an enthusiastic press.

By the end of 1926, Warner had produced 100 Vitaphone shorts, but the cost, and those of equipping cinemas for them, almost crippled the company. They had one last attempt, adapting Samson Raphaelson's play "The Jazz Singer," with Al Jolson reprising his stage role in the lead.

Intended only to have synchronized music and singing, audience reaction at the 6 October 1927 premiere was greatest to Jolson apparently talking to them from the screen. Ironically, his first words—"Wait a minute. Wait a minute. You ain't heard nothin' yet! Wait a minute, I tell you. You ain't heard nothing. You wanna hear toot-toot-tootsie? All right. Hold on"—were largely ad-libbed. Ignorant of this, the audience stood and cheered. Despite only containing 354 spoken words, The Jazz Singer's triumph firmly established talking pictures.

By cruel fate, Sam Warner died of a brain hemorrhage the night before the premiere. His surviving brothers—Harry, Albert, and Jack—were at his bedside, also missing their moment of glory. In the longer term, use of the Warners' synchronous Vitaphone system faded in favor of a sound-on-film system called Movietone, developed by the rival Fox Film Corporation.

See also **Audio Recording, Mechanical; Audio Recording, Electronic Methods; Audio Systems; Film and Cinema: High Fidelity to Surround Sound; Loudspeakers and Earphones**

PAUL COLLINS

Further Reading

Guy, R. *Warner Bros: Fifty Years of Film, 1923–1973.* LesLee Productions, Los Angeles, 1973.

Hirschhorn, C. *The Warner Bros. Story.* Octopus Books, London, 1979.

Millard, A. *America on Record: A History of Recorded Sound.* Cambridge University Press, Cambridge, 1995. An excellent idea of the interrelatedness of innovations in sound recording, radio and cinema.

Robertson, P. *The Guinness Book of Film Facts & Feats.* Guinness Superlatives, Enfield, Middlesex, 1980. The basic facts of who did what when.

Toulet, E. *Cinema is 100 Years Old.* Thames & Hudson, London, 1995. Covers the impact of cinema, and of the many innovations in production and presentation tried in its first decade or so.

Film and Cinema: High Fidelity to Surround Sound

While motion picture soundtracks typically generate less excitement than the visual content, the sound recording devices used in the motion picture industry have consistently been at the cutting edge of technological advancement. From the 1920s to the 1950s, these technologies reached the public in theaters long before anything similar was available from broadcasts or for home use. In fact, many of the early innovations in high fidelity recording and reproduction were created in the context of motion picture production and exhibition, while the phonograph, radio, and television lagged behind.

There were experimental linkings of sound and motion pictures from the earliest days of the cinema, and early designers of these systems struggled merely to provide sound at minimally acceptable levels of volume and quality. The nearly worldwide adoption of sound-on-film by the early 1930s corresponded to relatively favorable economic conditions in the motion picture industry, particularly in the U.S., and this encouraged even more experimentation with new audio techniques.

Some of the most notable achievements of the 1930s were related to stereophonic sound. Long before it was practical to introduce this technology to the public via the phonograph or radio, audiences in some places heard stereo soundtracks accompanying a small number of feature films. A landmark was *Fantasia*, the animated film by Walt Disney Studios (U.S.), which employed sound recorded on numerous separate optical tracks. These were then mixed down to three channels (left, right and center) of audio for exhibition along with a fourth "control" track, which was not audible but contained information that automatically controlled the volume of each of the three audio tracks. Only two theaters purchased the U.S.$85,000, 54-loudspeaker "Fantasound" system needed to reproduce these films on screen, but a traveling exhibition toured the U.S. when the film opened in 1940.

At the end of World War II, many motion picture producers adopted magnetic recording technology, which was known before the war but rarely used outside Germany. Magnetic recording was substituted for optical recording in the studios primarily because it was much less expensive to use; at a time when television was cutting deeply into theater attendance, cost cutting was imperative. However, that cost saving did not apply to the exhibition of films, and most theaters retained their optical-soundtrack projection equipment through the late 1980s. While studios repeatedly tried to introduce new theater systems using multichannel, high-fidelity sound, most exhibitors resisted. The theater was no longer at the forefront and many innovations in movie sound technology made after the 1950s were preceded by similar innovations in broadcasting or in home high-fidelity systems.

However, the experimental technologies of the 1950s are usually cited by film historians as great landmarks. One of the most notable examples was Cinerama, one of several widescreen formats that Hollywood studios believed would bring customers back to the theaters in the 1950s. Besides its remarkably wide screen, Cinerama featured a seven-track magnetic soundtrack, carried on a separate 35 mm film run on a player that was operated in parallel to multiple motion picture projectors. Like Fantasound before it, Cinerama was so expensive to exhibit that it saw only limited use. Somewhat more successful were systems based on a double-width, 70 mm film on a single projector. All of these used some variation of multichannel sound, and some used magnetic rather than optical soundtracks for theater reproduction. Perhaps the most commercially successful of these was Todd-AO (promoted by film producer Michael Todd and the American Optical Company), which used six audio channels.

A series of highly successful innovations was offered by Dolby Laboratories (U.K., later U.S.) beginning in 1965 with the introduction of what came to be known as Dolby A. This was a noise-reduction technology used to improve recordings made in the studio before they were released to theaters. Although used initially in the phonograph record industry, the first motion picture soundtrack made using Dolby A was *A Clockwork Orange*. Released in 1971, the movie was typical of the Dolby releases of the day in that it was originally recorded on multitrack magnetic recorders, mixed using Dolby noise reduction, but released in ordinary monophonic form, usually with an optical soundtrack. However, the next year Dolby introduced an improved optical soundtrack technology and the short film *A Quiet Revolution* was released to demonstrate to theater chain owners the value of using Dolby noise reduction equipment in the exhibition of these films. While this technology did not succeed, some theaters did begin to improve their audio equipment.

Sensurround, a multichannel system for theaters, was introduced as a sort of novelty with the film *Earthquake* in 1974. An optically recorded, inaudible control track triggered the reproduction of very low-frequency sounds, which were used to add emphasis to the soundtrack at key points (such as the rumbling of an earthquake). Sensurround-like systems would eventually evolve into the current "Surround Sound," but meanwhile the Dolby Laboratories once again introduced a new multitrack system. This one, called Dolby Stereo, electronically combined four soundtracks onto just two tracks for the final release print. The four tracks included left, right, and center channels plus a "surround" channel for special effects. Dolby Stereo could be reproduced by adding relatively inexpensive accessories to existing projectors. The first release in Dolby Stereo was *A Star is Born* in 1976. Following the advent of Dolby Stereo it became more common to advertise a film's sound technology along with its cast. This helped generate a popular interest in film sound, and along with the consolidation of exhibition and production companies, the rate of adoption of new theater technologies began to accelerate.

Digital recording techniques were tried by filmmakers from the early 1980s, although they were not in widespread use until the 1990s. An early optical digital playback system introduced in 1990 by the Eastman Kodak Company (U.S.) was not as successful as Dolby Digital, introduced in 1992. In this new system, a digitized version of the soundtrack was placed in the tiny spaces between the sprocket holes on the exhibition copy of the film, thus leaving room for a conventional analog soundtrack at the edge to be used as a backup or in theaters that had only the standard projectors. Under various names, the software algorithms developed for Dolby Digital have also been adapted for other formats, such as home theater and DVD discs. With the introduction of digital recording and playback systems, motion picture producers and movie theaters are once again acting as the channels for the introduction of new audio technologies to the public.

See also **Audio Recording, Compact Disc; Audio Recording, Electronic Methods; Audio Recording, Mechanical; Audio Recording, Stereo and Surround Sound; Audio Recording, Tape; Audio Recording, Wire; Audio Systems; Loudspeakers and Earphones**

DAVID MORTON

Further reading

Fielding, R., Ed. *A Technological History of Motion Pictures and Television*, University of California Press, Berkeley, 1967.

Weis E. and Belton, J. *Film Sound: Theory and Practice*. Columbia University Press, New York, 1985.

Film and Cinema: Sets and Techniques

During the early years of silent cinema until around 1909, interiors for film sets resembled theatrical and vaudeville stages with painted backdrops and limited three-dimensional representation. Action was depicted at long-shot distance immediately in front of the backdrop creating a shallow playing area. The earliest narrative films consisted of a single shot of actions on a stage and spectators were positioned in relation to the action as though viewing a traditional theatrical presentation. After 1911 the development of the American film industry in Hollywood as one based primarily on considerations of narrative storytelling over the esthetics of form, meant that a preoccupation of early studio set technicians and cinematographers, designers, producers and artistic directors was a drive toward greater narrative progression and the realistic representation of action on screen. Innovations in editing through the use of the fast film stocks characteristic of silent cinema had two profound cinematic effects.

First, continuity editing, as it came to be known, meant that scenes could be broken up into discrete shots, creating a break from theatrical conventions and the creation of a truly cinematic space. The standard continuity technique is shot-reverse shot and crosscutting. The former, used in filming two-way conversation, works in conjunction with editing and camera angles in order to produce the effect of seamless movement between speaking characters. Shots are taken from one character's point of view or over the shoulder and intercut with those from the other person. Crosscutting, pioneered and refined in the work of D. W. Griffith, refers to the alteration of shots, this time between scenes. It implies multiple actions in different locations that are occurring simultaneously. Techniques of continuity editing, in hiding the constructed nature of the action represented on screen, produces a second cinematic effect contributing to the establishment of the particular "look" or esthetic of American film. The codes and practices of editing have remained largely consistent with these early innovations and have been little affected by subsequent developments in technology. This emphasis on narrative continuity led to the early identification of American film with realism. Cinematic realism holds that what is represented on screen accurately reproduces that part of the real world to which it refers. Thus, American film through the use of continuity editing techniques introduced a realism of cinematic representation and positioned film spectators close to the action. This reversed technique that led to the distancing of the audience from the spectacle, common to European formalism and experimental cinema, offering audiences the illusion that they are watching a seamlessly coherent and wholly realistic representation of reality.

The realistic codes of American cinema also demanded the representation of three-dimensionality, the cinema frame as a window on the world extending beyond the limits of the frame into space off-screen. To this end developments intended to produce realistic cinematic effects occurred in liaison by cinematographers and art directors working with miniatures and techniques of composite photography such as glass shots and mattes. Glass shots worked to give cinematic depth through the technique of shooting through a clear pane of glass containing either a painted or photographic image. An early example of special effects, glass shots were used widely in the 1920s and 1930s and featured famously and extensively in the work of M. C. Cooper and E. B. Schoedsack in *King Kong* (1933).

By the 1920s, depth and three dimensionality was also being added to set design through modifications to stage architecture. The arrangements of props, increasing use of multiple room stage constructions, artificial lighting effects and careful construction of *mis-en-scéne* were all directed at creating the sensory illusion of depth and the participation of the audience in the screen space. Early attempts to represent the illusion of depth graphically were most convincing on location filming but proved incongruous when edited beside the shallow effects offered through traditional stage setting. However, developments as early as 1908 ushered forward techniques of deep focus cinematography through modifications to lens lengths and the use of wide angles, narrow apertures, and the manipulation of lighting, and provided for multiple planes of action area. As a technique of shooting, deep focus ensured that realism in depth could be achieved by allowing a number of actions to take place in multiple planes simultaneously during a single shot. Deep focus cinematography worked in combination with the development of techniques to enhance depth in set design. The illusion of depth offered through deep focus cinematography quickly became both the norm and the signature of American narrative cinema, though it was not fully realized until the pioneering work of Gregg Toland in the 1930s.

Toland, considered the greatest cinematographer of his age, experimented with arc lamps, wide angles, lens coating and fast film to produce a distinctive deep focus impression. The culmination of his work can be seen in *Citizen Kane* (1941) where Toland creates enormous and sumptuous spaces that achieve both great height as well as great depth. Toland's greatest innovation in *Citizen Kane* is considered to be his use of unusually long takes, using static cameras to achieve incredible shot depth of long duration. Toland's use of long takes added to the canon of deep focus techniques and was later developed in conjunction with the Garrett Brown's Steadicam to provide both an unusually long depth perception coupled with unerring rapid movement in *The Shining* (1980).

Contemporaneously, developments in special effects techniques from the pioneering work done in the 1920s and 1930s have become the modern signature of Hollywood productions. Common photographic techniques such as fades, wipes, and dissolves have been augmented by techniques such as rear projection in which live action is combined with painted backdrops or miniatures. Rear projection was used to fantastic effect by George Lucas in *Star Wars* (1977), the film considered to have reinvented the effects-led blockbuster film. Since then, digitization; that is, the use of electronically programmed motion control and computer graphics in which the fantastic is able to be represented realistically, has become ubiquitous in Hollywood productions in the wake of Steven Spielberg's breakthrough work of *Jurassic Park* in 1993.

Taken together, continuity editing, depth of field cinematography, composite special effects photography, increasingly sophisticated set design, and staging through mis-en-scéne, have each contributed to giving visual clarity and realism to multiple planes of action simultaneously. This has achieved the cinematic effect of greatly enhancing spectator perception of screen space by extending it forward toward the audience and approximating what the French theorist André Bazin, referring admiringly to American film production, called "total cinema."

See also **Cameras, Lens Designs: Wide Angle and Zoom; Film and Cinema, Early Sound Films; Film and Cinema, Widescreen Systems**

ANDREW PANAY

Further Reading

Allen, R.C., Gomery, D. *Film History: Theory and Practice.* McGraw-Hill, New York, 1985.

Bordwell, D., Staiger, J. and Thompson, K. *The Classical Hollywood Cinema: Film Style and Mode of Production to 1960.* Routledge, London, 1988.

Dick, B.F. *Anatomy of Film.* St. Martins Press, New York, 1998.

Katz, S.D. *Film Directing: Shot by Shot.* Michael Wiese Productions, Ann Arbor, MI, 1991.

Maltby, R. *Hollywood Cinema.* Blackwell, Oxford, 2000.

Nowel-Smith, G, Ed. *The Oxford History of World Cinema.* Oxford University Press, New York, 1996.

Film and Cinema: Widescreen Systems

The term "widescreen cinema" has been precisely defined by J. Belton as "a form of motion picture production and exhibition in which the width of the projected image is greater than its height, generally by a factor of at least 1.66:1." This restricted definition of widescreen excludes other large-screen formats such as the Imax and Omnimax formats, which produce a width to height ratio (or "aspect ratio") of 1.43:1.

Although widescreen formats vary, they all share in common the stretching of screen width far beyond the initial standard set by the film industry of 1.33:1. This standard later came to be called "narrow-screen" in contrast to the new widescreen industry standard. The previous format was based on 35 mm film gauge with a negative image of 1 inch wide by 0.75 inches high, producing a 4:3 image aspect ratio of 4 units wide by 3 units high. Although some experiments were undertaken with widescreen in the 1920s, "narrow-screen" endured with relatively minor variations as the industry standard for 64 years between 1889 and 1953 and remained the standard twentieth-century format for the television screen, 16 mm films and 8 mm home movies.

As with most developments in cinematic technology widescreen experiments in the 1920s like Magnascope in the U.S. and Polyvision in Europe were dismissed as short-term novelty gimmicks of no lasting value. That is, until the development of Cinerama by Fred Waller transformed established attitudes to screen dimensions. Waller was a ceaseless innovator, experimenting in a number of fields. His independent studies of depth perception established that the convincing illusion of three-dimensionality depended on peripheral as well as binocular vision. He was able to combine this insight with technical knowledge garnered from his work with Paramount's Special Effects Department. His Cinerama camera set three lenses at 48-degree angles to each other to produce a composite angle of view of 146 degrees by 55 degrees, closely matching the angle of view of human vision at 165 degrees by 60 degrees. Projected onto a deep curved screen by a three-sided multicamera system and supplemented by stereo sound, Cinerama seemed to surround the viewer at every point of visibility and heightened the viewing sensation as one of complete immersion in hugely inflated but realistically depicted moving images of unprecedented depth. The curved screen solved for Waller the problem of fitting the field of peripheral vision within theatrical space: a flat screen would have needed to be the length of a city block.

Cinerama was launched to huge acclaim on 30 September 1952 with the travelogue film *This is Cinerama*. Cinerama failed to follow up on its initial success and was only adopted by a major studio, MGM, ten years later to make two feature films, *How the West Was Won* and *The Wonderful World of the Brothers Grimm*. It was limited by its own expensive, specialized technology, designed more to showcase exhibitions than for standard feature films in conventional theaters. Cinerama films were shot and projected at 26 frames per second, rather than the standard 24, and the three-strip Cinerama production format used three and a half times more negative film stock as conventional processes. Moreover, clearly visible vertical lines on the screen were formed by three separately projected triptych images, badly distorting the desired optical illusion of seamless reality. Not until 1963 did Cinerama resolve the problems caused by multicamera perspective by converting to Ultra Panavision, a widefilm technology that stored images and sound on a single strip of 65 mm negative film, which could be projected in 70 mm Cinerama theaters or printed down to 35 mm CinemaScope for conventional theaters. This solution was at the cost of the uniqueness of Cinerama's full glory and the ambition of its original aspect ratio. Even innovatory productions like *2001: A Space Odyssey* could not halt Cinerama's steady decline and ultimate demise in May 1978.

Cinerama's lasting contribution is in the stimulation it provided for other, less expensive and technically complex widescreen processes like CinemaScope (1953) and Todd-AO (1955). Unlike Cinerama, CinemaScope was the product of the research department of a major film studio—Twentieth Century Fox. On 2 February 1953, Fox announced that it would make all of its subsequent movies in CinemaScope. Only seven months later, on 16 September 1953, Fox issued its first widescreen release, *The Robe*.

CinemaScope's rapid development and successful release crucially depended on combining innovations in computerized lens design, acetate film stock, and Eastman Color film, as well as in television innovations in magnetic recording equipment and screen materials. Critical to this was Fox's rights to the anamorphic optical system of cylindrical lenses, the Hypergonar, developed in France in 1927 by Henri Chretien. Attached to a standard camera lens, the Hypergonar compressed a wide horizontal image by a factor of 2 onto standard 35 mm film during recording without affecting the vertical image. In projection a reverse anamorphic lens "'stretched' the image for widescreen display, permitting an extreme panoramic image of 2.66:1. Fox's optical engineers worked with lens manufacturers Bausch & Lomb to modify Chretien's original square attachment to produce circular lens profiles, improving the overall consistency of projected light, resolution, depth of field and relative definition at the edges of the field.

With the screening of *Oklahoma!* on 10 October 1955, Fox's achievements in widescreen were eclipsed by Todd-AO, the process developed by Michael Todd and the engineering firm, American Optical. Todd-AO resolved the problem of magnification that beset other widescreen formats such as VistaVision and CinemaScope. By blowing up images to cover a screen 64 by 24 feet (19.5 by 17.3 meters), 35 mm film was magnified 330,000 times for CinemaScope, with some loss of image sharpness and resolution. Todd-AO recorded on 65 mm film and projected on 70 mm, the other 5 mm accommodated six magnetic soundtracks for stereo sound. The 65/70 mm format required a projected magnification of only 127,000 times to fill a curved 52 by 26 feet (15.8 by 7.9 meter) screen, ensuring sharper imagery. More expensive than 35 mm film, the 70 mm format acquired a prestigious reputation for cinematic quality. Todd assembled a talented team of researchers and contractors around American Optical, including the then current Director of the Institute of Optics, Brian O'Brien, to develop the process through a series of innovations. For example, a 128 degree "bug-eye" wide angle camera lens was developed and mounted on a single camera with three other lenses, 64, 48 and 37 degrees, to give a versatile range of coverage for long, medium and close-up shots.

In these few years in the 1950s widescreen developed out of past technical experiments and future expectations to meet market challenges and industry rivalries. It bequeathed a different sense of scale to cinema esthetics and provided an answer of sorts to the emergence of television as the leading mass medium. It lives on today in the form of 70 mm "blow-ups" where films such as *Star Wars* can be recorded on relatively cheap 35 mm stock and then "printed-up" to 70 mm for widescreen exhibition, shown in multiplex complexes with one or two 70 mm theaters. Widescreen's faint reflection can also be seen today in television's adoption of widescreen sets and video's use of "letterboxing."

See also **Cameras: Lens Designs, Wide Angle, and Zoom; Film and Cinema: High Fidelity to Surround Sound; Film and Cinema: Sets and Techniques; Television, Digital and High Definition Systems**

ALEX LAW

Further Reading

Bazin, A. Will CinemaScope save the Cinema? *Esprit*, 207–208, 672–683, 1953. Translated by Jones, C. and Neupert, N. and reprinted in *Velvet Light Trap, Review of Cinema*, 21, 1985. Bazin, A. *What is Cinema?* vol. 1. Editions du Cerf, Paris, 1958. Translated by Gray, H., University of California Press, Berkeley, 1967.

Belton, J. CinemaScope: the economics of technology. *Velvet Light Trap, Review of Cinema, Widescreen Issue*, 21, pp. 35–43, 1985.

Belton, J. *Widescreen Cinema*. Harvard University Press, Cambridge, MA, 1992.

Mitchell, R. History of widescreen formats. *American Cinematographer*, 68.5, 1987.

Spellberg, J. CinemaScope and ideology. *Velvet Light Trap, Review of Cinema, Widescreen Issue*, 21, 26–34, 1985.

Useful Websites

American Widescreen Museum website: http://www.widescreenmuseum.com/index.htm

Fire Engineering

The term "fire engineering" has gained growing acceptance in the construction industry only since the 1980s. However, the need for buildings that protected both the occupants and the structures themselves in case of fire has existed for 2000 years.

Major conflagrations (most famously, the Great Fire of London in 1666) often generated political pressures for legislation that required fireproof construction, though generally the object was saving buildings and their contents rather than lives. Until the late nineteenth century the word "fireproof" in the construction context was synonymous with "'incombustible." Throughout the nineteenth century an ever-increasing number of fireproof construction systems were patented for use in warehouses, factories and other large buildings. The earliest such system, from the 1790s, used brick or stone vaults (jack-arches) supported on cast-iron or, later, wrought iron beams and columns; this was used right up to the end of the nineteenth century, and many such buildings survive today. During the late nineteenth century systems generally incorporated iron beams in conjunction with concrete or hollow clay tiles.

However, the contents of factories and warehouses were often still highly flammable. Since much of the lighting in such buildings involved open flames, and cotton or wool fibers in the air formed a near explosive mixture, the conditions were nearly ideal for piloted ignition and spontaneous combustion. Severe fires and explosions were not uncommon and could lead to building collapse in two ways. An explosion could remove one or

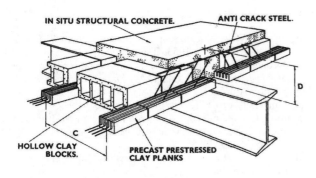

FIRE RESISTANCE

Fire resistance	Additional finish required
½ hour	Nil
1 hour	Refer to makers
2 hours	¾ inch gypsum plaster
4 hours	1 inch vermiculite plaster

Figure 3. Empirical approach to achieving fire protection. [*Source: British Constructional Steelwork Association.*]

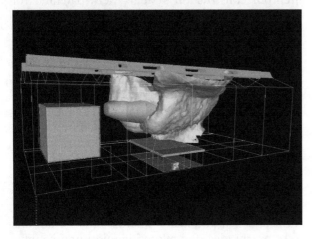

Figure 4. Two frames from a computational fluid dynamics (CFD) model of the spread of flames in a fire in a building.
[*Source: Buro Happold.*]

more vital structural members such as a column or load-bearing masonry wall. Alternatively, an intense fire could heat the exposed iron columns or beams to the point (about 550°C) where they lost virtually all their strength, and collapsed. Such was the technique of construction of these buildings that the removal of just one or two structural members could lead to a progressive and catastrophic collapse of a large part of the building, often causing many deaths. To such accidents were added a growing number of terrible fires in theaters in which the smoke from burning materials asphyxiated large numbers of the audience unable to escape through narrow corridors. By the end of the nineteenth century the whole idea of "fireproof" buildings had acquired a bad reputation and it was in this context that it was realized a new approach was needed to protecting buildings in fires – merely using incombustible construction materials was not sufficient.

Five main approaches were followed, almost simultaneously, during the early decades of the twentieth century. They all involved a logical, qualitative engineering approach to the problem and would be the necessary precursor to major quantitative developments later in the century. In contrast to the approach adopted in the previous century, these methods were tested experimentally in the growing number of fire research stations that were founded in many countries in the 1920s and 1930s:

1. The main cause of fire ignition was removed—electric lighting gradually replaced open flames.
2. Active means were installed to suppress or extinguish fires—portable fire extinguishers, an installed sprinkler system, or, in areas with electrical equipment, gas suppressants (usually carbon dioxide or Halon).
3. The idea of containment was introduced—ensuring that areas where a fire might start were isolated in a so-called "fire compartment," for instance through the use of self-closing doors or a fire safety curtain in theaters.
4. The exits from the occupied areas of buildings were also made larger, more accessible, and better signposted, for instance by providing fire escapes or isolated staircases.
5. Finally, steps were taken to retard the rate at which the iron or steel structural armature of the building heated up in a fire.

This latter idea was not new—structural timber had long been protected by plaster or even by iron

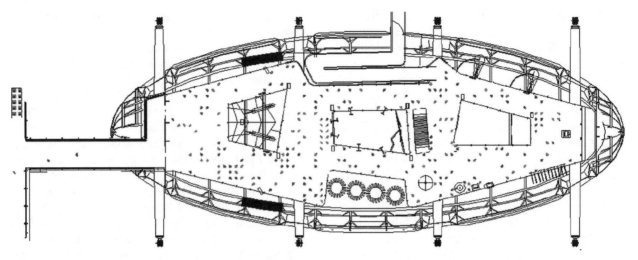

Figure 5.   A frame from a dynamic model of people escaping from a building in a fire.
Individual dots represent gender (male/female) and age (up to 30, 30–50 and > 50 years old).
[*Source: Buro Happold.*]

Figure 6.   Pompidou Centre, Paris (1969–1974; engineered by Ove Arup and Partners). The exposed steelwork inside the building is fire-protected using intumescent paint. The main, vertical tubular columns are fire-protected by being filled with water, which would be pumped around to remove heat in a fire.
[*Photo: Bill Addis.*]

sheets. It soon became a requirement to protect exposed steel and several methods were developed during the century—encasing columns in brickwork, covering or encasing beams in concrete, boxing the girders inside a case of insulating sheet made, for instance, from asbestos cement, and spraying a material containing a good insulator such as vermiculite to form an adhering layer up to 20 millimeters thick. The most recent development was intumescent paint which, when heated, foams and hardens to create a barrier of entrapped air bubbles which are highly insulating.

The aim of these different approaches was to achieve a certain fire resistance measured in units of time. A building element had to be made to survive the load of a standard fire test for a certain time, and the time depended upon the type of building (especially its height), and location of the element in the building. Typically the time might be 30 minutes for a two-story building or 60 minutes for a taller one. These times were deemed long enough for people to be evacuated and the fire itself to be suppressed. The whole issue of protecting the building structure thus usually became reduced to the relatively simple idea of providing sufficient "cover" to any load-bearing steel, whether a beam or the reinforcing bars in a reinforced concrete slab (Figure 3). However, as with many codified approaches to engineering design, this simplicity led to an unjustified feeling of comfort. Terrible fires still broke out and both collapses and deaths were still common. The matter was brought to a head in England by two terrible fires—the Summerland disaster in 1973 in

353

which 51 people, including many children, died in a fire in a leisure center, and a fire at Bradford football stadium in 1985. Both had been designed in accordance with current good practice.

These disasters provoked a fundamental rethink to achieving fire safety of building structures and, from the 1970s, an entirely new approach was developed. As already developed in other branches of engineering, it was based on mathematical models of the various phenomena involved. For the first time, attention was focused on the crucial parameter—the temperature of the metal in a building element subjected to a fire. The idea of a "fire load" was proposed in the 1920s by S.H. Ingberg and is typically measured in kilograms of wood; that is, the load on the building caused by burning so much timber. This, in many ways, is analogous to a gravity or wind load acting on a building. Knowing the thermal properties of the building materials, and with the newly available power of the computer, engineers were able to predict the thermal response of a structure in ways similar to predicting its structural response under gravity and wind loads. After much experimental testing to validate the mathematical models, it also soon became possible, for the first time, for engineers to model the behavior of fires themselves and thus understand and predict the behavior of buildings during fires. Computer

modeling, in the form of computational fluid dynamics (CFD), can now also be used to study the temperatures of the building structure in a fire as well as the flow of hot gases in fires (Figure 4). The computer modeling of fires in buildings is now even able to model the movement or flow of people escaping from fires so that escape times can be more precisely predicted (Figure 5).

These various modeling techniques, together with a full risk analysis of a fire situation, are now collectively called "fire engineering" and represent what has been, perhaps, a quiet revolution in building design. Yet without it, we would not have the dramatic, exposed-steel structures that are now a relatively common sight. The Pompidou Center in Paris, conceived in the early 1970s, was one of the first such buildings (Figure 6). The ability to model the fire load and the structural response to this load allowed the design engineers to adopt the unusual idea of achieving fire resistance by filling the main columns with water which, in a fire, would be pumped around to remove heat from the steel to prevent it heating up too quickly. More common nowadays are the many buildings in which exposed steel can be used in a rather understated way, and the fire engineering approach to design can mean that the need for applied fire protection can be avoided altogether (Figure 7).

Figure 7.   Bedfont Lakes near London (1993; engineered by Buro Happold). As a result of fire engineering design, none of the exposed steel needs fire protection.
[*Photo: Bill Addis.*]

See also **Concrete Shells; Constructed World**

W. A. ADDIS

Further Reading

Addis, W. Creativity and Innovation: The Structural Engineer's Contribution to Design. Architectural Press, Oxford, 2001.

Buchanan, A.H., Ed. *Fire Engineering Design Guide*. Centre for Advanced Engineering, University of Canterbury, New Zealand, 1994.

Drysdale, D. *Fire Dynamics*. Wiley, New York, 1998.

Hamilton S.B. *A Short History of the Structural Fire Protection of Buildings*. National Building Studies Special Report No. 27. Her Majesty's Stationery Office, London, 1958.

Ingberg, S.H. Tests of the severity of building fires. *National Fire Protection Association Quarterly, Boston*, 22, 1, 43–61, 1928.

Malhotra, H.L. *Fire Safety in Buildings*. Building Research Establishment, Garston, 1986.

Read, R.E.H. *A Short History of the Fire Research Station, Borehamwood*. Building Research Establishment, Garston, 1994.

Stollard, P. and Johnson, L. *Design Against Fire: An Introduction to Fire Safety Engineering Design*. Spon Press, London, 1994.

Fire Safety in Tall Buildings. Council on Tall Buildings and Urban Habitat, Lehigh University.

SFPE Handbook of Fire Protection Engineering, 2nd edn. SFPE & NFPA, New York, 1995.

Fish Farming

Controlled production, management, and harvesting of herbivorous and carnivorous fish has benefited from technology designed specifically for aquaculture. For centuries, humans have cultivated fish for dietary and economic benefits. Captive fish farming initially sustained local populations by supplementing wild fish harvests. Since the 1970s, aquaculture became a significant form of commercialized farming because wild fish populations declined due to overfishing and habitat deterioration. Growing human populations increased demand for reliable, consistent sources of fish suitable for consumption available throughout the year.

The United Nations Food and Agricultural Organization Food and Agriculture Organization (FAO) stated that fish farming productivity expanded more than 200 percent from 1985 to 1995. By the end of the twentieth century, fish farming provided approximately one third of fish eaten globally. Fish were a major protein source in developing countries where 85 percent of the world's fish farms exist. Farm-produced fish dominated agricultural production worldwide, outnumbering some types of livestock raised for meat. By 2000, the international aquaculture industry was estimated to value almost $50 billion.

In addition to contributing to markets, aquaculture created employment opportunities especially for women and unskilled laborers in Third World countries. Asia is the leading producer of farm-raised fish. China, where fish farming first occurred three centuries ago, grows 70 percent of farmed fish. In developing countries, fish farms offer people innovative means to enhance nutrition and to accrue profits. For example, South Pacific fish farmers drill holes in black-lipped oysters, *Pinctada margaritifera*, and submerge them on cords into water where the oysters form pearls around beads placed in their tissues. Because they offer alternative money sources, pearl farms reduce overfishing.

Technology appropriate for specific regions and species was crucial as fish farming industrialized. Aquaculture includes cultivation of freshwater and marine fish as well as mollusks, crustaceans, and aquatic plants. Fish farming relies on various engineering techniques to create functional artificial environments with suitable water quality to raise fish. Agricultural and civil engineers, chemists, and marine biologists collaborate, and the journal *Aquacultural Engineering* emphasizes this interdisciplinary approach.

Filtration, circulation, and oxygenation to provide adequate aeration and waste removal in tanks, man-made ponds, and pens are necessary to raise healthy fish. Some fish are cultivated in offshore floating pens. After they hatch, they are fed formula. Automated feeders and seal-scaring devices insure constant care. Fish farming protects fish from predators and minimizes exposure to diseases. Employees at the Dachigam National Park trout farm in Kashmir squeeze out eggs to hatch and raise fish in tanks prior to releasing them in the running waters of a contained stream.

Several fish farms have geothermally heated waters, which are warmed by power plant waste heat. Because the temperature of artificial fish habitats are regulated, fish can be raised indoors in facilities located in environments colder than their natural habitats. Indoor fish farming enables fish to be cultivated in urban areas. High yields can be produced in small spaces.

Initially, fish raised in captivity represented luxury species such as shrimp. In Europe and North and South America, salmon are raised in net pens. Other popular domesticated species include carp, cod, bass, and perch. Farm-raised fish are usually tastier and exhibit a greater consistency of size and quality than wild fish.

Because they are cold-killed in cool water, farm fish have a shelf life of at least two weeks. Fishing farms range in size and serve numerous purposes. Some backyard fish farms exist solely to feed owners' families, produce fish bait, generate extra revenues in local markets, or offer recreational fishing. Commercial fish farms raise fish for food, bait, aquarium pets, or for sport. Eggs, fry, and fingerlings are also sold to fisheries and to stock ponds. Oxygenated tanker trucks deliver fish to processors. Approximately 20 million trout are cultivated annually on U.K. fish farms. Catfish farming is widespread in the U.S.

Integrated fish farming (IFF) is a widely distributed technology. Farmers combine fish culture with other types of agriculture, such as raising rice or ducks either simultaneously or rotating cycles in the same water habitats as fish, especially when land and water resources are limited. In fields, fish live in furrows, pits, or ditches while rice grows on ridges. By utilizing farmland for multiple purposes, food and wastes are recycled and yields increase. Fish eat weeds and fertilize fields.

Biotechnology improves the quality of farm-raised fish by striving to produce hybrids and strains of fish with certain traits such as being disease-resistant and more resilient to cold climates. Researchers also apply technology to create genetically bigger fish that grow more quickly and need less oxygen to survive. Some transgenic fish produce more muscle from grain than their wild counterparts. The tilapia fish, indigenous to Africa, has been targeted for selective breeding experiments because it naturally thrives on wastes in poor habitats and requires minimal maintenance. The selectively bred strain reaches harvest maturity of more than 800 grams (1.75 pounds), has higher survival rates, and can reproduce when it is four months old, enabling as many as three harvests annually. The experimental channel catfish strain, NWAC-103, also attains maturity faster than other catfish. Conservation breeding is pursued to preserve endangered fish species. Fish breeding has demonstrated potential but lags behind genetic research and achievements applied to other livestock.

Computer technology has helped advance fish farming through the use of monitoring and security programs. Software such as FISHY tracks fish farm production. Electronic identification chips are used to tag farm fish. Satellite technology is used to determine climatic conditions such as weather patterns, atmospheric circulation, and oceanic tides that affect fish farms.

Agrichemicals serve as disinfectants, vaccines, and pharmaceuticals to combat bacterial infections, prevent disease, and kill parasites. Hormones are sometimes used to boost growth. Because carnivorous farm-raised fish consume large amounts of fish, researchers seek alternative feed sources to ease demands that have contributed to overfishing.

Fish farming technology can be problematic. If genetically engineered fish escape and mate with wild fish, the offspring might be unable to survive. Cultivated fish live in crowded tanks that sometimes cause suffocation, diseases, and immense amounts of waste and pollutants. Antibiotic use can sometimes result in resistant microorganisms. Coastal fish farms, especially those for shrimp, can be environmentally damaging if adjacent forests are razed.

See also **Breeding, Animal: Genetic Methods; Genetic Engineering, Applications**

ELIZABETH D. SCHAFER

Further Reading

Beveridge, MC.M., and McAndrew, B.J., Eds. *Tilapias: Biology and Exploitation*. Kluwer, Dordrecht, 2000.

Black, KD. and Pickering A.D., Eds. *Biology of Farmed Fish*. Sheffield Academic Press, Sheffield, 1998.

Ennion, SJ. and Goldspink, G., Eds. *Gene Expression and Manipulation in Aquatic Organisms*. Cambridge University Press, Cambridge, 1996.

Lahlou, B. and Vitiello, P., Eds. *Aquaculture: Fundamental and Applied Research*. American Geophysical Union, Washington D.C., 1993.

Mathias, J.A., Charles, A.T. and Bao-Tong, H., Eds. *Integrated Fish Farming*. CRC Press, Boca Raton, FL, 1998.

Pennell, W. and Barton, B.A., Eds. *Principles of Salmonid Culture*, Elsevier, New York, 1996.

Reinertsen, H., Dahle, L.A., Joergensen, L. and Tvinnereim, K., Eds. *Fish Farming Technology*. Balkema, Rotterdam, 1993.

Reinertsen, H. and Haaland, H., Eds. *Sustainable Fish Farming*. Balkema, Rotterdam, 1996.

Sedgwick, S.D. *Trout Farming Handbook*. Fishing News Books, Oxford, 1995.

Stickney, R.R. *Aquaculture in the United States: A Historical Survey*. Wiley, New York, 1996.

Treves-Brown, K.M. *Applied Fish Pharmacology*. Kluwer, Dordrecht, 2000.

Fission and Fusion Bombs

Fission weapons were developed first in the U.S., then in the Soviet Union, and later in Britain, France, China, India, and Pakistan. By the first decade of the twenty-first century, there were seven countries that announced that they had nuclear weapons, and another three or four suspected of developing them.

The first atomic weapon tested was Fat Man, the plutonium weapon designed at Los Alamos and detonated at the Trinity test, Alamogordo, New Mexico, 16 July 1945. A weapon on the Fat Man design was dropped 8 August 1945 on the city of Nagasaki Japan, two days after the dropping of the uranium-fueled weapon, Little Boy, on Hiroshima. The Soviets developed their first nuclear weapon utilizing information gathered through espionage on the American program, much of it supplied by the German-born British physicist, Klaus Fuchs. The Soviets tested a duplicate of Fat Man in a test dubbed "Joe 1" by the American press in August 1949. The British cabinet decided to build nuclear weapons in January 1947, and the first British test was held in Australia in 1952. In the British program of testing weapons above ground, altogether there were 21 devices detonated between 1952 and 1958, 12 in or near Australia and 9 at Christmas Island. In addition to announced tests, the British tested many weapons components at Maralinga, the Australian test range.

Several designs of hydrogen bombs or thermonuclear bombs that relied on nuclear fusion rather than on fission for the majority of their energy release were considered in the period 1946–1955 by both the U.S. and the Soviet Union. In some designs, tritium, an isotope of hydrogen, was used to create a fusion effect that would more thoroughly cause fission in the plutonium core of the weapon, an effect known as "boosting."

In the late 1940s, American weapons designers thought through and abandoned the Alarm Clock design in which a fission weapon would ignite a deuterium fusion fuel. The device could be known as a fission–fusion–fission weapon, with layers of fissionable material, including some uranium-238 that would fission in the intense neutron environment. The upper limit of such a weapon would be several hundred kilotons. That design was not pursued or built.

The U.S. pursued the design of a cooled device using liquid deuterium, in the Ivy-Mike test of October 26, 1952. Later, the Teller–Ulam design worked out by Edward Teller and Stansilaw Ulam was far smaller and could be transported by aircraft. Teller and some others called the later design the "Classical Super."

The Soviets pursued a weapon similar to the U.S. Alarm Clock design, called the "Sloyka" or "Layer Cake" design, first conceived by Yakov Zel'dovich as the "First Idea" and improved by Andrei Sakharov. Sakharov's "Second Idea" involved surrounding the deuterium with a uranium-238 shell, in the layers or "sloyka," as Sakharov called it, to increase the neutron flux during the detonation. In a Russian pun, colleagues referred to the Layer Cake design as having been "sugarized," as the name Sakharov means "of sugar."

The Soviet scientists tested the Layer Cake with the Joe 4 test of 12 August 1953 that yielded 400 kilotons, with 15 to 20 percent of its power derived from fusion. The U.S. already had larger boosted weapons, and American analysis of the Joe 4 test concluded that the device in the Joe 4 test was simply a variation on a boosted fission weapon. At the Russian weapons museum at Arzamas, the Joe 4 weapon is labeled the world's first hydrogen bomb, while U.S. analysts have continued to regard it as a boosted fission weapon. Thus the question of whether a boosted weapon can be regarded as a fusion weapon is crucial to the issue of priority of invention.

The U.S. tested a new design that could be delivered as a weapon, with the Teller–Ulam design in the Castle/Bravo test held in 1954. That device yielded an estimated 15 megatons. The "Third Idea" developed by Igor Tamm and Sakharov, like the Teller-Ulam device, could go much higher in yield than the limits imposed by the Layer Cake idea, and the later Soviet tests of hydrogen bombs followed the Third Idea, with true, two-stage detonations. The first Soviet test of the Third Idea weapon came 22 November 1955 with a yield estimated at 1.6 megatons. Later models of this weapon had a yield of 3 megatons. The accuracy of some of this information cannot be confirmed from official sources, as the material is derived from generally scholarly but unauthorized open literature.

Some observers believed that Joe 4 represented a thermonuclear weapon, just as the Soviets asserted. If that had been true, it would appear that the Soviets developed a deliverable nuclear weapon about seven months before the U.S. did so with the 1954 Castle/Bravo test of the huge low-temperature fusion device, too large to carry as a weapon aboard an aircraft. Careful evaluation of the Joe 4 test revealed that it should be regarded as a boosted fission weapon, somewhat along the lines of that detonated in the Greenhouse/Item test by the United States in 1952. By such logic, American scientists could claim that the Classical Super test of Ivy/Mike in 1952 was the first true thermonuclear detonation, and that the Castle/Bravo Teller–Ulam weapon tested in 1954 was the world's first deliverable thermonuclear weapon. That remains the American view of invention priority.

See also **Nuclear Reactors, Weapons Material; Warfare; Warfare, High Explosive Shells and Bombs**

RODNEY CARLISLE

Further Reading

Arnold, L. *A Very Special Relationship*. Her Majesty's Stationery Office, London, 1987.

Cochrane, T.B., *et al. Making the Russian Bomb:From Stalin to Yeltsin*. Westview Press, Boulder, CO, 1995.

Hewlett, R.G. and Anderson, O. The New World, 1939–1946, vol. 1: History of the United States Atomic Energy Commission. Atomic Energy Commission, Washington D.C., 1972.

Hewlett, R.G., and Duncan, F. *Atomic Shield, 1947–1952, vol. 2: History of the United States Atomic Energy Commission*. Atomic Energy Commission, Washington D.C., 1972.

Hewlett, R.G., and Holl, J.M. *Atoms for Peace and War, 1953–1961*. University of California Press, Berkeley, 1989.

Holloway, D. *Stalin and the Bomb*. Yale University Press, New Haven, 1994.

Rhodes, R. *The Making of the Atomic Bomb*. Simon & Schuster, New York, 1986.

Rhodes, R. *Dark Sun*. Simon & Schuster, New York, 1995.

Fly-by-Wire Systems

Most civilian and military aircraft are today controlled by "fly-by-wire" systems, in which controls are operated electrically, permitting greater maneuverability and safety than in conventional mechanically based control systems. Conventional flight control works as the pilot transmits directions through the control devices and the control surfaces are actuated according to the pilot's input. In larger airplanes, there is also a hydraulic system, like the power steering on an automobile, to actuate the control surfaces.

A fly-by-wire system breaks the mechanical lineages—a computer is added. Pilot input through the stick and rudder pedals are converted from a deflection to a voltage proportional to the deflection, and then the voltage is passed on to the computer, where it is further converted to binary numbers. Air data sensors are also connected to the computer. The computer then compares the pilot's desires to the actual state of the airplane and sends signals to reconcile them. These signals are converted back into voltages and used by electric servos at the hydraulics to move the control surfaces (see Figure 8).

Conventional systems depend on the pilot not doing something that is a bad idea, such as lifting the nose too much and causing a stall. In airplanes that are fly-by-wire, even if the pilot makes a mistake such as pulling up the nose too far or too fast, the computer can be programmed to prevent the airplane from responding in a dangerous way. This decoupling of the pilot from the airplane has to be lessened for fighter pilots, as they require increased maneuverability, not increased safety. Since many fly-by-wire airplanes are unstable in the pitch axis, it is simple to increase maneuverability using fly-by-wire.

During the first few decades of flight, airplanes were unstable. Originally there was a dispute about whether airplanes would be flown by "chauffeurs" or "pilots" The former would merely adjust the pointing of stable airframes, while the latter would have to successfully fight the instability of designs to keep the airplane in the air. At first, inventors who were would-be aviators adopted the chauffer model. When the Wright brothers realized that it was possible to pilot an unstable airplane, they flew. Their devices, plus most World War I fighters, were unstable and required great concentration to

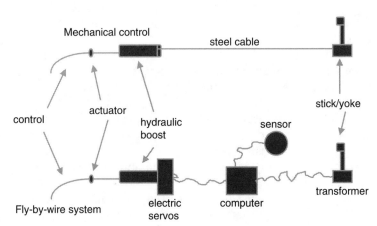

Figure 8. Fly-by-wire: how it works.

fly since the pilot's brain was the only "computer" and the seat of the pilot's pants the most important sensor on board. This was acceptable when fuel tanks were small and ranges limited. When airplanes were built to traverse long distances, they were designed to be stable in order to reduce the workload on their pilots. After decades of stable designs like these, once again the advantages of instability are available.

The Germans invented a sort of accidental fly-by-wire system to control part of their Mistel flying bomb during World War II. They filled worn-out Junkers-88 bombers with explosives, and a Messerschmidt-109 or Focke Wulf-190 fighter was then mounted on top. The fighter's pilot would steer this composite aircraft, a large three-engine biplane, to a target, aim it, and then release it to fly to the target. This was usually a dam or bridge with considerable concrete support. Junkers-88s were not commonly built with autopilots, but one was available. By placing potentiometers at the stick and rudder of the fighter, movements could be converted to electrical signals that were fed to the bomber's autopilot. Thus one man could fly both. The difference between the Mistel and a modern fly-by-wire aircraft is the presence of a computer.

Computers can be of two types: analog and digital. An analog computer operates on continuous functions, like voltage on a wire. A digital computer operates on discrete functions, so a voltage has to be converted to bits (ones and zeros) to be processed. In some ways, analog computers are better for airplane use, especially in that they accept continuous control inputs and do not need conversions. Digital computers are programmable with software and thus are more flexible. Analog computers have to be rebuilt to handle new flying characteristics; digital computers need only change programming. The first computers used in airplanes were analog. By the mid-1950s, most aeronautical engineers were convinced of the benefits of fly-by-wire. Only its practical implementation needed to be proved.

At about this time, Avro Canada was building a large, supersonic interceptor to help defend the country from Soviet nuclear attack by bombers. In American interceptors like the F-102 and the F-106, control passed to a ground-based site shortly after take-off. The ground controller would make the intercept, fire the weapons, and later pass control back to the pilot to complete the return to base and the landing. It would be far easier to make an interface with this type of system if the airplane's control system were electronic. Perhaps for this reason, or simply to achieve more

maneuverability in their unwieldy airframe, the Canadians made the CF-105 Arrow fly by wire. The five prototypes of the Arrow made 66 flights in 1958–1959, so even though the fly-by-wire system was not perfected, this airplane featured fly-by-wire as it has come to be used.

In the 1960s in the U.S., the control division at Wright-Patterson Air Force Base sponsored several low-cost research projects aimed at making fly-by-wire possible. About the middle of the decade, it salvaged the tail section from a crashed B-47 and experimented with fly-by-wire in the pitch axis on the ground. A flying B-47 lofted this same pitch controller for some test flights. Short on money, Wright-Patterson tried fly-by-wire as a survivability improvement. A study of fighter-bombers that were damaged in Vietnam but returned safely to base found that the damage was not in a certain limited area of the airplanes where a major junction of control cables for the mechanical flight control system was located. Presumably those planes that were brought down were damaged in that area. By emphasizing survival ability, engineers obtained funding and an aircraft similar to the ones studied in Vietnam. This F-4 was converted to an analog fly-by-wire system but retained the mechanical system for the first 27 flights as a backup. It first flew on 29 April 1972 under control of the mechanical system and then switched to fly-by-wire in flight.

Meantime, encouraged by positive results with computer control of the Gemini and Apollo spacecraft, the National Aeronautics and Space Administration (NASA) of the U.S., took a surplus Apollo computer and pioneered digital fly-by-wire by installing the computer on an F-8. This airplane first flew under fly-by-wire control about a month after the U.S. Air Force's F-4.

Even though Boeing had a fly-by-wire system on its YC-14 cargo plane prototype, its competitor, Airbus, built the first commercial aircraft with fly-by-wire, the A320. The entire family of derivatives: the A318, A319, A330, and A340, have similar systems. Therefore, a technology first pioneered in the 1970s was in common use less than 20 years later, a highly rapid pace of change.

Aside from reliability, fly-by-wire offers improvements in safety. At least three redundant systems or two dissimilar systems are now part of every fly-by-wire flight control system. For example, even though NASA's first digital computer-controlled airplane used a single-string (one computer) system, even it had three of the analog computers like those in its contemporary plane, the U.S. Air Force's modified F-4, as a backup. The second system installed on the NASA F-8 was triplex (three computers). This has been expanded to the point where the Boeing 777 has three strings each made up of three different computers, nine in all.

The Europeans have adopted a different method of redundancy. In Airbus fly-by-wire systems, there are two dissimilar strings. One controls the elevator and spoilers. The other controls the elevator and ailerons. The rudder is still mechanically controlled, as it is much less important to a commercial airplane. One of the systems has two computers; one has three. The machines contain software written by separate teams with dissimilar programming languages. The teams are even separated geographically. The basic idea is that two separate teams are unlikely to make the same errors, although this has not turned out to be the case. However, the dissimilar systems and redundant computers provide sufficient safety. Either method is a significant increase in overall safety.

See also **Aircraft Instrumentation; Warplanes, Fighters and Fighter Bombers**

JAMES E. TOMAYKO

Further Reading

Miller, F.L. and Emfinger, J.E. *Fly-by-Wire Techniques.* AFFDL-TR-67-53, U.S Air Force Flight Dynamics Laboratory, Wright-Patterson AFB, Ohio.

Morris, J. *Background Fly-by-Wire Historical Information.* U.S. Air Force Flight Dynamics Laboratory Memo, Wright-Patterson AFB, Ohio.

Tomayko, J. *History of NASA's Digital Fly-by-Wire Project.* SP-4224, NASA Washington D.C., 2000.

Tomayko, J. *Code Name Mistletoe.* American Heritage of Invention and Technology, , 2000, pp. 26–33.

Tomayko, J. Blind faith: The United States Air Force and the development of fly-by-wire technology, in *Technology and the Air Force.* Government Printing Office, Washington D.C., 1997, pp. 163–185.

Tomayko, J. The Fly-By-Wire Revolution in Flight Control. *American Heritage of Invention and Technology*, 1992.

Foods Additives and Substitutes

Advances in food and agricultural technology have improved food safety and availability. Food technology includes techniques to preserve food and develop new products. Substances to preserve and enhance the appeal of foods are called food additives, and colorings fit into this category of additives that are intentionally included in a processed food. All coloring agents must be proven to be safe and their use in terms of permitted quantity, type of food that can have enhanced coloring, and final level is carefully controlled.

Fat substitutes on the other hand are technically known as replacers in that they replace the saturated and/or unsaturated fats that would normally be found in processed food as an ingredient or that would be added in formulation of a processed food. Usually the purpose is to improve the perceived health benefit of the particular food substance. Technically speaking, substitutes are not additives but their efficacy and safety must be demonstrated.

The incorporation of a food coloring or fat substitute does not confer the food with the status synthetic food. That description would normally be restricted to a food that was made purely from chemically synthesized components. The regulation of what is permitted in food is complex and is governed in the U.S. by the Food and Drug Administration. In the U.K. the role of regulating food and enforcing the regulations is now the responsibility of the Food Standards Agency, which works with other government departments such as the Department of Health to ensure that food is safe and labeled appropriately. In the U.S. the law defines "Standards of Identity" covering exactly what ingredients can be used and where. At one time this exempted labeling, but now all ingredients must be labeled.

All synthetic colors must be demonstrated to be safe before addition to food. Safety tests are carried out in animal species to determine the highest level of a substance that has no deleterious effect. In the U.S. the amount that is permitted is 1 percent of this level. Other measures of safety include acceptable daily intake (ADI) level, tolerable limits for environmental and chemical contamination in food, and maximum residue levels (MRL). Since food colors are dispensable, they are the most heavily criticized group of food additive. They do not improve food safety in any way, but simply change or enhance its appearance. Therefore it is possible to require that their use entails no extra risk to the consumer. With other additives such as preservatives, for example, there may be a necessary compromise between the risk of using them and the risk of not using them.

Color additives are also used to help identify flavors, for example a lemon-colored sweet should taste lemony, black jelly beans should taste like licorice, the soft center of a chocolate should taste of oranges if it is orange but of raspberries or strawberries if it is pink. The use of a new synthetic coloring agent requires an application to be submitted that describes the chemical structure, processing route, range of intended use, and final concentrations. Prior to the discovery of synthetic dyes by Perkins (1856), only natural dyes were added to foods. Early synthetic dyes were metal based and included mercury and copper salts and arsenic poisons. Sometimes used to mask poor quality and even spoilage in food, many of these dyes were toxic. During the early part of the twentieth century a large number of cheaper dyes were synthesized, and many found their way into food products. Often these were complex chemicals and their physiological effect was not understood. These aniline or petroleum-based derivatives were covered by the U.S. Pure Food and Drug laws of 1906, which represented a landmark for food health and safety.

Once countries began to legislate what could be added to food, the number of natural and synthetic colors dropped markedly. By the late twentieth century there were major national differences in what is permitted. In Norway, the use of any synthetic coloring for food use has been banned since 1976. One source of great confusion is that natural dyes can be synthesized (the so-called nature-identical dyes). The coding for synthetic colors also underwent change; the E numbers for the European Union (EU) Code was being replaced by the Codex Alimentarius Commission system of coding, the so-called International Numbering System (INS). This largely follows the E numbers without the letter. In this system, tartrazine (C1 19140) is referred to as E102 in the U.K. and has an INS number of 102, but in the U.S. it is generally referred to as FD and C Yellow No. 5. It is a synthetic azo dye used in confectionery, fruit juices, canned fruits, brown sauces, pickles, pie fillings, and canned peas. It has an acceptable daily intake (ADI value) of 0 to 75 milligrams per kilogram of body weight. Tartrazine has been implicated in causing wakefulness in children and causing rhinitis, itching, and other allergenic symptoms. Individuals who wish to avoid tartrazine cannot simply avoid yellow-colored foods since tartrazine is also used to confer turquoise, green, and maroon colors. However U.S. and EU legislation requires that its use be listed on all food labels so that consumers can avoid it if they wish.

Amaranth E132 is a red food coloring made from synthetic coal-tar and azo dye. It was used in pudding mix, jelly crystals, and fruit fillings, but its use was banned in 1976 in the U.S. when a possible connection to malignant tumors was identified. However it is used in some countries with an average acceptable daily intake of 0 to 0.5 milligrams per kilogram of body weight as INS 123 (E123).

Approved food colorings are reviewed regularly and the ADI levels are amended accordingly. Reactions are rare (1 in 10,000 people are sensitive to tartrazine), but there is an Adverse Response Monitoring System (ARMS) in the U.S. and in most European countries.

Fat substitutes are compounds added in the formulation of complex foods that replace all or part of the fat and hence reduce the energy value and/or reduce the proportion of saturated fats. They also aid the digestion of foods. They must, by definition, function in the same way as the fat would in the body, for example act as binding agents or as humectants (giving food a moist texture). There are many fat substitutes available to the commercial food manufacturer. They are not required to be regulated as "novel foods" since the EU Novel Food Regulation (258/97) does not cover food processing aids. Low-fat spreads and cholesterol-lowering ingredients (containing phytosterols and phytostanols) are strictly speaking fat substitutes. Many of the low-fat spreads are whipped and have a high water content; hence they result in reduction in fat intake if they are replacing unmodified fat.

In addition to modifying food structure, providing more acceptable texture, imparting distinctive flavor, and adding to energy value, fats are often a source of fat-soluble vitamins (e.g., vitamins A and D). If an individual has an elevated cholesterol level, the use of a phytosterol or phytostanol spread to replace butter or margarine on bread will gradually reduce cholesterol levels. However, these products can cause gastrointestinal imbalance and are not recommended for babies, children, or those with conditions that might make them susceptible to vitamin A or D deficiency. One problem for consumers is that while it is clear what is in a package of these spreads, it is not always apparent which fats are incorporated in other foods such as yogurt, prepared baked goods, or restaurant meals. Hence it would be possible to unknowingly consume a daily intake above the recommended maximum. These issues remain of concern, and at the turn of the twenty-first century, the U.K. Food Standards Agency is looking at ways to monitor consumption of cholesterol-lowering spreads and prevent over consumption of specialized fats in diary goods by susceptible individuals.

See also **Food, Processed and Fast; Synthetic Foods, Mycoprotein and Hydrogenated Fats**

JANET BAINBRIDGE

Further Reading

Hochheiser, S. The evolution of U.S. food color standards, 1913–1919. *Agricult. Hist.*, 55, 4, 385–391, 1981.
Junod, SW. The chemogastric revolution and the regulation of food chemicals, in *Chemical Sciences in the Modern World*, Mauskopf, S., Ed. University of Pennsylvania Press, Philadelphia, 1993.

Useful Websites

U.S. Food and Drug Administration: Food Color Facts: http://vm.cfsam.fda.gov/-Ird/colorfac.html
FSA Website Homepage: www.food.gov.uk
FSA archive: www.archive.food.gov.uk

Food Preparation and Cooking

Twentieth century technological developments for preparing and cooking food consisted of both objects and techniques. Food engineers' primary objectives were to make kitchens more convenient and to reduce time and labor needed to produce meals. A variety of electric appliances were invented or their designs improved to supplement hand tools such as peelers, egg beaters, and grinders. By the close of the twentieth century, technological advancements transformed kitchens, the nucleus of many homes, into sophisticated centers of microchip-controlled devices. Cooking underwent a transition from being performed mainly for subsistence to often being an enjoyable hobby for many people.

Kitchen technology altered people's lives. The nineteenth-century Industrial Revolution had initiated the mechanization of homes. Cooks began to use precise measurements and temperatures to cook. Many people eagerly added gadgets to their kitchens, ranging from warming plates and toasters to tabletop cookers. Some architects designed kitchens with built-in cabinets, shelves, and convenient outlets to encourage appliance use. Because they usually cooked, women were the most directly affected by mechanical kitchen innovations. Their domestic roles were redefined as cooking required less time and was often accommodated by such amenities as built-in sinks and dishwashers. Ironically, machines often resulted in women receiving more demands to cook for events and activities because people no longer considered cooking to be an overwhelming chore.

Domestic technology contributed to home economics and dietetics becoming professional fields in the twentieth century. Home economists studied how household machines influenced cooking and advised people how to incorporate kitchen appli-ances into their lives. Guides instructed people how to utilize appliances to cook foods for entertaining groups. During the two world wars and economic depressions, people adjusted cooking techniques to cope with food shortages and rationing.

Throughout the twentieth century, inventors created new appliances to ease cooking burdens. When the century began, many kitchens in the U.S. still had wood-burning stoves or fireplaces. As electricity became available, some people invested in electric ranges. Costs limited mass acceptance, but gradually range costs became affordable. The standardization of electrical outlets, plugs, and currents in the 1920s and development of safety standards aided adoption of electric appliances. Electric stoves enabled cooks to bake goods without having to wait for a fire to warm sufficiently. These stoves were also cleaner than cooking on hearths. By the late 1970s, microwaves had replaced or supplemented stoves in many homes, altering how people prepared and cooked meals.

Cooks utilized a variety of appliances to prepare food for cooking. In the early twentieth century, engineers used small motors then magnetrons to create powerful kitchen appliances. Mixers quickly and smoothly combined dry goods with eggs, margarine, and other ingredients instead of people manually stirring dough. Crock pots and cookers enabled cooks to combine ingredients to cook unsupervised for a specified time. Automatic bread machines mixed, kneaded, raised, and baked breads. Coffee and tea makers brewed beverages timed for breakfast drinking and kept them warm. Espresso machines steamed frothy beverages. Cordless kettles heated liquids wherever people wanted to prepare hot drinks or soups. Juicers extracted liquid from fruits.

Some appliances were available only in certain geographical regions or met specific cultural needs such as rice steamers in Asia. As people traveled and encountered new devices, those technologies were often introduced to other countries. The Internet enabled people to become aware of and buy brands and types of appliances they might not find in local stores and are only available in specific countries or regions. Manufacturers such as Samsung and Toshiba produced appliances in Asia, while companies including DeLonghi and Bourgeois outfitted European homes. Innovators from many nations envisioned, adapted, and improved cooking tools.

Technologists worldwide created appliances to meet specific needs and local demand. The German manufacturer Miele produced the first electric dishwasher in Europe in 1929. Maurice Bourgeois invented the first European built-in oven, resulting in his company becoming the leader in the convection oven market. At Sunbeam, Ivar Jepson designed kitchen appliances between the World Wars. The Sunbeam Mixmaster patented in the late 1920s surpassed other mechanical mixers because it had two beaters with interlocking blades that could be detached from the machine. Previously, the popular mixer that L.H. Hamilton, Chester Beach, and Fred Osius patented in 1911 only had one beater. Attachments enabled the Mixmaster to perform other tasks, including peeling, grinding, shelling, and juicing. The Mixmaster also could polish and sharpen utensils and open cans.

Inventors devised various electric toasters designs during the twentieth century. Efforts to create a reusable, unbreakable heating element to produce radiant heat for toasting sliced bread stymied many people. Engineer Albert Marsh patented Nichrome, a nickel and chromium alloy, in 1905. His invention enabled toaster heating elements to be produced. By 1909, consumers could purchase electric toasters developed by General Electric. Ten years later, the first pop-up toaster was patented. By the 1980s, toasters were designed to accommodate bagels. Plastics were used in addition to metals for cases, and microchip controls monitored toasting options.

Modern appliances often had pre-twentieth century precedents. Denis Papin designed the first pressure cooker in 1679 France. Later engineers adapted his cooker to produce an airtight environment in which steam cooked food so that vitamins and minerals were retained. In 1939, the National Pressure Cooker Company first sold a cast iron saucepan pressure cooker called Presto. After World War II, pressure cookers were made from stainless steel. By the late 1950s, engineers designed electric cookware, which had removable heat controls so that the pans, griddles, skillets, and coffee makers could be immersed in water to clean.

From the 1970s, cooks also used appliances designed to produce small servings. Sandwich machines and small indoor grills quickly cooked meals for individuals. Electric deep fryers prepared single portions of onion rings, french fries, and other fried foods. In contrast, kitchen technology also offered healthier fare. Hot-air popcorn poppers did not use oil. By the 1980s, the electric SaladShooter sliced and shredded ingredients directly into bowls.

Cooking technology benefited from inventors' curiosity. In 1938 at a DuPont laboratory, Roy J. Plunkett discovered Teflon while investigating chemical reactions occurring in tetrafluoroethylene (TFE), a refrigerant gas. Gas molecules had bonded to form polytetrafluoroethylene (PTFE) resin, which had lubricating properties, a high melting point, and was inert to chemicals. The process to create this polymer was refined and patented. When cooking pots and pans are coated with this polymer, they have non-stick surface that makes foods such as eggs and batters easier to cook.

In 1946, Earl Tupper invented Tupperware, which transformed how people stored and prepared food. These plastic containers with airtight seals were light and unbreakable, inspiring food technologists worldwide to use plastic instead of glass and metal materials. Tupperware can be used for cooking in microwaves and was environmentally sounder than disposable plastic and aluminum foil wraps.

In the latter twentieth century, microprocessors and materials such as polymers were used to make appliances lighter and easier to use. Engineers strived to make appliances smaller, more versatile and stable, quieter, and requiring less energy to operate. Digital technology made cooking more convenient because appliances with timers and sensors could be programmed to perform certain functions at specific times.

Radio programs featured cooking programs that advised cooks. Television introduced people to such notable cooks as Nigella Lawson, Raymond Oliver, Catherine Langeais, and Julia Child. Through the medium of television and video, cooks could demonstrate preparation methods such as basting and stuffing and cooking techniques including sautéing and frying that cookbooks often insufficiently described for inexperienced cooks to follow adequately. Television personalities posted recipes on web sites. Restaurants and food-related industries used the Internet to inform consumers how to make favorite meals at home and use specific products.

See also **Food Preservation, Cooling and Freezing; Food, Processed and Fast; Microwave Ovens**

ELIZABETH D. SCHAFER

Further Reading

Anderson, J. *The American Century Cook Book: The Most Popular Recipes of the 20th Century*. C. Potter, New York, 1997.
Cowan, R.S. *More Work for Mother: The Ironies of Household Technology from the Open Hearth to the Microwave*. Basic Books, New York, 1983.
Davidson, A. *The Oxford Companion to Food*. Oxford University Press, Oxford, 1999.

Du Vall, N. *Domestic Technology: A Chronology of Developments.* G.K. Hall, Boston, 1988.

Kiple, KF. and Ornelas, K.C., Eds. *The Cambridge World History of Food*, 2 vols. Cambridge University Press, Cambridge, 2000.

Macdonald, A.L. *Feminine Ingenuity: Women and Invention in America.* Ballantine Books, New York, 1992.

MacKenzie, D. and Wajcman, J., Eds. *The Social Shaping of Technology: How the Refrigerator Got Its Hum.* Open University Press, Philadelphia, 1985.

McBride, T.M. *The Domestic Revolution: The Modernization of Household Service in England and France, 1820–1920.* Croom Helm, London, 1976.

Plante, E.M. *The American Kitchen, 1700 to the Present: From Hearth to Highrise.* Facts on File, New York, 1995.

Shapiro, L. *Perfection Salad: Women and Cooking at the Turn of the Century.* Farrar, Straus & Giroux, New York, 1986.

Symons, M. *A History of Cooks and Cooking.* University of Illinois Press, Urbana, 2000.

Trescott, M.M., Eds. *Dynamos and Virgins Revisited: Women and Technological Change in History.* Scarecrow, Metuchen, NJ, 1979.

Williams, C., Ed. *Williams-Sonoma Kitchen Companion: The A to Z Everyday Cooking, Equipment and Ingredients.* Time-Life Books, Alexandria, VA, 2000.

Wolke, R.L. *What Einstein Told His Cook: Kitchen Science Explained.* W.W. Norton, New York, 2002.

Yarwood, D. *Five Hundred Years of Technology in the Home.* B.T. Batsford, London, 1983.

Food Preservation: Cooling and Freezing

People have long recognized the benefits of cooling and freezing perishable foods to preserve them and prevent spoilage and deterioration. These cold storage techniques, which impede bacterial activity, are popular means to protect food and enhance food safety and hygiene. The food industry has benefited from chilled food technology advancements during the twentieth century based on earlier observations. For several centuries, humans realized that evaporating salt water removed heat from substances. As a result food was cooled by placing it in brine. Cold storage in ice- or snow-packed spaces such as cellars and ice houses foreshadowed the invention of refrigerators and freezers.

Before mechanical refrigeration became consistent, freezing was the preferred food preservation technique because ice inhibited microorganisms. Freezing technology advanced to preserve food more efficiently with several processes. Blast freezing uses high-velocity air to freeze food for several hours in a tunnel. Refrigerated plates press and freeze food for thirty to ninety minutes in plate freezing. Belt freezing quickly freezes food in five minutes with air forced through a mesh belt. Cryogenic freezing involves liquid nitrogen or

Freon absorbing food heat during several seconds of immersion.

In rural areas and small towns in the U.S., meat from butchering done by area residents was commonly frozen and stored in food lockers until after World War II. Refrigerators and freezers improved the quality of rural residents' lives and preserved milk from dairies, eggs from poultry, and meat from livestock. Worldwide, countries passed sanitation laws requiring specific refrigeration or freezing standards for foods. Refrigerators were miniaturized to become portable. Propane gas refrigerators were used to preserve food in recreational vehicles. By the end of the twentieth century, sophisticated refrigerators had several cooling zones within one unit to meet varying food needs such as chilling wine and crisping lettuce. Computer technology controlled these refrigerators' processes.

In the 1920s, government naturalist Clarence Birdseye (1886–1956) developed frozen foods by flash-freezing them. Inspired by watching Arctic people preserve meat in frozen seawater, Birdseye decided that quick freezing was why food remained edible. Fish cells were not damaged by ice crystals because freshly caught fish were frozen so quickly. Using ice, brine, and fans, Birdseye froze food rapidly in waxy boxes placed in a high-pressure environment. His freezing methods enabled foods to retain their flavors and textures. The foods' nutritional qualities were not altered. He established Birdseye Seafoods Inc., in 1924. Birdseye later sold his patents and trademarks for $22 million to the Goldman–Sachs Trading Corporation and Postum Company (later, the General Foods Corporation).

Birdseye's invention allowed people to consume a healthy diet of fresh vegetables and fruits out of season and at their convenience. He carefully marketed his frozen fare for retail sales. At Springfield, Massachusetts, consumers bought the first Birds Eye Frosted Foods. Despite the economic depression, consumers readily accepted buying frozen foods, including meat, vegetables, fruit, and seafood. Birdseye developed refrigerated display cases for grocery stores in 1930 and began manufacturing the cases four years later. By 1944, he shipped his frozen products in refrigerated boxcars, and they became popular throughout the U.S. His efforts served as a transportation and distribution model for other food companies.

Birdseye's inventiveness and marketing of frozen food inspired other food entrepreneurs globally. More types of food were frozen, including soups and pizza. Some frozen foods were packaged in foil. Frozen food production totaled 1.1 billion kilograms annually during the 1940s. Governmental agencies such as the U.S. Food and Drug Administration set frozen food standards. The National Association of Frozen Food Packers (NAFFP), later renamed the American Frozen Food Institute, was established.

By the 1950s, almost two thirds of American grocery stores had frozen food sections. Precooked and boil-in-the-bag frozen foods joined frozen raw foods in displays. In 1954, Gerry Thomas first sold Swanson TV dinners. These complete meals in a tray that were heated in ovens appealed to consumers. Approximately ten million TV dinners were sold within the first year of availability. Commercially popular, frozen foods injected billions of dollars into the global economy. Airlines and restaurants relied on frozen foods for customers' meals. The Cold War motivated the U.S. Federal Civil Defense Administration to include frozen food in an atomic bomb test called Operation Cue, which determined that radiation did not affect frozen foods.

Frozen foods gained new popularity in the late twentieth century when fast-food restaurants such as McDonald's and Burger King prepared frozen items fresh for each customer. Increased variety, smaller servings for one person, low-calorie foods, and microwave ovens encouraged people to rely on frozen foods as meal staples especially as more women join the workforce. Government agencies publicly declared that frozen foods were nutritious. Many stores expanded refrigerated sections to meet customers' demands. Cryogenic railcars were invented to distribute frozen foods. Luxuries such as ice cream and chilled drinks have become commonplace due to refrigeration and freezer technology.

See also **Food Preparation and Cooking; Food Preservation: Freeze-Drying, Irradiation, and Vacuum Packing; Food, Processed and Fast; Transport, Foodstuffs**

ELIZABETH D. SCHAFER

Further Reading

Cleland, A.C. Food Refrigeration Processes: *Analysis, Design, and Simulation.* Elsevier, New York, 1990.

Dellino, C.V.J., Ed. *Cold and Chilled Storage Technology.* Van Nostrand Reinhold, New York, 1990.

Gormley, T.R., Ed. *Chilled Foods: The State of the Art.* Elsevier, New York, 1990.

Gormley, T.R. and Zeuthen, P. Eds. *Chilled Foods: The Ongoing Debate.* Elsevier, New York, 1990.

Mallett, C.P., Ed. *Frozen Food Technology.* Blackie Academic & Professional, London, 1993.

Stringer, M., and Dennis, C., Eds. *Chilled Foods: A Comprehensive Guide,* 2nd edn. CRC Press, Boca Raton, FL, 2000.

Wiley, R.C., Eds. *Minimally Processed Refrigerated Fruits and Vegetables.* Chapman & Hall, New York, 1994.

Food Preservation: Freeze-Drying, Irradiation, and Vacuum Packing

Humans have used processes associated with freeze-drying for centuries by placing foods at cooler high altitudes with low atmospheric pressure where water content is naturally vaporized. Also called lyophilization, freeze-drying involves moisture being removed from objects through sublimation. Modern freeze-drying techniques dehydrate frozen foods in vacuum chambers, which apply low pressure and cause vaporization.

Freeze-drying reduces foods' weight for storage and transportation. Freeze-dried foods do not require refrigeration but do need dry storage spaces. Adding water to freeze-dried foods reconstitutes them. Unlike dehydration, freeze-drying does not remove components that give foods flavors. The nutritional qualities of foods are also retained. The process inhibits microorganisms and chemical reactions causing food to spoil because the water pathogens and enzymes need to thrive is absent.

Engineers developed freeze-dryers for specific tasks that use vacuum pumps and chambers to pull moisture out of food. Commercially, freeze-dried foods are popular because they require less storage space than other packaged food. Freeze-drying extends the shelf life of products. Freeze-drying technology has been applied to consumer food products since the 1930s. Coffee, first freeze-dried in 1938, is the most familiar commercial freeze-dried food product. The Nestle Company was the first to freeze-dry coffee because Brazil requested assistance to deal with coffee bean surpluses. The successful freeze-drying of coffee resulted in the creation of powdered drinks such as Tang, which contrary to popular belief was invented by General Foods not the National Aeronautics and Space Administration (NASA). Many types of soups and noodles are freeze-dried. Researchers develop and patent new freeze-drying processes to improve and vary foods for consumers.

Freeze-drying was appropriated as the best method to preserve food for astronauts in space flight. During his pioneering February 1962 orbit, John Glenn was the first human to eat food in space. When he expressed his dissatisfaction, food engineers attempted to improve the taste of space food while maintaining its nutritional qualities and minimizing its weight. As space technology advanced, more elaborate freeze-dried meals provided variety for longer duration space missions, which required compact, lightweight food cargo sufficient to feed crews at the international space station.

As early as 1963, radiation was used to control mold in wheat flour. The next year, white potatoes were irradiated to inhibit sprouting. By 1983, the Institute of Food Technologists released a study about the potential of radiation to sterilize and preserve food. Those experts declared that food processors might be reluctant to accept expensive irradiation technology unless they were confident that consumers would purchase irradiated goods. The Institute of Food Technologists warned that prices of irradiated food had to be affordably competitive with nonradiated foodstuffs and fulfill people's demands.

Irradiation is less successful than freeze-drying. Prior to irradiation, millions of people worldwide became ill annually due to contaminated foods with several thousand being hospitalized or dying due to food-borne pathogens. By exposing food to an electron beam, irradiation enhances food safety. Irradiated human and animal feed, especially grain, can be transported over distances and stored for a long duration without spoiling or posing contamination hazards. The radura is the international food packaging symbol for irradiation.

Small doses of radiation alters microbe DNA and kills approximately 99.9 percent of bacteria and parasites in meats. This exposure does not alter nutrients. Irradiation permits people to consume slightly cooked meats, including rare steaks and hamburgers. Most food irradiation uses cobalt-60 isotopes, but researchers developed alternative techniques such as gamma rays from cesium-137 and linear accelerators that transform electrons aimed at food into x-rays, which have greater penetration than electron beams. Those beams consist of high-energy electrons expelled from an electron gun and can only penetrate several centimeters compared to gamma and x-rays reaching depths of several feet. Irradiation sources are kept in water tanks that absorb the radiation until they are used to sterilize food in a thick concrete chamber.

Countries in North America, Europe, Asia, Africa, and the Middle East accepted irradiation. The World Health Organization endorsed irradiated food as safe for consumption. The U.S. Food and Drug Administration (FDA) approved irradiation of pork, fruit, vegetables, spices, and herbs in 1986 and poultry in 1990. Eight years later, the FDA approved the process of irradiating red meat to kill dangerous microorganisms and pests and slow meat spoilage. Despite federal approval, some states banned irradiation. During the 1990s, scientists improved irradiation methods to destroy toxins and bacteria, especially *Escherichia coli*, *Salmonella*, *Shigella*, and *Campylobacter*, in foods. The Institute of Food Technologists published a document that noted that the FDA's endorsement of irradiation legitimized that food preservation technology. Astronauts routinely eat irradiated food to prevent food-related sicknesses in space.

Despite irradiation's benefits, some consumers boycott purchasing or consuming irradiated foods that they consider dangerous and insist instead that facilities where agricultural goods are processed should be sanitized to reduce threats of contaminating toxins. Irradiation supporters assert that public opinion parallels how people initially reacted to milk pasteurization before accepting that process. Studies concerning how people and animals react to correctly irradiated food indicate that the foods are safe and not radioactive. Irradiation facilities are regulated by government licensing, and workers undergo rigorous training. Fatal accidents have occurred only when workers ignored rules regarding exposure to radioactive materials.

Irradiation has several significant limitations. Viruses are too small for irradiation dosages appropriate for safe food handling. Prions linked to bovine spongiform encephalopathy lack nucleic acid thus making irradiation ineffective.

Vacuum-packing food technologies involve a process that removes empty spaces around foods being packaged. Vacuum technology uses environments artificially modified to have atmospheric pressures that are lower than natural conditions. Vacuum packing extends the shelf life of food. The U.K. Advisory Committee on the Microbiological Safety of Foods warned that anaerobic pathogens such as *C. botulinum* can grow in vacuum-packed foods. Because vacuum packing often results in rubbery sliced cheese, some manufacturers use the modified atmosphere packaging (MAP) system, which utilizes gases to fill spaces so that cheese can mature to become tastier inside packaging.

See also **Food Preparation and Cooking; Food Preservation, Cooling and Freezing; Food, Processed and Fast; Transport, Foodstuffs**

ELIZABETH D. SCHAFER

Further Reading

Brennan, J.G. *Food Engineering Operations*, 3rd edn. Elsevier, London, 1990.

Dalgleish, J.M. *Freeze-Drying for the Food Industry*. Elsevier, London, 1990.

Jennings, T.A. *Lyophilization: Introduction and Basic Principles*. Interpharm Press, Englewood, CO, 1999.

Matz, S.A. *Technology of Food Product Development*. Pan-Tech International, McAllen, TX, 1994.

Rey, L. and May, J.C., Eds. *Freeze-Drying/Lyophilization of Pharmaceutical and Biological Products*. Marcel Dekker, New York, 1999.

Rotstein, E. Singh, R.P. and Valentas, K.J., Eds. *Handbook of Food Engineering Practice*. CRC Press, Boca Raton, FL, 1997.

Singh, R.P. and Heldman, D.R. *Introduction to Food Engineering*, 2nd edn. Academic Press, San Diego, 1993.

Wilkinson, V.M. and Gould, G.W. *Food Irradiation: A Reference Guide*. Butterworth-Heinemann, Oxford, 1996.

Food, Processed and Fast

Convenience, uniformity, predictability, affordability, and accessibility characterized twentieth-century processed and fast foods. Technology made mass-produced fast food possible by automating agricultural production and food processing. Globally, fast food provided a service for busy people who lacked time to buy groceries and cook their meals or could not afford the costs and time associated with eating traditional restaurant fare. As early as the nineteenth century, some cafeterias and restaurants, foreshadowing fast-food franchises, offered patrons self-service opportunities to select cooked and raw foods, such as meats and salads, from displays. Many modern cafeterias are affiliated with schools, businesses, and clubs to provide quick, cheap meals, often using processed foods and condiments, for students, employees, and members.

In the U.K., fish and chips shops introduced people to the possibilities of food being prepared and served quickly for restaurant dining or take-away. Sources credit the French with first preparing fried chips, also called fries, from potatoes, possibly in the seventeenth century. Simultaneously, the English bought fried fish at businesses that were so widespread that they were mentioned in nineteenth-century novels by writers including Charles Dickens. By the 1860s, English shopkeepers combined fish and chips for a nutritious and filling meal providing essential proteins and vitamins. Demand for fish and chips soared, with the approximately 8,500 modern shops, including the Harry Ramsden's chain, in the U.K. outnumbering

McDonald's eight to one, and extending to serve patrons in countries worldwide.

Food-processing technology is designed primarily to standardize the food industry and produce food that is more flavorful and palatable for consumers and manageable and inexpensive for restaurant personnel. Food technologists develop better devices to improve the processing of food from slaughter or harvesting to presentation to diners. They are concerned with making food edible while extending the time period it can be consumed. Flavor, texture, and temperature retention of these foods when they are prepared for consumers are also sought in these processes. Microwave and radio frequency ovens process food quickly, consistently, and affordably. Microwaves are used to precook meats before they are frozen for later frying in fast-food restaurants. Nitrogen-based freezing systems have proven useful to process seafood, particularly shrimp. Mechanical and cryogenic systems also are used. The dehydrating and sterilizing of foods remove contaminants and make them easier to package. Heating and thawing eliminate bacteria to meet health codes. These processes are limited by associated expenses and occasional damage to foods. Processing techniques have been adapted to produce a greater variety of products from basic foods and have been automated to make production and packaging, such as mixing and bottling, efficient enough to meet consumer demand.

McDonald's is the most recognized fast-food brand name in the world. Approximately 28,000 McDonald's restaurants operated worldwide by the end of the twentieth century, with 2,000 opening annually. That chain originated in 1948 at San Bernadino, California, when brothers Richard and Maurice McDonald created the Speedee Service System. The thriving post-World War II economy encouraged a materialistic culture and population expansion. People embraced such technological developments as automobiles and enjoyed traveling within their communities and to distant destinations. The interstate highway system encouraged travel, and new forms of businesses catered to motorists. Drive-in restaurants provided convenient, quick meal sources.

The McDonalds innovated the assembly-line production of food. They limited their menus to several popular meals that could be consumed without utensils and were easy for children to handle. A hamburger, fries, and soft drink or milkshake composed the ubiquitous fast-food meal. Instead of hiring carhops to wait on customers, the McDonalds created a system in which patrons served themselves. Employees did not need special skills to operate the McDonalds' food assembly line. As a result, the Speedee Service System reduced prices, establishing a family-friendly environment.

Multimixer milkshake machine salesman Ray Kroc bought rights to franchise the McDonalds' restaurant. He had traveled to the McDonalds' San Bernadino hamburger stand in 1954 because the brothers had bought eight milkshake machines. Because each machine had five spindles, 40 milkshakes could be produced at the same time. Curiously observing the stand, Kroc questioned customers about why they chose to patronize that restaurant. He then monitored kitchen activity and was especially intrigued by french fry preparations that created crispy yet soft fries. Kroc had an epiphany, deciding to establish a chain of identical hamburger stands where customers could expect food efficiently and consistently produced to taste the same regardless of their geographical location.

Technology was crucial for the spread of McDonald's. In addition to technical knowledge, tools, and materials to construct similar buildings which people would recognize as McDonald's anywhere, Kroc used technology to select suitable sites. He researched locations by flying above communities in airplanes or helicopters to determine where schools, shopping centers, and recreational places were located. Later, commercial satellites produced photographs for McDonald's representatives to identify where clusters of children and other possible customers were centralized. Computer software was utilized to analyze census and demographic information for optimal restaurant site selection.

Fast-food culture gradually altered eating habits in the latter twentieth century. Inspired by McDonalds' success, other major fast-food chains, including Burger King, Kentucky Fried Chicken, Wendy's, Domino's, Pizza Hut, Taco Bell, Dunkin' Donuts, and Carl's Jr. were created. Cake mixes, instant mashed potatoes, breakfast cereal, macaroni and cheese, and junk food such as potato chips, pretzels, canned cheese, and doughnuts also became popular processed and fast food available in grocery stores. Purchasing fast food became a routine activity for many people. The "drive-thru" enhanced fast food's convenience. In addition to restaurants, fast foods were served at airports, schools, gas stations, and large stores such as Wal-Mart.

McDonald's targeted children with marketing that promoted toys, often movie-related, in Happy Meals and featured children enjoying McDonald's playgrounds. McDonald's hired approximately one million workers annually. Most employees, many of them teenagers, were unskilled and paid minimum wages. Training was minimal, and the turnover rate was high. Fast-food corporations caused the demise of many independent food entrepreneurs. Conformity soon overshadowed regional cuisine.

Influencing agricultural distribution, McDonald's bought a large portion of the world's potatoes and meat. Slaughterhouses were industrialized to process hundreds of carcasses hourly on an assembly line system of conveyor belts and automated meat recovery (AMR) systems, which stripped all meat from bones. Because these machines often included bone and other contaminants in ground beef, European laws forbade their use. Meatpacking plants were cleaned with high-pressure hoses emitting a water and chlorine mixture. McDonald's implemented technological fast-food processes that altered how food was made. Only salad ingredients arrived at fast-food restaurants fresh. Everything else was reformulated and freeze-dried, dehydrated, frozen, or canned to prepare in fast-food kitchens designed by engineers. In laboratories, scientists manufactured chemicals to achieve desired flavors and smells in processed fast foods that were designed to please consumers.

Kroc hired people to assess water content of potato crops with hydrometers. He realized that potatoes needed to be stored in curing bins so that sugars converted into starches to prevent sugars from caramelizing during frying. Electrical engineer Louis Martino invented a potato computer for McDonald's to calculate how long fries should be cooked according to oil temperature. Fast-food chains used a system of blanching, drying, briefly frying, then freezing fries before they were deep fried for consumption. Sometimes fries were dipped in sugar or starch to achieve desired appearance and texture. At french fry factories, machines washed potatoes then blew off their skins. A water gun knife propelled potatoes 36 meters per second through a cutter that sliced them uniformly before they were frozen and shipped to restaurants.

Critics lambasted the high fat, sugar, salt, and calorie content of fast foods, which they linked to increased obesity, heart disease, and diabetes rates, especially in children. They demanded healthier options. Some menus were changed but were vulnerable to patrons' acceptance or rejection. Auburn University scientists replaced some fat in ground beef with carrageenan for moisture and flavor additives to create leaner beef that tasted like normal ground beef. Marketed by McDonald's as the McLean Deluxe in the early 1990s, this lean hamburger failed to attract consumers because it was described as health food.

Hindus and vegetarians were angered when McDonald's disclosed that french fries were fried in beef tallow. Some people protested at the placement of fast-food restaurants in their neighborhoods. In 1990, McDonald's sued David Morris and Helen Steel, members of London Greenpeace, for libel because they distributed leaflets critical of the chain. The defendants lost the McLibel trial, but The Court of Appeals overturned some of the initial judgment. McDonald's stopped purchasing genetically engineered potatoes in an effort to prevent European consumer protests spreading to the U.S.

Fast-food globalization occurred as fast-food chains were built worldwide. McDonald's owned the most retail property internationally. In addition to archetypal fast-food meals, restaurants often accommodated local tastes such as Wendy's in Seoul, Korea, selling noodle dishes. Fast-food symbols were seen throughout the world. Ronald McDonald statues appeared in areas of the former East Germany, and the first McDonald's Golden Arch Hotel opened in Zurich, Switzerland, in 2001. A large plastic Colonel Sanders greeted passengers at Don Muang airport, Bangkok, Thailand. Even St. Kitts, a small Caribbean island, has a Kentucky Fried Chicken restaurant. Fast food was the target of anti-American protests in China, which internationally is second in the number of fast-food restaurants, and other countries.

See also **Food Preparation and Cooking; Food Preservation, Cooling and Freezing; Food Preservation: Freeze-Drying, Irradiation, and Vacuum Packing; Transport, Foodstuffs**

ELIZABETH D. SCHAFER

Further Reading

Alfino, M., Caputo, J.S. and Wynyard, R., Eds. *McDonaldization Revisited: Critical Essays on Consumer Culture*. Praeger, Westport, CT, 1998.

Barbosa-Cánovas, and Gould, G.W., Eds. *Innovations in Food Processing*. Technomic, Lancaster, PA, 2000.

Fishwick, M.W., Ed. *Ronald Revisited: The World of Ronald McDonald*. Bowling Green University Popular Press, Bowling Green, OH, 1983.

Hightower, J. *Eat Your Heart Out: Food Profiteering in America*. Crown Publishers, New York, 1975.

Jakle, J.A. and Sculle, K.A. *Fast Food: Roadside Restaurants in the Automobile Age*. Johns Hopkins University Press, Baltimore, MD, 1999.

Kincheloe, J.L. *The Sign of the Burger: McDonald's and the Culture of Power*. Temple University Press, Philadelphia, PA, 2002.

Kroc, R. and Anderson, R. *Grinding It Out: The Making of McDonald's*. H. Regnery, Chicago, 1977.

Lee, T.-C. and Ho, C.-T., Eds. *Bioactive Compounds in Foods: Effects of Processing and Storage*. American Chemical Society, Washington D.C., 2002.

Lusas, E.W. and Rooney, L.W. *Snack Foods Processing*. Technomic, Lancaster, PA, 2001.

McLamore, J.W. *The Burger King: Jim McLamore and the Building of an Empire*. McGraw-Hill, New York, 1998.

Moreira, R.G. *Automatic Control for Food Processing Systems*. Aspen, Gaithersburg, ME, 2001.

Ritzer, G., Ed. *McDonaldization: The Reader*. Pine Forge Press, Thousand Oaks, CA, 2002.

Royle, T. and Towers, B. *Labour Relations in the Global Fast Food Industry*. Routledge, New York, 2002.

Rozin, E. *The Primal Cheeseburger*. Penguin, New York, 1994.

Schlosser, E. *Fast Food Nation: The Dark Side of the All-American Meal*. Houghton Mifflin, New York, 2001.

Smart, B., Ed. *Resisting McDonaldization*. Sage, London, 1999.

Vidal, J. *McLibel: Burger Culture on Trial*. New Press, New York, 1998.

Watson, J.L., Ed. *Golden Arches East: McDonald's in East Asia*. Stanford University Press, Stanford, CA, 1997.

Fossil Fuel Power Stations

Until the last third of the twentieth century, fossil fuels—coal, oil, natural gas—were the primary source of energy in the industrialized world. Large thermal power stations supplied from fossil fuel resources have capacities ranging up to 4000 to 5000 megawatts. Gas-fired combined-cycle power stations tend to be somewhat smaller, perhaps no larger than 1000 to 1500 megawatts in capacity.

Concerns in the 1970s over degradation of urban air quality due to particulate emissions and acid rain from sulfur dioxide emissions from fossil fuel power stations were joined from the 1990s by an awareness of the potential global warming effect of greenhouse gases such as carbon dioxide (CO_2), produced from the combustion of fossil fuels (see Electricity Generation and the Environment). However, despite a move towards carbon-free electricity generation, for example from nuclear power stations and wind and solar plants, fossil fuels remain the most significant source of electrical energy generation.

Basic Technology

Power station technology for converting fuel into useful electricity comprises two basic stages:

1. Combustion, either of fuel in a boiler to raise steam or gas or distillate in a gas turbine to produce direct mechanical power
2. Conversion of the energy in steam in a steam turbine to produce mechanical power

In both cases, the mechanical power is used to drive an electricity generator. The more efficient steam turbines developed by Charles Parsons and Charles Curtis at the end of the nineteenth century replaced the use of reciprocating (i.e. piston-driven) steam engines to drive the dynamo.

Efficiency in the fuel-to-electricity conversion process for large coal and oil-fired power stations is seldom greater than about 36 to 38 percent. Combined-cycle gas turbine stations on the other hand can have efficiencies around 60 percent.

Coal

Coal was the source of energy for the Industrial Revolution in the U.K. and for the rapid expansion of industry in the eighteenth and nineteenth centuries. Since the middle of the twentieth century, the introduction of other more convenient fuels than that of the labor-intensive coal industry has caused its decline throughout Western Europe. Coal as a fuel dominated the power industry in the first half of the twentieth century but in the U.K., its use for power generation has declined from 90 percent in 1950 to 15 percent in 2000. In the U.S., coal has become more important as other fossil fuel supplies decline and large coal reserves remain. In 1995, coal burning produced about 55 percent of the electricity generated in the U.S. The largest coal power station in Europe is situated at Drax in Yorkshire and has a capacity of 4000 megawatts.

In the 1920s, pulverized coal firing was developed, improving thermal efficiency at higher combustion temperatures. Powdered coal is suspended in air and blown directly into the burner. Cyclone furnaces, developed in the 1940s, used crushed coal and an injection of heated air, causing the coal-air mixture to swirl like a cyclone. This rotation produced higher heat densities and allowed the combustion of poorer grade of coal with less ash production and greater overall efficiency.

Electrostatic precipitators, filtration, or wet scrubbing (quenching the raw fuel gas with water) is used to reduce particulate emissions in flue gases. In the electrostatic precipitator, an ionizing field imparts an electric charge to the particles, allowing them to be collected on an oppositely charged surface.

Oil

From about 1945 onward, the rapid increase in the quantities of crude oil refined has given rise to large quantities of heavy fuel oil (HFO) being produced, which can be burned in large power stations to raise steam for turbine-generators. For many years, the cost of oil (on an equivalent heat basis) was lower than coal, and this gave rise to the construction of large oil-fired power stations, generally sited close to oil refineries on the coast. HFO can contain significant quantities of sulfur, which causes severe atmospheric pollution unless the sulfur oxide flue gases are subjected to chemical treatment (desulfurization). The price of HFO is prone to considerable fluctuation on the world market, and this can affect its competitive position as a power station fuel. Many power stations around the world, particularly medium-scale ones (up to 50 megawatt capacity), utilize distillate or "diesel" fuel to power large reciprocating engines which drive electricity generators. Such liquid fuel can be used also to power gas turbines to drive generators.

Natural Gas

Natural gas (methane) is generally associated with oil reserves and can be recovered for use as a domestic, industrial and power station fuel. While it can be burned in conventional power station boilers to raise steam for turbine-generators, it is more usually utilized and burned directly in gas turbines, which drive electrical generators (see Turbines, Gas). The efficiency of this method of power production can be raised significantly (to over 60 percent) by recovering the waste heat from the gas turbine exhaust and using this to raise steam to drive a separate turbine generator set. Most of the new power stations constructed in the developed world in the latter part of the twentieth century have been based on this so-called "combined-cycle" arrangement. Such power stations are generally cheaper, quicker to build, and require less land area than a conventional coal or oil-fired power station of equivalent capacity.

More efficient gas turbines have resulted in the energy-efficiency factor for gas increasing by 40 percent from the early 1980s to the late 1990s. New natural-gas-powered plants also have much reduced sulfur dioxide (SO_2) and nitrous oxides (NO_x) emissions compared to pulverized-coal-fired steam plant, even those with "scrubbers" that remove polluting gases and particulates.

Orimulsion and Petroleum Coke

Orimulsion is a bitumen-based fuel of which there are large reserves in North and South America and elsewhere. It is widely regarded by environmentalists as a "dirty" fuel, and thus it finds little application as a power station fuel at the beginning of the twenty-first century. It has enormous potential when used with appropriate emission-control equipment. Petroleum coke is a byproduct of oil refining and can be gasified to produce a synthetic gas (Syngas) for power generation purposes.

Delivery and Storage of Fuel
Continuous supplies of fuel are essential for any electricity generating plant. Coal tends to be delivered by railway trains or by ship. Gas or oil would be delivered by pipeline, either from a country's "gas grid" or from a marine terminal or refinery. Large quantities of coal or oil fuel are usually stored on a power station site to cover for periods when deliveries might be interrupted. An alternative or "back-up" fuel such as distillate would be stored at a gas-fired power station.

Siting of Power Stations
The three critical factors that determine the location of a large central power station are source of fuel, supplies of cooling water, and proximity of load.

In the case of a coal-fired power station, in Britain the site may be on or very close to a coal field or coal mine. Sometimes, large coal power stations might be situated on an estuary and have coal delivered via ships. Rail links might also be used to deliver the coal, using large 1000-ton capacity freight trains to a specially constructed rail head at the power station.

Oil-fired power stations tend to be sited adjacent to an oil refinery, the fuel being conveyed directly from the refinery to the power station by pipeline. Sometimes a location on an estuary might be selected and the fuel oil delivered to offshore jetties by large oil tankers.

Gas-fired power stations are built so as to gain access to the gas transmission network in a country and may also be located close to the point of gas production; for example, on the North Sea coast in the U.K.

Where the power station utilizes steam turbines to drive the electrical generators, large supplies of cooling water are required to condense the steam back to water after it leaves the steam turbine. This condensate is then circulated back to the boilers where it is converted into steam once again. Where rivers or the sea are not available or, for environmental reasons cannot be used as sources of cooling water, cooling towers may be used which serve the same condensing purposes, albeit at lower overall efficiency.

In the early days of the electricity industry, power stations were located in the middle of towns and cities and the electricity consumers were served by cables radiating from the power station. As stations became larger, the lack of suitable large sites tended to result in their siting in areas where either or both of fuel and cooling water were obtainable. The electricity was transmitted to the load center by a "grid" of high-voltage power lines. It is generally more economical to convey the electrical energy by high-voltage power lines from the power station over long distances than to transport the fuel by road, rail or sea over the same distance.

Where gas-fired power stations are established in heavily built-up regions like the U.K., the existence of a large gas transmission network over most of the country means that the stations, because of their relatively compact size, can be sited close to the load centers and thus reduce the requirements for high-voltage power lines.

See also **Electrical Energy Generation and Supply, Large Scale; Electricity Generation and the Environment; Energy and Power**

IAN BURDON

Further Reading

Edgerton, D. *Science, Technology and the British Industrial 'Decline' ca. 1870–1970*. Cambridge University Press/ Economic History Society, Cambridge, 1996.

Elliot, D. *Energy, Society and Environment*. Routledge, London 1997.

Leggett, J. *The Carbon War: Global Warming at the End of the Oil Era*. Penguin, London, 1999.

National Research Council. *Coal: Energy for the Future*. National Academy Press, Washington D.C., 1995.

Reddish, A. and Rand, M. The environmental effects of present energy policies, in *Energy, Resources and Environment*, Blunden, J., and Reddish, A., Eds. Hodder & Stoughten (in association with The Open University, London, 1991.

Wollenberg, B.F. *Power Generation, Operation and Control*. Wiley/Interscience, New York, 1996.

Useful Websites

U.S. Department for Energy, Office of Fossil Energy: http://www.fe.doe.gov/

Fuel Cells

The principle of the fuel cell (FC) is similar to that of the electrical storage battery. However, whereas the battery has a fixed stock of chemical reactants and can "run down," the fuel cell is continuously supplied (from a separate tank) with a stream of oxidizer and fuel from which it generates electricity. Electrolysis—in which passage of an electrical current through water decomposes it into its constituents, H_2 and O_2—was still novel in 1839 when a young Welsh lawyer–scientist, William Grove, demonstrated that it could be made to run in reverse. That is, if the H_2 and O_2 were not driven off but rather allowed to recombine in the presence of the electrodes, the result was water—and electrical current.

The mechanism of electrical generation was not fully understood until development of atomic theory in the twentieth century. Like a battery, an FC has two electrodes connected via an electrical load and physically separated by an electrolyte—a substance that will selectively pass either positive ions (acidic electrolyte) or negative ions (alkaline electrolyte). Oxidizer (e.g., O_2) enters at one electrode and fuel (e.g., H_2) at the other. At the cathode (the electrically positive electrode) atoms of fuel occasionally ionize, forming positively charged ions and free electrons. This is accelerated by a catalyst at the electrodes as well as suitable conditions of temperature and pressure. Similarly at the negative electrode (anode), oxidizer atoms spontaneously form negative ions. Then, depending on electrolyte, either positive or negative ions migrate through it to combine with ions of opposite polarity, while electrons (unable to pass through the electrolyte) flow through the electrical load from anode to cathode.

While the Second Law of Thermodynamics severely limits the efficiency of heat engines operating at practical temperatures, fuel cells can extract more power out of the same quantity of fuel compared to traditional combustion, since the hydrogen fuel is not converted to thermal energy, but used directly to produce mechanical energy. In principle, an FC consuming pure hydrogen and oxygen and producing liquid water can achieve an efficiency of 94.5 percent with an open-circuit potential of 1.185 volts. In practice a variety of losses cannot altogether be eliminated, but achievable efficiencies are considerably greater than those of most heat engines. Multiple cells connected in series supply higher voltages.

Practical challenges include electrolyte–electrode chemistry and physical properties, catalysts, fuel and oxidizer composition and purity, internal electrical losses, corrosion and other destructive chemical reactions, and a host of mechanical issues. Efficiency, silence, and lack of polluting emissions stimulated great interest, however. In searching for suitable systems, twentieth century researchers developed various approaches, most named for their electrolytes. Most of the developments in the field have come as a result of proprietary interests and the work of corporate teams of researchers.

Alkaline FCs (AFCs)

From the 1930s Francis Bacon of the U.K., followed later by U.S. researchers, turned from acidic electrolytes to more tractable potassium hydroxide (KOH). Bacon also pioneered use of porous electrodes through which gaseous reactants diffused. The first major application of FC technology was in the U.S. space program where AFCs provide power (and potable water) for the space shuttles. The need for extremely pure H_2 and O_2 made AFCs uneconomic for more mundane applications.

Molten carbonate FCs (MCFCs)

MCFCs grew out of SOFC research (see below). In the 1950s Dutch investigators G.H.J. Broers and J.A.A. Ketelaar turned to molten lithium-, sodium, and potassium carbonates as electrolytes, as did Bacon in the U.K. At typical 650°C operating temperatures, MCFCs produce waste heat in a form useful for industrial purposes or to power turbines for added electrical output. High temperatures relax the need for costly catalysts while their carbonate chemistry is tolerant of carbon monoxide (CO), which as an impurity is problematic for alkaline fuel cells. However chemical and mechanical problems have thus far impeded wide application.

Phosphoric Acid FCs (PAFCs)

Interest in phosphoric acid as an electrolyte emerged slowly until the mid-1960s, after which PAFCs rapidly became the first FCs to see significant commercialization. At around 200°C, PAFC waste heat may be used to reform (convert) hydrocarbons or coal to H_2 for fuel or power an auxiliary turbine. PAFCs are relatively tolerant of CO but sulfur must be separated. Units up to 250 kilowatts output are sold for fixed-site power applications and experimental units have shown promise in buses. Problems center on internal chemical reactions and corrosion.

Proton-Exchange Membrane FCs (PEMFCs)

In the early 1960s Thomas Grubb and Leonard Niedrach of General Electric in the U.S. developed a polymer membrane which, moistened with water, served as an effective and stable electrolyte. Initial space application attempts in the 1960s revealed reliability problems (and led to adoption of AFCs) but further development has held out strong promise for ground and marine vehicle applications as well as small fixed generators. Operating temperatures of less than 100°C and high power relative to size and weight, suit PEMFCs especially. Platinum catalysts are necessary and the H_2 fuel must be essentially free of CO. By century's end PEMFCs had shown some promise of being adaptable to liquid methanol in place of H_2 fuel.

Solid-Oxide FCs (SOFCs)

Beginning in the 1930s experimenters first in Switzerland and Russia sought high-temperature (around 1000°C) solid ceramic electrolytes. At such temperatures, reactions proceed rapidly without costly catalysts and many fuel stocks can be reformed to produce H_2 within the SOFC, while the waste heat can be used for many purposes. But physical and chemical problems of high-temperature operation have been difficult and it remained uncertain at century's end how well the promise of SOFCs could be realized.

Metal FCs (MFCs)

To avoid problems of H_2 supply or conversion some FC developers turned, late in the century, to metal fuels, usually zinc or aluminum. Electrolytes include liquid KOH and proton-exchange membranes. Waste products are metal oxides.

Conclusion

Over the twentieth century FCs moved from laboratory curiosity to practical application in limited roles and quantities. It is very possible that the twenty-first century will see them assume a major or even dominant position as power sources in a broad array of applications. Obstacles are largely economic and the outcome will be influenced by success in development of competing systems as well as FCs themselves.

See also **Batteries; Electrochemistry**

WILLIAM D. O'NEIL

Further Reading

The future of fuel cells. *Scientific American*, 1999.
Kartha, S. and Grimes, P. Fuel cells: energy conversion for the next century. *Physics Today*, 1994.
Montavalli, J. *Forward Drive: The Race to Build "Clean" Cars for the Future.* Sierra Club Books, 2000.
Riezenman, M.J. Metal fuel cells. *IEEE Spectrum*, 38, 2001.
Srinivasan, S. *et al.* Fuel cells: reaching the era of clean and efficient power generation in the twenty-first century. *Ann. Rev. of Energ. Environ.*, 24, 1999.

Useful Websites

Smithsonian Institution Fuel Cells History Project, http://fuelcells.si.edu/index.htm

Fusion, *see* **Nuclear Reactors: Fusion**

G

Gender and Technology

As a field of scholarship, the study of gender and technology evolved rapidly in the latter half of the twentieth century. The field began with the study of women in technology, focusing on the leading women in engineering or female inventors, who were found primarily in Western societies. These studies were often in a biographical and historical mode, leading to an appreciation of the barriers women faced and overcame to join technical professions. However, scholars came to understand that this privileged the conventional or common-sense definitions of technology, which was equated with large-scale, complex, and public enterprises such as aerospace, railway, and civil engineering projects. Cowan (1983) and McGaw (1996) argue that this realization led to scholarship that broadened the definition of technology to include what had often been stereotyped as "women's work" and which included such items as domestic appliances, typewriters and other tools of clerical labor, and artifacts of personal care. In these kinds of studies, the technical complexity of women's activities, including invention, adaptation, and the use of artifacts, was highlighted and understood as being interconnected with larger social and economic systems.

Concern with looking at women in technology and looking at women's technologies was informed by what is known as the "second wave" of feminism in the 1970s and later. This period reflected the dramatic changes in the status of women, starting first with the experiences of women in the workforce during both World War I and II and continuing through the radicalization of feminism in the context of the 1960s civil rights

era in the U.S. In these perspectives, issues of equal access and equal treatment in major social institutions were concerns of both scholarship and activism. This inclusive approach was also fueled by the launch of Sputnik in the USSR, which focused attention on U.S. competitiveness in space and science more generally, and spurred an interest in science and engineering for both young men and women. The historical studies were soon joined by sociological and anthropological inquiries such as those raised by Witz (1992) and Cockburn and Ormrod (1993) that asked questions about how it is that work is seen to be appropriate for people of one gender or the other and how technologies come to be associated with a particular gender. Sociologists such as McIlwee and Robinson (1992), Ranson and Reeves (1996) and Roos and Gatta (1999) have studied women in technological industries and workplaces and examined different patterns of compensation and promotion, both in terms of discrimination and success. This scholarship illustrated that while some gains had been made, promotion and compensation rates for women lagged those of men even when factoring in years of experience and child-rearing breaks in career paths.

Late twentieth century research also looked at cross-national comparisons of the professional engineering sector, as well as studies of how development projects differentially affect men and women in nonindustrialized countries. For example, while engineering is highly gender segregated in the U.S., other countries have more integrated engineering programs (the former Soviet Union and Italy) and computer science and telecommunications employment fields (as in Southeast Asia). International development projects often affect

375

women's livelihoods in nonindustrialized societies, sometimes improving their social position but often having a negative impact on work, property ownership, and family health.

This kind of complexity means that it is hard to generalize about exactly how technological change has influenced the roles of men and women. In some cases, technological change opened new avenues for work and economic mobility, as in the early days of clerical labor fostered by the typewriter or the early phases of the computer revolution. Frequently, however, the pattern of technological change followed existing lines of power in society and reinforced systems of stratification. For example, Ruth Schwartz Cowan argued that while "labor saving" household technologies such as washers and dryers or dishwashers save women from direct physical labor, women still spent nearly as much time as they did a century ago on housework, as well as handling the additional demands of working outside the home.

At the same time that scholars have made the definition of technology more inclusive and complex, there is increased understanding of the relationship between identity and work. Butler (1993) and Connell (1995) argue that how identities are defined by individuals and society, how they change over time, and how they vary with regard to race, class and ethnicity, have helped shape configurations of masculinity and femininity. The study of identity and difference has shown that gender is a way that cultures categorize the perceived differences between men and women. Thus the study of women and technology has been transformed into the study of gender and technology, the study of ideas about masculinity and femininity, and the inseparability of these concepts from cultural ideas about technology and work. This development of the understanding that gender is a constellation of values and ideas that any given culture associates with sexual differences is socially constructed and situationally produced and is not a natural category. Similarly, what is considered technology is a classification bounded by cultural ideas about gender, status, and skill. Croissant (2003) argues that in much the same ways that classical or "high" art is distinguished from crafts and hobbies, technology—particularly "high tech"—is a way of making distinctions of worth and value based on cultural stereotypes that overlap with gender systems. So, as scholars such as Oldenziel (1999) and Pacey (1983) have argued, ideas about technology in Western culture are strongly associated with masculinity, particularly a middle-class professional identity associated with engineering.

Several late twentieth century scholars such as McGaw (1996) and Croissant (2000), hoped that an understanding of how gender ideas help shape ideas about technology would influence people to look beyond stereotypical ideas about both technology and gender. This could result in opening employment opportunities for men and women and increase the diversity of engineering and related technical work. In addition, such an understanding could provide an important set of tools for critical thinking about the interconnections of our technological society.

JENNIFER CROISSANT

Further Reading

Ambrose, S.A. *Journeys of Women in Science and Engineering: No Universal Constants.* Temple University Press, Philadelphia, 1997.

Butler, J. *Bodies that Matter: On the Discursive Limits of Sex.* Routledge, New York, Routledge, 1993.

Cockburn, C. and Ormrod, S. *Gender and Technology in the Making.* Sage, London, 1993.

Connell, R.W. *Masculinities.* Polity Press, Cambridge, 1995.

Cowan, R.S. *More Work for Mother: The Ironies of Household Technology from the Open Hearth to the Microwave.* Basic Books, New York, 1983.

Croissant, J. Engendering technology: culture, gender, and work. *Knowledge and Society*, 12, 189–207, 2000.

McGaw, J. Reconceiving technology: why feminine technologies matter, in *Archaeology and Gender*, R.P. Wright, Ed. University of Pennsylvania Press, Philadelphia, 1996, pp. 52–75.

McIlwee, J.S. and Robinson J.G. *Women in Engineering: Gender, Power, and Workplace Culture.* SUNY Press, Albany, 1992.

Oldenziel, R. *Making Technology Masculine: Men, Women, and Modern Machines in America, 1870–1945.* Amsterdam University Press, Amsterdam, 1999.

Pacey, A. *The Culture of Technology.* Blackwell, Oxford, 1983.

Ranson, G. and Reeves, W. J. Gender, earnings, and proportions of women: lessons from a high-tech occupation. *Gender and Society*, 10, 2, 168–184, 1996.

Roos, P.A. and Gatta M.L. The gender gap in earnings: trends, explanations, and prospects, in *Handbook of Gender and Work*, Powell, G., Ed. Sage, Thousand Oaks, CA, 1999, pp. 95–125.

Stanley, A. *Mothers and Daughters of Invention: Notes for a Revised History of Technology.* Scarecrow Press, Metuchen, NJ, 1993.

Vare, E.A. and Ptacek, G. *Mothers of Invention: From the Bra to the Bomb : Forgotten Women & their Unforgettable Ideas.* Morrow, New York, 1988.

Vare, E.A. and Ptacek, G. *Patently Female: From AZT to TV Dinners: Stories of Women Inventors and their Breakthrough Ideas.* Wiley, New York, 2002.

Witz, A. *Professions and Patriarchy.* Routledge, London, 1992.

Gene Therapy

In 1971, Australian Nobel laureate Sir F. MacFarlane Burnet thought that gene therapy (introducing genes into body tissue, usually to treat an inherited genetic disorder) looked more and more like a case of the emperor's new clothes. Ethical issues aside, he believed that practical considerations forestalled possibilities for any beneficial gene strategy, then or probably ever. Bluntly, he wrote: "little further advance can be expected from laboratory science in the handling of 'intrinsic' types of disability and disease." Joshua Lederberg and Edward Tatum, 1958 Nobel laureates, theorized in the 1960s that genes might be altered or replaced using viral vectors to treat human diseases. Stanfield Rogers, working from the Oak Ridge National Laboratory in 1970, had tried but failed to cure argininemia (a genetic disorder of the urea cycle that causes neurological damage in the form of mental retardation, seizures, and eventually death) in two German girls using Swope papilloma virus. Martin Cline at the University of California in Los Angeles, made the second failed attempt a decade later. He tried to correct the bone marrow cells of two beta-thalassemia patients, one in Israel and the other in Italy. What Cline's failure revealed, however, was that many researchers who condemned his trial as unethical were by then working toward similar goals and targeting different diseases with various delivery methods. While Burnet's pessimism finally proved to be wrong, progress in gene therapy was much slower than antibiotic or anti-cancer chemotherapy developments over the same period of time.

In September 1990, National Institutes of Health investigators Michael Blaese, French Anderson, and Kenneth Culver carried out the first authorized gene therapy clinical trial. Their patient, 4-year-old Ashanthi DeSilva, suffered from a form of severe combined immune deficiency (SCID) caused by lack of an essential enzyme, adenosine deaminase (ADA). The protocol called for a recombinant virus—beginning with a denatured mouse leukemia virus, into which a human gene for adenosine deaminase and promoter/marker genes were spliced—to correct rather than destroy her circulating white blood cells. A blood specimen was taken from the child, and her cells were grown in the presence of the virus and later reinfused. In this and several subsequent cases, SCID patients also received a bovine-derived adenosine deaminase. Results were encouraging, but the addition of a replacement enzyme clouded proof that gene therapy alone was responsible for

improving the patient's health. In April 2000, Alain Fischer at the Hôpital Necker Enfants-Malades in Paris reported success following a similar strategy in treating five patients with a related type of SCID for which no replacement enzyme existed. Only then was there unchallenged proof for the principle of gene therapy.

During the 1990s, however, researchers in the U.S., Europe, and Japan conducted over 400 clinical gene therapy trials, with the largest group targeting cancers. Some strategies aimed at stimulating immune responses to tumor antigens; others sought to insert a functioning tumor suppressor gene, or, alternatively, to inhibit gene function to stop tumor growth. One protocol attempted to change the nature of a deadly brain cancer by inserting a retrovirus recombined with the thymidine kinase gene from the cold sore virus, herpes simplex. Because brain cells in the adult human grow very slowly if at all and retroviruses can only infect dividing cells, researchers believed they could alter only the tumor's cells. After surgery to implant virus-producing cells and allowing time for the virus to spread, the antiviral drug ganciclovir was administered to create a lethal toxin through its interaction with thymidine kinase. While some patients apparently benefited from this approach and adverse reactions were not significant, positive indications were hardly compelling, other than suggesting a need for more research.

As clinical experiments continued, more limitations became apparent. Several protocols relied on adenoviruses as vectors to deliver a functional gene to the lung tissue of cystic fibrosis patients, but the normal immune response rejected this approach in much the same way it fights a cold. No viral vector could be targeted, thus gene delivery carried a risk of disrupting some other essential genetic function. In the cancer trials, tumor cell heterogenicity and mutability defeated any stunning breakthrough. Many patients in these studies died, though in only one instance at the University of Pennsylvania in 1999 was the therapy itself responsible for a death. Ironically, had gene therapy proved to be more effective, it may also have been less safe.

In the late 1990s, investigators began pursuing another approach—gene repair. One protocol was directed at treating Crigler-Nijjar syndrome, a liver enzyme deficiency responsible for a fatal bilirubin clearance disorder. The approach relied on a synthesized oligonucleotide that intentionally mismatched the point mutation of the disease on chromosome 2. The aim was to provoke normal gene repair enzymes to correct the problem in enough cells (perhaps no more than 5 percent of the liver) so that a life-saving repair would result. The technique used liposomes for gene delivery. Because gene repair relied on nonviral delivery and aimed at generating self-repair in cells, its potential for efficacy and safety exceeded that of gene therapy. However, as of early 2002 no clinical trials had begun.

While gene therapy had limited success, it nevertheless remained an active area for research, particularly because the Human Genome Project, begun in 1990, had resulted in a "rough draft" of all human genes by 2001, and was completed in 2003. Gene mapping created the means for analyzing the expression patterns of hundreds of genes involved in biological pathways and for identifying single nucleotide polymorphisms (SNPs) that have diagnostic and therapeutic potential for treating specific diseases in individuals. In the future, gene therapies may prove effective at protecting patients from adverse drug reactions or changing the biochemical nature of a person's disease. They may also target blood vessel formation in order to prevent heart disease or blindness due to macular degeneration or diabetic retinopathy. One of the oldest ideas for use of gene therapy is to produce anticancer vaccines. One method involves inserting a granulocyte-macrophage colony-stimulating factor gene into prostate tumor cells removed in surgery. The cells then are irradiated to prevent any further cancer and injected back into the same patient to initiate an immune response against any remaining metastases. Whether or not such developments become a major treatment modality, no one now believes, as MacFarland Burnet did in 1970, that gene therapy science has reached an end in its potential to advance health.

See also **Genetic Engineering, Applications; Genetic Engineering, Methods**

G. TERRY SHARRER

Further Reading

Culver, K.W. *Gene Therapy: A Primer for Physicians*, 2nd edn. Mary Ann Liebert, Larchmont, NY, 1996.

Friedmann, T. *The Development of Human Gene Therapy*. Cold Spring Harbor Laboratory Press, Cold Spring Harbor, NY, 1999.

Krniec, E.B. Gene therapy. *American Scientist*, May 1999.

Lyon, J. and Gorner, P. *Altered Fates: Gene Therapy and the Retooling of Human Life*. W.W. Norton, New York, 1995.

Porter, R. *The Greatest Benefit to Mankind: A Medical History of Humanity*. W.W. Norton, New York, 1997.

Verma, I.M. and Somia, N. Gene therapy—promises, problems and prospects. *Nature*, 389, 239–242, 1997.

Walters, L. and Palmer, J.G. *The Ethics of Human Gene Therapy*. Oxford University Press, Oxford, 1997.

Useful Websites

Human Genome Project information, gene therapy: http://www.ornl.gov/hgmis/medicne/genetherapy.html

Genetic Engineering, Applications

For centuries, if not millennia, techniques have been employed to alter the genetic characteristics of animals and plants to enhance specifically desired traits. In a great many cases, breeds with which we are most familiar bear little resemblance to the wild varieties from which they are derived. Canine breeds, for instance, have been selectively tailored to changing esthetic tastes over many years, altering their appearance, behavior and temperament. Many of the species used in farming reflect long-term alterations to enhance meat, milk, and fleece yields. Likewise, in the case of agricultural varieties, hybridization and selective breeding have resulted in crops that are adapted to specific production conditions and regional demands.

Genetic engineering differs from these traditional methods of plant and animal breeding in some very important respects. First, genes from one organism can be extracted and recombined with those of another (using recombinant DNA, or rDNA, technology) without either organism having to be of the same species. Second, removing the requirement for species reproductive compatibility, new genetic combinations can be produced in a much more highly accelerated way than before. Since the development of the first rDNA organism by Stanley Cohen and Herbert Boyer in 1973, a number of techniques have been found to produce highly novel products derived from transgenic plants and animals.

At the same time, there has been an ongoing and ferocious political debate over the environmental and health risks to humans of genetically altered species. The rise of genetic engineering may be characterized by developments during the last three decades of the twentieth century.

1970 to 1979

The term genetic engineering was probably first coined by Edward L. Tatum during his 1963 Nobel Prize acceptance speech, but it was not until the following decade that many of the potential applications of gene transfer became more apparent to the emerging field of molecular biology. However, the period was witness to a clash between the new molecular biologists who were pioneering the techniques of genetic engineering and their more industrially related colleagues in microbiology. Whereas the microbiologists, with their roots in the fermentation industries, were more prepared to recognize the potential of genetic engineering, molecular biologists were initially much more concerned with the threats of environmental risk.

As a consequence of a letter to *Science* and *Nature* in 1974, known as the Berg letter after its first signatory Paul Berg, molecular biologists instigated a voluntary moratorium on gene transfer work until such time as the community was satisfied that necessary safety measures and procedures had been put in place. Two further key events were instrumental in moving the field forward. The first was the Asilomar Conference in February 1975, which addressed safety measures and, to a lesser extent, prospects for future applications. The second was the publication in 1996 of the first guidelines for gene transfer research released by the U.S. National Institutes for Health, which effectively lifted the moratorium. In 1977, Genentech reported the manufacture of the human hormone somatostatin in bacteria genetically engineered to contain a synthetic gene that produced a human protein. This step is widely considered to represent the opening moments of modern biotechnology production.

1980 to 1989
The early 1980s is the first period in which large-scale investments were made in the biotechnology industry against the anticipation of huge profits to follow. Indeed, the first biotechnology stocks to be floated on the markets in the 1980s rose in value far more rapidly than had any other sector up to that time. In 1980 the U.S. Supreme Court allowed patent protection for a genetically modified "oil-eating" bacterium, providing powerful financial incentives for biotechnology companies to expand research. The year 1980 also saw the introduction of the polymerase chain reaction (PCR) technique, through which DNA sequences are multiplied many times *in vitro*, which became the foundation of much of the work to follow. In 1981 a team at Ohio University produced the first transgenic animals, mice in this case. Shortly afterward, Harvard University released details of studies in which mice were engineered to carry a human gene that increased susceptibility to a form of human cancer. The "OncoMouse," as it became known, could be used to test the carcinogenicity of

different compounds and as a model for developing cures for cancer. The OncoMouse, for which Harvard filed a patent in 1984, focused much of the ensuing debate on the future health implications of genetic engineering. By 1983, the first patents had been granted on genetically engineered plants (actually only for the use of an antibiotic resistance "marker gene," that allowed researchers to select transformed plants by their ability to survive exposure to an otherwise lethal dose of an antibiotic). In 1985 the first U.S. field trials of genetically engineered crops (tomatoes with a gene for insect resistance and tobacco plants with herbicide resistance) took place, and in 1986, genetically engineered tobacco plants, modified with addition of a gene from the bacterium *Bacillus thuringiensis* (Bt) to produce a insecticidal toxin, making the hybrid resistant to the European corn borer and other pests, underwent field trials in the U.S. and France.

1990 to 2000
The 1990s saw considerable growth in a wide range and variety of biotechnological applications, though without necessarily fulfilling the huge

Figure 1. Tracy—PPL Therapeutics' transgenic ewe, modified in 1992 to produce a human protein in her milk. [*Courtesy of PPL Therapeutics.*]

expectations evident in the early 1980s. In respect to animal biotechnology products, a number of events can be seen to have defined the decade.

Sizeable resources were directed at the production of proteins and drug compounds in transgenic animals, resulting in over 50 varieties of genetically modified (GM) bioreactors. These methods had a number of advantages over traditional cell culture production including higher production volumes, particularly in respect to those proteins (such as human albumin) that cannot be produced in a sufficient volume using other available techniques. On a considerably smaller scale, research in the 1990s also focused on the production of transgenic animals as sources of transplant tissues and organs. However, the decade closed with little progress seen in either reducing tissue rejection or overcoming anxieties about transpecies disease.

By far the largest research activity was within the field of plant biotechnology. For instance, genetically modified herbicide-tolerant (GMHT) crops were intended to enable varieties to withstand chemical treatments that would normally damage them. The same concept was applied to the production of insect and virus-resistant plant varieties, in addition to altering the way fruits ripen so that they can withstand increased storage and travel stresses. In 1994, the Flavr Savr tomato, designed to delay ripening and resist rotting, became the first whole genetically engineered food to be approved for sale in the U.S. (China commercialized virus-resistant tobacco plants in the early 1990s). In 2000, a rice variety was genetically engineered to contain a gene that increases the vitamin A content of the grains. Similar improvements could be made to the composition of other important food staples. Another widespread application of genetic engineering prevented plants from pollinating in order to limit the chances of cross-fertilization with other species. A more controversial aspect of genetically modified plants was their inability to reproduce so that growers would be unable to collect seeds for replanting, and thus forced to purchase seed from the supplier each season.

The 1990s were also characterized by what became known as the "GM debate." Although the strength of the controversy varied considerably throughout the world, with much greater intensity in Europe than in the U.S., anxieties continued to focus on a number of potentially adverse environmental effects arising from GM foods. First, there were concerns that GM crops will indirectly reduce wild plant biodiversity through intensification of industrial agriculture (increased use of pesticides and herbicides) and potentially threaten species higher up the food chain, for example invertebrates that feed on the weeds, and their bird and mammal predators. Second, the debate has focused on the risk of cross-pollination and gene transfer between GM and non-GM plants. Contamination has implications for the labeling of foods, for organic methods, and for seed production. Finally, more intensive production methods, together with enhanced transportation tolerance, were seen to exacerbate other environmental problems, particularly pollution and global warming.

See also **Biotechnology, Breeding, Animal Genetic Methods; Breeding, Plant Genetic Methods; Gene Therapy; Genetic Engineering, Methods; Genetic Screening and Testing; Pesticides**

NIK BROWN

Further Reading

Boylan, M. and Brown, K.E. *Genetic Engineering: Science and Ethics on the New Frontier*. Prentice Hall, Upper Saddle River, NJ, 2001.

Bud, R. Molecular biology and the long-term history of biotechnology, in *Selection Indices and Prediction of Genetic Merit in Animal Breeding*, Cameron, N.D., Ed. CAB International, Oxford, 1997.

Reeve, E., Ed. *Encyclopedia of Genetics*. Fitzroy Dearborn, London and Chicago, 2000. See entries on Rice genetics: engineering vitamin A; Pharmaceutical proteins from milk of transgenic animals.

Thackray, A., Ed. *Private Science: Biotechnology and the Rise of the Molecular Sciences*. University of Pennsylvania Press, Philadelphia.

Useful Websites

Advisory Committee on Novel Foods and Processes. Annual Report 2000, FSA/0013/0301: http://www.foodstandards.gov.uk/multimedia/pdfs/acnfp2000

Genetic Engineering, Methods

The term "genetic engineering" describes molecular biology techniques that allow geneticists to analyze and manipulate deoxyribonucleic acid (DNA). At the close of the twentieth century, genetic engineering promised to revolutionize many industries, including microbial biotechnology, agriculture, and medicine. It also sparked controversy over potential health and ecological hazards due to the unprecedented ability to bypass traditional biological reproduction. This article describes common genetic engineering techniques, excluding gene therapy and organism cloning, which are covered in separate entries in this encyclopedia.

Recombinant DNA (rDNA) technology involves combining DNA from different organisms, often from radically different species. In 1970, Howard Temin and David Baltimore independently isolated an enzyme, reverse transcriptase, which cuts DNA molecules at specific sites. Paul Berg, who was awarded the Nobel Prize in Chemistry in 1980 for his work with DNA, in 1972 used a similar restriction enzyme, ligase, to paste two DNA strands together to form a recombinant DNA molecule. In 1973, Stanley Cohen at Stanford University and Herbert Boyer at the University of California in San Francisco, combined two plasmids (short, circular pieces of single-stranded DNA) that naturally exist in many bacteria. Each contained a different antibiotic resistance gene. The new "recombinant" plasmid was inserted into a bacterial cell, and the cell produced both types of antibiotic resistance proteins. This bacteria was the first recombinant DNA organism.

Rapid advances occurred in the late 1970s, including splicing genes from higher organisms such as humans into bacterial plasmids. By the early 1980s valuable proteins such as insulin, interferon, and human growth hormone were being synthesized using recombinant bacteria hosts (such as *Escherichia coli*), and there was anticipation that these proteins would soon be produced on an industrial scale.

During meiosis, the production of gametes from dividing sperm and egg cells, the two copies of each chromosome exchange DNA by a process called "genetic crossover." A crossover between two regions of DNA causes them to be separated and no longer co-inherited. Thus the frequency of co-inheritance is an indirect measure of the distance between DNA regions. This "genetic distance" is measured in "Morgans," after Thomas Hunt Morgan's studies of the phenomenon in fruit flies c.1915. The determination of genetic distance is called "genetic mapping."

Genetic mapping in humans was relatively uncommon prior to rDNA technology. In 1978, Yuet Wai Kan at the University of California, San Francisco, discovered a region of DNA genetically close to the sickle cell anemia gene using genetic mapping techniques. Genetic mapping in humans then expanded rapidly.

In 1975 Ed Southern at Oxford University in the U.K. invented the "Southern Blot." The procedure relies on two other technologies: gel electrophoresis and DNA probes. Gel electrophoresis was developed in the 1950s for separating molecules of different sizes on a gel subjected to an electric current. DNA probes, small strands of radioactive DNA with a known base sequence, bind to DNA with a base pair sequence that matches that of the probe, and the binding can be visualized using X-ray film. Using these two techniques, DNA strands containing a desired sequence can be isolated from a mixture of strands (see Figure 2).

These technologies can be used to produce a "physical map," a collection of overlapping DNA fragments arranged in their proper order. Two DNA fragments overlap if a probe that recognizes a unique sequence hybridizes to both fragments. A physical map is a useful tool when searching for genes. If a gene has been genetically mapped to a specific region of DNA, a physical map of this region can be used to locate and isolate this gene.

In 1977, methods for sequencing DNA were developed in two separate locations, by Walter Gilbert at Harvard University and by Fred Sanger at Cambridge University. These DNA sequencing processes were automated in the mid-1980s and marketed by Applied Biosystems, DuPont, and other companies. The "Sanger method" is the most common method; it uses dideoxynucleotides (ddNTPs), which are similar to deoxynucleotides (dNTPs), the normal components of DNA, but they lack an oxygen atom. There are four forms, corresponding to the four dNTP types found in DNA: adenine (A), thymine (T), guanine (G), and cytosine (C). When any of these is incorporated into a synthesizing DNA strand, DNA synthesis stops. Sequencing involves four separate DNA synthesis reactions, each containing one of the four ddNTPs and each using the original DNA strand as a template. DNA synthesis occurs in each reaction tube and ends when ddNTP gets incorporated into the strand. If the reaction is allowed to continue long enough, each of the four tubes will have a full assortment of DNA strands of varying

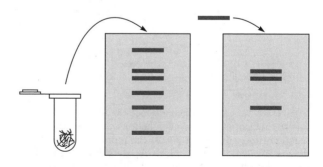

Figure 2. Southern Blot.
[*Source: Dr. John Medina, of the University of Washington, Seattle, Washington. Used with permission from Brinton, K. and Lieberman, K.-A. Basics of DNA Fingerprinting, University of Washington, 1994.*]

lengths, all ending with the ddNTP present in the reaction tube. When these strands are separated by gel electrophoresis, the sequence of the original DNA molecule can be determined by simply reading the sequencing gel from bottom to top (see Figure 3).

In 1985 Kary Mullis, a technician with the Cetus Corporation in the U.S., invented the polymerase chain reaction (PCR), a process that could rapidly identify and replicate a segment of DNA with a known sequence out of a complex mixture of DNA. The technique was time saving and simple compared with the Southern blot and

was used for a wide variety of applications, including genetic and physical mapping. Mullis received the 1993 Nobel Prize in Chemistry for this technique.

PCR involves successive rounds of DNA synthesis of double-stranded DNA (dsDNA). The two strands are first separated and used as a template to synthesize another strand each; these strands then form a new dsDNA. Both the original and the new dsDNA are then subjected to this same process. This cycle is performed from 20 to 40 times, and the result is an exponential amplification of the DNA of interest (see Figure 4).

Site-directed mutagenesis, invented in 1978 by Michael Smith (1932–2000) of the University of British Columbia in Canada, is a targeted alteration of DNA sequence. It involves inserting DNA to be mutated into a plasmid and exposing it to a short DNA fragment containing one changed base. This fragment then binds to the insert. A new DNA molecule is synthesized from this DNA fragment, and this results in a double-stranded plasmid. Multiplying this plasmid in bacteria will produce equal numbers of single-stranded plasmids with and without the mutation; the mutated DNA can then be isolated (see Figure 5). In 1982 Smith managed to produce the protein products of these induced mutations, which allowed for analysis of protein function. Smith was the co-recipient of the 1993 Nobel Prize in Chemistry for this technique and shared the prize with Mullis.

Introducing a recombinant DNA molecule into the host cell varies according to the host organism. Bacteria can take up DNA molecules naturally. In animal cells in culture and fertilized egg cells, DNA can be injected directly into the cell

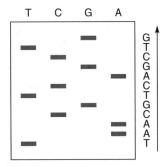

1. Sequencing reactions loaded onto polyacrylamide gel for fragment separation

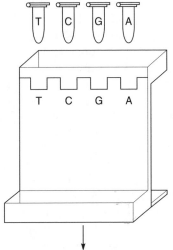

2. Sequence read (bottom to top) from gel autoradiogram

Figure 3.   Sanger sequencing. Once the four dideoxynucleotide sequencing reactions are concluded, each reaction tube, containing one of the four dideoxynucleotides (T, C, G, or A) are subjected to gel electrophoresis, which separates the different synthesized strands. The sequence of the original DNA strand is then ascertained by reading the gel from bottom to top. [*Source: Reprinted with permission from U.S. Department of Energy Human Genome Program, Primer on Molecular Genetics, 1992.*]

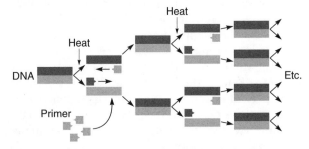

Polymerase Chain Reaction (PCR)

Figure 4.   Exponential amplification of DNA using the polymerase chain reaction (PCR). Heat is used to separate the DNA strands. Primers are short DNA strands used to initiate DNA synthesis. [*Source: Reprinted with permission from Wrobel, S. Serendipity, science, and a new hantavirus. FASEB Journal, 9, 1247–1254, 1995.*]

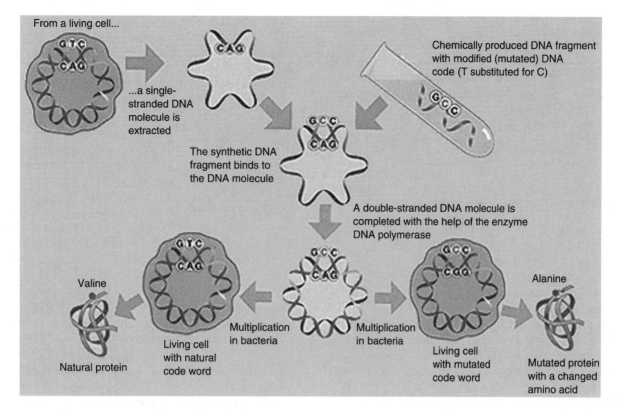

Figure 5. Site-directed mutagenesis.
[*Source: Reprinted with permission from the Royal Swedish Academy of Sciences from the Nobel E-Museum, Illustrated Presentation: The Nobel Prize in Chemistry, 1993.*]

nucleus, where it integrates with the host DNA. In plants, destruction of the cell wall permits DNA to be readily taken up by the plant cell, which then spontaneously reforms the cell wall. Alternatively, a vector or small carrier molecule can be derived from plasmids, bacteriophages, or plant and animal viruses. In 1977 it was shown that part of the tumor-inducing (Ti) plasmid of the soil-borne bacterial plant pathogen *Agrobacterium tumefaciens* (which causes galls or tumors on the plant near soil level) is transferred to the host plant cell during infection. If the tumor gene is removed and replaced with another gene, the bacteria can introduce the beneficial gene into plants. The Ti plasmid is now widely used as a vector system for transferring foreign genes into the plant genome.

More commonly, a retroviral vector that has been rDNA modified can be used to infect mammal cells: the viral DNA is integrated into the host cells.

Many modifications to these common techniques have increased their effectiveness and usefulness. Together, they allowed scientists to manipulate DNA in unprecedented ways and created major topics of interest and research for industry, government, and the general public.

See also **Cloning, Testing and Treatment Methods; Gene Therapy; Genetic Engineering, Applications**

TED EVERSON

Further Reading

Aldridge, S. *The Thread of Life: The Story of Genes and Genetic Engineering*. Cambridge University Press, New York, 1996.

Cantor, C.R. and Smith, C.L. *Genomics: The Science and Technology Behind the Human Genome Project*. Wiley, New York, 1999.

Cook-Deegan, R. *The Gene Wars: Science, Politics, and the Human Genome*. W.W. Norton, New York, 1996.

Fox Keller, E. *The Century of the Gene*. Harvard University Press, Cambridge, MA, 2000.

Green, J.J. and Rao, V.B., Eds. *Recombinant DNA Principles and Methodologies*. Marcel Dekker, New York, 1998.

Sheldon K. *Biotechnics and Society: The Rise of Industrial Genetics*. Praeger, New York, 1991.

Watson, J.D., Gilman, M., Witkowski, J. and Zoller, M. *Recombinant DNA: A Short Course*, W.H. Freeman, San Francisco, 1983.

Wright, S. Recombinant DNA technology and its social transformation, 1972–1982. *Osiris*, 2, 303–360, 1986.

Useful Websites

Southern blotting: http://www.biology.washington.edu/fingerprint/blot.html

Human Genome Project information: http://www.ornl.gov/hgmis

Nobel Prize in Chemistry, 1993: http://www.nobel.se/chemistry/laureates/1993/illpres/index.html

Genetic Screening and Testing

The menu of genetic screening and testing technologies now available in most developed countries increased rapidly in the closing years of the twentieth century. These technologies emerged within the context of rapidly changing social and legal contexts with regard to the medicalization of pregnancy and birth and the legalization of abortion. The earliest genetic screening tests detected inborn errors of metabolism and sex-linked disorders. Technological innovations in genomic mapping and DNA sequencing, together with an explosion in research on the genetic basis of disease which culminated in the Human Genome Project (HGP), led to a range of genetic screening and testing for diseases traditionally recognized as genetic in origin and for susceptibility to more common diseases such as certain types of familial cancer, cardiac conditions, and neurological disorders among others. Tests were also useful for forensic, or nonmedical, purposes.

The earliest genetic testing procedures began in the 1950s when amniocentesis was used to obtain fetal cells from amniotic fluid for analysis to ascertain the sex of fetuses being carried by women with a family history of sex-linked diseases such as hemophilia. However, this procedure was not rapidly adopted, not in the least part because of the nearly universal illegality of abortion until the late 1960s. A more precise form of genetic testing using amniocentesis emerged in the mid-1960s. The procedure followed the discovery in the late 1950s that the common form of the condition known then as mongolism and now Down's syndrome was caused by the presence of three copies of chromosome 21 (instead of the normal two). The condition came to be called trisomy 21 as other major chromosomal abnormalities associated with disease conditions were identified. Although it was initially difficult to detect abnormalities using cultures of fetal cells obtained via amniocentesis, these technological problems were overcome by the mid-1960s. Testing of pregnant women both due to family history but more widely to "advanced maternal age" (known to be associated with increased risk of trisomy 21) became common, particularly following the legalization of abortion and later with increased litigation against physicians, particularly in the U.S.

Also in the 1960s, widespread testing of newborns for the metabolic disease phenylketonuria (PKU), which results from inheritance of two copies of a mutatedgene responsible for making the enzyme phenylalanine hydroxylase, was made possible through the development of basic and economically efficient screening technologies. Newborn screening for PKU and other metabolic diseases involves taking a blood sample from the baby's heel and placing small blood spots on a piece of filter paper from which samples are punched out to test for metabolic errors. This method was developed by the U.S. microbiologist Robert Guthrie, whose name is now associated with the filter paper samples known as "Guthrie cards." Guthrie successfully lobbied U.S. state governments to require testing of all newborns, which he argued would reduce public health costs by allowing early dietary intervention and perhaps avoiding the disease's damaging early effects. The testing was not uncontroversial, however, since the diet was difficult to maintain and the rate of incorrect diagnoses (false positives and false negatives) was initially high.

A particularly notable early genetic screening program occurred in the U.S. in the early 1970s for sickle cell anemia. This condition affects approximately 1 in 400–600 African–Americans who have two copies of the sickle cell version of the β-hemoglobin gene, and 1 in 10–12 African–Americans are carriers for this gene. Mandatory screening programs were implemented, focusing on school-aged children and young adults, despite limited treatment options. These programs were widely criticized for problems with the accuracy of the test results, limited community consultation, and the stigmatization and discriminatory effects that often resulted, particularly for those who were carriers but who did not in fact have sickle cell anemia.

Few additional testing options became available until the 1980s and 1990s, when genetic technologies began to allow screening for many more disease conditions. The range of conditions detectable through prenatal testing expanded to include not only trisomy 21 and other major chromosomal abnormalities, but also conditions associated with point gene mutations such as cystic fibrosis. Screening procedures were initially targeted at those couples with a family history of the particular genetic disease and members of subpopulations where it was recognized that certain diseases were more common (e.g., testing for Tay–Sachs disease among those of Ashkenazi Jewish backgrounds). A third targeted group were women who had abnormal test results early in their pregnancies from procedures such as ultrasound or screening for maternal serum alpha-fetoprotein (MSAFP), a simple blood test that allows early detection of increased risk of neural tube defects and Down's syndrome. Prenatal testing is performed using amniocentesis or chorionic villus sampling (CVS), a newer technology that allows detection of genetic anomalies at an earlier stage of pregnancy (i.e., during the first trimester, or first three months). Although originally developed in the 1970s, CVS did not become commonly used until the 1980s. It was considered more desirable than amniocentesis by many consumers since it allows earlier decisions about pregnancy termination, although the procedure carries greater risks.

Adult presymptomatic testing also began for a limited range of inherited conditions, notably the late-onset neurological disorder Huntington's disease, which is a dominant condition transmitted via one copy of the affected gene so that every individual with one affected gene expresses the disorder (compared to recessive disorders, where two copies of the affected gene are needed). Many of these genetic testing programs, especially those associated with adult onset disorders, have not been popular for a number of reasons: lack of availability of treatment options, anticipated problems with discrimination with regard to life and health insurance, employment and other areas, concerns about predictive value (i.e., the lack of availability of information about age of onset, severity of disease), and the potential for possibly revealing the genetic status of other family members such as parents or twins.

The Human Genome Project, begun in 1990 and completed in 2003, refined existing mapping and sequencing technologies and provided more information about the function of genes and genetic pathways, resulting in an increased number of diseases that can be detected using genetic testing and screening. In some cases, genetic testing can now provide predictive information about the likely medical effects of particular genetic mutations or DNA sequence variations known as single nucleotide polymorphisms (SNPs), although for many diseases this information remains limited. Screening for predispositions to develop serious conditions such as certain types of familial cancer, neurological conditions, and cardiac disease also has become available, and such programs undoubtedly will expand to include many more common conditions.

Consideration of the ethical, legal, and social implications of clinical genetic testing raises concerns about privacy, potential misuse, and counseling. If genetic information is disclosed, how can we prevent discrimination by health insurers and employers? Knowledge of genetic differences may lead to anxiety about disease, relationship breakdown, and social stigmatization. A positive result for a predictive test for adult-onset diseases such as Huntington's disease does not guarantee a cure or treatment, and individuals in high-risk families may decide not to be tested. Screening for susceptibility to diseases is also likely to benefit few people in the developing world.

Genetic screening techniques are now available in conjunction with *in vitro* fertilization and other types of reproductive technologies, allowing the screening of fertilized embryos for certain genetic mutations before selection for implantation. At present selection is purely on disease grounds and selection for other traits (e.g., for eye or hair color, intelligence, height) cannot yet be done, though there are concerns for eugenics and "designer babies." Screening is available for an increasing number of metabolic diseases through tandem mass spectrometry, which uses less blood per test, allows testing for many conditions simultaneously, and has a very low false-positive rate as compared to conventional Guthrie testing. Finally, genetic technologies are being used in the judicial domain for determination of paternity, often associated with child support claims, and for forensic purposes in cases where DNA material is available for testing.

See also **Diagnostic Screening; Fertility, Human; Gene Therapy; Genetic Engineering, Applications; Genetic Engineering, Methods**

RACHEL A. ANKENY

Further Reading

Andrews, L. *Future Perfect: Confronting Decisions about Genetics.* Columbia University Press, New York, 2001.

Cowan, Ruth Schwartz. Aspects of the history of prenatal diagnosis. *Fetal Diagn. Ther.*, 8, 1, 10–17, 1993.

Holtzman, N.A. *Proceed with Caution: Predicting Genetic Risks in the Recombinant DNA Era.* Johns Hopkins University Press, Baltimore, 1989.

Kevles, D.J. and Hood, L., Eds. *Code of Codes: Scientific and Social Issues in the Human Genome Project.* Harvard University Press, Cambridge, 1992.

Lippman, A. Prenatal genetic testing and screening: constructing needs and reinforcing inequities. *Am. J. Law Med.*, 17, 15–50, 1991.

Lynch, M. and Jasanoff, S., Eds. Contested identities: science, law and forensic practice. *Soc. Stud. Sci.* 28, 5–6, 1998.

Millington, D.S. Newborn screening for metabolic diseases. *American Scientist*, 90, 1, 40–47, 2002.

Markel, H. The stigma of disease: implications of genetic screening. *Am. J. Med.*, 93, 2, 209–215, 1992.

Rothman, B.K. *The Tentative Pregnancy: Prenatal Diagnosis and the Future of Motherhood.* Viking, New York, 1986.

U.S. National Research Council. *The Evaluation of Forensic DNA Evidence.* National Academy Press, Washington D.C., 1996.

Useful Websites

Paul, D.B. The history of newborn phenylketonuria screening in the U.S. in *Promoting Safe and Effective Genetic Testing in the United States: Final Report of the Task Force on Genetic Testing*, Holtzman, N.A. and Watson M.S., Eds.

Markel, H. Scientific advances and social risks: historical perspectives of genetic screening programs for sickle cell disease, tay-sachs disease, neural tube defects and Down syndrome, 1970–1997, in *Promoting Safe and Effective Genetic Testing in the United States: Final Report of the Task Force on Genetic Testing*, Holtzman, N.A. and Watson M.S., Eds.

Both available at http://www.nhgri.nih.gov/ELSI/TFGT _final

Global Positioning System (GPS)

The use of radio signals for navigation began in the early twentieth century. Between 1912 and 1915, Reginald Fessenden of Boston, Massachusetts, transmitted radio waves from designated shore-based stations to correct ship chronometers and give mariners a sense of direction. The prospect of greater accuracy came in 1940 when Alfred Loomis of the National Defense Research Council suggested simultaneous transmission of pulsed signals from a pair of precisely surveyed ground stations to ship receivers and use of the difference in arrival times of the two signals to calculate a line of position. Developed during World War II by the Radiation Laboratory at the Massachusetts Institute of Technology, the system Loomis had proposed became known as Loran (LOng RAnge Navigation), and it supported convoys crossing the Atlantic. Subsequent improvements significantly increased Loran accuracy, but the technology was still limited to two dimensions—latitude and longitude. Not until 1960 did Ivan Getting of Raytheon Corporation in Lexington, Massachusetts, propose the first three-dimensional type of Loran system.

Intended to solve navigational problems associated with rail-based, mobile intercontinental ballistic missiles, Raytheon's system was called Mosaic (MObile System for Accurate ICBM Control). That system was never developed, however, because a new technology—artificial satellites—provided the basis for more precise, line-of-sight radio navigation.

The NAVSTAR (NAVigation System Timing And Ranging) Global Positioning System (GPS) provides an unlimited number of military and civilian users worldwide with continuous, highly accurate data on their position in four dimensions—latitude, longitude, altitude, and time—through all weather conditions. It includes space, control, and user segments (Figure 6). A constellation of 24 satellites in 10,900 nautical miles, nearly circular orbits—six orbital planes, equally spaced 60 degrees apart, inclined approximately 55 degrees relative to the equator, and each with four equidistant satellites—transmits microwave signals in two different L-band frequencies. From any point on earth, between five and eight satellites are "visible" to the user. Synchronized, extremely precise atomic clocks—rubidium and cesium—aboard the satellites render the constellation semiautonomous by alleviating the need to continuously control the satellites from the ground. The control segment consists of a master facility at Schriever Air Force Base, Colorado, and a global network of automated stations. It passively tracks the entire constellation and, via an S-band uplink, periodically sends updated orbital and clock data to each satellite to ensure that navigation signals received by users remain accurate. Finally, GPS users—on land, at sea, in the air or space—rely on commercially produced receivers to convert satellite signals into position, time, and velocity estimates.

Drawing the best concepts and technology from several predecessor navigation systems, engineers synthesized one that became known in the early 1970s as GPS. Those previous efforts began with Transit or the Naval Navigation Satellite System (NNSS), developed in the late 1950s by the Applied Physics Laboratory at Johns Hopkins University in Baltimore, Maryland, and used operationally by both the U.S. Navy and commercial mariners from 1962 to 1996. Also contributing to the mix was the U.S. Naval Research Laboratory's Timation satellite project that began during 1964 to provide very precise timing and time transfer between points on Earth. A third element originated with Aerospace Corporation's Project 57, which included aircraft navigation using satellite signals and which the

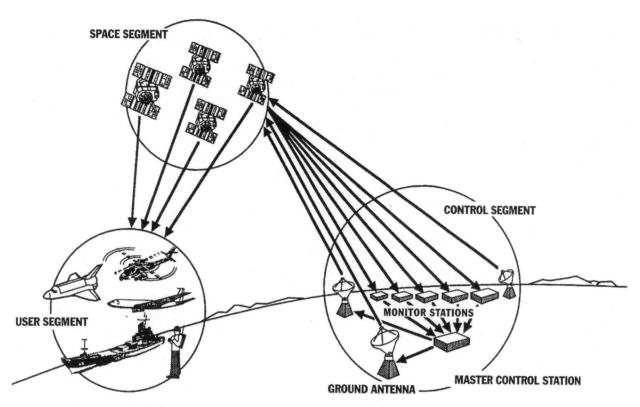

Figure 6. Major Global Position System (GPS) segments. [*Courtesy U.S. Air Force.*]

U.S. Air Force, under direction from the Advanced Research Projects Agency (ARPA), pursued as Project 621B beginning in October 1963. The formation in 1973 of a multiservice or joint program office headed by a U.S. Air Force colonel, Bradford Parkinson, drew together these various programmatic threads led to approval from the Defense System Acquisition and Review Council (DSARC) to proceed with GPS development. This led quickly to the launch of two Navigation Technology Satellites to explore technical capabilities. By the time the first Block I or Navigation Development Satellite was launched in February 1978, an initial control segment was functioning and several kinds of user equipment were being tested at Yuma Proving Ground on the Arizona–California border.

The GPS space segment evolved steadily from the late 1970s into the 1990s. Between February 1978 and October 1985, the U.S. Air Force successfully launched and operated ten of eleven Block I satellites (Figure 7). In February 1989, the first Block II fully operational GPS satellite went into orbit, and near the end of 1993, both military and civilian authorities had declared achievement of initial operational capability. With the launch of a Block IIA satellite on March 9, 1994, a fully

operational, 24-satellite GPS constellation was completed. To sustain the space segment, the U.S. Air Force began launching Block IIR satellites on July 23, 1997. The latter also enhanced performance through greater autonomy and increased protection against radiation—natural or man-made. Further improvements scheduled for 2005 with Block IIF and sometime after 2009 with Block III would make GPS less vulnerable to jamming by delivering a more powerful signal to military users. In addition, these future GPS satellites would be able to surge signals over a specific area by using spot beams and would employ new frequencies for civil "safety of life" applications.

Designed for use by both military and civilian parties at no charge, GPS originally offered two levels of service—precise and standard. With specially equipped receivers, authorized users could pinpoint their precise position and time with at least 22-meter horizontal, 27.7-meter vertical, and 200-nanosecond accuracy. Through a Selective Availability (SA) feature, the U.S. Defense Department could intentionally degrade GPS signals for the majority of users, who possessed standard receivers. The latter were accurate only to within 100 meters horizontally,

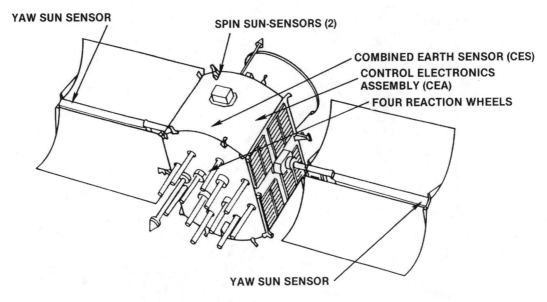

Figure 7.   Global Positioning System Block I satellite diagram indicating avionics equipment.
[*Source: Courtesy U.S. Air Force.*]

156 meters vertically, and 340 nanoseconds. By 1996, however, it had become evident that the exploding market for GPS among civilians demanded greater precision. Consequently, on May 1, 2000, President Clinton announced immediate discontinuation of SA. By giving everyone access to the more precise level of service, his administration hoped to encourage private sector investment in what was already a multibillion-dollar growth industry.

The uses for GPS data have become mind boggling in number and variety, as have the number of GPS devices. On the military side, the uses include aerial refueling, rendezvous operations, forward air control, location and recovery of downed pilots, targeting and delivery of precision weapons, and computer network integration. Civilians employ GPS to manage natural resources, control maritime and land-based traffic, survey and map terrain, conduct search and rescue operations, monitor earthquakes and storms, improve athletic performance, and link communication networks and disparate databases. In automobile GPS devices alone, there was a tenfold increase from 1,100,000 units in 1996 to 11,300,000 worldwide in 2001. As GPS receivers became smaller and cheaper, they could be found in everything from cellular telephones to wristwatches.

At the beginning of the twenty-first century, GPS signals were even being used for navigation in space, and it seemed likely that aviation authorities would soon rely on an augmented GPS capability for air traffic control. Nonetheless, skeptics doubted that GPS would someday govern people's lives to the extent enthusiasts predicted, and alarmists warned that jamming or hacking could render a GPS-dependent world militarily and economically helpless. Supporters worried about maintaining stable funding sources for the GPS space and control segments, but critics in the U.S. Congress accused the U.S. Air Force of poor program execution as well as budgeting too far ahead of what it needed simply to sustain the system. Meanwhile, engineers pondered future GPS designs that might improve positional accuracy to 30 centimeters and time transfer to 1 nanosecond before the end of the first decade of the new century.

See also **Gyrocompass and Inertial Guidance; Radionavigation**

RICK W. STURDEVANT

Further Reading

Adams, Thomas K. GPS vulnerabilities. *Mil. Rev.*, 81, 10–16, 2001.

Getting, I.A. The global positioning system. *IEEE Spectrum*, 30, 36–47, 1993.

Marquis, W. M is for modernization: Block IIr–m satellites improve on a classic. *GPS World*, 12, 36–42, 2001.

National Academy of Public Administration and National Research Council. *The Global Positioning System: Charting the Future*. National Academy Press, Washington D.C., 1995.

National Research Council. *The Global Positioning System: A Shared National Asset*, National Academy Press, Washington D.C., 1995.

Nordwall, B.D. Using pseudo-satellites to foil GPS jamming. *Aviation Week & Space Technology*, 155, 54–61, 2001.

Parkinson, B.W., and Spilker, J.J. Jr., Eds. *Global Positioning System: Theory and Applications*, 2 vols. American Institute of Aeronautics and Astronautics, Washington D.C., 1996.

Rip, M.R. and Hasik, J.M. *The Precision Revolution: GPS and Future of Aerial Warfare*. Naval Institute Press, Annapolis, 2002.

Useful Websites

Adams, T.K. GPS vulnerabilities. *Military Review*: http://www-cgsc.army.mil/milrev/English/MarApr01/adams.htm

Dana, P.H. *Global Positioning System Overview*: http://www.colorado.edu/geography/gcraft/notes/gps/gps.html

Federal Aviation Administration (FAA) Global Positioning System Product Team: http://gps.faa.gov/

Franson, P. GPS's Sweet Dreams: http://www.upside.com/texis/mvm/story?id=369123830

NAVSTAR GPS Joint Program Office (SMC/CZ): http://gps.losangeles.af.mil

Navtech, GPS Related Internet Sites: http://www.navtechgps.com/links.asp

Globalization

Technology was pivotal to globalization throughout the twentieth century. While the transfer and diffusion of technological innovations fueled economic growth, ever-greater global economic interdependence was enabled by new transport and communication technologies, which progressively tamed time and distance over the course of the century. Technology was thus both a direct agent of the economic growth and development underpinning globalization and, simultaneously, a medium of globalization facilitating the evergreater movement of materials, people, knowledge, and ideas on which economic interdependence depended. Daniele Archibugi and Jonathan Michie (1997) distinguished between the former concentration upon the *globalization of technology* and the latter concentration on *the technologies of globalization*. These two categories provide the focus for this entry. A common thread is provided by the mutual significance of the institutional, organizational, and broader socio-cultural systems of which technologies are part, as they are crucial to understanding the dynamics of their development, dissemination, and utilization. Railways, for example, not only facilitated the movement of people and goods but also inspired innovations in business organization, were intimately involved in the development of the national identities of various nations, and crucially shaped conceptions of time and timekeeping. While the post-World War II period of intense global economic growth and integration is most widely identified with the term globalization, analogous activity earlier in the century.

The Technologies of Globalization

The dawn of the twentieth century was preceded by some three decades of unprecedented and intense integration and consolidation in the world economy brought about by the traffic in materials, people, and ideas that resulted from railroads, steamships, and the telegraph. Newly emerging technologies including the telephone, the wireless, and the internal combustion engine were all soon destined to bring the world even closer together. Ongoing industrialization both facilitated and drove these developments with the early industrial, most notably northern European, nations relying heavily on this traffic to supply an escalating demand for both basic foodstuffs and industrial raw materials. This increasing demand stimulated a search for less expensive, and thus more remote, sources of supply and a concomitant outflow of capital, skilled labor, and technology to develop these resources. A number of countries, notably Argentina, Australia, Canada, New Zealand, South Africa, and Uruguay, experienced rapid development at this time as a result of the influx of overseas capital, labor, and technology resulting from this demand for their primary products. Perhaps the greatest single factor facilitating these developments was the improvement in transport, witnessed by increased investment in port facilities and railways both within industrialized nations and primary exporting countries. The development of these latter regions was, however, highly reliant on the diffusion of technological improvements, particularly in the mining and agricultural sectors.

The pre-World War I period paralleled the globalization that marked the closing decades of the twentieth century in a number of other ways. The average size of firms and business enterprises grew rapidly, sound forms of business organization increasingly became the norm, and dynamic multinational business activity emerged on a large scale. Changes in trade, manufacturing, and technology, such as the increasing use of electric motors, also resulted in the first significant cases of painful deindustrialization. Parts of northern England, for example, which were first to industrialize with steam and textiles, could not respond rapidly enough to these changes, and long-lasting pools of unemployment were created. The limit of this analogy to the closing decades of the century is underlined by the fact that when this period was abruptly halted by World War I, it was largely only Western Europe, North America, and Japan that had industrialized. Much of the rest of the globe, however, including some of the poorest nations such as those of Africa, were increasingly drawn upon as sources of raw materials.

The period after World War I saw the emergence of automobiles and air travel, which both vastly accelerated the taming of time and distance and extended these changes to far more people and further corners of the globe. However foreign trade declined in importance relative to the prewar period. As a result some contemporary economists argued that technological progress and the spread of industrialization would contribute to declining trade between nations. These arguments were never broadly accepted, and the surge in economic growth and accompanying expansion in trade that followed World War II effectively ended the discussion. The 1920s and 1930s did, nonetheless, lay a foundation for the coming surge in growth and globalization. The Middle East developed rapidly as a source of oil on the back of western investment. Russia's swift industrialization was facilitated by burgeoning domestic resources, while Japan set the scene for later developments by developing a regional economic hinterland, most notably in Korea.

In the years after 1945, there was not only development of the intimate global economic interdependence that inspired the term globalization but also a parallel and ultimately perhaps more profound political and socio-cultural interdependence. The most commonly perceived and perhaps most extensive of these latter developments was the global ascendency of American popular culture, the pervasiveness of which profoundly marked late twentieth century life. On the political front, the first truly global institutions came with the founding of the United Nations in 1945 and the development of the Bretton-Woods institutions (the World Bank and International Monetary Fund) that played particularly critical roles in late twentieth century economic globalization. Communications and reinforcing developments in entertainment technologies were key to the advance of American popular culture. The rise of television, the consolidation of Hollywood as a global focus for film making, and the emergence of rock and roll music and the youth culture that

accompanied it paralleled radical geopolitical changes external to the U.S. Such changes as the rise of welfare states in Europe, decolonization, and postwar rebuilding, left a vacuum that American popular culture was destined to fill. Technology played both an instrumental and a profoundly cultural role in these developments. Many technologies including automobiles and motorbikes, and for some politicians hydroelectric dams and nuclear power stations, took on an iconic status that was close to totemic. Analogous developments occurred on many fronts, from pop music and movies to fast food, soft drinks, and personal entertainment equipment and telephones.

The technological character of World War II itself signaled a fundamental shift in the importance and role assigned technology that would be borne out in increased research and development (R&D) spending in the postwar period. The development of radar, rockets, digital computers and the harnessing of science to technological imperatives, pioneered by the U.S. Manhattan Project to develop the first atomic bomb, all raised immense possibilities that were soon to be the focus of frenetic activity. R&D spending in the U.S. rose from 1 percent of gross domestic product (GDP) in 1950 to 2.8 percent in 1960. In the USSR over the same period, it grew from 1.2 percent to 2.5 percent, a trend reflected in most other leading industrial nations. By 1993 the U.S. figure was still at 2.8 percent, while other leading investors such as Germany at 2.5 percent, the U.K. at 2.15 percent, and South Korea at 2.1 percent reflected similar priorities and an increasing acknowledgement of the role of science and technology in wealth creation. The vast increases in trade accompanying these changes brought a significant growth in foreign direct investment (FDI) by which commercial interests in one country invest in activities in another. While U.S. multinational corporations initially dominated this growth, European multinationals were featured in the investment flows from the 1970s, with Japanese multinationals emerging during the 1980s. By the mid-1990s many other countries including Hong Kong, Korea, Singapore, Taiwan, China, Mexico, and Brazil were involved.

Railway construction continued in many developing regions after 1945, but in most industrialized countries, railways declined in significance in the face of competition from road transport and the growing importance of aviation. This led to rising oil consumption, reflected in a significant increase in the amount of oil traded globally and in oil tanker capacity. Increased air travel further promoted greater commercial interdependence by facilitating the face-to-face contact invaluable in business. From the late 1950s containerization vastly increased the speed and efficiency of the handling and transportation of goods, while emerging satellite links would revolutionize intercontinental telecommunications.

Much of the increase in economic interdependence after 1945 reflected an intensification of earlier trends and developments. For example, a significant element in Japan's rise to economic prominence was her skill in harnessing innovation and technological systems, notably in manufacturing and transport, in which innovation was as much organizational as technical. The success of Japan's postwar strategy to gain access to raw materials emulated the earlier success of the U.S. and northern Europe by harnessing improved technologies in shipping and bulk cargo handling across great distances and between countries. Throughout southern Asia and Australasia in particular, export-oriented primary commodity production and transportation systems were developed in which the size and scale of all elements were integrated to optimize efficiencies and economies of scale. Such elements included extractive technologies such as mining operations, transportation to port, port and shipping facilities, and Japanese domestic port and manufacturing operations. The success of Japanese innovation and production management in this period is highlighted by the way Japan's Ministry for International Trade and Industry (MITI) methods, such as "lean production" in the automobile industry, were so widely emulated by other leading industrial nations during the final decades of the century.

Technology was increasingly regarded as a key to competitive advantage, placing an ever-greater emphasis on information and knowledge and on their management rather than on technology *per se*. While this emphasis reflected the economic rationalism prevalent at this time, which closely resembled the *laissez-faire* perspective of the late nineteenth and early twentieth century, it was also an acknowledgment of the contemporary significance of knowledge in product innovation and development. This shift was reflected both in policy and in the nature of some of the most significant technologies of the closing decades of the century. The microprocessor-based information and communication technologies and genetic biotechnologies that emerged at this time were both centered on the encryption, transmission, and manipulation of knowledge and information.

The development of microprocessors in the 1970s unleashed unprecedented information processing and communication capabilities that became particularly marked with an exponential growth in their power and in the extent of the Internet in the 1980s. These technologies greatly facilitated both the economic deregulation of the leading industrial economies (given particular impetus by President Ronald Reagan in the U.S. and Prime Minister Margaret Thatcher in the U.K.) and the surge in transnational commerce and related rise of speculative global capital markets that were such a dominant feature of the late twentieth century. For example, the deregulation of electricity industries, a major concern of this period, was made possible by new systems able to coordinate transactions between numerous generators and consumers, scheduling and administering the transmission and the quality of supply. The scale of trading in speculative global capital markets facilitated by these technologies, including trading of such major economic indicators as currencies, was so great that it challenged the volume of the markets in material goods and industrial output by the close of the century.

These technologies soon permeated everyday life in the industrial world; electronic domestic monetary transactions were common and by the end of the century home computer use for work, entertainment, and information retrieval purposes were widespread. The significance of these changes was the subject of great debate as the century closed. What is clear is that there were significant differences with the scale and pace of past developments. A particularly notable example is the Internet, which achieved ubiquity within about a decade, many times faster than earlier comparable technologies such as the telephone. By the end of the century the management, transmission, and manipulation of information had become both a mainstay of wealth creation and of significance for lifestyles more generally. This was quite distinct from an earlier narrower emphasis on the production of material goods and from earlier "knowledge economies" in which stock market activity was tied directly to the production of material goods.

A similar emphasis on the manipulation of information is central to genetic biotechnology, which also emerged and developed rapidly in this period. The potential of this technology, a major product of some of the larger contemporary multinational agrichemical and pharmaceutical companies, was still largely untested and controversial as the century closed. However the implications of human understanding and intervention in the "coding of life" were widely viewed as being at least as profound as those of the information and communication technologies discussed above.

The "dematerialization" evident in these new information focused technologies and trends and the weakening of links to tradition and place characteristic of late twentieth century globalization were understood in a variety of ways. Some interpreted these changes in terms of the advent of a new postindustrial or postmodern society. However, while these notions flagged the significant changes, if not transformations, evident in leading industrial economies, some of the poorest nations, including many African ones, remained underdeveloped and poorly integrated with the global economy and broader global culture even by the close of the century.

Some of the most thoughtful insights into the nature of the changes marking late twentieth century life were made by those with a limited understanding of technology but a profound understanding of society and its dynamics. Anthony Giddens, a leading twentieth century social theorist, described how the implications of these technologies of globalization can be understood in terms of an increasing distancing of time and space involving a growing separation of social activity from localized contexts. The implications he draws from this include a novel requirement for trust in complex large-scale sociotechnical systems such as airlines or the Internet, a mounting ability to apply updated knowledge to personal circumstances that reinforces the dynamic complexity of contemporary life, and the way differentials in knowledge increasingly map into differentials in power. Giddens' analysis serves to underline how technology had become less a tool for human ends and more an integral element of human life and broader culture over the course of the twentieth century.

The Globalization of Technology

While late twentieth century globalization was both an effect and a cause of the large-scale development, exploitation, distribution, and transfer of scientific and technological innovation, mainstream economists such as Rosenberg (1982) regarded technology as a "black box" for most of the century. In this view, technology was simply a "public good" autonomous of broader economic and social factors. From the 1980s however, the economics of technology became more widely acknowledged and by the 1990s increasingly studied by policy makers. Inspired particularly by

Karl Marx and Joseph Schumpeter, the economics of technology concentrates on innovation as the key to technological change and economic growth. It encompasses a variety of approaches and ascribes a varying importance to firms, forces at the national level, global forces, or to particular technological systems or regimes. However, perhaps surprisingly in light of a widespread perception of the late twentieth century power of multinational corporations, the concept of "national systems of innovation" was particularly influential. Echoing the insights of the twentieth century German economist Friedrich List, these ideas underline the critical importance of education and training, investment in R&D, national integration and infrastructure, and the extent and quality of interaction between the many national players involved in innovation.

The increasing internationalization of business, witnessed by the twentieth century surge in FDI, underlined the fact that national technological competence was regarded as a key to harnessing technology. While direct national investment in R&D has the potential to benefit offshore interests in a globalized world, national technological competence was seen as more likely to encourage local innovation, inward investment, and an ability to utilize technologies developed elsewhere. Also, while education and training are crucial to national technological competence, the structuring of economic incentives and broader aspects of national culture such as attitudes to technology are also significant elements. The globalization of technology was thus regarded as being as much contingent on context and culture as it was on hard cash or engineering.

However, while business and governments universally promoted late twentieth century globalization as an incontestable good, in practice it advanced the interests of the most powerful industrial countries and did little to redress contemporary patterns of inequity and unequal development. These are complex and controversial matters in which technology played a small but not insignificant part. Technology transfer was generally acknowledged to be essential for more equitable economic growth and development, but commercial imperatives significantly constrained such transfer. For example, resistance by developed countries to effective technology transfer in the environmental domain, dating back to the very first multilateral environmental summit in Stockholm in 1972, was significant in impeding negotiation and agreement in both the Montreal Protocol (1987) and Kyoto Protocol (1997) processes. Another, rather different, example of the role of technology in these matters is provided by the speculative global capital markets. Widely implicated in the Asian "financial crisis" of 1997–1998, these markets were a target for regulation and reform as the century closed.

The complex and increasingly global relationships of technology and societies over the course of the twentieth century had many ramifications. Not only was the globe effectively a smaller, more closely knit place by the end of the century, but many aspects of life in the most developed parts of the globe were thoroughly technology-driven in ways quite distinct from earlier societies. In the industrialized world, technology formed part of the fabric of life in previously unimaginable ways. Electricity networks, airlines, the Internet, and highway systems fashioned what people did in the late twentieth century far more intensively than railroads, steamships, and the telegraph had done at the start of the century. Technological developments increasingly informed cultural developments. By the end of the century, digital technology was central to both the form and content of many types of visual art, music, and communication, with the latter opening up new forms of interpersonal and communal interactions that were evolving rapidly as the century closed.

While the quality of life enabled by these changes was immeasurably better in the developed countries at the end of the century, some developments, such as those in the environmental arena, were profoundly disturbing. What was becoming clearer as the century closed was that a thoroughly technologized globe would only be viable under conditions that addressed the complex interaction of technology and society. Although the twentieth century afforded the insight that this problem was as much a political, cultural, and economic matter as it was a technical one, it would be for the next century to unravel the repercussions of this.

See also **Communications; Computers, Uses and Consequences; Organization of Technology and Science; Research and Development in the Twentieth Century; Social and Political Determinants of Technological Change; Technology, Society, and the Environment; Telecommunications; Transport**

STEPHEN HEALY

Further Reading

Archibugi, D. and Michie, J. Technological globalization and national systems of innovation: an introduction, in *Technology, Globalization and Economic Performance*, Archibugi, D. and Michie, J., Eds. Cambridge University Press, Cambridge, 1997.

Archibugi, D. and Michie, J., Eds. *Trade, Growth and Technical Change*. Cambridge University Press, Cambridge, 1998.

Bunker, S. and Ciccantell, P. Restructuring space, time, and competitive advantage in the capitalist world economy: Japan and raw materials transport after World War II, in *A New World Order? Global Transformations in the Late Twentieth Century*, Smith, D. and Borocz, J., Eds. Greenwood Press, Westport, CT, 1995.

Cowan, R.S. *A Social History of American Technology*. Oxford University Press, New York, 1997.

Fichman, M. *Science, Technology and Society: A Historical Perspective*. Kendall/Hunt Publishing, Dubuque, IA, 1993.

Giddens, A. *The Consequences of Modernity*. Polity Press, Cambridge, 1990.

Kenwood, G. and Lougheed, A.L. *The Growth of the International Economy 1820–2000: An Introductory Text*. Routledge, London, 1999.

McMillan, C. Shifting technological paradigms: from the US to Japan, in *States Against Markets: The Limits of Globalization*, Boyer, R. and Drache, D., Eds. Routledge, New York, 1996.

Rosenberg, N. *Inside the Black Box: Technology and Economics*. Cambridge University Press, Cambridge, 1982.

Talalay, M., Farrands, C. and Tooze, R., Eds. *Technology, Culture and Competitiveness: Change and the World Political Economy*. Routledge, London, 1997.

Green Chemistry

The term "green chemistry," coined in 1991 by Paul T. Anastas, is defined as "the design of chemical products and processes that reduce or eliminate the use and generation of hazardous substances." This voluntary, nonregulatory approach to the protection of human health and the environment was a significant departure from the traditional methods previously used. While historically people tried to minimize exposure to chemicals, green chemistry emphasizes the design and creation of chemicals so that they do not possess intrinsic hazard.

Within the definition of green chemistry, the word chemical refers to all materials and matter. Therefore the application of green chemistry can affect all types of products, as well as the processes to make or use these products. Green chemistry has been applied to a wide range of industrial and consumer goods, including paints and dyes, fertilizers, pesticides, plastics, medicines, electronics, dry cleaning, energy generation, and water purification.

A fundamental aspect of green chemistry is the recognition that intrinsic hazard is simply another property of a chemical substance. Because properties of chemicals are caused by their molecular structure, the properties can be altered or modified by changing the chemical structure. Chemists and molecular scientists have been developing ways of manipulating chemical structure since the nineteenth century, and green chemistry uses the same expertise to design the property of minimal hazard into the molecular structure of chemicals.

The types of hazards that can be addressed by green chemistry vary and can include physical hazards such as explosiveness and flammability, toxicity including carcinogenicity (cancer-causing) and acute lethality, or global hazards such as climate change or stratospheric ozone depletion. Therefore, in the same way that a substance can be designed to be colored blue or to be flexible, it can also be designed to be nontoxic.

Principles of Green Chemistry

Paul T. Anastas and John E Warner's twelve principles of green chemistry outline a framework for chemists and chemical engineers to use in the quest to design more environmentally benign products and processes. These principles look at the entire life cycle of a product or process from the origins of the materials that go into its manufacture to the ultimate fate of the materials after they have finished their useful life. Through the use of the principles of green chemistry, scientists have been able to reduce the impact of chemicals in the environment. The principles are as follows:

1. It is better to prevent waste than to treat or clean up waste after it is formed.
2. Synthetic methodologies should be designed to maximize the incorporation of all materials used in the process into the final product.
3. Wherever practicable, synthetic methodologies should be designed to use and generate substances that possess little or no toxicity to human health and the environment.
4. Chemical products should be designed to achieve efficacy of function while reducing toxicity.
5. The use of auxiliary substances (e.g., solvents, separation agents, etc.) should be made unnecessary wherever possible and, innocuous when used.
6. Energy requirements should be recognized for their environmental and economic impacts and should be minimized. Synthetic methods should be conducted at ambient temperature and pressure.
7. A raw material or feedstock should be renewable rather than depleting wherever technically and economically practicable.
8. Unnecessary derivatization (e.g., blocking group, protection/deprotection, temporary modification of physical/chemical properties) should be avoided whenever possible.
9. Catalytic reagents (as selective as possible) are superior to stoichiometric reagents.
10. Chemical products should be designed so that at the end of their function they do not persist in the environment and break down into innocuous degradation products.
11. Analytical methodologies need to be further developed to allow for real-time, in-process monitoring and control prior to the formation of hazardous substances.
12. Substances and the form of a substance used in a chemical process should be chosen so as to minimize the potential for chemical accidents, including releases, explosions, and fires.

Green Chemistry Research and Development

Research and development in the field of green chemistry occurs in several different areas.

Alternative Feedstocks Historically, many of the materials that went into making the products we used often depleted finite resources, such as petroleum, or were toxic. Green chemistry is active in developing ways of making the products that we need from renewable and non-hazardous substances such as plants and agricultural wastes.

Benign Manufacturing Synthetic methods used to make chemical materials have often involved the use of toxic chemicals such as cyanide and chlorine. In addition, these methods have at times generated large quantities of hazardous and toxic wastes. Green chemistry research has developed new ways to make these synthetic methods more efficient and to minimize wastes while also ensuring that the chemicals used and generated by these methods are as nonhazardous as possible.

Designing Safer Chemicals Once it is ensured that the feedstocks and the methods to make a substance are environmentally benign, it is important to make certain that the end product is as nontoxic as possible. By understanding what makes a product harmful at the molecular level (as in the field of molecular toxicology), chemists and molecular scientists can design the molecular structure so that it cannot cause this harm.

Green Analytical Chemistry The detection, measurement, and monitoring of chemicals in the environment through analytical chemistry has been a part of the environmental movement since its beginning. Instead of measuring environmental problems after they occur, however, green chemistry seeks to prevent the formation of toxic substances and thus prevent problems before they arise. By putting sensors and instruments into industrial manufacturing processes, green analytical chemistry is able to detect trace amounts of a toxic substance and to adjust process controls to minimize or stop its formation altogether. In addition, while the traditional methods of analytical chemistry used substances such as hazardous solvents, green analytical methods are being developed to minimize the use and generation of these substances while carrying out the analysis.

The Need for Green Chemistry

The need for green chemistry has evolved with the increased use of chemicals. Many companies around the world, including those within the U.S., China, Germany, Italy, Japan, and the U.K., are adopting green chemistry, and there are several reasons for this. First, green chemistry can effectively reduce the adverse impact of chemicals on human health and the environment. The second reason is economic; many companies have found that it is less expensive, and even profitable, to meet environmental goals using green chemistry. The monetary savings come from the combined factors of higher efficiency, less waste, better product quality, and reduced liability. The third reason is legal. Many environmental laws and regulations target hazardous chemicals, and following all of the requirements can be a long and complex process. By using green chemistry, companies are finding they are able to comply with the law in much simpler and cheaper ways. Finally, green chemistry is a fundamental science-based approach. This means that because it deals with the problem of hazard at the molecular level, green chemistry approaches can be applied to all kinds of environmental issues.

Conclusion

Since 1991, there have been many advances in green chemistry in academic research and in industrial implementation. These advances, however, represent just a small fraction of the potential applications of green chemistry. Because the products and processes that form the basis of the economy and infrastructure are based on the design and utilization of chemicals and materials, the challenges facing green chemistry are vast.

See also **Biotechnology; Chemicals; Feedstocks; Environmental Monitoring**

PAUL T. ANASTAS

Further Reading

Anastas, P.T. and Farris, C.A., Eds. *Benign by Design: Alternative Synthetic Design for Pollution Prevention.* Oxford University Press, New York, 1994.

Anastas, P.T., Heine, L.G. and Williamson, T.C., Eds. *Green Chemical Syntheses and Processes.* Oxford University Press, New York, 2000.

Anastas, P.T., Heine, L.G. and Williamson, T.C., Eds. *Green Engineering.* ACS Symposium Series No. 766, Oxford University Press, New York, 2001.

Anastas, P.T. and Williamson, T.C., Eds. *Green Chemistry: Designing Chemistry for the Environment.* Oxford University Press, New York, 1996.

Anastas, P.T. and Williamson, T.C., Eds. *Green Chemistry: Frontiers in Benign Chemical Synthesis and Processes.* 1998.

Lankey, R.L. and Anastas, P.T., Eds. *Advancing Sustainability through Green Chemistry and Engineering.* ACS Symposium Series No. 823, Oxford University Press, New York, 2002.

Poliakoff, M.and Anastas, P. A principled stance. *Nature,* 413, 257–258, 2001.

Useful Websites

U.S. Environmental Protection Agency, Green Chemistry Program: http://www.epa.gov/greenchemistry/

Green Chemistry Institute at the American Chemical Society: http://www.acs.org/greenchemistryinstitute, 2002.

Gyrocompass and Inertial Guidance

Before the twentieth century, navigation at sea employed two complementary methods, astronomical and dead reckoning. The former involved direct measurements of celestial phenomena to ascertain position, while the latter required continuous monitoring of a ship's course, speed, and distance run. New navigational technology was required not only for iron ships in which traditional compasses required correction, but for aircraft and submarines in which magnetic compasses cannot be used. Owing to their rapid motion, aircraft presented challenges for near instantaneous navigation data collection and reduction. Electronics furnished the exploitation of radio and the adaptation of a gyroscope to direction finding through the invention of the nonmagnetic gyro-compass.

Jean Bernard Léon Foucault's invention of the gyroscope in the mid-nineteenth century spurred the idea of a spinning rotor within two gimbals aligned with north that might be adapted to direction finding. In 1908 Hermann Anschütz-Kaempfe introduced a gyrocompass for naval use, intending it for submarine guidance to reach the North Pole. His gyrocompass was installed in a German warship in 1910. Meanwhile, the American Elmer Ambrose Sperry, who began work on a seaworthy gyrocompass in 1896, established the Sperry Gyroscope Company in 1910 and installed his version in a warship in 1911. World War I furnished the first rigorous test of the gyrocompass. During the war years, gyroscopes enabled the installation of automatic pilots in ships and aircraft. After the war, Sperry Gyroscope Company became Sperry Corporation (and much later the Unisys Corporation), the leading world manufacturer and developer of gyrocompasses and other gyroscopic devices. Numerous American, German, and British inventors contributed modifications and improvements, although many of them remained unknown outside their industry because of military secrecy. Patent infringement claims have further obscured the lineage of gyrocompasses. Anschütz-Kaempfe's vision was not realized until 1958 with the voyages of the American submarines USS Nautilus and Skate under the polar ice, navigated by inertial guidance, the most important adaptation of the gyro in the last half of the century.

The gyrocompass relies on a key property of a gyro: its rotor or flywheel, mounted within gimbals, when set in motion can be oriented such that its axis points in any direction. The utility of a gyro used as a compass is its stability. Once oriented to north, the rotor axis maintains the plane of rotation unless acted upon by an external influence (known as gyroscopic inertia) and is unaffected by magnetism. Various influences, however, cause the axis in motion to precess, meaning that the rotation axis responds at right angles to the influence of an applied force. Gyrocompasses may wander due to the earth's rotation or when the gyro-controlled vehicle is stationary. A vehicle's motion may also influence the direction of the rotor axis (tilt and drift are common terms for axial wander). North is not assumed automatically by the rotor's axis.

Considerable sophistication in the application of gyroscopes to navigation had occurred before World War II. Wiley Post's 1933 solo air circum-

navigation involved multiple gyro applications including a Sperry autopilot. Figure 8 shows a two-gimballed directional gyroscope as installed in U.S. Navy aircraft in World War II. The compass card showing degrees is affixed to the vertical gimbal and can be viewed through a small window next to which a lubber's line has been marked. The rotor is powered by jets of air. The caging knob releases the vertical gimbal to rotate, allowing the device to be reset with the rotor's axis horizontal and aligned with north. The use of a magnetic compass to align the rotor, however, may introduce compass errors of variation and deviation, requiring the pilot to understand the limitations of both instruments. A magnetic sensor known as a flux valve, found in later aircraft, transformed the directional gyroscope into a gyrocompass. The most common marine gyrocompasses have featured not two but three gimbals. When the ship's motion introduces a tilt, marine gyrocompasses compensate by the action of a pendulous mass that applies a torque to the rotor, causing the spinning axis to return to the north–south meridian. In fact, the tilt is necessary to the determination of the meridian by allowing the pendulous mass to cause

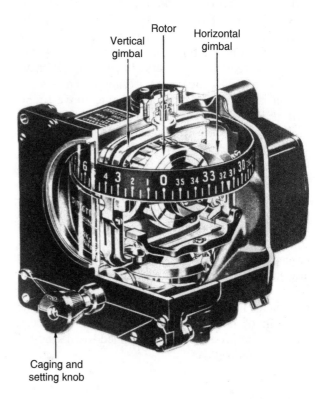

Figure 8. Gyroscopic direction finder used in U.S. Navy aircraft during World War II.
[*Source: U.S. Navy, Air Navigation Part Four: Navigation Instruments, Flight Preparation Training Series, McGraw-Hill, New York, 1944, p. 103.*]

the rotor to describe a diminishing circle about the pole, finally settling into the polar alignment. With no tilting, the gyrocompass ceases to work, a condition that occurs under some circumstances, precluding the marine gyrocompass from being applied in aircraft.

Although the Cold War arms race after World War II led to the development of inertial navigation, German manufacture of the V-2 rocket under the direction of Wernher von Braun during the war involved a proto-inertial system, a two-gimballed gyro with an integrator to determine speed. Inertial guidance combines a gyrocompass with accelerometers installed along orthogonal axes, devices that record all accelerations of the vehicle in which inertial guidance has been installed. With this system, if the initial position of the vehicle is known, then the vehicle's position at any moment is known because integrators record all directions and accelerations and calculate speeds and distance run. Inertial guidance devices can subtract accelerations due to gravity or other motions of the vehicle. Because inertial guidance does not depend on an outside reference, it is the ultimate dead reckoning system, ideal for the nuclear submarines for which they were invented and for ballistic missiles. Their self-contained nature makes them resistant to electronic countermeasures. Inertial systems were first installed in commercial aircraft during the 1960s. The expense of manufacturing inertial guidance mechanisms (and their necessary management by computer) has limited their application largely to military and some commercial purposes. Inertial systems accumulate errors, so their use at sea (except for submarines) has been as an adjunct to other navigational methods, unlike aircraft applications. Only the development of the global positioning system (GPS) at the end of the century promised to render all previous navigational technologies obsolete. Nevertheless, a range of technologies, some dating to the beginning of the century, remain in use in a variety of commercial and leisure applications.

See also **Global Positioning System (GPS); Radionavigation.**

ROBERT D. HICKS

Further Reading

Burger, W. and Corbet, A.G. *Marine Gyrocompasses and Automatic Pilots*, vol. 1. Pergamon Press, Oxford, 1963.

Defense Mapping Agency Hydrographic/Topographic Center. *The American Practical Navigator*. Pub. No. 9, DMAHTC, Bethesda, MD, 1995.

May, W.E. *A History of Marine Navigation*. W.W. Norton, New York, 1973.

Richardson, K.I.T. *The Gyroscope Applied*. Philosophical Library, New York, 1954.

Tetley, L. and Calcutt, D. *Electronic Navigation Systems*, 3rd edn. Butterworth-Heinemann, London, 2001.

Williams, J.E.D. *From Sails to Satellites: The Origin and Development of Navigational Science*. Oxford University Press, Oxford, 1994.

Hall Effect Devices

The "Hall effect," discovered in 1879 by American physicist Edwin H. Hall, is the electrical potential produced when a magnetic field is perpendicular to a conductor or semiconductor that is carrying current. This potential is a product of the buildup of charges in that conductor. The magnetic field makes a transverse force on the charge carriers, resulting in the charge being moved to one of the sides of the conductor. Between the sides of the conductor, measurable voltage is yielded from the interaction and balancing of the polarized charge and the magnetic influence.

Hall effect devices are commonly used as magnetic field sensors, or alternatively if a known magnetic field is applied, the sensor can be used to measure the current in a conductor, without actually plugging into it ("contactless potentiometers"). Hall sensors can also be used as magnetically controlled switches (see below), and as a contactless method of detecting rotation and position, sensing ferrous objects.

The first device employing the Hall effect was created in 1926 by Dr. Palmer H. Craig. The electromagnetic detector and amplifier were constructed of bismuth plates or films stacked together and wrapped with copper coils or wires. The device was able to supply radio reception without the use of batteries and vacuum tubes.

Hall effect switches combine Hall elements and a switching circuit. They are electronic switches operated by placing a magnet close to them. No direct physical contact is required, just close proximity to a magnet. The device can replace mechanical switches in various robotic and automotive applications. The use of a high-speed revolution counter can make a Hall effect switch operate at a high frequency. The advantages of Hall effect devices include their greater reliability compared to mechanical contact switches, the fact that, unlike optical sensors, they can function even when dirty (since they are contactless), and, because they are totally enclosed, they can operate in dirty environments, such as salt- and fuel-filled environments in automotive applications.

Some Hall effect devices have a binary output (on or off); others have a linear output. Devices with a linear output give voltage output proportional to the strength of the magnetic field and can provide an estimation of the distance from the magnet.

Hall effect devices have numerous electronic functions. They are used commonly as magnetic sensors such as a simple open or closed sensor. For example, in a seat belt sensor, the buckle, made of a ferrous material, interrupts the magnetic field between a magnet and the Hall effect device. When the field is interrupted, the device output switches on and when the buckle is removed the device switches off. This information is sent to the controller. Other automobile Hall sensors are electric windows and control systems for doors. Modern computerized engine control systems also use Hall effect sensors to sense crankshaft and camshaft speed and position. Again, the sensor detects a change in voltage when the magnetic field between a magnet and the Hall effect device is interrupted. A shutter rotating with the engine shaft passes through this opening, and with the change in voltage each time the shutter passes through the magnetic field, a rotation is counted.

The output Hall voltage is very small, and is usually amplified with several transistor circuits.

Chips based on the Hall effect became inexpensive and widely available in the 1970s, and eliminated the need for operators of the devices to design and manufacture their own magnetic circuits. In theory, Hall effect devices should have zero output voltage when no magnetic field is present, but in practice an "offset" voltage occurs due to material and mechanical defects. The offset voltage can be calibrated for, but may drift with temperature variations or mechanical stresses, masking low signals. Programmable linear Hall effect devices with active error correction provide better accuracy, linearity, and reliability with low thermal drift of the offset voltage.

The quantized Hall effect finds that the Hall resistance at low temperatures of a two-dimensional system (a thin sheet, such as in semiconductor transistors) exists as steps, rather than varying smoothly with voltage. The Hall resistance is said to be quantized. The quantum Hall effect, discovered by Klaus von Klitzing in 1980 is now the basis of very precise resistance measurements, needed to calibrate sensitive electronic test equipment. Quantum Hall devices produce very precise values of resistance of the order of 10 parts per billion.

See also **Computer Science; Electronics; Quantum Electronic Devices; Radio Receivers, Crystal Detectors and Receivers**

TIMOTHY S. BROWN

Further Reading

Kamimura, H. and Aoki, H. *The Physics of Interacting Electrons in Disordered Systems.* Clarendon Press, Oxford, 1989.

Lury, S., Xu, J. and Zaslavsky, A. *Further Trends in Microelectronics.* Institute of Electrical and Electronics Engineers, Wiley/Interscience, New York, 2002.

Putley, E.H. *The Hall Effect and Related Phenomena.* Butterworths, London, 1960.

Shrivastava, K.N. *Introduction to Quantum Hall Effect.* Nova Science Publishers, Hauppauge, NY, 2002.

Weisbuch, C. *Quantum Semiconductor Structures: Fundamentals and Applications.* Academic Press, Boston, 1991.

Health

The relationship between technology and health in the twentieth century has been multifaceted and complex. However, some general trends are observable. First, a variety of imaging techniques were pioneered during the course of the century, which permitted the *visual* representation of the living organism, including the ability to "see" the presence or absence of disease. Second, technologies were often "disease specific" in that they were designed with the intent of detecting or, in the case of vaccines, preventing particular diseases. Third, although most biomedical technologies have been beneficial, a few have come to be regarded as health threatening. Finally, the "information technologies" associated with the coming of the computer proved to be invaluable in recording and storing information about mortality and morbidity, which provided the foundation for public health policies. These trends are illustrated by examining some of the major health threats and biotechnological developments of the twentieth century. Although the focus will be on technology in the context of industrially developed Western societies, it is impossible to exclude other parts of the world when discussing these issues.

Perhaps the most famous twentieth century technology actually predated the emergence of the century by a little more than four years—the x-ray that was discovered by the German physicist Wilhelm Roentgen late in 1895. This was only the first in a series of "visual technologies" that permitted the examination of the inner workings of the human form. Two prominent examples from later in the century were computerized axial tomography (CAT) scan, developed by the British engineer Godfrey Newbold Hounsfield in the late 1960s and early 1970s, and the later technique of magnetic resonance imaging (MRI). Both techniques produced three-dimensional images of a human body on a screen. The 1970s also witnessed the introduction of positron emission tomography (PET) scanning. This technology used a radioactive tracer and permitted the observation of organ function (such as local blood flow) rather than focusing solely on organ structure as did the CAT and MRI scans. Although these technologies have become prominent symbols of the role of technology in understanding health and disease, they are technologies primarily available in the industrially developed countries of the world. According to one study by Tubiani in 1997, five sixths of the world's radiological equipment resides in countries that comprise one sixth of the world's population.

The x-ray soon proved its usefulness in screening for tuberculosis, one of the major health threats for much of the twentieth century. By the second decade of the century in the U.S., the newly formed National Association for the Study and Prevention of Tuberculosis urged that family and associates of a tubercular patient receive a precautionary chest x-ray. By the time of World War II, radiologists on both sides of the Atlantic anticipated the prospect of "mass radiology of the chest." In the U.S., this was required of all military recruits after January 1941.

Although the primary function of x-ray examinations remained screening for tuberculosis, other diseases could be detected as well. As the century wore on, lung cancer became increasingly prominent. Technology played a role in the emergence of this disease in the sense that the mass-production of cigarettes was a distinctively twentieth century technology. Cigarettes became readily available on a wide scale and contributed significantly to lung cancer as a serious health threat. Even though there has been a reduction in the amount of cigarette smoking in Western societies since the health risks became known in the 1950s and 1960s, it is still a widespread practice in much of the developing world.

In addition to their obvious uses for diagnostic purposes, technologies have also proved to be useful therapeutically. Soon after Pierre and Marie Curie discovered the radioactive substance radium in 1898, it was put to medical use. This set the stage for the eventual development of radiation therapy as a standard treatment for diseases, most notably certain types of cancer. In 1917, the physicist Albert Einstein showed that under certain circumstances molecules could be stimulated to produce directed electromagnetic radiation. This insight laid the foundation for what later became the technology of the laser (that is, light amplification by stimulated emission of radiation). Eventually laser technology proved useful for restoring detached retinas and clearing blocked coronary arteries. Although not without risks, these technologies were instrumental in restoring individuals to health.

Whereas x-rays and lasers were dependent on developments in physics, the technologies that relied on chemistry were also prominent in contributing to health in the twentieth century. Like the x-ray, innovations in the last decades of the nineteenth century provided a context for later developments. In the 1880s, the French scientist Louis Pasteur and his German counterpart Robert Koch ushered in the science of bacteriology, which held that microorganisms were the cause of infectious disease. These microorganisms could be observed under a microscope in a laboratory setting and, after being isolated in a pure culture, could produce the disease anew in laboratory animals.

As can be deduced etymologically, Pasteur's work led to the process of pasteurization, which helped to preserve liquids like milk and wine and thereby contributed greatly to the public health. Because milk and milk products were a source of such diseases as tuberculosis, brucellosis, typhoid, diphtheria, salmonellosis, and streptococcal infections, the technologies of heat processing of food proved to be very important in contributing to a decline in milk-borne diseases throughout the twentieth century.

Among Koch's discoveries was the cholera bacillus organism that caused cholera. Because this bacillus was water-borne, Koch's work proved to be a spur to ensure that, for the sake of public health, communities receive clean water supplies. However, the association of a clean water supply with improved public health actually predated Koch's work—it had already been developed by physicians and sanitary reformers who had been politically active in Europe and the U.S. from the second quarter of the nineteenth century onward. In the twentieth century, the efforts to ensure a clean and health-sustaining water supply have increasingly relied on chemical substances like chlorine and fluoride, which have been introduced into water used for swimming and drinking. Although these developments produced major improvements in public health (e.g., the reduction in dental decay), they have also caused concern because of fears that chlorinated disinfection byproducts might prove to be carcinogenic.

In addition to practical developments like providing a scientific rationale for a clean water supply, Koch is generally credited with giving bacteriology a theoretical foundation. Although some remained skeptical of Koch's claim that infectious diseases were caused by specific microorganisms, his work generated many followers on both sides of the Atlantic. By the twentieth century, the bacteriological laboratory became a symbol of a new "scientific" public health that targeted specific "at risk" populations because of their exposure to specific germs. In the decades leading up to World War II, the search for bacteria through the techniques that Koch had pioneered was extended throughout the globe. This period coincided with a time of European colonization of much of the rest of the world, and the extension of Koch's techniques by European medical overseers was part and parcel of this broader movement.

One early use of the diagnostic laboratory from the early twentieth century onward was to screen for syphilis. In 1901, Jules Bordet and Octave Gengou developed a method of using blood serum to determine the presence of microorganisms, and five years later August von Wasserman and his associates developed a blood serum test that detected the presence of syphilis. Within a very few years, local boards of health had added the Wasserman test to their list of standard procedures. In the U.S., many states required that couples undergo this test prior to marriage.

With the emergence of the modern pharmaceutical industry, the twentieth century witnessed the transformation of diabetes from a near fatal disease to a medically manageable condition. In the 1920s, Frederick Banting and Charles Best at the University of Toronto pioneered a method to develop insulin in a form that could be injected into the diabetic patient. When this fundamental scientific breakthrough was developed commercially by the Eli Lilly Company of Indianapolis, Indiana, it became possible for many diabetic individuals to lead comparatively normal lives. Because of the drug-induced potential to manage chronic conditions like diabetes, there were increasing calls for individuals to be screened for this condition on a regular basis.

By the end of World War II, major breakthroughs in industrially produced antibiotic drugs were occurring. Among the most prominent were the development of the drug streptomycin to treat tuberculosis and the multiple uses for the drug penicillin (e.g., to treat syphilis). Along with these success stories, there were also major health threats posed by the new and powerful drugs. In the latter category, probably the most famous example was the drug thalidomide. Produced commercially in Germany, it was widely prescribed in Europe in the late 1950s and early 1960s for women suffering from morning sickness. By 1961, pregnant women in 48 countries had taken the drug, and this resulted in the birth of over 8000 infants with deformities. This led to a ban on the drug in the U.S. and eventually to state testing and oversight of drugs sold by the pharmaceutical industry in most Western societies.

In the second half of the twentieth century, identification of chronic conditions such as hypertension and heart disease became more prominent. The technologies associated with detecting these health problems had been developed in the last years of the nineteenth century. In 1896, the Italian physician Scipione Riva-Rocci developed a method of measuring a patient's blood pressure by restricting blood flow in the arm through circular compression. Although there were later technical improvements, this approach to blood pressure measurement became the standard throughout Western medicine. In 1903, the Dutch physiologist Willem Einthoven developed the electrocardio-graph, a technology that permitted the electrical activity of the heart to be monitored in an effort to detect cardiac disease. After the 1940s, routine testing for high blood pressure became more widespread as the concept of multiphasic screening (the attempt to locate disease in seemingly well people by giving them a wide variety of tests) became common. By the last quarter of the century, a whole host of cardiovascular agents were developed in an attempt to make hypertension and other heart-related illnesses into manageable health problems.

This shift to a concern with heart disease vividly illustrates the key epidemiological transition of the twentieth century; that is, the shift from infectious to chronic diseases as the leading causes of mortality and morbidity. Chronic conditions take years to develop and often result from lifestyle choices regarding diet and exercise rather than exposure to an infecting agent. Both diet and exercise have produced their attendant technologies. In response to these changing health realities and consumer demand, food producers have created foods that are lower in fat and refined sugar, higher in fiber, and "fortified" with vitamins and minerals. Similarly, the concern with exercise has spawned the creation of health clubs with a variety of machines designed to produce a healthy body in addition to squash and tennis courts, swimming pools, exercise bicycles, and running/walking tracks.

The postwar era also saw the emergence of vaccines, which led to the eradication of many diseases. In 1960 American researcher John Enders attempted to develop a measles vaccine, which was licensed in 1963 and is credited with saving a large number of lives. Similar vaccines were developed for mumps and rubella, and it is now possible in most developed countries to receive a triple vaccine that immunizes the individual against all three conditions.

Polio is one of the most famous examples of disease eradication by means of vaccination. In the 1950s, American researchers Jonas Salk and Albert Sabin both developed polio vaccines. The Salk vaccine used an inactivated (killed) polio virus whereas Sabin used a live-attenuated virus that could be administered orally. With support from the National Foundation for Infantile Paralysis, industrial production of the Salk vaccine began in the U.S. immediately after its effectiveness had been established in a clinical trial conducted in 1955. In European countries, Salk's vaccine was either imported from the U.S. or developed in government laboratories. By the 1970s, however,

Salk's vaccine had been supplanted by Sabin's. United States pharmaceutical companies ceased producing the Salk vaccine, and the World Health Organization began to issue the Sabin vaccine to developing countries.

The development of vaccines was also of vital importance in the eradication of smallpox. As C.C. Booth has observed, the discovery of a method for producing a freeze-dried vaccine was an indispensable aspect of this endeavor. Following a worldwide vaccination campaign, smallpox was confined to India, Ethiopia, Somalia, and Botswana by 1973. By 1977, the disease had been removed from Ethiopia and Somalia, and the official eradication of the disease from the world was declared at a 1980 meeting of the World Health Organization. As the chairman of the commission overseeing the eradication campaign observed, this was an event that was "unique in the history of mankind". Despite this optimism, at the dawn of the twenty-first century concerns still linger that the few existing samples of the smallpox virus might be used for malevolent purposes should they fall into the wrong hands.

In the last two decades of the twentieth century, the new life-threatening disease of AIDS (acquired immune deficiency syndrome) emerged on the world stage. Although initially diagnosed in homosexual men and intravenous drug users in the U.S., it spread throughout the world with approximately 70 percent of the cases in sub-Saharan Africa; also, at the beginning of the twenty-first century, most infections result from heterosexual transmission. Based on whether one lives in the developed or developing world, AIDS is a different disease. In the industrially developed world, it has become a chronic condition that, although still life threatening, can be treated with drugs that interfere with viral replication. By contrast, in the developing world, AIDS remained a major killer with many countries lacking adequate financial resources to purchase the drugs that can prolong the life of infected individuals. To help ensure that all the world's people (and not just those in the developed parts) have access to vaccines, the International AIDS Vaccine Initiative (IAVI) was established in 1996. By setting up a Vaccine Purchase Fund, the IAVI hoped to establish that there will be a guaranteed market in the developing world and thereby create incentives for private sector vaccine companies to enter this market. However, all these developments are predicated on being able to develop an effective AIDS vaccine, and at the time of writing none of these efforts has been successful.

Although most biomedical technologies have been developed with the express intent of improving health, many technologies have also been perceived as health risks. From early on, the potential dangers of x-ray exposure were reported. Similarly, the technology of electronic fetal monitoring has had a varied history. Pioneered in the 1950s and 1960s by researchers in the U.S., Germany, and South America, the technology was heralded as a way to detect fetal distress during delivery. Even though this claim has been the subject of much debate within the medical community, the widespread use of the practice has had the effect of increasing the number of deliveries by caesarean section, which, like any surgical intervention, constitutes a health risk for the mother. Other technologies that have been alleged to be health risks (often in a court of law) include exposure to electromagnetic fields and silicone breast implants. Although many in the scientific community have questioned the empirical basis for these claims, the historically important fact is that the legal arena has become a prominent forum in which the health risks of technology are actively debated.

In addition to considering how technologies help to determine whether a particular individual is in a state of health or disease—as revealed by an x-ray image or blood pressure test—the more abstract concepts of "health" are dependent on technology as well. In particular, "information technologies" have been developed to retrieve and preserve data in individual health records and in aggregate data about populations. Of course, the collecting of aggregate data predates the twentieth century, but computer-based technologies have permitted this practice to be carried out on a much wider scale. By the 1990s there were calls on both sides of the Atlantic for health care providers to practice "evidence-based medicine," that is, practice based on empirical evidence derived from studying populations. In this respect, technology has gone beyond merely revealing the presence or absence of disease and has become ineluctably interwoven with how we conceive of health and disease in contemporary Western society.

See also **Antibacterial Chemotherapy; Biotechnology; Cardiovascular Disease, Diagnostic Screening; Genetic Screening and Testing; Immunization Technology; Medicine; Nuclear Magnetic Resonance (NMR, MRI); Positron Emission Tomography (PET); Tomography in Medicine**

J. ROSSER MATTHEWS

Further Reading

Annas, G.J. and Elias, S. Thalidomide and the *Titanic*: reconstructing the technological tragedies of the twentieth century. *Am. J. Pub. Health*, 89, 1, 98–101, 1999. This article is written as a policy piece and argues that, suitably regulated, the drug thalidomide could be reintroduced into the American market; however, it also provides the historical background to the thalidomide tragedy.

Banta, H.D. and Thacker, S.B. Historical controversy in health technology assessment: the case of electronic fetal monitoring. *Obst. Gynecol. Surv.*, 56, 11, 707–719, 2001. This article historically recounts how electronic fetal monitoring was rapidly diffused within obstetrical practice and discusses the debates that ensued when the authors published a 1979 study, which indicated that there was insufficient evidence to recommend its use as a standard procedure.

Berkley, S. HIV vaccine development for the world: an idea whose time has come? *AIDS Res. Hum. Retroviruses*, 14, s. 3, 191–196, 1998. The author advocates that an AIDS vaccine be developed and discusses possible financing arrangement like the International AIDS Vaccine Initiative.

Blume, S. and Geesink, I. A brief history of polio vaccines. *Science*, 288, 5471, 1593–1594, 2000. This article traces how the Sabin vaccine against polio eventually supplanted the Salk vaccine; the authors use this story as a case study of the process of technological "lock in," a view of technological change that they attribute to evolutionary economists.

Booth, C. Changing concepts of health and disease during the 20th century. *J. R. Coll. Phys. London*, 34, 1, 38–46, 2000. As the title (correctly) indicates, this article provides a nice overview of changing concepts of health and disease-including discussions of the role that biomedical technologies have played in this process.

Brandt, A.M. The cigarette, risk, and American culture. *Daedalus*, 119, 4, 155–176, 1990. Among other things, this article discusses how technological forces shaped "the rise of the cigarette" in twentieth century American culture, with its attendant health risks.

Caplan, A.L. The concepts of health, illness, and disease, in *Companion Encyclopedia of the History of Medicine*, Bynum, W.F. and Porter, R., Eds. Routledge, London, 1993. This article provides a philosophical overview of changing conceptions of health and disease, including a discussion of how technology can "objectify" the patient.

Daly, J. and McDonald, I. Cardiac disease construction on the Borderland. *Soc. Sci. Med.*, 44, 7, 1997, 1043–1049. Adopting a social constructionist perspective, the authors discuss the role of technology in defining whether an individual suffers from heart disease; the authors survey developments throughout the twentieth century.

Desrosières, A. *The Politics of Large Numbers: A History of Statistical Reasoning*. Harvard University Press, Cambridge, MA, 1998. Original French edition, 1993 As a work that attempts to synthesize a vast amount of the scholarly literature on the history of statistics, this book would be a good source for bibliographic leads on the history of medical and public health statistics in the twentieth century.

Higgs, E. The statistical big bang of 1911: ideology, technological innovation, and the production of medical statistics. *Soc. Hist. Med.*, 9, 3, 409–426, 1996. By discussing what he terms "the nuts and bolts of statistical creation," the author illuminates the pre-history that set the stage for the computer revolution in data collection later in the century; the geographical focus in on early twentieth century Britain.

Kennedy, R.G. Electronic fetal heart rate monitoring: retrospective reflections on a twentieth-century technology. *J. R. Soc. Med.*, 91, 244–250, 1998. The author discusses the history of how electronic fetal monitoring was introduced into British obstetric practice with a focus on why it disseminated so rapidly.

Marks, H.M. Medical technologies: social contexts and consequences, in *Companion Encyclopedia of the History of Medicine*, Bynum, W.F. and Porter, R., Eds. Routledge, London, 1993. A good overview with a focus on developments from 1850 to the present.

Porter, D. *Health, Civilization and the State: A History of Public Health from Ancient to Modern Times*. Routledge, London 1999. Although the title implies an encyclopedia overview, the focus is on developments in public health in the nineteenth and twentieth centuries. There is a discussion of "healthy" lifestyles and attendant developments; for example, body-building.

Porter, R. *The Greatest Benefit to Mankind: A Medical History of Humanity*. W.W. Norton, New York, 1999. A panoramic overview of the history of Western medicine, which includes discussions of developments in health and technology in the twentieth century.

Reiser, S.J. *Medicine and the Reign of Technology*. Cambridge University Press, Cambridge, 1978. This book surveys technological developments from early modern times (sixteenth to the eighteenth centuries) to contemporary concerns.

Reiser, S.J. The emergence of the concept of screening for disease. *The Milbank Memorial Fund Quarterly*, 56, 4, 403–425, 1978. This article provides a nice overview of how the concept of "screening" populations to determine their health status emerged in twentieth century American society.

Reiser, S.J. The science of diagnosis: diagnostic technology, in *Companion Encyclopedia of the History of Medicine*, Bynum, W.F. and Porter, P, Eds. Routledge, London, 1993. Although covering some of the same material as his book, Reiser brings the story up to more recent developments.

Sepkowitz, K.A. AIDS—the first 20 years. *New Engl. J. Med.*, 344, 23, 1764–1772, 2001. This article attempts to review the main features of the response to AIDS in the last 20 years.

Tubiana, M. From Bertha Roentgen's hand to current medical imaging: one century of radiological progress. *Eur. Radiol.*, 7, 9 1507–1513, 1997. This article provides a historical overview of the development of radiology and advocates that the technology be extended to the developing world.

Welch, R.W. and Mitchell, P.C. Food processing: a century of change. *Br. Med. Bull.*, 56, 1, 1–17, 2000. As the title states, this article surveys developments in food processing technology during the course of the twentieth century—drying, heat processing, freezing and chilling, controlled and modified atmospheres, radiation, and mechanization/standardization. Also, the health implications of changes in food processing are discussed.

Hearts, Artificial

Heart disease is a prominent cause of death in affluent nations. In the U.S., it is the number one killer and accounts for more than 39 percent of all deaths. Treatment for heart failure includes medical therapy, surgical intervention, cardiac transplantation, and the use of cardiac devices. Cardiac devices such as heart valves, pacemakers, defibrillators, and intra-aortic balloon pumps can assist the heart, and an artificial heart can replace it.

There are two groups of artificial hearts: total artificial hearts and partial artificial hearts. The total artificial heart (TAH) replaces the failing ventricles (pumping chambers) of a damaged human heart, which is excised or removed from the body. A partial artificial heart or ventricular assist device (VAD) attaches to a failing heart (which remains in the body) and serves to assist in the pumping function of the heart. The main components of both TAHs and VADs are the energy source, drive-control unit, actuator, and pump. Historically, the actuation of artificial hearts has included pneumatic, hydraulic, electrical (electrohydraulic, electromechanical, electromagnet) and nuclear (thermoenergy) power. Nuclear artificial heart development was halted due to individual and societal concerns relating to the plutonium radioisotope power source. The first artificial hearts were constructed out of polyvinylchloride, which tested well for endurance, and later silastic and polyurethane materials. Improved biomaterials, particularly compatible blood surface materials, continue to be developed. Over time, both total and partial artificial hearts have changed from large devices situated outside the body (paracorporeal and extracorporeal) to smaller devices intended to be placed inside the body (intracorporeal and fully implantable).

Willem Kolff and Michael DeBakey are among the pioneers in the field of artificial heart development. Kolff invented the artificial kidney machine in Nazi-occupied Holland in 1943. After the war, Kolff led an artificial organ research program at the Cleveland Clinic in Cleveland, Ohio that encouraged research into artificial kidneys, eyes, ears, arms, lungs, and hearts. In 1957, Kolff and Dr. Tetsuzo Akutsu successfully implanted a crude air-driven artificial heart, composed of two polyvinylchloride pumps, in a dog and maintained circulation for ninety minutes. The Cleveland Clinic group experimented with hydraulically-activated, electromechanical and pneumatic-driven TAHs, and later various types of VADs. At the Baylor College of Medicine in Houston, Texas,

DeBakey and his research team developed a different type of TAH. Their device consisted of two pneumatically-driven silastic sac-type pumps implanted for biventricular bypass. The Baylor College of Medicine group also expanded their research into VADs, heart valves and vascular prostheses.

DeBakey was instrumental in lobbying for the formation of the U.S. Artificial Heart program, established in 1964, at the National Heart Institute of the National Institutes of Health (NIH) in Bethesda, Maryland. The NIH became an important funding source and catalyst for artificial heart research. Under the initial direction of Frank Hastings, the NIH artificial heart program supported investigator research grants and industry contracts for the development of assist and replacement devices totaling more than $400 million. Outside of the U.S., artificial heart research programs were also established in Germany, the former Soviet Union, Japan, and other countries whose research teams experimented with various TAH and VAD prototypes.

By the late 1960s, there were four institutions in the U.S. working on developing TAHs for clinical application: Willem Kolff's team, now at the University of Utah in Salt Lake City, Utah; the Baylor College of Medicine group headed by Michael DeBakey; William Pierce and his research team at Pennsylvania State University; and the Cleveland Clinic under the direction of Yukihiko Nosé. These researchers contributed incremental improvements on device designs, materials, performance, durability and outcome measures through extensive animal experiments.

The device used in the first clinical case was the pneumatic Liotta TAH, developed by Domingo Liotta and tested in the Baylor College surgical laboratory on calves. At the Texas Heart Institute, Denton Cooley implanted the device as a bridge-to-transplantation in 47-year-old Haskell Karp who could not be weaned from cardiopulmonary bypass after heart surgery. The Liotta TAH sustained Karp for 64 hours, at which time the patient underwent cardiac transplantation. In 1981, Cooley implanted the pneumatic Akutsu III TAH, developed by Tetsuzo Akutsu, which provided a patient with 39 hours of support as a bridge-to-transplantation. Both device implant cases were technical successes, but both patients died shortly after the transplant surgery.

In 1982, William DeVries implanted the pneumatic Jarvik-7 TAH, developed by Robert Jarvik and fellow researchers at the University of Utah, in Barney Clark. Ineligible for cardiac transplantation, Clark consented to the TAH as a permanent implant. During his 112 days with the TAH, Clark experienced numerous complications including device problems, nasal bleeding and neurological incidents. In 1984 and 1985, three more patients in the U.S. and one in Sweden received permanent Jarvik-7 TAH implants, surviving from 10 to 620 days. In response to severe patient complications, the U.S. Food and Drug Administration suspended Jarvik-7 TAH production and its use as a permanent device in 1991. The Jarvik-7 TAH (renamed Symbion, later CardioWest) continued however to be implanted in patients as bridge-to-transplantation with moderate success rates.

Knowledge gained in TAH research contributed to VAD advancements. VADs experienced success as short-term mechanical support systems, bridge-to-transplantation, and possibly permanent cardiac devices. DeBakey's research group developed a pneumatic paracorporeal (outside the body) left ventricular assist device (LVAD) for short-term ventricular assistance. In 1967, the first clinical case was reported in which a patient, unable to be weaned from cardiopulmonary bypass, temporarily utilized the LVAD and then had it removed upon cardiac recovery. By the mid-1970s, clinical trials were underway at the Texas Heart Institute and Boston Children's Hospital to evaluate the intracorporeal pneumatic VAD—the TECO Model VII—as both a temporary recovery measure and

bridge-to-transplantation. In 1984, the electrically powered left ventricle assist system—the Novacor LVAS—emerged as both the first implantable system and successful bridge-to-transplant clinical case (Figure 1). In that same year, the paracorporeal, pneumatic Thoratec VAD also reported a successful bridge-to-transplantation case. Further refinements in the 1990s brought forth wearable ventricular assist systems, which allowed patients to resume basic activities. Clinical trials also began to evaluate VADs for permanent use in patients ineligible for cardiac transplantation.

In 2001, clinical trials with the electromechanical AbioCor TAH as a permanent device began in the U.S. The AbioCor TAH is the first fully implantable device; its components of the replacement heart (hydraulically-driven pump), controller and rechargeable internal battery are implanted in the patient's chest and abdomen. The AbioCor TAH utilizes a transcutaneous energy transmission (TET) device as a wireless energy transfer system without need for percutaneous lines risking infection.

Artificial heart research encompasses the development of total artificial hearts (TAHs) (see Figure 2) and partial artificial hearts (VADs). In both cases, these devices became more sophisticated, smaller in size, and fully implantable. Both devices were explored as temporary and permanent therapies. Whereas the TAH remains an experimental device, several ventricular assist systems are now

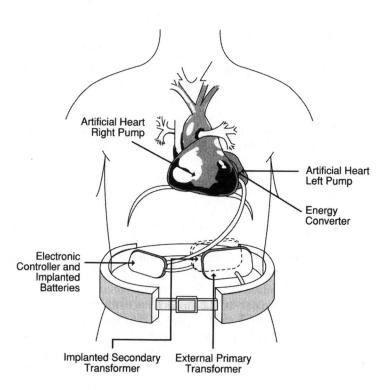

Figure 1.   The Novacor fully implantable ventricular assist device.
[*Used with permission from The Artificial Heart: Prototypes, Policies, and Patients, by the National Academy of Sciences, courtesy of the National Academy Press, Washington D.C., 1991.*]

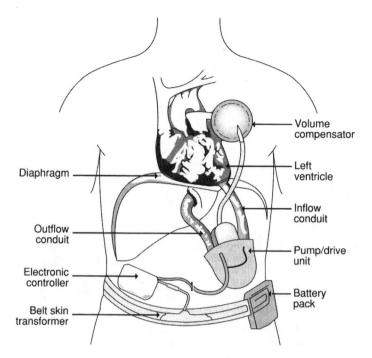

Figure 2.   A fully implantable total artificial heart. [*Used with permission from The Artificial Heart: Prototypes, Policies, and Patients, by the National Academy of Sciences, courtesy of the National Academy Press, Washington D.C., 1991.*]

FDA-approved for clinical use. Not without debate, artificial heart technologies have raised contentious issues of access, quality of life, cost-benefit and reimbursement. The broader context is the role that artificial heart technologies may play towards continuing or challenging the technological imperative in medicine.

See also **Biotechnology; Cardiovascular Surgery; Implants, Heart Pacemakers, and Heart Valves**

SHELLEY MCKELLAR

Further Reading

Cauwels, J.M. *The Body Shop: Bionic Revolutions in Medicine.* Mosby, St. Louis, MO, 1986.

Fox, R.C., Swazey, J.P. and Watkins, J.C. *Spare Parts: Organ Replacement in American Society.* Oxford University Press, New York, 1992.

Fox, R.C. and Swazey, J.P. *The Courage to Fail: A Social View of Organ Transplants and Dialysis*, 2nd edn. Transaction, New Brunsick, NJ, 2002.

Hogness, J.R., Van Antwerp, M. and National Heart Lung and Blood Institute, Eds. *The Artificial Heart: Prototypes, Policies, and Patients.* National Academy Press, Washington D.C., 1991.

Lubeck, D.P. and Bunker, J.P. *The Implications of Cost-Effectiveness Analysis of Medical Technology. Case Study Number 9: The Artificial Heart: Cost, Risks and Benefits. Background Paper Number 2: Case Studies of Medical Technologies.* U.S. Government Printing Office, Washington D.C., 1982.

Plough, A.L. *Borrowed Time: Artificial Organs and the Politics of Extending Lives.* Temple University Press, Philadelphia, 1986.

Preston, T.A. The artificial heart, in *Worse than the Disease: Pitfalls of Medical Progress*, Dutton, D.B., Ed. Cambridge University Press, New York, 1988, pp. 91–126.

Shaw, M.W. *After Barney Clark: Reflections on the Utah Artificial Heart Program.* University of Texas Press, Austin, 1984.

Helicopters

The term "helicopter" is derived from the ancient Greek "elikoeioas," which means winding, and "pteron" meaning feather. Its first documented use came in the 1860s when Frenchman Ponton d'Amécourt tested miniature steam-powered models. His experiments, however, were built on a notion dating back two millennia.

The notion of vertical flying and hovering probably originated in the manufacture of toys consisting of leaves or feathers attached to a small stick and spun between the hands before being released. The earliest evidence of such toys is found in the form of Chinese tops dating back to 400 BC. Such toys gained in sophistication beginning in the eighteenth century, when Enlightenment experimenters tried to attach spring devices to help the contraption rise.

Several paper projects also existed, most notably Leonardo da Vinci's 1483 sketch of an air screw pointed vertically. Though inspiring much speculation about its feasibility, more recent interpretations of the design suggest da Vinci, who may have tested a small model of his design, was not seriously considering building such a machine.

By the nineteenth century, however, several aviation pioneers had further focused their efforts on the necessary elements to build rotating wings. The lack of an effective power plant made any machine impossible to build, but such pioneers as Englishmen George Cayley and W. H. Phillips, through their articles and model experiments, lay the groundwork for practical tests in the early twentieth century.

Most practical experiments were short-lived. Indeed, it became clear that the vertical rotor produced torque, shifting the rest of the machine around its axis. Neutralizing this torque effect while installing directional control became a central concern of experimenters. Solutions ranged from Thomas Edison's testing of rotors in the 1880s, to Italian Enrico Forlanini's dual counter-rotating propellers; yet all attempts confirmed the need for both a high-powered engine and a better understanding of aerodynamics.

In France, Louis Bréguet and Paul Cornu both built a machine capable of hovering, but their 1907 flights failed to control their respective machines' direction.

Several helicopter pioneers were eventually successful in devising solutions to the problem of directional control. Spaniard Juan de la Cierva, suggested a hybrid airplane/helicopter solution known as the autogiro. Using the principle known as autorotation, which capitalized on the phenomenon affecting fixed wings, Cierva transferred the principle to rotating surfaces. By adding wings and a propeller engine, and perfecting later models with collective pitch control (which allowed the pilot to determine lift, thrust and rotor rotation speed), Cierva launched a craze. The "flying windmill" as it became nicknamed was sold as a kind of model T of the air, though its speed was limited and it depended on proper wind conditions to lift-off. The death of Cierva in a plane accident in 1936 also signaled the decline of his "autogiro" formula.

That same year, the French Bréguet–Dorand machine set several endurance and hovering records, but still failed to demonstrate effective control capacity. Meanwhile the German Focke Fa-61 with two counter-rotating rotors took to the air and broke several of the French records. In the hands of female pilot Hanna Reitsch, it became a remarkable propaganda machine during 1938, and paved the way for the development of model Fa-223. But true German success came with Anton Flettner's model Fl-282 Kolibri, lighter than the Focke machine, and maneuverable to the point

where it avoided simulated German fighter attacks, and became the first machine to reach a practical stage (and was ordered by the German navy).

The full breakthrough would come from the U.S., where Russian Igor Sikorsky had immigrated in 1919. The young inventor had been experimenting with helicopter formulas ever since returning from a trip to Paris in 1910, where he had learned of Bréguet's helicopter attempts. Through trial and error, Sikorsky began defining further the dynamics required to control lift-off. Finally in 1931, he applied for a basic helicopter patent that included a rotor with a small vertical propeller in the tail intended to counteract the torque effect created by the big rotor. The patent was granted in 1935 and Sikorsky's prototype, model VS-300, was rolled out in fall 1939. This seminal machine served as the basis for the design of the XR-4 model, which first flew in 1942 and was ordered into serial production for use in Burma in World War II, where the machines rescued downed pilots.

By the end of World War II several other helicopter manufacturers had entered the fray, including Bell helicopters, whose chief engineer Arthur Young helped design the highly successful Bell-47, which first appeared in 1949. By the time of the Korean War, helicopters began serving as medical evacuation machines, and soon after were used for commando drops. The performance of these machines, though steadily increasing, remained limited by the use of piston engines, which often experienced overheating. This changed with the introduction of turbine engines; that is, jet-power affixed to a propeller or a rotor. The French-built SE-3130 Alouette II was the first serial-built machine to use a jet engine, the turboshaft Artouste II.

Steady progress continued, yet helicopters did not evolve much with regard to the lift and control devices they use, which remained remarkably similar to those of Sikorsky's VS-300. A helicopter rotor includes two to five blades radiating symmetrically from a hub atop the machine. These function as both wings and propellers, which are driven by an engine. Because of a series of gears, the speed of the rotor blades is less than that of the engine. Unfortunately, this reduction on speed increases the phenomenon of rotational inertia, or torque, which pushes the helicopter in the direction opposite of that of rotation. This is usually countered through the installation of a smaller vertical rotor in the tail. Some models, however, incorporate a second rotor turning in the opposite direction, a solution adopted on several American and Russian models.

The lifting phenomenon functions in a manner similar to that of a plane's wings. However, to replace the rudders and ailerons, the blades (hinged to remain flexible) can rotate along their axis depending on which flight phase the helicopter is in. The rotor can modify the angle of the blades cyclically (each blade in turn as it completes a rotation), or simultaneously.

Piloting a helicopter relies on three primary controls. The "stick," similar to that in an airplane cockpit, is the cyclic pitch lever, which allows the aircraft to move forward, backward, or sideways; the collective pitch lever, on the left side of the pilot's seat, which controls the angle of descent and climb; and the rudder pedals.

Since the helicopter's flying depends entirely on the rotor, any engine shutdown can have catastrophic consequences. A safety feature that allows a rapid controlled descent is autorotation, where the gears are released, allowing the rotor to spin solely under the force of air, without engine input. This produces some lift, while safeguarding control functions.

The features of the modern-day helicopter have also limited its speed. Some combat machines, such as the Bell AH-1 Cobra gunship used in Vietnam, broke speed records, but their achievement remains an exception rather than the norm.

Soon after World War II, several airlines in the U.S. and Europe began applying for licenses to operate helicopter service, either from city to city or within a city, from downtown to the airport. Most of these scheduled services disappeared by the 1970s as a consequence of safety concerns (including several high-profile accidents in the U.S.), but also due to cost: the number of bookings for such flights failed to balance the high operational overhead that the maintenance of small heliports required. Instead, most helicopter lines rely on a combination of charter, sightseeing, and luxury services for high-end clientele to survive. Others specialize in specific hauling services, such as logging, heavy construction, and medical evacuation. The helicopter has also proven extremely useful to law enforcement in solving ground traffic problems and policing crime.

Though the overall shape of the helicopter has not changed much since Sikorsky's original patent, various innovations and materials have contributed to improving its performance. These range from the use of carbon composites, to placing the tail rotor within the axis of the tail to reduce noise. More recently, the use of directional outlets to blow air out of the tail (the NOTAR principle) has appeared as a way to replace the tail rotor.

Finally, newer combat models are displaying stealth capabilities. A recent innovation has involved the use of tilt rotor technology, whereby a vertical take-off machine tilts its engines and uses them in the manner of an airplane. The Bell Boeing V-22 Osprey is the latest incarnation of such an approach. Though touted as highly promising (especially in terms of speed gains and carrying capacity), it has thus far proven difficult to master operationally.

See also **Aircraft Design**

GUILLAUME DE SYON

Further Reading

Boulet, J. History of the Helicopter: As Told by its Pioneers. France-Empire, Paris, 1984.
Brooks, P.W. Cierva Autogiros: The Development of Rotary-Wing Flight. Smithsonian Press, Washington D.C., 1988.
Focke, H. German Thinking on Rotary Wing Development. J. R. Aeronaut. Soc., May 1965.
Leishmann, J.G. Principles of Helicopter Aerodynamics. Cambridge University Press, Cambridge, 2000.
Liberatore, E.K. Helicopters Before Helicopters. Krieger, Malabar, FL, 1998.
Spenser, J.P. Whirlybirds. A History of the U.S. Helicopter Pioneers. University of Washington Press, Seattle, 1998.

Hematology

Hematology is the medical specialty that deals with conditions of the blood and blood-forming organs. Plentiful in supply and relatively easy to access, the blood is a tissue well suited to manipulation and investigation. Nineteenth century innovations in microscopy, such as the introduction of new staining techniques and phase-contrast methods, brought great advances in analysis of the blood. Such advances, in combination with the mass-production of the relatively easy to use hemocytometer and hemoglobinometer (used for measuring the size and number of blood cells and hemoglobin concentration, respectively), meant that the morphological and quantitative analysis of the blood became a fashionable part of practice for many early twentieth century physicians, especially those wishing to demonstrate familiarity with the latest methods in "'scientific medicine," reflecting and stimulating changes in medical practice and research more widely. It was during this time in North America and Europe that new institutional and intellectual ties between clinical medicine and basic science were forged, dramatically affecting the nature of clinical research. The study of the form and functions of the blood in health and disease was a popular subject of research, attracting the attention of chemists and pathologists, research-oriented clinicians, and chemical physiologists. Similarly, as knowledge of the blood and its constituents increased, so routine diagnostic analysis of the blood became a central part of the work of the many new hospital laboratories springing up in this period across Europe and North America.

Between the mid-nineteenth and mid-twentieth centuries, research effort in hematology was particularly focused on the causes and treatment of the anemias. These were conditions commonly seen on the wards, which often provided great potential for analysis using the latest technologies. Sickle cell anemia among African–Americans, for instance, was formally identified by Chicago physician James Herrick in 1910, but as a rare disease found solely in members of an underprivileged minority population, it did not initially trigger much clinical or biological interest in the U.S., and still less in Europe, with its tiny black population. Nonetheless, by the 1930s, sickle cell anemia, along with thalassemia, another inherited hemoglobinopathy (first identified as a specific disease in 1925 by the Detroit pediatric physicians Thomas Cooley and Pearl Lee), was recognized as a genetically linked condition, and as such was fitted well in the growth of interest in genetic and molecular views of disease in the twentieth century. In 1945, chemist Linus Pauling and his new medically-trained graduate student Harvey Itano took up the problem of sickling in red blood cells as a tool through which to explore the chemical nature of hemoglobin, making sickle cell anemia the first disease to be fully described on a molecular level.

The link between iron deficiency and anemia, the root of one of the commonest of all nutritional deficiencies, was first made by Gustav von Bunge, Professor of Physiological Chemistry at the University of Basle in 1902. It was not until the 1930s, however, that British physicians Leslie Witts, D.T. Davies, and Helen MacKay each confirmed the role of nutritional iron in the replacement of iron contained in hemoglobin, and discussed the dietary lack or gastric malabsorption of this iron as leading to a specific disease process. In other anemia research, dietary treatment was also producing a great deal of research activity. In 1925 the Rochester physiologists George Whipple and Frieda Robscheit-Robbins published results showing that administration of beef liver to dogs with severe experimental anemia markedly increased the rate of blood regeneration, and they discussed the possible applications of this work to patients with pernicious anemia. Their

work was complemented on the clinical side by George Minot and William Murphy, two Harvard medical school physicians, who announced in 1926 a dietary treatment for pernicious anemia through the administration of raw liver. As had been the case for insulin treatment for diabetes (discovered a few years earlier), these findings were celebrated as a wonderful example of the power of combined laboratory and bedside research, and in 1934, Whipple, Minot and Murphy were awarded the Nobel Prize in Medicine for their work. The anti-anemic factor in liver was isolated and identified as vitamin B12 in 1948 by two teams working independently, one British, led by E. Lester Smith, head of Glaxo's biochemistry department, and one American, led by Merck's director of biochemistry, Karl Folkers. The vitamin's chemical structure was worked out during the mid-1950s by Dorothy Hodgkin, the Oxford x-ray crystallographer, for which work she received the Nobel Prize in Chemistry in 1964.

The absence of some "intrinsic factor," (i.e., a substance required to absorb the anti-anemic factor), was first proposed by William Castle to be the underlying cause of pernicious anemia arising from his work at the Boston City Hospital during the 1930s (Castle also contributed significantly to studies of hypochromic anemia and iron deficiency ongoing there at this time also). Castle's idea stimulated enormous amounts of further hematological research aimed at the isolation and identification of this mysterious entity, and then, it was hoped, the development of a cure for pernicious anemia. Today, the disease remains incurable but treatable, and is now understood as an autoimmune disease. The 1930s also saw the introduction of a pioneering method of quantitative analysis of the blood, devised by the U.S. physician Maxwell Wintrobe, making use of his new invention, the "Wintrobe tube," designed to measure the volume of packed red cells and erythrocyte sedimentation rate. From this analysis, Wintrobe produced a classification of the anemias that endures to the present day.

If anemia defined the study and practice of hematology in the first half of the twentieth century, then advances in the understanding of blood malignancy (the leukemias and lymphomas) and the introduction of effective chemotherapies firmly established hematology as a formal specialty in the post-World War II years. The first professional hematological society, the International Society of Hematology, was founded in 1946, followed by the American Society of Hematology in 1954, and then several national societies across

Europe and Asia. New journals such as *Blood* (1946), *The British Journal of Haematology* (1955) and *Progress in Hematology* (1956) became crucial professional vehicles for the nascent specialty. The new oncology and pediatric oncology bodies and journals emerging during the 1960s and 1970s further extended the professional interactions of hematologists.

By the end of the twentieth century, a career in hematology required first training as either a physician or a pathologist, and involved some or all of the following: the management of blood products and derivatives; the administration of immunosuppressives, chemotherapies, anticoagulants and antithrombotic agents; and supportive care for a range of systemic diseases. On the research side, hematologists are actively engaged in research on several fronts including anemia, cancer and chemotherapy, blood typing and blood products, pain management studies, and stem cell therapy.

See also **Blood Transfusion and Blood Products; Cancer, Chemotherapies; Diagnostic Screening**

HELEN K. VALIER

Further Reading

Foster, W.D. *A Short History of Clinical Pathology.* London, 1961.

Jaffe, E.R., The evolution of the American and International Societies of Hematology: a brief history. *Am. J. Hematol.*, 39, 67–69, 1992.

Loudon, I. The history of pernicious anaemia from 1822 to the present day, in *Health and Disease: A Reader*, 1st edn. Milton Keynes, 1984.

Robb-Smith, A.H.T. Osler's influence on haematology. *Blood Cells*, 7, 513–533, 1981.

Wintrobe, M.M., Ed. *Blood, Pure and Eloquent: A Story of Discovery, People and of Ideas.* McGraw-Hill, New York, 1980.

Wintrobe, M.M. *Hematology: The Blossoming of a Science.* Lea & Febiger, Philadelphia, 1985.

Wailoo, K. *Drawing Blood: Technology and Disease Identity in Twentieth-Century America.* Johns Hopkins University Press, Baltimore, 1997.

Highways

The planning and use of highways, or motorways in the U.K., have profoundly altered the speed, scope, and meaning of individual transport as well as the face of the environment in many developed countries, especially during the second half of the twentieth century. Cars, which were originally designed as intraurban alternatives to horse-drawn carriages, could only be transformed into major carriers of regional and long-distance transportation with thoroughfares connecting urban centers. A highway is generally defined as a multilane road with separate lanes for each direction, separated by a median or crash barrier, to which access is limited both technically through entry and exit ramps and legally by only allowing cars and trucks to use these roads. Many countries require minimum speeds on their highways; all countries, with the exception of Germany and some U.S. states, limit the maximum speed. Often equipped with their own police, service stations, rest areas, and road crews, these roads have become institutions of their own, especially in those countries where they are run by separate public or private bodies. Research into surfacing, soil treatment, and the planning of interchanges, has contributed to the growth of these networks.

In both Europe and the U.S., calls for the creation of "car-only" roads grew louder after World War I. The overwhelming majority of roads were unpaved, and early motorists disliked sharing them with farmers, carriages, and livestock. The exclusion of the nonmotorized was at the heart of proposals to build car-only roads as promoted by middle class and upper middle class lobbies and local chambers of commerce. But the number of cars was still low and the status of the automobile predominantly that of a luxury. The Milan engineer Piero Puricelli introduced the idea of an "Autostrada dei Laghi" in 1922 to connect the northern Italian industrial centers. Only with the support of Mussolini was this fantastic idea realized. The Italian Fascists saw the autostrada project as a chance to portray themselves as energetic and in the vanguard of technological modernity. The first of Puricelli's roads was opened in 1924, and by 1935, 485 kilometers of autostrada had been built. They lacked separate carriageways, and drivers had to pay tolls on these roads.

Inspired by the Italian example, a coalition of industrialists and upper-class car drivers pushed unsuccessfully for similar roads called "autobahnen" in Germany. The lobby called itself "Hafraba" since the road was to connect the Hanseatic cities, Frankfurt, and Basle, and the lobby created technical blueprints. Toll roads were illegal in Weimar Germany, so the group lobbied unsuccessfully for the removal of this provision. The Nazi and Communist parties also opposed this restriction. Yet, after his assumption of power in 1933, Hitler grasped the opportunity to present his new government as proactive in the face of high unemployment, and he pushed the autobahn project against the advice of the national railway, the department of transportation, and the army, which was opposed to building potential enemy targets. He created a separate government department for the Reichsautobahn, headed by the civil engineer Fritz Todt. The Nazi plans envisioned a network of 6,000 kilometers of roads—far more than demand—to be built by 600,000 laborers. When road building came to a halt in 1941, some 3,700 kilometers had been built by roughly 125,000 underpaid workers under harsh conditions. After the onset of World War II, Jews from Germany and the occupied countries were forced to work on these roads, shielded from public view. The exuberant propaganda of the autobahn, however, fashioned it into a symbiosis of technology and nature, thus belying the internal conflicts over road design between landscape architects and civil engineers. Through movies, radio, theater, paintings, and board games, Germans were told that "Adolf Hitler's roads" had been envisioned by the dictator during the 1920s. Another myth is the military and strategic role of the network during the war—motorways were of limited use to Nazi aggression.

When the Allied troops liberated Germany in 1945, civil engineers in the U.S. Army were astonished by the size of the autobahn network and were reminded of the lack of a similar system in the U.S. For the first four decades of the twentieth century, political conflict over taxes and financing had generally blocked interstate road construction in the highly federalized U.S. Local roads of varying standards did not extend into a transcontinental network. While "pleasure" roads such as the Merritt Parkway, the Mount Vernon Memorial Parkway, and New York City's Parkway System banned trucks, the Pennsylvania Turnpike was opened in 1940 as a labor-generating project using a railroad right-of-way. The federal Bureau of Public Roads, run by the technocrat Thomas MacDonald and aided by money from Congress, built or resurfaced over 145,000 kilometers of federally aided highways during the 1920s. The Bureau also sponsored and conducted road-building research, thus defining and homogenizing the standards for roads while ensuring the professional hegemony of the engineers. But not until the passage of the Interstate Highway Act in 1956 did these standards—including limited access and grade separation—become laws. After failing to pass the act the first time, the Eisenhower administration reintroduced the bill in Congress and advertised its ostensible military relevance. With remarkable speed, the U.S. set out to build the longest engineered structure ever, the Interstate Highway System. By 2000, some 89,400 kilometers

of these roads spanned the U.S. European countries extended or commenced their highway programs in the 1950s, yet not with the same vigor as the U.S.

During the first decades of the Cold War, optimism about these roads abounded in the U.S. They were extolled as embodiments of a specifically American freedom to move and roam; extending the network was often seen as directly contributing to economic growth. The federal and state engineers controlling the construction program had few incentives to consider issues of urban renewal and social regeneration in their planning. As a result, urban interstates were often built through minority and lower-class neighborhoods, causing social displacement and environmental problems. The "freeway revolt" of the 1960s and 1970s rallied against the destruction of inner cities and ended the sole authority of engineers over the design process. A number of projects came to a halt, and roads increasingly came to be identified with pollution, the despoliation of nature, and suburban sprawl. This reevaluation of highways is true for Europe as well. When the Newbury Bypass in the U.K. was opened to traffic in 1998, it occurred in secrecy at 1:25 A.M. in order to forestall protests.

See also **Environmental Monitoring**

THOMAS ZELLER

Further Reading

Bortolotti, L. and De Luca, G. *Fascismo e Autostrada. Un caso di sintesi: La Firenze-Mare [Fascism and Motorways. A Case of Synthesis: The Motorway from Florence to the Sea]*. Franco Angeli, Milan, 1994.

Davis, T.F. Mount Vernon Memorial Highway: Changing Conceptions of an American Commemorative Landscape, in *Places of Commemoration: Search for Identity and Landscape Design*, Wolschke-Bulmahn, J., Ed. Dumbarton Oaks, Washington D.C., 2001, pp. 123–177.

Lewis, T. *Divided Highways: Building the Interstate Highways, Transforming American Life*. Penguin, New York, 1997.

Netzwerk Autobahn [Network Autobahn]. *WerkstattGeschichte* [thematic issue], 7, 21, 1998.

McShane, C. *Down the Asphalt Path: The Automobile and the American City*. Columbia University Press, New York, 1994.

Rose, M. *Interstate: Express Highway Politics*, revised edn. University of Tennessee Press, Knoxville, 1990.

Schütz, E. and Gruber, E. *Mythos Reichsautobahn: Bau und Inszenierung der "Straßen des Führers" 1933–1941 [The Myth of the Reichsautobahn: Building and Representing the Reichsautobahn, 1933–1941]*. Ch. Links, Berlin, 1996.

Seely, B. *Building the American Highway System: Engineers as Policy Makers*. Temple University Press, Philadelphia, 1987.

West, G. *The Technical Development of Roads in Britain.* Ashgate, Aldershot, 2000.

Zeller, T. The landscape's crown: landscape, perceptions, and modernizing effects of the German autobahn system, 1934 to 1941, in *Technologies of Landscape*, Nye, D., Ed. University of Massachusetts Press, Amherst, 1999, pp. 218–238.

Hip Replacement, *see* **Implants, Joints, and Stents**

Histology

Histology, simply defined, is the study of tissues. The term comes from the Greek words "histos," meaning web (or tissue), and "logos," meaning study; and the word histology first appeared in 1819. Without the microscope, there would be no modern field of histology; however Marie Francois Bichat, an anatomist and surgeon in Montpellier, France, defined 21 types of tissues without this technology. Other technologies used in histology pertain to preparation, preservation, and visualization of tissue samples.

The basic techniques of sectioning and staining used in the early part of the twentieth century depended on the ability of a tissue or cell to retain enough of its morphological integrity to be useful after being processed. One problem with preparing specimens was obtaining cuts of tissue thin enough for visualization without destruction. The microtome, a device for cutting specimens, developed synchronously in France and Germany. A hand-held model by Nachet in 1890 was followed by a rotary microtome by Bausch and Lomb in 1901, and Leitz manufactured a sledge chain-driven machine in 1905. The basic principle involves the operation of a hand wheel that activates the advancement of a block of tissue embedded in wax toward a fixed knife blade that slices it into very thin pieces. Microtome knives were hand sharpened by histotechnologists until the 1960s when machines with disposable blades were introduced.

One challenge to early scientists was the death of cells after exposure to air and light: tissue scrapings placed on slides soon lost their shape and size. When tissues are removed from the body, they lose circulating nutrients and will deform unless treated with appropriate chemicals. Tissue had to be fixed, dehydrated, cleared, and infiltrated. During the first half of the century, each tissue was taken through a series of baths of formalin, alcohols, dioxane, and then paraffin. Formalin denatures the proteins so that they do not get damaged during the subsequent chemical baths. After rinsing, they were placed in 70 percent alcohol and successive baths of more concentrated alcohol until the application of xylene, a solvent for paraffin. Dioxane (diethyl dioxide alcohol) was used as the final dehydrant until the 1970s when its toxicity and pathogenicity were recognized.

After the tissue was cleared (alcohol removed) and made transparent, a material to support it was necessary. Liquid paraffin was embedded to infiltrate, support and enclose the specimens. In 1949, another embedding media, butyl metacrylate was introduced for the ultrathin sections used in electron microscopy. Celloidin, a nitrocellulose compound, was used in Europe instead of paraffin because it was considered superior with regard to support of tissues that were hard to infiltrate, such as bone or eyes. Carbowax, a water-soluble wax, was first used in the early 1960s. It took less time, but since it was hygroscopic, it required more care with regard to environmental moisture.

Until the 1970s, the technologist had to fabricate paper "boats" into which paraffin was poured to embed the prepared tissue. Some laboratories used plastic trays but until the introduction of an automated technology known as an embedding center (Tissue-Tek), the work was tedious and time consuming. After the paraffin cooled and solidified, the paraffin-containing specimen was cut on the microtome and mounted on a slide.

An alternative to the paraffin method of preparation is the cryostatic method, introduced in 1932 by Schultz-Brauns. This technology consisted of a refrigerated knife and microtome that could prepare the tissue at a temperature between -10 and $-20°C$. Its advantage was that so-called frozen specimens could be examined while the patient was still in surgery. If the tissue was pathological (usually cancerous), it could then be removed in the same surgical procedure. By the 1990s, the technique used was historadiography. Quickly-frozen tissues were dried, then prepared and photographed. The relative mass could be determined because there is a relationship between the film contrast and the various parts of the specimen.

Since cells are made from proteins, each subcellular organelle reacts to a dye by either staining to a color or not. The two most common stains are eosin, which stains the cytoplasm pink, and hematoxylin, a blue color used for nuclear material. The process of staining formerly required labor-intensive work in which a technologist applied stain to one slide at a time. The process took 12 hours from start to availability for examination in order to make a diagnosis. With the introduction of an automated system (Dako Autostainer) in the 1990s, 48 slides could be processed in a period of two hours and with more than one stain. This equipment applies the stain, advances the slides, and dries them in a uniform process.

Preservation of live material is only temporary, but in certain cases one can observe both structure and function. The phase contrast microscope allowed for optimal visualization of difficult materials, but a challenge to progress in histology has been the inverse relationship between magnification and light. The power of a microscope depends on the wavelength of the light and the light-gathering capacity (numerical aperture) of the objective. Most histological work is performed with lenses of numerical apertures of 1.0 or less. If the wavelength of light is reduced, it is possible to increase the resolving power of a microscope to 0.1 micrometers with ultraviolet light. Smaller particles can be seen in dark field, but shape and dimension are not accurate. The challenge to increase both the light and the resolving power of the microscope resulted in a phase contrast microscope developed by Zeiss in the 1930s that was able to film a cell division. Such microscopes were not widespread until the 1950s. In 1955 improvements were made in the prism design, known as differential interference contrast (DIC), and this allowed visualization of living cells without staining.

In the 1960s a technology for diagnosing cervical cancer was developed by George Papanicolaou. Known as the PAP smear, the technique consisted of lightly scraping the cervical mucosa, spreading the sample on the slide, fixing it, all in the doctor's office, and then sending it to a laboratory for analysis. In 1996, a ThinPrep (Cytyc) test was approved by the U.S. Food and Drug Administration that collected the cells, rinsed them into a vial of preservative solution, and after filtration, applied them to a microscope slice uniformly. This technique was extended to non-gynecologic specimens such as fine needle aspirations and endoscopic brushings.

The halogen lamp, powered by a circuit board that prolonged its life, solved the light versus resolution relationship. Microscopes with multiple eyepieces allowed more than one person to view a specimen. By adding a tube to the viewing chamber and attaching another set of eyepieces, it was possible for two pathologists to view the same slide simultaneously (Olympus split microscope). Variations on this technology allowed for a class arranged in a circle to observe the same material. By the 1990s, technology had arrived at infinity-corrected optics; optical microscopy allowed investigation at the micron and submicron level.

Perhaps the most revolutionary development in twentieth century medicine has been the emergence of biotechnology. Monoclonal antibodies, immunohistochemistry, tumor markers, and flow cytometry depend on the ability to work with DNA and fluorescent labeling of organelles such as microtubules and endoplasmic reticulum. New diseases such as AIDS required new technologies to detect antibodies. By the end of the twentieth century, automation technology had affected and improved every division of histology departments in hospitals, from cytology and microbiology to pathology. One microscope could be accessorized with DIC, fluorescence, polarized light, phase contrast, and photomicrography using several film formats and digital image capture (Olympus Provis AX-70). The entire paradigm of a lengthy fixation time and preservation was modified because immunohistochemistry, electron microscopy, and molecular biology techniques required change. An autostainer can produce 2000 slides per hour, 100 at a time. NASA technology led to the development of an automated cellular information system (ACIS) using automated microscopy and computerized image analysis. The slide software captures hundreds of fields and projects them on a screen, and the observer selects those of interest and adjusts the magnification to a higher order. A quantitative score is then computed with regard to staining intensity and other parameters. The data is converted to a format suitable for export to a spreadsheet or database program. At the close of the century, fiber optics, digital cameras, and camcorders as well as computer software used in microscopy were used in programs for both students and continuing professional education.

LANA THOMPSON

Further Reading

Fredenburgh, J. and Myers, R.B. *The Effects of Fixation and Antigen Retrieval on Immunohistochemical Staining of Paraffin Sections.* Monograph. Richard-Allan Scientific, Kalamazoo, MI, 2000.

Long, E.R. Recent trends in pathology 1929–1963, in *A History of Pathology.* Dover, New York, 1965. Reprint of Williams & Wilkins 1928 edition.

Maximow, A. and Bloom, W. *A Textbook of Histology.* W.B. Saunders, Philadelphia, 1953.

Preece, A. *A Manual for Histologic Technicians*, 2nd edn. Little, Brown & Company, Boston, 1965.

Hormone Therapy

Hormones are substances formed in one organ of the body and transported in the blood to another where they regulate and control physiological activity. The term hormone was coined in 1905 by William B. Hardy from the Greek meaning "arouse to activity." English physiologist Ernest H. Starling and Sir William Bayliss discovered the first hormone in 1902 and named it secretin. Chemically, hormones are divided into three classes: amino acid derivatives, peptides, and steroids. The purpose of hormone therapy is to provide or maintain physiological homeostasis with natural or synthetic hormone replacement.

Menopause

Perhaps the most familiar type of steroid hormone therapy is estrogen replacement therapy (ERT) for women whose ovaries have been removed surgically or women who have symptoms related to menopause. The first estrogen drug, under the trade name Premarin, was available in injectable form in 1942. The pill form gained popularity as the "medicalization" of menopause increased. The hormone was derived from the urine of pregnant mares. When the relationship between osteoporosis, hip fracture, and low estrogen levels was recognized, ERT was recommended for all women. The "baby-boomer" generation (those born between 1946 and 1964) was particularly receptive to ERT promotion, and by the end of the century one third of postmenopausal women in the U.S. were taking some kind of hormone replacement therapy.

In 1975, research indicated that unopposed estrogen raised the rate of endometrial cancers. Progesterone was added to estrogen, and the treatment became known as HRT (hormone replacement therapy). By the year 2000, esterified estrogens, synthetic estrogen from plant sources, and combinations with testosterone and progestins were prescribed as pills, transdermal skin patches, or intravaginal applicators. Studies early in the twenty-first century, however, determined that although these drugs were safer, they did not provide sufficient protection against osteoporosis or cardiovascular disease in women to warrant the risks of therapy.

Transgender Surgery

In 1930, Danish artist Einer Wegener underwent the first documented gender reassignment surgery and became Lily Elbe. Gender and sex, taboo topics for much of the century, became important research concerns in the 1960s. John Money began his longitudinal studies on children born with ambiguous genitalia (intersex individuals).

Understanding how male and female sex hormones worked on the brain paved the way for more sophisticated hormone therapies before and after gender reassignment surgery.

Androgens are administered to biological females, and estrogens, progesterone, and testosterone-blocking agents (anti-androgens) are given to biological males, either orally, transdermally, or by injection, to stimulate anatomical changes. Orchiectomy (removal of the testicles) may mean that feminizing hormone dosages can be reduced. Unfortunately, follow-up studies revealed high morbidity and mortality rates, and the reasons for this are unknown. Although one can correlate hormone therapy with these rates, it is difficult to isolate causal factors because of the complex relationship between physiological and psychological processes.

Prostate Cancer

Hormone therapy may be initiated to treat prostate cancer following, or in addition to, surgery or radiation therapy. The hormones used to treat prostate cancers are also anti-androgens, substances antagonistic to the male hormones. Estrogen and diethylstilbestrol (DES) were used until the late 1970s when some synthetic compounds were introduced. In 1987 Zoladex, a drug that interferes with the production of testosterone, was tested to replace DES. Cyproterone acetate (a type of anti-androgen) and Estracyt or Emcyt (a combination of estradiol and nitrogen mustard) were found to be effective when estrogen therapy failed. A synthetic hormone to treat prostate cancer, Lupron, suppresses the formation of steroids in the body. Paradoxically, it has also been prescribed for the treatment of endometriosis in women.

Human Growth Disorder

Human growth hormone (HGH, somatotropin) was discovered by Herbert Evans in 1921, but it was not used in the U.S. until 1958 to treat certain kinds of dwarfism in children related to hormone deficiency. Stature deficits were virtually eliminated with this hormone, produced from human pituitary glands from cadavers. Unfortunately, the deaths of four children were traced to cadavers in which Creutzfeldt–Jacob disease infectious tissue was present (a new variant of Creutzfeldt–Jacob disease became known in the 1990s as the human form of mad cow disease), and in April 1985 all hormone production using cadavers ceased. In October of that year, a biosynthetic hormone, somatrem, was produced using recombinant DNA technology. For thousands of children dependent on continued therapy for optimum growth, an unlimited supply of HGH could be produced. As with HRT for women, questions have been raised about the use and potential misuse of this hormone for human engineering.

Thyroid Deficiency

At the turn of the century, George Murray prepared an extract of thyroid from a sheep and injected it into a woman with myxedema, a severe form of thyroid deficiency, and the patient improved. Twenty-eight years later, he succeeded in making an oral preparation. Two hormones, thyroxine (T4), isolated in 1914, and triiodothyronine (T3), affect growth and metabolism, and both are controlled by thyrotropin, or thyroid stimulating hormone (TSH), produced by the pituitary gland. In 1910 the relationship between the thyroid gland and endemic goiter and iodine was discovered. When there is insufficient thyroid hormone, one cause of which is insufficient iodine in the diet, the body overproduces thyrotropin, causing enlargement of the thyroid gland; the condition is known as goiter. Replacement therapy suppresses thyrotropin, and the goiter shrinks. Despite the fact that thyroid and iodine supplements have been available for 50 years, endemic goiter and cretinism still occur worldwide.

Adrenal Hormones

Prior to the twentieth century, Addison's disease and its relationship with the adrenal glands were well known. But it was not until the 1930s that a great deal of interest was generated with regard to adrenocorticotropic hormone (ACTH). Deficiencies affected electrolyte balance, carbohydrate metabolism, hypoglycaemia, and sodium loss. It was postulated that two hormones were secreted by the adrenals: mineralocorticoids and glucocorticoids (salt and sugar hormones). Harvey Cushing described a syndrome in which too much ACTH was secreted. When the first extract of the adrenal gland was prepared in 1930, it was found to contain 28 different steroids, and five were biologically active. Adrenal cortex substance was of interest during World War II because of its effect in reducing the stress of oxygen deprivation. However, cortisone, the "anti-stress hormone," soon became the miracle drug of the twentieth century for treatment of arthritis and skin inflammations. In 1952, hydrocortisone, with fewer side effects, was developed using biosynthesis. By the 1960s, cortical steroids were being used for over 150 medical applications. Since the 1960s, technology has been directed at developing non-steroidal anti-inflammatory therapies.

Prostaglandins

In the mid-1930s, von Euler found that when the active substance of seminal fluid was injected into laboratory animals, it lowered their blood pressure. He named it prostaglandin. For the next thirty years his Swedish colleagues continued this work and discovered that prostaglandin was really four substances composed of fatty acids. This research led to the discovery of prostacyclin and thromboxane, hormones that have reciprocal actions on platelets. Prostacyclin has been used to treat circulatory problems. Flolan, a British product, is used to prevent blood from clotting in bypass machines.

Insulin

In 1921, Frederick Banting and Charles Best, two Canadian physiologists, were attempting to find a cure for diabetes mellitus. Their research on dogs involved removing the pancreas, waiting for symptoms of diabetes to develop, then injecting the animals with insulin. Previous experiments by others failed because the hormone degraded when taken orally. From the 1920s until the early 1980s, insulin was produced from the pancreases of pigs. In 1973, scientists at the Massachusetts Institute of Technology paved the way for synthetic insulin when they invented a technique for cloning DNA. In 1978 Genentech cloned the gene for synthetic insulin, and in 1982 the USDA approved an insulin derived from recombinant DNA called Humulin. Since that time, a disposable insulin pen has been developed to take the place of a syringe and vial. Other technologies include jet injectors, an external insulin pump, implantable insulin pumps, and insulin inhalers. Oral antidiabetic agents called sulfonylureas were developed in 1994. Another oral product, troglitazone, used concomitantly, was removed from the market in 1997 because of liver toxicity. In 2000, Eli Lilly, the original pharmaceutical company for the manufacture of insulin, announced a research partnership with Generex Biotechnology to develop another oral (buccal) form of insulin.

Negative Consequences

As previously noted, some of the technologies that have allayed suffering and prolonged life in the twentieth century have also been misused. In the 1990s, synthetic HGH was marketed to athletes to increase body strength and to adults as the "anti-aging hormone." Many of these non-medical applications had disastrous consequences. Controversies with HRT continued as some women believed the "hot flashes" and sleep disorders they experienced beginning in perimeno-pause could only be relieved with hormone ther-apy, while others simply considered estrogen useful to maintain a youthful appearance. On the other hand, for the millions of children and adults who have benefited from insulin, there is no controversy about hormone therapy.

See also **Diabetes Mellitus; Contraception, Hormonal Methods and Surgery; Fertility, Human; Genetic Engineering, Applications**

LANA THOMPSON

Further Reading

Lock, S., Last, J.M. and Dunea, G., Eds. *The Oxford Companion to Medicine.* Oxford University Press, Oxford, 2001.

Maisel, A.Q. *The Hormone Quest.* Random House, New York, 1965.

Money, J. *Discussion in Endocrinology and Human Behaviour.* Oxford University, London, 1968.

Money, J. *Sex Errors of the Body: Dilemmas, Education, Counseling.* Johns Hopkins University Press, Baltimore, 1968.

Money, J. and Ehrhardt, A. *Man and Woman, Boy and Girl: The Differentiation and Dimorphism from Conception to Maturity.* Johns Hopkins University Press, Baltimore, 1972.

Pediatrics, in *Medical and Health Annual.* Encyclopedia Brittanica, Chicago, 1987.

Seaman, B. and Seaman, G. *Women and the Crisis in Sex Hormones.* Rawson Associates, New York, 1977.

Shames, K. and Shames, R. *Thyroid Power.* Harper Information, New York, 2001.

Hovercraft, Hydrofoils, and Hydroplanes

These three vehicle types combine aspects of flying and floating in hybrid marine craft capable of far higher speeds than traditional boat hulls. They accomplish this feat, valuable in commercial ferries, military vehicles, and racing boats, in very different ways.

Hovercraft

Hovercraft fly a few feet over water or land on a cushion of high-pressure air forced between the vehicle's hull and the surface by motorized (often turbine) fans. Though the air-cushion idea dates back more than 200 years, practical achievement of a viable carrier of people and goods came only in the mid-twentieth century, thanks to the inventive genius of radio engineer Christopher Cockerell in Britain and Jean Bertin in France. Vital to their success was C. H. Latimer-Needham's develop-ment in the late 1950s of an inflated "skirt" of rubber-like material hanging down several feet from the rim of the vessel to hold a deeper air cushion in place. Air cushion vehicles (ACV) or hovercraft saw their greatest promise in the 1960s as a variety of larger and faster designs were introduced, primarily in Britain.

The first full-sized hovercraft, the Saunders Roe-designed SR-N1, traveled across the English Channel in 1959. Experimental hovercraft ferry services began in 1962, and regular cross-Channel passenger and car ferry services began in 1964. Improved models soon reached speeds of 50 knots and could handle 10-foot (3-meter) waves. The SR-N4 two-ship class was introduced into regular car and passenger-carrying service in 1968. Demand for more capacity led to both being lengthened by 17 meters to increase their carrying capability to nearly 400 passengers and more than 50 cars in 1978–1979. These vehicles served as the world's largest hovercraft for three decades. But the growing cost of their operation and maintenance finally ended cross-Channel ferry hovercraft opera-tions with the retirement of both vessels on 1 October 2000. They were replaced by other vessels including a growing fleet of Australian-built cata-maran *Seacat* ships, first introduced in 1990 and able to carry more passengers and vehicles, as well as the opening of the Channel tunnel in the mid-1990s.

Hovercraft also serve a variety of rescue and military roles. Marines in several nations have taken hovercraft potential furthest, with over 100 craft in use in America and 250 in the former Soviet Union. The Russian "Bora" class, which entered service in 2000, displace more than 1,000 tons and carry guided missiles and a crew of 70. Other military hovercraft carry just one person for reconnaissance.

Hydrofoils

A hydrofoil is a metal wing that "flies" in water rather than air. This almost always means a boat or ship with fins attached to the light hull that can travel as either a normal vessel or up on the fins (hydrofoils) at much greater speed (up to 113 kilometers per hour (km/h)) due to the lack of hull friction with the water. Because water is far denser than air, the hydrofoils can be much smaller than wings on an airplane yet still lift the hull when a minimum forward speed is achieved. Lift increases with speed, and a hydrofoil needs but half the power of a traditional ship. Most hydrofoils are ladder-like structures, enabling the marine craft to continue at high speed even in choppy seas as at least some of the hydrofoil structure remains under water. Hydrofoils are either surface piercing (intended to operate only partially submerged, and generally more stable) or fully submerged, with many variations of each. They can maneuver and take rough sea conditions more easily than traditional hull-borne vessels.

Though there were a number of nineteenth century hydrofoil ideas and experiments in both Europe and America, the first practical hydrofoils were designed in the late 1890s and early 1900s by Enrico Forlanini, an Italian engineer who worked on both advanced boats and airships. The first large-scale example, the HD-4, was designed by F. W. "Casey" Baldwin and a team backed by Alexander Graham Bell, and it achieved 113 km/h in tests in 1918. Parts of it, as well as a full-scale replica, are displayed in a Nova Scotia museum. Later developmental models were pursued into the late 1920s, but only HD-12, a 9-meter runabout, and HD-13, an outboard motor hydrofoil boat, were actually built, both in 1928

Hans von Schertel first began experimenting with hydrofoil craft in Germany in 1927 and had developed eight designs by 1936. Only in 1939 did the military first become interested in the potential of a hydrofoil boat. Various hydrofoils followed into late 1944, including one intended for torpedo attacks, one for coast defense, and another as a specialized landing craft. Schertel moved to Switzerland to continue his work after the war. In 1953 on Lake Maggiore connecting Switzerland and Italy, a 10-ton, 28-passenger von Schertal hydrofoil took 48 minutes to cross the lake (regular ferries took three hours). Using a similar design, in 1956 Carlo Rodriquez built several hydrofoils to carry passengers between Sicily and Italy and over the next four years more than a million people traveled by hydrofoil. In the 1950s and 1960s, the U.S. Army, Navy, and Marines all experimented with hydrofoil landing craft. Some of these were amphibians, and though successful, they were mechanically very complex and heavy for their limited payload capacities. The first U.S. Navy operational hydrofoil was the submarine chaser *High Point* of 1963, which served until 1978. A fleet of six (original plans called for a class of 30) Patrol, Hydrofoil, Missile (PHM) Pegasus-class navy vessels were built by Boeing Marine, based at

Key West, Florida, and placed in service from 1975 to 1982. Utilized in drug interdiction among other missions, each displaced more than 250 tons and could reach 89 km/h on their hydrofoils. They were withdrawn in mid-1993 due to their high cost of operation.

Hydrofoils have operated as commercial ferries in all parts of the world, including several major urban harbors. Seven boats were part of the State Transit fleet of hydrofoils that operated between Sydney and Manly, Australia from 1965 to 1991 before being replaced by fast catamarans. Hong Kong hydrofoils have provided fast connections with Macao since well before both colonies were ceded to China.

Hydroplanes

Sometimes called planing craft or skimmers, hydroplane hulls resemble the simple surfboard in their basic function. By varying the shape of the bottom of a hull form (sometimes termed "stepped hulls"), a hydrofoil at speed can ride partially out of the water, thus decreasing drag and gaining greater speed. Such hull forms were, by the early twentieth century, being applied to maritime vehicles with somewhat similar principles later used in the hulls of flying boat aircraft.

The first serious discussion of such hull forms, however, dates back to the mid-nineteenth century. Abraham Morrison of Pennsylvania in 1837 patented a planing boat (though he did not call it that) with a concave hull bottom. A paper delivered in London the same year by noted engineer John Scott Russell noted that at speed such a craft "emerges wholly from the fluid and skims its surface." In the early 1850s Joseph Apsey proposed a steamship using a planing hull although the inefficient power plants of the day could not accomplish the task. In 1872 Charles Ramus suggested a wedge-hulled planing craft to the British Admiralty, which led to unsuccessful trials of a model vehicle. The problem again was the inadequate power generated by steam engines of the time—indeed, even rocket power was suggested! Finally, a Swiss inventor, M. Raoul Pictet, tested a model vessel on Lake Geneva in the early 1880s with a very modern concave hull form.

Sir John Thornycraft was one of several experimenters to resurrect the planing craft idea early in the twentieth century. In 1908 he proposed what he termed a hydroplane, featuring a two-wedge concave hull bottom that, driven with sufficient power, might largely ride above the surface with the aid of an air cushion between hull and water. At about the same time, Comte de Lambert hit on an ideal combination when he developed a series of five floats with a structure to keep them aligned and featuring a blunt bow—with an engine driving a large airplane propeller mounted above the stern. This pioneering craft was slowly improved and one version entered passenger service on Lake Geneva in 1913, as did others on the River Nile. Modern versions of the light-hulled vehicles driven by huge caged propellers are now are widely used in the Florida Everglades and other swampy areas.

Hydroplane development in the postwar years centered primarily on constant attempts to achieve record-breaking speed rather than viable commercial services. Land vehicle racer Sir Malcolm Campbell developed a series of high-powered *Bluebird* hydroplane vessels that by 1939 had achieved better than 225 km/h in a closed course on England's Coniston Water. His son Donald died there in 1967 attempting to reach more than twice that speed when his jet-powered boat, *Bluebird K7*, broke up.

In the meantime in the U.S., Gold Cup Races beginning in Detroit in 1904 and later contests such as those running on Lake Washington near Seattle since 1950 and the Harmsworth International Trophy race, became the chief arenas for constant hydroplane hull and engine improvement. The *Slo-mo-shun* series of racing hydroplanes, for example, began development in the late 1930s. By 1950 the aircraft engine-powered *Slo-mo-shun IV* reached more than 257 km/h while the *Slo-mo-shun V* broke the 298 km/h barrier a year or so later.

Most racing hydroplanes since the late 1970s have been "propriders," so named as their driving propeller is partially submerged. The modern "three-point" (only two portions of the hull and part of its propeller are in the water at high speed) racing proprider largely rides on an air cushion and can be described as a kind of surface-effect vehicle.

CHRISTOPHER H. STERLING

Further Reading

Croome, A. *Hover Craft*, 4th edn. Hodder & Stoughton, London, 1984.

Cross, I. and O'Flaherty, C. *Hovercraft and Hoverports*. Pitman, London, 1975.

Gunston, W.T. *Hydrofoils and Hovercraft: New Vehicles for Sea and Land*. Doubleday, Garden City, NY, 1970.

Hynds, P. *Worldwide High Speed Ferries*. Conway Maritime Press, London, 1992.

Hogg, G. *The Hovercraft Story*. Abelard Schuman, London, 1970.

Jane's Surface Skimmers, Hovercraft and Hydrofoils. In 1994 became *Jane's High-Speed Marine Craft and Air-Cushion Vehicles*. Janes, London, 1968–date.

King, H.F. *Aeromarine Origins: The Beginnings of Marine Aircraft, Winged Hulls, Air-Cushion and Air-Lubricated Craft, Planing Boats and Hydrofoils*. Putnam, London, 1966.

McLeavy, R. *Hovercraft and Hydrofoils*. Blandford Press, London, 1976.

Parkin, J.H. *Bell and Baldwin: Their Development of Aerodromes and Hydrodromes at Baddeck, Nova Scotia*. University of Toronto Press, Toronto, 1964.

Watts, A.J. *A Sourcebook of Hydrofoils and Hovercraft*. Ward Lock, London, 1975.

Useful Websites

Classic Fast Ferries web magazine: http://www.classicfast ferries.com/cff/

Hovercraft History and Museum:

Hydroplane and Racing Boat Museum: http://www. thunderboats.org/

Hydroplane history: http://powerboat.about.com/gi/ dynamic/offsite.htm?site=http%3A%2F%2Fwww. lesliefield.com%2F%00

Hydroplane Racing: http://www.btinternet.com/~malcolm. mitchell/hydro/history.html

International Hydrofoil Society: http://www.foils.org/ index.html

U.S. Hovercraft Society: http://www.flash.net/~bla77/ ushs/TOC.htm

Hydroelectric Power Generation

It is estimated that about 50 percent of the economically exploitable hydroelectric resources, not including tidal resources, of North America and Western Europe have already been developed. Worldwide, however, the proportion is less than 15 percent.

The size of hydroelectric power plants covers an extremely wide range, from small plants of a few megawatts to large schemes such as Kariba in Zimbabwe, which comprises eight 125 megawatt generating sets. More recently, power stations such as Itaipu on the Parana River between Brazil and Paraguay in South America were built with a capacity of 12,600 megawatts, comprising eighteen generating sets each having a rated discharge of approximately 700 cubic meters per second.

Hydroelectric power has traditionally been regarded as an attractive option for power generation since fuel costs are zero; operating and maintenance costs are low; and plants have a long life—an economic life of 30 to 50 years for mechanical and electrical plant and 60 to 100 years for civil works is not unusual.

Small-scale hydropower schemes (typically less than 10 megawatts per site) utilize rivers, canals, and streams. Large-scale schemes generally include dams and storage reservoirs, with the option of pumped storage schemes to generate power to match demand. Pumped storage schemes are however a net energy consumer, and should not be considered as renewable projects. Small hydroelectric installations are numerous in countries such as Scotland, South America, and China, for example, and may be operated by power generation companies or privately. Although some plants have been in service since the turn of the century, a considerable number of developments took place after 1945 and up to the mid-1970s, with a few, small, run-of-river developments having taken place since then. It is likely that the investment criteria that were applied in the later years of the twentieth century were more onerous than those set previously, and this has meant that new developments became more difficult to justify. The capital cost of "green field" (i.e., undeveloped and particularly unpolluted land) hydroelectric developments are higher than most alternative power generation schemes. Environmental concerns, for example over the Chinese government's undertaking of building the Three Gorges Dam, the largest hydroelectric project ever undertaken, must also be considered. In terms of a straight financial comparison, small hydroelectric plants are difficult to justify where the "competition" is a generating plant on a developed nationwide grid system. The existence of the National Grid in the U.K., for example, has allowed the exploitation of significant economies of scale in conventional thermal and later the combined-cycle generating plant.

Manufacturers' Developments

Although the field of hydroelectric engineering does not lend itself readily to the application of "standardization" techniques, most plant manufacturers have managed to use this approach to cover plants in the low output range. Most can offer packaged, or factory-assembled units, or plants that are broken down into major components, which can virtually be self-contained power stations. The accurate determination of the net head available for establishing the plant rating is fundamental. This involves deducting frictional head losses for the water intake and penstock (the conduit which carries the water to the turbine) and consideration of maximum and minimum head and tailwater levels with a statistical assessment of their frequency of occurrence.

Small and micro hydroelectric plants will generally only be viable if the civil works are simple. Even for sites where the head has already been developed (an existing dam, for example) it is

unlikely that the civil engineering costs will be less than the cost of the plant.

The number, type and arrangement of generating sets will be decided upon at the feasibility study stage. The procedures for selecting the type of turbine and generator and for selecting synchronous speeds follow firmly established engineering principles, the objective ultimately being to employ the most cost-effective and appropriate overall arrangement to fulfill technical requirements. In collaboration with the civil engineer, the requirements for turbine setting (level of runner centerline relative to tailwater level to ensure cavitation-free operation), regulation issues (a function of the inertias of both the waterway and of the generating set), and the power station dimensions will be established.

A wide range of turbine types is available, each having distinct merits in certain fields and designed to cater for specific site conditions. The tubular turbine is a simple configuration of a propeller or Kaplan axial-flow reaction-type turbine suitable for low- to medium-head applications. A speed-increasing gear can be incorporated that enables a lower-cost high-speed generator to be employed. The cross-flow turbine is a partial admission radial impulse-type low-speed turbine. A speed-increasing gear or belt-driven arrangement is normally selected to permit cheaper high-speed generator designs to be used. The form of construction is simple, resulting in a lower-cost machine than the conventional types. The maximum efficiency is modest. The efficiency characteristic is, however, such that the machine can make efficient and therefore economical use of wide-ranging river flows.

The Francis turbine runner receives water under pressure from the spiral casing in a radial direction, discharging in an axial direction via a draft tube to the tailrace. For very low heads, the Francis turbine may be used in an "open flume" arrangement whereby the spiral casing is effectively replaced by a concrete forebay structure.

Pelton and Turgo turbines are impulse-type machines in which the penstock (potential energy) pressure is converted (via nozzles) into kinetic energy; the water jet impinges on the wheel and falls to the tailrace. Impulse-type turbines are relatively low speed and, therefore, except for small outputs, are not suited to the medium- to low-head range. Because the runner must be situated above maximum tailwater level to prevent "drowning," it is unavoidable that a proportion of the gross head cannot be utilized.

Two types of generator are used in hydroelectric installations: synchronous and asynchronous (or induction) type. Synchronous generators are capable of independent isolated operation, synchronous speed and voltage control being maintained by the action of a governor (speed controller) as the load varies. An asynchronous generator relies upon the system to which it is connected to provide its magnetization current, and it operates at a power factor of 0.8 or less. This can be improved by using capacitors for "power-factor correction." While the turbine output must be automatically adjusted according to the availability of water, there is no need for governing. Synchronizing is uncomplicated as the generator circuit breaker is closed at or near synchronous speed. There are speed and output limitations for asynchronous machines, but these would probably not apply for the small hydro output range. The cost advantage diminishes with increasing output. The method of construction of asynchronous machines means that their inertia is lower than for synchronous machines. If this is an important consideration, however, it can be overcome by the use of a flywheel.

See also **Dams; Energy and Power; Electricity Generation and the Environment**

IAN BURDON

Further Reading

Greer, C. *Water Management in the Yellow River Basin of China*. University of Texas Press, Austin, TX, 1979.

Hay, D. *Hydroelectric Development in the United States, 1880–1940*. Edison Electric Institute, Washington D.C., 1991.

Schnitter, N.J. *A History of Dams: The Useful Pyramids*. Balkema, Rotterdam, 1994.

Wolf, D.E., Wolf, D.D. and Lowitt, R. *Big Dams and Other Dreams: The Six Companies Story*. University of Oklahoma Press, Norman, OK, 1996.

Worster, D. *Rivers of Empire: Water, Aridity, and the Growth of the American West*. Oxford University Press, New York, 1985.

Iconoscope

The iconoscope, a vacuum electronic camera tube, was the first television camera tube to clearly image a picture by all-electronic means. It was invented by Vladimir K. Zworykin, of the Radio Corporation of America (RCA), developed by his research group from 1929, and first field tested by RCA in an experimental television system in 1933. A similar tube known as the emitron was independently engineered by James McGee and his research group at Electric and Musical Industries from about 1932. The iconoscope and emitron all-electronic systems marked a shift away from the electromechanical television scanners that dated from the 1880s. Though no longer used, the iconoscope is the precursor to all television camera tubes used today.

The iconoscope has a vacuum-tight glass envelope. Inside there is a photosensitive mosaic, onto which an image is focused by a lens, and a high velocity electron beam that scans the mosaic line by line. The electrode configuration is illustrated in Figure 1. The electron gun that produces the electron beam comprises an oxide-coated cathode (C) mounted within a controlling electrode (G), which has a small hole through which the electron beam passes. The first anode (A) accelerates the

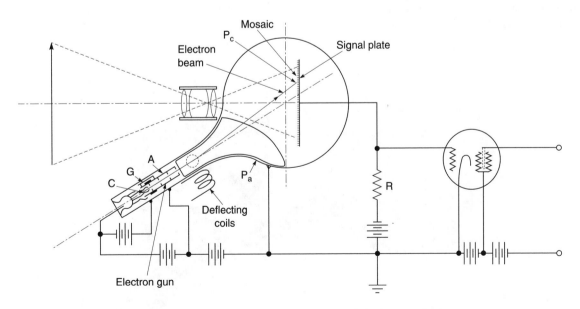

Figure 1.   The iconoscope camera tube showing cathode C, control grid G, first anode A, second anode P_a, and the mosaic elements P_c.
[*Source: Journal of the IEEE, 73, 1933, p. 441.*]

electrons. The second anode (Pa), consisting of a metallic coating on the inside of the glass bulb, further accelerates the electrons to about one tenth the speed of light. The additional important function of this second electrode is electrostatically to focus the beam into a sharp spot on the photosensitive mosaic layer (Pc), which coats one side of a thin sheet of mica. A metallic coating, the signal plate, is deposited on the other side of the mica sheet. The mosaic consists of a very large number of minute, insulated, photosensitized silver globules, each of which release electrons under the effect of illumination (the photoelectric effect). Both frame and line-scanning deflecting coils enable the electron beam to scan a raster, or series of horizontal lines, on the mosaic's surface.

In operation, when the scanning electron beam strikes an unilluminated element of the mosaic, the high energy of the beam leads to the emission of secondary electrons that, if the emission is saturated, exceed the number of primary electrons of the beam. Since the element is insulated, it acquires an equilibrium positive charge, with respect to the anode, when the rate at which the secondary electrons leave the surface is equal to the rate at which the primary electrons strike the surface. If the mosaic is then illuminated (by an optical image of a scene or object), the mosaic elements emit photoelectrons, which are collected by the positively charged electrode. The mosaic elements consequently accumulate positive charges. These are returned to the equilibrium condition by the scanning primary electron beam, which restores the electrons lost by photoemission. Thus, during each scanning period (typically one twenty-fifth of a second), all the mosaic elements follow a cycle of positive charge acquisition by photoemission, and equilibrium charge restoration by primary electron beam scanning. Since each mosaic element is capacitively coupled to the signal plate, any variation of electric charge on the elements (caused by variation in light intensity) induces a corresponding variation of charge on the signal plate. The rate of change of this charge gives rise to the signal current.

A noteworthy feature of the iconoscope is that it incorporates the principle of charge storage, possibly the most important principle of early television engineering. With a single scanning cell, as used in the Nipkow disk–photocell arrangement (an invention of Paul Nipkow) and also Philo T. Farnsworth's image dissector tube, the cell must respond to light changes in $1/(Nn)$ of a second where N is the frame rate and n is the number of scanned elements. For N to equal ten frames per second and n to equal 10,000 elements (corresponding to a square image having a 100-line definition), the cell must react to a change of light flux in less than ten millionths of a second. But if the scanned mosaic of cells is used, each cell has 100,000 millionths of a second in which to react, provided that each cell is associated with a charge storage element. Theoretically, the maximum increase in sensitivity with charge storage is n, although in practice the very early iconoscopes only had an efficiency of about 5 percent of n.

The iconoscope is not free from defects. Apart from keystone distortion, which arises because the mosaic screen is scanned obliquely by the primary electron beam, the early workers found that the

Figure 2. Iconoscope and deflecting electromagnets.
[*Courtesy of the David Sarnoff Library.*]

picture signals tended to become submerged in great waves of spurious signals associated with the secondary electrons emitted. Correcting signals, which became known as "tilt" and "bend," had to be electronically generated and added to both the line and frame signals to annul the unwanted signals.

Electric and Musical Industries' emitron camera was used in the world's first all-electronic, public, high-definition television station, the London Station, which opened in 1936 at Alexandra Palace. The emitron cameras were light, portable, noiseless, and reliable. Their sensitivity—about 25 percent of the ideal—was sufficiently high to permit them to be employed for outside television broadcasts, including the 1937 Coronation, as well as for studio productions. Moreover, because the emitron cameras produced practically no lag in the picture, moving objects were reproduced clearly.

In 1934 Alan D Blumlein and James D McGee invented the ultimate solution—cathode potential stabilization (CPS)—to the problem of the spurious signals. In a CPS emitron, the primary beam approaches the mosaic in a decelerating field and strikes the surface with substantially zero energy. Hence no secondary emission of electrons can occur. Cathode potential stabilization has several advantages. First, the utilization of the primary photoemission is almost 100 percent efficient, thus increasing the sensitivity by an order of magnitude; second, shading (or spurious) signals are eliminated; third, the signal generated is closely proportional to the image brightness at all points (i.e., it has a photographic gamma of unity); and fourth, the signal level generated during the scan return time corresponds to "picture black" in the image.

These advantages became of much importance in the operation of signal generating tubes such as the CPS emitron, the image orthicon, the vidicon, the plumbicon, and every one of the Japanese photoconductive tubes up to the advent of the charge-coupled devices. The first public outside broadcasts with the then new CPS emitron cameras were from the Wembley Stadium and the Empire Pool during the fourteenth Olympiad held in London in 1948. The cameras were about 50 times more sensitive than the existing cameras and enabled what was proclaimed a wealth of detail and remarkable depth of field to be obtained.

See also **Television, Digital and High Definition Systems; Television, Electromechanical Systems; Television, Late Nineteenth and Early Twentieth Century Ideas**

RUSSELL W. BURNS

Further Reading

Abramson, A. *Zworykin, Pioneer of Television*. University of Illinois Press, Chicago, 1995.
Abramson, A. *The History of Television, 1880 to 1941*. McFarland, Jefferson, 1987.
Burns, R.W. *Television: An International History of the Formative Years*. Institution of Electrical Engineers, London, 1998.
Everson, G. *The Story of Television: The Life of Philo T. Farnsworth*. W.W. Norton, New York, 1949.
Fink, D.G. *Principles of Television Engineering*. McGraw-Hill, New York, 1940.
Maloff, I.G. and Epstein, D.W. *Electron Optics in Television*. McGraw-Hill, New York, 1938.
Zworykin, V.K., and Morton, G.A. *Television: The Electronics of Image Transmission*. Wiley, New York, 1940.

Immunological Technology

Immunization

At the start of the twentieth century, it was known that bacteria and some mysterious filterable agents called viruses could cause infectious diseases, and that people who recovered from a given infection often become specifically immune to that particular infection. There were only two vaccines known to be useful in preventing human disease, namely the widely used smallpox vaccine and the rarely used rabies vaccine of Louis Pasteur. It had been known for about 10 years that one molecular mechanism of specific immunity is the production of antibody, a protein produced usually in response to a foreign agent and able to bind specifically to that agent.

The smallpox vaccine evolved originally from the cowpox virus, which was used by Edward Jenner to immunize humans in 1796. During serial transmission from human to human, and later from rabbit to rabbit, it evolved into a distinctly different virus now called vaccinia. In the 1970s it played a central role in the first ever global eradication of a human disease: smallpox was officially declared eliminated from the human race in 1979.

Vaccinia was the prototype of many live viral vaccines, whose immunizing effects depend on their being able to cause an infection in the patient, albeit a very mild infection. Other live viral vaccines in common use today include those that immunize against measles, mumps, rubella, polio, varicella, and yellow fever.

These viral vaccines are all said to be attenuated, or of reduced virulence; that is, the infection that they cause is very mild compared with that caused by their wild viral ancestors. Attenuation of infective agents is usually achieved by very simple methods, such as maintaining them for a long time in artificial culture or in an unusual animal host. Most of the attenuated viral vaccines produce strong, long-lasting protective immunity in most people after a single dose.

The only important attenuated bacterial vaccine is the BCG vaccine (Bacille Calmette–Guérin), which is a live, attenuated form of *Mycobacterius bovis*. It provides about 80 percent protection against tuberculosis in the U.K. but does not provide any demonstrable benefits when used in India.

Some vaccines are created merely by killing the infective agent by exposure to heat, alcohol, formaldehyde, or thiomersal. The usual pertussis (whooping cough) vaccines are of this sort, as is the killed polio vaccine. Others consist of purified single components of the infective agent. Examples are the vaccines against pneumococci, *Haemophilus influenzae* B, and meningitis C. The important vaccines against tetanus and diphtheria are made by purifying the very potent protein toxins that the bacterial agents produce and then exposing them to formaldehyde, which slightly alters the chemical structure of the toxin, making it nontoxic.

Some pathogens; for example, hepatitis B virus, are difficult to culture in a laboratory, and so the usual methods of creating a vaccine cannot be used. However, if only one protein is important for induction of the immunity, it is usually possible to transfer the gene that codes for it into a yeast or a bacterium that can easily be cultured. The required protein can then be recovered from the cultures and used for immunization. The current hepatitis B vaccine is of this type.

The net effect of vaccines in the twentieth century has been one of the most celebrated success stories of medicine. Beginning with the diphtheria vaccine in the early 1940s, many new vaccines have come into widespread use and the incidence of most childhood infections has fallen about 1000-fold. Smallpox has disappeared and polio is also well on its way towards extinction. On the other hand, there are still no affordable and effective vaccines against some of the world's biggest killers, such as malaria, schistosomiasis, trypanosomiasis, amoebic dysentery, and HIV.

Allergy

The same mechanisms that protect against infection sometimes swing into action in response to relatively harmless substances such as pollen, house dust mites, penicillin, aspirin, latex, wheat proteins, peanuts, fruits, shellfish, dyes, and nickel salts. These are known as allergens. Upon a second exposure to the same allergen, the body's immune system can overreact, causing disease rather than preventing it. This phenomenon is called allergy, a word that is usually restricted to adverse reactions known to be mediated by the immune system. When these results occur naturally, they can cause conditions such as hay fever, rhinitis, asthma, eczema, urticaria, celiac disease, and the life-threatening anaphylactic shock, in which blood pressure can fall dramatically and the bronchi can become severely constricted.

One of the great mysteries of the twentieth century is that allergies and asthma in particular were rare at the start of the century and progressively increased in prevalence in developed countries. By the end of the century about 20 percent of the population suffered from an allergy and about 14 percent of children in some areas suffered from asthma.

By about 1961 it was widely accepted that the prick test was one of the best methods of identifying which allergens were responsible for the most acute allergic reactions. A drop of the allergen solution is placed on the skin and a trace is introduced into the epidermis by pricking it gently with a lancet. If the patient is allergic, a swollen, pale, edematous weal appears within 15 minutes at the site of pricking, surrounded by a reddened flare region that itches intensely. This test helps to decide which allergens the patient ought to avoid.

Management of allergic patients has been greatly improved by the use of drugs such as sodium cromoglycate that prevent histamine from being released from mast cells, and by antihistamines that prevent the histamine from exerting its inflammatory effects. For asthma, a slow-acting but long-lasting benefit is conferred by inhaled steroids such as Beclomethasone, also known as Vanceril or Beclovent, while an acute attack can be treated with β-adrenergic agonists such as Salbutamol, which relaxes the smooth muscle of the bronchi. The main therapy for anaphylactic shock is injected adrenalin, also known as epinephrine.

Desensitization to insect venoms and inhaled allergens such as pollen can often be achieved by giving the patient a long course of immunotherapy with very small but increasing doses of the allergen, but this is seldom successful with food allergens.

Antibodies

Injection of antitetanus antibodies into patients at risk of tetanus was introduced in 1914 and had an immediate, major impact on the number of deaths from tetanus among those wounded on the battlefield. Antibodies produced by a patient during an infection are sometimes used to identify the infective agent, especially in virus infections, and they have been used since 1900 to detect differences among different strains of bacteria and different blood cells. Many species of salmonella are distinguished largely by their reaction with antibodies, and blood typing for transfusion purposes depends largely on observing the reaction of the red cells with antibody.

Since antibodies can be made that react specifically with almost any soluble substance of our choice, they provide a convenient type of reagent that can be used to assay many different substances or to locate them in living tissues. Popular techniques based on this idea include radioimmunoassays and enzyme-linked immunosorbent assays (ELISA). The simplest, direct immunofluorescence techniques involve covalently coupling an antibody with a fluorochrome to produce a staining reagent that has the extraordinary specificity of antibodies as well as the extraordinary ease of detection of fluorochromes. It may be used to demonstrate the precise location of a particular substance within a microscopic section of a biological tissue.

All these techniques originally suffered from the fact that it was impossible to make exactly the same antibody twice, and so any assay had to be revalidated and recalibrated when a given batch of antibody was exhausted. But in 1975 the immunologists Georges Kohler and Cesar Milstein showed how to create a potentially immortal clone of cells that would continue to make the same antibody for as long as required, thereby allowing many different researchers to use an identical product. This monoclonal antibody also had some advantages arising from the fact that all of its molecules had very nearly the same structure.

Autoimmunity

In 1945 a case of autoimmune hemolytic anemia was described by Coombs, Mourant, and Race. They showed that the patient's anemia was caused by an antibody that reacted with his own red blood cells, triggering their destruction. Since then many human diseases have been found to be associated with immunity to one's own tissues. These diseases include rheumatoid arthritis, pernicious anemia, insulin-dependent diabetes, Hashimoto's thyroiditis, thyrotoxicosis, idiopathic Addison's disease, myasthenia gravis, psoriasis, vitiligo, systemic lupus erythematosus, and many others.

Steroids and immunosuppressive drugs have helped to control these conditions, and some therapeutic antibodies against inflammatory mediators have initially showed some promise, but a cure is not yet available.

Transplantation Immunology

In 1900 the ABO blood groups were discovered by Karl Landsteiner, and this made it possible to match blood donors with recipients and therefore transfuse blood safely. This may be regarded as the start of transplantation. Unfortunately, almost all human cells other than red blood cells bear on their surface not only the ABO substances but also a set of molecules known as histocompatibility molecules, which differ greatly between individuals and result in strong immune reactions against most transplanted tissues.

The problem can be partly alleviated by matching donor with recipient, first tried in 1966, but a complete match is almost impossible except between monozygotic twins.

Organ transplantation, however, became an important and useful treatment with the development of cytotoxic drugs that suppress the immune response. The main such drug is cyclosporin A, first used in 1978. The prevention of immune-mediated rejection, however, is not quite complete. A low-grade immune reaction still may limit the survival of the graft, and the drugs have several unfortunate side effects, including the suppression of desirable protective immune responses. Some see therapeutic cloning as a potential long-term solution to the problem of transplant rejection. Others hope to find an effective method of inducing specific tolerance to the donor tissue.

See also **Organ Transplantation**

IAN C. McKAY

Further Reading

Elliman, D.A.C. and Bedford, H.E. MMR vaccine—worries are not justified. *Arch. Dis. Childhood*, 85, 71–274, 2001. A concise review of the evidence concerning safety of the MMR vaccine.

Voltaire (*Lettres*. 1733). Describes the history of variolation and the incidence and effects of smallpox in the early eighteenth century.

Useful Websites

Centers for Disease Control and Prevention. General recommendations on immunization: recommendations of the Advisory Committee on Immunization Practices and the American Academy of Family Physicians.

Morbidity and Mortality Weekly Report (MMWR), 51, RR-2), 2002. All *MMWR* references are available on the Internet at: http://www.cdc.gov/mmwr

Information on general recommendations on immunization, and information about the National Immunization Program: http://www.cdc.gov/

World Health Organisation, Information on Vaccines and Immunizations: http://www.who.int/m/topics/vaccines _immunizations/en/index.html

Public Health Laboratory Service, U.K.: http://www.phls.co.uk/

Web pages on immunization maintained by the author of the above article: http://www.immunol.com

Implants, Joints, and Stents

Joint replacement, particularly for the hip and knee, has become a common surgical procedure for the treatment of joints affected by arthritis. Joint replacement replaces the surfaces of the natural joint and leads to restoration of joint function and pain relief.

The hip consists of the articulation between the spherical femoral head and the cup-shaped acetabulum. The first total hip replacement was performed in 1938 by Philip Wiles at the Middlesex Hospital in London. The design, made from stainless steel, consisted of a spherical femoral component attached to a bolt passing down the femur and a cup-shaped acetabular part that was secured by screws. However, it was not until the 1950s and 1960s that the idea of joint replacement became possible through the pioneering work of Kenneth McKee (Norwich, U.K.) and John Charnley (Wrightington, U.K.). McKee had designed a hip replacement, similar to the one used by Wiles, with a stainless steel ball and socket secured by screws. However, the results of replacements were poor because of inadequate fixation.

Charnley developed the idea of using dissimilar materials to create a low friction implant. He used a stainless steel femoral component and plastic socket made from polytetrafluorethylene (PTFE), more commonly know as Teflon. Charnley also introduced the idea of using large amounts of acrylic cement to fix the implant. The cement was used as a grout, relying on the mechanical fit rather than a glue. Over 300 hip replacements were undertaken by Charnley before it was realized in 1962 that PTFE was not a suitable material for hip replacement. Studies found that high wear rates resulted in a severe tissue reaction. Later in 1962, Harry Craven, the technician at Charnley's biomechanical laboratory, obtained a new plastic known as high-molecular weight polyethylene (HMWPE). At first Charnley was dismissive of the new material, but Craven tested it on a wear machine and found that it had much lower wear rates than PTFE and appeared to be a better material for hip replacement. Charnley began implanting the new metal-on-plastic hips in November 1962, and this is the basis of the metal-on-plastic hip implant that was most commonly used up to the twenty-first century. The metal component, however, was often a cobalt chrome molybdenum alloy (CoCrMo), and the plastic was ultra-high molecular weight polyethylene (UHMWPE) (see Figure 3).

Metal-on-polymer hip replacements can be expected to last for at least 15 years, but the wear particles developed from the metal articulating against the polymer leads to wear debris, which can cause osteolysis. Osteolysis is a tissue reaction that causes bone resorption and loosening of the implant, and the problem has led to growing interest in alternative biomaterials for hip replacements. Metal-on-metal (CoCrMo) and ceramic-on-ceramic (alumina) are considered to lead to less wear debris and to avoid the problem of osteolysis.

The knee consists of the articulation between the femur and the tibia; the two bones being separated by fibrous cartilage called the meniscus. In addition

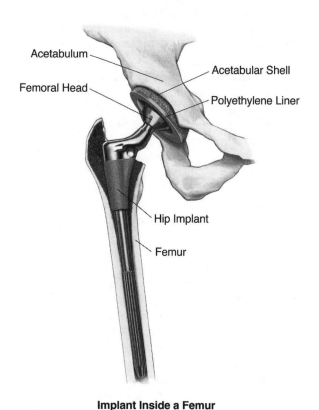

Implant Inside a Femur

Figure 3. A total hip replacement showing the femoral and acetabular components.
[*Reproduced with kind permission from Zimmer Inc., U.S.*]

the patella (knee cap) articulates with the femur. The development of knee replacements followed a similar trend to that of hip replacements in which early attempts failed, mainly because of loosening. In the early 1950s, hinged knee replacements were developed by Wallidus (1951) and Shiers (1953). These two implants consisted of a stainless steel hinge, but they required large bone resection, and there were high failure rates associated with them. Modern-day knee replacement took off after the success of the Charnley metal-on-polymer hip replacement. In 1969, Gunston, who worked with Charnley at Wrightington, developed a metal-on-polymer knee replacement that was fixed with acrylic cement. The implant consisted of a stainless steel femoral component and a HMWPE tibial component.

By the end of the twentieth century, there were various designs of knee replacement available that consisted of femoral, tibial, and/or patellar components. The implant can be: (1) unconstrained; (2) semi-constrained; or (3) fully constrained, depending on whether the knee ligaments are present. Implants are also divided according to the proportion of the knee replaced: (1) unicompartmental to replace either the medial or lateral compartment; (2) bicompartmental to replace both medial and lateral components; (3) tricompartmental to replace medial and lateral components and the patella. Present-day knee replacement (see Figure 4) consists of a metal femoral component (CoCrMo) that has a stem cemented into the femur, a tibial stem of titanium alloy cemented into the tibia, a tibial tray (UHMWPE) that fits to the tibial stem, and a patellar component (also of UHMWPE).

The fixation, or setting, of fractures has been aided by the use of screws and pins. The use of screws in bone began around the late 1840s, and today bone screws are commonly used either alone to treat fractures or in conjunction with plates or intramedullary nails. They are generally made from stainless steel or titanium alloy. The screws can either be self-tapping (the screw cuts a thread as it is inserted) or non-self-tapping (requires a tapped hole). Two main types of screws are used to get good purchase in the two types of bone: cortical (small threads) and cancellous (large threads). The holding power of a bone screw depends on a number of factors including diameter, length of thread engagement, and thread geometry. Hence, there are many different designs, lengths, and diameters of screw available for fracture fixation. Screws can also be cannulated (have a hole down the center) for use with wires and pins.

Pins or wires are also used in fracture fixation. The use of wires dates to the 1770s. Straight wires are known as Steinmann pins, while wires of diameter less than 2.4 millimeters are known as Kirschner wires. Wires and pins are primarily used to hold bone fragments together, either temporarily or permanently. They can also be used to guide large screws or intramedullary nails during insertion. The pins have a sharpened tip designed to penetrate easily through the bone.

A stent is a device used to maintain an orifice or cavity in the body. Stents are mainly used in vascular diseases to maintain blood flow, although they can be used to maintain a passage in other sites such as the urinary duct or for a bronchial obstruction. The most common type of stent used for coronary artery disease is made from fine stainless steel wire; the device has a lattice appearance resembling chicken wire. In angioplasty procedures, the stent is inserted into the

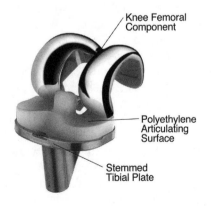
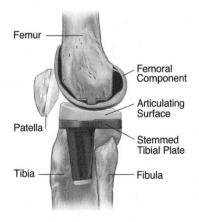

Figure 4. A total knee replacement showing the femoral, tibial, and patellar components. [*Reproduced with kind permission from Zimmer Inc., U.S.*]

artery on a balloon catheter. The balloon is inflated, causing the stent to expand and press against the inside of the artery. Once the balloon has been deflated and removed, the stent remains in place keeping the artery open. Other stents can be made from a variety of materials: nitinol (a nickel–titanium alloy), polymers, elastomers, and biodegradable materials.

See also **Cardiovascular Surgery and Implants**
DUNCAN E.T. SHEPHERD

Further Reading

Chandran, K.B., Burg K.J.L. and Shalaby S.W. Soft tissue replacements, in *The Biomedical Engineering Handbook*, Bronzino, J.D., Ed. CRC Press, Boca Raton, FL, 2000.

Dowson, D. New joints for the millennium: wear control in total replacement hip joints. *Proc. Inst. Mech. Eng: Part H, J. of Eng. Med.*, 215, 335–358, 2001.

Dowson, D. and Wright, V. *An Introduction to the Biomechanics of Joints and Joint Replacement*. Mechanical Engineering Publications, London, 1981.

Park S.-H., Llinás, G.V.K. and Keller J.C. Hard tissue replacements. in *The Biomedical Engineering Handbook*, Bronzino, J.D., Ed. CRC Press, Boca Raton, FL, 2000.

Waugh, W. *John Charnley: the Man and the Hip*. Springer Verlag, London, 1990.

Park, J. and Lakes, R. *Biomaterials: an Introduction*. Plenum, New York, 1992.

Industrial Gases

While the gases that are now commonly referred to as "industrial," namely oxygen, hydrogen, carbon dioxide, and nitrogen, were not fully understood until the nineteenth century, scientists in the twentieth century moved rapidly to utilize the knowledge. Driven largely by the demands of manufacturing industries in North America and Western Europe, rapid improvements in the technology of production and storage of industrial gases drove what has become a multibillion dollar business, valued at $34 billion in 2000. At the start of the twenty-first century, industrial gases underpin nearly every aspect of the global economy, from agriculture, welding, metal manufacturing and processing, refrigerants, enhanced oil recovery, food and beverage processing, electronic component manufacturing, to rocket propulsion. Oxygen for metal manufacturing is the largest volume market, with chemical processing and electronics using significant volumes of hydrogen and lower volumes of specialty gases such as argon.

Until at least the fifteenth century, gases were thought of inclusively as "air," part of the alchemical construction of the world. It was not until the seventeenth and eighteenth centuries that properties and characteristics unique to individual gases were recognized. Once one gas was finally isolated, a cavalcade of similar discoveries followed. In 1754, for example, English aristocrat and chemist Henry Cavendish "discovered" hydrogen, in 1756 English chemist Joseph Black discovered that carbon dioxide was a constituent of carbonate rocks, in 1772 Swede Carl Wilhelm Scheele discovered oxygen's properties, Englishman Joseph Priestley isolated oxygen (which he called dephlogisticated air) by 1774, and by 1784, Frenchman Antoine Laurent Lavoisier came up with a method of decomposing water into constituent elements, which he named hydrogen and oxygen, by passing water vapor over hot charcoal. Lavoisier was one of the first quantitative chemists, and showed that water was composed of two-thirds hydrogen and one-third oxygen.

These earliest discoveries highlighted the challenges ahead for harnessing the benefits of these gases: how to isolate and store them. Hydrogen, known for its "lighter than air" qualities as early as 1774, was first extracted from water using electrolysis by William Nicholson and Anthony Carlisle in 1800, but the high cost of electricity proved to make this an expensive method. As demand increased for hydrogen's use in airships, from zeppelins to military observation balloons in the 1930s, more economic means of extraction appeared. Since the 1920s, hydrogen has been produced by liquefaction of natural gas, partial oxidation of heavy oil, or gasification of coal.

Oxygen was first used in medicine (as an anesthetic or ventilator) and for "limelight," a theatrical lighting method from burning oxygen and hydrogen together. Early oxygen-using equipment could function with low-purity oxygen in small amounts, acquired by using several different chemical and heating processes that would break the oxygen into molecules, which was then compressed and sold in cylinders. By 1902, a system of rectifying "liquid air" pioneered by German Carl von Linde produced oxygen up to 99 percent pure.

In the early nineteenth century, nitrogen fertilizers were obtained from discoveries of immense bat guano deposits and caliche (nitrate-bearing rocks) in South America. As threats of famine loomed at the close of the nineteenth century, the need for agricultural fertilizers drove early production of nitrogen from the atmosphere. While "liquid air" contained nitrogen, it was difficult to separate from oxygen. Three solutions were developed in Europe: the cyanamide process (c.1900), which involved passing steam over certain carbides to form calcium cyanamide; the electric arc process (c.1903), which imitated lightning discharges to isolate nitrogen from air; and the Haber process, created by German Fritz Haber in 1904 and later developed into an industrial process by Carl Bosch, in which nitrogen is reacted with hydrogen to form ammonia.

By the twentieth century, carbon dioxide was extracted from many natural sources, especially cracks in the earth's crust due to volcanic activity, and as a byproduct of limekiln operations and synthetic ammonia production. Carbonic acid was used in bottled soft drinks, cooling, and in dry ice. Combined with sodium bicarbonate and ammonium sulfate it also created a foam, which deployed from a pressurized canister, became the fire extinguisher (carbon dioxide will not support combustion, and foam application ensures the gas does not quickly disperse).

Acetylene, discovered in 1836 and used in home and street lighting, was not produced industrially from calcium carbide until 1892. In 1901 Charles Picard and Edmond Fouché separately invented the oxyacetylene lamp, now widely used in arc welding, by combining acetylene with oxygen to produce an intense heat. Argon was used as an inert gas for electric lights, neon was used in lighted signs by 1913, and helium was used in balloons, dirigibles, welding, and medicine, and also mixed with oxygen for compression into cylinders for divers (the first practical compressed air diving apparatus was produced in 1925).

After World War II, applications for industrial gases expanded, using combinations of the basic four with lesser-known gases, as with the oxyacetylene lamp. Because acetylene was highly volatile in its compressed (and saleable) form, several innovations in transportation of the gas became necessary, with acetylene transported in pressurized steel cylinders.

However it was the liquefaction of gases—bringing a gas to a liquid state by intense pressure followed by cooling expansion—that was critical to the transport and application of these gases. Building on compression technology invented in the late nineteenth century, the science of cryogenics was the major contribution to liquefying gases. The method of cryogenic separation or distillation developed by German Carl von Linde in 1895 for liquefied air, involved dropping the temperature of the air to below $-140°C$ and increasing pressure to eight to ten times that of the atmosphere. As the liquid air is boiled, gases are boiled off at different boiling temperatures: vaporized nitrogen first, then argon, then oxygen. By 1908, all the gases had been separated using several different cryogenic machines, leading to a

boost in the availability of these liquid gases, and increasing demand as subsequent uses for the gases developed. In the post-World War II years, oxygen was produced in large quantities, thanks to the Linde–Frankl process for liquefying air (developed in 1928–1929), which ultimately gave way in the 1960s to the molecular sieve or pressure swing adsorption—a method that removes carbon dioxide from air. The abundance of oxygen also contributed to great advances in medicine—namely anesthesia and respiratory support, as well as facilitating high-altitude flying.

The growth of industrial use of oxygen and acetylene meant that pressure vessels were needed for transportation and storage. One of the earliest storage devices for liquefied gas was a double-walled vessel with a vacuum in the annular space, known as the "Dewar flask," invented by Sir James Dewar in 1892. Building on this early technology, ultrahigh vacuum pumps, as well as aluminum and foil matting used as insulation, contributed to advances in the storage and transport of highly pressured gases. In 1930, the first vacuum-insulated trains and trucks carried refrigerated, liquefied gases in the U.S., and by the 1950s, pipelines carried gas—with the largest in France—transporting oxygen, nitrogen and hydrogen from several different plants. The first generation of compressed gas cylinders (1902–1930) used carbon steel cylinders. Problems with rupturing led to the development of quenched and tempered alloy steel cylinders.

By the 1960s, oxygen was being used by steel manufacturers to enhance combustion in furnaces, and by the 1970s, nitrogen was widely employed as inert packaging for food preservation, as well as freeze-drying of food and heat treatment of metals. By the 1960s liquid hydrogen was used in rocket fuel and as a coolant for superconductors. By the 1980s, semiconductors were big customers for bulk industrial gases such as oxygen, argon, and hydrogen used as carrier gases for epitaxial growth, with specialist applications demanding higher quality gas (for example high-purity silane as a dopant, chlorine as an etchant). Cryosurgery using liquid nitrogen contributed to advances in medicine, including fertility treatments with frozen embryos, blood bank storage, and organ transplantation.

Industries from the automotive industry to steel making rely heavily on industrial gases. Hydrogen is used to cool alternators in power plants, and high-pressure cylinder gases are used in hospitals, small welding businesses, and in fire extinguishers. Liquefied carbon dioxide is used for dry cleaning textiles and as an industrial solvent, for example degreasing machine parts, coffee and tea decaffeination, and extracting essential oils and medicinal compounds from plants. While cryogenic distillation is still the favored form of production for industrial gases, some industries have shifted to less expensive noncryogenic methods. Pressure swing adsorption, for example, pumps pressurized air into either a molecular sieve, or an adsorptive agent, which removes the "unwanted" gas from the air stream, leaving the desired gas behind (for example, removing carbon dioxide from air). While the end product is not as pure as cryogenic products, engineers are perfecting the method. Noncryogenic separation techniques such as pressure swing adsorption have made on-site production more affordable. On-site production, primarily of oxygen, reduces distribution costs for remote locations, and guarantees essential supply, for example for hospitals.

One of the byproducts of the expansive use of industrial gas is an increase in undesirable environmental pollutants—contributing to the "greenhouse effect" and an overabundance of nitrates from agriculture application. Subsequently, government controls worldwide have led the gas industry to revamp some of its distribution and application. Hydrogen is likely to be the gas of the future, employed in "green" fuel cell technology, and glass and steel manufacturers are reducing nitrous dioxide emissions by mixing oxygen with coal.

See also **Cryogenics: Liquefaction of Gases; Nitrogen Fixation**

LOLLY MERRELL

Further Reading

Almquivst, E. *History of Industrial Gases.* Kluwer Academic/Plenum, New York, 2003.
Anonymous. Something in the air. *Process Engineering*, 32, Jan. 2000.
Downie, N. *Industrial Gases.* Kluwer Academic, Dordrecht, 1996.
Gardner, J. One hundred years of commercial oxygen production. *BOC Technology Magazine*, 5, 3–15, 1986.
Kroesche, A. Out of thin air: due to technology and environmental concerns, industrial gases are seeing good growth opportunities ahead. *Chem. Mark. Rep.*, 243, 17, 1993.
Scurlock, R.G., Ed. *History and Origins of Cryogenics.* Oxford University Press, New York, 1992.
Wett, Ted. Technology galore. *Chem. Mark. Rep.*, 240, 24, 13–14, 1991.

Information Theory

Information theory, also known originally as the mathematical theory of communication, was first explicitly formulated during the mid-twentieth century. Almost immediately it became a foundation; first, for the more systematic design and utilization of numerous telecommunication and information technologies; and second, for resolving a paradox in thermodynamics. Finally, information theory has contributed to new interpretations of a wide range of biological and cultural phenomena, from organic physiology and genetics to cognitive behavior, human language, economics, and political decision making.

Reflecting the symbiosis between theory and practice typical of twentieth century technology, technical issues in early telegraphy and telephony gave rise to a proto-information theory developed by Harry Nyquist at Bell Labs in 1924 and Ralph Hartley, also at Bell Labs, in 1928. This theory in turn contributed to advances in telecommunications, which stimulated the development of information theory *per se* by Claude Shannon and Warren Weaver, in their book *The Mathematical Theory of Communication* published in 1949. As articulated by Claude Shannon, a Bell Labs researcher, the technical concept of information is defined by the probability of a specific message or signal being picked out from a number of possibilities and transmitted from A to B. Information in this sense is mathematically quantifiable. The amount of information, I, conveyed by signal, S, is inversely related to its probability, P. That is, the more improbable a message, the more information it contains. To facilitate the mathematical analysis of messages, the measure is conveniently defined as $I = \log^2 1/P(S)$, and is named a binary digit or "bit" for short. Thus in the simplest case of a two-state signal (1 or 0, corresponding to on or off in electronic circuits), with equal probability for each state, the transmission of either state as the code for a message would convey one bit of information. The theory of information opened up by this conceptual analysis has become the basis for constructing and analyzing digital computational devices and a whole range of information technologies (i.e., technologies including telecommunications and data processing), from telephones to computer networks.

As was noticed early on by Leo Szilard in his classic 1929 paper on Maxwell's Demon and then later by Denis Gabor in ideas on light and information (1951), information in this technical sense is the opposite of entropy. According to the Second Law of Thermodynamics, the entropy or disorder of any closed system tends naturally to increase, so that the probability of energy differentials within it approaches 0. A heat engine, for instance, depends on the existence of an improbable or anti-entropic energy differential, created typically by energy input from a confined heating source, so that energy then flows in accord with the Second Law from one part of the system to another (as in the Carnot cycle of a steam engine). But in a late nineteenth century thought experiment, the physicist James Clerk Maxwell posed a paradox. Imagine, Maxwell suggested, a closed container with a gas at equilibrium temperature or high entropy; that is, without any detectable energy differential. In such a case, the standard distribution curve will nevertheless dictate that some gas molecules possess slightly higher energy states than others. Introduce into the container a partition with a small door operated by a "demon" that lets randomly higher energy molecules move one way and randomly lower energy molecules move the other. The demon will thus tend gradually to create an energy differential in the system, thus reducing entropy and creating a potential heat engine, without any energy input. Before the formulation of information theory, Maxwell's demon seemed to be getting something for nothing, whereas in fact it is introducing information (about the energy of molecules) into the system.

Information thus functions like negative entropy or "negentropy" (a term coined by Erwin Schrödinger). Expanding on the connections between information theory and thermodynamics, Norbert Wiener developed the closely related theory of cybernetics (1948) to analyze information as a means for control and communication in both animals and machines. In Wiener's theory, Maxwell's demon becomes a *kubernetes* (the Greek word for steersman) who utilizes some small portion of a system output as information and energy feedback to regulate itself in order to attain or maintain a predefined state.

To no machine has information theory been applied with more intensity than the digital computer. Indeed, as one computer scientist, Goldstine, noted in 1972, despite the implications of its name, a computer "does not just operate on numbers; rather, it transforms information and communicates it." With a computer the simple transmission of encoded information, as in telecommunications networks, is transformed into a complex storing and processing of the code so as to sort, compound, interrelate, or unpack messages in ways useful to machine or human operations. One may think of advanced computers as composed of millions of differentially programmed Maxwell demons embedded in the silicon of integrated circuits so as to yield structured outputs which, in appropriate contextual configurations, display what has been

called (after the lead of computer scientist Marvin Minsky) artificial intelligence. To deal with such complexities, information theory itself has been redefined as a mathematical analysis of both the transmission and manipulation of information.

According to Wiener, however, it is also the case that organisms, functioning as local anti-entropic systems, depend on Maxwell demon-like entities: "Indeed, it may well be that enzymes are metastable Maxwell demons, decreasing entropy, perhaps not by the separation between fast and slow particles but by some equivalent process." For Weiner the very essence of a living organism is that its input–output physiological functions (e.g., eating producing energy) are complemented by multiple feedback loops in which output becomes information regulating further output. In place of the static organization of nonliving matter, this creates a dynamic homeostasis through which an organism remains the same in form even while it undergoes continuous changes in the material out of which its form is constituted. This cybernetic view of biology is deepened with the discovery of DNA and the genetic code, to which information theory is again applied, and DNA is conceived as the means for transmitting an anatomy and physiology from one generation to another. The attractiveness of the information theory metaphor of a "book of life" in molecular biology has been extensively documented by Lily Kay (2000).

Cognitive psychologists were especially attracted by the idea that information theory and cybernetics could serve as the basis for a general interpretation of not only organic physiology but for much overt human behavior. In the last third of the twentieth century and with the help of computer scientists, psychologists developed increasingly sophisticated models of the mind—some attempting to replicate mental outputs by any technical means, others focused on replicating known or hypothesized brain structures and thinking processes. Independent of the extent to which the brain itself actually functions as a technical information processor, this dual research strategy proved remarkably fruitful for the investigation of human cognition and the design of advanced information processing machines. Examples in the latter case include theories of parallel information processing, neural networks, and emergent complexities.

In contrast to information theory and the technical concept of information is the concept of information in a semantic or natural linguistic sense. According to Shannon, although messages certainly have meaning, the "semantic aspects of communication are irrelevant to the engineering problem" of efficient message transmission. Information theory, in its concern to develop a mathematical theory of technical communication, takes the linguistic meaning of any message as unproblematic and seeks to determine the most efficient way to encode and then transmit that code, which may then once again be decoded and understood. Semantic information, however, is not a two-term relation—that is, a signal being transmitted between A and B—but a three- or four-term relation: a signal being transmitted between A and B and saying something about C (possibly to D). Nevertheless, despite the difficulty of establishing an unambiguous relation between the technical and semantic notions of information (no matter what its probability, some particular signal or message may possess any number of different semantic meanings) information theory sense has had a strong tendency to influence ideas about natural language. The best background to this research program is Cherry (1978), with one representative achievement being Fred I. Dretske (1981).

Finally information theory has influenced the rational reconstruction of economic and political decision making. In this field, information theory also tends to merge with game and decision theory. After first being rejected as "bourgeois ideology," information theory and cybernetics appealed to leaders in the Soviet Union during the 1960s and 1970s as a nonideological means to reform an increasingly inefficient state planning system. Ironically enough, the same appeal has also been operative among some business and government management theorists in Europe and North America.

See also **Error Correction**

CARL MITCHAM

Further Reading

Cherry, C. *On Human Communication: A Review, a Survey, and a Criticism*, 3rd edn. MIT Press, Cambridge, MA, 1978.

Dretske, F.I. *Knowledge and the Flow of Information*. MIT Press, Cambridge, MA, 1981.

Goldstine, H.H. *The Computer: From Pascal to Von Neumann*. Princeton University Press, Princeton, NJ, 1972.

Hartley, R.V.L Transmission of information. *Bell Syst. Techn. J.*, 7, 535–563, 1928.

Kay, L.E. *Who Wrote the Book of Life? A History of the Genetic Code*. Stanford University Press, Stanford, CA, 2000.

Nyquist, H. Certain factors affecting telegraph speed. *Bell Syst. Techn. J.*, 3, 324–346, 1924.

Shannon, CE. and Weaver, W. *The Mathematical Theory of Communication*. University of Illinois Press, Urbana, IL, 1949. Collates two articles: an expanded version of Weaver, W. Recent contributions to the mathematical theory of communication. *Scientific American*, July 1949; and Shannon, C.E. The mathematical theory of communication. *Bell Syst. Techn. J.* 1948.

Wiener, N. *Cybernetics: Or, Control and Communication in the Animal and the Machine*. MIT Press, Cambridge, MA, 1948.

Wiener, N. *The Human Use of Human Beings: Cybernetics and Society*. Houghton Mifflin, Boston, 1950.

Infrared Detectors

Infrared detectors rely on the change of a physical characteristic to sense illumination by infrared radiation (i.e., radiation having a wavelength longer than that of visible light). The origins of such detectors lie in the nineteenth century, although their development, variety and applications exploded during the twentieth century. William Herschel (c. 1800) employed a thermometer to detect this "radiant heat"; Macedonio Melloni, (c. 1850) invented the "thermochrose" to display spatial differences of irradiation as color patterns on a temperature-sensitive surface; and in 1882 William Abney found that photographic film could be sensitized to respond to wavelengths beyond the red end of the spectrum. Most infrared detectors, however, convert infrared radiation into an electrical signal via a variety of physical effects. Here, too, nineteenth century innovations continued in use well into the twentieth century.

Electrical photodetectors can be classed as either thermal detectors or quantum detectors. The first infrared detectors were thermal detectors: they responded to infrared radiation by the relatively indirect physical process of an increase in temperature. A thermal detector having a blackened surface is sensitive to radiation of any wavelength, a characteristic that was to become valuable to spectroscopists. The discovery of new physical principles facilitated the development of new thermal detectors. Thomas J. Seebeck reported a new "thermoelectric effect" in 1821 and then demonstrated the first "thermocouple," consisting of junctions of two metals that produced a small potential difference (voltage) when at different temperatures. In 1829 Leopoldo Nobili constructed the first "thermopile" by connecting thermocouples in series, and it was soon adapted by Melloni for radiant heat measurements—in modern parlance, for detecting infrared radiation—rather than for temperature changes produced by contact and conduction. In 1880, Samuel

P. Langley announced the "bolometer," a temperature-sensitive electrical resistance device to detect weak sources of radiant heat.

Such detectors were quickly adopted by physicists for studying optical radiation of increasingly long wavelength. Early twentieth century research in spectroscopy was largely detector-centered. This program sought to show the connections—indeed, to bridge the perceived gap—between infrared "optical" radiation and electrically generated "radio" waves. Infrared methods, for the most part, developed as an analog of visible methods while relying implicitly on electrical detectors.

During the twentieth century a variety of quantum detectors were developed and applied to a growing range of detection and measurement devices. Relying on a direct link between photons of infrared radiation and the electrical properties of the detecting material, they proved dramatically more sensitive than thermal detectors in restricted wavelength regions. As with thermal detectors, quantum detectors rely on a variety of principles. They may exhibit increased conductivity when illuminated with infrared radiation (examples of such "photoconductive" materials being pure crystals such as selenium, and compound semiconductors such as lead sulfide or lead selenide). Alternatively, quantum detectors may generate electrical current directly from infrared illumination. Examples of these "photovoltaic" detectors include semiconductor compounds such as indium antimonide or gallium arsenide.

Physical research on infrared detectors soon attracted military sponsors. Military interest centered initially on the generation and detection of invisible radiation for signaling. During the World War I, Theodore W. Case found that sulfide salts were photoconductive and developed thallous sulfide (Tl2S) cells. Supported by the U.S. Army, Case adapted these unreliable "thalofide" detectors for use as sensors in an infrared signaling device consisting of a searchlight as the source of radiation, which would be alternately blocked and uncovered to send messages (similar to smoke signals or early optical telegraphs) and a thalofide detector at the focus of a receiving mirror. With this system messages were successfully sent several miles.

During the 1930s, British infrared research focused on aircraft detection via infrared radiation as an alternative to radar; and, during World War II, relatively large-scale development programs in Germany and America generated a number of infrared-based prototypes and limited production devices.

Edgar W. Kutzscher developed the lead sulfide (PbS) photoconductive detector in Germany in 1932. This became the basis of a major wartime program during the following decade, studying the basic physics of detectors and materials, as well as production techniques and applications of infrared detection. Like the other combatants, the German military managed to deploy only limited production runs of infrared detectors and devices during World War II, for example using the radiation reflected from targets such as tanks to direct guns and developing the "lichtsprecher," or optical telephone.

In the U.S., successful developments during World War II included an infrared-guided bomb that used a bolometer as sensor, heat-sensitive phosphors for night vision "metascopes," and scanning systems used for the detection of infrared-radiating targets.

In the years following World War II, German detector technology was rapidly disseminated to British and American firms. Some of this information was recognized as having considerable military potential and was therefore classified. Infrared detectors were of great interest for locating the new jet aircraft and rockets; for their ability to be used "passively" (i.e., by measuring the radiation emitted by warm bodies rather than having to illuminate the targets with another source, as in radar); and for their increasing reliability. The potential military applications promoted intensive postwar research on the sensitivity of infrared detectors.

Whilst largely a product of military funding, these detectors gradually became available to academic spectroscopists. Improved sensitivity to infrared radiation was the postwar goal both of military designers and research scientists. The concurrent rise of infrared spectroscopy provided an impetus to improve laboratory-based detectors and led to developments such as the "Golay cell" by Marcel Golay in 1947. While this hybrid device, essentially an optically monitored pneumatic expansion cell, was a thermal detector such as the commonly used thermopile or bolometer, it was a reliable and sensitive alternative for use in spectrometers. Another new thermal detector was the thermistor bolometer, based on a blackened semiconductor (generally an oxide of a transition metal) having a narrow band-gap. Spectroscopists were also eager to discover the ultimate limitations of the newer quantum infrared detectors, and work by Peter B. Fellgett and by R. C. Jones in the early 1950s demonstrated the poor practical performance and theoretical potential of contemporary detectors.

During this period, further developments in Germany included the thallous sulfide and lead sulfide (PbS) detectors; Americans added the lead selenide (PbSe), lead telluride (PbTe), and indium antimonide (InSb) detectors; and British workers introduced mercury–cadmium–telluride (HgCdTe) infrared detectors. The military uses found rapid application. A guided aircraft rocket (the American GAR-2) was in production by 1956, and missile guidance systems, fire control systems, bomber-defense devices, and thermal reconnaissance equipment, all employing infrared measurement devices, were available to many countries by the mid-1960s.

By the late 1970s the military technology of infrared detection was increasingly available in the commercial sphere. Further military research and development during the 1980s extended capabilities dramatically, especially for detector arrays sensitive to the mid-infrared and high background temperatures. This technology was also adapted by civilian astronomers throughout the 1980s for high-sensitivity, far-infrared use. Modern infrared detection systems fall into three distinct classes:

1. Thermal imaging devices, operating analogously to visible-light cameras
2. Low-cost single-element thermal detectors
3. Radiometric or spectrometric devices, employed for precise quantification of energy.

Detectors adapted to new, low-cost markets included the pyroelectric sensor, reliant upon the change in electrical polarization produced by the thermal expansion of a ferroelectric crystal. This is considerably more sensitive than traditional bolometers and thermopiles.

While many infrared detector principles were patented, their commercial exploitation was seldom determined in this way. Nevertheless, individual firms were able to dominate market sectors partly because of in-house manufacturing expertise, particularly for the post-World War II semiconductor detectors, or the small markets for specific detector types.

Detectors, both as single element devices and in arrays, are now increasingly packaged as components. Hence thermopiles have benefited from silicon micromachining technology, and can be manufactured by photolithography. Radiometric instruments typically are designed around such single-detector systems, although infrared arrays have been adopted for special applications, parti-

cularly astronomical observation. Such arrays, capable of precise spectroscopic, radiometric, and spatial measurement, now match thermal imagers in spatial resolution. From the Infrared Astronomy Satellite (IRAS), launched in 1983 and having a mere 62 detector elements, arrays had grown in complexity to exceed one million detector elements by 1995.

See also **Spectroscopy, Infrared**

SEAN F. JOHNSTON

Further Reading

Hudson, R.D. Jr. *Infrared System Engineering*. Wiley, New York, 1969.
Hudson, R.D. Jr. and Hudson, J.W., *Infrared Detectors*. Dowden, Hutchison & Ross, Stroudsburg, 1975.
Jha, H.R. *Infrared Technology: Applications to Electro-optics, Photonic Devices, and Sensors*. Wiley, New York, 2000.
Jamieson, J.A., McFee, R.H., Plass, G.N., Grube, R.H., and Richards, R.G. *Infrared Physics and Engineering*. McGraw-Hill, New York, 1963.
Johnston, S.F. *Fourier Transform Infrared: A Constantly Evolving Technology*. Ellis Horwood, Chichester, 1991.
Johnston, S.F. *A History of Light and Colour Measurement: Science in the Shadows*. Institute of Physics Press, Bristol, 2001.
Smith, R.E., Jones, F.E., and Chasmar, R.P. *Detection and Measurement of Infrared Radiation*. Clarendon, Oxford, 1957.
Wolfe, W.L. *Handbook of Military Infrared Technology*. Office of Naval Research, Washington D.C., 1965.
Wolfe, W.L. and Zissis, G.J. *The Infrared Handbook*. Office of Naval Research, Washington D.C., 1978.

Integrated Circuits, Design and Use

Integrated circuits (ICs) are electronic devices designed to integrate a large number of microscopic electronic components, normally connected by wires in circuits, within the same substrate material. According to the American engineer Jack S. Kilby, they are the realization of the so-called "monolithic idea": building an entire circuit out of silicon or germanium. ICs are made out of these materials because of their properties as semiconductors—materials that have a degree of electrical conductivity between that of a conductor such as metal and that of an insulator (having almost no conductivity at low temperatures). A piece of silicon containing one circuit is called a die or chip. Thus, ICs are known also as microchips. Advances in semiconductor technology in the 1960s (the miniaturization revolution) meant that the number of transistors on a single chip doubled every two years, and led to lowered microprocessor costs and the introduction of consumer products such as handheld calculators.

In 1952, the British engineer G. W. A. Dummer of the Royal Radar Establishment in England, observed during a conference in Washington that the step following the development of transistors would be the manufacture of "electronic equipment in a solid block with no connecting wires." At the time, the American military establishment had already recognized the need to miniaturize the available electrical circuits for airplanes and missiles. Those circuits used several discrete modules such as transistors, capacitors, and resistors wired into masonite boards. Although the replacement of vacuum tubes (valves) with transistors had improved miniaturization, the circuit's manufacture was expensive and cumbersome. Thus, the military financed research programs to develop molecular electronics; that is, micromodules able to perform electronic functions on a miniaturized platform. In 1958, Kilby designed a micromodule for the firm Texas Instruments within a military-funded program. He reproduced the property of a circuit that converts direct current (DC) into alternating current (AC) into a thin wafer of germanium attached to four electrical contacts. Kilby's invention was the object of a legal controversy because in the same period, the American chemical engineers Robert Noyce and Gordon Moore, of the firm Fairchild Semiconductors, developed and patented a similar device using silicon rather than germanium. The legal controversy marked the birth of integrated circuits, and silicon eventually became the material used in the IC industry.

In the 1960s, the introduction of new types of transistors—metal-oxide semiconductors (MOS) and metal-oxide semiconductors field-effect transistors (MOSFET)—greatly improved the performance of ICs and ensured their commercial viability. In 1963, Fairchild's engineers designed ICs capable of performing resistor–transistor logic (RTL). This implied that ICs were now able to contain logic gates and perform Boolean function of the type NOT/OR/AND. The logic properties embodied in ICs made computer memory storage their main field of application. When ICs began storing computational information, their capacity was calculated in bits (binary digits). During the 1960s, ICs passed from a memory storage space capacity of 8 bits to 1024 bits. The need for greater memory space made IC design more complicated, and new techniques for computer-aided design (CAD) were thus introduced. Fairchild produced the first CAD microchip (called Micromosaic) in

1967. In the 1960s, ICs also constituted the main component in the guidance systems for spacecraft (e.g., *U.S. Freedom*, *U.S. Apollo*) and ballistic missiles' navigation systems (e.g., *U.S. Minuteman*).

In 1968, Noyce and Moore left Fairchild to establish a new company: the Integrated Electronics Corporation, or Intel. From the 1970s, their company became a leader in IC technology. In 1970, the engineers Joel Karp and Bill Regitz designed the model 1103 for Intel, the first 1024-bit dynamic random access memory (RAM) for computers. The great innovation developed by Intel, however, was the development of the first microprocessor. In 1971, Intel engineers Marcian E. Hoff, Stan Mazor, and Federico Faggin developed an IC that could be programmed with proper software and thus perform virtually any electronic function. Before 1971, each logic or arithmetic function of a computer was assigned to a specific IC, while microprocessors would work eventually as miniaturized general-purpose computers. The microprocessor was commercialized by Intel in the models 4004 (4 bits) and 8008 (8 bits). Model 4004 executed 60,000 operations in a second and deployed 2300 transistors. The introduction of microprocessors ensured an extension of IC capabilities and the development of large-scale integration (LSI), the integration of ever-increasing logic operations within the same IC.

During the 1970s, the microchip industry grew enormously, providing integrated circuits for a number of electronic devices. These commercial applications helped the IC manufacturers gain independence from military funding and to open new markets. Research and development was driven by a few companies such as Intel, mainly located in the Silicon Valley area near San Francisco. Although old electronics companies such as Texas Instruments or IBM proved strong competitors, the introduction of the model 8086 in 1978 ensured Intel's long-term success as the provider of microprocessors for computers. More generally, the development of ICs during the 1970s came to be seen as the technological basis for the so-called microelectronics revolution. According to Robert Noyce, the technological change brought by ICs would bring about a qualitative change in human capabilities. He observed that the increased number of IC components (10 in the 1960s; 1,000 in the 1970s; 100,000 and more in the 1980s) would bring about a rapid decline in the cost of given electronic functions. Furthermore, the increasing miniaturization would ease IC application in several electronic devices. Thus, ICs would become a cheap, very reliable, and widespread technological tool in use in almost all electronic equipment.

At the end of the 1970s the American IC industry found a strong competitor in Japan, which could produce more efficient microscopic devices at a cheaper price and with more convincing design. Companies such as Nippon Electronics Corporation (NEC), Fujitsu, and Hitachi soon gained success through programs of very large scale of integration (VLSI), with ICs containing more than 1000 logic gates. Between 1977 and 1984, the American companies lost nearly 20 percent of their market to Japanese competitors. Chip development implied the definition of new designs and architectures capable of compressing functions into simpler circuits and reduced dimensions. New microprocessors also embodied the idea of reduced instruction-set computing (RISC), simplified architectures capable of operating subsets of common instructions with standard parts of microprocessors. RISC improved the speed of operation and miniaturization.

The crisis of American IC companies was overcome in the late 1980s through a new government policy that restored the traditional link between the military and the IC industry. The Pentagon funded new research for the very high-speed integrated circuit (VHSIC or vee-sick) to develop its program called the Strategic Defense Initiative (SDI) or Star Wars. Meanwhile, new materials such as gallium arsenide were introduced to replace silicon and provide faster and more reliable ICs. In the 1990s, Intel strengthened its monopoly in microprocessors for computers with the development of Pentium, the most complex processor ever built by the company after the 8086, with 3 million transistors and a speed of 100 million instructions per second.

See also **Computer Memory, Early; Integrated Circuits, Fabrication; Transistors**

SIMONE TURCHETTI

Further Reading

Augarten S. *State of the Art. A Photographic History of the Integrated Circuit.* Ticknor & Fields, New York, 1983.

Braun E. and Macdonald S. *Revolution in Miniature: The History and Impact of Semiconductor Electronics.* Cambridge University Press, Cambridge, 1982.

Jackson, T. *Inside Intel: How Andy Grove Built the World's Most Successful Chip Company.* Harper Collins, London: 1998.

Noyce, R.N. Microelectronics. *Scientific American*, 237, 3, 1977. Reprinted in *The Microelectronics Revolution*, Forester, T. Ed. Blackwell, Oxford, 1980.

Queisser H. *The Conquest of the Microchip.* Harvard University Press, Cambridge, MA, 1988.

Integrated Circuits, Fabrication

The fabrication of integrated circuits (ICs) is a complicated process that consists primarily of the transfer of a circuit design onto a piece of silicon (the silicon wafer). Using a photolithographic technique, the areas of the silicon wafer to be imprinted with electric circuitry are covered with glass plates (photomasks), irradiated with ultraviolet light, and treated with chemicals in order to shape a circuit's pattern. On the whole, IC manufacture consists of four main stages:

1. Preparation of a design
2. Preparation of photomasks and silicon wafers
3. Production
4. Testing and packaging

Preparing an IC design consists of drafting the circuit's electronic functions within the silicon board. This process has radically changed over the years due to the increasing complexity of design and the number of electronic components contained within the same IC. For example, in 1971, the Intel 4004 microprocessor was designed by just three engineers, while in the 1990s the Intel Pentium was designed by a team of 100 engineers. Moreover, the early designs were produced with traditional drafting techniques, while from the late 1970s onward the introduction of computer-aided design (CAD) techniques completely changed the design stage. Computers are used to check the design and simulate the operations of perspective ICs in order to optimize their performance. Thus, the IC drafted design can be modified up to 400 times before going into production.

Once the IC design is thoroughly checked, computers prepare the drawings of the circuit's layers. Each circuit may contain up to 15 layers and each layer defines an essential part of the circuit (gates, connections, and contacts). The drawings are then reproduced by a pattern generator in the form of sets of optical reticles that are used to create the photomasks. The reticles are up to ten times bigger than the actual chip size; hence the photomasks are produced through a photoreductive process known as step-and-repeat. The photomasks are finally cut with lithographic techniques that exploit laser beams, light, or x-rays. From the 1980s, electron-beam lithography has allowed the production of photomasks directly from computer memory, eliminating the intermediate stages.

While photomasks are manufactured, other factories are involved in the production of silicon;

the substrate material on which circuits are installed. Cylinders of raw crystalline silicon are obtained from vats of molten silicon at the temperature of 1400°C in a process that resembles the making of candles. Silicon crystals are then cut in ultra-thin wafers with a diamond saw, polished, and inspected.

At this point the manufacture of ICs begins. The method of IC's manufacture derives largely from the planar process used in the production of transistors and conceived in the late 1950s by the Swiss physicist Jean Hoerni and the American chemical engineer Robert Noyce of the American firm Fairchild Semiconductors. The planar process was a method of batch production that allowed metal connections to be evaporated onto the oxidized surface of a semiconductor. Thus the circuit areas were engraved into the substrate material. Hoerni and Noyce observed that the planar process could be easily applied only on silicon batches (rather than germanium) because in its oxidized form (silicon oxide) it allowed the introduction of insulating parts between different areas of the circuit. Thus, the isolating and conducting parts of a single component could be easily deployed and divided on the substrate material. The introduction of the planar process was one of the keys to the commercial success of ICs.

Building on the principles of the planar process, silicon wafers are covered with a layer of insulating silicon dioxide that forestalls the circuits. At this stage the photomasks come into use. The silicon wafer is exposed to ultraviolet light through the photomask. This exposure hardens some selected areas of the silicon wafer and leaves the others soft. The soft areas are then etched away with an acid bath. Recent research has allowed the definition of new dry-etching techniques without the use of corrosive acids.

The process of irradiation, treatment with chemicals, and etching is repeated at least five times. A first mask isolates the individual chips in the wafer. A second provides some areas with photoresistant material (photoresist) that defines the gates into individual chips. A third ensures the definition of the circuit's contacts by the removal of oxide parts. A fourth defines the interconnections through the introduction of metal (this is either aluminum or an alloy of aluminum and copper). Finally, a fifth mask strengthens the bonding pads through a protective glass-like substance called vapox. During some of these operations the wafers are heated to temperatures of 1000°C. The process also requires great accuracy as

each photomask has to be aligned to the wafer consistently with the previous ones.

The final stage consists of testing and packaging the ICs. A first check occurs after the final overcoat on the silicon wafer passing electric current into the several ICs. Then, the wafer is cut again with the diamond saw into individual chips. At this point each chip is packaged and the IC looks like a caterpillar. The IC is tested again and the defective ones are identified and reworked.

The early stages of the IC manufacturing process are heavily automated (design, production of photomasks, silicon wafers, and ICs), while the later ones are labor-intensive (packaging and testing) and usually carried out by affiliated factories. An IC plant looks like something between a factory and a hospital. Although workers and machines fabricate chips in a highly automated context, the rooms must be clean and the workers must wear white gloves, caps, and shoe covers because even the smallest dust particle can make the circuits defective.

See also **Integrated Circuits, Design and Use; Semiconductors: Crystal Growing, Purification; Transistors**

SIMONE TURCHETTI

Further Reading

Augarten S. *State of the Art. A Photographic History of the Integrated Circuit.* Ticknor & Fields, New York, 1983.
Braun E. and Macdonald S. *Revolution in Miniature. The History and Impact of Semiconductor Electronics.* Cambridge University Press, Cambridge, 1982.
Oldham, W.G. The fabrication of microelectronic circuits. *Scientific American,* 237, 3, 1977. Reprinted in *The Microelectronics Revolution,* Forester, T., Ed. Blackwell, Oxford, 1980.
Queisser H. *The Conquest of the Microchip.* Harvard University Press, Cambridge, MA, 1988.

Intensive Care and Life Support

Intensive care is defined as the medical treatment of patients with serious reversible conditions, generally with a significant element of respiratory failure, who are judged to be capable of recovery.

The Crimean War (1853–1856) saw the start of modern nursing, pioneered by Florence Nightingale. The traditional hospital ward, with beds in rows alongside the walls was (and still is) known as a Nightingale ward. A system of progressive patient care was established in which the sickest patients were segregated and placed in the area immediately adjacent to the nursing station. In the 1940s and 1950s special areas were developed for the care of, for example, post-operative thoracic surgical or neurosurgical patients. By the 1960s, the concept of general intensive care, or therapy, units evolved, often from anesthetist-led postoperative recovery units. The first general intensive therapy units (ITUs) were opened in the U.K. in 1966. At about the same time, coronary care units (CCUs), generally for postoperative heart surgery patients, were opened in the U.S. and the U.K. Since hospitals had been traditionally organized under a somewhat rigid hierarchical structure, the concept of multidisciplinary care with physicians, anesthetists, nurses, physiotherapists and biochemists working together was a novel one.

The impetus for these units included:

1. The increasing ability to keep patients alive using ventilators and other mechanical devices
2. Increasing understanding of the metabolic changes occurring in critical illness and the ability to monitor these changes
3. The means to treat them
4. The improvement in anesthesiology techniques that allowed complex surgery to be performed on very ill patients

Management of the systemic inflammatory response syndrome, previously referred to under the generic term "shock," gradually became better understood. Treatment of acidosis, fluid and blood replacement, and cardiovascular support with medication became the basis of intensive care. Some of the advances made possible by the isolation and treatment of patients in these specialized high technology environments define intensive care and life support.

Ventilators

The poliomyelitis epidemics of the 1930s had seen wards of patients "breathing" with the help of "iron lungs." The polio virus affected the neurological system and caused paralysis of the muscles (thus the early name of the disease, infantile paralysis), including those that make it possible to breathe. The iron lung was a bulky and cumbersome device: the patient's upper body was inside an airtight chamber in which alternate negative and positive pressures were created by a system of bellows, enabling the influx and expulsion of air from the lungs. The machine was developed by Harvard University engineer Philip Drinker and Louis Shaw in 1928. Because of iron lungs, innumerable lives were saved during the

polio epidemics of the 1930s and 1940s. The supervision of these patients was almost entirely in the hands of medical students who worked a shift system. Although outdated by the development of positive pressure ventilators after World War II, 300 of these devices were amazingly still in use in the U.S. in 1985. The success of this rather primitive equipment was all the more remarkable when it is remembered that the modern, noninvasive techniques of oximetry (the measurement of blood oxygen levels) had not been invented. In 1946 a "pneumatic balance" breathing machine was described, and the Bird and Cape respirators of 1955 onward were based on this concept. Inspiratory gas flow is terminated when a preset pressure or flow is reached. Flow is generated by a mechanically driven piston, which delivers a volume of gas to the patient. Positive pressure respirators of varying degrees of complexity became the norm in hospitals although ventilation modes became much more sophisticated. By the last decade of the century it was possible to vary the pressure throughout the respiratory cycle using different waveforms and frequencies to enable patients to interact much more with the ventilator depending on their individual needs.

Cardiopulmonary Bypass

As long ago as 1812, French physician Julien Jean-Cesar LeGallois suggested that it would be theoretically possible to replace the heart with a pump. This technique required the blood to be oxygenated (receive oxygen) while it was being pumped outside the body. By the end of the nineteenth century, primitive film and bubble oxygenators had been developed. In 1916 Jay McLean, a medical student at Johns Hopkins University in Baltimore, discovered heparin, a blood component that prevents the clotting. This discovery solved the problem of blood clotting in extracorporeal circulation. In 1934, heart surgeon Michael DeBakey described the roller pump as a perfusion device. In 1944, Willem J. Kolff developed hemodialysis using a semipermeable membrane and extracorporeal blood circulation to remove impurities from the blood when the kidneys were unable to do so. John H. Gibbon Jr., performed the first "open heart" surgery using a heart–lung machine in 1953. By 1994 it was estimated that over 650,000 open-heart operations were performed worldwide annually using cardiopulmonary bypass. These machines are not without problems, which include the amount of blood required during procedures and possible damage to the patient's blood cells with subsequent inflammatory response syndrome. To avoid these complications, there has been a trend to perform "off-pump" surgery.

Aortic Balloon Counterpulsation

During severe cardiac ischemia (lack of oxygenated blood in the tissues) or after cardiac surgery, the heart muscle can fail and lack sufficient pumping action to maintain the circulation. While awaiting definitive treatment such as a transplant or the reversal of complications as in severe disorders of cardiac rhythm, aortic balloon counterpulsation can help stabilize the patient. A balloon is positioned in the arch of the aorta, which, by precise timing of the cycle, can deflate while the left ventricle contracts (systole) forcing blood out through the arterial system, and inflate while the ventricle relaxes (diastole). This allows the maximum blood volume into circulation with each contraction and augments the coronary blood during relaxation. The device was introduced into clinical practice in 1976.

Acute Renal Failure

Modern understanding of acute renal failure (ARF) developed during the World War II years when large numbers of civilians developed renal failure following crush injuries due to collapsing buildings. As renal function failed and urinary output fell, it became apparent that too much fluid in the system was lethal. By restricting fluids to the "volume obligatoire" (i.e., just the replacement of fluid lost in perspiration and breathing), water intoxication could be avoided and renal function could recover. Following the development of dialysis techniques, ARF became somewhat more manageable.

Monitoring

Maintenance of the normal pH, or acid–base balance, of the blood is essential to health. As early as 1928, Laurence J. Henderson wrote "The condition known as acidosis is today hardly less familiar than anemia." The recognition of the importance of metabolic balance and the steady progress in resuscitation from this time culminated in the automated intensive care and critical care units of the 1990s. The patients admitted to an ICU frequently have metabolic reasons for acidosis, such as low blood pressure and lack of blood in the tissues secondary to shock, systemic infection, diabetic ketosis, or renal failure. In addition, these conditions are frequently complicated by respiratory problems, which exert a profound influence on maintenance of a normal pH because of the amount of carbon dioxide (CO_2) normally expired. The movements in pH initiated by these factors can often be in different directions. The biochemical basis for the understanding of acid–base balance was laid down between 1908 and 1920 when Henderson and K. A. Hasselbalch reported that measurement of the three variables, pH, pCO_2, and base deficit, was necessary to define acid–base status. Portable, easy-to-use blood gas machines that used blood from an arterial puncture gave rapid answers. In-dwelling arterial or venous catheters enabled frequent monitoring of other electrolytes, notably sodium, potassium, and creatinine. Continuous oximetry and simultaneous pulmonary artery pressure recording by means of disposable fiber-optic catheters had become routine by 1970. Continuous monitoring of cardiac output could be achieved by placing a Doppler probe in the esophagus. All these devices provide valuable information relevant to fluid balance and cardiovascular drug support. In addition, the status of the central nervous system can be obtained by monitoring intracranial pressure and cerebral artery blood flow.

Antibiotics

Infection, and particularly septicemia in which infection has spread through the bloodstream, either as a cause of the presenting illness or as a secondary infection, has always been a problem in ICUs. It is also among the most serious problems. Since the introduction of penicillin in 1941, innumerable increasingly potent antibiotics have been developed, only to be overused and therefore made less and less effective because of acquired resistance by microorganisms.

Conclusion

Intensive care units are not cost-effective. The technology is only manageable by dedicated one-to-one nursing by highly trained staff and the availability of an equally well-trained team of medical and other specialists. By the 1980s it had become apparent that the unrelenting stress of providing intensive care could cause psychological problems in the care givers. Research also looked at the effects on patients of sensory deprivation amid the array of technological equipment. The ability to sustain life in such an environment seemingly indefinitely had also raised questions about end-of-life care and decisions about resuscitation and life support that continued beyond the twentieth century.

See also **Dialysis**

JOHN J. HAMBLIN AND CATHERINE BROSNAN

Further Reading

Atkinson, R., Hamblin, J., and Wright, J. *Handbook of Intensive Care*. Chapman & Hall, 1981. Of historic interest only.

Fairman, J. and Lynaugh, J. *Critical Care Nursing: A History*. University of Pennsylvania Press, Philadelphia, 2000.

Galley, E. *Renal Failure: Critical Care Focus*. BMJ Books, 1999.

Hillman, K. and Bishop, G. *Clinical Intensive Care*. Cambridge University Press, Cambridge, 1996. Comprehensive up-to-date textbook.

Sasada, M and Smith, S. *Drugs in Anaesthesia and Intensive Care*. Oxford University Press, Oxford, 1997.

Sykes, K. and Young, D., Eds. *Respiratory Support in Intensive Care*. BMJ Books, 1999.

Internal Combustion Piston Engine

Born in the nineteenth century, the internal combustion piston engine did not come of age until the twentieth century. Through a long stream of refinements and expansion into a variety of sizes, shapes, and types, this device must be considered among the most influential technological developments in human history. Internal combustion piston engines, such as those used to power model aircraft, can be small enough to fit in the palm of the hand. At the other end of the spectrum, engines used to operate power generators, ships, or rail locomotives can weigh several tons and take up the space of an entire room.

As its name implies, the internal combustion engine burns fuel inside the engine itself. The most common type of internal combustion piston engine uses the four-stroke cycle (see Figure 5), also known as the "Otto" cycle after Nikolaus Otto, a German who patented a four-stroke engine design in 1876.

Most four-stroke engines employ poppet valves, which open and close by a camshaft to allow air and fuel to enter and exhaust to exit. Engine operation is through a sequence of four piston strokes, identified as intake, compression, power,

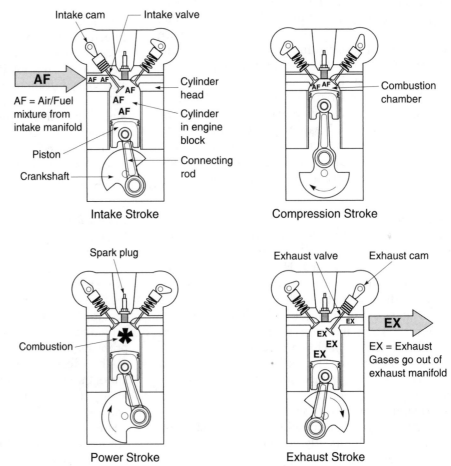

Figure 5. 4-stroke cycle engine.
[*Reproduced with permission of 2 Stroke International, c 2002.*]

and exhaust. The power stroke is the point at which thermal energy is converted to kinetic energy as a result of the air/fuel mixture burning rapidly, creating high pressure in the cylinder, and pushing with great force on the head of the piston.

A two-stroke internal combustion piston engine (see Figure 6) must accomplish all of the same tasks as the four-stroke, but in half the number of strokes, identified as intake/compression stroke and power/exhaust stroke. In place of poppet valves, two-stroke engines control the flow of gases by the piston covering and uncovering ports in the cylinder, assisted by reed valves in some engines.

This arrangement gives the two-stroke engine the advantage of mechanical simplicity, and allows for smaller size and lighter weight. Early development work on modern two-stroke engine design took place in the Britain during the 1880s, prominent contributors to that work being Sir Dugald Clerk, a Scot, and Englishman James Robson.

Internal combustion piston engines are also classified based on the method of igniting the fuel. The most widely used variety is the four-stroke petrol, or gasoline, engine, in which electric spark plugs ignite the fuel, thus the designation "spark ignition." Cars and trucks are the most prolific user of this type of engine and were a primary source of motivation for engine development and improvement throughout the twentieth century. Other common applications include agricultural machinery, aircraft, and military vehicles, which are discussed in detail in separate entries.

Many of the machines powered by spark-ignition gasoline engines can also operate using compression-ignition engines. This other major classification of engines is also known as the diesel engine after Rudolph Diesel, a German engineer who patented a design for a compression-ignition engine in 1893. In lieu of an electric spark, a diesel engine applies greater compression to air in the cylinder, generating enough heat to ignite the fuel as it is injected. Some diesel engines operate on the

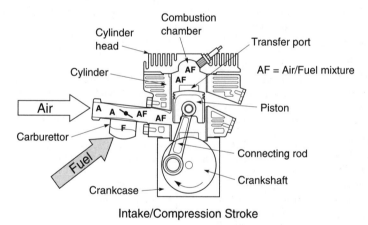

Intake/Compression Stroke

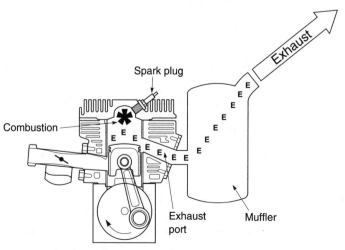

Power/Exhaust Stroke

Figure 6. 2-stroke cycle engine.
[*Reproduced with permission of 2 Stroke International, c 2002.*]

443

two-stroke cycle; however, the four-stroke cycle is more common. Diesel fuels are composed of heavier distillates of petroleum and have a greater energy density than gasoline.

Mobile uses of diesel engines include large over-the-road trucks, earthmoving and construction equipment, watercraft such as ferries and tugboats, and rail locomotives. The infusion of the internal combustion piston engine into rail transport clearly revolutionized the industry and has provided a good example of the influence of the engine as emerging technology. As the twentieth century opened, steam locomotives powered all rail transport. Their operation required that a large boiler be stoked with wood or coal, an example of an external combustion engine. A considerable time delay was involved in building up steam pressure before the locomotive could be set into motion. Frequent stops along the way were required to refill the boiler with water and the hopper car with fuel. By contrast, a diesel locomotive starts almost instantaneously, achieves greater speeds, and can travel many hundreds of miles before refueling. By the middle of the twentieth century, the steam locomotive had all but disappeared, and the diesel and diesel/electric locomotive ruled the rails.

Diesel engines are used in many stationary applications, such as water pumping stations and electrical power generators. Diesel generators are frequently used as a supplementary power source, in conjunction with electrical generating stations operating on wind or ocean tides. Backup systems for hydroelectric and even nuclear power plants often utilize diesel generators, which compared with turbine and nuclear generators are more easily transportable and operate on commonly available fuels. For these reasons, diesel generators have brought electrical power to remote regions of the world that otherwise would have none.

Diesel engines and four-stroke gasoline engines are generally larger and bulkier than two-stroke engines. Most four-strokes require an oil sump for lubrication, which means that the engine cannot be inverted or operated at differing angles of movement. Lubrication of the two-stroke engine is accomplished by mixing oil into the fuel, therefore lightweight two-stroke engines are used to power the majority of small hand-held equipment, such as chain saws and weed trimmers. Slightly larger machines such as lawnmowers, garden tillers, portable pumps and generators, jet ski-type watercraft, and outboard boat motors may be powered by either a two-stroke engine or a small four-stroke. Light motorcycles known as "dirt bikes" run on two-stroke engines, as do millions of small

motor scooters common in Europe and Asia. Some compact automobiles, snowmobiles, and even small aircraft known as "ultralights" use two-stroke engines.

Many varieties of internal combustion piston engines can be fitted with minor modifications to run on natural gas, liquid petroleum gas, or hydrogen gas. As alternatives to petroleum-based fuels, renewable organic engine fuels such as alcohols (methanol and ethanol) and "biodiesel" fuels are increasing in availability and usage.

See also **Automobiles, Internal Combustion; Civil Aircraft, Propeller Driven; Farming, Mechanization; Helicopters; Motorcycles; Rail, Diesel and Diesel Electric Locomotives; Wright Flyers**

REESE E. GRIFFIN, JR.

Further Reading

Cummins, L. *Diesel's Engine*. Society of Automotive Engineers, Carnot Press, Warrendale, PA, 1993.
Cummins, L. *Internal Fire*, 3rd edn. Society of Automotive Engineers, Carnot Press, Warrendale, PA, 2000.
Griffin, Reese, Ed. *Theory Manuals on CD*. 2 Stroke International, Beaufort, SC, 2001.
Hardenberg, H.O. *The Antiquity of the Internal Combustion Engine: 1509–1688*. Society of Automotive Engineers, Warrendale, PA, 1993.
Hardenberg, H.O. *The Middle Ages of the Internal Combustion Engine: 1794–1886*. Society of Automotive Engineers, Warrendale, PA, 1999.
Heywood, JB. *Internal Combustion Engine Fundamentals*. McGraw Hill, Columbus, OH, 1988.
Heywood, J.B. *The Two Stroke Cycle Engine*. Society of Automotive Engineers, Warrendale, PA, 1999.
Obert, E.F. *Internal Combustion Engines and Air Pollution*. Addison-Wesley, Boston, MA, 1973.
Stone, R. *Introduction to Internal Combustion Engines*. Society of Automotive Engineers and MacMillan, Warrendale, PA, 1999.
Suzuki, T. *The Romance of Engines*. Society of Automotive Engineers, Warrendale, PA, 1997.

Useful Websites

2 Stroke International: http://www.2si.com
Animated Engines: http://www.keveney.comJEngines.html
DieselNet: http://www.dieselnet.com
How Stuff Works: www.howstuffworks.com
Science Museum: www.sciencemuseum.org.uk
Society of Automotive Engineers, International: www.sae.org/servlets/index

Internet

The Internet is a global computer network of networks whose origins are found in U.S. military efforts. In response to Sputnik and the emerging space race, the Advanced Research Projects Agency (ARPA) was formed in 1958 as an agency of the Pentagon. The researchers at ARPA were given a generous mandate to develop innovative technologies such as communications.

In 1962, psychologist J.C.R. Licklider from the Massachusetts Institute of Technology's Lincoln Laboratory joined ARPA to take charge of the Information Processing Techniques Office (IPTO). In 1963 Licklider wrote a memo proposing an interactive network allowing people to communicate via computer. This project did not materialize. In 1966, Bob Taylor, then head of the IPTO, noted that he needed three different computer terminals to connect to three different machines in different locations around the nation. Taylor also recognized that universities working with IPTO needed more computing resources. Instead of the government buying machines for each university, why not share machines? Taylor revitalized Licklider's idea, securing $1 million in funding, and hired 29-year-old Larry Roberts to direct the creation of ARPAnet.

Universities were reluctant to share their precious computing resources and concerned about the processing load of a network. Wes Clark of Washington University in St. Louis, Missouri, proposed an interface message processor (IMP), a separate smaller computer for each main computer on the network that would handle the network communication. Another important idea came from Paul Baran of the RAND Corporation, who had been concerned about the vulnerability of the U.S. telephone communication system since 1960 but had yet to convince the telephone monopoly AT&T of the virtues of his ideas on distributed communications. He devised a scheme of breaking signals into blocks of information to be reassembled after reaching their destination. These blocks of information traveled through a "distributed network" where each "node," or communication point, could independently decide which path the block of information took to the next node. This allowed data to automatically flow around blockages in the network. Donald Davies at the British National Physical Laboratory (NPL) independently developed a similar concept in 1965 that he termed "packet switching," each packet being a block of data. While Baran was interested in a communications system that could continue to function during a nuclear war, ARPAnet was purely a research tool, not a command and control system.

Bolt, Beranek and Newman (BBN), a small consulting firm in Cambridge, Massachusetts, got the contract to construct ARPA's IMP in December 1968. They decided that the IMP would only handle the routing, not the transmitted data content. As an analogy, the IMP looked only at the addresses on the envelope, not at the letter inside. Faculty and graduate students at the host universities created host-to-host protocols and software to enable the computers to understand each other. They called themselves the Network Working Group (NWG) and began to share information via request for comment (RFC) papers. Steve Crocker at the University of California at Los Angeles wrote RFC Number 1, entitled "Host Software," on April 7, 1969. Because the machines did not know how to talk to each other as peers, the researchers wrote programs that fooled the computers into thinking they were talking to preexisting dumb terminals. By the summer of 1969 the NWG agreed on an overall protocol called network control protocol (NCP). RFCs are still used today to communicate about issues of Internet protocol, a versatile system for creating technical standards that allows technical excellence to dominate.

ARPAnet began with the installation of the first IMP in the fall of 1969 at UCLA, a Honeywell minicomputer built by BBN with 12 kilobyte of memory and weighing 400 kilograms, followed by three more nodes at the Stanford Research Institute (SRI), University of California at Santa Barbara, and the University of Utah. The first message transmitted between UCLA and SRI was "L," "O," "G," the first three letters of "LOGIN," then the system crashed. Initial bugs were overcome, and ARPAnet grew a node every month in 1970. In 1973 the first international nodes were added in the U.K. and Norway. More improvements and new protocols quickly followed. Electronic mail, file transfer (FTP), and remote login (Telnet) became the dominant applications. BBN invented remote diagnostics in 1970.

In 1974, Robert Kahn and Vincent Cerf proposed the first internetworking protocol, a way for datagrams (packets) to be communicated between disparate networks, and they called it an "internet." Their efforts created transmission control protocol/internet protocol (TCP/IP). In 1982, TCP/IP replaced NCP on ARPAnet. Other networks adopted TCP/IP and it became the dominant standard for all networking by the late 1990s.

In 1981 the U.S. National Science Foundation (NSF) created Computer Science Network (CSNET) to provide universities that did not have access to ARPAnet with their own network. In 1986, the NSF sponsored the NSFNET "backbone" to connect five supercomputing centers. The backbone also connected ARPAnet and CSNET together, and the idea of a network of networks became firmly entrenched. The open technical architecture of the Internet allowed numerous innovations to be grafted easily onto the whole. When ARPAnet was dismantled in 1990, the Internet was thriving at universities and technology-oriented companies. The NSF backbone was dismantled in 1995 when the NSF realized that commercial entities could keep the Internet running and growing on their own, without government subsidy. Commercial network providers worked through the Commercial Internet Exchange to manage network traffic.

Other western industrialized nations also built computer networks in the mid-1970s, and gradually these also joined the Internet. The introduction in 1991 of the hypertext transport protocol (HTTP), the basis of the World Wide Web, took advantage of the modular architecture of the Internet and became widely successful. What began with four nodes as a creation of the Cold War in 1969 became a worldwide network of networks, forming a single whole. By early 2001, an estimated 120 million computers were connected to the Internet in every country of the world.

See also **Computer Networks; Electronic Communications; Packet Switching; World Wide Web**

DAVID L. FERRO AND ERIC G. SWEDIN

Further Reading

Abbate, J. *Inventing the Internet*. MIT Press, Cambridge, MA, 2000.
Hafner, K. and Lyon, M. *Where Wizards Stay Up Late: The Origins of the Internet*. Simon & Schuster, New York, 1996.
Moschovitis, C.J.P Poole, H., Schuyler, T., Senft, T.M. *History of the Internet: A Chronology, 1943 to the Present*. ABC-CLIO, Santa Barbara, CA, 1999.
Segaller, S. *Nerds 2.0.1: A Brief History of the Internet*. TV Books, New York, 1988.

Useful Websites

RFC 2555, 30 Years of RFCs: ftp://ftp.rfc-editor.org/in-notes/rfc2555.txt.
Internet Society (ISOC), histories of the Internet: http://www.isoc.org/internet/history/February 2002.

Iron and Steel Manufacture

Iron and steel manufacture is possibly the most basic industry of the twentieth century industrial economy. For much of the century, iron and steel tonnage was the most commonly used comparative measurement of a nation's economic strength.

The first step in steel manufacture is to extract iron from iron ore, then process it into steel by adding carbon and alloying elements. The predominant method of iron extraction uses a blast furnace, simply a huge shell of over 25 meters high lined with a ceramic refractory material and with holes, or taps, through which molten iron and waste products can be drawn off. The name comes from the blast of air blown into the furnace that provides oxygen for combustion. Three materials are required: iron ore, coke, and limestone. Iron ore is any of several minerals such as hematite (Fe_2O_3), limonite ($Fe_2O_3 \cdot 3H_2O$), or magnetite (Fe_3O_4), while coke, made from refining coal by heating, is the fuel for the furnace. When these materials are combined inside the furnace, the burning of coke gives off carbon monoxide, which reacts with the iron oxide as:

$$Fe_2O_3 + 3CO = 3CO_2 + 2Fe$$

The limestone serves as a scavenger to remove silica and other impurities present in the iron ore—for example, the limestone's calcium carbonate combines with silicon dioxide to form calcium silicate. The impurities float on top of the furnace melt as "slag" which is drawn off as waste through the "cinder notch." The limestone reaction also provides some amount of extra carbon monoxide that further assists in the primary reaction. Molten iron is "tapped" from the furnace and cast in molds in the form of pig iron, named for the appearance of the molds as piglets gathered around a sow.

Blast furnace technology improved slowly and steadily through the twentieth century. One significant development was the pressurizing of furnaces, which began around the middle of the century and allowed more efficient combustion and increased yield. Although wrought iron, created through a process called puddling, is useful for some limited applications, pig iron is usually further strengthened by further processing into steel by the removal of impurities such as unwanted silicon, phosphorus, sulfur, and manganese and the homogenization of carbon through the solution.

The classic process for making steel is the Bessemer process developed in the nineteenth century. Air or oxygen is blown through molten pig iron in a pear-shaped, tiltable steel container called a Bessemer converter to oxidize impurities into slag. After removing the slag, the desired alloying materials are added. The Bessemer process was highly efficient and capable of massive economic production of steel, providing steel as the basic material for the rapid expansion of the world's economy. Its disadvantage was that it required high-quality iron ore.

The basic oxygen process, or Linz–Donawitz process named for the Austrian mills that first developed it, gradually replaced the Bessemer process from the 1950s. The principle is the same; the primary difference is simply the use of pure oxygen for combustion instead of air. The basic oxygen furnace has the ability to function with lower quality ore and with up too 35 percent scrap.

Other methods of steel production were also used, including the open-hearth process. This system, which used a flat, rectangular brick hearth for oxidation of impurities, could achieve a purer metal than other processes. The use of the open-hearth process peaked in the 1950s. Because of the serious economic consequences to the industry and the often unionized laborers, controversy attended the changes in steel manufacturing at the time with intense debates over the best processes and furnaces.

The 1960s saw the introduction of the "mini-mill," in which an electric arc furnace was used to produce steel, usually of a specialized variety. The advantages of the electric furnace are tighter temperature control, no contamination from fuel, and desulfurization, while using even higher levels of scrap iron.

The production of raw steel in ingot form is only an intermediate step toward a finished product. After casting, steel is typically hot-rolled or cold-rolled into coiled sheet or strip, or into various other forms suitable for even further process: billets, skelp, slag, or bloom. There are various types of steel rolling mills, the most basic being the two-stand mill in which an ingot passes back and forth between too reversing rollers, becoming thinner and longer with each pass. A later development was the continuous rolling mill, in which stands of rollers operated in series, each at a faster rotation and smaller gap, allowing an ingot to pass through once in a single direction, eliminating the reversing and reducing cycle time. An important development in the finishing of steel was the move in the last decade of the twentieth century toward continuous casting of steel into thin coils, which reduced the number of subsequent rolling passes required.

See also **Alloys, Light and Ferrous; Nuclear Reactor Materials; Prospecting, Minerals**

FRANK WATSON

Further Reading

Chandler, H. *Metallurgy for the Non-Metallurgist*. ASM International, Materials Park, OH, 1998.
Hall, C.G.L. *Steel Phoenix: The Rise and Fall of the U.S. Steel Industry*. St. Martin's Press, New York, 1997.
Hempstead, C.A., Ed. *Cleveland Iron and Steel*. The British Steel Corporation, 1979.
McGannon, HE., Ed. *The Making, Shaping and Treating of Steel*. United States Steel Corporation, Pittsburgh, PA, 1971.
Vukmir, R.B. *The Mill*. University Press of America, Lanham, MD, 1999.

Irrigation Systems

Since the onset of human civilization, the manipulation of water through irrigation systems has allowed for the creation of agricultural bounty and the presence of ornamental landscaping, often in the most arid regions of the planet. These systems have undergone a widespread transformation during the twentieth century with the introduction of massive dams, canals, aqueducts, and new water delivery technology. In 1900 there were approximately 480,000 square kilometers of land under irrigation; by 2000 that total had surged to 2,710,000 square kilometers, with India and China as the world leaders in irrigated acreage. Globally, the agriculture industry uses about 69 percent of the available fresh water supplies, producing 40 percent of the world's food on just about 18 percent of the world's cropland. (It takes 1000 tons of water to produce 1 ton of grain.)

Traditionally flood irrigation, practiced as far back as Ancient Egypt and Sumeria, has been the standard irrigation delivery system. Flood irrigation requires relatively flat land, the construction of retaining dikes and canals, and the laborious task of breaking and filling gaps in the dikes to control the flood of water onto the field, or manually lifting the water from plot to plot.

Furrow irrigation also requires a relatively flat surface. The water from central ditches is used to fill furrows or gated pipes between rows of crops. Gated pipes are made of aluminum and have sliding, or screw-open, gates to release the water into the irrigation furrow. Siphon hoses are generally employed to deliver the water from the main ditches to the furrows. Flood and furrow irrigation are still widely employed throughout the world, but the systems require significant amounts of labor, capital, and readily available water.

With the construction of large dams, waterworks, and wells equipped with centrifugal pumps powered by electricity, diesel, gas, or steam around the globe in the twentieth century, there were also

Figure 7. Agricultural Research Service scientists download data about the movement of a center-pivot irrigation system to reconstruct the amount of water and time it took to irrigate an area.
[*Photo by Scott Bauer. ARS/USDA.*]

new delivery systems for moving irrigation water to the fields. These included gated portable piping, sprinkler, and center pivot sprinkler systems, and drip irrigation systems. Massive government water projects, such as those built by the U.S. Bureau of Reclamation in the American west, led to a dramatic increase in irrigated agriculture, which quadrupled in scale from 1900 to 1990. With gas turbine pumps and deep wells, it became possible to recover, or "mine," water from underground aquifers as deep as 300 meters. Older windmills and water wheels that could provide a limited volume of water were replaced. By the year 2000, the U.S. alone had more than 500,000 such irrigation wells.

In the world's most productive farm region, the Central Valley of California where it generally rains less than 250 millimeters per year, irrigation is used on 97 percent of the acreage to produce

high value crops. Irrigated by massive governmental projects that date back to the 1930s, the Central Valley is one of the most "transformed" places in the world due to irrigation. Depletion of groundwater supplies has actually led to the "sinking" of the Central Valley by several feet in some areas, although in recent years recharge basins have been refilling underground aquifers during times of water surplus. Pesticide concentration, soil salinization, and competition from urban and environmental sources have contributed to a reduction of irrigated acreage in the Central Valley since 1980.

The introduction of gated aluminum piping allowed for more efficient delivery of water to the fields where it could then be used for flood or furrow irrigation. Less water evaporates or "percolates" into the soil when aluminum irrigation pipes are used; however, the pipes are expensive and require upkeep or constant transportation from field to field. When connected to turbine pumps and improved sprinkler heads, a gated pipe system may also be used to spray a crop with water from gun-type sprinklers. Sprinkler irrigation expanded dramatically in the twentieth century in spite of problems with evaporation, salinization of fields, and high energy and equipment and maintenance costs. "Gun" sprinklers evolved from nineteenth century British firefighting technology, and by the 1870s mechanized portable, linear lateral and boom sprinkler systems had emerged, which saved labor costs. Sprinkler irrigation made it easier to control the amount of water on a crop, and it allowed irrigation on irregular surfaces.

After the 1940s in some parts of the world, such as the Great Plains of the U.S., center pivot sprinkler systems helped bring new areas under cultivation. These systems are driven by electric motors and manipulated by hydraulic systems. A giant boom with nozzles attached, the center pivot system rotates on wheels in a circle that can irrigate an entire 640-acre (2.6-square kilometer) section, thus eliminating the costly job of moving pipe. In general, sprinkler systems are expensive and result in a large amount of evaporation of water into the air, essentially wasting a resource in limited supply.

Since the 1970s drip irrigation systems have become increasingly popular throughout the world. With drip systems, underground polyvinyl chloride pipes are connected to polyethylene hoses placed above the crops. Applicators in the hoses slowly drip filtered water onto individual plants, allowing precision watering, simultaneous chemical applications, and lower water usage than other

Figure 8. Because the water level of the river varies, farmers along the Missouri River use floating pumps like this one to collect irrigation water without the pump clogging. Another concern is streambank stability, which is now more manageable thanks to an online guide developed by scientists.
[*Photo by USDA-NRCS.*]

irrigation systems. Drip irrigation systems are relatively inexpensive and easy to maintain, and they allow less evaporation than other irrigation systems.

The expansion of irrigation in the twentieth century created more productive farmland for a burgeoning world population and more forage and grain for livestock production. Water analysts predict that by 2025, 3 billion people in 48 countries will be living in conditions of water scarcity. While dam projects continue to be funded in China, Turkey, and India, the economic and ecological costs of the projects has led to a decline in development of large-scale water storage facilities.

New technologies to monitor evaporation, plant transpiration, and soil moisture levels have helped increase the efficiency of irrigation systems. The US is the world leader in irrigation technology, exporting upward of $800 million of irrigation equipment to the rest of the world each year, with the sales of drip irrigation equipment increasing 15 to 20 percent per annum in the 1990s. Golf course and landscape irrigation are also an increasing part of the irrigation technology market. Intense competition for water from cities and for environmental restoration projects might mean a reduction in irrigated agriculture in future years. At the same time, salinization of fields, infiltration of aquifers by sea water, and depleted water availability could lead to a reduction in land under irrigation worldwide.

See also **Agriculture and Food; Farming, Agricultural Methods.**

RANDAL BEEMAN

Further Reading

Fiege, M. and Cronon, W. *Irrigated Eden: The Making of an Agricultural Landscape in the American West*. Seattle, 2000.

Finkel, H.J. *Handbook of Irrigation Technology*. Boca Raton, FL, 1983.

Hurt, R.D. *Agricultural Technology in the Twentieth Century*. Manhattan, 1991.

Morgan, R.M. *Water and the Land: A History of American Irrigation*. Fairfax, 1993.

Postel, S. *Last Oasis: Facing Water Scarcity*. New York, 1992.

Reisner, M. *Cadillac Desert: The American West and Its Disappearing Water*. New York, 1986.

Useful Websites

U.S. Bureau of Reclamation Official Site: www.usbr.gov/historyborhist.htm 11.01.2003

Working Group on the History of Irrigation, Drainage and Flood Control: www.wg-hist.icidonline.org 11.01.93

(Includes references to international irrigation projects in the twentieth century.)
Irrigation Association/World Water Council: www.world watercouncil.org

Isotopic Analysis

Beyond the analysis of the chemical elements in a sample of matter, it is possible to determine the isotopic content of the individual chemical elements. The chemical analysis of a substance generally takes its isotopic composition to be a "standard" that represents terrestrial composition, because for most purposes the isotopic ratios are more or less fixed, allowing chemical weight to be a useful laboratory parameter for most elements. Deviations from the standard composition occur because of differences in:

1. Nuclear synthesis
2. Radioactive decay
3. Geological, biological, and artificial fractionation
4. Exposure to various sources of radiation

Applications of isotopic analysis make use of these sources of variation. Our knowledge results from analytical techniques developed during the twentieth century that allow precisions of 10^{-5} and that have led to significant improvements in the understanding of the Earth and solar system and even of archaeology.

Synthesis

Current cosmological theory holds that the chemical elements were formed in the original creation of the universe (the Big Bang), in synthesis by stars, and by the interaction of cosmic rays with interstellar matter. Within minutes after the Big Bang, all matter was composed of isotopes of hydrogen, helium, and trace amounts of lithium. Stars formed from this gas began burning hydrogen by stages to elements as heavy as iron. At this point further burning is not possible because energy is no longer released in the fusion of lighter to heavier elements, and the star undergoes a transformation dependent on its mass, initial composition, and relationship to possible companions. For stars of mass much greater than the sun, this can lead to the collapse of the iron core with the release of an enormous flux of neutrons, together with the explosion of the outer layers from nuclear reactions. The neutrons allow the synthesis of elements to uranium and beyond. Knowledge of nuclear reactions combined with theoretical models of stellar evolution allows prediction of the composition of the matter that is ejected into space. This isotopic signature varies according to the nature of the explosion, thus the solar system carries the signature of a supernova thought to have created its elements. The light elements beryllium, boron, and some additional lithium are produced by cosmic rays interacting with dust in interstellar space. The assumption that the isotopes forming the Earth were well mixed in the initial melted state cannot be relied on in detail, as special "inhomogeneities" are found within the accuracy of modern mass spectrometry.

Radioactive Decay

Most elements had radioactive isotopes in their original nucleosynthesis, but in the few million years during which the solar system formed, all but a few decayed with the daughter products joining in the composition. Some have half-lives comparable to the existence time of the solar system—most importantly uranium and thorium—which decay through a series of intermediate products ending in lead. There are about a dozen other radioactive elements scattered through the periodic table. Measuring the ratio of the parent to its daughter is the basis of determining the age when a rock solidified, but complications entailed in the history of the rock require a high level of experimental and geological skill for these determinations. These studies have attached definite ages to the epochs and periods that geologists had identified from the comparison of fossils in sedimentary rock strata. The history of the solar system during the first few million years of its formation has been established through dating based on radioactive elements, later extinct, whose daughter isotopes have been identified.

Fractionation

The isotopic composition of a chemical element can be altered through diffusion and transmutation by irradiation. Heat causes the evaporation of atoms and molecules from a sample, and the process proceeds preferentially with the lighter isotopes leaving at a greater rate than the heavy, in proportion to the ratio of the square roots of the masses. This is an important phenomenon in depleting ocean water of the isotope ^{16}O, making the ocean isotopically heavier than atmospheric oxygen. This fact is used in the principal method of determining temperatures in past ages by measuring the $^{18}O/^{16}O$ ratios at various depths in ice cores and in certain kinds of sedimentary deposits. Biochemical reactions also cause isotopic fractionation, which is strongly expressed in the two stable isotopes of carbon and nitrogen and the four of sulfur. Differences as large as a few percent exist among the reservoirs that range from marine carbonates through a variety of plants and bacteria. Ratios of $^{34}S/^{32}S$ vary by 5 percent in coal and petroleum and 9 percent in sedimentary rocks. Diffusion is used on an industrial scale to separate isotopes with application to uranium being the best known.

Exposure to Radiation

Cosmic rays—protons with extremely high energies—can induce nuclear reactions, forming both stable and radioactive isotopes. These cosmogenic isotopes are produced on the Earth's surface and in its atmosphere and on meteorites. They figure importantly in investigations as tracers of atmospheric and geological surface processes and are important in determining time sequences in the formation of the solar system.

Star Dust

To this point, this entry has been concerned with matter in the solar system whose isotopic composition has been altered in some way. In 1987 isotopic ratios found in mineral grains of meteorites were so unusual that they forced the conclusion that the grains, which are very rare but recognizable optically, were from interstellar dust derived from the nucleosynthesis of a different star or stars than the one that formed our solar system. Measurements of the isotopic composition of these grains of "star dust" allowed the prediction of models of stellar nucleosynthesis to be compared with observation.

Archaeology

The isotopic composition of lead varies significantly according to the source from which the metal was extracted. Early metal workers introduced lead into their products either as a principal ingredient or an impurity, and measurements of their compositions offer the archaeologist clues to the origin and history of metal objects. It is a difficult subject because what is observed is the result not only of the initial metals but also of subsequent reworking by generations of craftsmen. To this end the isotopes of tin, whose primordial composition is fixed, are proving valuable. These isotopes undergo a significant fractionation with each new melting, owing to tin's volatility. Archaeologists can trace food patterns of peoples and animals with measurements of $^{13}C/^{12}C$.

See also **Mass Spectrometry**

LOUIS BROWN

Further Reading

Basu, A. and Hart, S., Eds. *Earth Processes: Reading the Isotopic Code.* American Geophysical Union, Washington D.C., 1996.

Clayton, R.N. Isotopic anomalies in the early solar system/ *Ann. Rev. of Nucl. Part. Sci.*, 28 501–522, 1981.

Cowan, J.J., Thielemann, F.-K. and Truran, J.W. Radioactive dating of the elements. *Ann. Rev. Astron. Astrophys.*, 29 447–497, 1991. Discusses the various processes of nuclear synthesis in stars.

Heaman, L. and Ludden, J.N. Applications of radiogenic isotope systems to problems in geology, in *Short Course Handbook*, 19, 1991.

Valley, J.W., and Cole, D.R., Eds. Stable isotope geochemistry. *Rev. Mineral. Geochem.*, 43, 2001.

Wetherill, G.W. Formation of the Earth. *Ann. Rev. Earth Planet. Sci.*, 18, 205–256, 1990.

Zindler, A., and Hart, S. Chemical geodynamics, in *Ann. Rev. Earth Planet. Sci.*, 14, 493–571, 1986.

J

Josephson Junction Devices

One of the most important implications of quantum physics is the existence of so-called tunneling phenomena in which elementary particles are able to cross an energy barrier on subatomic scales that it would not be possible for them to traverse were they subject to the laws of classical mechanics. In 1973 the Nobel Prize in Physics was awarded to Brian Josephson, Ivan Giaever and Leo Esaki for their work in this field. Josephson's contribution consisted of a number of important theoretical predictions made while a doctoral student at Cambridge University. His work was confirmed experimentally within a year of its publication in 1961, and practical applications were commercialized within ten years.

The device that has commonly become known as a Josephson junction consists of a thin piece of insulating material between two superconductors. The thickness of the insulator is generally of the order of 25 Å (1 Å = 10^{-10} meters) or less. The electrons from the superconducting material tunnel through the nonsuperconducting insulator enabling a current to flow across the junction.

Josephson predicted two different theoretical effects for this sandwich: the first, commonly known as the DC Josephson effect, shows that a current can flow through the insulating region even when no voltage is applied across the junction; the second, the so-called AC Josephson effect, occurs when a constant voltage is applied across the junction and results in an alternating current flowing across the insulator with a high frequency in the microwave range. In the latter effect the junction acts much like an atom in which quantum mechanical transitions are taking place because electrons in superconducting solids are able to drop into a lower energy state as a result of their pairing.

When Josephson junctions are subjected to magnetic fields the current that flows across the insulator is dependent on the intensity of the field on the junction. This has important implications because when two Josephson junctions are combined in parallel as part of a superconducting loop it is possible for interference effects to occur between the junctions in an analog of the famous Young's slit experiment with light waves. In another variant only one Josephson junction is used in the superconducting loop where it acts as a parametric amplifier. Both arrangements are commonly known as superconducting quantum interference devices or SQUIDs. Modern commercially available SQUIDs have reached a high level of reliability in the manufacture of high-quality thin layers, and there has been a huge amount of research into SQUIDs. Aside from the Josephson effect they rely on another property of quantum systems, namely the quantization of magnetic flux flowing through a closed loop.

One of the most important uses of SQUIDs is as highly sensitive magnetometers, and the first SQUID magnetometer was built in 1972. It is possible to measure the miniscule magnetic fields caused by the human brain, and even the effects of thinking. As virtually all biological systems involve the movement of charged ions around the body, a corresponding magnetic field is caused. These fields are commonly of the magnitude of 10^{-13} Tesla (far smaller than the Earth's magnetic field which is of the order of 10^{-4} Tesla). A modern SQUID system for medical use may consist of up to 200 SQUIDs arranged in different axes and connected to an advanced computer control and display system, which is capable of automatically compensating for background magnetic effects. The imaging, in real time, is interpreted as a functional map rather than a structural map. By 2002 approximately 120 hospitals in the world used low-temperature superconductor SQUID systems for cardiac (magnetocardiography) or neural imaging (magnetoencephalography), and as the technology continues to develop, this number is certain to increase.

Another interesting use of SQUIDs is in nondestructive testing of materials through measurement of magnetic fields caused by eddy currents in conducting samples. It is possible using SQUIDs to measure defects in materials nondestructively to a far greater depth than is possible using conventional techniques. As with medical imaging uses, it is common for the SQUID to be used in conjunction with a pick-up coil, which is placed adjacent to the sample. Although this technology is of great use in a laboratory setting, there are practical difficulties with using it to test materials *in situ*, such as in aerospace and civil engineering structures. These generally arise because of the requirement that the Josephson junctions must operate below a superconducting transition temperature (T_c) in order for the magnetic susceptibility to be established. For low temperature superconductors, this is of the order of $9°$Kelvin ($-264°$C). Commercialization of portable nondestructive evaluation (NDE) systems could be as early as 2005.

Despite the use of Josephson junctions in SQUIDs, possibly the most important use of the technology will lie in future developments of the digital computer. Because Josephson junctions can operate in two different voltage states, it is possible for each junction to act as a digital switch. The switches rely on quantum transitions, and therefore it is possible for them to operate at phenomenally high speeds of a few picoseconds. When combined in a large array, this may lead to exciting developments in ultrafast digital switching and ultimately quantum computing. There are, however, a number of unfortunate practical complications in the technology as it currently stands. The most important of these is what has commonly been called the "reset" problem. Once the Josephson junction switch has been set, it is necessary to reset it by removing the background bias current, and this operation is, relatively speaking, very slow.

Another issue concerns the cost of refrigeration to superconducting temperatures, which is commonly achieved using very costly liquid helium apparatus (4 to $5°$Kelvin), though high-T_c materials discovered in the 1980s require cooling only to $77°$Kelvin (liquid nitrogen).

Another interesting use of Josephson junctions arises when a junction is subjected to a microwave field. It is then possible for it to work in reverse so that a precise voltage is generated across the junction. When many junctions are combined it is possible to produce extremely accurate standard voltage sources, as long as the frequency of the microwaves driving the junctions is precisely known. Currently sources that generate up to 12 volts are commercially available, and it appears likely that this technology will be of increasing importance for calibration in future years. Many national metrological laboratories have adopted the Josephson effect voltage standard.

Josephson's theoretical predictions have spawned a huge field of technology since 1961 and one that seems set to become increasingly important in biomedical applications, in the testing of materials, and in the development of a quantum computer. However, perhaps more significantly, the existence and reliability of Josephson junctions provide direct confirmation of some of the most philosophically paradoxical postulates of quantum mechanics.

See also **Quantum Electronic Devices; Superconductivity, Applications**

NICHOLAS SAUNDERS

Further Reading

Barone, A. and Paternò, G. *The Physics and Applications of the Josephson Effect*. Wiley, New York, 1982.

Barone, A., Ed. *The Josephson Effect: Achievements and Trends*. World Scientific, Singapore, 1986.

Gallop, J.C. *SQUIDS: The Josephson Effects and Superconducting Electronics*. Adam Hilger, Philadelphia.

Jenks W.G, Sadeghi, S.S.H. and Wikswo, J. P. Review Article: SQUIDs for nondestructive evaluation. *J. Phys. D: Appl. Phys.* 30 293–323, 1997.

Josephson, B.D. Possible new effects in superconductive tunnelling. *Phys. Lett.*, 1 251–3, 1961.

Pagano, S. and Barone, A. Josephson junctions. *Superconductor Sci. Technol.*, 10 904–908, 1997.

Pizzella, V., Penna, S.D., Del Gratta, C. and Romani, G.L. Review Article: SQUID systems for biomagnetic imaging. *Superconductor Sci. Technol.*, 14 R79–R114, 2001.

L

Lasers, Applications

More than 1000 different lasers (an acronym for light amplification by stimulated emission of radiation) exist and can be classified according to the type of lasing material employed, which may be a solid, liquid, gas, or semiconductor. The characteristic wavelength and power output of each laser type determines its application. The first true lasers developed in the 1960s were solid-state lasers (e.g., ruby, or neodymium: yttrium–aluminum garnet "Nd:YAG" lasers). Ruby lasers were used as early as 1961 in retinal surgery and are today used mainly in surgery and scientific research, and increasingly for micromachining. Gas lasers (helium and helium–neon being the most common, but there are also carbon dioxide lasers), developed in 1964, were soon investigated for surgical uses. Gas lasers were first used in industry in 1969 and are still heavily used in high-power applications in manufacturing for drilling, cutting, and welding. Excimer lasers use the noble gas compounds for lasing. Dye lasers, which became available in 1969, use solutions of organic dyes, such as rhodamine 6G, which can be stimulated by ultraviolet light or fast electrons. Semiconductor lasers (sometimes called diode lasers) are generally small and low power. Emitting in either the infrared or visible range, semiconductor lasers produce light based on free electrons in the conduction band, which are stimulated by an electrical current to combine with others in the valance band of the material. Covering the wavelength range used in optical fiber communications (see Lasers in Optoelectronics), semiconductor lasers are today the most important and widespread type of laser.

Lasers are employed in virtually every sector of the modern world including industry, commerce, transportation, medicine, education, science, and in many consumer devices such as CD players and laser printers. The intensity of lasers makes them ideal cutting tools since their highly focused beam cuts more accurately than machined instruments and leaves surrounding materials unaffected. Surgeons, for example, have employed carbon dioxide or argon lasers in soft tissue surgery since the early 1970s. These lasers produce infrared wavelengths of energy that are absorbed by water. Water in tissues is rapidly heated and vaporized, resulting in disintegration of the tissue. Visible wavelengths (argon ion laser) coagulate tissue. Far-ultraviolet wavelengths (higher photon energy, as produced by excimer lasers) break down molecular bonds in target tissue and "ablate" tissue without heating. Excimer lasers have been used in corneal surgery since 1984. Short pulses only affect the surface area of interest and not deeper tissues. The extremely small size of the beam, coupled with optical fibers, enables today's surgeons to conduct surgery deep inside the human body often without a single cut on the exterior. Blue lasers, developed in 1994 by Shuji Nakamura of Nichia Chemical Industries of Japan, promise even more precision than the dominant red lasers currently used and will further revolutionize surgical cutting techniques.

Commerce throughout the world has been profoundly affected by laser applications, including the widely used barcode scanning systems used in supermarkets, warehouse inventories, libraries, universities, and schools. The red "scan line" is a laser spot rapidly moving across at 30 or 40 times per second. A photo diode measures the intensity of light reflected back from the object: since dark bars absorb light, the bar and space pattern in the barcode can be determined.

Lasers are widely employed for micromachining and automated cutting in industrial applications. Laser welding is routinely used in large industrial applications such as automotive assembly lines, providing much cheaper, better, and dramatically quicker welds than those possible with traditional techniques. Laser alloying utilizes the precise and powerful application of lasers to melt metal coatings and a portion of the underlying substrate to create surfaces that have unique and highly desirable qualities. Alloying produces the desired effect at the precise location where it is needed, meaning that less expensive materials can be utilized for the remainder of the instrument. New alloys with unique properties have been developed using this technology over the past few years with resultant new applications.

Laser diagnostic instruments have revolutionized studies across the entire range of the sciences and engineering. These include applications such as laser-induced fluorescence to measure tiny amounts of trace materials and laser Doppler anemometry, which enables fluid flow to be precisely monitored. Laser spectroscopy has revolutionized the study of very fast chemical reactions, the study of structural changes in complex molecules, and other areas of biology and chemistry. Laser photobiology and photochemistry are large and growing subdisciplines with their own conferences, journals, theories, and nomenclature.

Lasers are widely used in telecommunications, especially in fiber-optic cables. These systems employ low-powered, computer-controlled, semiconductor lasers that transmit encoded information in rapid infrared pulses. Regular light cannot perform suitably because its waves are not in parallel and therefore become too weakened over long distances resulting in an unacceptable loss of essential information. Semiconductor lasers can also "read" the pits and lands on the surface of a compact disk or DVD (see Audio Recording, Compact Disk). Because the wavelength of blue light is shorter than red light, the blue semiconductor lasers developed in the 1990s will be able to form much smaller spots on the recording layer of the disc, increasing the density of optical data storage.

The precise, unchanging nature of a laser-generated light beam makes it ideal for a wide range of applications involving measurement. Surveyors, construction personnel, oceanographers, geologists, and astronomers use laser ranging between a source and a reflector some significant distance away to measure distances or to ensure proper alignment of objects. Transit time can be used to calculate distance to an extremely high level of accuracy. Geologists routinely employ lasers, for example, to measure regional deformation of the Earth's crust to aid in understanding and predicting earthquakes. This same precision over long distances forms the core of modern laser-guided weapons, ranging from hand-held sniper rifles to long-range missiles and other "smart" weapons. Lasers are used to "paint" a target that is then precisely honed in by the weapon "tuned" to that particular wavelength. Efforts to create laser-based missile defense systems are under development in the U.S, but as of the early twenty-first century, results suggest that success is many years away (see Missiles, Defensive).

Holography is a widely employed and enjoyed aspect of the modern application of lasers. In addition to being employed for artistic and esthetic purposes, holography is used in the manufacture of optical instruments, in analyzing materials without harming them, and in storing data in extremely compact form. Two absolutely identical light beams are essential to forming a holographic image, and lasers ideally perform this function.

See also **Audio Recording, Compact Disk; Computer Memory; Lasers, Theory and Operation; Lasers in Optoelectronics; Semiconductors, Compound; Spectroscopy, Raman**

DENNIS W. CHEEK AND CAROL A. CHEEK

Further Reading

Bunkin, A. and Voliak, K. *Laser Remote Sensing of the Ocean: Methods and Applications.* Wiley, New York, 2001.

Committee on Optical Science and Engineering, National Research Council. *Harnessing Light: Optical Science and Engineering for the 21st Century.* National Academy Press, Washington, D.C., 1998.

Crafer, R.C., Ed. *Laser Processing in Manufacturing.* Chapman & Hall, New York, 1993.

Diehl, R. *High-Power Diode Lasers.* Springer Verlag, New York, 2000.

Duley, W.W. *Laser Welding.* Wiley, New York, 1998.

Fotakis, C, Papazoglou, T. and Kalpouzos, C. *Optics and Lasers in Biomedicine and Culture.* Springer Verlag, New York, 2000.

Hecht, J. and Teresi, D. *Laser: Light of a Million Uses.* Dover, New York, 1998.

Nakamura, S., Pearton, S. and Fasol, G. *The Blue Laser Diode: The Complete Story*, 2nd edn. Springer Verlag, Berlin.

Puliafito, C.A. *Laser Surgery and Medicine: Principles and Practices.* Wiley, New York, 1996.

Ready, J.F. *Industrial Applications of Lasers*, 2nd edn. Academic Press, New York, 1997.

Rubahn, H.-G. *Laser Applications in Surface Science and Technology.* Wiley, New York, 1999.

Lasers in Optoelectronics

Optoelectronics, the field combining optics and electronics, is dependent on semiconductor (diode) lasers for its existence. Mass use of semiconductor lasers has emerged with the advent of CD and DVD technologies, but it is the telecommunications sector that has primarily driven the development of lasers for optoelectronic systems. Lasers are used to transmit voice, data, or video signals down fiber-optic cables.

Theodore Maiman, of Hughes Aircraft Company in the U.S., demonstrated the first laser in 1960, some four decades after the prediction of the lasing phenomenon by Einstein in 1917. Following this demonstration, in synthetic ruby crystals, it was inevitable that lasing action in other media would be investigated. It was two years later in 1962 that the Massachusetts Institute of Technology (MIT), IBM and General Electric simultaneously built the first semiconductor lasers using gallium arsenide (GaAs) as the lasing medium. Initial lasers were severely limited by heating problems and could only operate for short periods. Teams at Bell Labs, U.S. and the Ioffe Physical Institute in Russia succeeded with continuous operation at room temperature in 1970, with commercial production emerging from 1975 onwards. The development of silica optical fiber, through the 1960s and into the 1970s, has provided the complementary technology to drive and dominate the development of lasers for optoelectronics.

The basis of the semiconductor laser is a junction between *p*-type and *n*-type semiconductor materials, familiar from transistor technology. By applying a current across the junction (i.e. electrical pumping), concentrations in holes and free electrons are altered in such a way as to produce a population inversion. The injected electrons recombine with holes with the result being emission of a photon of light. Through this process of recombination in the "active region," light is emitted and, through diffusion of the carriers through the material, this light exits at the surface of the semiconductor. Natural cleavage planes of the semiconductor material form the laser end faces and define the resonant cavity within which the emission of photons of the same wavelength can build up. Once the input current reaches a certain level (usually a few milliamps, known as the "threshold current"), the optical output increases, almost linearly, with electrical input. Below the threshold, any optical emission is spontaneous, but above the threshold, the stimulated emission that defines lasing is dominant. In order to transmit information for communications purposes, the laser has to be turned on and off (modulated) and this is achieved by varying the input current by small amounts around a fixed operating current.

Devices are produced by growing semiconductor crystals, layer by layer, using epitaxial and chemical vapor deposition techniques. The wavelength of light from the laser is dependent on the band-gap energy of the semiconductor material. Thus, by selecting particular semiconductor materials and doping with various impurities it is possible to grow a range of lasers with different band-gaps, and therefore differing wavelengths of laser emission. Gallium aluminum arsenide (AlGaAs) lasers emit around 850 nanometers ($1 \, nm = 10^{-9}$ meters) and were used in the first U.K. telephone field trials in 1977 at the British Post Office Research Laboratories facility at Martlesham Heath. The use of compound semiconductors from groups III and V of the periodic table continued to dominate (e.g., indium gallium arsenide phosphide (InGaAsP) which emits around 1300 nanometers). Due to low power losses (attenuation) in optical fibers at this wavelength, it remained the standard for most optical fiber systems for some time. Later generations of systems operated at 1550 nanometers (e.g., InGaAsP/InP), where lower fiber losses and the availability of optical amplifiers (erbium-doped optical fibers) at that wavelength to overcome transmission loss meant that longer spans were obtainable.

An optoelectronic communication system comprises of an electronic signal transmitted to a laser, which produces an intensity-modulated pulse stream. This modulated light signal is transmitted along optical fiber to a detector, where the optical signal is converted to electrical and amplified. When optical fibers were first proposed in the 1960s, high attenuation was a problem for long-distance communication. Removal of metal impurities achieved by Corning Glass Works (now Corning Inc.) researchers in the U.S. in 1970, kick-started a revolution in communications by fiber optic. The first commercial fiber-optic telephone systems were installed in 1977 by AT&T and GTE in Chicago and Boston respectively.

As the propagation characteristics of optical fiber (transmission losses, dispersion) were improved, so the performance characteristics of the lasers that served as the transmitters in these communication systems were refined. By adjusting the mix of semiconductor ingredients, the peak emission wavelengths have been tuned to match transmission windows (a wavelength region that offers low optical loss) in the optical fiber; thus the greatest effort continued to be directed at improving semiconductor lasers that emit at the windows of 850, 1300, and 1550 nanometers.

With the explosion of data transmission in the 1990s, there came a need to utilize the communications infrastructure to its greatest possible capacity. This was achieved in two main ways, both made possible by developments in laser technology. First, the architecture of the semiconductor crystals evolved to enable increased modulation speeds. From 2 megabytes per second (Mb/s) in the 1970s, through 140 Mb/s in the 1980s, to 40 gigabytes per second (Gb/s) at the end of the century, there was a relentless rise in the amount of information that could be transmitted, and semiconductor laser source operating lifetimes improved greatly. Second, more optical channels were squeezed into a single fiber using dense wavelength division multiplexing (DWDM) technology. Each wavelength channel requires a separate signal-generating transmission laser. These are generally spaced at 100 or 200 gigahertz intervals across the 1530- to 1570-nanometer gain bandwidth. A wavelength multiplexer combines the separate wavelength laser signals into a single transmission fiber. Multiple wavelengths demanded increased laser stability, narrower linewidths (wavelength spread about the center wavelength) and refinement in the central wavelengths emitted. At the receiving end, a wavelength demultiplexer separates the signals into separate wavelengths of light, which are each detected at a photodiode to interface the signal into the computer or electronic network.

For a long time, semiconductor lasers emitted light parallel to the boundaries between the layered structure of the devices and were known as edge-emitters. However, surface-emitting lasers (SELs) had the advantage of being easier—and therefore more economic—to couple to other optoelectronic components. In particular, vertical cavity surface-emitting lasers (VCSELs) emerged in the late 1980s after theoretical postulation in late 1970s. Cheaper to manufacture in quantity, this laser architecture held the potential to reduce the cost of high-speed fiber optics connections, with the added benefits of low threshold current and almost circular beam symmetry (rather than elliptical), leading to improved coupling to fibers and thus efficient operation. The first VCSELs emitted at 850 nanometers, and since the 1990s lasers at 780,

859 and 980 nanometers have been commercialized into optical systems. In 1996 research started into green–blue ultraviolet devices, and at the other end of the spectrum, a VCSEL operating at 1300 nanometers was demonstrated at Sandia Laboratories in the U.S. in 2000.

While the success of lasers within telecommunication systems seems unquestioned thanks to their utility in long-distance large-capacity, point-to-point links, these lasers also find use in many other applications and are ubiquitous in the developed world. Their small physical size, low power operation, ease of modulation (via simple input current variation) and small beam size mean that these lasers are now part of our everyday world, from CDs and DVDs, to supermarket checkouts and cosmetic medicine (see Lasers, Applications).

See also **Optical Amplifiers; Optical Materials; Optoelectronics, Dense Wavelength Division Multiplexing; Semiconductors, Compound; Telecommunications**

Susan M. Hodgson

Further Reading

Bertolotti M. *The History of the Laser*. Institute of Physics Press, Bristol, 2004.

Bhattacharya, P. *Semiconductor Optoelectronic Devices*. Prentice Hall, Upper Saddle River, NJ, 1997.

Brinkman, W.F., Koch, T.L., Lang, D.V. and Wilt, D.P. The lasers behind the communications revolution. *Bell Lab. Tech. J.*, 5, 1, 150–167, 2000.

Dakin, J. and Brown, R. *The Handbook of Optoelectronics*. Institute of Physics Press, Bristol, 2004.

Hecht, J. *City of Light: The Story of Fiber Optics*. Oxford University Press, Oxford, 1999.

Holonyak, N. Jr. The semiconductor laser: A thirty-five year perspective. *Proceedings of the IEEE*, 85, 11, 1678–1693, 1997.

Wilson, J. and Hawkes, J. *Optoelectronics: An Introduction*. Prentice Hall, London, 1998.

Lasers, Theory and Operation

Lasers (an acronym for light amplification by stimulated emission of radiation) provide intense, focused beams of light whose unique properties enable them to be employed in a wide range of applications in the modern world (see Lasers, Applications). The key idea underlying lasers originated with Albert Einstein who published a paper in 1916 on Planck's distribution law, within which he described what happens when additional energy is introduced into an atom. Atoms have a heavy and positively charged nucleus surrounded by groups of extremely light and negatively charged electrons. Electrons orbit the atom in a series of "fixed" levels based upon the degree of electromagnetic attraction between each single electron and the nucleus. Various orbital levels also represent different energy levels. Normally electrons remain as close to the nucleus as their energy level permits, with the consequence that an atom's overall energy level is minimized. Einstein realized that when energy is introduced to an atom; for example, through an atomic collision or through electrical stimulation, one or more electrons become excited and move to a higher energy level. This condition exists temporarily before the electron returns to its former energy level. When this decay phenomenon occurs, a photon of light is emitted. Einstein understood that since the energy transitions within the atom are always identical, the energy and the wavelength of the stimulated photon of light are also predictable; that is, a specific type of transition within an atom will yield a photon of light of a specific wavelength. Hendrick Kramers and Werner Heisenberg obtained a series of more extensive calculations of the effects of these stimulated emissions over the next decade. The first empirical evidence supporting these theoretical calculations occurred between 1926 and 1930 in a series of experiments involving electrical discharges in neon.

After World War II, a large amount of radar equipment was obtained by universities throughout the U.S. and Europe. Experiments utilizing this equipment began to lay the groundwork for the laser. By the mid-1950s, Felix Bloch, Edward Purcell, Robert V. Pound, and others had demonstrated that a significant proportion of electrons within various substances could be stimulated to raise their energy levels and then naturally decay, releasing large quantities of photons (a process known as transient population inversions). Simultaneous work on optical "pumping" techniques by Jean Brossel and Alfred Kastler demonstrated that this phenomenon could be greatly increased so that a substantially larger numbers of electrons within a substance are excited rather than unexcited. Joseph Weber described at an Electron Tube Conference in Canada in 1952 how these processes might be combined to create and maintain a large population inversion, but he was not able to produce one in the laboratory. Charles Townes at Columbia University in the U.S., working with James P. Gordon and Herbert J. Zeiger, managed such a sustained inversion process in ammonia vapor utilizing a microwave pump in 1951. In 1954 Townes formally announced the operation of an oscillator that could make this

occur. Aleksandr Prokhorov and Nicolay Basov from the Lebedev Institute in Moscow published a paper in the same year about similar results using microwave spectroscopes. Both devices involved microwave amplification by stimulated emission of radiation (MASER). Townes' maser emitted at a wavelength of 1.25 centimeters and therefore did not result in visible light. In 1964 the three scientists (Townes, Prokhorov, and Basov) would all share the Nobel Prize in Physics for their work on the development of the maser-laser principle.

By 1958, Townes and Arthur Schawlow at Bell Laboratories (where Townes was also a consultant) determined that a carefully crafted optical cavity using optical mirrors would result in photons released by the pumping process being reflected back into the medium to strike other electrons. Their resultant decay from excitation would release yet further photons, the process repeating itself many times over until an intense visible beam of light would be created (all these photons would be of the same wavelength and in phase with one another). The result, light amplification by stimulated emission of radiation (LASER), would then exit the cavity through the semitransparent mirror on the one end once it had achieved sufficient intensity. The observer would see a thin beam of concentrated light. Theodore Maiman of Hughes Research Laboratories in the U.S. succeeded in turning Townes and Schawlow's ideas into reality, demonstrating in 1960 the first true laser using a ruby crystal as the laser source and a flash lamp as the source of energy "pumped" into the system. In this particular case, the two flat ends of the ruby crystal itself were coated with a reflective substance. When the flash lamp was turned on, electrons within the crystal were raised to an excited state and released photons in their decay, which were in turn reflected by the two ends of the crystal back into the medium and so on, until very quickly an intense red beam emerged from one end of the crystal. A year later, Ali Javan, William Bennett, Donald Herriott, L. F. Johnson, and K. Nassau at Bell Laboratories in the U.S., built lasers using a mixture of helium and neon gases, and a solid-state laser from neodymium (see Figure 1.) In 1962 the first semiconductor (diode) lasers using gallium arsenide were achieved, by IBM Research Laboratory in New York, General Electric in New York, and the Massachusetts Institute of Technology Lincoln Laboratory.

In 1962, Robert Hall of the General Electric Research Laboratories in the U.S. created the first infrared carbon dioxide laser, one of a new class of gas lasers whose power was unprecedented.

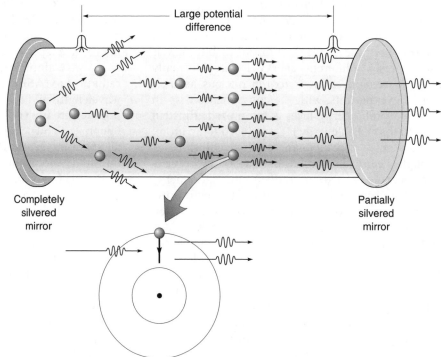

Figure 1. Schematic drawing of a helium/neon laser. [*Source: Cutnell, J.D. and Johnson, K.W. Physics, 4th edn. Wiley, New York, 1998, p. 936.*]

Creation of new types of lasers proceeded quickly over the next few years with the invention of ion lasers using mercury vapor (1963), argon (1964), and a special type of ion laser employing blue helium and cadmium metal vapor (1966). In 1966 researchers Peter Sorokin and John Lankard of the IBM Research Laboratories in the U.S. developed a liquid laser utilizing fluorescent dye that allowed investigators to "tune" certain lasers to a desired wavelength. By the mid-1970s rare gas excimer lasers such as xenon fluoride (1975) and a free-electron laser amplifier in the infrared zone (1976) were introduced. The first soft x-ray laser was successfully demonstrated by D. Matthews and colleagues at the Lawrence Livermore Laboratories in California in 1985. By the end of the twentieth century, over 1000 types of laser had been developed. Recent astronomical observations have confirmed that natural laser action also occurs in interstellar medium associated with emerging stars.

See also **Lasers, Applications; Quantum Electronic Devices**

DENNIS W. CHEEK AND CAROL A. CHEEK

Further Reading

Bertolotti, M. *Masers and Lasers: An Historical Approach*. Adam Hilger, Bristol, 1983.

Bromberg, J.L. *The Laser in America*. MIT Press, Cambridge, MA, 1991.

Committee on Optical Science and Engineering, National Research Council. *Harnessing Light: Optical Science and Engineering for the 21st Century*. National Academy Press, Washington D.C., 1998. A government-sponsored study by experts of the present and future directions of optical science and engineering, including further developments and applications of lasers.

Brown, R.G.W. and Pike, E.R. A history of optical and optoelectronic physics in the twentieth century, in *Twentieth Century Physics*, Brown, L., Pais, A. and Pippard, B., Eds. Institute of Physics Press, Bristol, 1995.

Harbison, J.P. and Nahory, R.E. *Lasers: Harnessing the Atom's Light*. W.H. Freeman, New York, 1997.

Hecht, J. *Laser Pioneers*. Academic Press, New York, 1992. Interviews with 12 scientists who were instrumental in developing masers and various kinds of lasers, including three Nobel Prize winners in Physics.

Hitz, C.B., Ewing, J.J. and Hecht, J. *Introduction to Laser Technology*, 3rd edn. Wiley, New York, 2001.

Schawlow, A.L. and Townes, C.H. Infrared and optical masers. *Phys. Rev.*, 112 1940–1949, 1958. A seminal article in the field by two of its early pioneers.

Siegman, A.E. *Lasers*, 5th revised edn. Springer Verlag, New York, 1999.

Silvast, W.T. *Laser Fundamentals*. Cambridge University Press, Cambridge, 1996.

Svelto, O. *Principles of Lasers*. 4th edn. Plenum Press, New York, 1998.

Townes, C.H. *How the Laser Happened: Adventures of a Scientist*. Oxford University Press, New York, 1999. A Nobel Prize winner's personal account of the invention of the maser and the laser and his interactions with many key scientists in the fields of applied physics and optical engineering.

Laundry Machines and Chemicals

In the nineteenth century, American home economist Catherine Beecher described laundry as the "housekeeper's hardest problem." Thousands of patents for washing machines attested to the tedium of "blue Monday," as the day commonly allocated to that task was known. The task involved hauling, boiling, and pouring hundreds of gallons of water, exposure to caustic chemicals, and rubbing, pounding, wringing, and lifting the heavy, wet linens and clothing. The automation of washing and drying machines in the twentieth century and improvements in detergents resulted in a considerable reduction in the intensity of labor required for this task. At the same time, historians of technology claim that rising standards of cleanliness and the cultural turn away from commercial laundry and domestic laundresses ensured that time spent in domestic laundering remained significant. Laundry technology never fully eliminated human effort, even in Western countries, where a person was still required to gather, sort, load, unload, fold, and iron the laundry. In less developed countries, many women continued to perform the full range of laundering tasks without machines at the end of the twentieth century.

Washing Machines

Sir John Hoskins' first British patent for a washing machine dates to 1677, and the first American patent of 1805 reflects early work on industrial machines. The British and French textile industries' use of washing machines in the bleaching process was noted in the *Grande Encyclopedie* of 1780. In fact, until the twentieth century when laundering was individualized and permanently allocated to the home, innovations in washing machines were developed principally for large commercial laundries. American James T. King's industrial cylinder washing machine of 1851 incorporated technological principles that would be applied in twentieth century automated industrial and domestic washing machines. It consisted of an outer, fixed cylinder and an inner, perforated rotating cylinder. A coal or wood fire under the outer cylinder heated soapy water that circulated through the clothes contained in the inner cylinder, which revolved by huge belts attached to an industrial steam engine. Other machines, such as Montgomery Ward's "Old Faithful" of 1927 produced well into the twentieth century, imitated the motion of the human hand rubbing against a ribbed washboard. Hand wringing to remove excess water was replaced with a hand-cranked and later electric wringer that appeared on washing machines such as the Westinghouse A4TP-2 as late as the 1960s.

By 1900, semi-automated washing machines followed two designs that would be refined and improved over the next century. In the first configuration, an exterior cylindrical shell contained a perforated inner cylindrical shell that held the clothing and rotated on a horizontal axis. In the second vertical arrangement, an outer cylinder contained an inner, perforated cylinder, or basket, in which an agitator or gyrator moved the clothing through water. The cylindrical tubs could be made of wood, copper, cast iron, and later porcelain. Later in the century, the inner shells of both designs would be configured to spin their contents at high speed in order to expel most of the water after the wash cycle was complete.

By 1939, washing machines that could agitate clothing, pipe water in and out of their tubs, and spin-dry at speed- were being marketed directly and most heavily to women. Several technological and cultural developments necessarily preceded the post-World War II mass consumption of these machines in North America and Western Europe. The electric motor, available from around the turn of the century, was initially installed in washing machines in a form that exposed users to electric shocks. It made hand cranking obsolete, although by no means in the short term. Electrical appliances in general could not become widely popular until a significant proportion of households had current delivered through an electrical network. Electrical grids by which electrical current was distributed across wide geographic areas were set up in Britain by 1926 and over a decade earlier in the U.S. Fully plumbed houses and access to piped hot water did not become the norm until after the middle of the twentieth century. Little wonder then, that a British market study of 1945 found that washing boards, dollies (hand agitators), and wringers were among the staple laundry technologies in British homes at that time.

Laundry technologies were refined at the end of the twentieth century in response to popular interest in energy conservation and environmental issues. "Smart" or fuzzy logic technologies emerged from Japan in the 1980s and were subsequently applied to major appliances, including washing machines. These technologies involved sensors that anticipated the fabric type, the turbidity (soil level), and even the hardness of the water. Total washing time, power and water level, the amount of detergent, and the number of rinses could be regulated in the most economical manner. An example of a fuzzy rule might be: if the wash is very dirty and the dirt has a muddy quality, increase the wash time.

Chemicals

By the turn of the century, soap had become the most widespread among many chemicals used for laundering. Other products commonly used well into the century included lye, salt soda, lime, borax, and even urine and manure. Soap is made by combining animal fats or vegetable oils (glycerides) and alkalis such as sodium or potassium (once derived from ashes). Its discovery is thought to date back to Roman times. Until the twentieth century, when soap flakes and powders were introduced to the market, soaps manufactured for laundry were sold in bar or tablet form and had to be cut up or flaked by the consumer. In 1918 the British manufacturer Lever Brothers began marketing Rinso, the first granulated general laundry soap. In all these cases, the reaction between the magnesium and calcium content in hard water and soap left residual scum and was a less than satisfactory cleanser, especially in light of the rise in standards of cleanliness. Manufacturers responded by developing synthetic cleaning agents based on two-part molecules they called synthetic surfactants. The results of research at the U.S. company Proctor & Gamble were the products Dreft, followed in 1946 by Tide, which set a new standard for laundry detergents. In the final decades of the twentieth century, in response to consumer concern over the environmental impact of synthetic detergents, manufacturers developed biological and biodegradable detergents. In the former, enzymes work on biological stains. Biodegradable detergents were designed so they would not resist the biological products used to purify sewage, as had synthetic surfactants and phosphates. It was hoped that this would avoid the release of foam into rivers, which has not proven to be the case.

Among the last major innovations of the twentieth century in laundering were "detergent-less" washing machines. Korean electronics company Daewoo's Midas model and others like it were designed to use an electrolysis device that filters a number components out of tap water to produce ionized water that would clean the clothes.

See also **Detergents; Electric Motors**

JENNIFER COUSINEAU

Further Reading

Cowan, R.S. *More Work for Mother: The Ironies of Household Technology from the Open Hearth to the Microwave.* Basic Books, New York, 1983.

Davidson, C. *A Woman's Work is Never Done: A History of Housework in the British Isles 1650–1950.* Chatto & Windus, London, 1982.

Forty, A. *Objects of Desire: Design and Society from Wedgwood to IBM.* Pantheon, New York, 1986.

Gideon, S. *Mechanization Takes Command: a Contribution to Anonymous History.* W.W. Norton, London, 1948.

Hardyment, C. *From Mangle to Microwave: The Mechanization of Household Work.* Polity Press, Oxford, 1988.

Lupton, E. and Miller, J.A. *Mechanical Brides: Women and Machines From Home to Office.* Princeton Architectural Press, Princeton, 1993.

Lupton, E. and Miller, J.A. *The Bathroom The Kitchen and the Aesthetics of Waste: A Process of Elimination.* Princeton Architectural Press, New York, 1992.

Nye, D. *Electrifying America: Social Meanings of A New Technology.* MIT Press, Cambridge, MA, 1990.

Strasser, S. *Never Done: A History of American Housework.* Pantheon, New York, 1982.

Yarwood, D. *Five Hundred Years of Technology in the Home.* B.T. Batsford, London, 1983.

Useful Websites

Badami, V.V. and Chbat, N.W. *Home Appliances Get Smart*: http://www.hjtu.edu.cn/depart/xydzzxx/ec/spectrum/homeapp/app.html

Life Support, *see* **Intensive Care and Life Support**

Light Emitting Diodes

Light emitting diodes, or LEDs, are semiconductor devices that emit monochromatic light once an electric current passes through it. The color of light emitted from LEDs depends not on the color of the bulb, but on the emission's wavelength. Typically made of inorganic materials like gallium or silicon, LEDs have found frequent use as "pilot," or indicator, lights for electronic devices. Unlike incandescent light bulbs, which generate light from "heat glow," LEDs create light more efficiently and are generally more durable than traditional light sources. Despite having some evidence that Ernest Glitch produced luminescence during mid-nineteenth century experiments with crystals made from silicon and carbon, Henry Joseph Round is credited with the first published observation of electroluminescence from a semiconductor in 1907. Round, a radio engineer and inventor, created the first LED when he touched a silicon carbide (SiC) crystal with electrodes to produce a glowing yellow light. The SiC crystals, or carborundum, were widely used as abrasives in sandpaper at the time, but the crystals were difficult to work with—and the light produced was too dim—to be of any practical use as a source of lighting. Round soon abandoned his research. Although O.V. Lossev later confirmed in 1928 that the source of the crystal's light was from "cool" luminescence, work on LEDs was largely forgotten.

Research on LEDs resumed in the 1950s, but unlike Round's use of SiC crystals, work centered on the use of gallium arsenide (GaAs) as a semiconductor. The first commercially viable LED, invented by Nick Holonyak, Jr., became available in 1962. Holonyak, who received his PhD in electrical engineering from the University of Illinois at Urbana in 1954, was the first graduate student of John Bardeen who, along with Walter Brattain, invented the transistor in 1947. After graduation, Holonyak joined Bell Labs and, in 1957, moved to General Electric (GE) laboratories in Syracuse, NY. At GE, Holonyak joined the race against other teams from GE, IBM, and the Massachusetts Institute of Technology (MIT) to develop the first infrared semiconductor laser. Robert Hall, also at GE, was the first to create a working infrared laser diode; however, Holonyak did make an interesting discovery during his research. Unlike most of his colleagues, Holonyak made his semiconductor with an alloy of gallium arsenide phosphide (GaAsP). By adding phosphorus, Holonyak increased the band gap, which reduced the wavelength of the emitted light from infrared to red. By widening the band gap, researchers discovered that LEDs could eventually produce light across the spectrum. In 1963, Holonyak returned to the University of Illinois to teach, and his former student, M. George Craford, developed a yellow LED in 1970.

From Holonyak's work, GE offered the first commercially available LED in 1962, but at a price of $260, the device proved too expensive for all but the most exclusive functions. Anticipating wider use, the Monsanto Corporation began mass-producing LEDs in 1968 and enjoyed steadily increasing sales into the next decade. In the late 1960s, electronics manufacturers began using LEDs to produce numeric displays for calculators and later wristwatches with the introduction of the Hamilton Watch Corporation's Pulsar digital watch in 1972. With time, researchers at Monsanto were eventually able to produce red, orange, yellow, and green light from LEDs, but contemporary LED technology still produced light that was too dim and consumed too much power for wider application. By the late 1970s, most manufacturers of

calculators and digital watches replaced LEDs with liquid crystal displays (LCDs). Still, advances in LED technology offered much for use as indicators in devices such as telephones. The popularity of AT&T's "Princess" telephone in the 1960s, with its light-up dial, and the increasing use of multiline telephones in large corporate offices, created significant problems for the company. Existing indicators in these telephones required installation near 110-volt outlets, and the bulbs burned out often, leading to costly service calls. AT&T had much to gain from indicator lights that were both more durable and more energy efficient than current technologies. Given that LEDs had an anticipated life of over fifty years when used in telephones, the continued development of the technology offered significant cost savings for the company. Moreover, as LEDs required only about 2 volts of current they could be powered through telephone lines, which are typically around 40 volts, without affecting call quality, thereby negating the need for a separate outlet.

By the early 1980s, researchers began to develop the next generation of super high brightness LEDs. The first significant development used aluminum gallium arsenide phosphide (GaAlAsP) for LEDs that were ten times brighter than previously existing LEDs, but were only available in red and tended to have a shorter life. Research continued through the rest of the decade to achieve more colors and extra brightness without adversely effecting longevity. At the end of the 1980s, LED designers borrowed techniques from the recent developments in laser diode technology, which were gaining popular use in barcode readers, to produce visible light from indium gallium aluminum phosphide (InGaAlP) LEDs. In addition to gaining extra brightness, efficiency, and durability, the new material allowed more flexibility in changing the band gap to produce red, orange, yellow, and green LEDs. In the early 1990s, commercial applications for LEDs expanded to include traffic control devices, variable message signs, and indicator lights for the automotive industry. Also in the 1990s, researchers succeeded in creating the first blue LEDs, which made full color applications and the creation of white light from LEDs possible. By 2003, LEDs were a multibillion dollar industry. With traffic signals in the U.S. scheduled to be replaced with LEDs by 2006, the anticipated universal application of LEDs in automobiles by the end of the decade, and new uses for LEDs in medical phototherapy, this figure is sure to increase exponentially. Current designers expect that the development of LEDs made from organic materials will further reduce the cost of production, and, as brightness and efficiency increase, some anticipate that LEDs will soon serve as a general light source that could potentially produce huge savings in annual energy costs by 2025.

See also **Computer Displays; Lighting Techniques; Semiconductors, Compound**

AARON L. ALCORN

Further Reading

Schubert, E.F. *Light-Emitting Diodes.* Cambridge University Press, New York, 2003.

Shinar, J., Ed. *Organic Light-Emitting Devices.* Springer Verlag, New York, 2003.

Zukauskas, A., Shur, M.S. and Caska, R. *Introduction to Solid-State Lighting.* Wiley, New York, 2002.

Useful websites

Hall, A. A laser in your light bulb? *Business Week Online,* 27 October 2000: http://www.businessweek.com/bwdaily/dnflash/oct2000/nf20001027_054.htm.

Martin, P. The Future Looks Bright for Solid-State Lighting. *Laser Focus World* 37, 9, September 2001: http://lfw.pennnet.com/Articles/Article_Display.cfm?Section=Articles&Subsection=Display&ARTICLE_ID =119946&KEYWORD=leds

Perry, T.S. Red hot. *IEEE Spectrum Online,* June 2003: http://www.spectrum.ieee.org/WEBONLY/public feature/ jun03/med.html

Rotsky, G. LEDs cast Monsanto in unfamiliar role. *Electronic Engineering Times,* 944, 10 March 1997: http://eetimes.com/anniversary/designclassics/monsanto.html

Lighting, Public and Private

At the turn of the twentieth century, lighting was in a state of flux. In technical terms, a number of emerging lighting technologies jostled for economic dominance. In social terms, changing standards of illumination began to transform cities, the workplace, and the home. In design terms, the study of illumination as a science, as an engineering profession, and as an applied art was becoming firmly established.

Overt competition between lighting methods was a relatively new phenomenon exacerbated by the spread of electric lighting during the 1880s (Edison's Lighting System and its many competitors), which rivaled the firmly entrenched gas lighting networks. During the late Victorian period, competitors marshaled science to justify the superiority of their technologies. Gas lighting, revitalized by efficient burners, incandescent mantles, and high-pressure operation, vied with the new filament electric lamps and, in larger spaces, with electric arc lighting. Between 1890 and 1910, the difficulties of incandescent lamp manufacture and the potential profits to be made from more efficient technologies motivated engineers to seek alternatives. New, more reliable, and more economical sources were continually being developed, such as the Nernst glower lamp. During this 20-year period, both innovation and technical development blossomed. The great illuminating efficiency of the firefly was much discussed, for example, and an electrochemical or luminescent analog was actively sought.

The enthusiasts who were developing lighting systems were faced with marketing, physiological, and economic questions. How were they to convince purchasers of the need for more or better lighting? How could they meaningfully compare the competing light sources in terms of brightness, color, and efficiency? How much light was needed for various tasks, and how should lighting systems best be installed and employed? Thus characteristics such as cost, intensity, and uniformity of light distribution increasingly were measured and compared. Firms devoted attention to developing more efficient (and competitive) incandescent gas mantles and electric lamp enclosures, notably the Holophane Company in New York, which produced "scientifically designed" prismatic glass globes. Yet, as more than one engineer of the period complained, the term "illumination" was more closely associated in the popular mind with medieval manuscripts or fireworks than with lighting. Indeed, perhaps the most significant alteration in the understanding of lighting during this period was the rising interest in attending to the illumination of surfaces rather than the brightness of the light source itself.

By the turn of the twentieth century, the identification of such problems was common and contributed to a growing self-awareness by practitioners of "illuminating engineering." The gas lighting industry, via local, regional and national societies, and exhibitions, was an initial nucleus for the movement. Legislators and administrators, too, had a long-standing concern with the industry owing to regulation of gas standards from the 1860s. William Preece, for example, Engineer in Chief of the Post Office in Britain, attempted to organize interest in the wider problems of lighting schemes and illumination, and the German national laboratory, the Physikalisch-Technische Reichsanstalt, undertook research for its country's gas industry during the 1890s. The PTR, in fact, effectively marshaled scientific research to measure and regulate gas and electric lighting: the Lummer–Brodhun photometer became the standard visual device internationally. The German laboratory also undertook certification and standardization of lamps, especially after the introduction of an "illuminant tax" in 1909. Similarly, the new National Physical Laboratory in Britain (1899) and the Bureau of Standards in America (1901) made lighting standards and research an important part of their activities. The first decade of the new century saw a proliferation of interest in questions of lighting. Industrially funded lighting laboratories also appeared, such as the research arm of the National Electric Lamp Association which was founded in Cleveland in 1908. Practitioners of the lighting art in New York and London, still very much divided along the lines of their favored technologies, formed "Illuminating Engineering Societies" in 1906 and 1909, respectively; other countries followed over the next two decades. These societies, led by proselytizers Van R. Lansingh in America and Leon Gaster in Britain, sought to unite the "better educated" electrical engineers, gas engineers, physiologists, scientists, architects, and designers of lighting fixtures. An international organization, the Commission Internationale de l'Éclairage (CIE), was formed in 1913 from what had begun at a gas congress in 1900. By 1935 most European countries had similar national or regional illuminating societies contributing to the international CIE.

Such organizations promoted illumination research, standardization of lighting products, technical and public education about good lighting, and increasingly, international competition for nationally important lighting industries. A flurry of books for practitioners appeared in the years between the world wars, during which time the subject was reoriented firmly as a specialty, drawing upon electrical engineering, physics, physiology, and the psychology of perception.

An important aspect of such activities was defining of standards of illumination in public and private spaces. The interaction of safety, quality, and cost were the principal issues encouraging government involvement. In Britain, five government studies of factory and workshop lighting were carried out between 1909 and 1922. In America, the Illuminating Engineering Society published a lighting code in 1910, which led to regulations for factory lighting in five states. During the World War I, the U.S. National Defense Advisory Council issued a similar nationwide code. Purpose-designed factories of the period commonly employed large windows to maximize daylight illumination, but were inadequately supplemented by gas flames, which were often unshielded or diffused by mantles. Such dim working environments were increasingly labeled as unproductive, and an inverse correlation between illumination level and the rate of accidents was demonstrated. Moreover, workplaces frequently suffered from extremes of illumination and bright reflections. Such high contrast, or "glare," was counteracted by the employment of diffusers, shields, and reflectors over the light sources themselves and by matt finishes on walls.

Public lighting also came under increasing scrutiny. Glare in road lighting became a major concern of illuminating engineering during the 1920s and was systematically improved by careful design of the placement and elevation of road lighting fixtures and of the road surfaces themselves. During the interwar years too, signaling for vehicles, railways, roads and airports was standardized by international agreement.

The lighting of offices and homes was similarly transformed. Tables of recommended illumination had been determined empirically from the early 1890s using makeshift portable illumination photometers. Lighting levels were increasingly specified for different purposes; 136 environments ranging from coal washing to school sewing rooms to bowling alleys were specified by one source. The recommended level of lighting for offices, later correlated with working speed and accuracy, increased 25-fold over a half-century period in America: from 3 to 4 foot-candles in 1910 to 100 foot-candles in 1959. Lighting levels fell from the 1970s to 300 to 500 lux in 1993 (500 lux, or lumens per square meter, is about 47 foot-candles). This decrease in recommended lighting levels was a consequence both of higher energy costs and refinements of models of human vision, especially for lighting in conjunction with visual display units and computer usage. Other national standards and practices have risen and fallen similarly. For example, office lighting standards in Finland were 225 lux (21 foot-candles) in 1985 and 400 lux (37 foot-candles) in the Netherlands in 1991. From the 1960s, European countries began to augment such illumination standards by defining criteria for acceptable glare and color balance from lighting installations, aspects later adopted in Japan and Russia.

One reason for the rapid mid-century rise in illumination levels was the introduction of fluorescent fixtures, particularly in offices after World War II. These cylindrical sources of light have the advantage of reducing shadows and controlling contrast in the working area and of providing high illumination levels at low electrical cost. Their disadvantages—faint hum, flicker, and a different rendition of color than incandescent lights—has limited their use in the home.

In the last decades of the twentieth century, the technological and social choices in lighting attained considerable stability both technically and socially. Newer forms of compact fluorescent lighting, despite their greater efficiency, have not significantly replaced incandescent bulbs in homes owing to higher initial cost. Low-pressure sodium lamps, on the other hand, have been adopted increasingly for street and architectural lighting owing to lower replacement and maintenance costs. As with fluorescent lighting in the 1950s, recent lighting technologies have found niche markets rather than displacing incandescents, which have now been the dominant lighting system for well over a century.

See also **Lighting Techniques**

SEAN F. JOHNSTON

Further Reading

Anon. Recommended values of illumination. *Trans. Illum. Eng. Soc. of London*, 1, 42–44, 1936.

Cady, F.E. and Dates, H.B., Eds. *Illuminating Engineering*. Wiley, New York, 1928.

Clewell, C.E. *Factory Lighting*. McGraw-Hill, New York, 1913.

Gaster, L. and Dow, J.S. *Modern Illuminants and Illuminating Engineering*. Pitman, London, 1920.

Hughes, T.P. *Networks of Power: Electrification of Western Society, 1880–1930*. MIT Press, Baltimore, 1983.

Johnston, S.F. *A History of Light and Colour Measurement: Science in the Shadows*. Institute of Physics Press, Bristol, 2001.

Trotter, A.P. *Illumination: Its Distribution and Measurement*. Macmillan, London, 1911.

Walsh, J.W.T. *Photometry*. Constable, London, 1926.

Lighting Techniques

In 1900 electric lighting in the home was a rarity. Carbon filament incandescent lamps had been around for 20 years, but few households had electricity. Arc lamps were used in streets and large buildings such as railway stations. Domestic lighting was by candle, oil and gas.

Non-Electric Lighting

Victorian gas lighting was just a burning jet of gas from a small hole, the flame little brighter than a few candles. In the 1890s the efficiency was greatly increased by the gas mantle, a fine web of rare earth oxides, which are "selective emitters," giving light only at specific wavelengths when heated. Early mantles were upright, held above the flame on a fireclay support. The "inverted" mantle, fully developed by 1905, sent light downward where it was needed. It hung from a porcelain unit that provided thermal insulation between flame and gas pipe and also deflected the combustion products away from the metal pipe.

A further advance was the high-pressure gas lamp, the subject of a practical trial in 1930. Two consecutive lengths of road in Lewisham, south London, were lit, one by high-pressure gas and the other by electric light. The local authority concluded that gas was preferable, but within a couple of years new lamps gave electricity a clear advantage.

Three other types of lighting were important where there was no electricity or gas. Most important was acetylene, known earlier but not available in any great quantity until the end of the nineteenth century. A colorless, nonpoisonous gas, it burns with a very bright flame in which colors appear as in daylight. For private houses acetylene was produced by the reaction of calcium carbide and water. Only a little gas was stored, in a gas bell floating in a tank of water. Water dropped through a control valve on to a tray of carbide. The gas passed through a pipe into the bell, which rose, operating a linkage that closed the control valve. Sometimes acetylene was used with gas mantles.

Another "home-made" gas for domestic lighting was prepared from vaporized petrol mixed with air to produce "air gas" or "petrol gas." The mixture was only explosive if the proportion of petrol was between 2 and 5 percent. Inside the drum was a rotating impeller. It forced a stream of air across gasoline stored in a separate tank. The air picked up droplets of gas.

In a simple oil lamp, the wick carries the fuel to the flame where the oil is vaporized by the heat. It is the vapor rather than the liquid that burns. Some nineteenth-century oil lamps preheated the fuel, but this only became important when incandescent mantles could exploit the increased heat of the flame. These new oil vapor lamps needed heating on starting, often using a small methylated spirit burner. In some, a small hand pump in the fuel tank helped maintain the flow. Such lamps normally had mantles and were more economical than the simple oil lamp.

Electric Lighting: Filament Lamps

The efficiency of a filament lamp increases with the temperature. In 1900 all filaments were carbon, and the practical limit was about 1600°C. At higher temperatures material evaporating from the filament blackened the glass, reducing the light output even though the lamp still functioned. The General Electric Company in America made an improved lamp, the GEM (General Electric Metallized) lamp, in 1904. During manufacture, the carbon filament was heated to 3500°C in carbon vapor. This formed a very hard outer coating and changed the electrical properties to give a positive temperature coefficient of resistance like a metal. GEM lamps operated 200°C hotter than ordinary carbon lamps and sold widely in the U.S. between 1905 and 1918. Their popularity declined as tungsten lamps became available. They were never a success in Britain or Europe where metal filaments directly succeeded the earlier carbon types.

Several metals with very high melting points were tried, including vanadium (melting point 1680°C) and niobium (1950°C). Because metals have a lower electrical resistance than carbon, metal filaments have to be longer and thinner. Furthermore, the refractory metals are brittle and difficult to draw into fine wires using conventional techniques. The initial solution was to mix powdered metal with a binder, squirt the mixture through a die, and heat the resulting thread to drive off the binder and sinter the metal particles. The first metal used successfully was osmium, a rare metal melting at about 3000°C. Osmium lamps were fragile and expensive because the long filament (typically 700 millimeters) required an elaborate "squirrel cage" support. Some lamps used an alloy of osmium and tungsten, under the name Osram, from *os*mium and wolf*ram*, the German for tungsten. These lamps were not a success, but the name was kept. In 1905 Siemens and Halske in Germany began making filament lamps with tantalum, which melts at 2996°C, and can be drawn easily. Tantalum filaments were stronger than osmium, but on alternating current circuits the metal recrystallized and became brittle.

Tungsten has the highest melting point of any metal at 3410°C. Tungsten filament lamps were first produced in Vienna by Alexander Just and Franz Hanaman, using a sintering, or heating, process. When Hugo Hirst heard of this, he set up the Osram Lamp Works at Hammersmith, which made both osmium and tungsten lamps from 1909. Manufacturing filaments by sintering was complex. After much research William Coolidge, in the General Electric research laboratory in Cleveland, Ohio, developed a process in which tungsten powder was heated and hammered, after which the metal could be drawn.

Early lamps had the bulb evacuated as completely as possible, until it was found that a little gas discouraged evaporation of the filament, increasing the life (or allowing the lamp to be run hotter for the same life). The gas, however, tended to cool the filament, reducing overall efficiency. Irving Langmuir studied the cooling of a hot wire in gas and found that although the rate of heat loss is proportional to the length of the wire, it is not much affected by the diameter. This led to winding the filament into a tight coil, which reduced the heat losses. Nitrogen-filled, coiled-filament lamps were marketed in 1913. Argon, denser and with a lower thermal conductivity, was subsequently adopted, usually with a little nitrogen (about 10 percent) because if the filament broke in pure argon an arc could form and draw a heavy current. For this reason a fuse was normally included within the lamp.

The drawn tungsten filament in a vacuum gave an efficiency of about 9 lumens per watt (1.25 watts per candle power), and the coiled filament and gas filling increased this. In 1914, Osram marketed the "Half-watt" lamp, claiming an efficiency of half a watt per candle power. This had a filament temperature of about 2800°C—some 600°C hotter than earlier tungsten lamps—and gave a whiter light.

The high working temperature necessitated changes in lamp design. Because material evaporated from the filament was carried upward by convection currents in the filling gas, the bulb was given a long neck in which evaporated material could be deposited without obscuring the light. A mica disc across the neck discouraged hot gases from overheating the cap and lamp holder.

Until the early 1920s lamps were evacuated through the end opposite the cap, leaving a characteristic, fragile "pip." Lamps are now evacuated through a tube at the cap end, protecting the sealing point within the cap. From 1934 the coiled filament was often coiled again upon itself, resulting in the modern "coiled coil" filament with an efficiency of up to fifteen lumens per watt.

To reduce glare, lamps are often made with "pearl" glass, produced by etching the surface with hydrofluoric acid. Initially, etching the glass on its inner surface resulted in a very brittle bulb, so the process was usually applied to the outside. The rough surface bulb showed finger marks and was difficult to clean. John Aldington modified the process to give a smoother finish, making internally etched bulbs as strong as clear glass.

Attempts were made to prevent the blackening of carbon filament lamps by including chlorine in the bulb to react with the evaporated carbon and form a transparent compound. The idea was not successful with carbon lamps but was found to extend the life of tungsten lamps. Any halogen will react with tungsten vapour, and the resulting compound will break down in the region near the filament. The ultimate mechanism of failure is the same as when no halogen is present: metal evaporates from the filament leaving a thin portion which eventually overheats and then melts, but the glass is not blackened by evaporated metal. These lamps have a bulb of harder glass, with a higher softening temperature, or of fused silica (quartz). The bulb, usually much smaller than that of a conventional lamp, is enclosed in an outer bulb of ordinary glass.

Tubular halogen lamps rated at 1.5 kilowatts with a life of 2000 hours and an efficiency of 22 lumens per watt were on sale in the U.S. for floodlighting in 1960. The first halogen lamps used commercially in Europe were 200-watt lamps for airfield lighting installed at Luton Airport in the U.K. in 1961. They were also adopted for vehicle headlights where the small size of the light source was particularly advantageous. Using iodine as the halogen and quartz bulbs, they were known as quartz iodine lamps. Within a few years the quartz was superseded by hard glass, and other halogens, usually bromine, were used; the term "tungsten halogen lamps" was adopted. In the 1970s manufacturers began to use fluorine as the halogen. Tungsten hexafluoride only breaks down above 3000°C, and this temperature is achieved only at the filament itself. Consequently all the evaporated tungsten is redeposited on the filament, providing a self-healing action. The quantity of fluorine is extremely small, typically about 25 micrograms. Since the early 1990s, tungsten halogen lamps have been marketed for domestic use, both as general purpose, main-voltage lamps for higher brightness and a longer life and also as low-voltage spotlights.

Much of the energy in a filament lamp is dissipated as infrared radiation. If the heat could be returned to the filament in some way, the overall efficiency would be improved. In 1912, S.O. Hoffman obtained a patent in the U.S. for a system of prisms and mirrors designed to separate the infrared radiation and reflect it back to the filament. In 1922 he proposed the use of special coatings to achieve the effect, but it was not until the late 1970s that a potentially practical infrared

reflecting layer was demonstrated. This used a so-called "silver sandwich," layers of silver and titanium dioxide arranged to reflect the heat backward while light was transmitted forward. Such lamps are ideal for display purposes because the objects being illuminated are not unduly heated. Similar layers may be used as dichroic filters to transmit some colors while reflecting others, giving a high-efficiency color filter.

Electric Lighting: Discharge Lamps

Mercury Discharge Lamps. By the end of World War I, it must have seemed that the development of electric lighting had reached its peak. Tungsten filament lamps were established as best for most purposes. Other ways of making light were being investigated, though none were as simple and convenient. Most promising were gas discharge lamps.

Several British lamp manufacturers established research laboratories. The General Electric Company (GE) opened a laboratory at Wembley in 1919, headed by Clifford Paterson. Before the outbreak of war in 1914, GE and Osram had obtained all their research support from Germany. GE's chairman, Hugo Hirst, resolved that they should never again be dependent on foreign science. Most of the lighting manufacturers cooperated because there was so much to study. The nature of the "glass" envelope, the electrodes, the filling gas, and problems of sealing the lead-in wires all needed investigation. Research in Britain concentrated on the mercury discharge lamp, although Philips in Holland favored the sodium lamp. No one could find a combination of gases in a discharge tube that would both produce white light and produce it more efficiently than a filament lamp. All early discharge lamps were more efficient and therefore cheaper to run than carbon filament lamps, but their bulky nature, very high operating voltage, and poor color were such serious disadvantages that they were never used on a large scale.

Paterson became president of the Institution of Electrical Engineers in 1930. His inaugural address was on the importance of research in electrical engineering, illustrated from two areas, heating and lighting. He thought the tungsten filament lamp had been developed to its limit, which was not quite correct, although subsequent improvements have been relatively modest. He discussed how the various gas discharge lamps might be improved, but he gave no hint of the high-pressure mercury lamp, which his own company introduced commercially less than two years later.

The low-pressure discharge lamps of 1930 were typically glass tubes up to a meter long, which produced light of various colors for floodlighting. They were started with a high-voltage circuit employing a Tesla coil but could then run from the 230-volt mains. The color depended on the gas filling. Neon gave an orange–red light and an efficiency of up to 35 lumens per watt. Mercury gave a blue light and had a similar efficiency. In discussing "the colors most easily obtained," Paterson mentioned helium for tubes giving a yellow light. He also said that sodium vapor offered the possibility of a lamp giving up to 50 lumens per watt. At that time the chemical problems posed by the high reactivity of hot sodium vapor had not been solved.

The practical high-pressure mercury vapor lamp dates from July 1932, when GE installed them on the road outside their laboratories. The first were rated at 400 watts and had an arc 160 millimeters long in an aluminosilicate glass discharge tube 34 millimeters in diameter. The discharge tube contained a little argon and sufficient mercury to give a vapor pressure of 1 atmosphere at the operating temperature of 600°C. To reduce heat losses, the discharge tube was mounted in an evacuated outer bulb, giving an efficiency of about 36 lumens per watt and a life of 1200 hours.

Detailed improvements between 1932 and 1939 led to a range of lamps rated between 150 and 750 watts, with a life of up to 7000 hours and an efficiency of up to 42 lumens per watt. The arc tube was often made of quartz, which required new ways of sealing in the connecting wires. Platinum wires had been used in the earliest filament lamps because platinum and soda glass have almost the same coefficient of expansion, but a platinum wire sealed through quartz shrinks on cooling while the quartz maintains its dimensions. Similarly the dumet wire that replaced platinum in ordinary tungsten filament lamps cannot be used with quartz. Molybdenum was eventually adopted: a thin molybdenum wire could be sealed satisfactorily through quartz. A thicker wire did not make a good seal until Aldington found that if part of the wire were rolled flat, a good seal could be obtained around the flattened portion.

Much effort was devoted to improving the color of the light. An early approach was the "blended" lamp, a mercury lamp incorporating a tungsten filament run at a low temperature. The filament acted as the ballast resistance to control the current in the circuit and contributed red light to improve the overall color. Other lamps incorporated the vapor of metals such as zinc and cadmium to improve the color spectrum of the output. These "high brightness" lamps concentrated the discharge in a small volume between massive tungsten electrodes. The arc length of less than 1 centimeter was ideal for projectors. In 1947 they were introduced in film studios where previously the carbon arc had been the only suitable light.

Even more compact mercury discharge lamps with the arc in a narrow, water-cooled, quartz tube were developed in World War II as a substitute for the carbon arc in searchlights, but the complication of water cooling made them impractical.

Sodium Discharge Lamps. The simple, low-pressure, sodium lamp is the most efficient way known of producing light. Made in sizes from 10 to 180 watts, with efficiencies ranging from 100 to 180 lumens per watt, it has proved popular for street lighting, where the operating cost is a dominant consideration, even though the yellow light is unattractive. The spectrum of a low-pressure sodium lamp contains two lines, at 589.0 and 589.6 nanometers, close together in the yellow part of the spectrum so that for practical purposes the light is monochromatic, and colors cannot be seen by it.

The sodium lamp works in a similar way to the mercury lamp. Sodium is solid at ordinary temperatures, so the lamps contain a little neon in which a discharge starts with a red glow when the lamp is first switched on. As the lamp warms, which takes several minutes, the sodium vaporizes and takes over the arc. The neon ceases to play any part. To reduce heat losses and improve efficiency, sodium lamps are enclosed in a vacuum jacket, which in early lamps was detachable and reusable.

Sodium is highly chemically reactive. Glasses containing silica are attacked and blackened by hot sodium. Silica-free aluminoborate glass will withstand hot sodium, but it is attacked by atmospheric moisture and difficult to work. Low-pressure sodium lamps have an envelope of soda-lime glass coated on the inside with a very thin layer of aluminoborate glass. The electrodes are of oxide-coated tungsten similar to those in the high-pressure mercury lamp.

The high-pressure sodium lamp, introduced commercially in the 1970s, is nearly as efficient as the low-pressure lamp and gives a wide-spectrum, "salmon pink" light in which most colors are readily seen. The arc tube is made of alumina (Al_2O_3), a translucent ceramic, which is ground to a fine powder, pressed, and sintered in the required tubular shape. Alumina melts at 2050°C and is one of the most stable substances known, occurring naturally as sapphire and ruby; it is unaffected by hot sodium. The alumina tube can be cut to length but cannot otherwise be worked. In one construction method, the lead-in wires supporting the electrodes are sealed through an end cap, also made of alumina, and sealed to the tube with a sodium-resistant glass. The electrodes are oxide-coated tungsten connected by a niobium lead-in wire. The filling gas is sodium with a little mercury and either argon or xenon to assist in starting.

Electric Lighting: Fluorescent Lamps

The electric light most commonly found in shops and offices, and increasingly in the home, is the fluorescent tube, available in many shapes, sizes, and colors. These also exploit an electric discharge in mercury but use the ultraviolet light that is produced. A phosphor coating on the inside of the glass tube converts this ultraviolet into visible light. The glass is opaque to ultraviolet light so none escapes.

Although efficient, early fluorescent lamps were not popular because of poor color rendering and tendency to flicker, especially when old. They needed special control equipment, making their initial cost greater than that of filament lamps, a further disadvantage. The best modern fluorescents have largely overcome these problems and, with their low operating costs, are rapidly increasing their market share.

By 1930 it was known that a mixture of mercury vapor and another gas could convert as much as 60 percent of the electrical energy in a discharge into radiation at a single ultraviolet wavelength, 253.7 nanometers. Cold cathode fluorescent tubes, operating at several thousand volts, were introduced for lighting large spaces in 1938, quickly followed by the familiar hot cathode tube operating at mains voltage. Commercial exploitation of fluorescent lighting was delayed by the onset of World War II. Although it was used in factories, it was not commonly seen by the general public until the late 1940s.

Nearly all fluorescent tubes were made with a diameter of 38 millimeters until the early 1970s because that gave the greatest efficiency. The most widely used tube was the 40-watt, 1200-millimeter-long, 38-millimeter-diameter lamp. Continuing improvements in phosphors and electrodes steadily improved efficiency and life. Most of the mineral-based phosphors used in the first fluorescent tubes were sulfides such as cadmium sulfide and zinc

sulfide. From the late 1940s they were replaced by halophosphates, a British discovery of 1942. From about 1970 alkaline-earth silicate phosphors have been developed. These are more efficient than the halophosphates, but, whereas the halophosphates give a broad spectrum, the newer materials usually give light with a single line spectrum. In most fluorescent lamps a mixture of fluorescent materials is used to obtain the desired color and efficiency. Most modern fluorescent lamps employ several narrow-band phosphors of distinct colors as may be seen by viewing their light through a prism.

From 1975 the "slimline" fluorescent tubes, 26 millimeters in diameter, gradually superseded the 38-millimeter ones. The 26-millimeter tubes have krypton rather than argon for the inert gas filling and use 10 percent less electricity. Ever narrower lamps continue to be developed, using more efficient phosphors and operating at higher power densities than earlier phosphors.

The compact fluorescent lamp, launched by Philips in 1980, was designed to be a direct replacement of the incandescent lamp. The control circuitry was contained within its cap. It had four times the efficiency and five times the life of a filament lamp. A variant introduced in 1982 by Thorn Lighting in Britain was their two-dimensional lamp, which was plugged into the control gear, a separate, reusable unit that was then placed into a conventional lamp. Early compact fluorescents were heavy because the current was controlled by an iron-cored inductance, which has now been replaced by the electronic ballast. Electronic control has other advantages: the lamp starts without flicker after a half-second delay, during which the electrodes are prewarmed before the full voltage is applied, and then runs at a more efficient frequency of about 25,000 Hertz (compared with the mains frequency of 50 Hertz). Although a direct descendant of the long fluorescent tube, the compact fluorescent is virtually a new lamp and has been marketed vigorously for its "energy saving" qualities.

Electric Lighting: Other Lamps

In addition to mercury or sodium, many other elements can provide the medium for a light-giving discharge. The metal halide lamp (not to be confused with the tungsten halogen lamp) introduced in the 1960s was a logical development from high-pressure mercury lamps, containing additional metal vapors to improve the color. Several other metals are potentially suitable, but in the high-temperature and high-pressure conditions within the arc tube they would react chemically either with the silica of the tube or with the electrodes and lead-in wires. The successful metal halide lamp depends on the discovery that such chemical attack can be avoided by including the metals as their halides, usually the iodide. Metals used include dysprosium, gallium, indium, lithium, scandium, sodium, thallium and thorium. When the lamp is cold, the metal in its halide form is condensed on the silica tube. When switched on, it starts as a high-pressure mercury lamp, but the metal halides quickly vaporize, and in the intense heat they dissociate. The metal vapor then produces its own characteristic emission in the arc, but any metal atoms that diffuse toward the silica wall or the electrodes, which are cool compared with the arc, recombine with the halogen before they can cause damage by chemical attack.

Metal halide lamps are similar in construction to high-pressure mercury lamps, though usually smaller. They require similar starting and control circuits, and are widely used in large lighting schemes where bright light of good color is required, such as sports fields or arenas and television studios. Their efficiency (typically 70 to 80 lumens per watt) is better than the high-pressure mercury lamp (52 lumens per watt) though not as high as the high-pressure sodium lamp (115 lumens per watt), but their color is better than either.

The "induction" lamp, also introduced in the 1990s, is a development of the fluorescent lamp. Conventional fluorescent lamps eventually fail either because the electrodes wear out or because the seal between the glass and the lead-in wires fails. The induction lamp has no electrodes and no lead-in wires; the discharge is created by a very high-frequency electromagnetic field outside the bulb.

Completely new at the time was the "sulfur" lamp, an induction lamp invented in 1990 by Fusion Lighting of Maryland. A glass bulb the size of a golf ball contains sulfur and argon. When energized from a high frequency source, it gives a bright, very white light whose spectral composition is similar to sunlight. It produces very little ultraviolet. Each sulfur lamp absorbs 5,900 watts but gives as much light as 25,000 watts of incandescent lighting. The high frequency supply comes from a magnetron. The lamp itself is environmentally friendly because it contains no mercury or any other toxic substances.

Some semiconductor devices emit light. These light emitting diodes, or LEDs, are often used for small signs and indicator lamps. They are very efficient light sources but were long limited to low powers and a restricted color range. At the start of the century, bright green LEDs became available, and LEDs are appearing in traffic lights. LEDs seem likely to be an important light source in the twenty-first century.

See also **Lighting, Public and Private; Light Emitting Diodes**

BRIAN BOWERS

Further Reading

Cady, F.E. and Dates, H.B., Eds. *Illuminating Engineering.* Wiley, New York, 1928.

Gaster, L. and Dow, J.S. *Modern Illuminants and Illuminating Engineering.* Pitman, London, 1920.

Hughes, T.P. *Networks of Power: Electrification of Western Society, 1880–1930.* MIT Press, Boston, 1983.

Johnston, S.F. *A History of Light and Colour Measurement: Science in the Shadows.* Institute of Physics Press, Bristol, 2001.

Liquid Crystals

The term "liquid crystals" contains a contradiction in terms in that crystals are solid, not liquid. But a number of materials can be in a phase that is in between the solid and the liquid phase, and the term liquid crystals refers to this phase, called the mesomorphic phase, rather than to a specific class of materials. Meso refers to middle, as this state in is the middle between solid and liquid. On the other hand, liquid crystals combine characteristics of both the solid crystalline phase and the liquid phase. They have the mobility of a liquid, but also the anisotropy (anisotropy means that the properties mentioned depend on the direction in which they are measured) in optical and electrical properties that otherwise is only found in crystals. This makes them suitable for a number of applications.

The mesomorphic phase was discovered in 1888 by an Austrian botanist named F. Reinitzer. He found that when cholesterol in plants was being melted, it passed not one but two transition points. The first one marked the transition from solid into a hazy liquid at a certain temperature (the melting point), and the second one marked the transition from a hazy into a transparent liquid at a higher temperature (the clearing point). Moreover he observed the phenomenon of birefringe: the liquid crystal differentiates in diffraction index between light with different polarization directions. In 1922 G. Friedel developed a classification that is still used today and which is discussed below. D. Vorländer contributed to the interest in liquid crystals by synthesizing numerous organic compounds that showed the mesomorphic phase properties. In 1933 the Royal Society held a conference on liquid crystals. It was so successful in suggesting that all possible knowledge about liquid crystals had been acquired that scientific interest in liquid crystals decreased rapidly. In 1957 there was a revival following the publication of a review article on liquid crystals by Glenn Brown, an American chemist, which nearly coincided with the finding of Westinghouse researchers that liquid crystals could be used to make temperature sensors. In the 1960s, research into liquid crystals greatly intensified because a number of other applications were envisioned, and again several new compounds were developed. The main purpose of searching for these new compounds was that they could be brought into the mesomorphic state at lower temperatures (in the room temperature range), which made them much more usable for practical applications. A major breakthrough came in 1973 with the introduction of cyanobiphenyl liquid crystals by George Gray because these were stable over a large temperature range.

The molecules in the liquid crystals usually have a rod-like structure. Liquid crystals can be in three phases: the nematic, cholesteric, and smectic phases (see Figure 2).

In the nematic phase the rod-like molecules are all aligned in the same direction, along a common axis, which is called the director. The term nematic refers to the Greek word νημα, which means wire or filament.

In the cholesteric phase there are layers of nematic phases, whereby the director rotates when moving from one layer to the next. The distance through the layers over which the director makes a full turn is called the pitch (just as with a screw). As the cholesteric phase is a variant of the nematic phase, it is also called the chiral-nematic phase.

The third or smectic phase was derived from the Greek word σμηγμα, which means soap or salve.

The term was chosen because in this phase the liquid crystal has some mechanical properties that make it behave like soap. In the smectic phase the rod-like molecules are ordered not only in a certain direction (indicated by the director) but also in equidistant planes.

A particular type of phase that is used in displays is the twisted nematic phase (which closely resembles the cholesteric phase), first described by C. Maugin in 1911. This situation can be created because the molecules can be directed along a rubbed glass plate. When the direction in the bottom plate is different from that of the top plate, the director is twisted between the bottom and top glass plates. In displays this is combined with the phenomenon that an electric field can change the direction of the molecules due to their dipolar nature. Switching on and off an electric field disturbs the twisted nematic, and the display pixel changes from transparent to black and vice versa. This disturbance is called the Freedericsz transition, named after a Russian scientist Vsevolod K. Freedericksz, who found the effect in the 1930s. This effect is the basis of liquid crystal displays.

In the last decades of the twentieth century, numerous applications of liquid crystals have been developed. The first applications were the simple displays that we can still find in calculators and watches. For this application the nematic phase is used. Later, more sophisticated versions were developed for use in portable computers (laptops and notebooks. In this application the influence of an electric field on the molecules is used. In other applications the influence of temperature is used. This is done in liquid crystal thermometers where the cholerestic phase is required. As the reflected color depends on the pitch of the cholerestic phase, and the pitch changes with temperature, the reflected color is an indicator for the temperature when there is a mixture of different compounds with different pitches. By the end of the twentieth

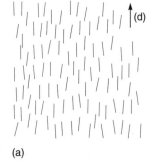 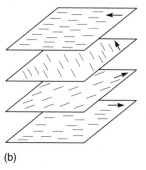 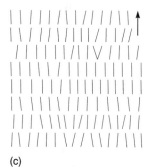

Figure 2. The phases of a liquid crystal: (a) nematic, (b) cholesteric, and (c) smectic with (d) director. [*Source: Philips Tech. Rev., 37, 5/6, p. 132, 1977.*]

(a) (b) (c)

century, devices for almost any temperature range were possible. Liquid crystal thermometers are used, for instance, in medical applications to measure skin temperature and in electronics to find bad connections on a circuit board by identifying places of higher temperature. Other applications include nondestructive mechanical testing of materials under stress and measuring the pressure of walking feet on the ground. Another application being explored at the turn of the twenty-first century was optical imaging. This application used the influence of an electric field, as in the display application. The liquid crystals are placed between layers of photoconductive material. At places where the photoconductor is hit by light, an electric field is produced, and this changes the state of the liquid crystal. Thus the light signal can be detected as an electric signal and then recorded.

See also **Computer Displays**

MARC J. DE VRIES

Further Reading

Bahadur, B. *Liquid Crystals:Applications and Uses*, vol. 1. World Scientific, 1990.

Collings, P.J. and Hird, M. *Introduction to Liquid Crystals: Chemistry and Physics*. Taylor & Francis, London, 1997.

Collings, P.J. *Liquid Crystals: Nature's Delicate Phase of Matter*. Adam Hilger, Bristol, 1990.

Gray, G.W. Introduction and historical development, in *Physical Properties of Liquid Crystals*. Gray, G.W., Ed. Wiley, New York, 1999.

Sluckin, Tim. The liquid crystal phases: physics and technology. *Contemp. Phys.*, 41, 37–56, 2000.

Loudspeakers and Earphones

The loudspeaker is the final element of several technological systems that transmit or reproduce sounds. Once the source has been transformed into an electrical signal, the loudspeaker finally converts it back into sound waves. The earliest research on these transducers was part of the development of the telephone in the late nineteenth century. Ernst Siemens and Alexander Graham Bell received patents on dynamic or moving coil transducers in which a coil of wire moved axially within a magnetic field, driving a diaphragm to produce sound. This was the basic idea of the modern direct radiator loudspeaker, but in the era of early telephony it could only handle a small portion of the frequency range of sounds—the range that extends from the low-frequency bass notes to the high-frequency treble notes.

The introduction of Western Electric's system of electrical recording in the 1920s meant that new loudspeakers had to be devised to handle the heavier loads and extended frequency range, especially in theaters presenting the new talking pictures. Chester W. Rice and Edward Kellog of Western Electric designed a moving coil transducer with a small coil-driven cone acting as a diaphragm. This driver design was influenced by the research carried out by E.C. Wente to improve public address loudspeakers. The Rice–Kellog loudspeaker proved to have uniform response to the much wider range of sound frequencies employed in electrical recording. It was widely adopted and so completely dominated the field for loudspeakers that most other types disappeared. Rice and Kellog published the specifications of their loudspeaker in a seminal paper in 1925 and this remained the blueprint for speaker design for the next 20 years.

Loudspeakers were employed in a variety of uses in radio sets, record players, and theater sound systems. The large business organizations responsible for these products employed research laboratories to improve all elements of the system: driver, enclosure, and acoustic dampening and insulation. Bell Laboratories produced the key innovations which determined the path of development: the bass reflex principle (a vented enclosure to radiate low-frequency sounds emerging from the rear of the diaphragm) and the divided range concept, in which separate drivers dealt with different parts of the frequency range in a two-way speaker.

In the 1940s, several important designs for two-way speakers—in which the frequencies were divided between a high-frequency driver (the horn-like "tweeter") and a low-frequency driver (the large bass "woofer")—were produced for film sound reproduction in theaters. James Lansing, John Blackburn, and John Hilliard were all involved in the research sponsored by the MGM film studio, which led to the widely used two-way "Voice of the Theater" loudspeaker. Lansing and Hilliard were also involved in developing co-axial speakers, which mated small high-frequency drivers to bass woofers. These ideas were later applied to loudspeakers like the Altec Lansing Duplex 604, which established the standard for monitors in recording studios and broadcast stations and remained in production from 1943 to 1998.

During World War II, emergency research programs in electronics and acoustics produced valuable new information that was eventually applied to audio systems. The development of more powerful and responsive magnetic materials had important consequences in loudspeaker design. The new nickel-based alloys, such as alnico, made the permanent magnet at the heart of the loudspeaker much more efficient. Wartime technological advances were applied to a broad range of products, but equally important in this progress were the young men who were trained in electronics. After the war several of these technicians founded their own innovative companies to introduce new technologies into consumer goods. Startup companies in the audio field, such as the loudspeaker manufacturer Altec Lansing, soon challenged the technological supremacy of the large organizations.

In the 1950s, the growing interest in high-fidelity home stereo players diverted attention from movie theater sound systems, which had always been the first to incorporate new ideas in loudspeakers. The Rice–Kellog specifications determined that a very large speaker was necessary to reproduce the frequency range required by the discerning home listener. The invention of a dome-shaped tweeter by Roy Allison and the acoustic suspension enclosure by Edgar Villchur and Henry Koss removed this size constraint. Villchur used a cushion of enclosed air in the speaker to act as a stiffener, doing away with the large bass woofers needed to reproduce low-frequency sounds. The new "bookshelf" speakers, less than 60 centimeters high, set new standards for bass response. This innovation was marketed by several small companies including Acoustic Research (founded by Koss and Villchur), Advent, and Allison Acoustics.

The market for loudspeakers broadened dramatically in the 1960s: audiophiles demanded bookshelf speakers with uniform frequency response and low distortion; guitarists needed large, heavy-duty speakers to deal with the unprecedented increase of output from their amplifiers; and teenagers wanted to hear clearly the music coming from their pocket-sized transistor radios. Every user had a different concept of high fidelity, and the merits of loudspeakers were so often debated that there could be no commonly accepted standard of performance.

Earphones were only a small part of this market as they were primarily a low-fidelity alternative to loudspeakers. They were scaled down versions of the dynamic moving coil speaker, with all the same elements of the driver enclosed in a device that channeled the sound directly into the ear. The unexpected success of the personal stereo in the 1980s directed research into improving the frequency response of the earphone and establishing true stereo sound. Injecting the sound into the ear takes away the indications, such as reverberation, which give the impression of listening in a space such as a large room or concert hall. This was accomplished by programming delays into the dispersion of sound going to each ear so that the listener gets an impression of the movement of sound.

The introduction of digital sound reproducing systems placed an even greater burden on loudspeakers and earphones because the background noise and distortion was reduced and frequency response increased: listeners could hear more of the source sound and became more discriminating in their evaluation of loudspeakers. Subsequently more innovations were produced in cone and enclosure design, and in electronic manipulation and filtering of the signals in the crossover network.

The introduction of digital surround sound theater systems accelerated the pace of innovation in the 1990s, and the fruits of this research were immediately applied to home systems that employed multiple speakers and complex crossover networks that fed selected frequencies into the appropriate drivers. Innovations such as the subwoofer, which was powered by its own amplifier to give powerful bass response, brought even more fidelity and volume to the home listener.

See also **Audio Recording, Electronic Methods; Audio Recording, Stereo and Surround Sound; Audio Systems, Personal Stereo**

ANDRE MILLARD

Further Reading

Colloms, M. *High Performance Loudspeakers*. Wiley, New York, 1997.

Millard, A. *America on Record: A History of Recorded Sound*. Cambridge University Press, Cambridge, 1995.

Milton, P. Headphones: history and measurement. *Audio*, 26, 90–94, 1978.

Read, O. and Welch, W.L. *From Tin Foil to Stereo: Evolution of the Phonograph*. SAMS, Indianapolis, 1976.

Rice, C.W. and Kellog, E.W. Notes on the development of a new type of hornless loudspeaker. *J. Aud. Eng. Soc.*, 30 512–521, 1982. A reprint of the original 1925 paper.

Useful Websites

A history of the loudspeaker: http://history.acusd.edu/gen/recording/loudspeaker.html